Chrysler
Rear-wheel drive
Automotive Repair Manual

by Robert Maddox and John H Haynes

Member of the Guild of Motoring Writers

Models covered:

All Dodge, Plymouth and Chrysler rear-wheel drive models with a six-cylinder or V8 engine and a single carburetor
1971 through 1989

(30050-2K10)

J H Haynes & Co. Ltd.
Haynes North America, Inc.
www.haynes.com

Disclaimer

There are risks associated with automotive repairs. The ability to make repairs depends on the individual's skill, experience and proper tools. Individuals should act with due care and acknowledge and assume the risk of performing automotive repairs.

The purpose of this manual is to provide comprehensive, useful and accessible automotive repair information, to help you get the best value from your vehicle. However, this manual is not a substitute for a professional certified technician or mechanic.

This repair manual is produced by a third party and is not associated with an individual vehicle manufacturer. If there is any doubt or discrepancy between this manual and the owner's manual or the factory service manual, please refer to the factory service manual or seek assistance from a professional certified technician or mechanic.

Even though we have prepared this manual with extreme care and every attempt is made to ensure that the information in this manual is correct, neither the publisher nor the author can accept responsibility for loss, damage or injury caused by any errors in, or omissions from, the information given.

Acknowledgements

We are grateful to the Chrysler Corporation for providing technical information and certain illustrations. Technical writers who contributed to this project include Mark Ryan, Mike Stubblefield and Larry Warren.

A book in the Haynes Automotive Repair Manual Series

ISBN 13: 978-1-56392-098-1
ISBN 10: 1-56392-098-0

Library of Congress Catalog Card Number 94-75678

98-384

Contents

Haynes mechanic with 1986 Dodge Diplomat

About this manual

Its purpose

The purpose of this manual is to help you get the best value from your vehicle. It can do so in several ways. It can help you decide what work must be done, even if you choose to have it done by a dealer service department or a repair shop; it provides information and procedures for routine maintenance and servicing; and it offers diagnostic and repair procedures to follow when trouble occurs.

We hope you use the manual to tackle the work yourself. For many simpler jobs, doing it yourself may be quicker than arranging an appointment to get the vehicle into a shop and making the trips to leave it and pick it up. More importantly, a lot of money can be saved by avoiding the expense the shop must pass on to you to cover its labor and overhead costs. An added benefit is the sense of satisfaction and accomplishment that you feel after doing the job yourself.

Using the manual

The manual is divided into Chapters. Each Chapter is divided into numbered Sections, which are headed in bold type between horizontal lines. Each Section consists of consecutively numbered paragraphs.

At the beginning of each numbered Section you will be referred to any illustrations which apply to the procedures in that Section. The reference numbers used in illustration captions pinpoint the pertinent Section and the Step within that Section. That is, illustration 3.2 means the illustration refers to Section 3 and Step (or paragraph) 2 within that Section.

Procedures, once described in the text, are not normally repeated. When it's necessary to refer to another Chapter, the reference will be given as Chapter and Section number. Cross references given without use of the word "Chapter" apply to Sections and/or paragraphs in the same Chapter. For example, "see Section 8" means in the same Chapter.

References to the left or right side of the vehicle assume you are sitting in the driver's seat, facing forward.

Even though we have prepared this manual with extreme care, neither the publisher nor the author can accept responsibility for any errors in, or omissions from, the information given.

NOTE

A **Note** provides information necessary to properly complete a procedure or information which will make the procedure easier to understand.

CAUTION

A **Caution** provides a special procedure or special steps which must be taken while completing the procedure where the Caution is found. Not heeding a Caution can result in damage to the assembly being worked on.

WARNING

A **Warning** provides a special procedure or special steps which must be taken while completing the procedure where the Warning is found. Not heeding a Warning can result in personal injury.

Introduction to the Chrysler rear-wheel drive models

The Chrysler rear-wheel drive models feature a front engine/rear-wheel drive unit-body chassis.

The inline six-cylinder and V8 engines used in these vehicles are equipped with single carburetors.

The engine drives the rear wheels through either a three- or four-speed manual or automatic transmission via a driveshaft. A solid rear axle is used.

Front suspension is independent, featuring torsion bars and conventional shock absorbers. Leaf springs and shock absorbers are used at the rear. The conventional gearbox-type steering system is power assisted on most models.

Most models are equipped with front disc brakes, although some models through 1976 have drum front brakes. Virtually all models use drum brakes at the rear. A few 1974 and 1975 Imperial models use disc brakes at the rear also. Most models have power-assisted brakes. Some 1972 and 1973 Imperial models are equipped with a four-wheel Anti-lock Brake System (ABS).

Body type identification

Although the vehicles covered in this manual share many common chassis design features, differences by body style will be identified in many repair procedures. Since so many model names were used for the various body styles, we have chosen to identify body styles within this manual by their common industry identifier. The chart below shows which identifier letter is used for each model covered by this manual **(see illustration)**. **Note:** *Chrysler, Dodge and Plymouth factory service manuals use different body identifiers (called "carline" identifiers). When writing this manual, we discovered these carline identifier letters were re-used for different body styles over the long range of years we chose to cover, and, therefore, we chose not to use them. So, if you open a factory service manual for your year of vehicle, or talk to someone at a dealer service department, you may find they are using a different letter to identify your body style.*

	A	E	F	M	J	B	R	C
Dodge	Dart,Demon ('71-'76)	Challenger ('71-'74)	Aspen ('76-'80)	Diplomat ('77-'89)	Mirada ('80-'83)	Coronet ('71-'76); Charger, Magnum ('71-'79); Monaco ('77-'78)	St.Regis ('79-'81)	Polara ('71-'73); Monaco ('71-'76); Royal Monaco/Monaco Brougham ('77-'78)
Plymouth	Valiant, Duster ('71-'76)	Barracuda ('71-'74)	Volar'e, Road Runner ('76-'80)	GranFury ('82-'89); Caravelle ('77-'81)		Satellite, Sebring, Road Runner, GTX ('71-'74); Fury ('75-'78)	GranFury ('79-'81)	Fury ('71-'74); GranFury ('75-'77)
Chrysler				LeBaron ('77-'81); New Yorker ('82); Fifth Avenue ('83-'89); Newport ('84-'89)	Cordoba ('80-'83); 300 ('80)	Cordoba, 300 ('75-'79)	Newport, NewYorker ('79-'81)	Newport, NewYorker, Town and Country ('71-'78); 300 ('71)
Imperial								LeBaron ('71-'75)

This chart shows the common industry letter identifier for each of the models covered in this manual

Vehicle identification numbers

Modifications are a continuing and unpublicized process in vehicle manufacturing. Since spare parts manuals and lists are compiled on a numerical basis, the individual vehicle numbers are essential to correctly identify the component required.

Vehicle Identification Number (VIN)

This very important identification number is stamped on a plate attached to the left side of the dashboard just inside the windshield on the driver's side of the vehicle **(see illustration)**. The VIN also appears on the Vehicle Certificate of Title and Registration. It contains information such as where and when the vehicle was manufactured, the model year and the body style.

VIN year and engine codes

Two particularly important pieces of information located in the VIN are the model year and engine codes. On models through 1980, counting from the left, the engine code is the fifth digit and the model year code is the sixth digit. On 1981 and later models, the 8th digit is the engine code and the 10th digit is the model year code. After obtaining the engine code from the VIN, use the accompanying chart to identify which engine was installed in your vehicle at the factory **(see illustration)**. On earlier models, it's possible the original engine may have been "swapped" for a different engine; see *Engine sequence number* below to determine if this has happened.

Data plate

This number-stamped metal plate is located on the left front fender side shield, the wheel housing or left upper radiator support, depending on model. It contains valuable information concerning the production of the vehicle as well as information about the way in which the vehicle is equipped. This

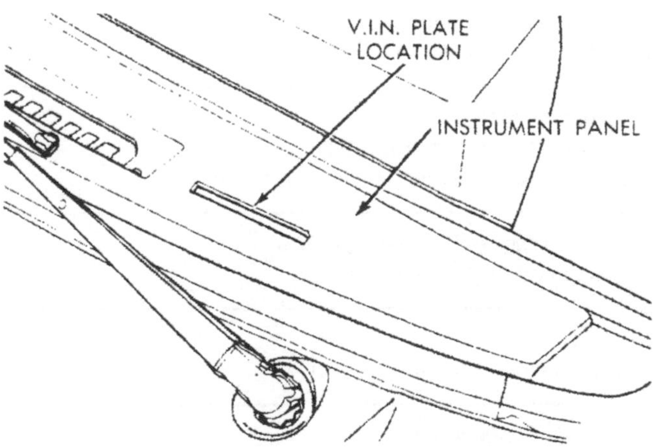

The Vehicle Identification Number (VIN) is visible from outside the vehicle, through the driver's side of the windshield

Example (1971 through 1980 models):

C L 41 G 9 C 100001

Engine code (5th digit) — G
Model year code (6th digit) — 9

Example (1981 through 1989 models):

1 B 3 B K 5 1 D 3 E C 100001

Engine code (8th digit) — D
Model year code (10th digit) — E

Model Year Codes

1 = 1971	8 = 1978	E = 1984
2 = 1972	9 = 1979	F = 1985
3 = 1973	0 = 1980	G = 1986
4 = 1974		H = 1987
5 = 1975	B = 1981	J = 1988
6 = 1976	C = 1982	K = 1989
7 = 1977	D = 1983	

B	198 CID six-cylinder engine (1971 through 1974)
C	225 CID six-cylinder engine (1971 and later)
D	225 CID six-cylinder engine, high-performance (1977 through 1979)
E	Special order six-cylinder engine (1971 through 1980)
F	225 CID six-cylinder engine, high-performance (1981 and 1982)
G	318 CID V8, standard engine (1971 through 1980) 225 CID six-cylinder engine, heavy-duty (1981 and 1982)
H	340 CID V8, high-performance engine (1971 through 1973) 318 CID V8, four-barrel carburetor (1978 through 1980) 225 CID six-cylinder engine, heavy duty (1981 and 1982) 225 CID six-cylinder engine, standard duty (1983)
J	360 CID V8 engine, four-barrel carburetor (1973 through 1980) 318 CID V8 engine, fuel injection (1981 and 1982) 225 CID six-cylinder engine, heavy duty (1983)
K	360 CID V8, standard engine (1971 through 1980) 318 CID V8, two-barrel carburetor, standard engine (1981 and 1982)
L	383 CID V8, standard engine (1971) 360 CID V8, high-performance engine (1974 through 1980) 318 CID V8, two-barrel carburetor, heavy-duty engine (1981 and 1982)
M	400 CID V8, standard engine (1972 through 1976) 318 CID V8, four-barrel carburetor, standard engine (1981 and 1982)
N	383 CID V8, high-performance engine (1971) 400 CID V8, four-barrel carburetor (1974 through 1980) 318 CID V8, four-barrel carburetor, heavy-duty engine (1981 and 1982)
P	400 CID V8, high-performance engine (1972 through 1978) 318 CID V8, two-barrel carburetor (1984 through 1989)
R	426 CID "Hemi" V8 engine (1971) 318 CID V8, four-barrel carburetor (1984 through 1988)
S	318 CID V8, police engine (1984 through 1988) 318 CID V8, standard engine (1989)
T	440 CID V8, standard engine (1971 through 1978)
U	440 CID V8, high-performance engine (1971 through 1978)
V	440 CID V8, "Six-pack" engine (1971)
Z	Special order V8 engine (1971 through 1980) 318 CID V8 engine (1987 and 1988)
4	318 CID V8, heavy-duty engine (1984 through 1989)

Engine identification codes for the models covered in this manual

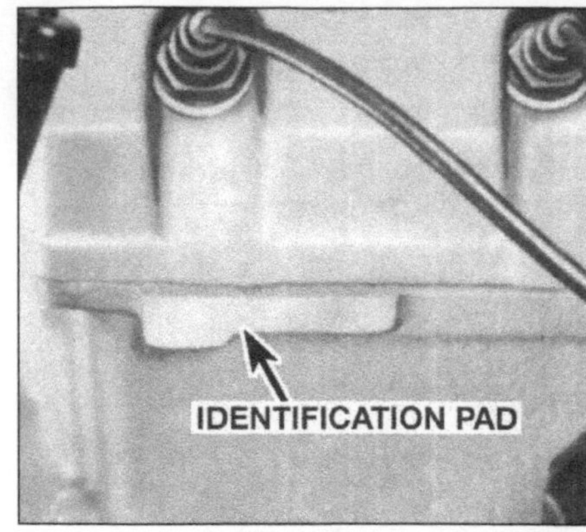

On the six-cylinder engine, the engine number is located on the identification pad below the right rear corner of the cylinder head

plate is especially useful for matching the color and type of paint during repair work.

Safety Certification label

The Safety Certification label is affixed to the left front door pillar. The plate contains the name of the manufacturer, the month and year of production, the Gross Vehicle Weight Rating (GVWR) and the certification statement.

Engine identification number

The engine ID number on the six-cylinder engine is stamped into the rear of the block, just below the cylinder head and in front of the bellhousing **(see illustration)**. On the small-block V8 engine, the number is on the front of the left cylinder bank, just below the cylinder head **(see illustration)**. On the big-block V8 engine, the number is on a machined pad at the front of the engine **(see illustrations)**. On the 383 and 400 engines, the pad is on the right (passenger's side) of the distributor. On the 440 engine, the pad is just in front of the intake manifold on the engine block (on air-conditioned 440 models, the air conditioning compressor often obscures the pad).

Engine sequence number

In addition to the engine identification number, a sequence number, which may be helpful when buying replacement parts, is used. On six-cylinder engines, the number is directly below the engine identification number. On V8 engines, the number is on the right side of the block, just above the oil pan **(see illustration)**. If the original engine is still installed in your vehicle, the last six digits of this number will match the last six digits of the VIN number.

Transmission identification number

The ID number on manual transmissions is located on a pad on the right side of the case **(see illustration)**. On automatic transmissions, the number is stamped on the transmission case near the fluid dipstick tube **(see illustration)**.

On small-block V8 engines, the number is located just below the driver's side cylinder head

On 383 and 400 engines, the identification number is just below the distributor

On 440 engines, the identification number is located in front of the intake manifold (it's often obscured by the air-conditioning compressor)

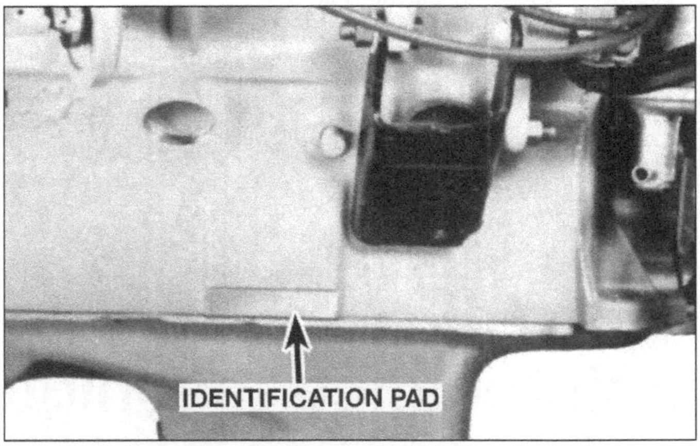

On V8 engines, the sequence number is stamped onto this identification pad, just above the oil pan on the right (passenger's) side of the engine

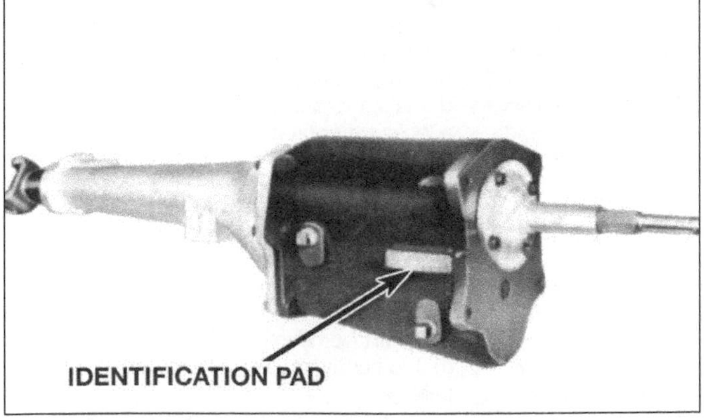

Typical manual transmission identification number location

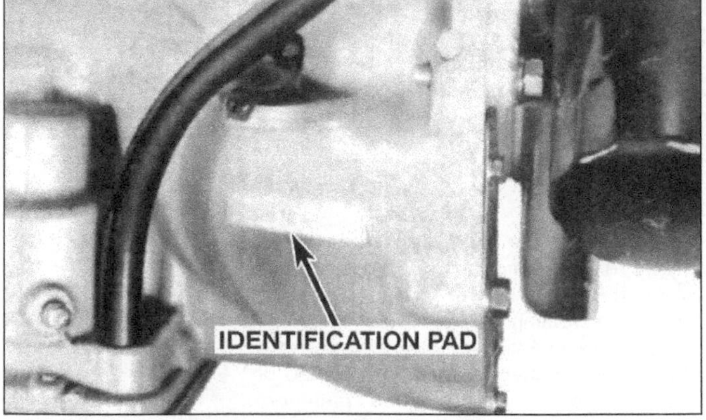

Automatic transmission number

Buying parts

Replacement parts are available from many sources, which generally fall into one of two categories - authorized dealer parts departments and independent retail auto parts stores. Our advice concerning these parts is as follows:

Retail auto parts stores: Good auto parts stores will stock frequently needed components which wear out relatively fast, such as clutch components, exhaust systems, brake parts, tune-up parts, etc. These stores often supply new or reconditioned parts on an exchange basis, which can save a considerable amount of money. Discount auto parts stores are often very good places to buy materials and parts needed for general vehicle maintenance such as oil, grease, filters, spark plugs, belts, touch-up paint, bulbs, etc. They also usually sell tools and general accessories, have convenient hours, charge lower prices and can often be found not far from home.

Authorized dealer parts department: This is the best source for parts which are unique to the vehicle and not generally available elsewhere (such as major engine parts, transmission parts, trim pieces, etc.).

Warranty information: If the vehicle is still covered under warranty, be sure that any replacement parts purchased - regardless of the source - do not invalidate the warranty!

To be sure of obtaining the correct parts, have engine and chassis numbers available and, if possible, take the old parts along for positive identification.

Maintenance techniques, tools and working facilities

Maintenance techniques

There are a number of techniques involved in maintenance and repair that will be referred to throughout this manual. Application of these techniques will enable the home mechanic to be more efficient, better organized and capable of performing the various tasks properly, which will ensure that the repair job is thorough and complete.

Fasteners

Fasteners are nuts, bolts, studs and screws used to hold two or more parts together. There are a few things to keep in mind when working with fasteners. Almost all of them use a locking device of some type, either a lockwasher, locknut, locking tab or thread adhesive. All threaded fasteners should be clean and straight, with undamaged threads and undamaged corners on the hex head where the wrench fits. Develop the habit of replacing all damaged nuts and bolts with new ones. Special locknuts with nylon or fiber inserts can only be used once. If they are removed, they lose their locking ability and must be replaced with new ones.

Rusted nuts and bolts should be treated with a penetrating fluid to ease removal and prevent breakage. Some mechanics use turpentine in a spout-type oil can, which works quite well. After applying the rust penetrant, let it work for a few minutes before trying to loosen the nut or bolt. Badly rusted fasteners may have to be chiseled or sawed off or removed with a special nut breaker, available at tool stores.

If a bolt or stud breaks off in an assembly, it can be drilled and removed with a special tool commonly available for this purpose. Most automotive machine shops can perform this task, as well as other repair procedures, such as the repair of threaded holes that have been stripped out.

Flat washers and lockwashers, when removed from an assembly, should always be replaced exactly as removed. Replace any damaged washers with new ones. Never use a lockwasher on any soft metal surface (such as aluminum), thin sheet metal or plastic.

Fastener sizes

For a number of reasons, automobile manufacturers are making wider and wider use of metric fasteners. Therefore, it is important to be able to tell the difference between standard (sometimes called U.S. or SAE) and metric hardware, since they cannot be interchanged.

All bolts, whether standard or metric, are sized according to diameter, thread pitch and length. For example, a standard 1/2 - 13 x 1 bolt is 1/2 inch in diameter, has 13 threads per inch and is 1 inch long. An M12 - 1.75 x 25 metric bolt is 12 mm in diameter, has a thread pitch of 1.75 mm (the distance between threads) and is 25 mm long. The two bolts are nearly identical, and easily confused, but they are not interchangeable.

In addition to the differences in diameter, thread pitch and length, metric and standard bolts can also be distinguished by examining the bolt heads. To begin with, the distance across the flats on a standard bolt head is measured in inches, while the same dimension on a metric bolt is sized in millimeters (the same is true for nuts). As a result, a standard wrench should not be used on a metric bolt and a metric wrench should not be used on a standard bolt. Also, most standard bolts have slashes radiating out from the center of the head to denote the grade or strength of the bolt, which is an indication of the amount of torque that can be applied to it. The greater the number of slashes, the greater the strength of the bolt. Grades 0 through 5 are commonly used on automobiles. Metric bolts have a property class (grade) number, rather than a slash, molded into their heads to indicate bolt strength. In this case, the higher the number, the stronger the bolt. Property class numbers 8.8, 9.8 and 10.9 are commonly used on automobiles.

Strength markings can also be used to distinguish standard hex nuts from metric hex nuts. Many standard nuts have dots stamped into one side, while metric nuts are marked with a number. The greater the number of dots, or the higher the number, the greater the strength of the nut.

Metric studs are also marked on their ends according to property class (grade). Larger studs are numbered (the same as metric bolts), while smaller studs carry a geometric code to denote grade.

It should be noted that many fasteners, especially Grades 0 through 2, have no distinguishing marks on them. When such is the case, the only way to determine whether it is standard or metric is to measure the thread pitch or compare it to a known fastener of the same size.

Standard fasteners are often referred to as SAE, as opposed to metric. However, it should be noted that SAE technically refers to a non-metric fine thread fastener only. Coarse thread non-metric fasteners are referred to as USS sizes.

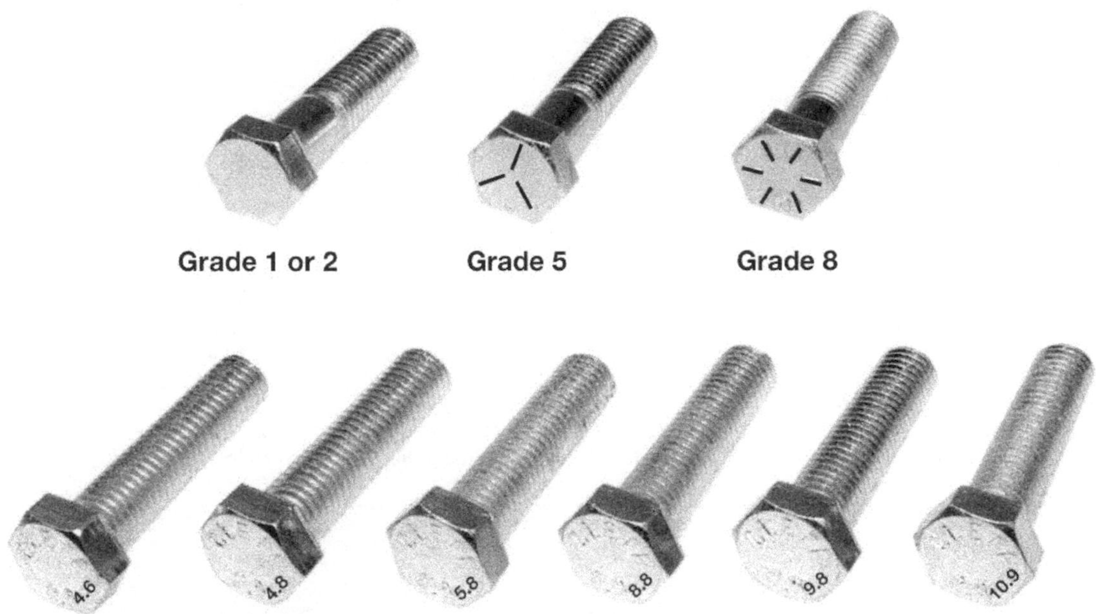

Bolt strength marking (standard/SAE/USS; bottom - metric)

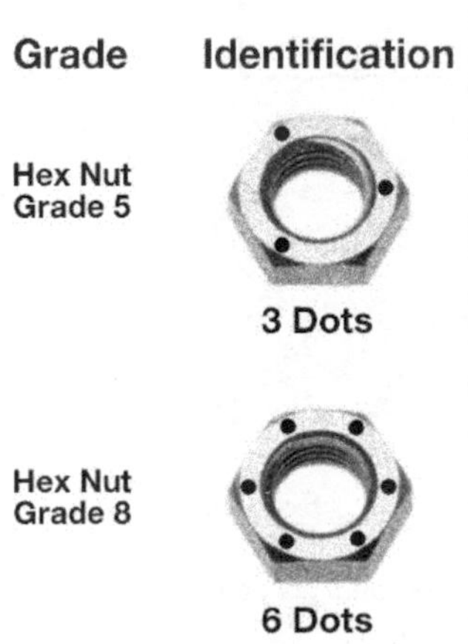

Standard hex nut strength markings

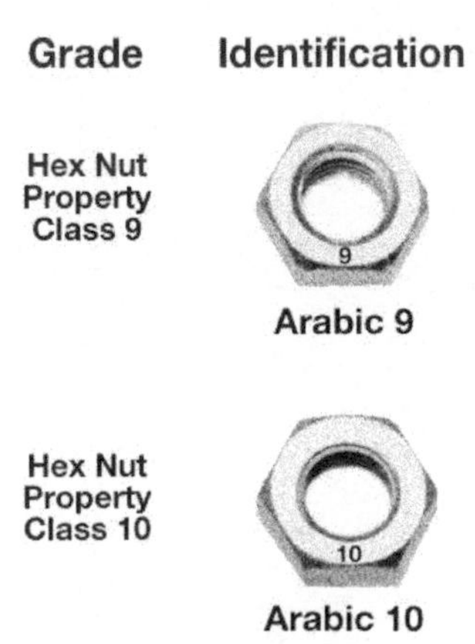

Metric hex nut strength markings

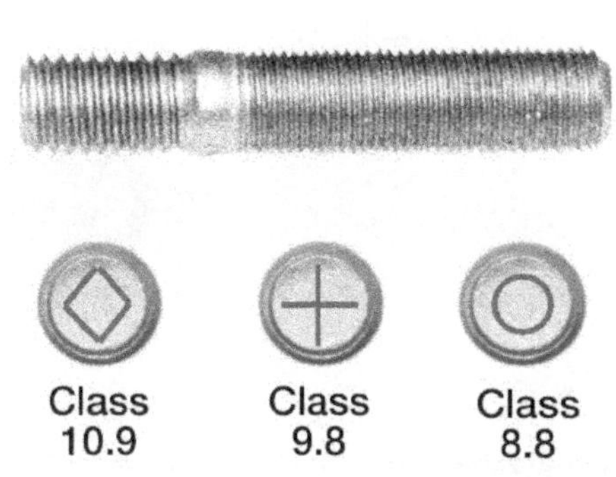

Metric stud strength markings

00-1 HAYNES

Since fasteners of the same size (both standard and metric) may have different strength ratings, be sure to reinstall any bolts, studs or nuts removed from your vehicle in their original locations. Also, when replacing a fastener with a new one, make sure that the new one has a strength rating equal to or greater than the original.

Tightening sequences and procedures

Most threaded fasteners should be tightened to a specific torque value (torque is the twisting force applied to a threaded component such as a nut or bolt). Overtightening the fastener can weaken it and cause it to break, while undertightening can cause it to eventually come loose. Bolts, screws and studs, depending on the material they are made of and their thread diameters, have specific torque values, many of which are noted in the Specifications at the beginning of each Chapter. Be sure to follow the torque recommendations closely. For fasteners not assigned a specific torque, a general torque value chart is presented here as a guide. These torque values are for dry (unlubricated) fasteners threaded into steel or cast iron (not aluminum). As was previously mentioned, the size and grade of a fastener determine the amount of torque that can safely be applied to it. The figures listed here are approximate for Grade 2 and Grade 3 fasteners. Higher grades can tolerate higher torque values.

Fasteners laid out in a pattern, such as cylinder head bolts, oil pan bolts, differential cover bolts, etc., must be loosened or tightened in sequence to avoid warping the component. This sequence will normally be shown in the appropriate Chapter. If a specific pattern is not given, the following procedures can be used to prevent warping.

Metric thread sizes	**Ft-lbs**	**Nm**
M-6	6 to 9	9 to 12
M-8	14 to 21	19 to 28
M-10	28 to 40	38 to 54
M-12	50 to 71	68 to 96
M-14	80 to 140	109 to 154
Pipe thread sizes		
1/8	5 to 8	7 to 10
1/4	12 to 18	17 to 24
3/8	22 to 33	30 to 44
1/2	25 to 35	34 to 47
U.S. thread sizes		
1/4 - 20	6 to 9	9 to 12
5/16 - 18	12 to 18	17 to 24
5/16 - 24	14 to 20	19 to 27
3/8 - 16	22 to 32	30 to 43
3/8 - 24	27 to 38	37 to 51
7/16 - 14	40 to 55	55 to 74
7/16 - 20	40 to 60	55 to 81
1/2 - 13	55 to 80	75 to 108

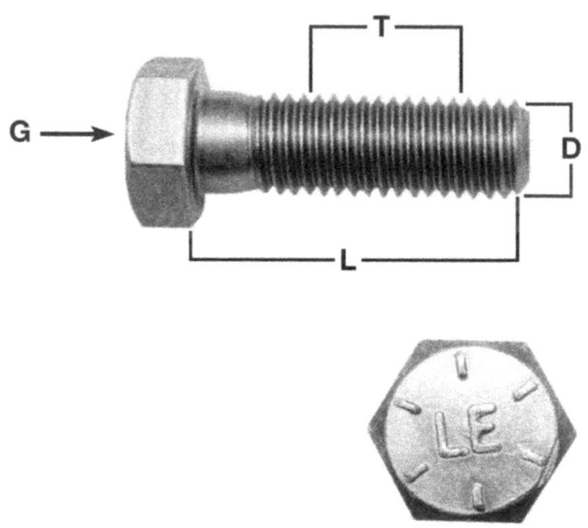

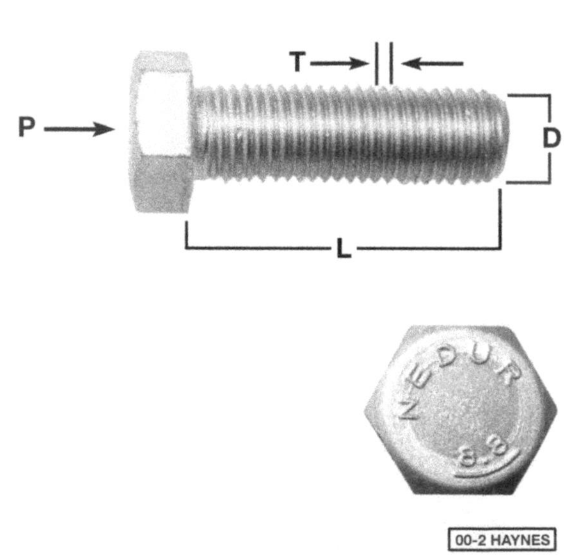

Standard (SAE and USS) bolt dimensions/grade marks

- *G* *Grade marks (bolt strength)*
- *L* *Length (in inches)*
- *T* *Thread pitch (number of threads per inch)*
- *D* *Nominal diameter (in inches)*

Metric bolt dimensions/grade marks

- *P* *Property class (bolt strength)*
- *L* *Length (in millimeters)*
- *T* *Thread pitch (distance between threads in millimeters)*
- *D* *Diameter*

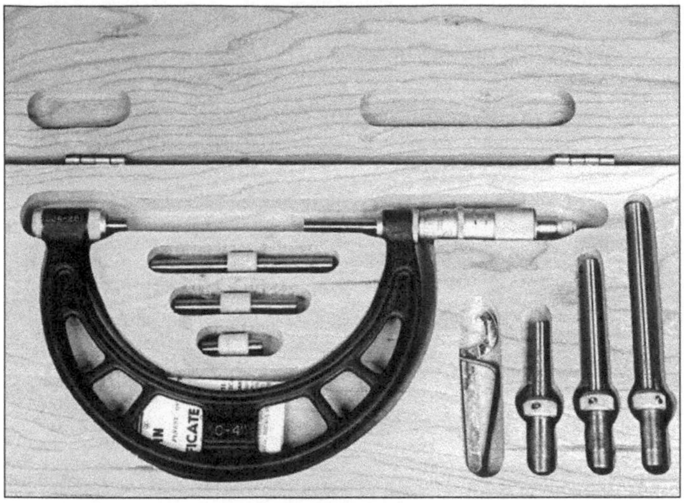

Micrometer set

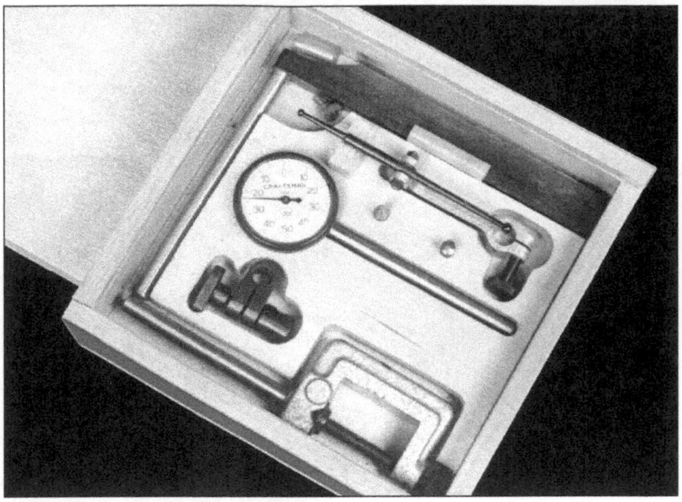

Dial indicator set

Initially, the bolts or nuts should be assembled finger-tight only. Next, they should be tightened one full turn each, in a criss-cross or diagonal pattern. After each one has been tightened one full turn, return to the first one and tighten them all one-half turn, following the same pattern. Finally, tighten each of them one-quarter turn at a time until each fastener has been tightened to the proper torque. To loosen and remove the fasteners, the procedure would be reversed.

Component disassembly

Component disassembly should be done with care and purpose to help ensure that the parts go back together properly. Always keep track of the sequence in which parts are removed. Make note of special characteristics or marks on parts that can be installed more than one way, such as a grooved thrust washer on a shaft. It is a good idea to lay the disassembled parts out on a clean surface in the order that they were removed. It may also be helpful to make sketches or take instant photos of components before removal.

When removing fasteners from a component, keep track of their locations. Sometimes threading a bolt back in a part, or putting the washers and nut back on a stud, can prevent mix-ups later. If nuts and bolts cannot be returned to their original locations, they should be kept in a compartmented box or a series of small boxes. A cupcake or muffin tin is ideal for this purpose, since each cavity can hold the bolts and nuts from a particular area (i.e. oil pan bolts, valve cover bolts, engine mount bolts, etc.). A pan of this type is especially helpful when working on assemblies with very small parts, such as the carburetor, alternator, valve train or interior dash and trim pieces. The cavities can be marked with paint or tape to identify the contents.

Whenever wiring looms, harnesses or connectors are separated, it is a good idea to identify the two halves with numbered pieces of masking tape so they can be easily reconnected.

Gasket sealing surfaces

Throughout any vehicle, gaskets are used to seal the mating surfaces between two parts and keep lubricants, fluids, vacuum or pressure contained in an assembly.

Many times these gaskets are coated with a liquid or paste-type gasket sealing compound before assembly. Age, heat and pressure can sometimes cause the two parts to stick together so tightly that they are very difficult to separate. Often, the assembly can be loosened by striking it with a soft-face hammer near the mating surfaces. A regular hammer can be used if a block of wood is placed between the hammer and the part. Do not hammer on cast parts or parts that could be easily damaged. With any particularly stubborn part, always recheck to make sure that every fastener has been removed.

Avoid using a screwdriver or bar to pry apart an assembly, as they can easily mar the gasket sealing surfaces of the parts, which must remain smooth. If prying is absolutely necessary, use an old broom handle, but keep in mind that extra clean up will be necessary if the wood splinters.

After the parts are separated, the old gasket must be carefully scraped off and the gasket surfaces cleaned. Stubborn gasket material can be soaked with rust penetrant or treated with a special chemical to soften it so it can be easily scraped off. A scraper can be fashioned from a piece of copper tubing by flattening and sharpening one end. Copper is recommended because it is usually softer than the surfaces to be scraped, which reduces the chance of gouging the part. Some gaskets can be removed with a wire brush, but regardless of the method used, the mating surfaces must be left clean and smooth. If for some reason the gasket surface is gouged, then a gasket sealer thick enough to fill scratches will have to be used during reassembly of the components. For most applications, a non-drying (or semi-drying) gasket sealer should be used.

Hose removal tips

Warning: *If the vehicle is equipped with air conditioning, do not disconnect any of the A/C hoses without first having the system depressurized by a dealer service department or a service station.*

Hose removal precautions closely parallel gasket removal precautions. Avoid scratching or gouging the surface that the hose mates against or the connection may leak. This is especially true for radiator hoses. Because of various chemical reactions, the rubber in hoses can bond itself to the metal spigot that the hose fits over. To remove a hose, first loosen the hose clamps that secure it to the spigot. Then, with slip-joint pliers, grab the hose at the clamp and rotate it around the spigot. Work it back and forth until it is completely free, then pull it off. Silicone or other lubricants will ease removal if they can be applied between the hose and the outside of the spigot. Apply the same lubricant to the inside of the hose and the outside of the spigot to simplify installation.

As a last resort (and if the hose is to be replaced with a new one anyway), the rubber can be slit with a knife and the hose peeled from the spigot. If this must be done, be careful that the metal connection is not damaged.

If a hose clamp is broken or damaged, do not reuse it. Wire-type clamps usually weaken with age, so it is a good idea to replace them with screw-type clamps whenever a hose is removed.

Tools

A selection of good tools is a basic requirement for anyone who plans to maintain and repair his or her own vehicle. For the owner who has few tools, the initial investment might seem high, but when compared to the spiraling costs of professional auto maintenance and repair, it is a wise one.

To help the owner decide which tools are needed to perform the tasks detailed in this manual, the following tool lists are offered: *Maintenance and minor repair,*

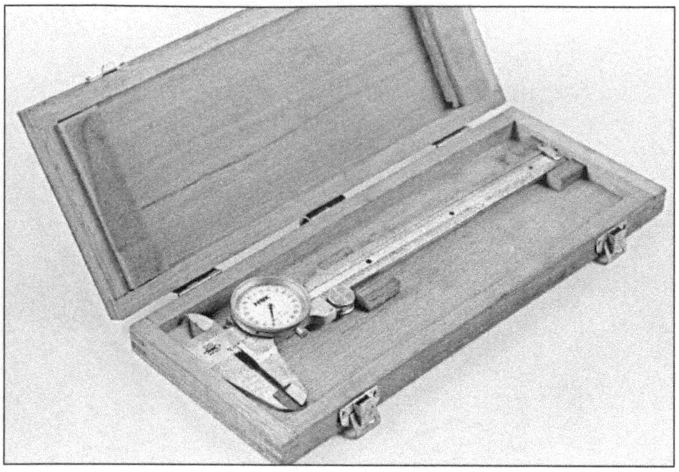
Dial caliper

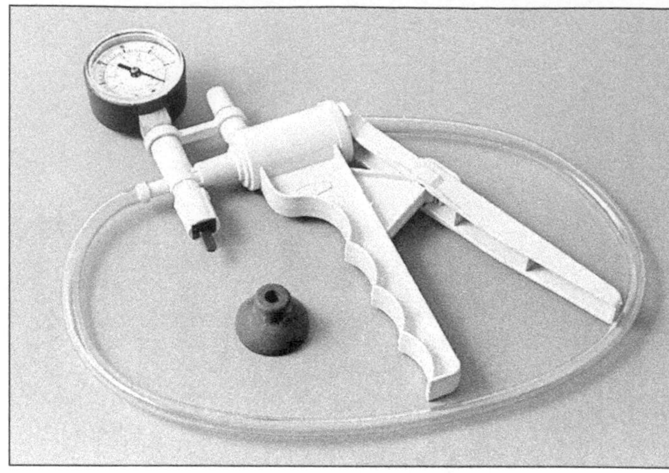
Hand-operated vacuum pump

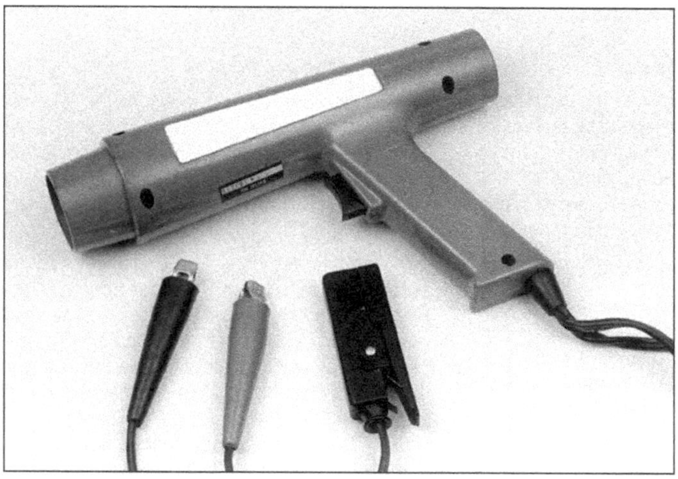
Timing light

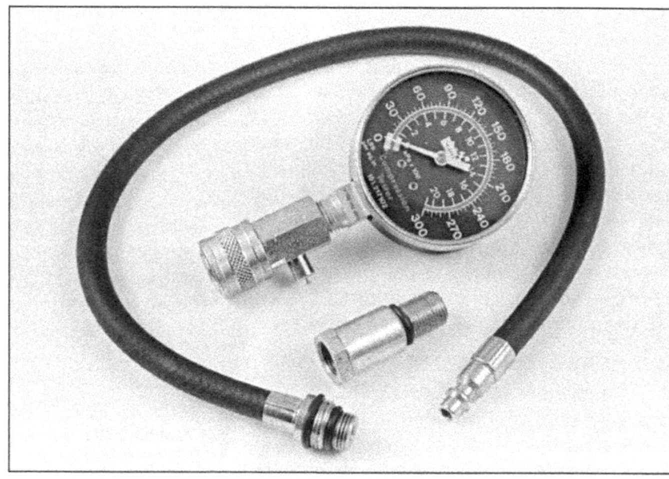
Compression gauge with spark plug hole adapter

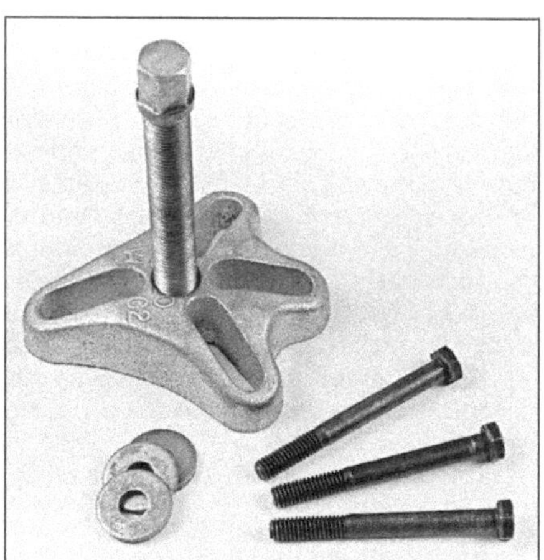
Damper/steering wheel puller

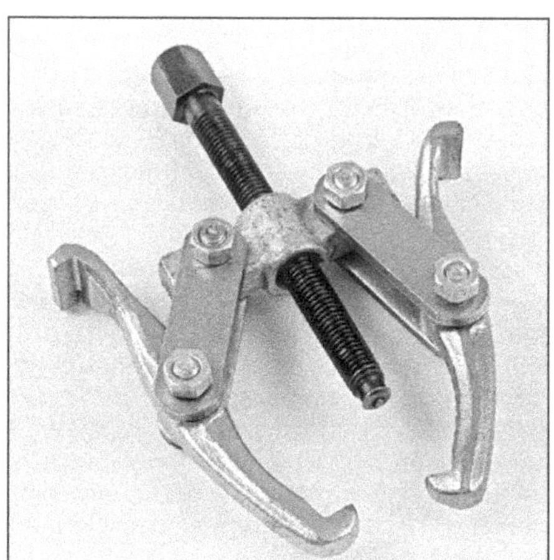
General purpose puller

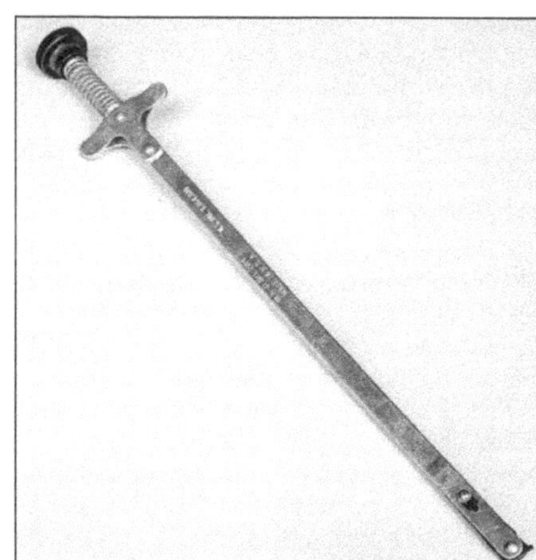
Hydraulic lifter removal tool

Repair/overhaul and *Special.*

The newcomer to practical mechanics should start off with the *maintenance and minor repair* tool kit, which is adequate for the simpler jobs performed on a vehicle. Then, as confidence and experience grow, the owner can tackle more difficult tasks, buying additional tools as they are needed. Eventually the basic kit will be expanded into the *repair and overhaul* tool set. Over a period of time, the experienced do-it-yourselfer will assemble a tool set complete enough for most repair and overhaul procedures and will add tools from the special category when it is felt that the expense is justified by the frequency of use.

Maintenance and minor repair tool kit

The tools in this list should be considered the minimum required for performance of routine maintenance, servicing and minor repair work. We recommend the purchase of combination wrenches (box-end and open-

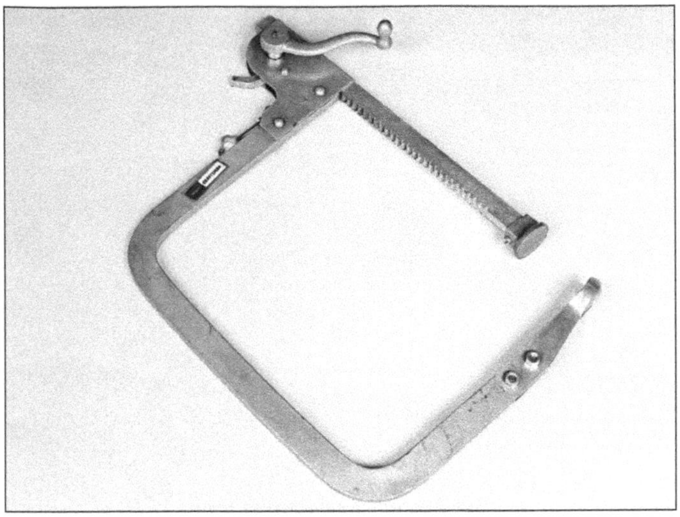

Valve spring compressor

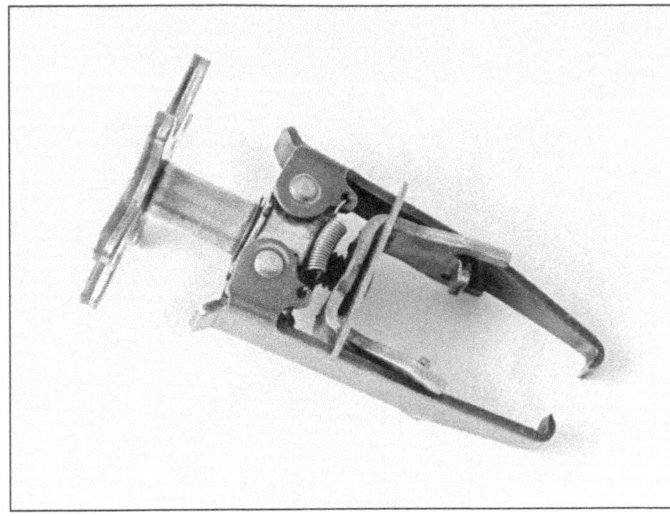

Valve spring compressor

Ridge reamer

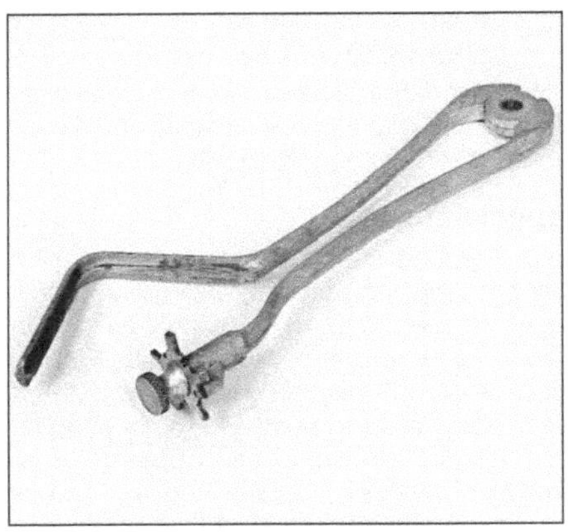

Piston ring groove cleaning tool

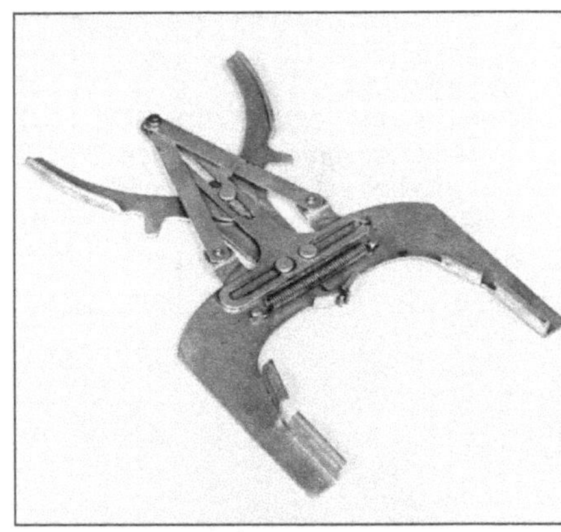

Ring removal/installation tool

end combined in one wrench). While more expensive than open end wrenches, they offer the advantages of both types of wrench.

Combination wrench set (1/4-inch to 1 inch or 6 mm to 19 mm)
Adjustable wrench, 8 inch
Spark plug wrench with rubber insert
Spark plug gap adjusting tool
Feeler gauge set
Brake bleeder wrench
Standard screwdriver (5/16-inch x 6 inch)
Phillips screwdriver (No. 2 x 6 inch)
Combination pliers - 6 inch
Hacksaw and assortment of blades
Tire pressure gauge
Grease gun
Oil can
Fine emery cloth
Wire brush
Battery post and cable cleaning tool
Oil filter wrench
Funnel (medium size)
Safety goggles
Jackstands (2)
Drain pan

Note: *If basic tune-ups are going to be part of routine maintenance, it will be necessary to purchase a good quality stroboscopic timing light and combination tachometer/dwell meter. Although they are included in the list of special tools, it is mentioned here because they are absolutely necessary for tuning most vehicles properly.*

Repair and overhaul tool set

These tools are essential for anyone who plans to perform major repairs and are in addition to those in the maintenance and minor repair tool kit. Included is a comprehensive set of sockets which, though expensive, are invaluable because of their versatility, especially when various extensions and drives are available. We recommend the 1/2-inch drive over the 3/8-inch drive. Although the larger drive is bulky and more expensive, it has the capacity of accepting a very wide range of large sockets. Ideally, however, the mechanic should have a 3/8-inch drive set and a 1/2-inch drive set.

Socket set(s)
Reversible ratchet
Extension - 10 inch
Universal joint
Torque wrench (same size drive as sockets)
Ball peen hammer - 8 ounce
Soft-face hammer (plastic/rubber)

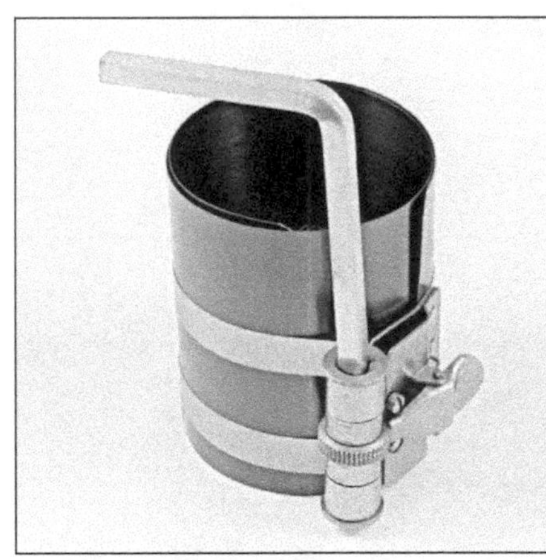

Ring compressor

Standard screwdriver (1/4-inch x 6 inch)
Standard screwdriver (stubby - 5/16-inch)
Phillips screwdriver (No. 3 x 8 inch)
Phillips screwdriver (stubby - No. 2)
Pliers - vise grip
Pliers - lineman's
Pliers - needle nose
Pliers - snap-ring (internal and external)
Cold chisel - 1/2-inch

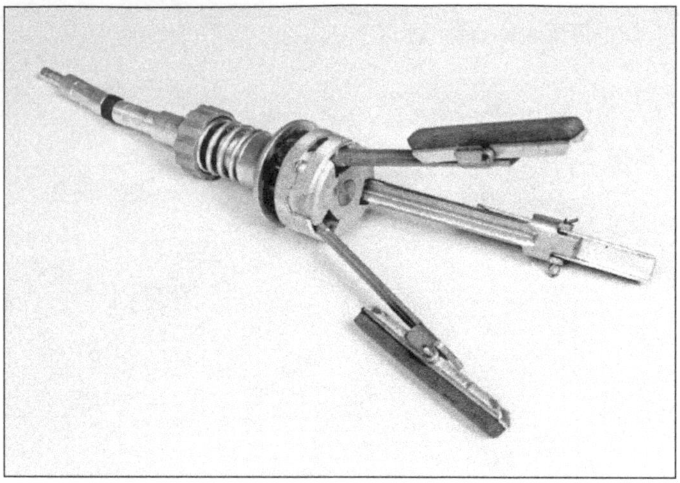

Cylinder hone

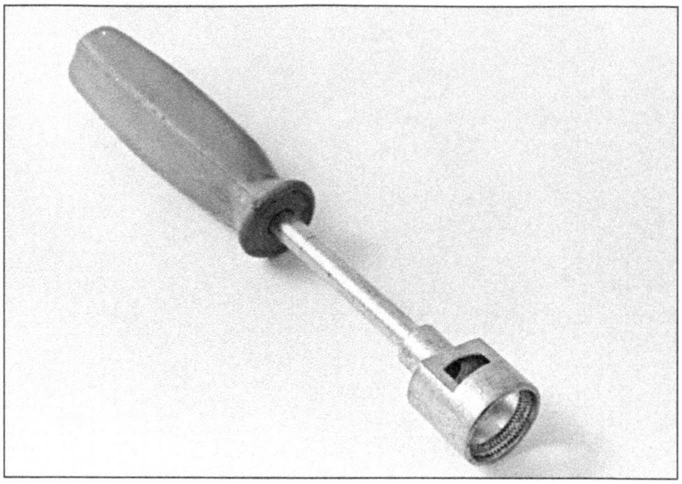

Brake hold-down spring tool

Scribe
Scraper (made from flattened copper tubing)
Centerpunch
Pin punches (1/16, 1/8, 3/16-inch)
Steel rule/straightedge - 12 inch
Allen wrench set (1/8 to 3/8-inch or 4 mm to 10 mm)
A selection of files
Wire brush (large)
Jackstands (second set)
Jack (scissor or hydraulic type)

Note: *Another tool which is often useful is an electric drill with a chuck capacity of 3/8-inch and a set of good quality drill bits.*

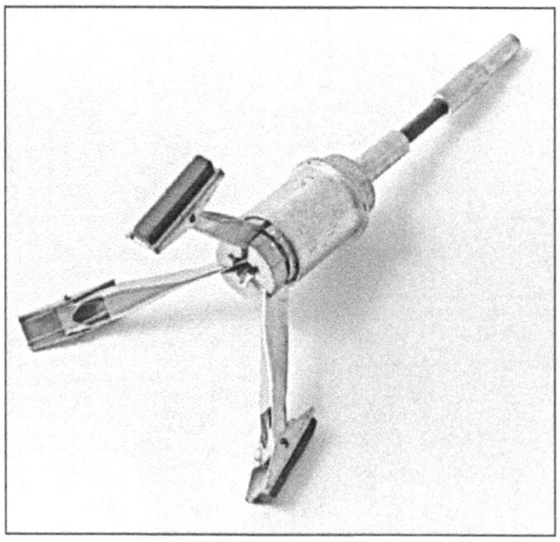

Brake cylinder hone

Special tools

The tools in this list include those which are not used regularly, are expensive to buy, or which need to be used in accordance with their manufacturer's instructions. Unless these tools will be used frequently, it is not very economical to purchase many of them. A consideration would be to split the cost and use between yourself and a friend or friends. In addition, most of these tools can be obtained from a tool rental shop on a temporary basis.

This list primarily contains only those tools and instruments widely available to the public, and not those special tools produced by the vehicle manufacturer for distribution to dealer service departments. Occasionally, references to the manufacturer's special tools are included in the text of this manual. Generally, an alternative method of doing the job without the special tool is offered. However, sometimes there is no alternative to their use. Where this is the case, and the tool cannot be purchased or borrowed, the work should be turned over to the dealer service department or an automotive repair shop.

Valve spring compressor
Piston ring groove cleaning tool
Piston ring compressor
Piston ring installation tool
Cylinder compression gauge
Cylinder ridge reamer
Cylinder surfacing hone
Cylinder bore gauge
Micrometers and/or dial calipers
Hydraulic lifter removal tool
Balljoint separator
Universal-type puller
Impact screwdriver
Dial indicator set
Stroboscopic timing light (inductive pick-up)
Hand operated vacuum/pressure pump
Tachometer/dwell meter
Universal electrical multimeter
Cable hoist
Brake spring removal and installation tools
Floor jack

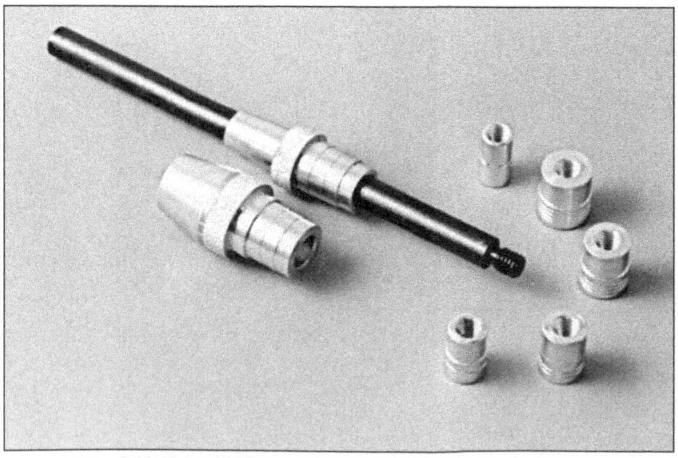

Clutch plate alignment tool

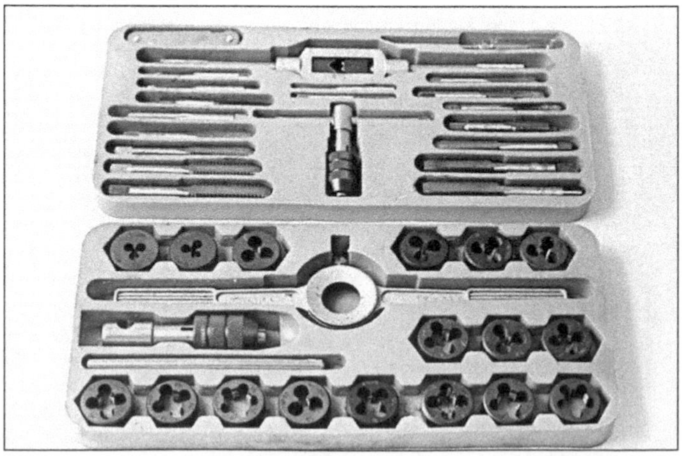

Tap and die set

Buying tools

For the do-it-yourselfer who is just starting to get involved in vehicle maintenance and repair, there are a number of options available when purchasing tools. If maintenance and minor repair is the extent of the work to be done, the purchase of individual tools is satisfactory. If, on the other hand, extensive work is planned, it would be a good idea to purchase a modest tool set from one of the large retail chain stores. A set can usually be bought at a substantial savings over the individual tool prices, and they often come with a tool box. As additional tools are needed, add-on sets, individual tools and a larger tool box can be purchased to expand the tool selection. Building a tool set gradually allows the cost of the tools to be spread over a longer period of time and gives the mechanic the freedom to choose only those tools that will actually be used.

Tool stores will often be the only source of some of the special tools that are needed, but regardless of where tools are bought, try to avoid cheap ones, especially when buying screwdrivers and sockets, because they won't last very long. The expense involved in replacing cheap tools will eventually be greater than the initial cost of quality tools.

Care and maintenance of tools

Good tools are expensive, so it makes sense to treat them with respect. Keep them clean and in usable condition and store them properly when not in use. Always wipe off any dirt, grease or metal chips before putting them away. Never leave tools lying around in the work area. Upon completion of a job, always check closely under the hood for tools that may have been left there so they won't get lost during a test drive.

Some tools, such as screwdrivers, pliers, wrenches and sockets, can be hung on a panel mounted on the garage or workshop wall, while others should be kept in a tool box or tray. Measuring instruments, gauges, meters, etc. must be carefully stored where they cannot be damaged by weather or impact from other tools.

When tools are used with care and stored properly, they will last a very long time. Even with the best of care, though, tools will wear out if used frequently. When a tool is damaged or worn out, replace it. Subsequent jobs will be safer and more enjoyable if you do.

How to repair damaged threads

Sometimes, the internal threads of a nut or bolt hole can become stripped, usually from overtightening. Stripping threads is an all-too-common occurrence, especially when working with aluminum parts, because aluminum is so soft that it easily strips out.

Usually, external or internal threads are only partially stripped. After they've been cleaned up with a tap or die, they'll still work. Sometimes, however, threads are badly damaged. When this happens, you've got three choices:

1) *Drill and tap the hole to the next suitable oversize and install a larger diameter bolt, screw or stud.*
2) *Drill and tap the hole to accept a threaded plug, then drill and tap the plug to the original screw size. You can also buy a plug already threaded to the original size. Then you simply drill a hole to the specified size, then run the threaded plug into the hole with a bolt and jam nut. Once the plug is fully seated, remove the jam nut and bolt.*
3) *The third method uses a patented thread repair kit like Heli-Coil or Slimsert. These easy-to-use kits are designed to repair damaged threads in straight-through holes and blind holes. Both are available as kits which can handle a variety of sizes and thread patterns. Drill the hole, then tap it with the special included tap. Install the Heli-Coil and the hole is back to its original diameter and thread pitch.*

Regardless of which method you use, be sure to proceed calmly and carefully. A little impatience or carelessness during one of these relatively simple procedures can ruin your whole day's work and cost you a bundle if you wreck an expensive part.

Working facilities

Not to be overlooked when discussing tools is the workshop. If anything more than routine maintenance is to be carried out, some sort of suitable work area is essential.

It is understood, and appreciated, that many home mechanics do not have a good workshop or garage available, and end up removing an engine or doing major repairs outside. It is recommended, however, that the overhaul or repair be completed under the cover of a roof.

A clean, flat workbench or table of comfortable working height is an absolute necessity. The workbench should be equipped with a vise that has a jaw opening of at least four inches.

As mentioned previously, some clean, dry storage space is also required for tools, as well as the lubricants, fluids, cleaning solvents, etc. which soon become necessary.

Sometimes waste oil and fluids, drained from the engine or cooling system during normal maintenance or repairs, present a disposal problem. To avoid pouring them on the ground or into a sewage system, pour the used fluids into large containers, seal them with caps and take them to an authorized disposal site or recycling center. Plastic jugs, such as old antifreeze containers, are ideal for this purpose.

Always keep a supply of old newspapers and clean rags available. Old towels are excellent for mopping up spills. Many mechanics use rolls of paper towels for most work because they are readily available and disposable. To help keep the area under the vehicle clean, a large cardboard box can be cut open and flattened to protect the garage or shop floor.

Whenever working over a painted surface, such as when leaning over a fender to service something under the hood, always cover it with an old blanket or bedspread to protect the finish. Vinyl covered pads, made especially for this purpose, are available at auto parts stores.

Booster battery (jump) starting

Observe the following precautions when using a booster battery to start a vehicle:

a) *Before connecting the booster battery, make sure the ignition switch is in the Off position.*
b) *Ensure that all electrical equipment (lights, heater, wipers etc.) are switched off.*
c) *Make sure that the booster battery is the same voltage as the discharged battery in the vehicle.*
d) *If the battery is being jump started from the battery in another vehicle, the two vehicles MUST NOT TOUCH each other.*
e) *Make sure the transaxle is in Neutral (manual transaxle) or Park (automatic transaxle).*
f) *Wear eye protection when jump starting a vehicle.*

Connect one jumper lead between the positive (+) terminals of the two batteries. Connect the other jumper lead first to the negative (-) terminal of the booster battery, then to a good engine ground on the vehicle to be started **(see illustration)**. Attach the lead at least 18 inches from the battery, if possible. Make sure that the jumper leads will not contact the fan, drivebelt of other moving parts of the engine.

Start the engine using the booster battery and allow the engine idle speed to stabilize. Disconnect the jumper leads in the reverse order of connection.

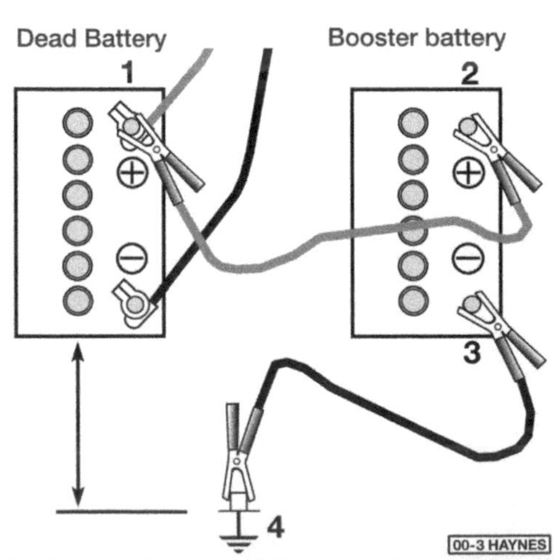

Make the booster battery cable connections in the numerical order shown (note that the negative cable of the booster battery is NOT attached to the negative terminal of the dead battery)

Jacking and towing

Jacking

Warning: *The jack supplied with the vehicle should only be used for raising the vehicle when changing a tire or placing jackstands under the frame. Never work under the vehicle or start the engine while this jack is being used as the only means of support.*

1 The vehicle should be on level ground with the hazard flashers on, the wheels blocked, the parking brake applied and the transmission in Park (automatic) or Reverse (manual). If a tire is being changed, remove the wheel cover (if equipped) and loosen the lug nuts one-half turn, leaving them in place until the wheel is raised off the ground.

2 If using a bumper-type jack, hook the jack head under the bumper, near the wheel or place the hook through the slot in the bumper; make sure the jack post is vertical and secure. If using another type of jack, place the jack under the vehicle in the indicated position **(see illustration)**. Operate the jack with a slow, smooth motion until the wheel is raised off the ground. Remove the lug nuts, pull off the wheel, install the spare and thread the lug nuts back on with the beveled sides facing in. Tighten them snugly, but wait until the vehicle is lowered to tighten them completely. Note that some spare tires are designed for temporary use only - don't exceed the recommended speed, mileage or other restrictions accompanying the spare.

3 Lower the vehicle, remove the jack and tighten the nuts (if loosened or removed) in a criss-cross pattern.

Towing

4 As a general rule, vehicles can be towed with all four wheels on the ground for 15 miles or less at speeds below 35 mph. If the vehicle must be towed further, the driveshaft must be removed (see Chapter 8).

5 Equipment specifically designed for towing should be used and should be attached to the main structural members of the vehicle, not the bumper or brackets.

6 Safety is a major consideration when towing and all applicable state and local laws must be obeyed. A safety chain must be used for all towing (in addition to the tow bar).

7 While towing, the parking brake must be released and the transmission must be in Neutral. The steering must be unlocked (ignition switch in the Off position). Remember that power steering and power brakes will not work with the engine off.

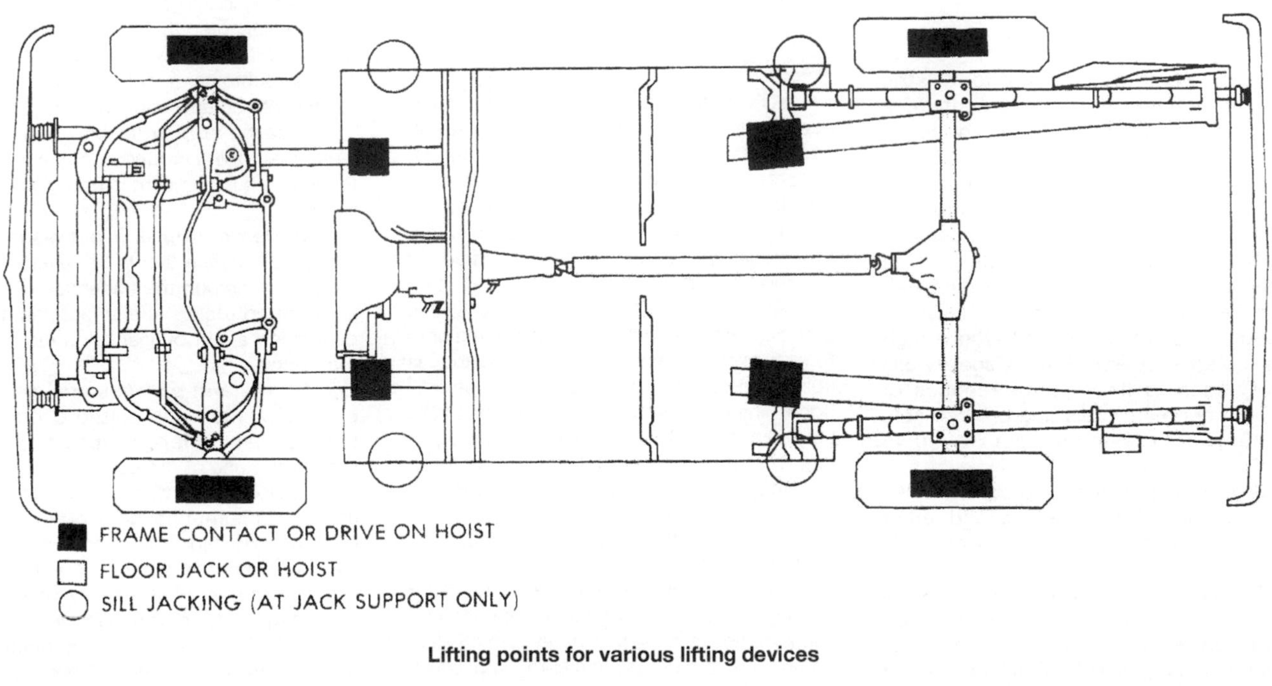

Lifting points for various lifting devices

Automotive chemicals and lubricants

A number of automotive chemicals and lubricants are available for use during vehicle maintenance and repair. They include a wide variety of products ranging from cleaning solvents and degreasers to lubricants and protective sprays for rubber, plastic and vinyl.

Cleaners

Carburetor cleaner and choke cleaner is a strong solvent for gum, varnish and carbon. Most carburetor cleaners leave a dry-type lubricant film which will not harden or gum up. Because of this film it is not recommended for use on electrical components.

Brake system cleaner is used to remove grease and brake fluid from the brake system, where clean surfaces are absolutely necessary. It leaves no residue and often eliminates brake squeal caused by contaminants.

Electrical cleaner removes oxidation, corrosion and carbon deposits from electrical contacts, restoring full current flow. It can also be used to clean spark plugs, carburetor jets, voltage regulators and other parts where an oil-free surface is desired.

Demoisturants remove water and moisture from electrical components such as alternators, voltage regulators, electrical connectors and fuse blocks. They are non-conductive, non-corrosive and non-flammable.

Degreasers are heavy-duty solvents used to remove grease from the outside of the engine and from chassis components. They can be sprayed or brushed on and, depending on the type, are rinsed off either with water or solvent.

Lubricants

Motor oil is the lubricant formulated for use in engines. It normally contains a wide variety of additives to prevent corrosion and reduce foaming and wear. Motor oil comes in various weights (viscosity ratings) from 0 to 50. The recommended weight of the oil depends on the season, temperature and the demands on the engine. Light oil is used in cold climates and under light load conditions. Heavy oil is used in hot climates and where high loads are encountered. Multi-viscosity oils are designed to have characteristics of both light and heavy oils and are available in a number of weights from 5W-20 to 20W-50.

Gear oil is designed to be used in differentials, manual transmissions and other areas where high-temperature lubrication is required.

Chassis and wheel bearing grease is a heavy grease used where increased loads and friction are encountered, such as for wheel bearings, balljoints, tie-rod ends and universal joints.

High-temperature wheel bearing grease is designed to withstand the extreme temperatures encountered by wheel bearings in disc brake equipped vehicles. It usually contains molybdenum disulfide (moly), which is a dry-type lubricant.

White grease is a heavy grease for metal-to-metal applications where water is a problem. White grease stays soft under both low and high temperatures (usually from -100 to +190-degrees F), and will not wash off or dilute in the presence of water.

Assembly lube is a special extreme pressure lubricant, usually containing moly, used to lubricate high-load parts (such as main and rod bearings and cam lobes) for initial start-up of a new engine. The assembly lube lubricates the parts without being squeezed out or washed away until the engine oiling system begins to function.

Silicone lubricants are used to protect rubber, plastic, vinyl and nylon parts.

Graphite lubricants are used where oils cannot be used due to contamination problems, such as in locks. The dry graphite will lubricate metal parts while remaining uncontaminated by dirt, water, oil or acids. It is electrically conductive and will not foul electrical contacts in locks such as the ignition switch.

Moly penetrants loosen and lubricate frozen, rusted and corroded fasteners and prevent future rusting or freezing.

Heat-sink grease is a special electrically non-conductive grease that is used for mounting electronic ignition modules where it is essential that heat is transferred away from the module.

Sealants

RTV sealant is one of the most widely used gasket compounds. Made from silicone, RTV is air curing, it seals, bonds, waterproofs, fills surface irregularities, remains flexible, doesn't shrink, is relatively easy to remove, and is used as a supplementary sealer with almost all low and medium temperature gaskets.

Anaerobic sealant is much like RTV in that it can be used either to seal gaskets or to form gaskets by itself. It remains flexible, is solvent resistant and fills surface imperfections. The difference between an anaerobic sealant and an RTV-type sealant is in the curing. RTV cures when exposed to air, while an anaerobic sealant cures only in the absence of air. This means that an anaerobic sealant cures only after the assembly of parts, sealing them together.

Thread and pipe sealant is used for sealing hydraulic and pneumatic fittings and vacuum lines. It is usually made from a Teflon compound, and comes in a spray, a paint-on liquid and as a wrap-around tape.

Chemicals

Anti-seize compound prevents seizing, galling, cold welding, rust and corrosion in fasteners. High-temperature ant-seize, usually made with copper and graphite lubricants, is used for exhaust system and exhaust manifold bolts.

Anaerobic locking compounds are used to keep fasteners from vibrating or working loose and cure only after installation, in the absence of air. Medium strength locking compound is used for small nuts, bolts and screws that may be removed later. High-strength locking compound is for large nuts, bolts and studs which aren't removed on a regular basis.

Oil additives range from viscosity index improvers to chemical treatments that claim to reduce internal engine friction. It should be noted that most oil manufacturers caution against using additives with their oils.

Gas additives perform several functions, depending on their chemical makeup. They usually contain solvents that help dissolve gum and varnish that build up on carburetor, fuel injection and intake parts. They also serve to break down carbon deposits that form on the inside surfaces of the combustion chambers. Some additives contain upper cylinder lubricants for valves and piston rings, and others contain chemicals to remove condensation from the gas tank.

Miscellaneous

Brake fluid is specially formulated hydraulic fluid that can withstand the heat and pressure encountered in brake systems. Care must be taken so this fluid does not come in contact with painted surfaces or plastics. An opened container should always be resealed to prevent contamination by water or dirt.

Weatherstrip adhesive is used to bond weatherstripping around doors, windows and trunk lids. It is sometimes used to attach trim pieces.

Undercoating is a petroleum-based, tar-like substance that is designed to protect metal surfaces on the underside of the vehicle from corrosion. It also acts as a sound-deadening agent by insulating the bottom of the vehicle.

Waxes and polishes are used to help protect painted and plated surfaces from the weather. Different types of paint may require the use of different types of wax and polish. Some polishes utilize a chemical or abrasive cleaner to help remove the top layer of oxidized (dull) paint on older vehicles. In recent years many non-wax polishes that contain a wide variety of chemicals such as polymers and silicones have been introduced. These non-wax polishes are usually easier to apply and last longer than conventional waxes and polishes.

Conversion factors

Length (distance)						
Inches (in)	X	25.4	= Millimetres (mm)	X	0.0394	= Inches (in)
Feet (ft)	X	0.305	= Metres (m)	X	3.281	= Feet (ft)
Miles	X	1.609	= Kilometres (km)	X	0.621	= Miles
Volume (capacity)						
Cubic inches (cu in; in^3)	X	16.387	= Cubic centimetres (cc; cm^3)	X	0.061	= Cubic inches (cu in; in^3)
Imperial pints (Imp pt)	X	0.568	= Litres (l)	X	1.76	= Imperial pints (Imp pt)
Imperial quarts (Imp qt)	X	1.137	= Litres (l)	X	0.88	= Imperial quarts (Imp qt)
Imperial quarts (Imp qt)	X	1.201	= US quarts (US qt)	X	0.833	= Imperial quarts (Imp qt)
US quarts (US qt)	X	0.946	= Litres (l)	X	1.057	= US quarts (US qt)
Imperial gallons (Imp gal)	X	4.546	= Litres (l)	X	0.22	= Imperial gallons (Imp gal)
Imperial gallons (Imp gal)	X	1.201	= US gallons (US gal)	X	0.833	= Imperial gallons (Imp gal)
US gallons (US gal)	X	3.785	= Litres (l)	X	0.264	= US gallons (US gal)
Mass (weight)						
Ounces (oz)	X	28.35	= Grams (g)	X	0.035	= Ounces (oz)
Pounds (lb)	X	0.454	= Kilograms (kg)	X	2.205	= Pounds (lb)
Force						
Ounces-force (ozf; oz)	X	0.278	= Newtons (N)	X	3.6	= Ounces-force (ozf; oz)
Pounds-force (lbf; lb)	X	4.448	= Newtons (N)	X	0.225	= Pounds-force (lbf; lb)
Newtons (N)	X	0.1	= Kilograms-force (kgf; kg)	X	9.81	= Newtons (N)
Pressure						
Pounds-force per square inch (psi; lbf/in^2; lb/in^2)	X	0.070	= Kilograms-force per square centimetre (kgf/cm^2; kg/cm^2)	X	14.223	= Pounds-force per square inch (psi; lbf/in^2; lb/in^2)
Pounds-force per square inch (psi; lbf/in^2; lb/in^2)	X	0.068	= Atmospheres (atm)	X	14.696	= Pounds-force per square inch (psi; lbf/in^2; lb/in^2)
Pounds-force per square inch (psi; lbf/in^2; lb/in^2)	X	0.069	= Bars	X	14.5	= Pounds-force per square inch (psi; lbf/in^2; lb/in^2)
Pounds-force per square inch (psi; lbf/in^2; lb/in^2)	X	6.895	= Kilopascals (kPa)	X	0.145	= Pounds-force per square inch (psi; lbf/in^2; lb/in^2)
Kilopascals (kPa)	X	0.01	= Kilograms-force per square centimetre (kgf/cm^2; kg/cm^2)	X	98.1	= Kilopascals (kPa)
Torque (moment of force)						
Pounds-force inches (lbf in; lb in)	X	1.152	= Kilograms-force centimetre (kgf cm; kg cm)	X	0.868	= Pounds-force inches (lbf in; lb in)
Pounds-force inches (lbf in; lb in)	X	0.113	= Newton metres (Nm)	X	8.85	= Pounds-force inches (lbf in; lb in)
Pounds-force inches (lbf in; lb in)	X	0.083	= Pounds-force feet (lbf ft; lb ft)	X	12	= Pounds-force inches (lbf in; lb in)
Pounds-force feet (lbf ft; lb ft)	X	0.138	= Kilograms-force metres (kgf m; kg m)	X	7.233	= Pounds-force feet (lbf ft; lb ft)
Pounds-force feet (lbf ft; lb ft)	X	1.356	= Newton metres (Nm)	X	0.738	= Pounds-force feet (lbf ft; lb ft)
Newton metres (Nm)	X	0.102	= Kilograms-force metres (kgf m; kg m)	X	9.804	= Newton metres (Nm)
Vacuum						
Inches mercury (in. Hg)	X	3.377	= Kilopascals (kPa)	X	0.2961	= Inches mercury
Inches mercury (in. Hg)	X	25.4	= Millimeters mercury (mm Hg)	X	0.0394	= Inches mercury
Power						
Horsepower (hp)	X	745.7	= Watts (W)	X	0.0013	= Horsepower (hp)
Velocity (speed)						
Miles per hour (miles/hr; mph)	X	1.609	= Kilometres per hour (km/hr; kph)	X	0.621	= Miles per hour (miles/hr; mph)
Fuel consumption*						
Miles per gallon, Imperial (mpg)	X	0.354	= Kilometres per litre (km/l)	X	2.825	= Miles per gallon, Imperial (mpg)
Miles per gallon, US (mpg)	X	0.425	= Kilometres per litre (km/l)	X	2.352	= Miles per gallon, US (mpg)

Temperature

Degrees Fahrenheit = (°C x 1.8) + 32

Degrees Celsius (Degrees Centigrade; °C) = (°F - 32) x 0.56

**It is common practice to convert from miles per gallon (mpg) to litres/100 kilometres (l/100km), where mpg (Imperial) x l/100 km = 282 and mpg (US) x l/100 km = 235*

Safety first!

Regardless of how enthusiastic you may be about getting on with the job at hand, take the time to ensure that your safety is not jeopardized. A moment's lack of attention can result in an accident, as can failure to observe certain simple safety precautions. The possibility of an accident will always exist, and the following points should not be considered a comprehensive list of all dangers. Rather, they are intended to make you aware of the risks and to encourage a safety conscious approach to all work you carry out on your vehicle.

Essential DOs and DON'Ts

DON'T rely on a jack when working under the vehicle. Always use approved jackstands to support the weight of the vehicle and place them under the recommended lift or support points.

DON'T attempt to loosen extremely tight fasteners (i.e. wheel lug nuts) while the vehicle is on a jack - it may fall.

DON'T start the engine without first making sure that the transmission is in Neutral (or Park where applicable) and the parking brake is set.

DON'T remove the radiator cap from a hot cooling system - let it cool or cover it with a cloth and release the pressure gradually.

DON'T attempt to drain the engine oil until you are sure it has cooled to the point that it will not burn you.

DON'T touch any part of the engine or exhaust system until it has cooled sufficiently to avoid burns.

DON'T siphon toxic liquids such as gasoline, antifreeze and brake fluid by mouth, or allow them to remain on your skin.

DON'T inhale brake lining dust - it is potentially hazardous (see *Asbestos* below).

DON'T allow spilled oil or grease to remain on the floor - wipe it up before someone slips on it.

DON'T use loose fitting wrenches or other tools which may slip and cause injury.

DON'T push on wrenches when loosening or tightening nuts or bolts. Always try to pull the wrench toward you. If the situation calls for pushing the wrench away, push with an open hand to avoid scraped knuckles if the wrench should slip.

DON'T attempt to lift a heavy component alone - get someone to help you.

DON'T rush or take unsafe shortcuts to finish a job.

DON'T allow children or animals in or around the vehicle while you are working on it.

DO wear eye protection when using power tools such as a drill, sander, bench grinder, etc. and when working under a vehicle.

DO keep loose clothing and long hair well out of the way of moving parts.

DO make sure that any hoist used has a safe working load rating adequate for the job.

DO get someone to check on you periodically when working alone on a vehicle.

DO carry out work in a logical sequence and make sure that everything is correctly assembled and tightened.

DO keep chemicals and fluids tightly capped and out of the reach of children and pets.

DO remember that your vehicle's safety affects that of yourself and others. If in doubt on any point, get professional advice.

Asbestos

Certain friction, insulating, sealing, and other products - such as brake linings, brake bands, clutch linings, torque converters, gaskets, etc. - may contain asbestos. Extreme care must be taken to avoid inhalation of dust from such products, since it is hazardous to health. If in doubt, assume that they do contain asbestos.

Fire

Remember at all times that gasoline is highly flammable. Never smoke or have any kind of open flame around when working on a vehicle. But the risk does not end there. A spark caused by an electrical short circuit, by two metal surfaces contacting each other, or even by static electricity built up in your body under certain conditions, can ignite gasoline vapors, which in a confined space are highly explosive. Do not, under any circumstances, use gasoline for cleaning parts. Use an approved safety solvent.

Always disconnect the battery ground (-) cable at the battery before working on any part of the fuel system or electrical system. Never risk spilling fuel on a hot engine or exhaust component. It is strongly recommended that a fire extinguisher suitable for use on fuel and electrical fires be kept handy in the garage or workshop at all times. Never try to extinguish a fuel or electrical fire with water.

Fumes

Certain fumes are highly toxic and can quickly cause unconsciousness and even death if inhaled to any extent. Gasoline vapor falls into this category, as do the vapors from some cleaning solvents. Any draining or pouring of such volatile fluids should be done in a well ventilated area.

When using cleaning fluids and solvents, read the instructions on the container carefully. Never use materials from unmarked containers.

Never run the engine in an enclosed space, such as a garage. Exhaust fumes contain carbon monoxide, which is extremely poisonous. If you need to run the engine, always do so in the open air, or at least have the rear of the vehicle outside the work area.

If you are fortunate enough to have the use of an inspection pit, never drain or pour gasoline and never run the engine while the vehicle is over the pit. The fumes, being heavier than air, will concentrate in the pit with possibly lethal results.

The battery

Never create a spark or allow a bare light bulb near a battery. They normally give off a certain amount of hydrogen gas, which is highly explosive.

Always disconnect the battery ground (-) cable at the battery before working on the fuel or electrical systems.

If possible, loosen the filler caps or cover when charging the battery from an external source (this does not apply to sealed or maintenance-free batteries). Do not charge at an excessive rate or the battery may burst.

Take care when adding water to a non maintenance-free battery and when carrying a battery. The electrolyte, even when diluted, is very corrosive and should not be allowed to contact clothing or skin.

Always wear eye protection when cleaning the battery to prevent the caustic deposits from entering your eyes.

Household current

When using an electric power tool, inspection light, etc., which operates on household current, always make sure that the tool is correctly connected to its plug and that, where necessary, it is properly grounded. Do not use such items in damp conditions and, again, do not create a spark or apply excessive heat in the vicinity of fuel or fuel vapor.

Secondary ignition system voltage

A severe electric shock can result from touching certain parts of the ignition system (such as the spark plug wires) when the engine is running or being cranked, particularly if components are damp or the insulation is defective. In the case of an electronic ignition system, the secondary system voltage is much higher and could prove fatal.

Troubleshooting

Contents

This Section provides an easy reference guide to the more common problems that may occur during the operation of your vehicle. Various symptoms and their probable causes are grouped under headings denoting components or systems, such as Engine, Cooling system, etc. They also refer to the Chapter and/or Section that deals with the problem.

Remember that successful troubleshooting isn't a mysterious 'black art' practiced only by professional mechanics, it's simply the result of knowledge combined with an intelligent, systematic approach to a problem. Always use a process of elimination starting, with the simplest solution and working through to the most complex - and never overlook the obvious. Anyone can run the gas tank dry or leave the lights on overnight, so don't assume that you're exempt from such oversights.

Finally, always establish a clear idea why a problem has occurred and take steps to ensure that it doesn't happen again. If the electrical system fails because of a poor connection, check all other connections in the system to make sure they don't fail as well. If a particular fuse continues to blow, find out why - don't just go on replacing fuses. Remember, failure of a small component can often be indicative of potential failure or incorrect functioning of a more important component or system.

Engine and performance

1 Engine will not rotate when attempting to start

1 Battery terminal connections loose or corroded. Check the cable terminals at the battery; tighten cable clamp and/or clean off corrosion as necessary (see Chapter 1).
2 Battery discharged or faulty. If the cable ends are clean and tight on the battery posts, turn the key to the On position and switch on the headlights or windshield wipers. If they won't run, the battery is discharged.
3 Automatic transmission not engaged in park (P) or Neutral (N).
4 Broken, loose or disconnected wires in the starting circuit. Inspect all wires and connectors at the battery, starter solenoid and ignition switch (in steering column).
5 Starter motor pinion jammed in flywheel ring gear. If manual transmission, place transmission in gear and rock the vehicle to manually turn the engine. Remove starter (Chapter 5) and inspect pinion and flywheel (Chapter 2) at earliest convenience.
6 Starter solenoid faulty (Chapter 5).
7 Starter relay faulty (Chapter 5)
8 Starter motor faulty (Chapter 5).
9 Neutral start switch faulty (automatic transmission models) (Chapter 7B).
10 Ignition switch faulty (Chapter 12).
11 Engine seized. Try to turn the crankshaft with a large socket and breaker bar on the pulley bolt.
12 Although not very likely, sometimes the connections to the ammeter, behind the instrument cluster, become loose or disconnected, causing a discharged battery or no-start condition. Occasionally, a bad connection inside the ammeter will also cause this problem. Usually, this will cause most of the electrical system to also not operate. The vehicle will act like it has a dead battery. Bypass the ammeter with a heavy-gauge jumper wire, then try to start the vehicle again.

2 Engine rotates but will not start

1 Fuel tank empty.
2 Battery discharged (engine rotates slowly). Check the operation of electrical components as described in previous Section.
3 Battery terminal connections loose or corroded. See previous Section.
4 Choke not operating properly (Chapter 1).
5 Water inside the distributor cap (particularly in foul weather). Remove the cap and dry it (see Chapter 1).
6 Fouled spark plugs or bad spark-plug wires (Chapter 1).
7 Faulty distributor components. Check the cap and rotor (Chapter 1).
8 Fuel not reaching the carburetor. With the engine off, remove the air cleaner top plate, open the choke plate by hand and have an assistant press the accelerator while you look down the carburetor bore - if you can see small streams of gasoline squirting into the carburetor, fuel is reaching the carburetor; otherwise, check for a clogged fuel filter or lines and a defective fuel pump. Also make sure the tank vent lines aren't clogged (Chapter 4).
9 Faulty distributor pick-up coil or ignition module (1973 and later models) (Chapter 5).
10 Burned or otherwise damaged ignition points (1971 and 1972 models) (Chapter 1).
11 Low cylinder compression. Check as described in Chapter 2.
12 Severe vacuum leak. Check for a loose carburetor or intake manifold or a disconnected PCV or power brake booster vacuum hose.
13 Water in fuel. Drain tank and fill with new fuel.
14 Defective ignition coil (Chapter 5).
15 Dirty or clogged carburetor jets. Carburetor out of adjustment. Check the float level (Chapter 4).
16 Broken, loose or disconnected wires in the starting circuit (see previous Section).
17 Loose distributor (changing ignition timing). Turn the distributor body as necessary to start the engine, then adjust the ignition timing as soon as possible (Chapter 1).
18 Broken, loose or disconnected wires at the ignition coil or faulty coil (Chapter 5).
19 Timing chain failure or wear affecting valve timing (Chapter 2).

3 Starter motor operates without turning engine

1 Starter pinion sticking. Remove the starter (Chapter 5) and inspect.
2 Starter pinion or flywheel/driveplate teeth worn or broken. Remove the inspection cover and inspect.

4 Engine hard to start when cold

1 Battery discharged or low. Check as described in Chapter 1.
2 Fuel not reaching the carburetor. Check the fuel filter, lines and fuel pump (Chapters 1 and 4).
3 Choke inoperative (Chapters 1 and 4).
4 Defective spark plugs (Chapter 1).

5 Engine hard to start when hot

1 Air filter dirty (Chapter 1).
2 Fuel not reaching the carburetor (see Section 4). Check for a vapor lock situation, brought about by clogged fuel tank vent lines.
3 Bad engine ground connection.
4 Choke sticking (Chapter 1).
5 Defective pick-up coil in the distributor (Chapter 5).
6 Float level too high (Chapter 4).

6 Starter motor noisy or engages roughly

1 Pinion or flywheel/driveplate teeth worn or broken. Remove the inspection cover on the left side of the engine and inspect.
2 Starter motor mounting bolts loose or missing.

7 Engine starts but stops immediately

1 Loose or damaged wire harness connections at distributor, coil or alternator.
2 If the engine stalls immediately after the ignition key is turned away from Start, check the ignition ballast resistor (Chapter 5).
3 Intake manifold vacuum leaks. Make sure all mounting bolts/nuts are tight and all vacuum hoses connected to the manifold are attached properly and in good condition.
4 Insufficient fuel flow (see Chapter 4).

8 Engine 'lopes' while idling or idles erratically

1 Vacuum leaks. Check the mounting bolts at the intake manifold and carburetor (see Chapter 1) for tightness. Make sure that all vacuum hoses are connected and in good condition. Use a length of fuel hose held against your ear to listen for vacuum leaks while the engine is running. A hissing sound will be heard. A soapy water solution will also detect leaks - spray it around suspected vacuum leak areas (for example, the intake manifold-to-cylinder head mating point, the base of the carburetor, the base of the EGR valve, etc). If the idle becomes smooth when the solution is sprayed at a particular point, you've found the area of the vacuum leak.
2 Leaking EGR valve or plugged PCV valve (see Chapters 1 and 6).
3 Air filter clogged (Chapter 1).
4 Fuel pump not delivering sufficient fuel (Chapter 4).
5 Leaking head gasket. Perform a cylinder compression check (Chapter 2).
6 Timing chain worn (Chapter 2).
7 Camshaft lobes worn (Chapter 2).
8 Valves burned or otherwise leaking (Chapter 2).
9 Ignition timing out of adjustment (Chapter 1).
10 Defective ignition points or loose distributor shaft, causing dwell to change (1971 and 1972 models) (Chapter 1).
11 Faulty spark plugs, spark plug wires, distributor cap or rotor (Chapter 1).
12 Thermostatic air cleaner not operating properly (Chapter 1).
13 Choke not operating properly (Chapters 1 and 4).
14 Carburetor dirty, clogged or out of adjustment. Check the float level (Chapter 4).
15 Idle speed out of adjustment (Chapter 1).

9 Engine misses at idle speed

1 Spark plugs fouled, faulty or not gapped properly (Chapter 1).
2 Faulty spark plug wires (Chapter 1).
3 Wet or damaged distributor components (Chapter 1).
4 Short circuits in ignition, coil or spark-plug wires.
5 Sticking or faulty EGR valve (see Chapters 1 and 6).
6 Clogged fuel filter and/or foreign matter in fuel. Remove the fuel filter (Chapter 1) and inspect.
7 Vacuum leaks at intake manifold or hose connections. Check as described in Section 8 of *Troubleshooting*.
8 Defective ignition points or loose distributor shaft, causing dwell to change (1971 and 1972 models) (Chapter 1).
9 Incorrect idle speed (Chapter 4) or idle mixture (Chapter 4).
10 Incorrect ignition timing (Chapter 1).
11 Low or uneven cylinder compression. Check as described in Chapter 2.
12 Choke not operating properly (Chapter 1).

10 Excessively high idle speed

1 Vacuum leak. Check as described in Section 8 of *Troubleshooting*.
2 Idle speed incorrectly adjusted (Chapter 1).
3 Sticking throttle linkage (Chapter 4).
4 Choke opened excessively at idle (Chapter 4).

11 Battery will not hold a charge

1 Alternator drivebelt defective or not adjusted properly (Chapter 1).
2 Battery cables loose or corroded (Chapter 1).
3 Alternator not charging properly or voltage regulator faulty (Chapter 5).
4 Loose, broken or faulty wires in the charging circuit
(Chapter 5). Sometimes, the connections to the ammeter, behind the instrument cluster, become loose, causing a discharge condition. Bypass the ammeter with a heavy-gauge jumper wire to check for this.
5 Short circuit causing a continuous drain on the battery.
6 Battery defective internally.

12 Ammeter registers a discharge

1 Fault in alternator, voltage regulator or charging circuit (Chapter 5).
2 Alternator drivebelt defective or not properly adjusted (Chapter 1).
3 Loose, broken or faulty wires in the charging circuit
(Chapter 5). Sometimes, the connections to the ammeter, behind the instrument cluster, become loose, causing a discharge condition. Bypass the ammeter with a heavy-gauge jumper wire to check for this.

13 Ammeter does not function (needle never moves)

1 Fault in the instrument cluster printed circuit or dash wiring (Chapter 12).
2 Short circuit that bypasses the ammeter.
3 Faulty ammeter.

14 Engine misses throughout driving speed range

1 Fuel filter clogged and/or impurities in the fuel system. Check fuel filter (Chapter 1) or clean system (Chapter 4).
2 Fouled, faulty or incorrectly gapped spark plugs (Chapter 1).
3 Incorrect ignition timing (Chapter 1).
4 Cracked distributor cap, disconnected distributor wires or damaged distributor components (Chapter 1).
5 Defective spark plug wires (Chapter 1).
6 Defective ignition points or loose distributor shaft, causing dwell to change (see Chapter 1).
7 Low or uneven cylinder compression pressures. Check as described in Chapter 2.
8 Weak or faulty ignition coil or loose connections at the coil (Chapter 5).
9 Weak or faulty ignition system (Chapter 5).
10 Vacuum leaks (see Section 8 of *Troubleshooting*).
11 Dirty or clogged carburetor (Chapter 4).
12 Leaky EGR valve (Chapter 6).
13 Carburetor out of adjustment (Chapter 4).

15 Hesitation or stumble during acceleration

1 Ignition timing incorrect (Chapter 1).
2 Faulty spark-plug wires, distributor cap or ignition coil (Chapters 1 and 5).
3 Dirty or clogged carburetor (Chapter 4).
4 Low fuel pressure. Check for proper operation of the fuel pump and for restrictions in the fuel filter and lines (Chapter 4).
5 Carburetor out of adjustment (Chapter 4).

16 Engine stalls

1 Idle speed incorrect (Chapter 4).
2 Fuel filter clogged and/or water and impurities in the fuel system (Chapter 1).
3 Choke not operating properly (Chapter 1).
4 Damaged or wet distributor cap and wires.
5 Emissions system components faulty (Chapter 6).
6 Faulty or incorrectly gapped spark plugs (Chapter 1). Also check the spark plug wires (Chapter 1).
7 Vacuum leak at the carburetor, intake manifold or vacuum hoses. Check as described in Section 8 of *Troubleshooting*.

17 Engine lacks power

1 Incorrect ignition timing (Chapter 1).
2 Excessive play in distributor shaft, causing dwell to change (1971 and 1972 models) (Chapter 1). At the same time, check for faulty distributor cap, wires, etc. (Chapter 1).
3 Faulty or incorrectly gapped spark plugs (Chapter 1).

4 Air filter dirty (Chapter 1).
5 Faulty ignition coil (Chapter 5).
6 Incorrectly adjusted throttle cable (not allowing throttle plates to open completely) (Chapter 4).
7 Choke not opening completely (Chapter 1).
8 Brakes binding (Chapters 1 and 10).
9 Automatic transmission fluid level incorrect, causing slippage (Chapter 1).
10 Clutch slipping (Chapter 8).
11 Fuel filter clogged and/or impurities in the fuel system (Chapters 1 and 4).
12 EGR system not functioning properly (Chapters 1 and 6).
13 Use of sub-standard fuel. Fill tank with proper octane fuel.
14 Low or uneven cylinder compression pressures. Check as described in Chapter 2.
15 Vacuum leak at the carburetor or intake manifold (check as described in Section 8 of *Troubleshooting*).
16 Dirty or clogged carburetor jets or malfunctioning choke (Chapters 1 and 4).

18 Engine backfires

1 EGR system not functioning properly (Chapter 6).
2 Ignition timing incorrect (Chapter 1).
3 Thermostatic air cleaner system not operating properly (Chapter 6).
4 Vacuum leak (refer to Section 8 of *Troubleshooting*).
5 Damaged valve springs or sticking or burned valves - a vacuum-gauge check will often reveal this problem (Chapter 2C).
6 Carburetor float level out of adjustment (Chapter 4).

19 Engine surges while holding accelerator steady

1 Vacuum leak (see Section 8 of *Troubleshooting*).
2 Fuel pump not working properly (Chapter 4).

20 Pinging or knocking engine sounds when engine is under load

1 Incorrect grade of fuel. Fill tank with fuel of the proper octane rating.
2 Ignition timing incorrect (Chapter 1).
3 Carbon build-up in combustion chambers. Try using a top-engine cleaning solvent that's poured through the carburetor. These are commonly available from auto parts stores.
4 Incorrect spark plugs (Chapter 1).

21 Engine diesels (continues to run) after being turned off

1 Idle speed too high (Chapter 4).
2 Ignition timing incorrect (Chapter 1).
3 Incorrect spark plug heat range (Chapter 1).
4 Vacuum leak (refer to Section 8 of *Troubleshooting*).
5 Carbon build-up in combustion chambers. Try using a top-engine cleaning solvent that's poured through the carburetor. These are commonly available from auto parts stores.
6 Valves sticking (Chapter 2).
7 EGR system not operating properly (Chapter 6).
8 Fuel shut-off system not operating properly (Chapter 6).
9 Check for causes of overheating (Section 27).

22 Low oil pressure

1 Improper grade of oil (viscosity too low) - if you're using the correct grade, try a higher grade to raise the pressure.
2 Oil pump worn or damaged (Chapter 2).
3 Engine overheating (refer to Section 27).
4 Clogged oil filter (Chapter 1).
5 Clogged oil strainer (Chapter 2).
6 Oil pressure gauge not working properly (Chapter 2).

23 Excessive oil consumption

1 Loose oil pan drain plug.
2 Loose bolts or damaged oil pan gasket (Chapter 2).
3 Loose bolts or damaged front cover gasket (Chapter 2).
4 Front or rear crankshaft oil seal leaking (Chapter 2).
5 Loose bolts or damaged valve cover gasket (Chapter 2).
6 Loose oil filter (Chapter 1).
7 Loose or damaged oil pressure switch (Chapter 2).
8 Pistons and cylinders excessively worn (Chapter 2).
9 Piston rings not installed correctly on pistons (Chapter 2).
10 Worn or damaged piston rings (Chapter 2).
11 Intake and/or exhaust valve oil seals worn or damaged (Chapter 2).
12 Valve stem oil seals missing or damaged.
13 Worn or damaged valves/guides (Chapter 2).

24 Excessive fuel consumption

1 Dirty or clogged air filter element (Chapter 1).
2 Incorrect ignition timing (Chapter 1).
3 Incorrect idle speed (Chapter 4).
4 Idle mixture not correctly set (Chapter 4).
5 Low tire pressure or incorrect tire size (Chapter 10).
6 Fuel leakage. Check all connections, lines and components in the fuel system (Chapters 1 and 4).
7 Choke sticking closed (Chapter 1).
8 Dirty or clogged carburetor jets (Chapter 4).

25 Fuel odor

1 Fuel leakage. Check all connections, lines and components in the fuel system (Chapter 4).
2 Fuel tank overfilled. Fill only to automatic shut-off.
3 Charcoal canister filter in Evaporative Emissions Control system clogged (Chapter 1).
4 Vapor leaks from Evaporative Emissions Control system lines (Chapter 6).

26 Miscellaneous engine noises

1 A strong dull noise that becomes more rapid as the engine accelerates indicates worn or damaged crankshaft bearings or an unevenly worn crankshaft. To pinpoint the trouble spot, remove the spark plug wire from one plug at a time and crank the engine over. If the noise stops, the cylinder with the removed plug wire indicates the problem area. Replace the bearing and/or service or replace the crankshaft (Chapter 2).
2 A similar (yet slightly higher pitched) noise to the crankshaft knocking described in the previous paragraph, that becomes more rapid as the engine accelerates, indicates worn or damaged connecting rod bearings (Chapter 2). The procedure for locating the problem cylinder is the same as described in Paragraph 1.
3 An overlapping metallic noise that increases in intensity as the engine speed increases, yet diminishes as the engine warms up indicates abnormal piston and cylinder wear (Chapter 2). To locate the problem cylinder, use the procedure described in Paragraph 1.
4 A rapid clicking noise that becomes faster as the engine accelerates indicates a worn piston pin or piston pin hole. This sound will happen each time the piston hits the highest and lowest points in the stroke (Chapter 2). The procedure for locating the problem piston is described in Paragraph 1.
5 A metallic clicking or groaning noise coming from the water pump indicates worn or damaged water pump bearings or pump. Replace the water pump with a new one (Chapter 3).
6 A rapid tapping sound or clicking sound that becomes faster as the engine speed

increases indicates "valve tapping." This can be identified by holding one end of a section of hose to your ear and placing the other end at different spots along the valve cover. The point where the sound is loudest indicates the problem valve. If the pushrod and rocker arm components are in good shape, you likely have a collapsed valve lifter. Changing the engine oil and adding a high-viscosity oil treatment (such as STP) will sometimes cure a stuck lifter problem. If the problem persists, the lifters, pushrods and rocker arms must be removed for inspection (see Chapter 2).

7 A steady metallic rattling or rapping sound coming from the area of the timing chain cover indicates a worn, damaged or out-of-adjustment timing chain. Service or replace the chain and related components (Chapter 2).

Cooling system

27 Overheating

1 Insufficient coolant in system (Chapter 1).
2 Drivebelt defective or not adjusted properly (Chapter 1).
3 Radiator core blocked by corrosion or radiator grille dirty and restricted (Chapter 3).
4 Thermostat faulty (Chapter 3).
5 Fan not functioning properly (Chapter 3).
6 Radiator cap not maintaining proper pressure. Have cap pressure tested by gas station or repair shop.
7 Ignition timing incorrect (Chapter 1).
8 Defective water pump (Chapter 3).
9 Improper grade of engine oil.
10 Inaccurate temperature gauge (Chapter 12).

28 Overcooling

1 Thermostat faulty, not installed or of too low a temperature (Chapter 3).
2 Inaccurate temperature gauge (Chapter 12).

29 External coolant leakage

1 Deteriorated or damaged hoses. Loose clamps at hose connections (Chapter 1).
2 Water pump seals defective. If this is the case, water will drip from the weep hole in the water pump body (Chapter 3).
3 Leakage from radiator core or header tank. This will require the radiator to be professionally repaired (see Chapter 3 for removal procedures).
4 Engine drain plugs or water jacket freeze plugs leaking (see Chapters 1 and 2C).
5 Leak from coolant temperature switch (Chapter 3).
6 Leak from damaged gaskets or small cracks (Chapter 2).
7 Damaged head gasket. This can be verified by checking the condition of the engine oil, as noted in Section 30.

30 Internal coolant leakage

Note: *Internal coolant leaks can usually be detected by examining the oil. Check the dipstick and inside the valve cover for water deposits and an oil consistency like that of a milkshake.*

1 Leaking cylinder head gasket. Have the system pressure tested or remove the cylinder head (Chapter 2) and inspect.
2 Cracked cylinder bore or cylinder head. Dismantle engine and inspect (Chapter 2).
3 Loose cylinder head bolts (tighten as described in Chapter 2).

31 Abnormal coolant loss

1 Overfilling system (Chapter 1).
2 Coolant boiling away due to overheating (see causes in Section 27).
3 Internal or external leakage (see Sections 29 and 30).
4 Faulty radiator cap. Have the cap pressure tested.
5 Cooling system being pressurized by engine compression. This could be due to a cracked head or block or leaking head gasket(s).

32 Poor coolant circulation

1 Inoperative water pump. A quick test is to pinch the top radiator hose closed with your hand while the engine is idling, then release it. You should feel a surge of coolant if the pump is working properly (Chapter 3).
2 Restriction in the cooling system. Drain, flush and refill the system (Chapter 1). If necessary, remove the radiator (Chapter 3) and have it reverse flushed or professionally cleaned.
3 Loose water pump drivebelt (Chapter 1).
4 Thermostat sticking (Chapter 3).
5 Insufficient coolant (Chapter 1).

33 Corrosion

1 Excessive impurities in the water. Soft, clean water is recommended. Distilled or rainwater is satisfactory.
2 Insufficient antifreeze solution (refer to Chapter 1 for the proper ratio of water to antifreeze).
3 Infrequent flushing and draining of system. Regular flushing of the cooling system should be carried out at the specified intervals as described in (Chapter 1).

Clutch

Note: *All clutch related service information is located in Chapter 8, unless otherwise noted.*

34 Fails to release (pedal pressed to the floor - shift lever does not move freely in and out of Reverse)

1 Clutch contaminated with oil. Remove clutch plate and inspect.
2 Clutch plate warped, distorted or otherwise damaged.
3 Diaphragm spring fatigued. Remove clutch cover/pressure plate assembly and inspect.
4 Insufficient pedal stroke. Check and adjust as necessary.
5 Lack of grease on pilot bushing.

35 Clutch slips (engine speed increases with no increase in vehicle speed)

1 Worn or oil soaked clutch plate.
2 Clutch plate not broken in. It may take 30 or 40 normal starts for a new clutch to seat.
3 Diaphragm spring weak or damaged. Remove clutch cover/pressure plate assembly and inspect.
4 Clutch adjusted too tight (see Chapter 1).
5 Flywheel warped (Chapter 2).

36 Grabbing (chattering) as clutch is engaged

1 Oil on clutch plate. Remove and inspect. Repair any leaks.
2 Worn or loose engine or transmission mounts. They may move slightly when clutch is released. Inspect mounts and bolts.
3 Worn splines on transmission input shaft. Remove clutch components and inspect.
4 Warped pressure plate or flywheel. Remove clutch components and inspect.
5 Diaphragm spring fatigued. Remove clutch cover/pressure plate assembly and inspect.
6 Clutch linings hardened or warped.
7 Clutch lining rivets loose.

37 Squeal or rumble with clutch engaged (pedal released)

1 Improper pedal adjustment. Adjust pedal freeplay (Chapter 1).
2 Release bearing binding on transmission shaft. Remove clutch components and check bearing. Remove any burrs or other damage on the shaft.
3 Pilot bushing worn or damaged.

4 Clutch rivets loose.
5 Clutch plate cracked.
6 Fatigued clutch plate torsion springs. Replace clutch plate.

38 Squeal or rumble with clutch disengaged (pedal depressed)

1 Worn or damaged release bearing.
2 Worn or broken pressure plate diaphragm fingers.

39 Clutch pedal stays on floor when disengaged

Binding linkage or release bearing. Inspect linkage or remove clutch components as necessary.

Manual transmission

Note: *All manual transmission service information is located in Chap-ter 7, unless otherwise noted.*

40 Noisy in Neutral with engine running

1 Input shaft bearing worn.
2 Damaged main drive gear bearing.
3 Insufficient transmission lubricant (see Chapter 1).
4 Transmission lubricant in poor condition. Drain and fill with proper grade lubricant. Check old lubricant for water and debris (Chapter 1).
5 Noise can be caused by variations in engine torque. Change the idle speed and see if noise disappears.

41 Noisy in all gears

1 Any of the above causes, and/or:
2 Worn or damaged output gear bearings or shaft.

42 Noisy in one particular gear

1 Worn, damaged or chipped gear teeth.
2 Worn or damaged synchronizer.

43 Slips out of gear

1 Transmission loose on clutch housing.
2 Stiff shift lever seal.
3 Shift linkage binding.
4 Broken or loose input gear bearing retainer.
5 Dirt between clutch lever and engine housing.
6 Worn linkage.
7 Damaged or worn check balls, fork rod ball grooves or check springs.
8 Worn mainshaft or countershaft bearings.
9 Loose engine mounts (Chapter 2).
10 Excessive gear endplay.
11 Worn synchronizers.

44 Oil leaks

1 Excessive amount of lubricant in transmission (see Chapter 1 for correct checking procedures). Drain lubricant as required.
2 Rear oil seal or speedometer oil seal damaged.
3 To pinpoint a leak, first remove all built-up dirt and grime from the transmission. Degreasing agents and/or steam cleaning will achieve this. With the underside clean, drive the vehicle at low speeds so the air flow will not blow the leak far from its source. Raise the vehicle and determine where the leak is located.

45 Difficulty engaging gears

1 Clutch not releasing completely.
2 Loose or damaged shift linkage. Make a thorough inspection, replacing parts as necessary.
3 Insufficient transmission lubricant (Chapter 1).
4 Transmission lubricant in poor condition. Drain and fill with proper grade lubricant. Check lubricant for water and debris (Chapter 1).
5 Worn or damaged shift rod.
6 Sticking or jamming gears.

46 Noise occurs while shifting gears

1 Check for proper operation of the clutch (Chapter 8).
2 Faulty synchronizer assemblies. Measure baulk ring-to-gear clearance. Also, check for wear or damage to baulk rings or any parts of the synchromesh assemblies.

Automatic transmission

Note: *Due to the complexity of the automatic transmission, it's difficult for the home mechanic to properly diagnose and service. For problems other than the following, the vehicle should be taken to a reputable mechanic.*

47 Fluid leakage

Note: *For further leak diagnosis and seal replacement information, refer to Chapter 7B.*
1 Automatic transmission fluid is a deep red color, and fluid leaks should not be confused with engine oil which can easily be blown by air flow to the transmission.
2 To pinpoint a leak, first remove all built-up dirt and grime from the transmission. Degreasing agents and/or steam cleaning will achieve this. With the underside clean, drive the vehicle at low speeds so the air flow will not blow the leak far from its source. Raise the vehicle and determine where the leak is located. Common areas of leakage are:

- *a)* ***Fluid pan:*** *tighten mounting bolts and/or replace pan gasket as necessary (Chapter 1).*
- *b)* ***Rear extension:*** *tighten bolts and/or replace oil seal as necessary.*
- *c)* ***Filler pipe:*** *replace the rubber oil seal where pipe enters transmission case.*
- *d)* ***Transmission oil lines:*** *tighten fittings where lines enter transmission case and/or replace lines.*
- *e)* ***Vent pipe:*** *transmission overfilled and/or water in fluid (see checking procedures, Chapter 1).*
- *f)* ***Speedometer connector****: replace the O-ring where speedometer cable enters transmission case.*

48 General shift mechanism problems

Chapter 7 deals with checking and adjusting the shift linkage on automatic transmissions. Common problems which may be caused by out of adjustment linkage are:

- *a) Engine starting in gears other than P (park) or N (Neutral).*
- *b) Indicator pointing to a gear other than the one actually engaged.*
- *c) Vehicle moves with transmission in P (Park) position.*

49 Transmission will not downshift with the accelerator pedal pressed to the floor

Chapter 7 deals with adjusting the throttle rod to enable the transmission to downshift properly.

50 Engine will start in gears other than Park or Neutral

Chapter 7 deals with adjusting the Neutral start switch installed on automatic transmissions.

51 Transmission slips, shifts rough, is noisy or has no drive in forward or Reverse gears

1 There are many probable causes for the above problems, but the home mechanic should concern himself only with one possibility; fluid level.

2 Before taking the vehicle to a shop, check the fluid level and condition as described in Chapter 1. Add fluid, if necessary, or change the fluid and filter if needed. If problems persist, have a professional diagnose the transmission.

Driveshaft

Note: *Refer to Chapter 8, unless otherwise specified, for service information.*

52 Leaks at front of driveshaft

Defective transmission rear seal and/or bushing. See Chapter 7B for replacement procedure. As this is done, check the splined yoke for burrs or roughness that could damage the new seal. Remove burrs with a fine file or whetstone.

53 Knock or clunk when transmission is under initial load (just after transmission is put into gear)

1 Loose or disconnected rear suspension components. Check all mounting bolts and bushings (Chapters 7 and 10).
2 Loose driveshaft bolts. Inspect all bolts and nuts and tighten them securely.
3 Worn or damaged universal joint bearings (Chapter 8).
4 Worn sleeve yoke and mainshaft spline.

54 Metallic grating sound consistent with vehicle speed

Pronounced wear in the universal joint bearings. Replace U-joints or driveshafts, as necessary.

55 Vibration

Note: *Before blaming the driveshaft, make sure the tires are perfectly balanced and perform the following test.*
1 Install a tachometer inside the vehicle to monitor engine speed as the vehicle is driven. Drive the vehicle and note the engine speed at which the vibration (roughness) is most pronounced. Now shift the transmission to a different gear and bring the engine speed to the same point.
2 If the vibration occurs at the same engine speed (rpm) regardless of which gear the transmission is in, the driveshaft is NOT at fault since the driveshaft speed varies.
3 If the vibration decreases or is eliminated when the transmission is in a different gear at the same engine speed, refer to the following probable causes.
4 Bent or dented driveshaft. Inspect and replace as necessary.
5 Undercoating or built-up dirt, etc. on the driveshaft. Clean the shaft thoroughly.
6 Worn universal joint bearings. Replace the U-joints or driveshaft as necessary.
7 Driveshaft and/or companion flange out of balance. Check for missing weights on the shaft. Remove driveshaft and reinstall 180-degrees from original position, then recheck. Have the driveshaft balanced if problem persists.
8 Loose driveshaft mounting bolts/nuts.
9 Defective center bearing, if so equipped.
10 Worn transmission rear bushing (Chapter 7).

56 Scraping noise

Look for scrape marks around the driveshaft that will help you locate the source of the scraping.

57 Whining or whistling noise

Defective center bearing, if so equipped.

Rear axle and differential

Note: *For differential servicing information, refer to Chapter 8, unless otherwise specified.*

58 Noise - same when in drive as when vehicle is coasting

1 Road noise. No corrective action available.
2 Tire noise. Inspect tires and check tire pressures (Chapter 1).
3 Front wheel bearings loose, worn or damaged (Chapter 1).
4 Insufficient differential oil (Chapter 1).
5 Defective differential.

59 Knocking sound when starting or shifting gears

Defective or incorrectly adjusted differential.

60 Noise when turning

Defective differential.

61 Vibration

See probable causes under Driveshaft. Proceed under the guidelines listed for the driveshaft. If the problem persists, check the rear wheel bearings by raising the rear of the vehicle and spinning the wheels by hand. Listen for evidence of rough (noisy) bearings. Remove and inspect (Chapter 8).

62 Oil leaks

1 Pinion oil seal damaged (Chapter 8).
2 Axleshaft oil seals damaged (Chapter 8).
3 Differential cover leaking. Tighten mounting bolts or replace the gasket as required.
4 Loose filler plug on differential (Chapter 1).
5 Clogged or damaged breather on differential.

Brakes

Note: *Before assuming a brake problem exists, make sure the tires are in good condition and inflated properly, the front end alignment is correct and the vehicle is not loaded with weight in an unequal manner. All service procedures for the brakes are included in Chapter 9, unless otherwise noted.*

63 Vehicle pulls to one side during braking

1 Defective, damaged or oil contaminated brake pad or shoe on one side. Inspect as described in Chapter 1. Refer to Chapter 9 if replacement is required.
2 Excessive wear of brake pad/shoe material or disc/drum on one side. Inspect and repair as necessary.
3 Loose or disconnected front suspension components. Inspect and tighten all bolts securely (Chapters 1 and 10).
4 Defective caliper or wheel cylinder assembly. Remove the caliper or wheel cylinder and inspect for a stuck piston or damage.
5 Brake pad to rotor adjustment needed. Inspect automatic adjusting mechanism for proper operation.
6 Scored or out-of-round rotor.
7 Loose caliper mounting bolts.
8 Incorrect wheel bearing adjustment.
9 On models with front drum brakes, the automatic adjusters may not be functioning correctly (see Chapter 9).

64 Noise (high-pitched squeal)

Note: *Squeal coming from front disc brakes is common and sometimes hard to fix. Anti-squeal compounds that are applied to the back of the brake pads often help. These are available from auto parts stores. Also available are anti-squeal shims that help as well. Semi-metallic brake pads last longer than conventional organic linings, but they also*

contribute to squeal and cause more rotor wear. Also make sure the caliper mounting bolts are tight and all the caliper sliding surfaces are well lubricated (see Chapter 9).
1 Front brake pads worn out. Replace pads with new ones immediately!
2 Glazed or contaminated pads.
3 Dirty or scored rotor.
4 Bent support plate.

65 Excessive brake pedal travel

1 Partial brake system failure. Inspect entire system (Chapter 1) and correct as required.
2 Insufficient fluid in master cylinder. Check (Chapter 1) and add fluid - bleed system if necessary.
3 Air in system. Bleed system.
4 Excessive lateral rotor play.
5 Brakes out of adjustment. Check the operation of the automatic adjusters.
6 Defective proportioning valve. Replace valve and bleed system.

66 Brake pedal feels spongy when depressed

1 Air in brake lines. Bleed the brake system.
2 Deteriorated rubber brake hoses. Inspect all system hoses and lines. Replace parts as necessary.
3 Master cylinder mounting nuts loose. Inspect master cylinder bolts (nuts) and tighten them securely.
4 Master cylinder faulty.
5 Incorrect shoe or pad clearance.
6 Defective check valve. Replace valve and bleed system.
7 Clogged reservoir cap vent hole.
8 Deformed rubber brake lines.
9 Soft or swollen caliper seals.
10 Poor quality brake fluid. Bleed entire system and fill with new approved fluid.

67 Excessive effort required to stop vehicle

1 Power brake booster not operating properly.
2 Excessively worn linings or pads. Check and replace if necessary.
3 One or more caliper pistons seized or sticking. Inspect and rebuild as required.
4 Brake pads or linings contaminated with oil or grease. Inspect and replace as required.
5 New pads or linings installed and not yet seated. It'll take a while for the new material to seat against the rotor or drum.
6 Worn or damaged master cylinder or caliper assemblies. Check particularly for frozen pistons.
7 Also see causes listed under Section 66.

68 Pedal travels to the floor with little resistance

Little or no fluid in the master cylinder reservoir caused by leaking caliper piston(s) or loose, damaged or disconnected brake lines. Inspect entire system and repair as necessary.

69 Brake pedal pulsates during brake application

1 Wheel bearings damaged, worn or out of adjustment (Chapter 1).
2 Caliper not sliding properly due to improper installation or obstructions. Remove and inspect.
3 Rotor not within specifications. Remove the rotor and check for excessive lateral runout and parallelism. Have the rotors resurfaced or replace them with new ones. Also make sure that all rotors are the same thickness.
4 Out of round rear brake drums. Remove the drums and have them turned or replace them with new ones.

70 Brakes drag (indicated by sluggish engine performance or wheels being very hot after driving)

1 Output rod adjustment incorrect at the brake pedal or between the brake booster and master cylinder.
2 Obstructed master cylinder compensator. Disassemble master cylinder and clean.
3 Master cylinder piston seized in bore. Overhaul master cylinder.
4 Caliper assembly in need of overhaul.
5 Brake pads or shoes worn out.
6 Piston cups in master cylinder or caliper assembly deformed. Overhaul master cylinder.
7 Rotor not within specifications (Section 69).
8 Parking brake assembly will not release.
9 Clogged brake lines.
10 Wheel bearings out of adjustment (Chapter 1).
11 Brake pedal height improperly adjusted.
12 Wheel cylinder needs overhaul.
13 Improper shoe-to-drum clearance. Adjust as necessary.

71 Rear brakes lock up under light brake application

1 Tire pressures too high.
2 Tires excessively worn (Chapter 1).
3 Defective power brake booster.

72 Rear brakes lock up under heavy brake application

1 Tire pressures too high.
2 Tires excessively worn (Chapter 1).
3 Front brake pads or shoes contaminated with oil, mud or water. Replace the pads.
4 Front brake pads or shoes excessively worn.
5 Defective master cylinder or caliper assembly.

Suspension and steering

Note: *All service procedures for the suspension and steering systems are included in Chapter 10, unless otherwise noted.*

73 Vehicle pulls to one side

1 Tire pressures uneven (Chapter 1).
2 Defective tire (Chapter 1).
3 Excessive wear in suspension or steering components
(Chapters 1 and 10).
4 Front end alignment incorrect.
5 Front brakes dragging. Inspect as described in Section 71.
6 Wheel bearings improperly adjusted (Chapter 1).
7 Wheel lug nuts loose.

74 Shimmy, shake or vibration

1 Tire or wheel out of balance or out of round. Have them balanced on the vehicle.
2 Loose, worn or out of adjustment wheel bearings (Chapter 1).
3 Shock absorbers and/or suspension components worn or damaged (see Chapter 10).

75 Excessive pitching and/or rolling around corners or during braking

1 Defective shock absorbers. Replace as a set.
2 Broken or weak leaf springs and/or suspension components.
3 Worn or damaged stabilizer bar or bushings.

76 Wandering or general instability

1 Improper tire pressures.
2 Incorrect front end alignment.
3 Worn or damaged steering linkage or suspension components.
4 Improperly adjusted steering gear.
5 Out-of-balance wheels.

6 Loose wheel lug nuts.
7 Worn rear shock absorbers.
8 Fatigued or damaged rear leaf springs.

77 Excessively stiff steering

1 Lack of fluid in the power steering fluid reservoir, where appropriate (Chapter 1).
2 Incorrect tire pressures (Chapter 1).
3 Lack of lubrication at balljoints (Chapter 1).
4 Front end out of alignment.
5 Steering gear out of adjustment or lacking lubrication.
6 Improperly adjusted wheel bearings.
7 Worn or damaged steering gear.
8 Interference of steering column with turn signal switch.
9 Low tire pressures.
10 Worn or damaged balljoints.
11 Worn or damaged steering linkage.
12 See also Section 76.

78 Excessive play in steering

1 Loose wheel bearings (Chapter 1).
2 Excessive wear in suspension bushings (Chapter 1).
3 Steering gear improperly adjusted.
4 Incorrect front end alignment.
5 Steering gear mounting bolts loose.
6 Worn steering linkage.

79 Lack of power assistance

1 Steering pump drivebelt faulty or not adjusted properly (Chapter 1).
2 Fluid level low (Chapter 1).
3 Hoses or pipes restricting the flow. Inspect and replace parts as necessary.
4 Air in power steering system. Bleed system.
5 Defective power steering pump.

80 Steering wheel fails to return to straight-ahead position

1 Incorrect front end alignment.
2 Tire pressures low.
3 Steering gears improperly engaged.
4 Steering column out of alignment.
5 Worn or damaged balljoint.
6 Worn or damaged steering linkage.
7 Improperly lubricated idler arm.
8 Insufficient oil in steering gear.
9 Lack of fluid in power steering pump.

81 Steering effort not the same in both directions (power system)

1 Leaks in steering gear.
2 Clogged fluid passage in steering gear.

82 Noisy power steering pump

1 Insufficient oil in pump.
2 Clogged hoses or oil filter in pump.
3 Loose pulley.
4 Improperly adjusted drivebelt (Chapter 1).
5 Defective pump.

83 Miscellaneous noises

1 Improper tire pressures.
2 Insufficiently lubricated balljoint or steering linkage.
3 Loose or worn steering gear, steering linkage or suspension components.
4 Defective shock absorber.
5 Defective wheel bearing.
6 Worn or damaged suspension bushings.
7 Damaged leaf spring.
8 Loose wheel lug nuts.
9 Worn or damaged rear axleshaft spline.
10 Worn or damaged rear shock absorber mounting bushing.
11 Incorrect rear axle endplay.
12 See also causes of noises at the rear axle and driveshaft.

84 Excessive tire wear (not specific to one area)

1 Incorrect tire pressures.
2 Tires out of balance. Have them balanced on the vehicle.
3 Wheels damaged. Inspect and replace as necessary.
4 Suspension or steering components worn (Chapter 1).

85 Excessive tire wear on outside edge

1 Incorrect tire pressure.
2 Excessive speed in turns.
3 Front end alignment incorrect (excessive toe-in).

86 Excessive tire wear on inside edge

1 Incorrect tire pressure.
2 Front end alignment incorrect (toe-out).
3 Loose or damaged steering components (Chapter 1).

87 Tire tread worn in one place

1 Tires out of balance. Have them balanced on the vehicle.
2 Damaged or buckled wheel. Inspect and replace if necessary.
3 Defective tire.

Notes

Chapter 1
Tune-up and routine maintenance

Contents

Specifications

Recommended lubricants and fluids

Note: *Listed here are manufacturer recommendations at the time this manual was written. Manufacturers occasionally upgrade their fluid and lubricant specifications, so check with your local auto parts store for current recommendations.*

Engine oil type	API grade SG or SG/CC multigrade and fuel-efficient oil
Engine oil viscosity	See accompanying chart
Automatic transmission fluid type	
1974 through 1977	Dexron II or Mercon automatic transmission fluid
1978 through 1984	
With lockup torque converter	Mopar type 7176 or equivalent fluid meeting this specification
Without lockup torque converter	Dexron II or Mercon automatic transmission fluid
1985 and later models	Mopar type 7176 or equivalent fluid meeting this specification
Manual transmission lubricant type	Dexron II or Mercon automatic transmission fluid*
Differential lubricant type	SAE 80W-90 GL-5 gear lubricant

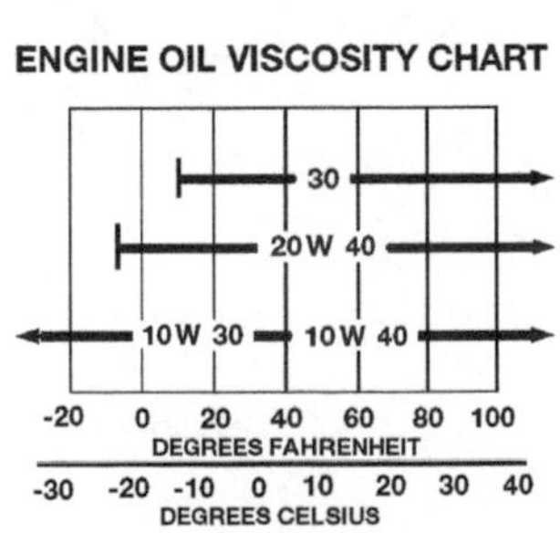

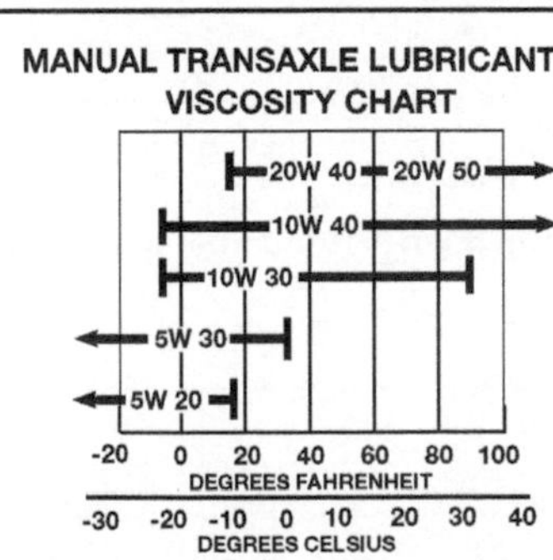

Engine oil viscosity chart - for best fuel economy and cold starting, select the lowest SAE viscosity grade for the expected temperature range

1-A1 HAYNES

Recommended lubricants and fluids (continued)

Limited slip differential	Add Chrysler Friction Modifier No. 4318060, or equivalent, to the specified lubricant
Brake fluid type	DOT 3 brake fluid
Power steering fluid	Chrysler power steering fluid or equivalent
Manual steering gear lubricant type	SAE 90 GL-5 hypoid gear lubricant
Chassis grease type	NLGI no. 2 EP chassis grease
Front wheel bearing grease	NLGI no. 2 EP high-temperature wheel bearing grease

* **Note:** *All manual transmissions covered by this manual came from the factory filled with Dexron ATF; however, a worn or noisy transmission may function better with SAE 75W-90 GL-5 gear lubricant. If switching to gear lubricant, be sure all ATF is drained from the transmission before refilling.*

Capacities

Engine oil (with filter change - approximate)	5 qts
Cooling system (approximate)	
Six-cylinder engine	11.5 to 14.5 qts
V8 engines	15 to 17.5 qts
Automatic transmission (approximate)	
When draining pan and replacing filter	4 qts
When filling a new or rebuilt transmission from dry	8 qts
Manual transmission	2 qts

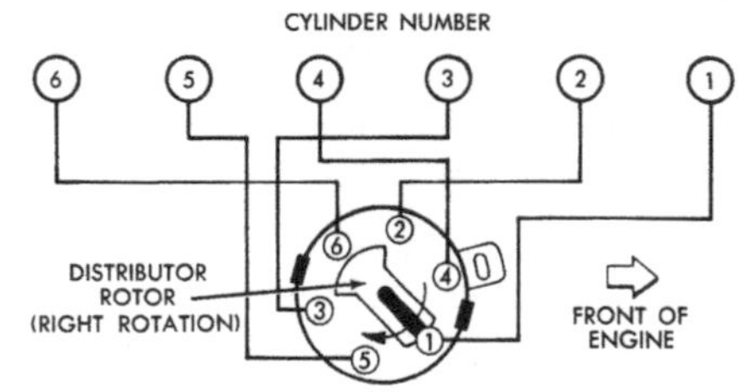

Cylinder locations and numbering - note the arrow that shows the direction the distributor rotates

Ignition system

Spark plug type	
Six-cylinder engine	
1974 and earlier	Champion no. RN14YC
1975 and later	Champion no. RV17YC
Small-block V8 engine - 318 CID (5.2L), 340 CID (5.6L) and 360 CID (5.9L)	
1971 and 1972	
With two-barrel carburetor	Champion no. RN14YC
With four-barrel carburetor	Champion no. RN12YC
1973 through 1985 (all)	Champion no. RN14YC
1986 and later (all)	Champion no. RN12YC
Big-block V8 engine - 383 CID (6.5L), 400 CID (6.6L) and 440 CID (7.2L)	
1971 through 1976	
Standard engines	Champion no. RJ14YC
High-performance engines	Champion no. RJ12YC
1977 and later (all)	Champion no. RJ12YC
Spark plug gap (all engines, all years)	0.035 inch
Ignition timing	Refer to the *Vehicle Emissions Control Information* label in the engine compartment
Ignition point dwell and gap (1971 and 1972 models)	
Six-cylinder engine	
Dwell	41 to 46 degrees
Gap	0.017 to 0.023 inch
Small-block V8 engine - 318 CID (5.2L), 340 CID (5.6L) and 360 CID (5.9L)	
Dwell	30 to 34 degrees
Gap	0.014 to 0.019 inch
Big-block V8 engine - 383 CID (6.5L), 400 CID (6.6L) and 440 CID (7.2L)	
Dwell	28.5 to 32.5 degrees
Gap	0.016 to 0.021 inch
Firing order	
Six-cylinder engine	1-5-3-6-2-4
V8 engine	1-8-4-3-6-5-7-2

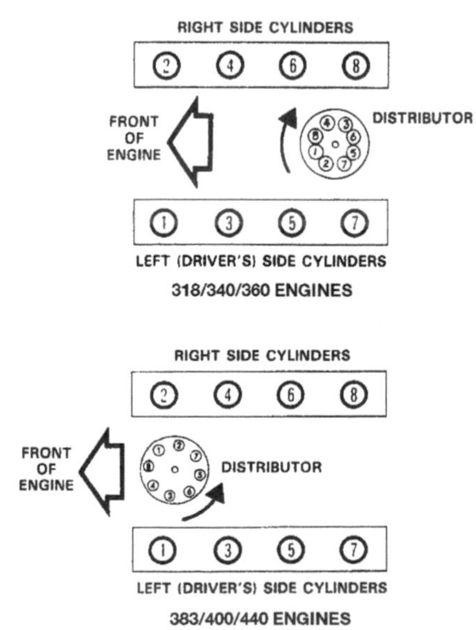

Cylinder location and distributor rotation

General

Valve clearance - engine hot (1980 and earlier six-cylinder engine)	
Intake	0.010 inch
Exhaust	0.020 inch
Clutch pedal freeplay	1 inch
Disc brake pad lining thickness (minimum)	5/16 inch
Drum brake shoe lining thickness (minimum)	1/16 inch

Automatic transmission band adjustment (number of turns backed off)

Kickdown band*	
A-904 transmission	
1971 through 1979	2
1980 on	2-1/2
A-727 transmission	
1971 through 1978 440 dual-exhaust	2
All others	2-1/2

Low-Reverse band*	
A-904 transmission	
1971 through 1983	
198 and 225 six-cylinder engine	
1971 through 1973	3-1/4
1974 through 1983	7
318 V8 engine	4
360 V8 engine (1976 through 1980)	4
1984 through 1989 (all models)	4
A-727 transmission (all years, all models)	2

*Note:

(1) 1971 through 1973, backed off from 72 in-lbs
(2) 1974 through 1983, backed off from 41 in-lbs
(3) 1984 and later, backed off from 72 in-lbs

Torque specifications

	Ft-lbs (unless otherwise indicated)
Kickdown band	
Adjusting screw (torque during adjustment procedure)	
1971 through 1973	72 in-lbs
1974 through 1983	41 in-lbs
1984 on	72-in-lbs
Adjusting screw locknut	
1971	29
1972 and 1973	35
Reverse band	
Adjusting screw (torque during adjustment procedure)	
1971 through 1973	72 in-lbs
1974 through 1983	41 in-lbs
1984 on	72 in-lbs
Adjusting screw locknut	
1971	35
1972 and 1973	30
Automatic transmission pan bolts	96 in-lbs
Carburetor mounting nuts/bolts	96 in-lbs
Differential cover bolts	10 to 20
EGR valve bolts	17
Spark plugs	
Six-cylinder engine	10
V8 engine	30
Engine oil drain plug	20
Oxygen sensor	20
Wheel lug nuts	85

1 Introduction

This Chapter is designed to help the home mechanic maintain Chrysler rear-wheel drive models with the goals of maximum performance, economy, safety and reliability in mind.

Included is a master maintenance schedule (page 1-5), followed by procedures dealing specifically with each item on the schedule. Visual checks, adjustments, component replacement and other helpful items are included. Refer to the accompanying illustrations of the engine compartment and the underside of the vehicle for the locations of various components.

Servicing your vehicle in accordance with the mileage/time maintenance schedule and the step-by-step procedures will result in a planned maintenance program that should produce a long and reliable service life. Keep in mind that it's a comprehensive plan, so maintaining some items but not others at the specified intervals will not produce the same results.

As you service your vehicle, you will discover that many of the procedures can - and should - be grouped together because of the nature of the particular procedure you're performing or because of the close proximity of two otherwise unrelated components to one another.

For example, if the vehicle is raised for chassis lubrication, you should inspect the exhaust, suspension, steering and fuel systems while you're under the vehicle. When you're rotating the tires, it makes good sense to check the brakes since the wheels are already removed. Finally, let's suppose you have to borrow or rent a torque wrench. Even if you only need it to tighten the spark plugs, you might as well check the torque of as many critical fasteners as time allows.

The first step in this maintenance program is to prepare yourself before the actual work begins. Read through all the procedures you're planning to do, then gather up all the parts and tools needed. If it looks like you might run into problems during a particular job, seek advice from a mechanic or an experienced do-it-yourselfer.

The following maintenance intervals are based on the assumption the vehicle owner will be doing the maintenance or service work, as opposed to having a dealer service department do the work. Although the time/mileage intervals are loosely based on factory recommendations, most have been shortened to ensure, for example, that such items as lubricants and fluids are checked/changed at intervals that promote maximum engine/driveline service life. Also, subject to the preference of the individual owner interested in keeping the vehicle in peak condition at all times, and with the vehicle's ultimate resale in mind, many of the maintenance procedures may be performed more often than recommended in the following schedule. We encourage such owner initiative.

When the vehicle is new it should be serviced initially by a factory authorized dealer service department to protect the factory warranty. In many cases the initial maintenance check is done at no cost to the owner (check with your dealer service department for additional information).

Engine compartment components (typical six-cylinder model)

1 *Brake fluid reservoir*
2 *Engine oil dipstick*
3 *Windshield washer fluid reservoir*
4 *Power steering fluid reservoir*
5 *Battery*
6 *Upper radiator hose*
7 *Engine oil filler cap*
8 *Coolant reservoir*
9 *Spark plug wires*
10 *Air cleaner housing*
11 *Automatic transmission fluid dipstick*

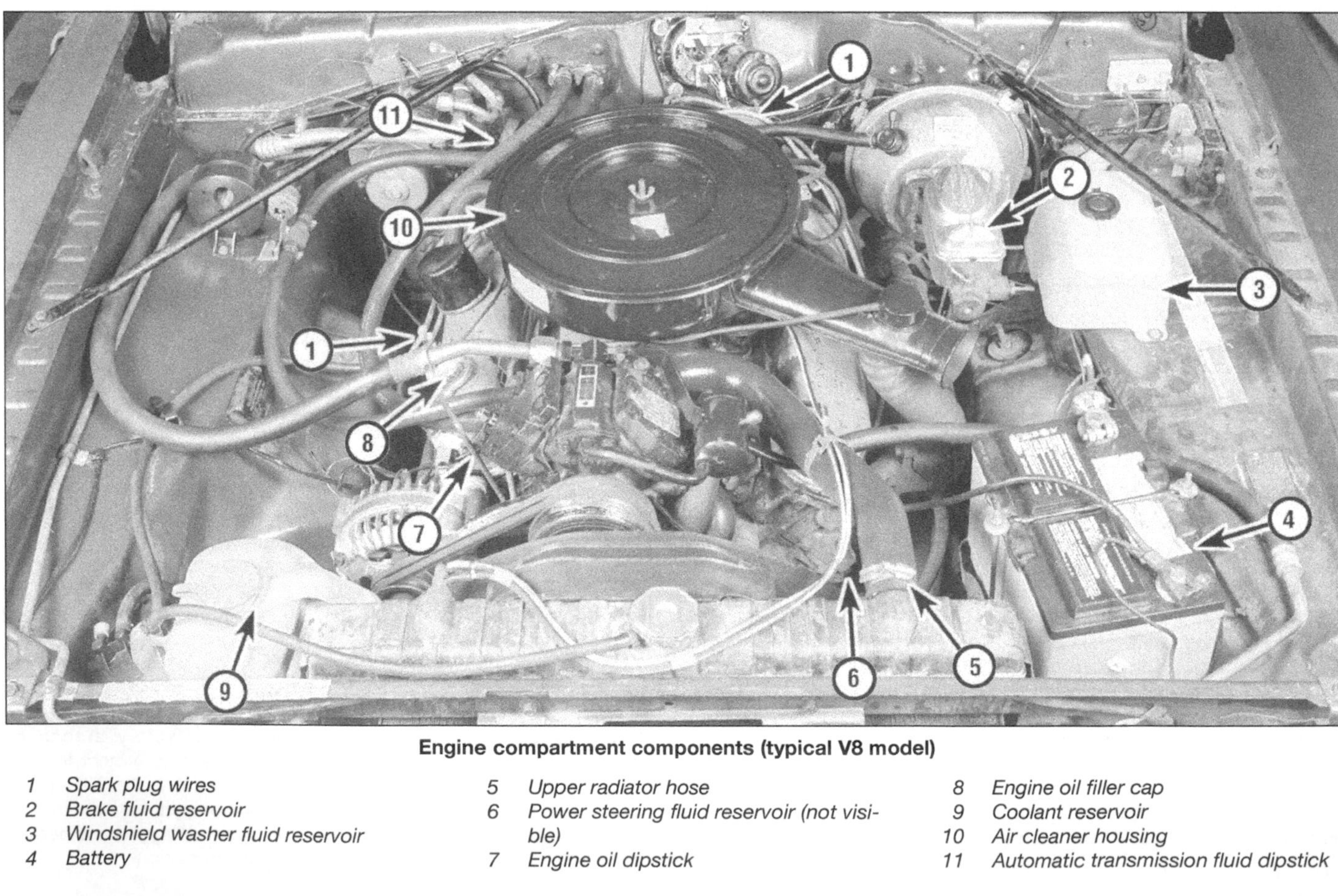

Engine compartment components (typical V8 model)

1 *Spark plug wires*
2 *Brake fluid reservoir*
3 *Windshield washer fluid reservoir*
4 *Battery*
5 *Upper radiator hose*
6 *Power steering fluid reservoir (not visible)*
7 *Engine oil dipstick*
8 *Engine oil filler cap*
9 *Coolant reservoir*
10 *Air cleaner housing*
11 *Automatic transmission fluid dipstick*

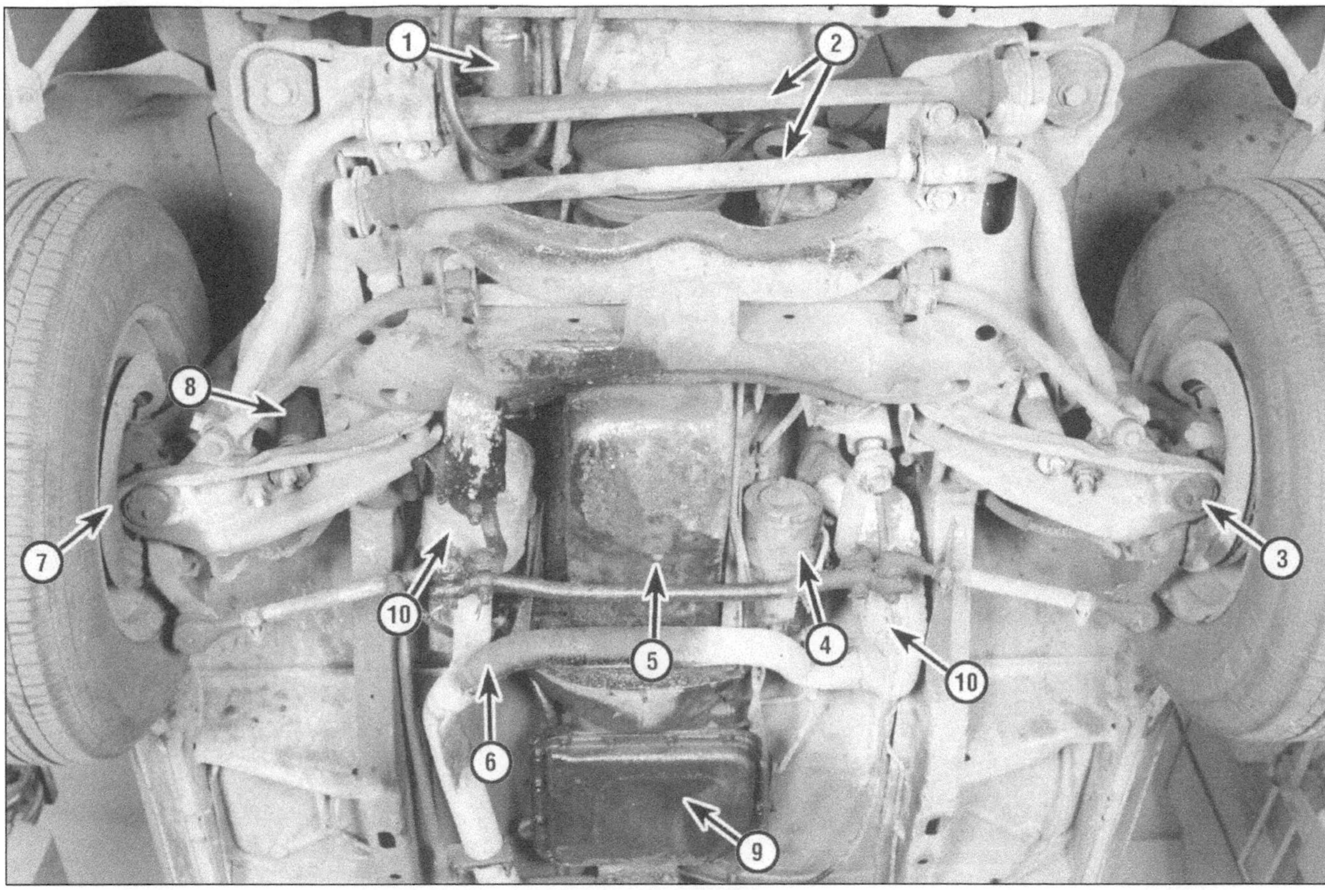

Typical engine compartment underside components (1986 Diplomat with V8 engine shown, others similar)

1. *Lower radiator hose*
2. *Torsion bars*
3. *Suspension balljoint*
4. *Starter motor*
5. *Engine oil drain plug*
6. *Exhaust pipe*
7. *Front brake disc*
8. *Shock absorber*
9. *Automatic transmission fluid pan*
10. *Catalytic converters (***Warning:** *They're very hot when the engine's warm!)*

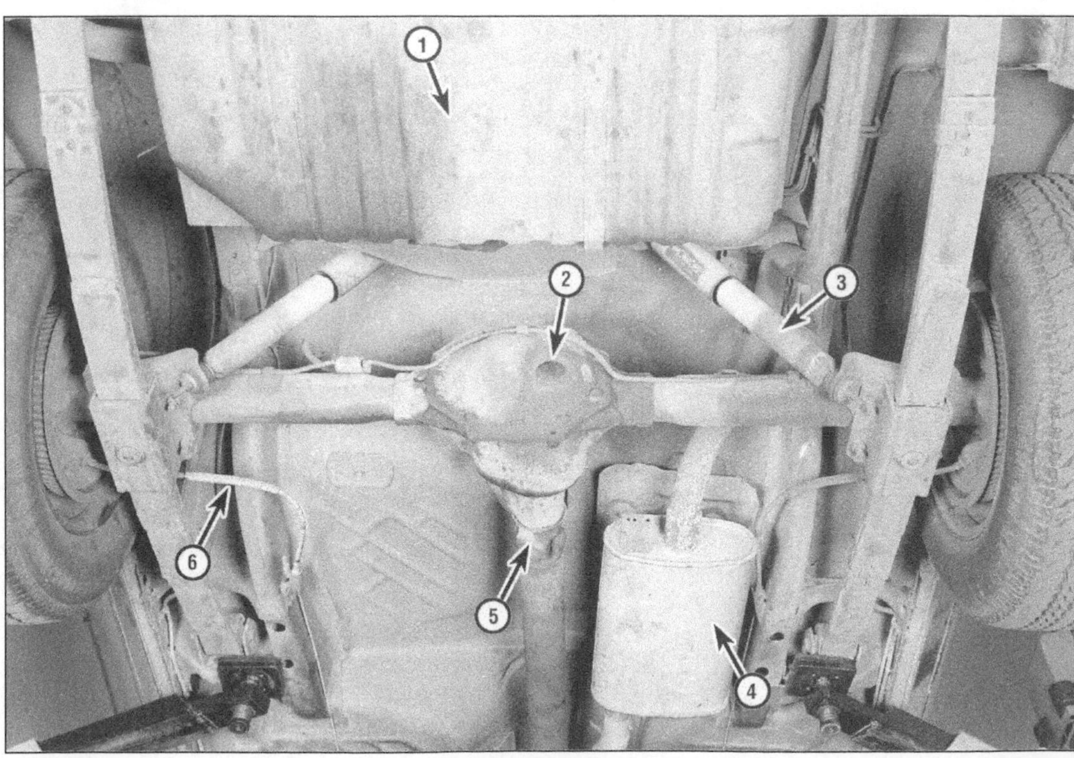

Typical rear underside components

1. *Fuel tank*
2. *Differential check/fill plug*
3. *Shock absorber*
4. *Muffler*
5. *Driveshaft rear universal joint*
6. *Parking brake cable*

Maintenance Schedule

Every 250 miles or weekly, whichever comes first

Check the engine oil level (Section 4)
Check the engine coolant level (Section 4)
Check the windshield washer fluid level (Section 4)
Check the brake and clutch fluid levels (Section 4)
Check the tires and tire pressures (Section 5)

Every 3000 miles or 3 months, whichever comes first

All items listed above, plus . . .
Check the automatic transmission fluid level (Section 6)
Check the power steering fluid level (Section 7)
Change the engine oil and filter (Section 8)
Check and service the battery (Section 9)
Check the cooling system (Section 10)
Inspect and replace, if necessary, all underhood hoses (Section 11)
Inspect and replace, if necessary, the windshield wiper blades (Section 12)
Inspect the suspension and steering components (Section 13)
Inspect the exhaust system (Section 14)
Check the clutch pedal freeplay (Section 15)
Check the manual transmission lubricant (Section 16)
Check the differential lubricant level (Section 17)

Every 7500 miles or 6 months, whichever comes first

Rotate the tires (Section 18)
Check the brakes (Section 19) *
Inspect the fuel system (Section 20)
Check the carburetor mounting bolt/nut torque (Section 21)
Check the throttle linkage (Section 22)
Check the engine drivebelts (Section 23)
Check the seat belts (Section 24)
Check the neutral start switch (Section 25)

Every 15,000 miles or 12 months, whichever comes first

All items listed above, plus . . .
Lubricate the chassis components (Section 26) *
Check and replace if necessary, the ignition points (1971 and 1972 models only) (Section 27)
Check and adjust if necessary, the valve clearances (1980 and earlier six-cylinder models) (Section 28)

Every 30,000 miles or 24 months, whichever comes first

All items listed above, plus . . .
Replace the air filter (Section 29)
Change the automatic transmission fluid and filter (Section 30)**
Adjust the automatic transmission bands (Section 31)**
Change the manual transmission lubricant (Section 32)
Change the differential lubricant (Section 33)
Check and repack the front wheel bearings (Section 34)*
Service the cooling system (drain, flush and refill) (Section 35)
Check the Positive Crankcase Ventilation (PCV) system (Section 36)
Check the evaporative emissions control system (Section 37)
Check the Exhaust Gas Recirculation (EGR) system (Section 38)
Check the thermostatically-controlled air cleaner (Section 39)
Check the crankcase inlet filter (Section 40)
Replace the fuel filter (Section 41)
Check and clean, if necessary, the exhaust manifold heat control valve (Section 42)
Clean the carburetor choke shaft (Section 43)
Replace the spark plugs (Section 44)
Inspect the spark plug wires, distributor cap and rotor (Section 45)
Check and adjust, if necessary, the idle speed (Section 46)
Check and adjust, if necessary, the ignition timing (Section 47)

Every 52,500 miles or 36 months, whichever comes first

Replace the oxygen sensor (Electronic Fuel Control system-equipped models) (Section 48)

* This item is affected by "severe" operating conditions, as described below. If the vehicle is operated under severe conditions, perform all maintenance indicated with an asterisk (*) at 3000 mile/three-month intervals. Severe conditions exist if you mainly operate the vehicle . . .
in dusty areas
towing a trailer
idling for extended periods and/or driving at low speeds when outside temperatures remain below freezing and most trips are less than four miles long

** If operated under one or more of the following conditions, change the automatic transmission fluid and adjust the bands every 15,000 miles:
in heavy city traffic where the outside temperature regularly reaches 90-degrees F or higher
in hilly or mountainous terrain
frequent trailer pulling

4.2 The engine oil dipstick (arrow) is located at the front of the engine, behind the radiator on small-block V8 engines, as shown here. On big-block V8 and six-cylinder engines, it's located on the left (driver's side) of the engine

3 Tune-up general information

The term tune-up is used in this manual to represent a combination of individual operations rather than one specific procedure.

If, from the time the vehicle is new, the routine maintenance schedule is followed closely and frequent checks are made of fluid levels and high wear items, as suggested throughout this manual, the engine will be kept in relatively good running condition and the need for additional work will be minimized.

More likely than not, however, there will be times when the engine is running poorly due to lack of regular maintenance. This is even more likely if a used vehicle, which has not received regular and frequent maintenance checks, is purchased. In such cases, an engine tune-up will be needed outside of the regular routine maintenance intervals.

The first step in any tune-up or diagnostic procedure to help correct a poor running engine is a cylinder compression check. A compression check (see Chapter 2, Part C) will help determine the condition of internal engine components and should be used as a guide for tune-up and repair procedures. If, for instance, the compression check indicates serious internal engine wear, a conventional tune-up won't improve the performance of the engine and would be a waste of time and money. Because of its importance, the compression check should be done by someone with the right equipment and the knowledge to use it properly.

The following procedures are those most often needed to bring a generally poor running engine back into a proper state of tune.

Minor tune-up

Check all engine related fluids (Section 4)
Clean, inspect and test the battery (Section 9)

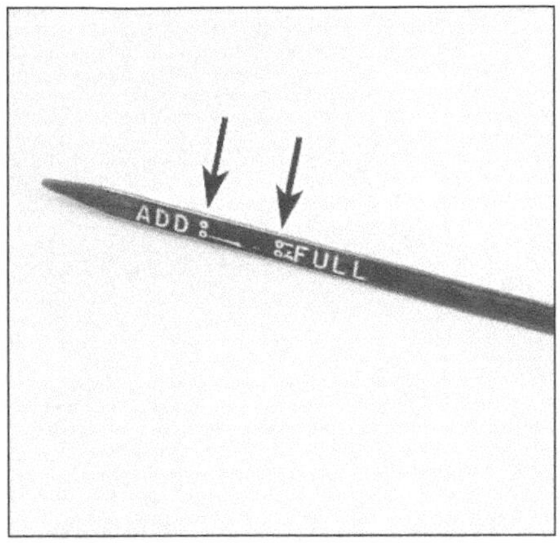

4.4 The oil level must be maintained between the marks at all times - it takes one quart of oil to raise the level from the ADD to the FULL mark

Check the cooling system (Section 10)
Check all underhood hoses (Section 11)
Check the air filter (Section 29
Check and adjust the drivebelts (Section 23)
Check the PCV valve (Section 36)
Replace the spark plugs (Section 44)
Replace the ignition points (1971 and 1972 models only) (Section 27)
Inspect the spark plug and coil wires (Section 45)
Check and adjust the idle speed (Section 46)
Inspect the distributor cap and rotor (Section 45)
Check and adjust the ignition timing (Section 47)

Major tune-up

All items listed under Minor tune-up, plus . . .
Check the fuel system (Section 20)
Replace the air filter (Section 29)
Check the EGR system (Section 38)
Replace the spark plug wires (Section 45)
Replace the distributor cap and rotor (Section 45)
Check the ignition system (Chapter 5)
Check the charging system (Chapter 5)

4 Fluid level checks (every 250 miles or weekly)

Note: *The following are fluid level checks to be done on a 250 mile or weekly basis. Additional fluid level checks can be found in specific maintenance procedures which follow. Regardless of intervals, be alert to fluid leaks under the vehicle which would indicate a fault to be corrected immediately.*

1 Fluids are an essential part of the lubrication, cooling, brake, clutch and windshield washer systems. Because the fluids gradually become depleted and/or contaminated during normal operation of the vehicle, they must be periodically replenished. See *Recommended lubricants and fluids* at the beginning of this Chapter before adding fluid to any of the following components. **Note:** *The vehicle must be on level ground when fluid levels are checked.*

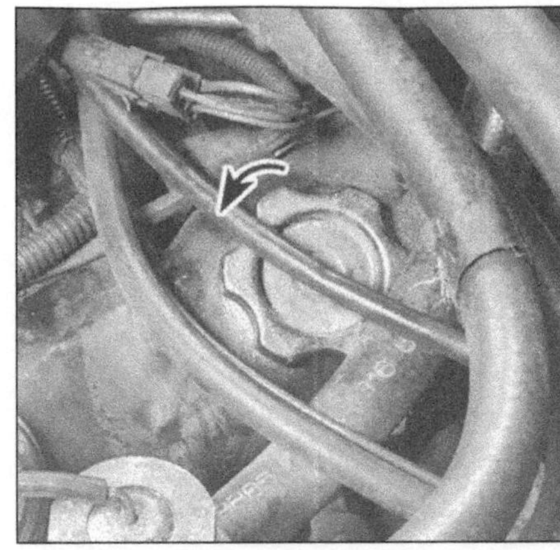

4.6 Rotate the oil filler cap on the valve cover 1/4-turn counterclockwise, then lift it off and add oil through the opening with a funnel - always make sure the area around the opening is clean before removing the cap to prevent dirt from contaminating the engine

Engine oil

Refer to illustrations 4.2, 4.4 and 4.6

2 The engine oil level is checked with a dipstick that extends through a tube and into the oil pan at the bottom of the engine **(see illustration)**.

3 The oil level should be checked before the vehicle has been driven, or about 15 minutes after the engine has been shut off. If the oil is checked immediately after driving the vehicle, some of the oil will remain in the upper engine components, resulting in an inaccurate reading on the dipstick.

4 Pull the dipstick out of the tube and wipe all the oil from the end with a clean rag or paper towel. Insert the clean dipstick all the way back into the tube, then pull it out again. Note the oil at the end of the dipstick. Add oil as necessary to keep the level between the ADD and FULL marks on the dipstick **(see illustration)**.

5 Do not overfill the engine by adding too much oil since this may result in oil fouled spark plugs, oil leaks or oil seal failures.

6 Oil is added to the engine after unscrewing a cap from the valve cover **(see illustration)**. A funnel may help to reduce spills.

7 Checking the oil level is an important preventive maintenance step. A consistently low oil level indicates oil leakage through damaged seals, defective gaskets or past worn rings or valve guides. If the oil looks milky or has water droplets in it, the cylinder head gasket(s) may be blown or the head(s) or block may be cracked. The engine should be checked immediately. The condition of the oil should also be checked. Whenever you check the oil level, slide your thumb and index finger up the dipstick before wiping off

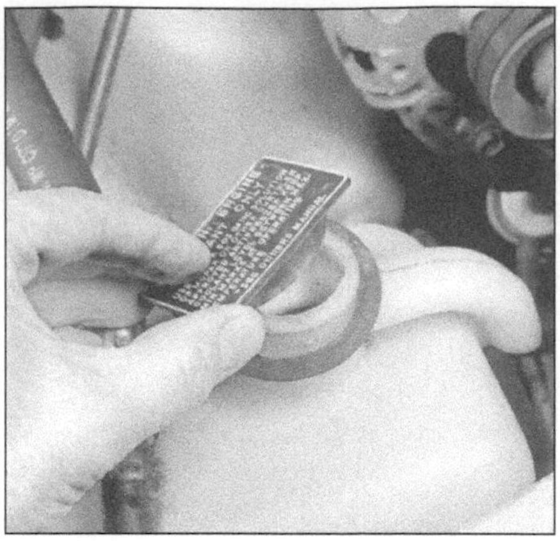

4.8 The coolant reservoir is located in the right front corner of the engine compartment on most models - keep the level between the MAX and COOL marks on the side of the reservoir

the oil. If you see small dirt or metal particles clinging to the dipstick, the oil should be changed (see Section 8).

Engine coolant

Refer to illustration 4.8

Warning: *Do not allow antifreeze to come in contact with your skin or painted surfaces of the vehicle. Flush contaminated areas immediately with plenty of water. Don't store new coolant or leave old coolant lying around where it's accessible to children or pets - they're attracted by its sweet smell. Ingestion of even a small amount of coolant can be fatal! Wipe up garage floor and drip pan coolant spills immediately. Keep antifreeze containers covered and repair leaks in the cooling system as soon as they are noted.*

8 Most vehicles covered by this manual are equipped with a pressurized coolant recovery system. A white plastic coolant reservoir located in the engine compartment is connected by a hose to the radiator filler neck **(see illustration)**. If the engine overheats, coolant escapes through a valve in the radiator cap and travels through the hose into the reservoir. As the engine cools, the coolant is automatically drawn back into the cooling system to maintain the correct level. **Warning:** *Do not remove the radiator cap to check the coolant level when the engine is warm.*

9 The coolant level in the reservoir should be checked regularly. The level in the reservoir varies with the temperature of the engine. When the engine is cold, the coolant level should be at or slightly above the MIN mark on the reservoir. Once the engine has warmed up, the level should be at or near the MAX mark. If it isn't, allow the engine to cool, then remove the cap from the reservoir and add a 50/50 mixture of ethylene glycol-based antifreeze and water.

10 Drive the vehicle and recheck the coolant level. If only a small amount of coolant is required to bring the system up to the proper level, water can be used. However, repeated additions of water will dilute the antifreeze and water solution. In order to maintain the proper ratio of antifreeze and water, always top up the coolant level with the correct mixture. An empty plastic milk jug or bleach bottle makes an excellent container for mixing coolant. Do not use rust inhibitors or additives.

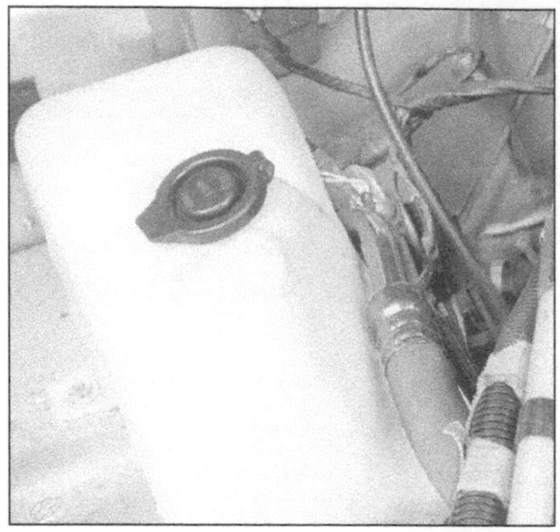

4.14 This windshield washer fluid reservoir can be easily confused with the coolant reservoir if you're not careful - remember that the coolant reservoir is connected to the radiator, just below the radiator cap, by a small-diameter rubber hose

11 If the coolant level drops consistently, there may be a leak in the system. Inspect the radiator, hoses, filler cap, drain plugs and water pump (see Section 10). If no leaks are noted, have the radiator cap pressure tested by a service station.

12 If you have to remove the radiator cap, wait until the engine has cooled, then wrap a thick cloth around the cap and turn it to the first stop. If coolant or steam escapes, let the engine cool down longer, then remove the cap.

13 Check the condition of the coolant as well. It should be relatively clear. If it's brown or rust colored, the system should be drained, flushed and refilled. Even if the coolant appears to be normal, the corrosion inhibitors wear out, so it must be replaced at the specified intervals.

Windshield washer fluid

Refer to illustration 4.14

14 Fluid for the windshield washer system is located in a plastic reservoir in the engine compartment **(see illustration)**.

15 In milder climates, plain water can be used in the reservoir, but it should be kept no more than 2/3 full to allow for expansion if the water freezes. In colder climates, use windshield washer system antifreeze, available at any auto parts store, to lower the freezing point of the fluid. Mix the antifreeze with water in accordance with the manufacturer's directions on the container. **Caution:** *Don't use cooling system antifreeze - it will damage the vehicle's paint.*

16 To help prevent icing in cold weather, warm the windshield with the defroster before using the washer.

4.17 Remove the cell caps to check the electrolyte level in the battery - if the level is low, add distilled water only

Battery electrolyte

Refer to illustration 4.17

17 Most vehicles with which this manual is concerned are equipped with a battery which is permanently sealed (except for vent holes) and has no filler caps. Water doesn't have to be added to these batteries at any time. If a maintenance-type battery is installed, the caps on the top of the battery should be removed periodically to check for a low electrolyte level **(see illustration)**. This check is most critical during the warm summer months.

Brake fluid

Refer to illustrations 4.19a and 4.19b

18 The brake master cylinder is mounted on the upper left of the engine compartment firewall.

19 The fluid inside can be checked after removing the cover or cap **(see illustrations)**. Be sure to wipe the top of the reservoir cover

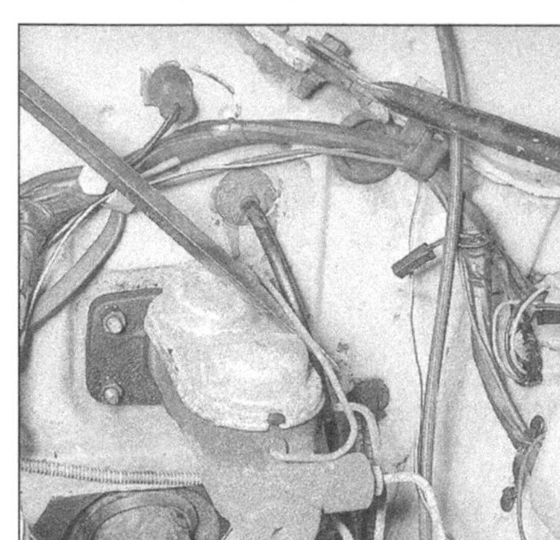

4.19a On earlier models, use a screwdriver to detach the spring retainer, then remove the cover to check the brake fluid level - it should be about 1/4-inch from the top (but not completely full)

14.19b The brake fluid level on later models is easily checked after unscrewing the reservoir caps - the level should be even with the ring at the base of the filler neck

with a clean rag to prevent contamination of the brake and/or clutch system before removing the cover.

20 When adding fluid, pour it carefully into the reservoir to avoid spilling it on surrounding painted surfaces. Be sure the specified fluid is used, since mixing different types of brake fluid can cause damage to the system. See *Recommended lubricants and fluids* at the front of this Chapter or your owner's manual. **Warning:** *Brake fluid can harm your eyes and damage painted surfaces, so use extreme caution when handling or pouring it. Do not use brake fluid that has been standing open or is more than one year old. Brake fluid absorbs moisture from the air. Excess moisture can cause a dangerous loss of brake performance.*

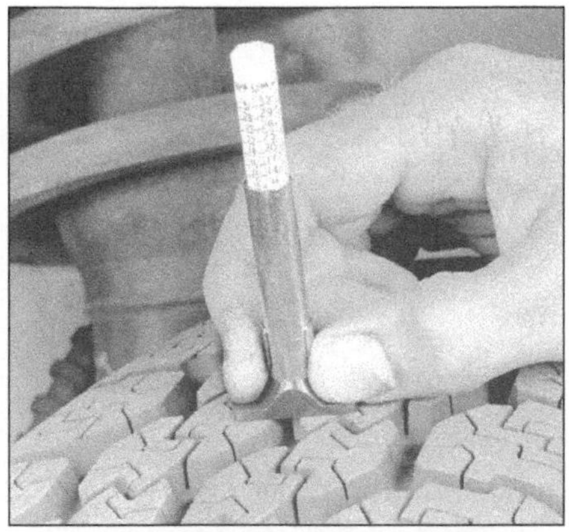

5.2 Use a tire tread depth indicator to monitor tire wear - they are available at auto parts stores and service stations and cost very little

21 At this time, the fluid and master cylinder can be inspected for contamination. The system should be drained and refilled if deposits, dirt particles or water droplets are seen in the fluid.

22 After filling the reservoir to the proper level, make sure the cover or cap is on tight to prevent fluid leakage.

23 The brake fluid level in the master cylinder will drop slightly as the pads at the front wheels wear down during normal operation. If the master cylinder requires repeated additions to keep it at the proper level, it's an indication of leakage in the brake system, which should be corrected immediately. Check all brake lines and connections (see Section 19 for more information).

24 If, upon checking the master cylinder fluid level, you discover one or both reservoirs empty or nearly empty, the brake system should be bled (Chapter 9).

5 Tire and tire pressure checks (every 250 miles or weekly)

Refer to illustrations 5.2, 5.3, 5.4a, 5.4b and 5.8

1 Periodic inspection of the tires may spare you the inconvenience of being stranded with a flat tire. It can also provide you with vital information regarding possible problems in the steering and suspension systems before major damage occurs.

2 The original tires on this vehicle are equipped with 1/2-inch wear bands that will appear when tread depth reaches 1/16-inch, but they don't appear until the tires are worn out. Tread wear can be monitored with a simple, inexpensive device known as a tread depth indicator **(see illustration)**.

3 Note any abnormal tread wear **(see illustration)**. Tread pattern irregularities such as cupping, flat spots and more wear on one

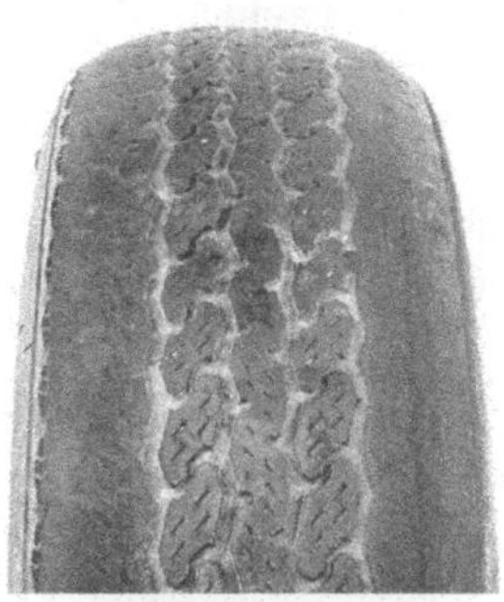

UNDERINFLATION

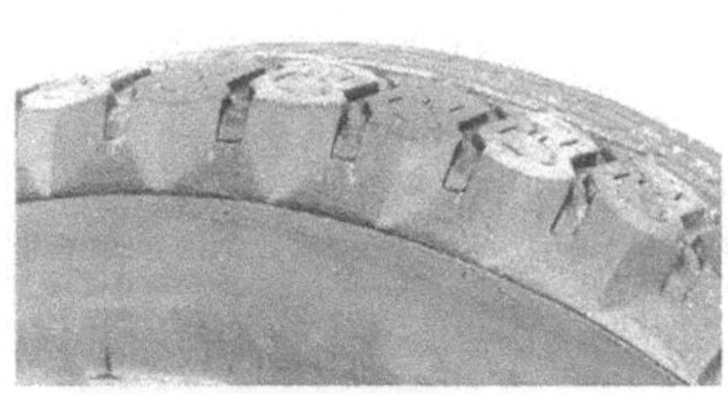

CUPPING

Cupping may be caused by:

- Underinflation and/or mechanical irregularities such as out-of-balance condition of wheel and/or tire, and bent or damaged wheel.
- Loose or worn steering tie-rod or steering idler arm.
- Loose, damaged or worn front suspension parts.

OVERINFLATION

INCORRECT TOE-IN OR EXTREME CAMBER

FEATHERING DUE TO MISALIGNMENT

5.3 This chart will help you determine the condition of the tires, the probable cause(s) of abnormal wear and the corrective action necessary

5.4a If a tire loses air on a steady basis, check the valve stem core first to make sure it's snug (special inexpensive wrenches are commonly available at auto parts stores)

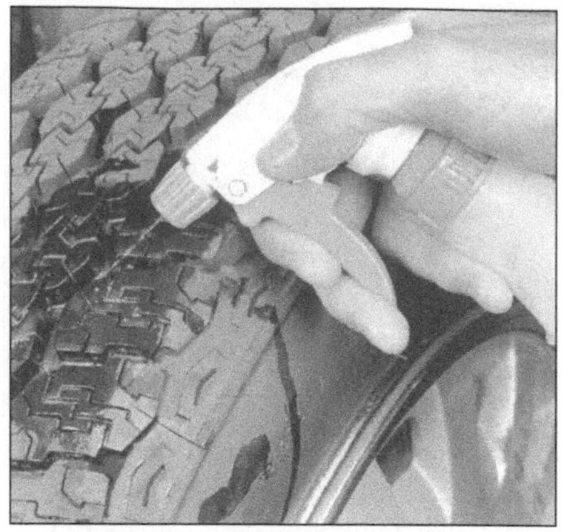

5.4b If the valve stem core is tight, raise the corner of the vehicle with the low tire and spray a soapy water solution onto the tread as the tire is turned slowly - leaks will cause small bubbles to appear

5.8 To extend the life of the tires, check the air pressure at least once a week with an accurate gauge (don't forget the spare!)

6.3 The automatic transmission dipstick (arrow) is located on the rear of the engine compartment, on the right (passenger's) side

side than the other are indications of front end alignment and/or balance problems. If any of these conditions are noted, take the vehicle to a tire shop or service station to correct the problem.

4 Look closely for cuts, punctures and embedded nails or tacks. Sometimes a tire will hold air pressure for a short time or leak down very slowly after a nail has embedded itself in the tread. If a slow leak persists, check the valve stem core to make sure it's tight **(see illustration)**. Examine the tread for an object that may have embedded itself in the tire or for a "plug" that may have begun to leak (radial tire punctures are repaired with a plug that's installed in a puncture). If a puncture is suspected, it can be easily verified by spraying a solution of soapy water onto the puncture area **(see illustration)**. The soapy solution will bubble if there's a leak. Unless the puncture is unusually large, a tire shop or service station can usually repair the tire.

5 Carefully inspect the inner sidewall of each tire for evidence of brake fluid leakage. If you see any, inspect the brakes immediately.

6 Correct air pressure adds miles to the lifespan of the tires, improves mileage and enhances overall ride quality. Tire pressure cannot be accurately estimated by looking at a tire, especially if it's a radial. A tire pressure gauge is essential. Keep an accurate gauge in the vehicle. The pressure gauges attached to the nozzles of air hoses at gas stations are often inaccurate.

7 Always check tire pressure when the tires are cold. Cold, in this case, means the vehicle has not been driven over a mile in the three hours preceding a tire pressure check. A pressure rise of four to eight pounds is not uncommon once the tires are warm.

8 Unscrew the valve cap protruding from the wheel or hubcap and push the gauge firmly onto the valve stem **(see illustration)**. Note the reading on the gauge and compare the figure to the recommended tire pressure shown on the placard on the driver's side door pillar. Be sure to reinstall the valve cap to keep dirt and moisture out of the valve stem mechanism. Check all four tires and, if necessary, add enough air to bring them up to the recommended pressure.

9 Don't forget to keep the spare tire inflated to the specified pressure (refer to your owner's manual or the tire sidewall).

6 Automatic transmission fluid level check (every 250 miles or weekly)

Refer to illustrations 6.3 and 6.6

1 The automatic transmission fluid level should be carefully maintained. Low fluid level can lead to slipping or loss of drive, while overfilling can cause foaming and loss of fluid.

2 With the parking brake set, start the engine, then move the shift lever through all the gear ranges, ending in Neutral. The fluid level must be checked with the vehicle level and the engine running at idle. **Note:** *Incorrect fluid level readings will result if the vehicle has just been driven at high speeds for an extended period, in hot weather in city traffic, or if it has been pulling a trailer. If any of these conditions apply, wait until the fluid has cooled (about 30 minutes).*

3 With the transmission at normal operating temperature, remove the dipstick from the filler tube. The dipstick is located at the rear of the engine compartment on the passenger's side **(see illustration)**.

4 Carefully touch the fluid at the end of the dipstick to determine if it's warm or hot. Wipe the fluid from the dipstick with a clean rag and push it back into the filler tube until the cap seats.

5 Pull the dipstick out again and note the fluid level.

6 If the fluid felt warm, the level should be between the two dimples **(see illustration)**. If

6.6 Check the fluid with the transmission at normal operating temperature - the level should be kept in the OK range

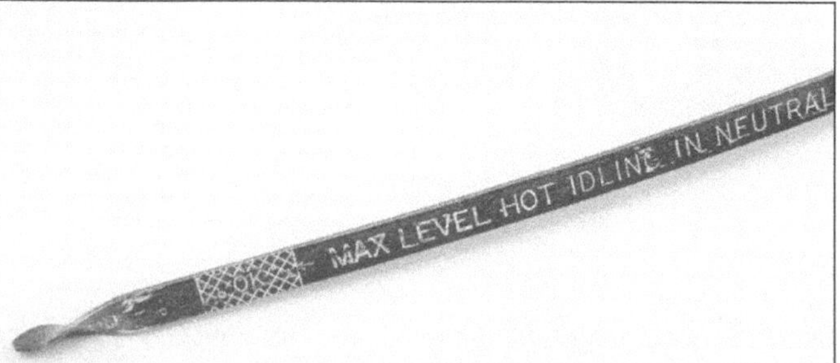

7.2 The power steering fluid reservoir is located near the front of the engine - turn the cap counterclockwise to remove it

it felt hot, the level should be in the crosshatched area, near the MAX line. If additional fluid is required, add it directly into the tube using a funnel. It takes about one pint to raise the level from the bottom of the crosshatched area to the MAX line with a hot transmission, so add the fluid a little at a time and keep checking the level until it's correct.

7 The condition of the fluid should also be checked along with the level. If the fluid at the end of the dipstick is a dark reddish-brown color, or if it smells burned, it should be changed. If you are in doubt about the condition of the fluid, purchase some new fluid and compare the two for color and smell.

7 Power steering fluid level check (every 3000 miles or 3 months)

Refer to illustrations 7.2, 7.6a and 7.6b

1 Unlike manual steering, the power steering system relies on fluid which may, over a period of time, require replenishing.

2 The fluid reservoir for the power steering pump is located on the pump body at the front of the engine **(see illustration)**.

3 For the check, the front wheels should be pointed straight ahead and the engine should be off.

4 Use a clean rag to wipe off the reservoir cap and the area around the cap. This will help prevent any foreign matter from entering the reservoir during the check.

5 Twist off the cap and check the temperature of the fluid at the end of the dipstick with your finger.

6 Wipe off the fluid with a clean rag, reinsert the dipstick, then withdraw it and read the fluid level. The fluid should be at the proper level, depending on whether it was checked hot or cold **(see illustrations)**. Never allow the fluid level to drop below the lower mark on the dipstick.

7 If additional fluid is required, pour the specified type directly into the reservoir, using a funnel to prevent spills.

8 If the reservoir requires frequent fluid additions, all power steering hoses, hose connections and the power steering pump should be carefully checked for leaks.

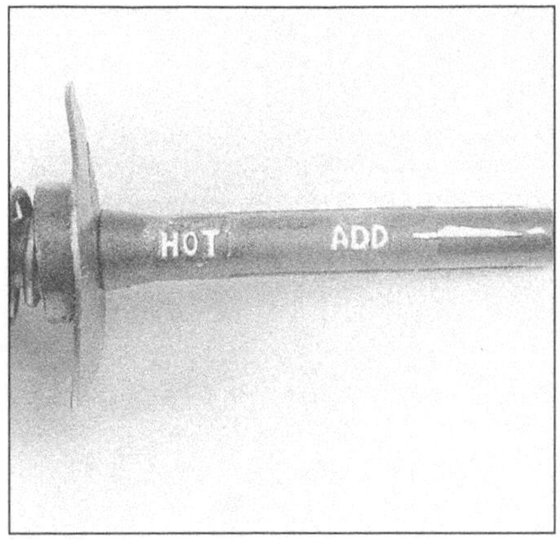

7.6a The power steering fluid dipstick has marks on both sides so the fluid level can be checked when hot . . .

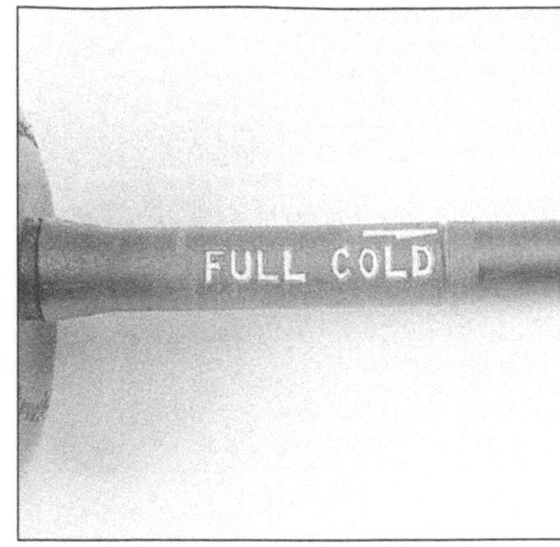

7.6b . . . or cold

8 Engine oil and filter change (every 3000 miles or 3 months)

Refer to illustrations 8.3, 8.9, 8.14 and 8.18

1 Frequent oil changes are the most important preventive maintenance procedures that can be done by the home mechanic. As engine oil ages, it becomes diluted and contaminated, which leads to premature engine wear.

2 Although some sources recommend oil filter changes every other oil change, we feel that the minimal cost of an oil filter and the relative ease with which it is installed dictate that a new filter be installed every time the oil is changed.

3 Gather together all necessary tools and materials before beginning this procedure **(see illustration)**.

4 You should have plenty of clean rags and newspapers handy to mop up any spills. Access to the under side of the vehicle may be improved if the vehicle can be lifted on a hoist, driven onto ramps or supported by jackstands. **Warning:** *Do not work under a vehicle which is supported only by a bumper, hydraulic or scissors-type jack.*

5 If this is your first oil change, get under the vehicle and familiarize yourself with the locations of the oil drain plug and the oil filter. The engine and exhaust components will be warm during the actual work, so note how they are situated to avoid touching them when working under the vehicle.

6 Warm the engine to normal operating temperature. If the new oil or any tools are needed, use this warm-up time to gather everything necessary for the job. The correct type of oil for your application can be found in *Recommended lubricants and fluids* at the beginning of this Chapter.

7 With the engine oil warm (warm engine oil will drain better and more built-up sludge will be removed with it), raise and support the vehicle. Make sure it's safely supported!

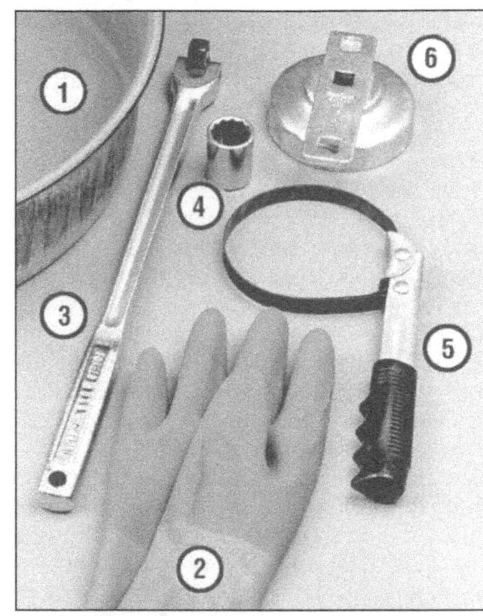

8.3 These tools are required when changing the engine oil and filter

1 **Drain pan** - *It should be fairly shallow in depth, but wide to prevent spills*
2 **Rubber gloves** - *When removing the drain plug and filter, you will get oil on your hands (the gloves will prevent burns)*
3 **Breaker bar** - *Sometimes the oil drain plug is tight, and a long breaker bar is needed to loosen it*
4 **Socket** – *To be used with the breaker bar or a ratchet (must be the correct size to fit the drain plug - six-point preferred)*
5 **Filter wrench** - *This is a metal band-type wrench, which requires clearance around the filter to be effective*
6 **Filter wrench** - *This type fits on the bottom of the filter and can be turned with a ratchet or breaker bar (different-size wrenches are available for different types of filters)*

8 Move all necessary tools, rags and newspapers under the vehicle. Set the drain pan under the drain plug. Keep in mind that the oil will initially flow from the pan with some force; position the pan accordingly.

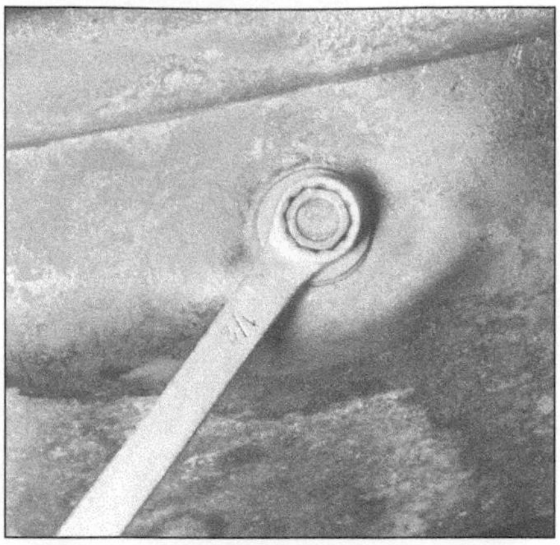

8.9 Use a box-end wrench or socket to remove the oil drain plug (this prevents rounding off the corners of the plug hex)

9 Being careful not to touch any of the hot exhaust components, use a wrench to remove the drain plug near the bottom of the oil pan **(see illustration)**. Depending on how hot the oil is, you may want to wear gloves while unscrewing the plug the final few turns.

10 Allow the old oil to drain into the pan. It may be necessary to move the pan as the oil flow slows to a trickle.

11 After all the oil has drained, wipe off the drain plug with a clean rag. Small metal particles may cling to the plug and would immediately contaminate the new oil.

12 Clean the area around the drain plug opening and reinstall the plug. Tighten the plug securely with the wrench. If a torque wrench is available, use it to tighten the plug to the torque listed in this Chapter's Specifications.

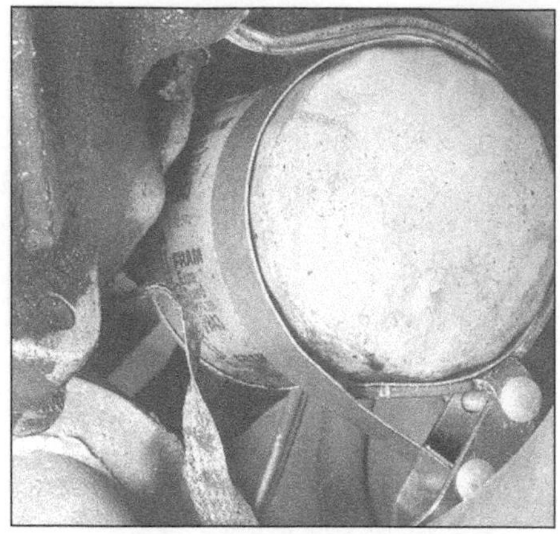

8.14 Since the oil filter is on very tight, you'll need a special wrench for removal - DO NOT use the wrench to tighten the new filter. This photo shows removing the filter on a small-block V8 engine (the filter is at the right rear of the engine). On six-cylinder and big-block V8 engines, the filter is also on the right (passenger's) side, but near the front of the engine.

13 Move the drain pan into position under the oil filter.

14 Use the filter wrench to loosen the oil filter **(see illustration)**. Chain or metal-band filter wrenches may distort the filter canister, but it doesn't matter, since the filter will be discarded anyway.

15 Completely unscrew the old filter. Be careful; it's full of oil. Empty the oil inside the filter into the drain pan, then lower the filter (V8 engines) or lift the filter out from above (six-cylinder engines).

16 Compare the old filter with the new one to make sure they're the same type.

8.18 Lubricate the oil filter gasket with clean engine oil before installing the filter on the engine

17 Use a clean rag to remove all oil, dirt and sludge from the area where the oil filter mounts to the engine. Check the old filter to make sure the rubber gasket isn't stuck to the engine. If the gasket is stuck to the engine (use a flashlight if necessary), remove it.

18 Apply a light coat of clean oil to the rubber gasket on the new oil filter **(see illustration)**.

19 Attach the new filter to the engine, following the tightening directions printed on the filter canister or packing box. Most filter manufacturers recommend against using a filter wrench due to the possibility of overtightening and damage to the seal.

20 Remove all tools, rags, etc. from under the vehicle, being careful not to spill the oil in the drain pan, then lower the vehicle.

21 Move to the engine compartment and locate the oil filler cap.

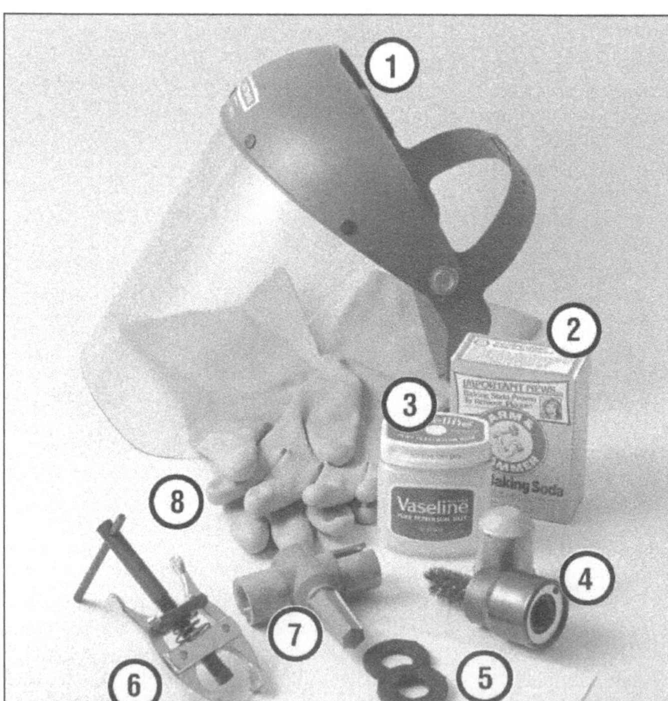

9.1 Tools and materials required for battery maintenance

1 ***Face shield/safety goggles*** *- When removing corrosion with a brush, the acidic particles can easily fly up into your eyes*

2 ***Baking soda*** *- A solution of baking soda and water can be used to neutralize corrosion*

3 ***Petroleum jelly*** *- A layer of this on the battery posts will help prevent corrosion*

4 ***Battery post/cable cleaner*** *- This wire brush cleaning tool will remove all traces of corrosion from the battery posts and cable clamps*

5 ***Treated felt washers*** *- Placing one of these on each post, directly under the cable clamps, will help prevent corrosion*

6 ***Puller*** *- Sometimes the cable clamps are very difficult to pull off the posts, even after the nut/bolt has been completely loosened. This tool pulls the clamp straight up and off the post without damage*

7 ***Battery post/cable cleaner*** *- Here is another cleaning tool which is a slightly different version of Number 4 above, but it does the same thing*

8 ***Rubber gloves*** *- Another safety item to consider when servicing the battery; remember that's acid inside the battery!*

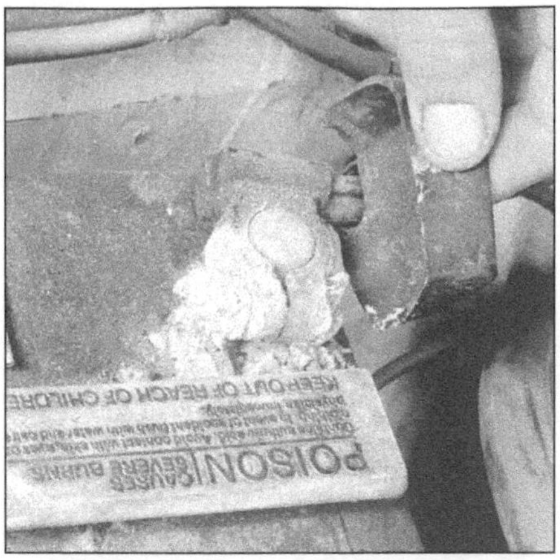

9.6a Battery terminal corrosion usually appears as light, fluffy powder

22 Pour the fresh oil through the filler opening. A funnel may be helpful.

23 Pour four quarts of fresh oil into the engine. Wait a few minutes to allow the oil to drain into the pan, then check the level on the oil dipstick (see Section 4 if necessary). If the oil level is above the ADD mark, start the engine and allow the new oil to circulate.

24 Run the engine for only about a minute and then shut it off. Immediately look under the vehicle and check for leaks at the oil pan drain plug and around the oil filter. If either is leaking, tighten with a bit more force.

25 With the new oil circulated and the filter now completely full, recheck the level on the dipstick and add more oil as necessary.

26 During the first few trips after an oil change, make it a point to check frequently for leaks and proper oil level.

27 The old oil drained from the engine cannot be reused in its present state and should be disposed of. Oil reclamation centers, auto repair shops and gas stations will normally accept the oil, which can be refined and used again. After the oil has cooled it can be drained into a container (capped plastic jugs, topped bottles, milk cartons, etc.) for transport to one of these disposal sites. Don't dispose of the oil by pouring it on the ground or down a drain!

9 Battery check, maintenance and charging (every 7500 miles or 6 months)

Refer to illustrations 9.1, 9.6a, 9.6b, 9.7a and 9.7b

Warning: *Certain precautions must be followed when checking and servicing the battery. Hydrogen gas, which is highly flammable, is always present in the battery cells, so keep lighted tobacco and all other open flames and sparks away from the battery. The electrolyte inside the battery is actually dilute sulfuric acid, which will cause injury if splashed on your skin or in your eyes. It will also ruin clothes and painted surfaces. When removing the battery cables, always detach the negative cable first and hook it up last!*

9.6b Removing the cable from a battery post with a wrench - sometimes special battery pliers are required for this procedure if corrosion has caused deterioration of the nut hex (always remove the ground cable first and hook it up last!)

Check and maintenance

1 A routine preventive maintenance program for the battery in your vehicle is the only way to ensure quick and reliable starts. But before performing any battery maintenance, make sure that you have the proper equipment necessary to work safely around the battery **(see illustration)**.

2 There are also several precautions that should be taken whenever battery maintenance is performed. Before servicing the battery, always turn the engine and all accessories off and disconnect the cable from the negative terminal of the battery.

3 The battery produces hydrogen gas, which is both flammable and explosive. Never create a spark, smoke or light a match around the battery. Always charge the battery in a ventilated area.

4 Electrolyte contains poisonous and corrosive sulfuric acid. Do not allow it to get in your eyes, on your skin on your clothes. Never ingest it. Wear protective safety glasses when working near the battery. Keep children away from the battery.

5 Note the external condition of the battery. If the positive terminal and cable clamp on your vehicle's battery is equipped with a rubber protector, make sure that it's not torn or damaged. It should completely cover the terminal. Look for any corroded or loose connections, cracks in the case or cover or loose hold-down clamps. Also check the entire length of each cable for cracks and frayed conductors.

6 If corrosion, which looks like white, fluffy deposits **(see illustration)** is evident, particularly around the terminals, the battery should be removed for cleaning. Loosen the cable clamp bolts with a wrench, being careful to remove the ground cable first, and slide them off the terminals **(see illustration)**. Then disconnect the hold-down clamp bolt and nut, remove the

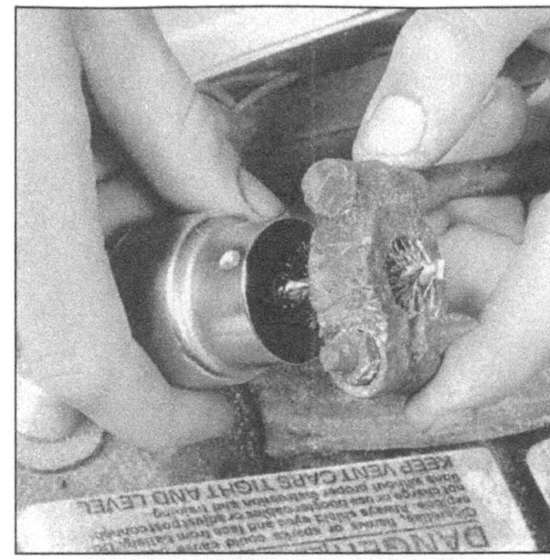

9.7a When cleaning the cable clamps, all corrosion must be removed (the inside of the clamp is tapered to match the taper on the post, so don't remove too much material)

9.7b Regardless of the type of tool used on the battery posts, a clean, shiny surface should be the result

clamp and lift the battery from the engine compartment.

7 Clean the cable clamps thoroughly with a battery brush or a terminal cleaner and a solution of warm water and baking soda **(see illustration)**. Wash the terminals and the top of the battery case with the same solution but make sure that the solution doesn't get into the battery. When cleaning the cables, terminals and battery top, wear safety goggles and rubber gloves to prevent any solution from coming in contact with your eyes or hands. Wear old clothes too - even diluted, sulfuric acid splashed onto clothes will burn holes in them. If the terminals have been extensively corroded, clean them up with a terminal cleaner **(see illustration)**. Thoroughly wash all cleaned areas with plain water.

8 Make sure that the battery tray is in good condition and the hold-down clamp bolts are tight. If the battery is removed from the tray, make sure no parts remain in the bottom of the tray when the battery is rein-

stalled. When reinstalling the hold-down clamp bolts, do not overtighten them.

9 Any metal parts of the vehicle damaged by corrosion should be covered with a zinc-based primer, then painted.

10 Information on removing and installing the battery can be found in Chapter 5. Information on jump starting can be found at the front of this manual. For more detailed battery checking procedures, refer to the *Haynes Automotive Electrical Manual*.

Charging

Warning: *When batteries are being charged, hydrogen gas, which is very explosive and flammable, is produced. Do not smoke or allow open flames near a charging or a recently charged battery. Wear eye protection when near the battery during charging. Also, make sure the charger is unplugged before connecting or disconnecting the battery from the charger.*

Note: *The manufacturer recommends the battery be removed from the vehicle for charging because the gas that escapes during this procedure can damage the paint or interior, depending on the model. Fast charging with the battery cables connected can result in damage to the electrical system.*

11 Slow-rate charging is the best way to restore a battery that's discharged to the point where it will not start the engine. It's also a good way to maintain the battery charge in a vehicle that's only driven a few miles between starts. Maintaining the battery charge is particularly important in the winter when the battery must work harder to start the engine and electrical accessories that drain the battery are in greater use.

12 It's best to use a one or two-amp battery charger (sometimes called a "trickle" charger). They are the safest and put the least strain on the battery. They are also the least expensive. For a faster charge, you can use a higher amperage charger, but don't use one rated more than 1/10th the amp/hour rating of the battery. Rapid boost charges that claim to restore the power of the battery in one to two hours are hardest on the battery and can damage batteries not in good condition. This type of charging should only be used in emergency situations.

13 The average time necessary to charge a battery should be listed in the instructions that come with the charger. As a general rule, a trickle charger will charge a battery in 12 to 16 hours.

14 Remove all the cell caps (if equipped) and cover the holes with a clean cloth to prevent spattering electrolyte. Disconnect the negative battery cable and hook the battery charger cable clamps up to the battery posts (positive to positive, negative to negative), then plug in the charger. Make sure it is set at 12-volts if it has a selector switch.

15 If you're using a charger with a rate higher than two amps, check the battery regularly during charging to make sure it doesn't overheat. If you're using a trickle charger, you can safely let the battery charge overnight after you've checked it regularly for the first couple of hours.

16 If the battery has removable cell caps, measure the specific gravity with a hydrometer every hour during the last few hours of the charging cycle. Hydrometers are available inexpensively from auto parts stores - follow the instructions that come with the hydrometer. Consider the battery charged when there's no change in the specific gravity reading for two hours and the electrolyte in the cells is gassing (bubbling) freely. The specific gravity reading from each cell should be very close to the others. If not, the battery probably has a bad cell(s).

17 Some batteries with sealed tops have built-in hydrometers on the top that indicate the state of charge by the color displayed in the hydrometer window. Normally, a bright-colored hydrometer indicates a full charge and a dark hydrometer indicates the battery still needs charging.

18 If the battery has a sealed top and no built-in hydrometer, you can hook up a digital voltmeter across the battery terminals to check the charge. A fully charged battery should read 12.6 volts or higher.

19 Further information on the battery and jump starting can be found in Chapter 5 and at the front of this manual.

10 Cooling system check (every 7500 miles or 6 months)

Refer to illustration 10.4

1 Many major engine failures can be attributed to a faulty cooling system. If the vehicle is equipped with an automatic transmission, the cooling system also cools the transmission fluid and thus plays an important role in prolonging transmission life.

2 The cooling system should be checked with the engine cold. Do this before the vehicle is driven for the day or after it has been shut off for at least three hours.

3 Remove the radiator cap by turning it to the left until it reaches a stop. If you hear a hissing sound (indicating there is still pressure in the system), wait until this stops. Now press down on the cap with the palm of your hand and continue turning to the left until the cap can be removed. Thoroughly clean the cap, inside and out, with clean water. Also clean the filler neck on the radiator. All traces of corrosion should be removed. The coolant inside the radiator should be relatively transparent. If it is rust colored, the system should be drained and refilled (see Section 35). If the coolant level is not up to the top, add additional antifreeze/coolant mixture (see Section 4).

4 Carefully check the large upper and lower radiator hoses along with the smaller diameter heater hoses which run from the engine to the firewall. Inspect each hose along its entire length, replacing any hose which is cracked, swollen or shows signs of deterioration. Cracks may become more apparent if the hose is squeezed **(see illustration)**. Regardless of condition, it's a good idea to replace hoses with new ones every two years.

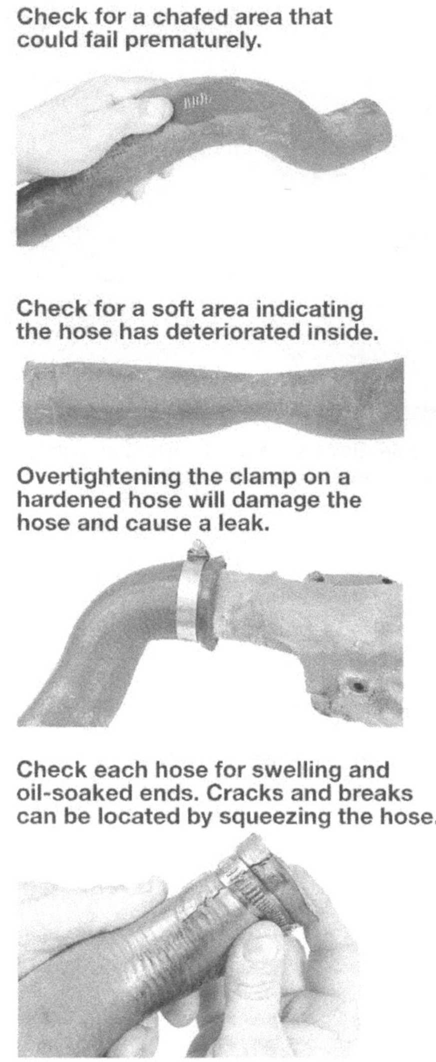

10.4 Hoses, like drivebelts, have a habit of failing at the worst possible time - to prevent the inconvenience of a blown radiator or heater hose, inspect them carefully as shown here

5 Make sure all hose connections are tight. A leak in the cooling system will usually show up as white or rust colored deposits on the areas adjoining the leak. If wire-type clamps are used at the ends of the hoses, it may be a good idea to replace them with more secure screw-type clamps.

6 Use compressed air or a soft brush to remove bugs, leaves, etc. from the front of the radiator or air conditioning condenser. Be careful not to damage the delicate cooling fins or cut yourself on them.

7 Every other inspection, or at the first indication of cooling system problems, have the cap and system pressure tested. If you don't have a pressure tester, most gas stations and repair shops will do this for a minimal charge.

11.1 Air conditioning hoses are identified by the metal tubes used at connections (arrow) - DO NOT disconnect or accidentally damage the air conditioning hoses (the system is under high pressure)

11 Underhood hose check and replacement (every 7500 miles or 6 months)

Refer to illustration 11.1

General

1 **Caution:** *Replacement of air conditioning hoses must be left to a dealer service department or air conditioning shop that has the equipment to depressurize the system safely. Never remove air conditioning components or hoses* **(see illustration)** *until the system has been depressurized.*

2 High temperatures in the engine compartment can cause the deterioration of the rubber and plastic hoses used for engine, accessory and emission systems operation. Periodic inspection should be made for cracks, loose clamps, material hardening and leaks. Information specific to the cooling system hoses can be found in Section 10.

3 Some, but not all, hoses are secured to the fittings with clamps. Where clamps are used, check to be sure they haven't lost their tension, allowing the hose to leak. If clamps aren't used, make sure the hose has not expanded and/or hardened where it slips over the fitting, allowing it to leak.

Vacuum hoses

4 It's quite common for vacuum hoses, especially those in the emissions system, to be color coded or identified by colored stripes molded into them. Various systems require hoses with different wall thicknesses, collapse resistance and temperature resistance. When replacing hoses, be sure the new ones are made of the same material.

5 Often the only effective way to check a hose is to remove it completely from the vehicle. If more than one hose is removed, be sure to label the hoses and fittings to ensure correct installation.

6 When checking vacuum hoses, be sure to include any plastic T-fittings in the check.

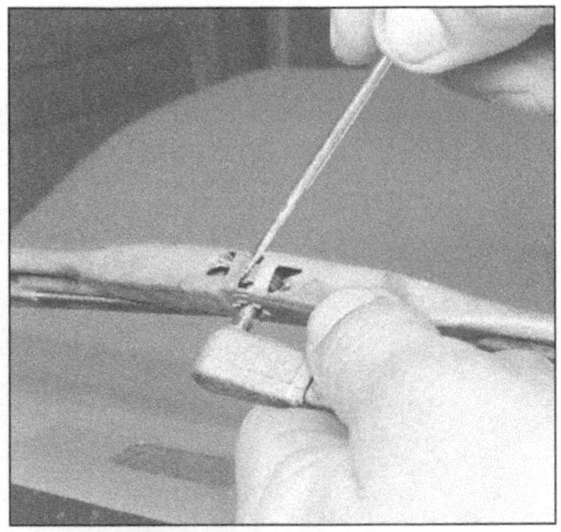

12.6 Use a small screwdriver to pry the release lever up (on some models, you will have to press down on a tab)

Inspect the fittings for cracks and the hose where it fits over the fitting for distortion, which could cause leakage.

7 A small piece of vacuum hose (1/4-inch inside diameter) can be used as a stethoscope to detect vacuum leaks. Hold one end of the hose to your ear and probe around vacuum hoses and fittings, listening for the "hissing" sound characteristic of a vacuum leak. **Warning:** *When probing with the vacuum hose stethoscope, be very careful not to come into contact with moving engine components such as the drivebelt, cooling fan, etc.*

Fuel hose

Warning: *There are certain precautions which must be taken when inspecting or servicing fuel system components. Work in a well ventilated area and do not allow open flames (cigarettes, appliance pilot lights, etc.) or bare light bulbs near the work area. Mop up any spills immediately and do not store fuel soaked rags where they could ignite. On fuel-injected models, the fuel system is under high pressure, so if any fuel lines are to be disconnected, the pressure in the system must be relieved first (see Chapter 4 for more information).*

8 Check all rubber fuel lines for deterioration and chafing. Check especially for cracks in areas where the hose bends and just before fittings, such as where a hose attaches to the fuel filter.

9 High quality fuel line, usually identified by the word *Fluoroelastomer* printed on the hose, should be used for fuel line replacement. Never, under any circumstances, use unreinforced vacuum line, clear plastic tubing or water hose for fuel lines.

10 Spring-type clamps are commonly used on fuel lines. These clamps often lose their tension over a period of time, and can be "sprung" during removal. Replace all spring-type clamps with screw clamps whenever a hose is replaced.

Metal lines

11 Sections of metal line are often used for fuel line between the fuel pump and carburetor. Check carefully to be sure the line has not been bent or crimped and that cracks have not started in the line.

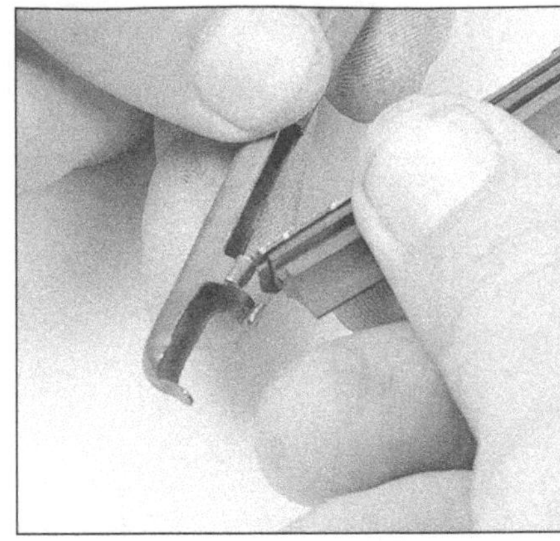

12.7a On some models, you'll have to squeeze the tabs on the end of the blade, then slide the element out of the blade

12 If a section of metal fuel line must be replaced, only seamless steel tubing should be used, since copper and aluminum tubing don't have the strength necessary to withstand normal engine vibration.

13 Check the metal brake lines where they enter the master cylinder and brake proportioning unit (if used) for cracks in the lines or loose fittings. Any sign of brake fluid leakage calls for an immediate thorough inspection of the brake system.

12 Wiper blade inspection and replacement (every 7500 miles or 6 months)

Refer to illustrations 12.6, 12.7a and 12.7b

1 The windshield wiper and blade assembly should be inspected periodically for damage, loose components and cracked or worn blade elements.

2 Road film can build up on the wiper blades and affect their efficiency, so they should be washed regularly with a mild detergent solution.

3 The action of the wiping mechanism can loosen the bolts, nuts and fasteners, so they should be checked and tightened, as necessary, at the same time the wiper blades are checked.

4 If the wiper blade elements (sometimes called inserts) are cracked, worn or warped, they should be replaced with new ones.

5 Pull the wiper blade/arm assembly away from the glass.

6 Insert a small screwdriver into the release slot and pry up or depress the tab (depending on model) to detach the blade assembly. Slide it off the wiper arm and over the retaining stud **(see illustration)**.

7 On some models, squeeze the tabs on the end, pull the blade out of the bridge claw, then slide the element out of the blade assembly bridge **(see illustration)**. On other

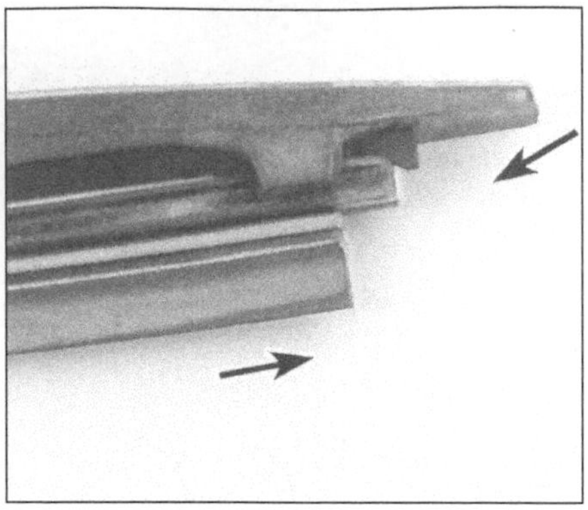

12.7b On other models, lift the end of the element up over the tang on the end of the blade, then slide the element out of the blade

models, lift the end of the assembly up and over the tang in the end of the bridge **(see illustration)**.

8 Compare the new element with the old for length, design, etc.

9 Slide the new element into place. It will automatically lock at the correct location.

10 Reinstall the blade assembly on the arm, wet the windshield and check for proper operation.

13 Suspension and steering check (every 7500 miles or 6 months)

Refer to illustration 13.4

1 Indications of a fault in these systems are excessive play in the steering wheel before the front wheels react, excessive sway around corners, body movement over rough roads or binding at some point as the steering wheel is turned.

2 Raise the front of the vehicle periodically and visually check the suspension and steering components for wear. Because of the work to be done, make sure the vehicle cannot fall from the stands.

3 Check the wheel bearings. Do this by spinning the front wheels. Listen for any abnormal noises and watch to make sure the wheel spins true (doesn't wobble). Grab the top and bottom of the tire and pull in-and-out on it. Notice any movement which would indicate a loose wheel bearing assembly. If the bearings are suspect, refer to Section 34 and Chapter 10 for more information.

4 From under the vehicle, check for loose bolts, broken or disconnected parts and deteriorated rubber bushings on all suspension and steering components. Look for fluid leaking from the steering gear assembly. Check the power steering hoses, belts and connections for leaks. Inspect the shock absorbers for fluid leaks, indicating the need for replacement **(see illustration)**.

5 Have an assistant turn the steering wheel from side-to-side and check the steering components for free movement, chafing and binding. If the steering doesn't react with the movement of the steering wheel, try to determine where the slack is located.

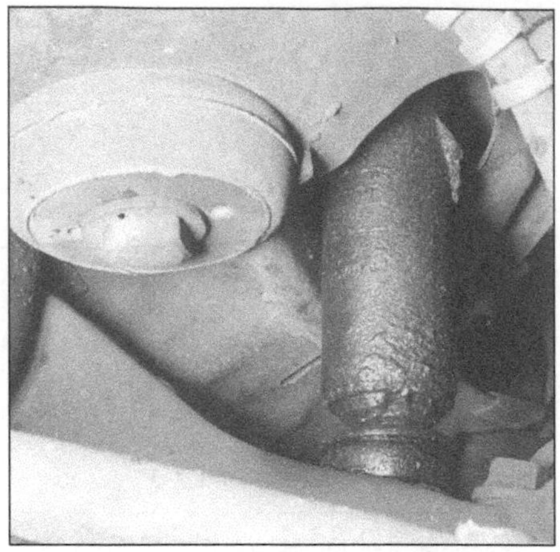

13.4 Inspect the shock absorbers for leaking fluid - this one is leaking badly and should be replaced (always replace shock absorbers in pairs)

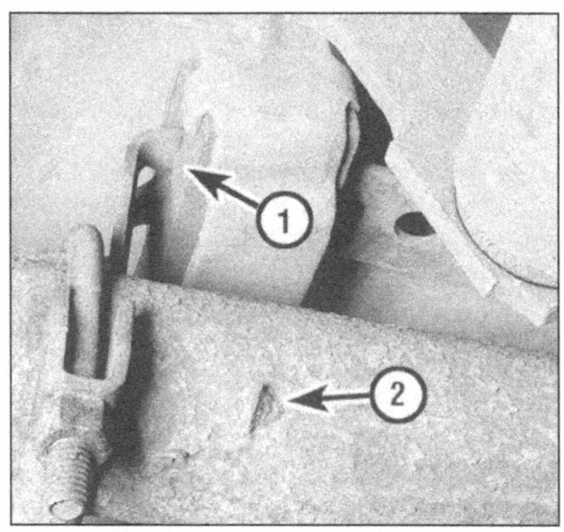

14.4b Inspect the rubber mounts (1) for damage, deterioration and sagging. Also check for rusted holes in the exhaust pipe and muffler (2); this pipe should be repaired or replaced

14 Exhaust system check (every 7500 miles or 6 months)

Refer to illustrations 14.4a and 14.4b

1 With the engine cold (at least three hours after the vehicle has been driven), check the complete exhaust system from the manifold to the end of the tailpipe. Be careful around the catalytic converter, which may be hot even after three hours. The inspection should be done with the vehicle on a hoist to permit unrestricted access. If a hoist isn't available, raise the vehicle and support it securely on jackstands.

2 Check the exhaust pipes and connections for signs of leakage and/or corrosion indicating a potential failure. Make sure that all brackets and hangers are in good condition and tight.

3 Inspect the underside of the body for holes, corrosion, open seams, etc. which may allow exhaust gasses to enter the passenger compartment. Seal all body openings with silicone sealant or body putty.

4 Rattles and other noises can often be traced to the exhaust system, especially the hangers, mounts and heat shields **(see illustrations)**. Try to move the pipes, mufflers and catalytic converter. If the components can come in contact with the body or suspension parts, secure the exhaust system with new brackets and hangers.

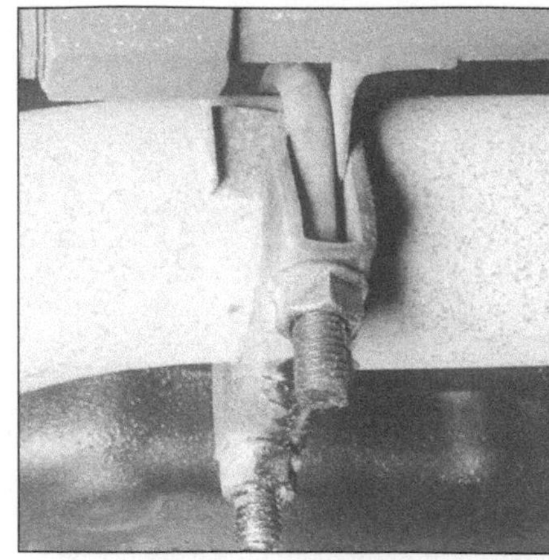

14.4a Check the exhaust system clamp nuts to make sure they are tight

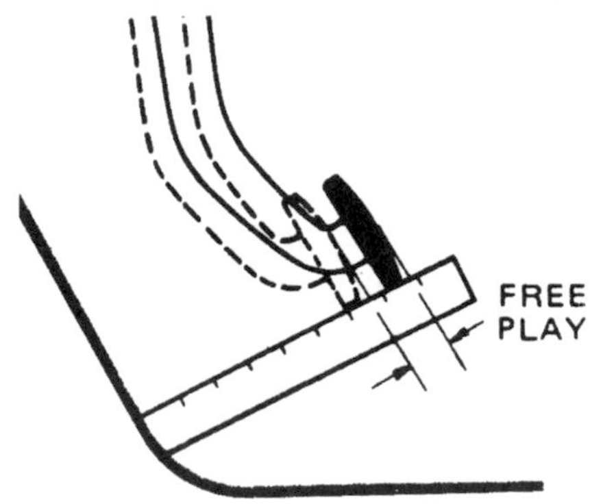

15.1 To determine the clutch pedal freeplay, depress the pedal, stop the moment clutch resistance is felt, then measure the distance (freeplay)

15 Clutch pedal freeplay check and adjustment (every 7500 miles or 6 months)

Refer to illustrations 15.1, 15.2 and 15.3

1 Press down lightly on the clutch pedal and, with a small steel ruler, measure the distance that it moves freely before the clutch resistance is felt **(see illustration)**. The freeplay should be within the specified limits. If it isn't, it must be adjusted.

2 On gearshift interlock-equipped models (some 1974 and earlier six-cylinder models), disconnect the device at the interlock rod swivel clamp screw before adjusting freeplay **(see illustration)**.

3 On all models, turn the self-locking adjusting nut on the clutch fork rod to achieve the specified freeplay at the pedal **(see illustration)**.

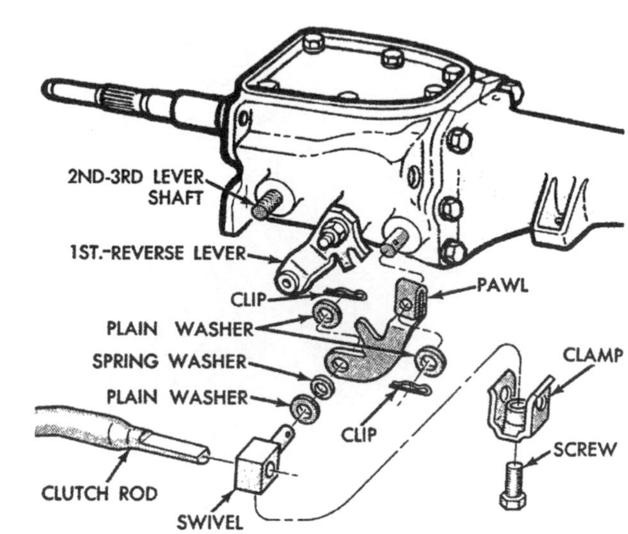

15.2 On interlock-equipped models, loosen the clamp screw on the clutch rod swivel before adjusting the clutch pedal freeplay

16 Manual transmission lubricant level check (every 7500 miles or 6 months)

Refer to illustration 16.2

1 The manual transmission has a filler plug which must be removed to check the lubricant level. If the vehicle is raised to gain access to the plug, be sure to support it safely on jackstands - DO NOT crawl under a vehicle which is supported only by a jack! Be sure the vehicle is level or the check may be inaccurate.

2 Using an open-end wrench, unscrew the plug from the transmission **(see illustration)** and use your little finger to reach inside the housing to feel the lubricant level. The level should be at or near the bottom of the plug hole.

3 If it isn't, add the recommended lubricant through the plug hole with a syringe or squeeze bottle.

4 Install and tighten the plug and check for leaks after the first few miles of driving.

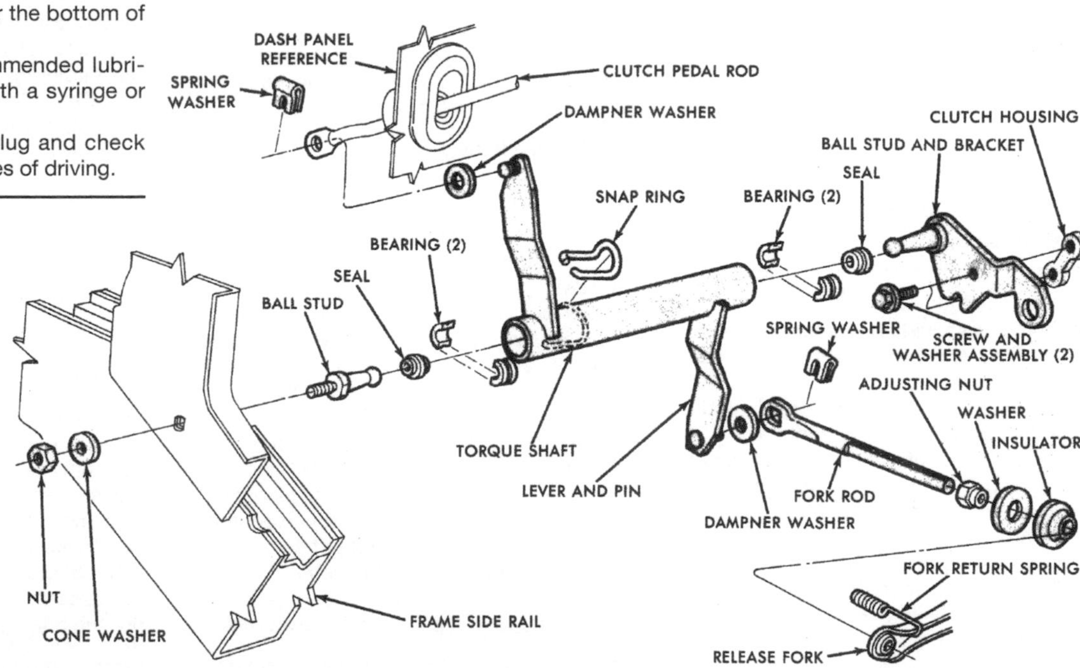

15.3 Adjust clutch pedal freeplay at the fork rod

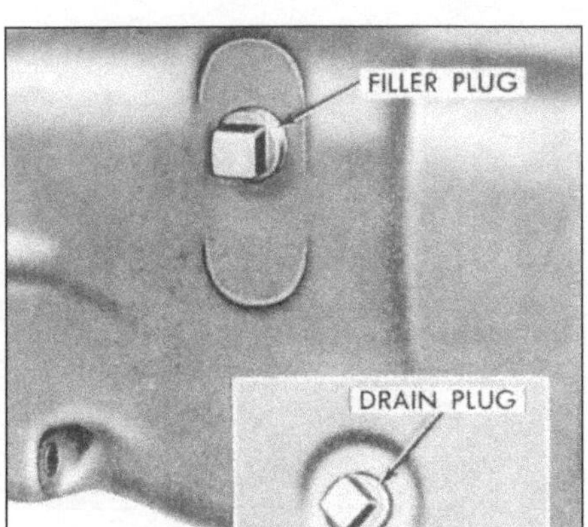

16.2 Typical filler plug and drain plug locations on manual transmissions - use an open-end wrench to unscrew the plugs

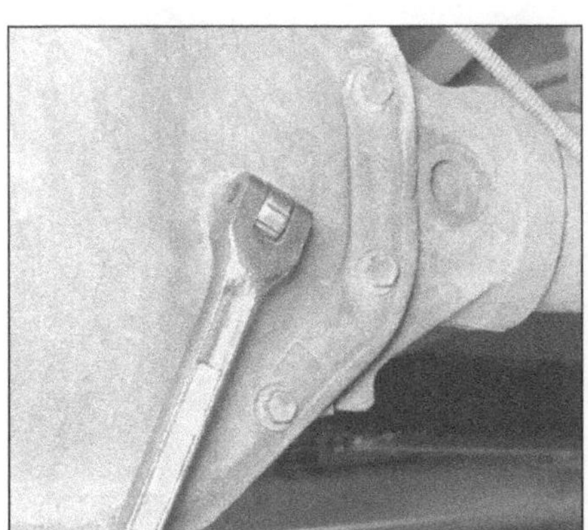

17.2a On some models, the inspection plug is removed by unscrewing it from the differential case

17 Differential lubricant level check (every 7500 miles or 6 months)

Refer to illustrations 17.2a, 17.2b and 17.3

1 The rear differential has a threaded metal or rubber press-in filler plug which must be removed to check the lubricant level. If the vehicle is raised to gain access to the plug, be sure to support it safely on jackstands - DO NOT crawl under the vehicle when it's supported only by the jack. Be sure the vehicle is level or the check may not be accurate.

2 Remove the plug from the filler hole in the differential housing or cover **(see illustrations)**.

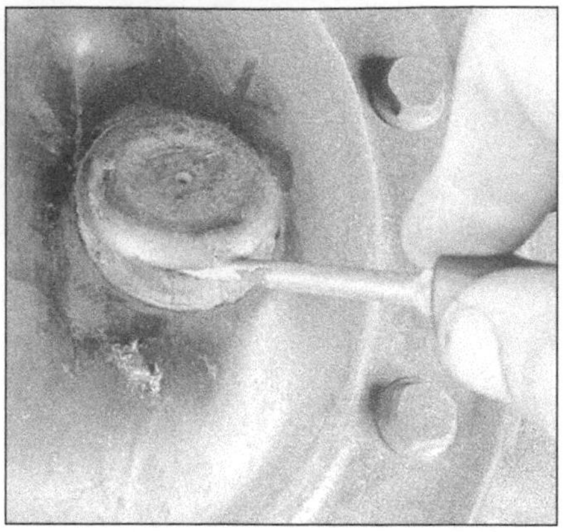

17.2b On later models, use a small screwdriver to pry out the rubber plug

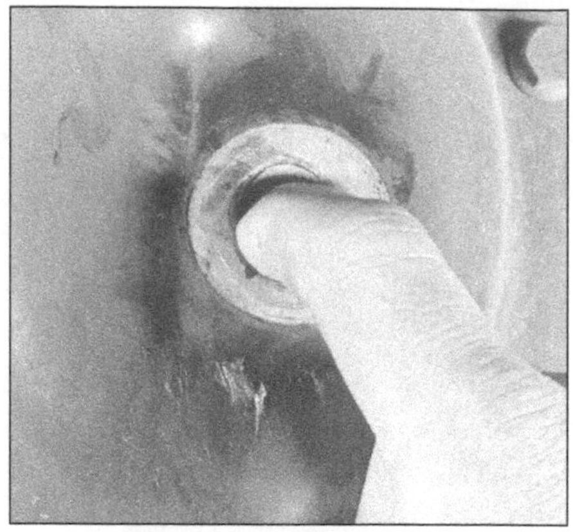

17.3 Use your finger as a dipstick to check the lubricant level

3 The lubricant level should be at the bottom of the filler hole **(see illustration)**. If not, use a syringe to add the recommended lubricant until it just starts to run out of the opening. On some models a tag is located in the area of the plug which gives information regarding lubricant type, particularly on models equipped with Sure Grip.

4 Install the plug securely into the filler hole.

18 Tire rotation (every 7500 miles or 6 months)

Refer to illustration 18.2

1 The tires should be rotated at the specified intervals and whenever uneven wear is noticed.

2 Refer to the accompanying illustration for the preferred tire rotation pattern.

3 Refer to the information in *Jacking and towing* at the front of this manual for the proper procedures to follow when raising the vehicle and changing a tire. If the brakes are to be checked, don't apply the parking brake as stated. Make sure the tires are blocked to prevent the vehicle from rolling as it's raised.

4 Preferably, the entire vehicle should be raised at the same time. This can be done on a hoist or by jacking up each corner and then lowering the vehicle onto jackstands placed under the frame rails. Always use four jackstands and make sure the vehicle is safely supported.

5 After rotation, check and adjust the tire pressures as necessary and be sure to check the lug nut tightness.

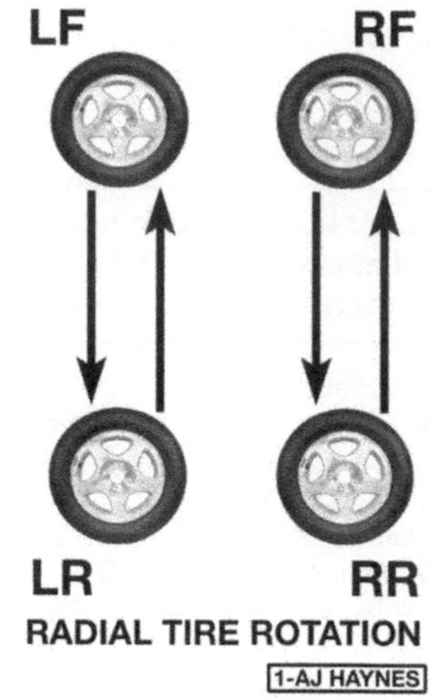

18.2 The recommended tire rotation pattern for these vehicles

19 Brake check (every 7500 miles or 6 months)

Note: *For detailed photographs of the brake system, refer to Chapter 9.*

Warning: *Brake system dust may contain asbestos, which is hazardous to your health. DO NOT blow it out with compressed air, inhale it or use gasoline or solvents to remove it. Use brake system cleaner only.*

1 In addition to the specified intervals, the brakes should be inspected every time the wheels are removed or whenever a defect is suspected.

2 To check the brakes, raise the vehicle and place it securely on jackstands. Remove the wheels (see *Jacking and towing* at the front of the manual, if necessary).

Disc brakes

Refer to illustrations 19.5a and 19.5b

3 Disc brakes are used on the front wheels on most models. Extensive rotor damage can occur if the pads are not replaced when needed.

4 The disc brake calipers, which contain the pads, are visible with the wheels removed. There is an outer pad and an inner pad in each caliper. All pads should be inspected.

5 Each caliper has a "window" or opening to inspect the pads. Check the thickness of the pad lining by looking into the caliper at each end and down through the inspection window at the top of the housing **(see illustrations)**. If the pad material has worn to about 5/16-inch or less, the pads should be replaced.

6 If you're unsure about the exact thickness of the remaining lining material, remove the pads for further inspection or replacement (refer to Chapter 9).

7 Before installing the wheels, check for leakage and/or damage (cracks, splitting, etc.) around the brake hose connections. Replace the hose or fittings as necessary, referring to Chapter 9.

8 Check the condition of the rotor. Look for score marks, deep scratches and burned spots. If these conditions exist, the hub/rotor assembly should be removed for servicing (see Section 34).

Drum brakes

Refer to illustrations 19.11, 19.13 and 19.15

9 On front brakes, remove the drum/hub assembly as described in Section 34. On rear brakes, remove the drum by pulling it off the axle and brake assembly. If this proves difficult, make sure the parking brake is released, then squirt penetrating oil around the center hub areas. Allow the oil to soak in and try to pull the drum off again.

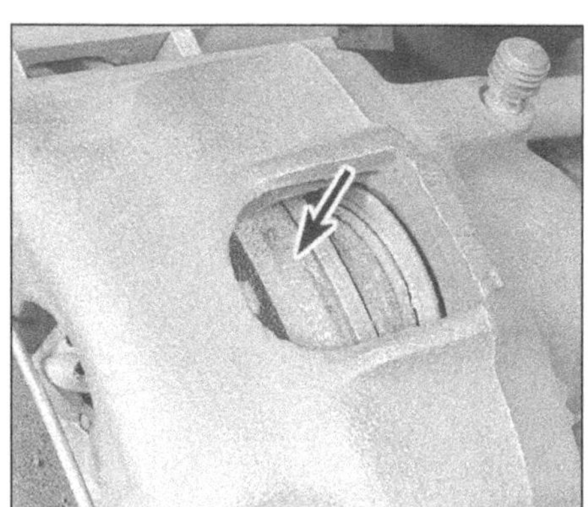

19.5a You will find an inspection hole like this in most calipers - placing a ruler across the hole should enable you to determine the thickness of remaining pad material

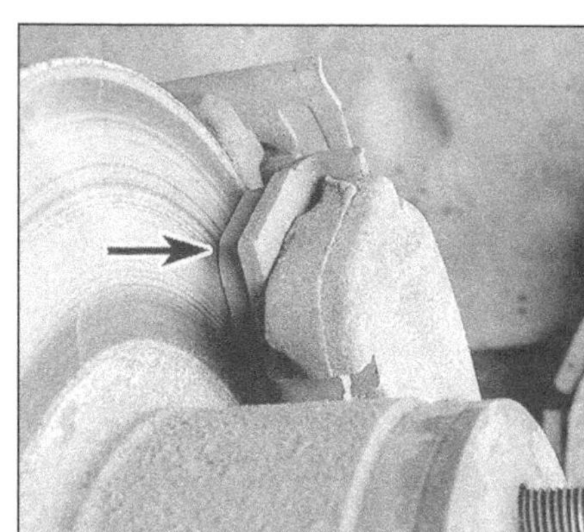

19.5b Some models don't have an inspection hole, but you can check the pad thickness by looking at the end of the caliper

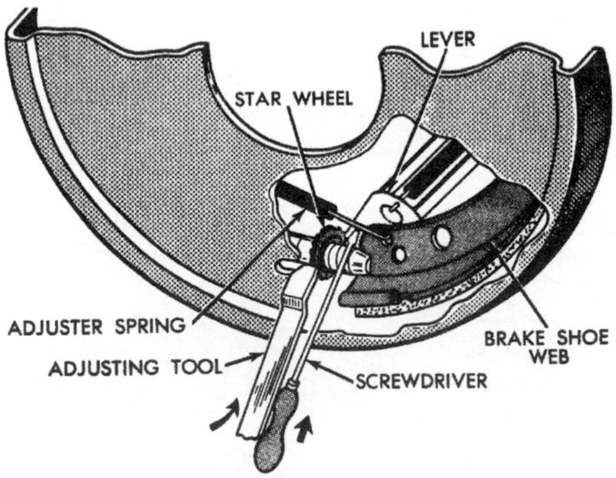

19.11 If the brake drum won't pull off the hub, use a thin screwdriver to push the lever away, then use an adjusting tool or another screwdriver to back off the star wheel

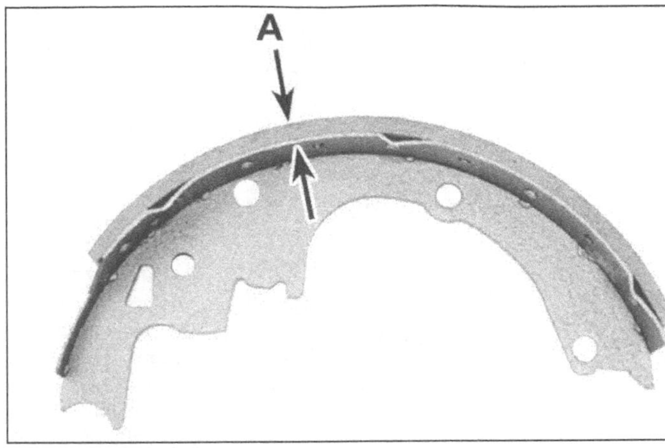

19.13 If the lining is bonded to the brake shoe, measure the lining thickness from the outer surface to the metal shoe, as shown here; if the lining is riveted to the shoe, measure from the lining outer surface to the rivet head

10 If the drum still cannot be pulled off, the parking brake lever will have to be lifted slightly off its stop. This is done by first removing the small plug from the backing plate.

11 With the plug removed, insert a thin screwdriver and lift the adjusting lever off the star wheel, then use an adjusting tool or screwdriver to back off the star wheel several turns **(see illustration)**. This will move the brake shoes away from the drum. If the drum still won't pull off, tap around its inner circumference with a soft-face hammer.

12 With the drum removed, do not touch any brake dust (see the Warning at the beginning of this Section).

13 Note the thickness of the lining material on both the front and rear brake shoes. If the material has worn away to within 1/16-inch of the recessed rivets or metal backing, the shoes should be replaced **(see illustration)**. The shoes should also be replaced if they're cracked, glazed (shiny surface) or contaminated with brake fluid.

14 Make sure that all the brake assembly springs are connected and in good condition.

15 Check the brake components for any signs of fluid leakage. Carefully pry back the rubber cups on the wheel cylinders located at the top of the brake shoes **(see illustration)**. Any leakage is an indication that the wheel cylinders should be overhauled immediately (see Chapter 9). Also check brake hoses and connections for signs of leakage.

16 Wipe the inside of the drum with a clean rag and brake cleaner or denatured alcohol. Again, be careful not to breathe the dangerous asbestos dust.

17 Check the inside of the drum for cracks, score marks, deep scratches and hard spots, which will appear as small discolorations. If these imperfections cannot be removed with fine emery cloth, the drum must be taken to a machine shop equipped to turn the drums.

18 If after the inspection process all parts are in good working condition, reinstall the brake drum.

19 Install the wheels and lower the vehicle.

19.15 To check for wheel cylinder leakage, use a small screwdriver to pry the boot away from the cylinder

Parking brake

20 The parking brake is operated by a hand lever or a foot pedal and locks the rear brake system. The easiest, and perhaps most obvious method of periodically checking the operation of the parking brake assembly is to stop the vehicle on a steep hill with the parking brake set and the transmission in Neutral. If the parking brake cannot prevent the vehicle from rolling, adjust it (see Chapter 9).

20 Fuel system check (every 7500 miles or 6 months)

Refer to illustration 20.5

Warning: *Gasoline is extremely flammable, so take extra precautions when working on any part of the fuel system. Don't smoke or allow open flames or bare light bulbs in or near the work area, and don't work in a garage where a natural gas-type appliance (such as a water heater or clothes dryer) with a pilot light is present. Since gasoline is carcinogenic, wear latex gloves when there's a possibility of being exposed to fuel, and, if you spill fuel on your skin, rinse it off immediately with soap and water. Have a Class B fire extinguisher on hand.*

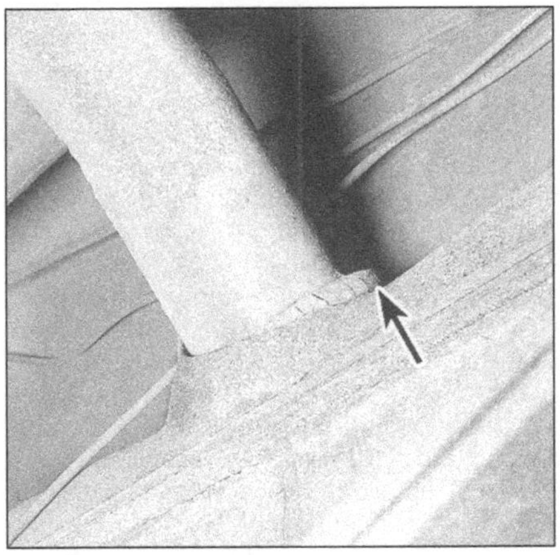

20.5 Check the fuel tank fill connections, as well as all hoses connected to the tank, for cracks or leaks. Leaks can often be found by looking for clean, damp areas around the hoses

1 The fuel tank is located under the rear of the vehicle.

2 The fuel system is most easily checked with the vehicle raised on a hoist so the components underneath the vehicle are readily visible and accessible.

3 If the smell of gasoline is noticed while driving or after the vehicle has been in the sun, the system should be thoroughly inspected immediately.

4 Remove the gas tank cap and check for damage, corrosion and an unbroken sealing imprint on the gasket. Replace the cap with a new one if necessary.

5 With the vehicle raised, check the fuel tank and filler neck for punctures, cracks and other damage **(see illustration)**. The connection between the filler neck and the tank is especially critical. Sometimes a rubber filler neck will leak due to loose clamps or deterio-

21.4 There are usually four carburetor mounting nuts/bolts (arrows) on V8 models, but many six-cylinder models only have two

rated rubber; problems a home mechanic can usually rectify. **Warning:** *Do not, under any circumstances, try to repair a fuel tank yourself (except rubber components). A welding torch or any open flame can easily cause the fuel vapors to explode if the proper precautions are not taken!*

6 Carefully check all rubber hoses and metal lines leading away from the fuel tank. Look for loose connections, deteriorated hoses, crimped lines and other damage. Follow the lines to the front of the vehicle, carefully inspecting them all the way. Repair or replace damaged sections as necessary.

7 If a fuel odor is still evident after the inspection, refer to Section 37.

21 Carburetor mounting bolt/nut torque check

Refer to illustration 21.4

1 The carburetor is attached to the top of the intake manifold by bolts and/or nuts. These fasteners can sometimes work loose from vibration and temperature changes during normal engine operation and cause a vacuum leak.

2 If you suspect that a vacuum leak exists at the bottom of the carburetor, obtain a length of hose. Start the engine and place one end of the hose next to your ear as you probe around the base with the other end. You will hear a hissing sound if a leak exists (be careful of hot or moving engine components).

3 Remove the air cleaner assembly (see Chapter 4), tagging each hose to be disconnected with a piece of numbered tape to make reassembly easier.

4 Locate the mounting bolts at the base of the carburetor. Decide what special tools or adapters will be necessary, if any, to tighten the fasteners **(see illustration)**.

5 Tighten the bolts or nuts to the torque listed in this Chapter's Specifications. Don't overtighten them, as the threads could strip.

6 If, after the bolts are properly tightened, a vacuum leak still exists, the carburetor must be removed and a new gasket installed. See Chapter 4 for more information.

7 After tightening the fasteners, reinstall the air cleaner and return all hoses to their original positions.

22 Throttle linkage inspection (every 7500 miles or 6 months)

1 Inspect the throttle linkage for damage and missing parts and for binding and interference when the accelerator pedal is depressed.

2 Lubricate the various linkage pivot points with chassis grease. Do not lubricate carburetor or throttle body linkage balljoint type pivots, throttle control cables or transmission control cables.

23 Drivebelt check, adjustment and replacement (every 7500 miles or 6 months)

Refer to illustrations 23.3, 23.4 and 23.6

1 The drivebelts, or V-belts as they are often called, are located at the front of the engine and play an important role in the overall operation of the vehicle and its components. Due to their function and material make-up, the belts are prone to failure after a period of time and should be inspected and adjusted periodically to prevent major engine damage.

2 The number of belts used on a particular vehicle depends on the accessories installed. Drivebelts are used to turn the alternator, power steering pump, water pump and air-conditioning compressor. Depending on the pulley arrangement, more than one of these components may be driven by a single belt.

Check

3 With the engine off, open the hood and locate the various belts at the front of the engine. Using your fingers (and a flashlight, if necessary), move along the belts checking for cracks and separation of the belt plies. Also check for fraying and glazing, which gives the belt a shiny appearance **(see illustration)**. Both sides of each belt should be inspected, which means you will have to twist the belt to check the underside.

4 The tension of each V-belt is checked by pushing on the belt at a distance halfway between the pulleys. Push firmly with your thumb and see how much the belt moves (deflects) **(see illustration)**. A rule of thumb is that if the distance from pulley center-to-pulley center is between 7 and 11 inches, the belt should deflect 1/4-inch. If the belt travels between pulleys spaced 12 to 16 inches apart, the belt should deflect 1/2-inch. **Note:** *On serpentine belts, the belt tension is controlled by a tensioner and the belt does not require periodic adjustment.*

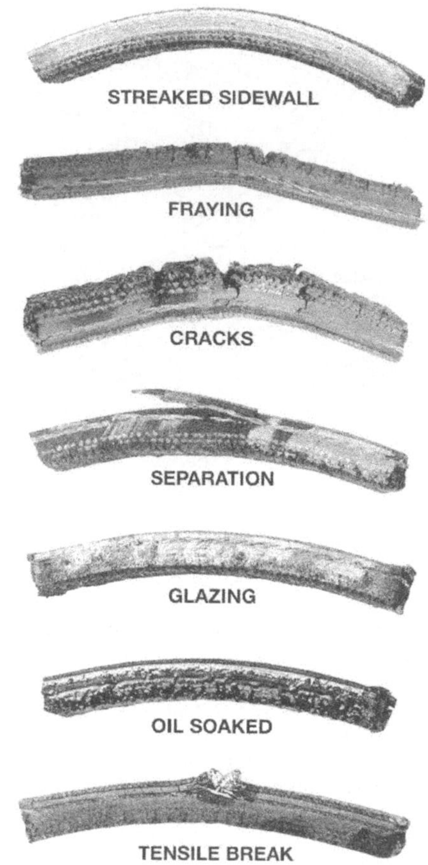

23.3 Here are some of the more common problems associated with drivebelts (check the belts very carefully to prevent an untimely breakdown)

Adjustment

5 If it is necessary to adjust the belt tension, either to make the belt tighter or looser, it is done by moving the belt-driven accessory on the bracket.

6 For each component there will be an adjusting bolt and a pivot bolt. Both bolts must be loosened slightly to enable you to move the component **(see illustration)**.

7 After the two bolts have been loosened, move the component away from the engine to tighten the belt or toward the engine to loosen the belt. Hold the accessory in position and check the belt tension. If it is correct, tighten the two bolts until just snug, then recheck the tension. If the tension is all right, tighten the bolts.

8 To adjust the alternator drivebelt on some later models, loosen the pivot bolt and locknut, then turn the adjusting bolt to tension the belt. After the belt is tensioned, tighten the pivot bolt and locknut.

9 It will often be necessary to use some sort of prybar to move the accessory while the belt is adjusted. If this must be done to gain the proper leverage, be very careful not to damage the component being moved or the part being pried against.

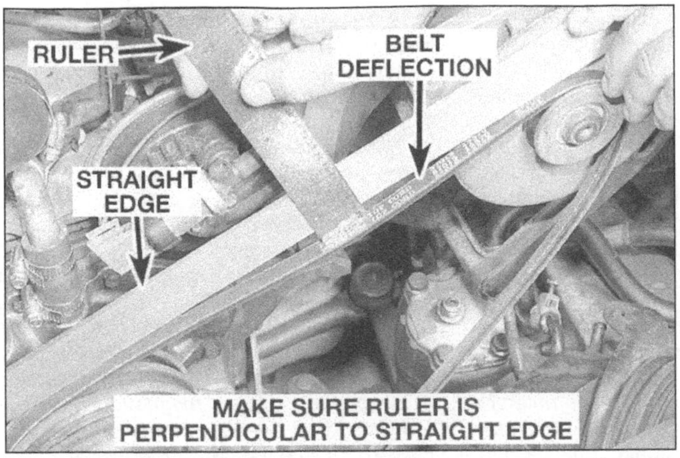

23.4 Measuring drivebelt deflection with a straightedge and ruler

23.6 With the pivot bolt and adjusting bolt (arrow) loose, pivot the alternator up to increase the drivebelt tension, then tighten the pivot bolt and adjusting bolts

Replacement

10 To replace a belt, follow the above procedures for drivebelt adjustment but slip the belt off the pulleys and remove it. Since belts tend to wear out more or less at the same time, it's a good idea to replace all of them at the same time. Mark each belt and the corresponding pulley grooves so the replacement belts can be installed properly.
11 Take the old belts with you when purchasing new ones in order to make a direct comparison for length, width and design.
12 Adjust the belts as described earlier in this Section.

26.1 Materials required for chassis and body lubrication

1 ***Engine oil*** *- Light engine oil in a can like this can be used for door and hood hinges*
2 ***Graphite spray*** *- Used to lubricate lock cylinders*
3 ***Grease*** *- Grease, in a variety of types and weights, is available for use in a grease gun. Check the Specifications for your requirements*
4 ***Grease gun*** *- A common grease gun, shown here with a detachable hose and nozzle, is needed for chassis lubrication. After use, clean it thoroughly!*

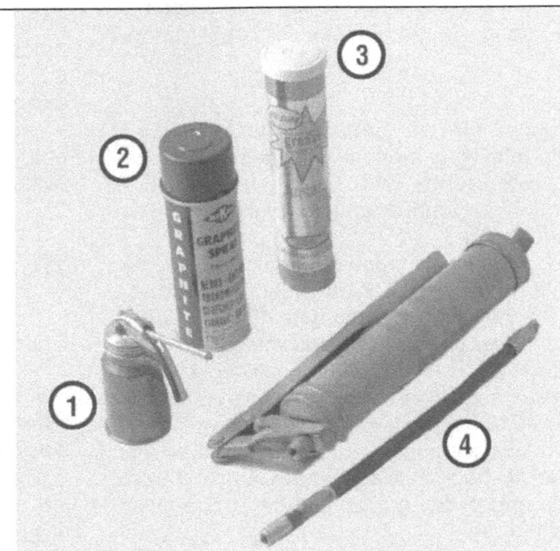

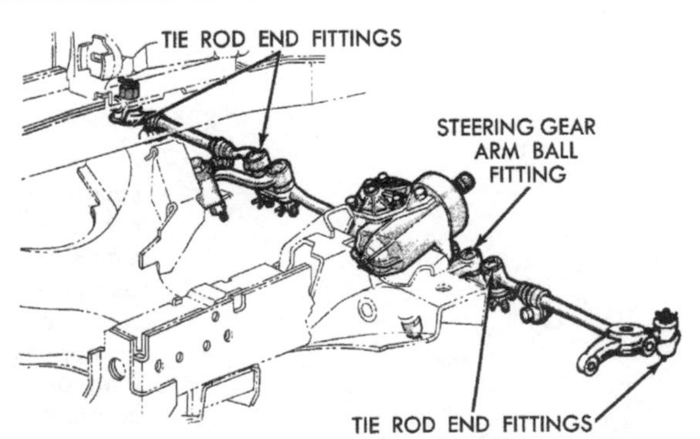

26.2 Here are the typical points on the steering linkage where you'll find grease fittings - there are also four suspension balljoints that need lubrication - one at the top and bottom, just to the inside of each front wheel. The suspension balljoints usually require more grease than the steering-linkage components

24 Seat belt check (every 7500 miles or 6 months)

1 Check the seat belts, buckles, latch plates and guide loops for any obvious damage or signs of wear.
2 On later models, make sure the seat belt reminder light comes on when the key is turned on.
3 The seat belts are designed to lock up during a sudden stop or impact, yet allow free movement during normal driving. The retractors should hold the belt against your chest while driving and rewind the belt when the buckle is unlatched.
4 If any of the above checks reveal problems with the seat belt system, replace parts as necessary.

25 Neutral start switch check (models with automatic transmission) (every 7500 miles or 6 months)

Warning: *During the following checks there is a chance the vehicle could lunge forward, possibly causing damage or injuries. Allow plenty of room around the vehicle, apply the parking brake firmly and hold-down the regular brake pedal during the checks.*

1 The neutral start switch prevents the engine from being cranked unless the gear selector is in Park or Neutral.
2 Try to start the vehicle in each gear. The engine should crank only in Park or Neutral. If it cranks in any other gear, the neutral start switch is faulty or in need of adjustment (see Chapter 7 Part B).
3 Make sure the steering column lock allows the key to go into the Lock position only when the shift lever is in Park.
4 The ignition key should come out only in the Lock position.

26 Chassis lubrication (every 15,000 miles or 12 months)

Refer to illustrations 26.1, 26.2 and 26.6

1 Refer to *Recommended lubricants and fluids* at the front of this Chapter to obtain the necessary grease, etc. You'll also need a grease gun **(see illustration)**. Occasionally plugs will be installed rather than grease fittings. If so, grease fittings will have to be purchased and installed.
2 Look under the vehicle and locate the

26.6 Pump the grease into each suspension balljoint fitting until the rubber seal (arrow) is firm to the touch; don't pump in too much grease or you'll rupture the seal!

grease fittings **(see illustration)**. Some models with manual transmissions have grease fittings on the shift linkage which should be lubricated whenever the chassis is lubricated.

3 For easier access under the vehicle, raise it with a jack and place jackstands under the frame. Make sure it's safely supported by the stands. If the wheels are to be removed at this interval for tire rotation or brake inspection, loosen the lug nuts slightly while the vehicle is still on the ground.

4 Before beginning, force a little grease out of the nozzle to remove any dirt from the end of the gun. Wipe the nozzle clean with a rag.

5 With the grease gun and plenty of clean rags, crawl under the vehicle and begin lubricating the components.

6 Wipe one of the grease fitting nipples clean and push the nozzle firmly over it. Pump the gun until the component is completely lubricated. On balljoints, stop pumping when the rubber seal is firm to the touch **(see illustration)**. Do not pump too much grease into the fitting as it could rupture the seal. For all other suspension and steering components, continue pumping grease into the fitting until it oozes out of the joint between the two components. If it escapes around the grease gun nozzle, the nipple is clogged or the nozzle is not completely seated on the fitting. Resecure the gun nozzle to the fitting and try again. If necessary, replace the fitting with a new one.

7 Wipe the excess grease from the components and the grease fitting. Repeat the procedure for the remaining fittings.

8 Also clean and lubricate the parking brake cable, along with the cable guides and levers. This can be done by smearing some of the chassis grease onto the cable and its related parts with your fingers.

27 Ignition point check and replacement (1971 and 1972 models) (every 15,000 miles or 12 months)

Refer to illustrations 27.1 and 27.16

Caution: *The ignition key should be in the OFF position during this procedure.*

1 The ignition points must be replaced at regular intervals on vehicles not equipped with electronic ignition. Occasionally, the rubbing block on the points will wear sufficiently to require readjustment. Several special tools are required to replace the points **(see illustration)**.

2 After removing the distributor cap and rotor, the points and condenser are plainly visible. The points may be examined by gently prying them open to reveal the condition of the contact surfaces. If they are rough, pitted or dirty, they should be replaced. **Caution:** *The following procedure requires the removal and installation of small screws which can easily fall into the distributor. To retrieve them, the distributor would have to be removed and disassembled. Use a magnetic or spring-loaded screwdriver and exercise caution.*

3 To replace the condenser, which should be replaced along with the points, remove the screw that secures the condenser to the point plate. Loosen the nut or screw retaining the condenser wire and the primary wire to the point assembly. Remove the condenser and mounting bracket.

4 Remove the point assembly mounting screw.

5 Attach the new point assembly to the distributor plate with the mounting screw.

6 Install the new condenser and tighten the mounting screw.

7 Attach the primary ignition wire and the condenser lead to the point assembly. Make sure that the forked connectors for the primary ignition and condenser leads do not touch the distributor plate or any other grounded surface.

8 Two adjusting methods are available for setting the points. The first and most effective method involves an instrument called a dwell meter.

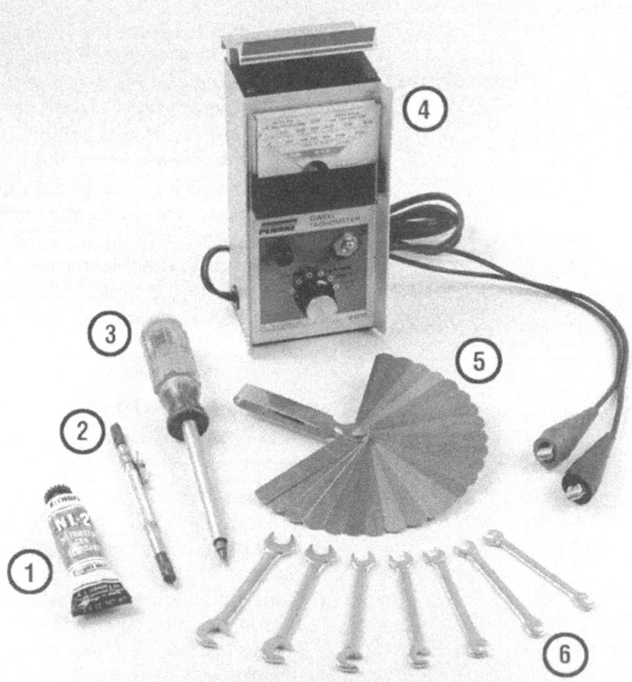

27.1 Tools and materials needed for ignition point replacement and dwell angle adjustment

1 ***Distributor cam lube*** - *Sometimes this special lubricant comes with the new points; however, its a good idea to buy a tube and have it on hand*

2 ***Screw starter*** - *This tool has special claws which hold the screw securely as it is started, which helps prevent accidental dropping of the screw*

3 ***Magnetic screwdriver*** - *Serves the same purpose as 2 above. If you do not have one of these special screwdrivers, you risk dropping the point mounting screws down into the distributor body*

4 ***Dwell meter*** - *A dwell meter is the only accurate way to determine the point setting (gap). Connect the meter according to the instructions supplied with it*

5 ***Blade-type feeler gauges*** - *These are required to set the initial point gap (space between the points when they are open)*

6 ***Ignition wrenches*** - *These special wrenches are made to work within the tight confines of the distributor. Specifically, they are needed to loosen the nut/bolt which secures the leads to the points*

27.16 If a dwell meter is not available, a feeler gauge can be used to adjust the point gap (which is equivalent to setting the dwell angle); however, you must be very precise when using this method, and, even then, you probably won't get the accuracy of using a dwell meter (dwell meters are quite inexpensive)

9 Connect one dwell meter lead to the primary ignition terminal of the point assembly or to the distributor terminal of the coil. Connect the other lead of the dwell meter to an engine ground. Some dwell meters may have different connecting instructions, so always follow the instrument manufacturer's directions.

10 Have an assistant crank the engine over or use a remote starter and crank the engine with the ignition switch turned to On.

11 Note the dwell reading on the meter and compare it to the Specifications found at the front of this chapter or on the engine tune-up decal. If the dwell reading is incorrect, adjust it by first loosening the point assembly mounting screw a small amount.

12 Move the point assembly plate with a screwdriver inserted into the slot provided next to the points. Closing the gap on the points will increase the dwell reading, while opening the gap will decrease the dwell.

13 Tighten the screw after the correct reading is obtained and recheck the setting.

14 If a dwell meter is unavailable, a feeler gauge can be used to set the points.

15 Have a helper crank the engine in short bursts until the rubbing block of the points rests on a high point of the cam assembly. The rubbing block must be exactly on the apex of one of the cam lobes for correct point adjustment. It may be necessary to rotate the front pulley of the engine with a socket and breaker bar to position the cam lobe exactly.

16 Measure the gap between the contact points with the correct size feeler gauge **(see illustration)**. If the gap is incorrect, loosen the mounting screw and move the point assembly until the correct gap is achieved.

17 Tighten the screw and check the gap one more time.

18 Apply a small amount of distributor cam lube to the point cam on the distributor shaft.

28.4 Be sure to use the correct feeler gauge when adjusting the valves - intake and exhaust valves require different clearances

28 Valve clearance check and adjustment (1980 and earlier six-cylinder models) (every 15,000 miles or 12 months)

Refer to illustration 28.4

Note: *Although the factory-recommended procedure requires the engine to be running while the valve clearances are checked/-adjusted, the following procedure is easier, safer and less messy and will produce the same results.*

1 Make sure the engine is at normal operating temperature before beginning this procedure.

2 Refer to Chapter 2A and remove the valve cover, then position the number one piston at TDC on the compression stroke (the procedure is in Chapter 2, Part C).

3 Slip the specified size feeler gauge between the number one cylinder intake valve stem and the rocker arm (the intake valves are directly opposite the intake manifold runners, while the exhaust valves are opposite the exhaust manifold runners). It should fit between them with a slight amount of drag if the clearance is correct.

4 If the feeler gauge will not fit, or if it fits loosely, turn the adjusting screw with a box-end wrench until you can feel a slight drag on the feeler gauge as it is withdrawn **(see illustration)**. **Note:** *The rocker arm adjusting screws have interference threads and no locknut, so always use the correct size box-end wrench to turn them (a six-point wrench is preferred).*

5 Repeat the procedure for the number one cylinder exhaust valve. Be sure to use the correct size feeler gauge - intake and exhaust valves require different clearances (see the Specifications at the front of this Chapter).

6 Turn the crankshaft 120-degrees to position the next piston in the firing order at TDC on the compression stroke. The ignition rotor in the distributor will be pointing at the appropriate terminal in the cap. Refer to the Specifications at the front of this chapter for the firing order.

7 Adjust the valves, then repeat the procedure to position the next piston in the firing order at TDC.

8 Continue working through the firing order until all valves for all cylinders have been adjusted, then reinstall the rocker arm cover.

9 Start the engine and check for oil leaks at the valve cover-to-cylinder head junction

29 Air filter replacement (every 30,000 miles or 24 months)

Refer to illustrations 29.2 and 29.4

1 At the specified intervals, the air filter element should be replaced with a new one.

2 The filter is located on top of, or adjacent to the carburetor, and is replaced by unscrewing the wing nut from the top of the filter housing and lifting off the cover **(see illustration)**.

3 While the top plate is off, be careful not to drop anything down into the carburetor.

4 Lift the air filter element out of the housing **(see illustration)** and wipe out the inside of the air cleaner housing with a clean rag.

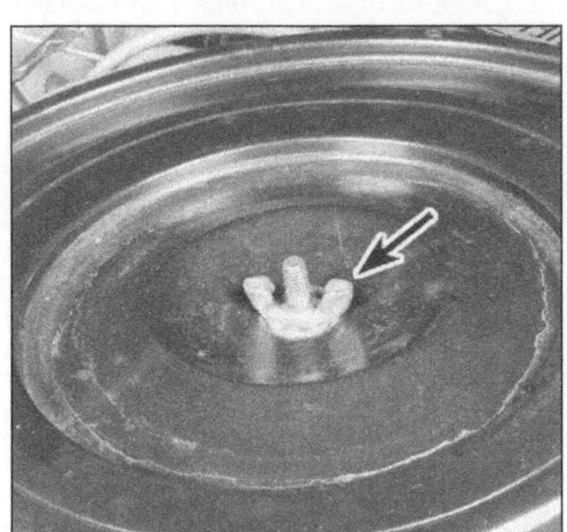

29.2 Remove the air cleaner wing nut (arrow)

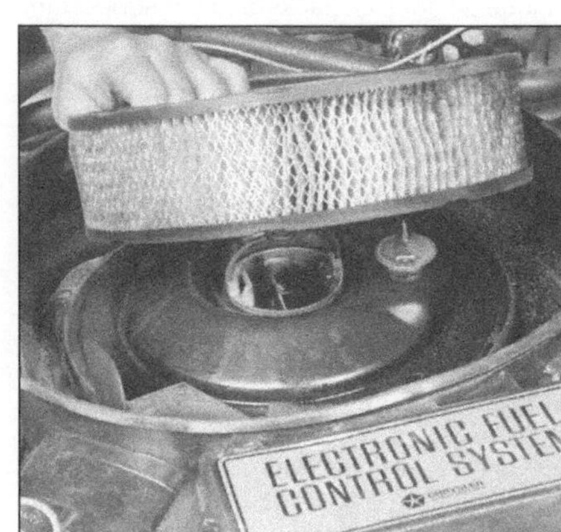

29.4 Lift the element out of the air cleaner housing

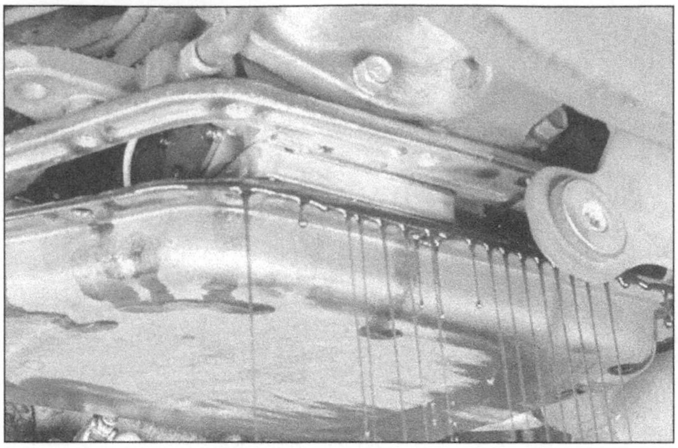
30.6 With the front bolts in place but loose, pull the rear of the pan down to drain the fluid

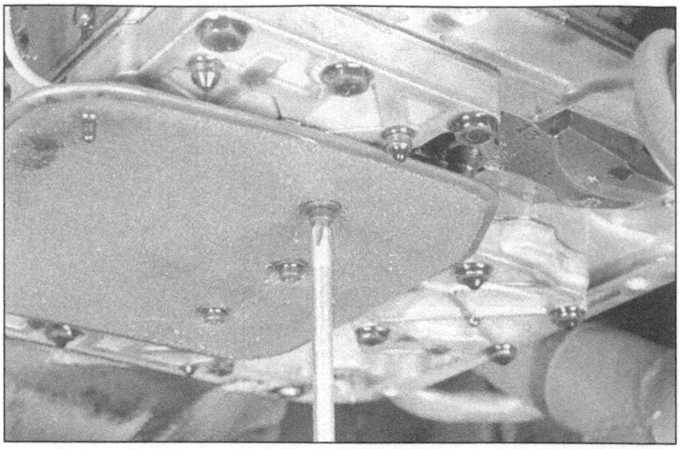
30.9 Remove the three filter screws and lower the filter from the transmission valve body - some more fluid will drain when the filter is removed

5 Place the new filter element in the air cleaner housing. Make sure it seats properly in the bottom of the housing.
6 Install the top plate and tighten the wing nut securely.

30 Automatic transmission fluid and filter change (every 30,000 miles or 24 months)

Refer to illustrations 30.6, 30.9, 30.11a and 30.11b

1 At the specified intervals, the transmission fluid should be drained and replaced. Since the fluid will remain hot long after driving, perform this procedure only after the engine has cooled down completely. The manufacturer also recommends adjusting the transmission bands at this time, since this procedure requires removing the fluid pan (see Section 31).
2 Before beginning work, purchase the specified transmission fluid (see *Recommended lubricants and fluids* at the front of this Chapter) and a new filter.
3 Other tools necessary for this job include a floor jack, jackstands to support the vehicle in a raised position, a drain pan capable of holding at least eight pints, newspapers and clean rags.
4 Raise the vehicle and support it securely on jackstands.
5 Place the drain pan underneath the transmission pan. Remove the rear and side pan mounting bolts, but only loosen the front pan bolts approximately four turns.
6 Carefully pry the transmission pan loose with a screwdriver, allowing the fluid to drain **(see illustration)**.
7 Remove the remaining bolts, pan and gasket. Carefully clean the gasket surface of the transmission to remove all traces of the old gasket and sealant.
8 Drain the fluid from the transmission pan, clean the pan with solvent and dry it with compressed air, if available.

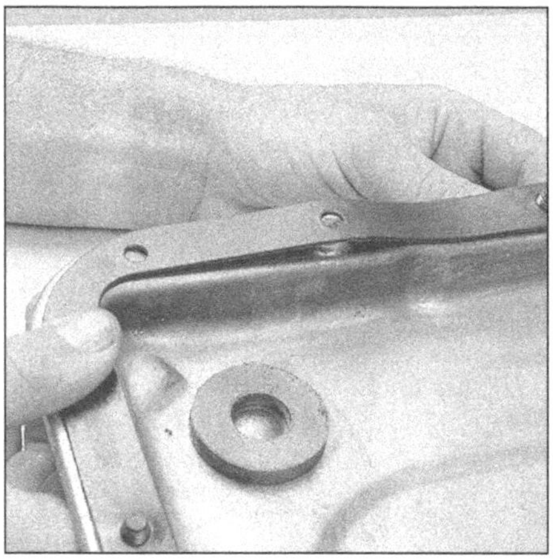
30.11a Place a new gasket in position on the pan and install two bolts to hold it in place - some later gaskets, like the one shown here, have two holes slightly smaller than the others so the bolts can grip the gasket securely

9 Remove the filter from the valve body inside the transmission **(see illustration)**. Use a gasket scraper to remove any traces of old gasket material that remain on the valve body. **Note:** *Be very careful not to gouge the delicate aluminum gasket surface on the valve body.*
10 Install a new gasket and filter. On many replacement filters, the gasket is attached to the filter to simplify installation.
11 Make sure the gasket surface on the transmission pan is clean, then install a new gasket on the pan **(see illustration)**. Put the pan in place against the transmission and, working around the pan, tighten each bolt a little at a time until the final torque figure is reached **(see illustration)**.
12 Lower the vehicle and add approximately seven pints of the specified type of automatic transmission fluid through the filler tube (see Section 6).
13 With the transmission in Park and the parking brake set, run the engine at a fast idle, but don't race it.

30.11b Hold the pan in place and install all of the bolts snugly before tightening them fully

14 Move the gear selector through each range and back to Park. Check the fluid level. It will probably be low. Add enough fluid to bring the level between the two dimples on the dipstick.
15 Check under the vehicle for leaks during the first few trips. Check the fluid level again when the transmission is hot (see Section 6)

31 Automatic transmission band adjustment (every 30,000 miles or 24 months)

Refer to illustrations 31.2 and 31.7

1 The transmission bands should be adjusted at the specified interval when the transmission fluid and filter are being replaced (see Section 30).

Kickdown band

2 The kickdown band adjusting screw is located on the left side of the transmission **(see illustration)**.

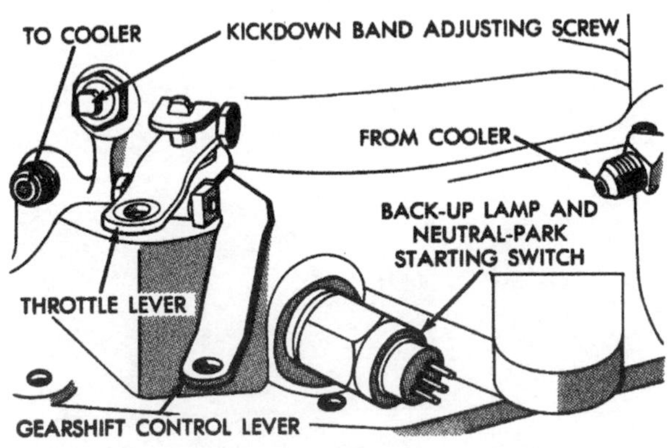

31.2 Location of the kickdown band adjusting screw

31.7 Loosen the Low-Reverse band locknut (arrow)

3 Raise the front of the vehicle and support it securely on jackstands.

4 Loosen the adjusting screw locknut approximately five turns, then loosen the adjusting screw a few turns. Make sure the adjusting screw turns freely, with no binding; lubricate it with penetrating oil if necessary.

5 Tighten the adjusting screw to the torque listed in the Specifications section at the beginning of this Chapter, then back it off the number of turns listed in the Specifications. Hold the screw from turning, then tighten the locknut securely.

Low-Reverse (rear) band

6 To gain access to the Low-Reverse band, the fluid pan must be removed (see Section 30).

7 Loosen the adjusting screw locknut and back it off four turns **(see illustration)**. Make sure the screw turns freely in the lever.

8 Tighten the adjusting screw to the torque listed in the Specifications section at the beginning of this Chapter, then back it off the number of turns listed in the Specifications. Hold the screw from turning, then tighten the locknut securely.

9 Install the transmission fluid pan.

32 Manual transmission lubricant change (every 30,000 miles or 24 months)

1 Raise the vehicle and support it securely on jackstands.

2 Move a drain pan, rags, newspapers and wrenches under the transmission.

3 Remove the transmission drain plug at the bottom of the case and allow the lubricant to drain into the pan **(see illustration 16.2)**.

4 After the lubricant has drained completely, reinstall the plug and tighten it securely.

5 Remove the fill plug from the side of the transmission case. Using a hand pump, syringe or funnel, fill the transmission with the specified lubricant until it begins to leak out through the hole. Reinstall the fill plug and tighten it securely.

6 Lower the vehicle.

7 Drive the vehicle for a short distance, then check the drain and fill plugs for leakage.

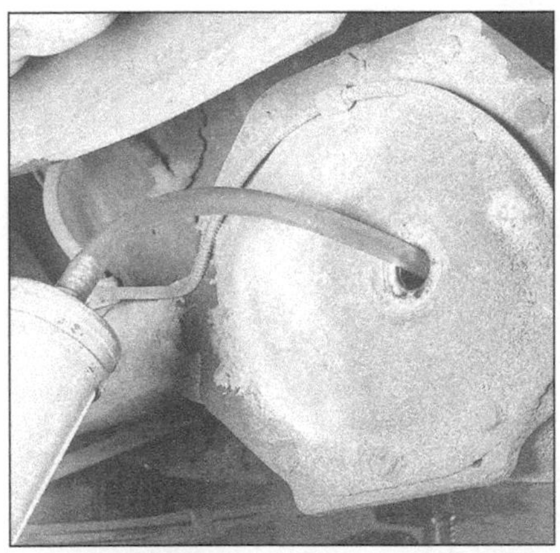

33.3 This is the easiest way to remove the lubricant: Work the end of the hose to the bottom of the differential housing and draw out the old lubricant with a suction pump. Since this differential has a bolt-on cover, the preferred method is to unbolt and remove the cover, as shown in the following illustrations

33 Differential lubricant change (every 30,000 miles or 24 months)

All models

Refer to illustration 33.3

1 This procedure should be performed after the vehicle has been driven so the lubricant will be warm and therefore will flow out of the differential more easily.

2 Raise the vehicle and support it securely on jackstands. If the differential has a bolt-on cover at the rear, it is usually easiest to remove the cover to drain the lubricant (which will also allow you to inspect the differential. If there's no bolt-on pan, look for a drain plug at the bottom of the differential housing. If there's not a drain plug and no cover, you'll have to remove the lubricant through t;he filler plug hole with a suction pump. If you'll be draining the lubricant by removing the cover or a drain plug, move a drain pan, rags, newspapers and wrenches under the vehicle.

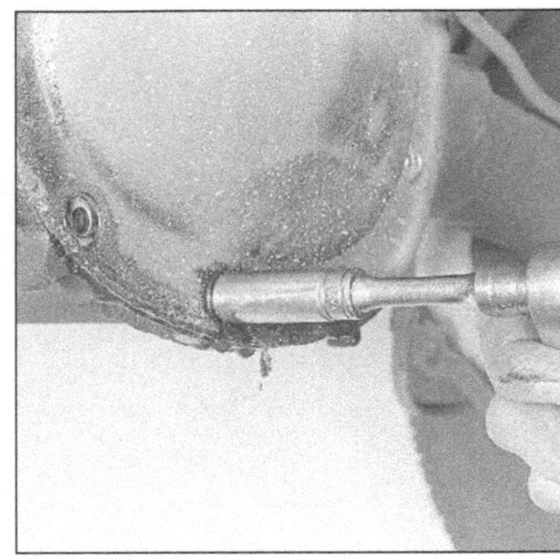

33.4a Remove the bolts from the lower edge of the cover . . .

3 Remove the filler plug from the differential (see Section 17). If a suction pump is being used, insert the flexible hose. Work the hose down to the bottom of the differential housing and pump the lubricant out **(see illustration)**. If you'll be draining the lubricant through a drain plug, remove the plug and allow the lubricant to drain into the pan, then reinstall the drain plug.

Models where the rear cover must be removed

Refer to illustrations 33.4a, 33.4b, 33.4c and 33.6

4 If the differential is being drained by removing the cover plate, remove the bolts on the lower half of the plate **(see illustration)**. Loosen the bolts on the upper half and use them to keep the cover loosely attached

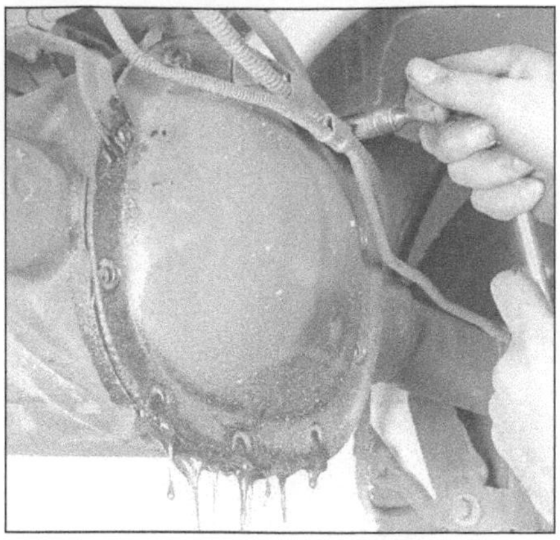

33.4b . . . then loosen the top bolts and let the lubricant drain out

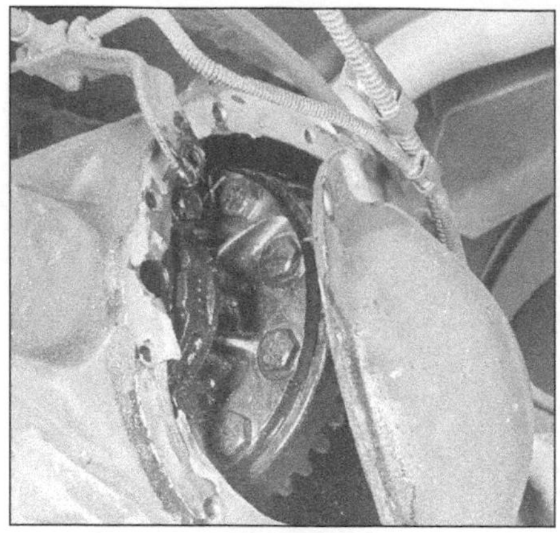

33.4c After the lubricant has drained, remove the bolts and the cover

33.6 Carefully scrape the old gasket material off to ensure a leak-free seal

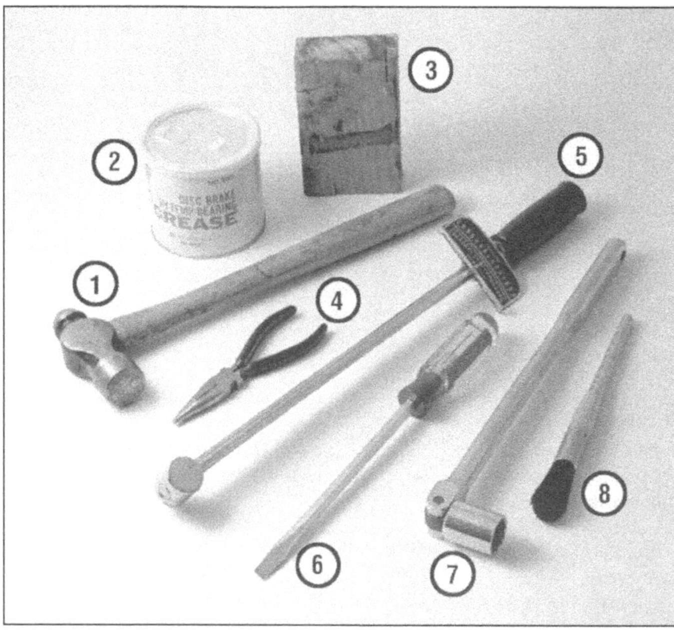

34.1 Tools and materials needed for front wheel bearing maintenance

1 ***Hammer*** *- A common hammer will do just fine*
2 ***Grease*** *- High-temperature grease that is formulated for front wheel bearings should be used*
3 ***Wood block*** *- If you have a scrap piece of 2x4, it can be used to drive the new seal into the hub*
4 ***Needle-nose pliers*** *- Used to straighten and remove the cotter pin in the spindle*
5 ***Torque wrench*** *- This is very important in this procedure; if the bearing is too tight, the wheel won't turn freely - if it's too loose, the wheel will "wobble" on the spindle. Either way, it could mean extensive damage*
6 ***Screwdriver*** *- Used to remove the seal from the hub (a long screwdriver is preferred)*
7 ***Socket/breaker bar*** *- Needed to loosen the nut on the spindle if it's extremely tight*
8 ***Brush*** *- Together with some clean solvent, this will be used to remove old grease from the hub and spindle*

(see illustration). Allow the oil to drain into the pan, then completely remove the cover **(see illustration)**.

5 Using a lint-free rag, clean the inside of the cover and the accessible areas of the differential housing. As this is done, check for chipped gears and metal particles in the lubricant, indicating that the differential should be more thoroughly inspected and/or repaired.

6 Thoroughly clean the gasket mating surfaces of the differential housing and the cover plate. Use a gasket scraper or putty knife to remove all traces of the old gasket **(see illustration)**.

7 Apply a thin layer of RTV sealant to the cover flange, then press a new gasket into position on the cover. Make sure the bolt holes align properly.

8 Place the cover on the differential housing and install the bolts. Tighten the bolts securely.

All models

9 Use a hand pump, syringe or funnel to fill the differential housing with the specified lubricant until it's level with the bottom of the plug hole.

10 Install the filler plug and make sure it is secure.

34 Front wheel bearing check, repack and adjustment (every 30,000 miles or 24 months)

Refer to illustrations 34.1, 34.6, 34.7, 34.8, 34.11 and 34.15

1 In most cases the front wheel bearings will not need servicing until the brake pads are changed. However, the bearings should be checked whenever the front of the vehicle is raised for any reason. Several items, including a torque wrench and special grease, are required for this procedure **(see illustration)**.

2 With the vehicle securely supported on jackstands, spin each wheel and check for noise, rolling resistance and freeplay.

3 Grasp the top of each tire with one hand and the bottom with the other. Move the wheel in-and-out on the spindle. If there's any noticeable movement, the bearings should be checked and then repacked with

34.6 Dislodge the dust cap by working around the outer circumference with a hammer and chisel

34.7 **Remove the cotter pin and discard it - use a new one when the hub is reinstalled**

34.11 **Use a large screwdriver to pry the grease seal out of the rear of the hub**

grease or replaced if necessary.

4 Remove the wheel.

5 Remove the brake caliper (see Chapter 9) and hang it out of the way on a piece of wire. A wood block can be slid between the brake pads to keep them separated, if necessary.

6 Pry the dust cap out of the hub using a screwdriver or hammer and chisel **(see illustration)**.

7 Straighten the bent ends of the cotter pin, then pull the cotter pin out of the nut lock **(see illustration)**. Discard the cotter pin and use a new one during reassembly.

8 Remove the nut lock, nut and washer from the end of the spindle **(see illustration)**.

9 Pull the hub/disc assembly out slightly, then push it back into its original position. This should force the outer bearing off the spindle enough so it can be removed.

10 Pull the hub/disc assembly off the spindle.

11 Use a screwdriver to pry the seal out of the rear of the hub **(see illustration)**. As this is done, note how the seal is installed.

12 Remove the inner wheel bearing from the hub.

13 Use solvent to remove all traces of the old grease from the bearings, hub and spindle. A small brush may prove helpful; however make sure no bristles from the brush embed themselves inside the bearing rollers. Allow the parts to air dry.

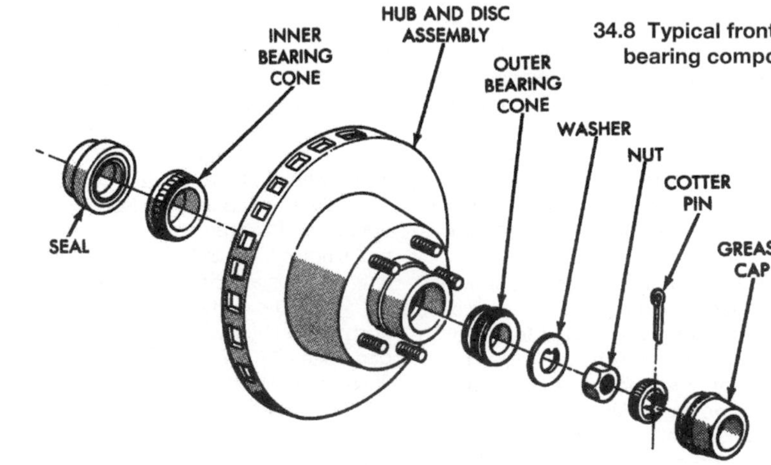

34.8 **Typical front hub and bearing components**

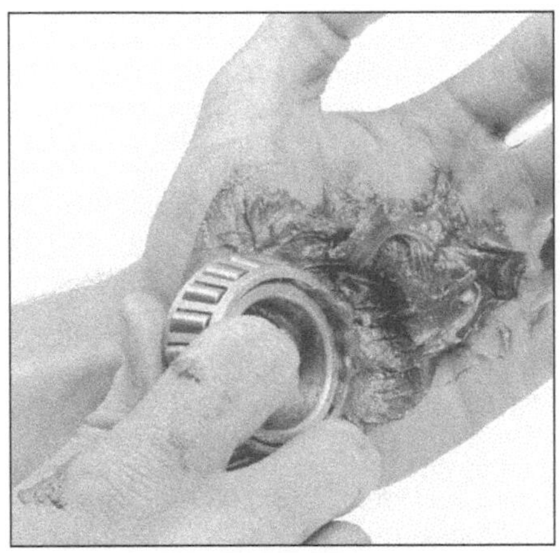

34.15 **Work the grease completely into the bearing rollers - if you don't like getting greasy, special bearing packing tools that work with a common grease gun are available inexpensively from auto parts stores**

14 Carefully inspect the bearings for cracks, heat discoloration, worn rollers, etc. Check the bearing races inside the hub for wear and damage. If the bearing races are defective, the hubs should be taken to a machine shop with the facilities to remove the old races and press new ones in. Note that the bearings and races come as matched sets and old bearings should never be installed on new races.

15 Use high-temperature front wheel bearing grease to pack the bearings. Work the grease completely into the bearings, forcing it between the rollers, cone and cage from the back side **(see illustration)**.

16 Apply a thin coat of grease to the spindle at the outer bearing seat, inner bearing seat, shoulder and seal seat.

17 Put a small quantity of grease inboard of each bearing race inside the hub. Using your finger, form a dam at these points to provide extra grease availability and to keep thinned grease from flowing out of the bearing.

18 Place the grease-packed inner bearing into the rear of the hub and put a little more grease outboard of the bearing.

19 Place a new seal over the inner bearing and tap the seal evenly into place with a hammer and blunt punch until it's flush with the hub.

20 Carefully place the hub assembly onto the spindle and push the grease-packed outer bearing into position.

21 Install the washer and spindle nut. Tighten the nut only slightly (no more than 12 ft-lbs of torque).

22 Spin the hub in a forward direction while tightening the spindle nut to approximately 20 ft-lbs to seat the bearings and remove any grease or burrs which could cause excessive bearing play later.

23 Loosen the spindle nut 1/4-turn, then using your hand (not a wrench of any kind), tighten the nut until it's snug. Install the nut lock and a new cotter pin through the hole in the spindle and the slots in the nut lock. If the nut lock slots don't line up, remove the nut lock and turn it slightly until they do.

24 Bend the ends of the cotter pin until they're flat against the nut. Cut off any extra length which could interfere with the dust cap.

25 Install the dust cap, tapping it into place with a hammer.

26 Place the brake caliper near the rotor and carefully remove the wood spacer. Install the caliper (see Chapter 9).

27 Install the wheel on the hub and tighten the lug nuts.

28 Grasp the top and bottom of the tire and check the bearings in the manner described earlier in this Section.

29 Lower the vehicle.

35 Cooling system servicing (draining, flushing and refilling) (every 30,000 miles or 24 months)

Refer to illustrations 35.3 and 35.4

Warning: *Do not allow antifreeze to come in*

contact with your skin or painted surfaces of the vehicle. Rinse off spills immediately with plenty of water. Antifreeze is highly toxic if ingested. Never leave antifreeze lying around in an open container or in puddles on the floor; children and pets are attracted by it's sweet smell and may drink it. Check with local authorities about disposing of used antifreeze. Many communities have collection centers which will see that antifreeze is disposed of safely.

1 Periodically, the cooling system should be drained, flushed and refilled to replenish the antifreeze mixture and prevent formation of rust and corrosion, which can impair the performance of the cooling system and cause engine damage. When the cooling system is serviced, all hoses and the radiator cap should be checked and replaced if necessary.

2 Apply the parking brake and block the wheels. **Warning:** *If the vehicle has just been driven, wait several hours to allow the engine to cool down before beginning this procedure.*

3 Move a large container under the radiator drain to catch the coolant. Attach a 3/8-inch diameter hose to the drain fitting (if possible) to direct the coolant into the container, then open the drain fitting **(see illustration)** (a pair of pliers may be required to turn it).

4 After coolant stops flowing out of the coolant reservoir, remove the radiator cap, allow the radiator to drain, then, move the container under the engine block drain plugs - there's one on each side of the block on V8 engines and one on the right side only on six-cylinder engines **(see illustration)**. Remove the plugs and allow the coolant in the block to drain. **Note:** *Frequently, the coolant will not drain from the block after the plug is removed. This is due to a rust layer that has built up behind the plug. Insert a Phillips screwdriver into the hole to break the rust barrier.*

5 While the coolant is draining, check the condition of the radiator hoses, heater hoses and clamps (refer to Section 11 if necessary).

6 Replace any damaged clamps or hoses.

7 Once the system is completely drained, flush the radiator with fresh water from a garden hose until it runs clear at the drain. The flushing action of the water will remove sediments from the radiator but will not remove rust and scale from the engine and cooling tube surfaces.

8 These deposits can be removed with a chemical cleaner. Follow the procedure outlined in the manufacturer's instructions. If the radiator is severely corroded, damaged or leaking, it should be removed (see Chapter 3) and taken to a radiator repair shop.

9 Remove the overflow hose from the coolant reservoir and flush the reservoir with clean water, then reconnect the hose.

10 Close and tighten the radiator drain fitting. Install and tighten the block drain plugs.

11 Place the heater temperature control in the maximum heat position.

12 Slowly add new coolant (a 50/50 mixture of water and antifreeze) to the radiator until it's full. Add coolant to the reservoir up to the lower mark.

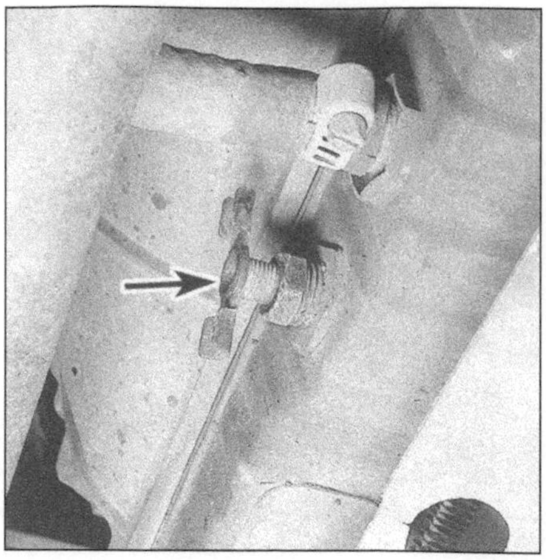

35.3 The drain fitting (arrow) is located at the bottom of the radiator - turn this "wing bolt" counterclockwise to drain the coolant

13 Leave the radiator cap off and run the engine in a well-ventilated area until the thermostat opens (coolant will begin flowing through the radiator and the upper radiator hose will become hot).

14 Turn the engine off and let it cool. Add more coolant mixture to bring the level back up to the lip on the radiator filler neck.

15 Squeeze the upper radiator hose to expel air, then add more coolant mixture if necessary. Replace the radiator cap.

16 Start the engine, allow it to reach normal operating temperature and check for leaks.

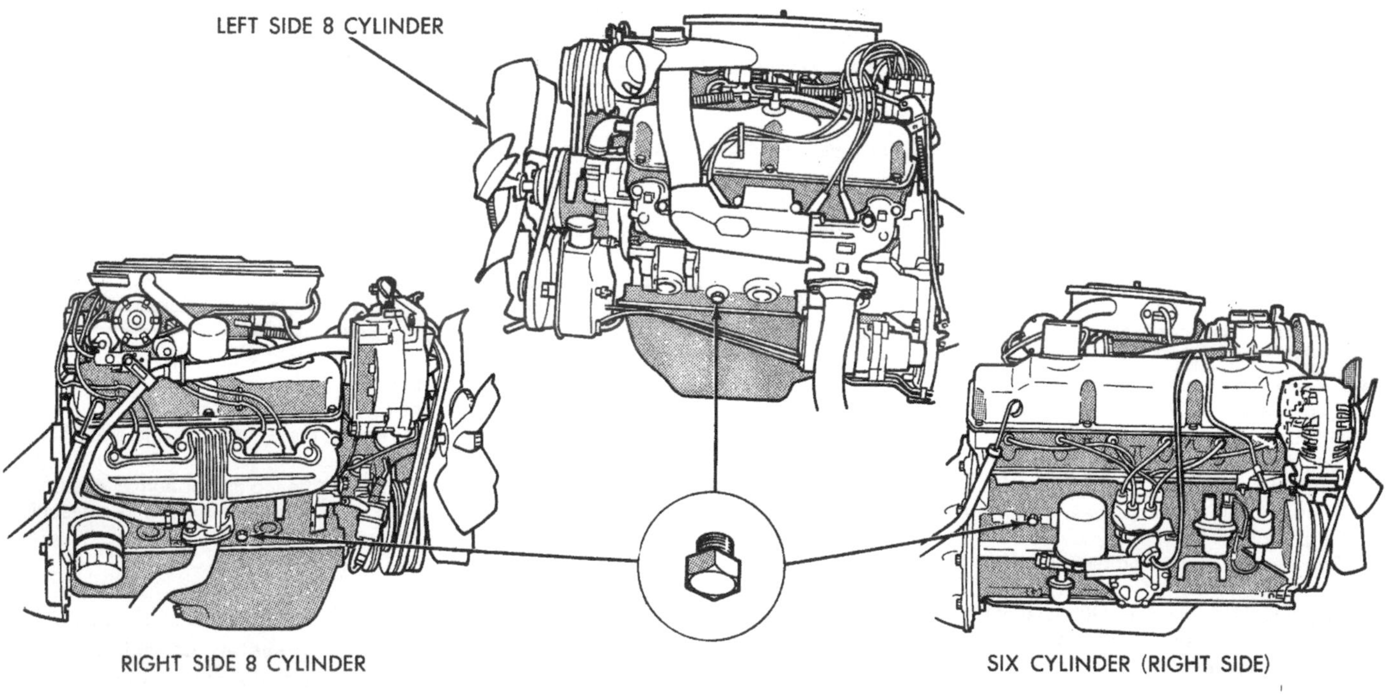

35.4 Cylinder block drain plug locations - the V8 engine shown is a small-block, but big-block engines also have a drain plug on each side of the block

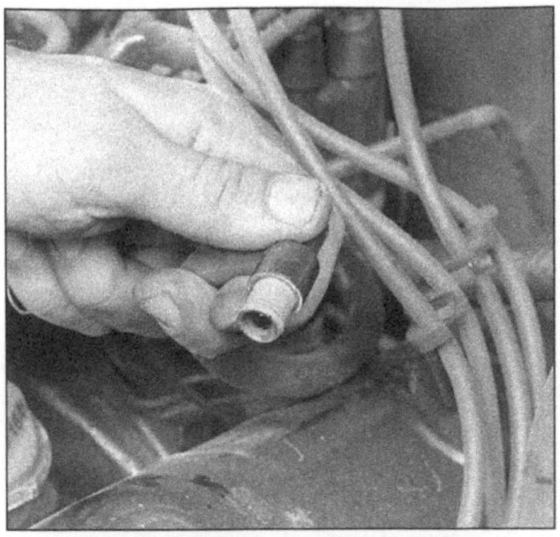

36.2 The PCV valve fits into the valve cover

36 Positive Crankcase Ventilation (PCV) valve check and replacement (every 30,000 miles or 24 months)

Refer to illustration 36.2

1 The PCV valve is located in the valve cover.

2 With the engine idling at normal operating temperature, pull the valve (with hose attached) from the rubber grommet in the cover **(see illustration)**.

3 Place your finger over the valve opening. If there's no vacuum at the valve, check for a plugged hose, manifold port, or the valve itself. Replace any plugged or deteriorated hoses.

4 Turn off the engine and shake the PCV valve, listening for a rattle. If the valve doesn't rattle, replace it with a new one.

5 To replace the valve, pull it from the end of the hose, noting its installed position.

6 When purchasing a replacement PCV valve, make sure it's for your particular vehicle and engine size. Compare the old valve with the new one to make sure they're the same.

7 Push the valve into the end of the hose until it's seated.

8 Inspect the rubber grommet for damage and hardening. Replace it with a new one if necessary.

9 Push the PCV valve and hose securely into position.

37 Evaporative emissions control system check (every 30,000 miles or 24 months)

Refer to illustration 37.2

1 The function of the evaporative emissions control system is to draw fuel vapors from the gas tank and fuel system, store them in a charcoal canister and route them to the intake manifold during normal engine operation.

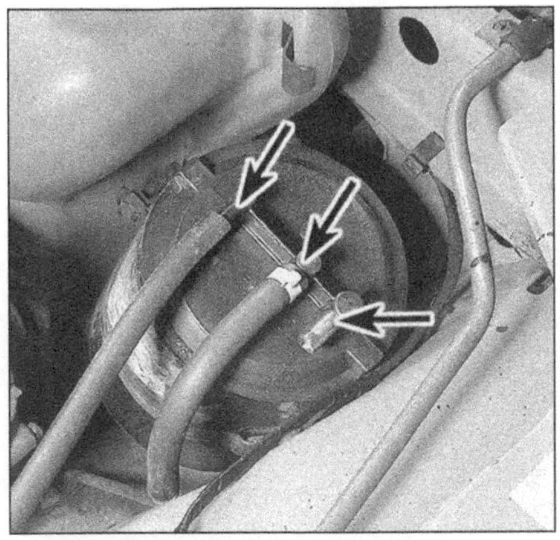

37.2 The evaporative emissions canister is located in the right front corner of the engine compartment - check the hoses and connections (arrows) for damage

2 The most common symptom of a fault in the evaporative emissions system is a strong fuel odor in the engine compartment. If a fuel odor is detected, inspect the charcoal canister, located in the engine compartment or under the vehicle **(see illustration)**. Check the canister and all hoses for damage and deterioration.

3 The evaporative emissions control system is explained in more detail in Chapter 6.

38 Exhaust Gas Recirculation (EGR) system check (every 30,000 miles or 24 months)

Refer to illustration 38.2

1 The EGR valve is usually located on the intake manifold, adjacent to the carburetor. The EGR valve is operated by engine vacuum. Most of the time when a problem develops in the emissions system, it's due to a stuck or corroded EGR valve or a damaged or leaking hose or vacuum diaphragm.

2 Warm the engine up to operating temperature and allow it to idle with the transmission in Neutral. Accelerate the engine abruptly up to about 2000 rpm and look under the valve housing to make sure the movement marker groove in the stem moves **(see illustration)**.

3 If the stem doesn't move, check all the vacuum hoses for leaks or damage. If the hoses are not damaged, check the EGR valve operation. Disconnect the vacuum hose, connect a hand-vacuum pump and apply 10 inches Hg of vacuum. If the valve stem still does not move, the valve itself is faulty and should be replaced.

4 If the valve stem moves, clamp the hose and make sure the valve stays open (diaphragm holds vacuum) 30 seconds or more. If it doesn't, the diaphragm is faulty and the valve should be replaced.

5 Refer to Chapter 6 for more information on the EGR system.

38.2 Look under the EGR valve housing (arrow) to make sure the stem moves

39 Thermostatic air cleaner check (every 30,000 miles or 24 months)

Refer to illustration 39.3

1 These models are equipped with a thermostatically controlled air cleaner which draws air to the carburetor/throttle body from different locations, depending upon engine temperature.

2 This is a visual check. If access is limited, a small mirror may have to be used.

3 Open the hood and locate the damper door inside the air cleaner assembly **(see illustration)**.

4 If there is a flexible air duct attached to the end of the air cleaner, disconnect it at the air cleaner. This will enable you to see the damper inside.

5 The check should be done when the engine is cold. Start the engine and look at the damper, which should move to a closed position. With the damper closed, air cannot enter through the air cleaner opening, but instead enters the air cleaner through a flexible duct attached to a heat stove on the exhaust manifold.

6 As the engine warms up to operating temperature, the damper should open to

39.3 The damper door (arrow) is located in the air cleaner housing snorkel

40.3 Lubricate the filter by pouring clean engine oil into the large opening until it drains out the smaller opening

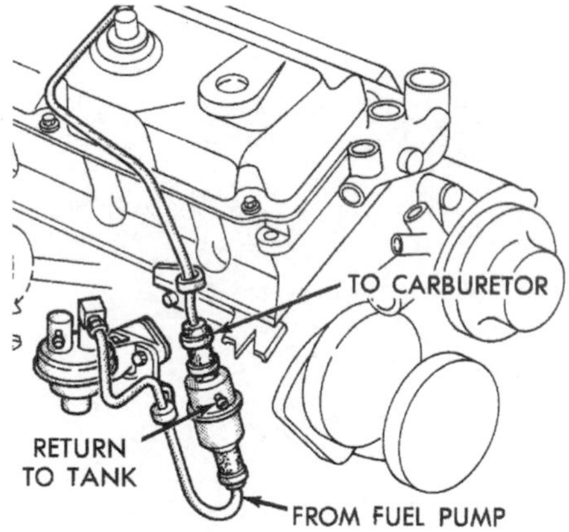

41.1a Typical six-cylinder fuel filter location

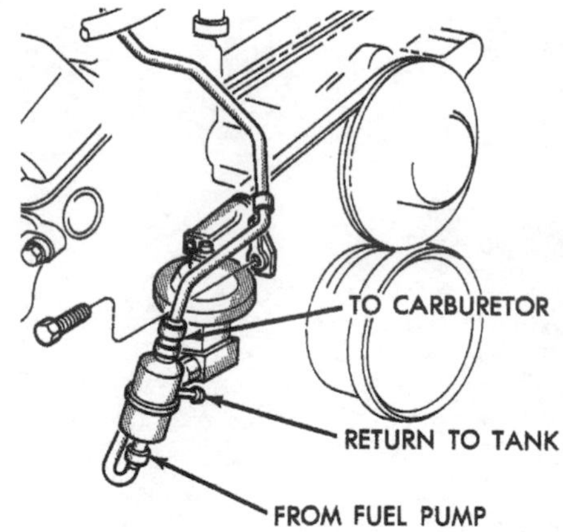

41.1b Typical V8 engine fuel filter location

allow air through the air cleaner opening. Depending on outside temperature, this may take 10 to 15 minutes. To speed up this check, you can reconnect the air duct, drive the vehicle, then check to see if the damper is completely open.

7 If the thermo-controlled air cleaner is not operating properly, see Chapter 6 for more information.

40 Crankcase inlet filter - cleaning (every 30,000 miles or 24 months)

Refer to illustration 40.3

1 Disconnect the hose and pull the crankcase inlet filter out of the valve cover.

2 Wash the inside of the filter with solvent.

3 Lubricate the filter by pouring clean engine oil into the large opening and allowing into to drain out through the smaller (inlet) opening **(see illustration)**.

4 Reinstall the filter in the valve cover and connect the hose.

41 Fuel filter replacement (every 30,000 miles or 24 months)

Refer to illustrations 41.1a and 41.1b

Warning: *Gasoline is extremely flammable, so take extra precautions when working on any part of the fuel system. Don't smoke or allow open flames or bare light bulbs in or near the work area, and don't work in a garage where a natural gas-type appliance (such as a water heater or clothes dryer) with a pilot light is present. Since gasoline is carcinogenic, wear latex gloves when there's a possibility of being exposed to fuel, and, if you spill fuel on your skin, rinse it off immediately with soap and water. Have a Class B fire extinguisher on hand.*

1 The fuel filter is mounted in the line, between the carburetor and the fuel pump **(see illustrations)**.

2 Place rags or newspapers under the filter to absorb any spilled fuel. If the original fuel filter is still installed, use wire cutters to cut the Keystone clamps securing the fuel and vapor return hoses to the filter, then detach the hoses from the filter. Loosen aftermarket screw-type clamps with a screwdriver or the correct-size six-point socket.

3 Installation is the reverse of removal. Use new screw-type clamps to replace the Keystone clamps. Make sure the vapor return line is routed correctly **(see illustrations 41.1a and 41.1b)**.

42 Exhaust manifold heat control valve - check and servicing (every 30,000 miles or 24 months)

Warning: *The heat control valve, as well as the exhaust manifold and pipe, are very hot. Never touch these parts with bare hands until the engine has cooled completely. If you must touch exhaust-system parts when they're hot, wear heavy, heat-resistant gloves, such as welding gloves.*

1 The manifold heat control valve is located on the exhaust manifold.

Check

2 Allow the engine to warm up to normal operating temperature, then have an assistant accelerate the engine from idle. The valve counterweight should move counterclockwise when the engine accelerates.

3 If it doesn't, deposits have probably built up on the valve shaft which inhibit its movement.

Servicing

Refer to illustration 42.4

4 With the engine cold, spray MOPAR Manifold Heat Control Valve solvent, or equivalent, to each end of the shaft **(see illustration)**.

5 Allow the solvent to soak in for a few minutes, then work the valve shaft back and forth by hand until it moves freely.

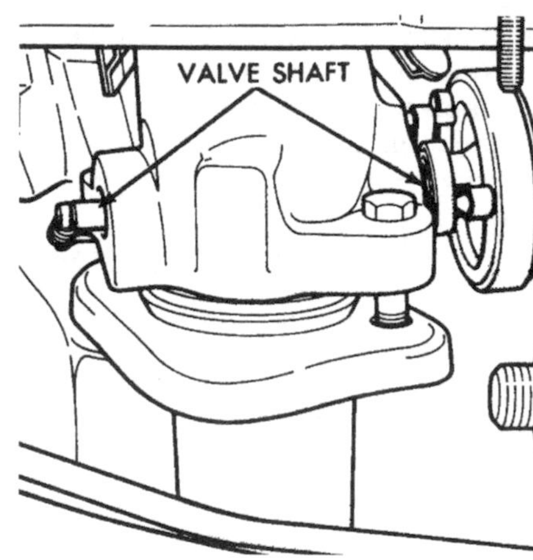

42.4 Spray the solvent at both ends of the valve shaft, let it penetrate, then work the valve back and forth to loosen any deposits which will cause binding

43 Carburetor choke check and shaft cleaning (every 30,000 miles or 24 months)

Refer to illustrations 43.3 and 43.9

1 The choke operates when the engine is cold, so this check can only be performed before the vehicle has been started for the day.

2 Open the hood and remove the top plate of the air cleaner assembly (see Section 29). If any vacuum hoses must be disconnected, make sure you tag the hoses for reinstallation in their original positions. Place the top plate and wing nut aside, out of the way of moving engine components.

3 Look at the center of the air cleaner housing. You will notice a flat plate at the carburetor opening **(see illustration)**.

4 Press the accelerator pedal to the floor. The plate should close completely. Start the engine while you watch the plate at the carburetor. **Warning:** *Do not position your face directly over the carburetor, as the engine could backfire, causing serious burns.* When

43.3 The choke plate (arrow) is located at the top of the carburetor air horn (shown in the open position)

43.9 Work the choke shaft back-and-forth while spraying carburetor cleaner into the contact areas

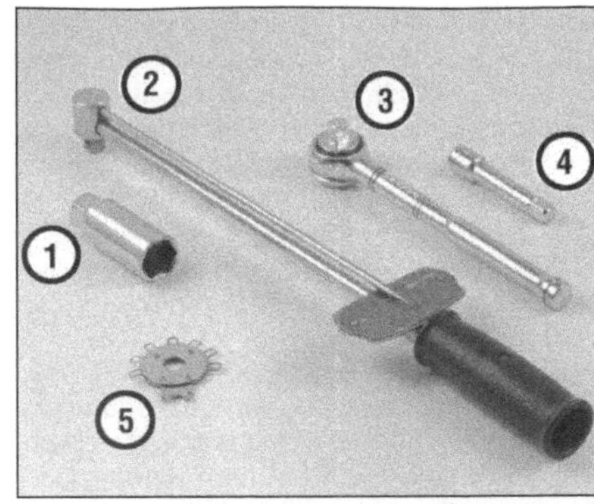

44.2 Tools required for changing spark plugs

1 ***Spark plug socket*** *- This will have special padding inside to protect the spark plug's porcelain insulator*
2 ***Torque wrench*** *- Although not mandatory, using this tool is the best way to ensure the plugs are tightened properly*
3 ***Ratchet*** *- Standard hand tool to fit the spark plug socket*
4 ***Extension*** *- Depending on model and accessories, you may need special extensions and universal joints to reach one or more of the plugs*
5 ***Spark plug gap gauge*** *- This gauge for checking the gap comes in a variety of styles. Make sure the gap for your engine is included*

the engine starts, the choke plate should open slightly.

5 Allow the engine to continue running at an idle speed. As the engine warms up to operating temperature, the plate should slowly open, allowing more air to enter through the top of the carburetor.

6 After a few minutes, the choke plate should be fully open to the vertical position. Lightly tap the accelerator to make sure the fast-idle cam disengages.

7 You will notice that the engine speed corresponds with the plate opening. With the plate fully closed, the engine should run at a fast idle speed. As the plate opens and the throttle is moved to disengage the fast idle cam, the engine speed will decrease.

8 Refer to Chapter 4 for specific information on adjusting and servicing the choke components.

9 At the specified intervals, or when the choke shaft binds, apply MOPAR Combustion Chamber Conditioner 4318001 or equivalent to the choke shaft and fast idle cam contact surfaces to remove deposits and ensure free movement **(see illustration)**.

44 Spark plug replacement (every 30,000 miles or 24 months)

Refer to illustrations 44.2, 44.5a, 44.5b, 44.6 and 44.10

1 Open the hood and label each spark plug wire to ensure proper installation.

2 In most cases, the tools necessary for spark plug replacement include a spark plug socket which fits onto a ratchet (spark plug sockets are padded inside to prevent damage to the porcelain insulators on the new plugs), various extensions and a gap gauge to check and adjust the gaps on the new plugs **(see illustration)**. A special plug wire removal tool is available for separating the wire boots from the spark plugs, but it isn't absolutely necessary. A torque wrench should be used to tighten the new plugs.

3 The best approach when replacing the spark plugs is to purchase the new ones in advance, adjust them to the proper gap and replace them one at a time. When buying the new spark plugs, be sure to obtain the correct plug type for your particular engine. This information can be found on the *Emission Control Information* label located under the hood, in the factory owner's manual and the Specifications at the front of this Chapter. If differences exist between the plug specified on the emissions label and in the owner's manual, assume that the emissions label is correct.

4 Allow the engine to cool completely before attempting to remove any of the plugs. While you're waiting for the engine to cool, check the new plugs for defects and adjust the gaps.

5 The gap is checked by inserting the proper-thickness gauge between the electrodes at the tip of the plug **(see illustration)**. The gap between the electrodes should be the same as the one specified on the *Emissions Control Information* label or in this Chapter's Specifications. The wire should just slide between the electrodes with a slight amount of drag. If the gap is incorrect, use the adjuster on the gauge body to bend the curved side electrode slightly until the proper gap is obtained **(see illustration)**. If the side electrode is not exactly over the center electrode, bend it with the adjuster until it is. Check for cracks in the porcelain insulator (if any are found, the plug should not be used).

6 With the engine cool, remove the spark plug wire from one spark plug. Pull only on

44.5a Spark plug manufacturers recommend using a wire-type gauge when checking the gap - if the wire does not slide between the electrodes with a slight drag, adjustment is required

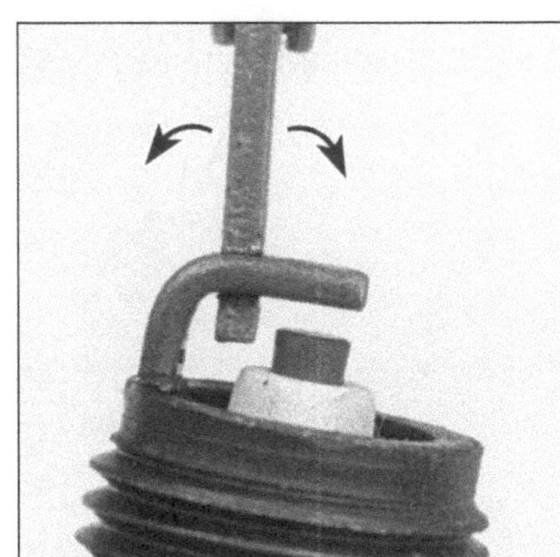

44.5b To change the gap, bend the side electrode only, as indicated by the arrows, and be very careful not to crack or chip the porcelain insulator surrounding the center electrode

the boot at the end of the wire - do not pull on the wire. A plug wire removal tool should be used if available **(see illustration)**.

7 If compressed air is available, use it to blow any dirt or foreign material away from the spark plug hole. A common bicycle pump will also work. The idea here is to eliminate the possibility of debris falling into the cylinder as the spark plug is removed.

8 Place the spark plug socket over the plug and remove it from the engine by turning it in a counterclockwise direction. On six-cylinder engines, also remove the stamped-steel spark plug cup from the cylinder head and replace the rubber oil seal on installation. **Warning:** *On some engines, particularly big-block engines, it may be easier to remove the rear spark plugs from below. If this is necessary, be sure to support the vehicle securely on jackstands - never get under a vehicle that is only supported by a jack!*

9 Compare the spark plug to those shown in the color photos inside the back cover of this manual to get an indication of the general running condition of the engine.

10 Thread one of the new plugs into the hole until you can no longer turn it with your fingers, then tighten it with a torque wrench (if available) or the ratchet. It might be a good idea to slip a short length of rubber hose over the end of the plug to use as a tool to thread it into place **(see illustration)**. The hose will grip the plug well enough to turn it, but will start to slip if the plug begins to cross-thread in the hole - this will prevent damaged threads and the accompanying repair costs.

11 Before pushing the spark plug wire onto the end of the plug, inspect it following the procedures outlined in Section 45.

12 Attach the plug wire to the new spark plug, again using a twisting motion on the boot until it's seated on the spark plug.

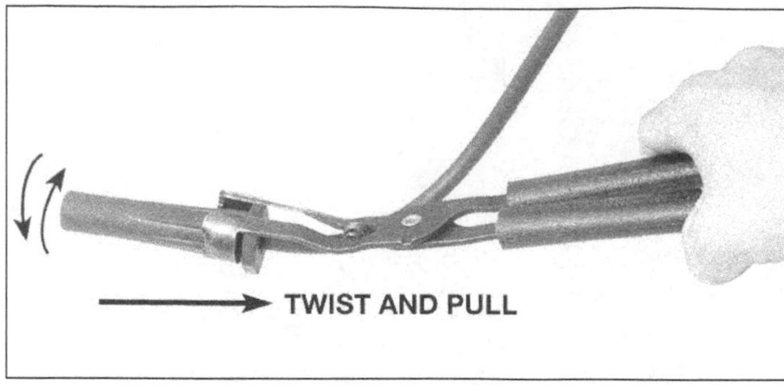

44.6 When removing the spark plug wires, pull only on the boot and twist it back-and-forth

13 Repeat the procedure for the remaining spark plugs, replacing them one at a time to prevent mixing up the spark plug wires.

45 Spark plug wires and distributor cap and rotor - check and replacement (every 30,000 miles or 24 months)

Refer to illustrations 45.10 and 45.13

1 The spark plug wires should be checked at the recommended intervals and whenever new spark plugs are installed in the engine.

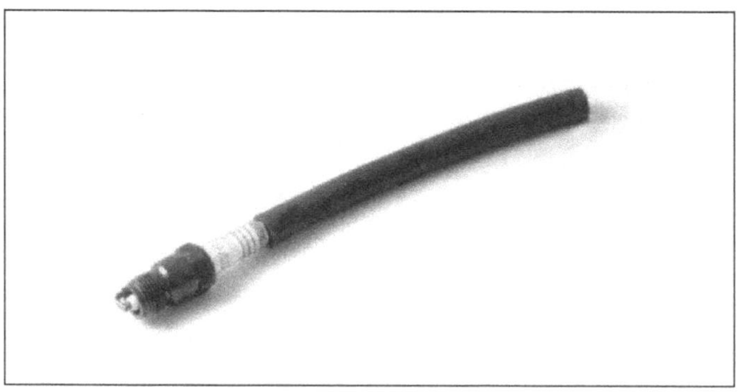

44.10 A length of 3/8-inch ID rubber hose will save time and prevent damaged threads when installing the spark plugs

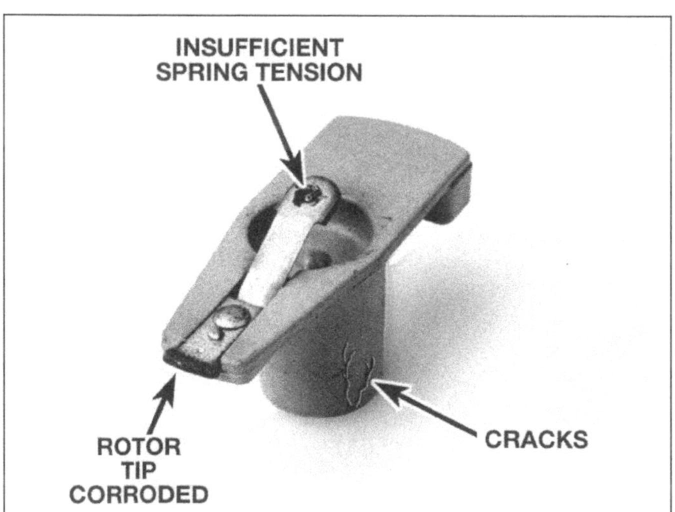

45.10 The ignition rotor should be checked for wear and corrosion as indicated here (if in doubt about its condition, buy a new one)

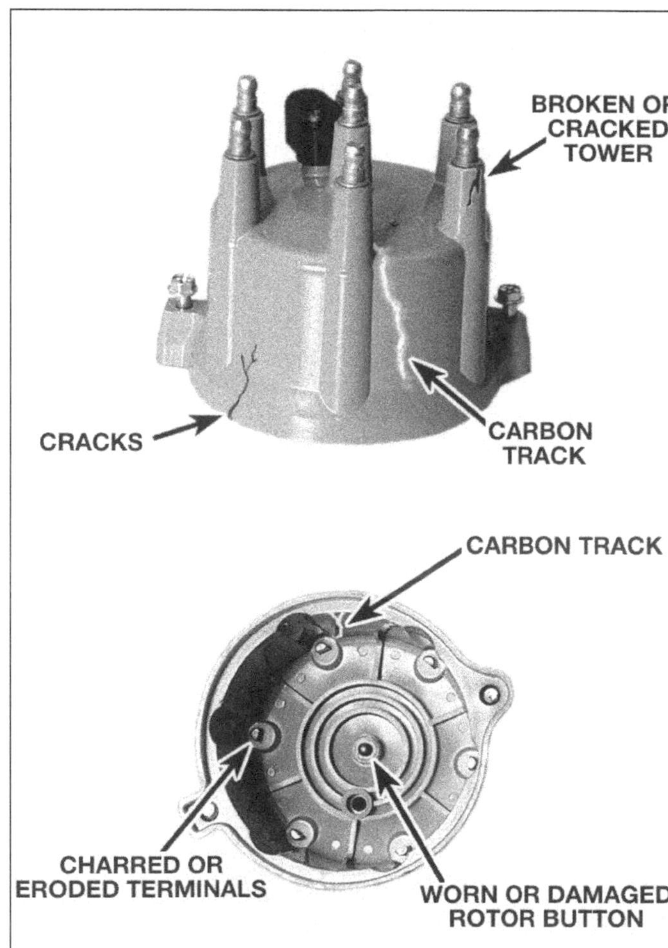

45.13 Shown here are some of the common defects to look for when inspecting the distributor cap (if in doubt about its condition, install a new one)

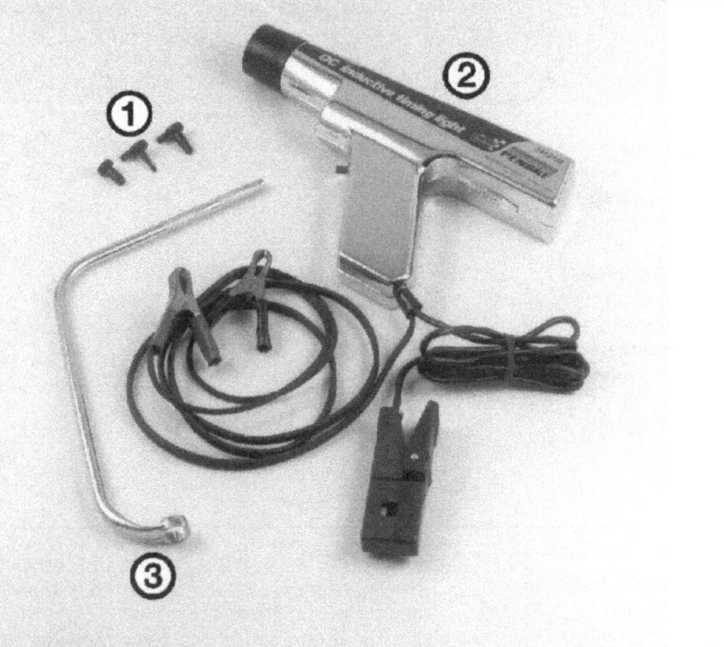

47.2 Tools needed to check and adjust the ignition timing

1 ***Vacuum plugs*** *- Vacuum hoses will, in most cases, have to be disconnected and plugged. Molded plugs in various shapes and sizes are available for this*

2 ***Inductive pick-up timing light*** *- Flashes a bright concentrated beam of light when the number one spark plug fires. Connect the leads according to the instructions supplied with the light*

3 ***Distributor wrench*** *- On some models, the hold-down bolt for the distributor is difficult to reach and turn with a conventional wrench or socket. A special wrench like this must be used*

2 The wires should be inspected one at a time to prevent mixing up the order, which is essential for proper engine operation. On some models it will be necessary to remove the two screws and detach the distributor splash shield for access.

3 Disconnect the plug wire from one spark plug. To do this, grab the rubber boot, twist slightly and pull the wire free. Do not pull on the wire itself, only on the rubber boot.

4 Check inside the boot for corrosion, which will look like a white crusty powder. Push the wire and boot back onto the end of the spark plug. It should be a tight fit on the plug. If it isn't, remove the wire and use a pair of pliers to carefully crimp the metal connector inside the boot until it fits securely on the end of the spark plug.

5 Using a clean rag, wipe the entire length of the wire to remove any built-up dirt and grease. Once the wire is clean, check for holes, burned areas, cracks and other damage. Don't bend the wire excessively or the conductor inside might break.

6 Disconnect the wire from the distributor cap. Pull the wire straight out of the cap. Pull only on the rubber boot during removal. Check for corrosion and a tight fit in the same manner as the spark plug end. Reattach the wire to the distributor cap.

7 Check the remaining spark plug wires one at a time, making sure they are securely fastened at the distributor and the spark plug when the check is complete.

8 If new spark plug wires are required, purchase a new set for your specific engine model. Wire sets are available pre-cut, with the rubber boots already installed. Remove and replace the wires one at a time to avoid mix-ups in the firing order. The wire routing is extremely important, so be sure to note exactly how each wire is situated before removing it.

9 Release the distributor cap clips. Pull up on the cap, with the wires attached, to separate it from the distributor, then position it to one side.

10 The rotor is now visible on the end of the distributor shaft. Check it carefully for cracks and carbon tracks. Make sure the center terminal spring tension is adequate and look for corrosion and wear on the rotor tip **(see illustration)**. If in doubt about its condition, replace it with a new one.

11 If replacement is required, detach the rotor from the shaft and install a new one. The rotor is a press fit on the shaft and can be pried or pulled off.

12 The rotor is indexed to the shaft so it can only be installed one way. It has an internal key that must line up with a slot in the end of the shaft (or vice versa).

13 Check the distributor cap for carbon tracks, cracks and other damage. Closely examine the terminals on the inside of the cap for excessive corrosion and damage **(see illustration)**. Slight deposits are normal. Again, if in doubt about the condition of the cap, replace it with a new one. Be sure to apply a small dab of silicone dielectric grease to each terminal before installing the cap. Also, make sure the carbon brush (center terminal) is correctly installed in the cap - a wide gap between the brush and rotor will result in rotor burn-through and/or damage to the distributor cap.

14 To replace the cap, simply separate it from the distributor and transfer the spark plug wires, one at a time, to the new cap. Be very careful not to mix up the wires!

15 Reattach the cap to the distributor, then reposition the clips or tighten the screw latches to hold it in place.

46 Idle speed check and adjustment (every 30,000 miles or 24 months)

This procedure is covered under *Carburetor adjustments* in Chapter 4.

47 Ignition timing check and adjustment) (every 30,000 miles or 24 months)

Refer to illustrations 47.2 and 47.5

1 All vehicles are equipped with an *Emissions Control Information* label inside the engine compartment. The label contains important ignition timing specifications and the proper timing procedure for your specific vehicle. If the information on the emissions label is different from the information included in this Section, follow the procedure on the label.

2 At the specified intervals, or when the distributor has been removed, the ignition timing must be checked and adjusted if necessary. Tools required for this procedure include an inductive pick-up timing light, a tachometer, a distributor wrench and, in some cases, a means of plugging vacuum hoses **(see illustration)**.

3 Before you check the timing, make sure the idle speed is correct (see Chapter 4) and the engine is at normal operating temperature.

4 With the engine off, connect a timing light in accordance with the manufacturer's instructions. Usually, the light must be connected to the battery and the number one spark plug wire in some fashion. The number one spark plug is the one at the very front on six-cylinder engines and the one at the front of the left (driver's side) of the engine on V8 engines. Attach the timing light lead near the plug.

5 Locate the numbered timing scale on the front cover of the engine **(see illustration)**. On small-block V8 engines, it's cast into the front cover on the left (driver's side) of the engine. On six-cylinder and big-block V8 engines, the scale is a stamped-steel tab on the right (passenger's side) of the engine front cover.

6 Locate the notched groove across the crankshaft pulley or the flywheel. It may be necessary to have an assistant temporarily turn the ignition on and off in short bursts without starting the engine in order to bring the groove into a position where it can easily be cleaned and marked. **Warning:** *Stay clear of all moving engine components when the engine is turned over in this manner.*

7 Use white soap-stone, chalk or paint to mark the groove in the pulley or flywheel. Also, put a mark on the timing scale corresponding to the number of degrees specified on the *Vehicle Emission Control Information* label in the engine compartment.

8 Aim the timing light at the marks, again being careful not to come into contact with moving parts. The marks made should appear stationary. If the marks are in alignment, the timing is correct. If the marks are not aligned, turn off the engine.

9 Loosen the hold-down bolt or nut at the base of the distributor. Loosen the bolt/nut only slightly, just enough to turn the distributor (see Chapter 5).

10 Now restart the engine and turn the distributor very slowly until the timing marks are aligned.

11 Shut off the engine and tighten the distributor bolt/nut, being careful not to move the distributor.

12 Start the engine and recheck the timing to make sure the marks are still in alignment.

13 Disconnect the timing light and tachometer and reconnect any components which were disconnected for this procedure.

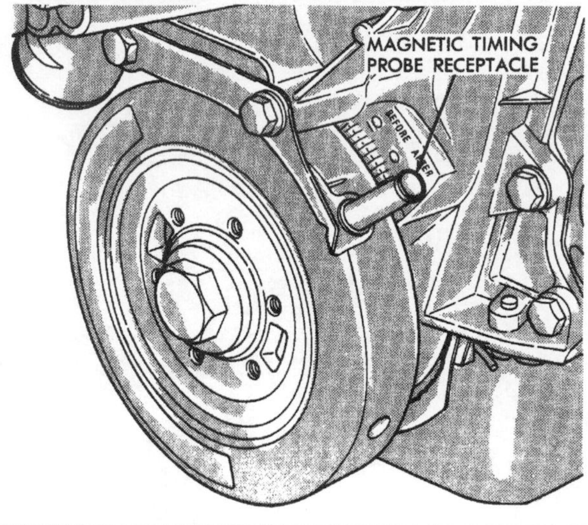

47.5 The timing scale is just to the rear of the damper and is marked to indicate the number of degrees before and after TDC - the receptacle, used on some later models, is where dealer mechanics can hook up a special magnetic probe to monitor ignition timing (small-block V8 engine shown)

48 Oxygen sensor replacement (Electronic Fuel Control system-equipped models) (every 52,000 miles or 36 months)

Refer to illustration 48.2

1 The oxygen sensor must be replaced at the specified interval. The sensor is threaded into the exhaust pipe or manifold. It may be necessary on some models to raise the front of the vehicle and support it securely on jackstands for access to the underside of the engine compartment.

2 Disconnect the oxygen sensor wire by unplugging the connector **(see illustration).**

3 Unscrew the sensor with a box-end wrench.

4 Use a tap to clean the threads in the exhaust pipe or manifold.

5 If the old sensor is to be reinstalled, apply anti-seize compound to the threads. New sensors will already have the anti-seize compound on the threads.

6 Install the sensor, tighten it to the torque listed in this Chapter's Specifications and plug in the connector.

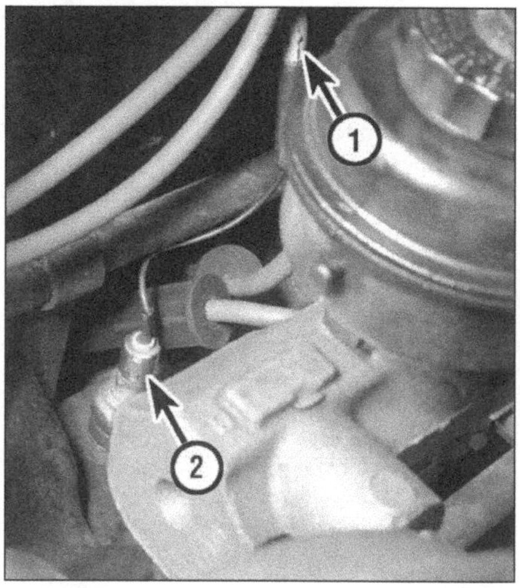

48.2 Unplug the connector (1), then unscrew the oxygen sensor (2)

Chapter 2 Part A
Six-cylinder engines

Contents

Specifications

General

Cylinder numbers (front-to-rear)	1-2-3-4-5-6
Displacement	198 or 225 cubic inches
Bore (all)	3.40 inches
Stroke	
198 cubic inches	3.640 inches
225 cubic inches	4.125 inches
Distributor rotation	Clockwise, as viewed from top
Firing order	1-5-3-6-2-4
Minimum compression	100 psi
Maximum compression variation between cylinders	25 psi

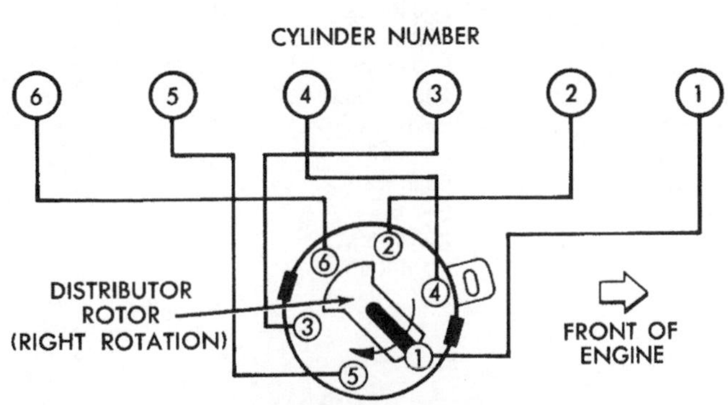

Cylinder locations and numbering - note the arrow that shows the direction the distributor rotates

Camshaft

Bearing journal diameter	
No. 1	1.998 to 1.999 inches
No. 2	1.982 to 1.983 inches
No. 3	1.967 to 1.968 inches
No. 4	1.951 to 1.952 inches
Bearing oil clearance	0.001 to 0.003 inch

Oil pump

Minimum oil pressure at curb idle speed	10 psi
Normal operating oil pressure	30-to-70 psi at 2000 rpm
Pressure drop from a clogged filter	7-to-9 psi
Oil pump cover surface variation limit	0.0015 inch maximum
Outer rotor thickness limit	
Through 1980	0.649 inch minimum
1981 and later	0.825 inch minimum
Outer rotor diameter limit	2.469 inches minimum
Inner rotor thickness limit	
Through 1980	0.649 inch minimum
1981 and later	0.825 inch minimum
Clearance over rotors	0.004 inch maximum
Outer rotor clearance limit	0.014 inch maximum
Rotor tip clearance limit	0.010 inch maximum
Relief valve spring free length	2-1/4 inches

Valve lash/clearance - solid lifters (1980 and earlier models)

Intake	0.010 inch
Exhaust	0.020 inch

Valve lift

1971 through 1980	
Intake	0.406 inch
Exhaust	0.414 inch
1981 through 1983	
Intake	0.378 inch
Exhaust	0.378 inch

Torque specifications **Ft-lbs*** (unless otherwise indicated)

Camshaft sprocket bolt	50
Cooling fan-to-water pump bolts or nuts	200 in-lbs
Crankshaft damper/pulley	Press fit
Crankshaft rear main seal retainer bolts	30
Cylinder head bolts	
1971	65
1972 through 1983	70
Engine mounts	
Mount-to-engine bolts	65
Mount-to-bracket through-bolt	75
Exhaust manifold-to-exhaust pipe bolts/nuts	35
Flywheel/driveplate-to-crankshaft bolts	55
Intake-to-exhaust manifold	
Stud nut	20
Bolts	200 in-lbs
Manifold assembly-to-cylinder head nuts	120 in-lbs
Oil pan screws	200 in-lbs
Oil pump attaching bolts	200 in-lbs
Oil pump cover bolts	95 in-lbs
Oil filter attaching stud	120 in-lbs
Oil pressure gauge sending unit	60 in-lbs
Rocker arm shaft bolts	24
Timing chain cover bolts	200 in-lbs
Valve cover bolts	40 in-lbs

***Note:** *Refer to Part C for additional specifications*

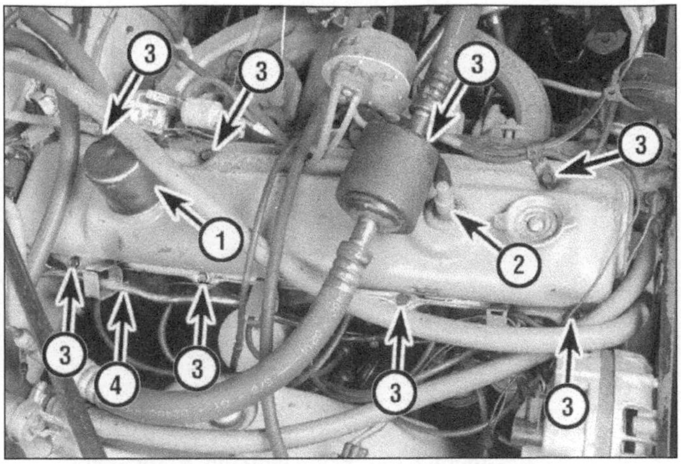

3.3 Disconnect anything attached to the valve cover

1 PCV breather (lift to remove)
2 PVC valve (lift to remove)
3 Valve cover bolts (two are hidden by hoses and wiring in this photo)

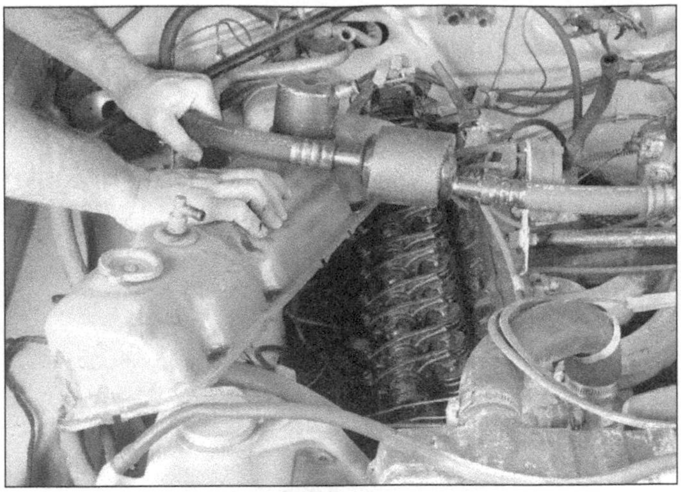

3.5 Carefully lift the air conditioning hose (if equipped) and remove the valve cover from the cylinder head

1 General information

This part of Chapter 2 is devoted to in-vehicle repair procedures for the six-cylinder engine. All information concerning engine removal and installation and engine block and cylinder head overhaul can be found in Part C of this Chapter.

Since the repair procedures included in this Part are based on the assumption that the engine is still installed in the vehicle, if they are being used during a complete engine overhaul (with the engine already out of the vehicle and on a stand) many of the steps included here will not apply.

The specifications included in this Part of Chapter 2 apply only to the procedures found here. The specifications necessary for rebuilding the block and cylinder heads are included in Part 2C.

2 Repair operations possible with the engine in the vehicle

Many major repair operations can be accomplished without removing the engine from the vehicle.

Clean the engine compartment and the exterior of the engine with some type of pressure washer before any work is done. A clean engine will make the job easier and will help keep dirt out of the internal areas of the engine.

Depending on the components involved, it may be a good idea to remove the hood to improve access to the engine as repairs are performed (refer to Chapter 11 if necessary).

If oil or coolant leaks develop, indicating a need for gasket or seal replacement, the repairs can generally be made with the engine in the vehicle. The oil pan gasket, the cylinder head gaskets, intake and exhaust manifold gaskets, timing cover gaskets and the crankshaft oil seals are accessible with the engine in place.

Exterior engine components, such as the water pump, the starter motor, the alternator, the distributor, oil pump and the carburetor as well as the intake and exhaust manifolds, can be removed for repair with the engine in place.

Since the cylinder heads can be removed without pulling the engine, valve component servicing can also be accomplished with the engine in the vehicle.

Replacement of, repairs to or inspection of the timing chain and sprockets and the oil pump pick-up tube with the screen are all possible with the engine in place.

In extreme cases caused by a lack of necessary equipment, repair or replacement of piston rings, pistons, connecting rods and rod bearings is possible with the engine in the vehicle. However, this practice is not recommended because of the cleaning and preparation work that must be done to the components involved.

3 Valve cover - removal and installation

Removal

Refer to illustrations 3.3 and 3.5

1 Disconnect the cable from the negative battery terminal.

2 Remove the air cleaner assembly (see Chapter 4).

3 Remove the breather tube, PCV valve and hoses, vacuum line and electrical wiring clips from the valve cover **(see illustration)**.

4 Remove the valve cover mounting bolts **(see illustration 3.3)**.

5 Remove the valve cover **(see illustration)** while lifting up gently on the air conditioning hose, as shown, and moving the valve cover up and then forward until it has room to be removed. **Note:** *If the cover is stuck to the head, bump the cover with a block of wood and a hammer to release it. If it still will not come loose, try to slip a flexible putty knife between the head and cover to break the seal. Don't pry at the cover-to-head joint, as damage to the sealing surface and cover flange will result and oil leaks will develop.*

Installation

6 The mating surfaces of the cylinder head and valve cover must be perfectly clean when the cover is installed. Use a gasket scraper to remove all traces of sealant or old gasket, then wipe the mating surfaces with a cloth saturated with lacquer thinner or acetone. If there is sealant or oil on the mating surfaces when the cover is installed, oil leaks may develop. **Note:** *Gasket removal solvents are available from auto parts stores and may prove helpful.*

7 Inspect the valve cover gasket flange for distortion. Carefully straighten it with a wooden block and the rounded end of a ball-peen hammer, if necessary.

8 Make sure any threaded holes are clean. Run a tap into them to remove corrosion and restore damaged threads.

9 Mate the new gaskets to the cover before the cover is installed. Apply a thin coat of RTV sealant to the cover flange, then position the gasket inside the cover lip and allow the sealant to set up so the gasket adheres to the cover (if the sealant is not allowed to set, the gasket may fall out of the cover as it is installed on the engine).

10 Carefully position the cover on the head and install the bolts.

11 Tighten the bolts to the torque listed in this Chapter's Specifications.

12 The remaining installation steps are the reverse of removal.

13 Start the engine and check carefully for oil leaks as the engine warms up.

4.2 Remove the rocker arm shaft bolts and curved washers (arrows), working from the end bolts towards the center, loosening them a little at a time

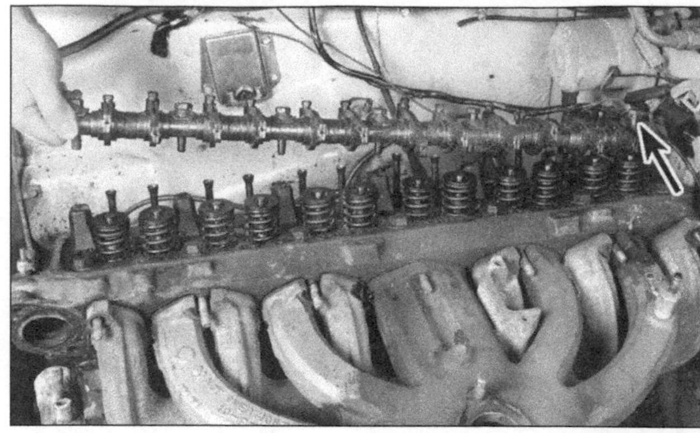

4.3 Lift the shaft assembly from the head without disassembling it; notice the location of the one long bolt at the rear of the shaft (arrow)

4 Rocker arms and shaft - removal, inspection and installation

Removal

Refer to illustrations 4.2 and 4.3

1 Remove the valve cover and gasket (see Section 3).

2 Loosen the rocker shaft bolts 1/2-turn at a time, moving from bolt to bolt **(see illustration)**.

3 When all the bolts are loose, remove the rocker arm shaft as an assembly with the bolts still installed **(see illustration)**. The bolts will keep the assembly together.

4 Remove the pushrods and store them in a cardboard box with numbered holes so they can be reinstalled in their original locations.

Inspection

Refer to illustrations 4.5 and 4.6

5 Once the shaft assembly is removed, disassemble the components for cleaning and inspection **(see illustration)**. **Caution:** *All parts MUST be reinstalled in their original locations. Store or lay the parts out in order so they can be put back on the rocker shaft in their original positions. One strategy is to tie a heavy piece of wire (such as an unbent coat hanger) around a short wood dowel or similar object. Label the dowel to indicate the front of the engine, then remove each rocker arm and spacer from the engine, starting at the front of the engine, and slide them over the wire (the dowel will keep them from falling off at the other end).*

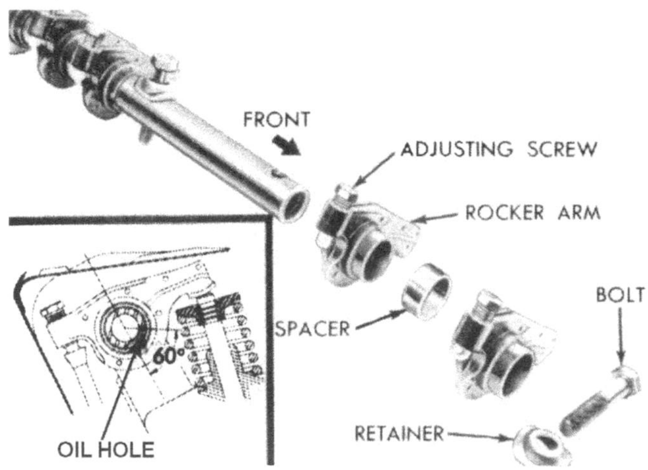

4.5 Disassemble the components for cleaning and inspection, being very careful to keep everything in order so all the parts can be reassembled in exactly the same locations

6 Inspect the rocker arms (especially the surfaces where they contact the valve stems and rocker arms) and rocker arm shaft for signs of excessive wear, galling or damage. If any of these conditions exist, replace the rocker arm and shaft assembly with new parts. **Note:** *Normal wear will appear as shiny areas where there is metal-to-metal contact* **(see illustration)** *and, unless there is not excessive wear, scoring, damage or overheating, the parts are reusable*

7 Inspect the pushrods to see if they are bent, worn or cracked. Make sure the cup at the top of each rocker arm is not loose. Roll each pushrod across a piece of plate glass - it it wobbles, it's bent. If any of these conditions exist, replace the pushrods with new ones.

Installation

Refer to illustration 4.8

8 Reassemble the rocker shaft, rocker arms, spacers, retainers and bolts, applying

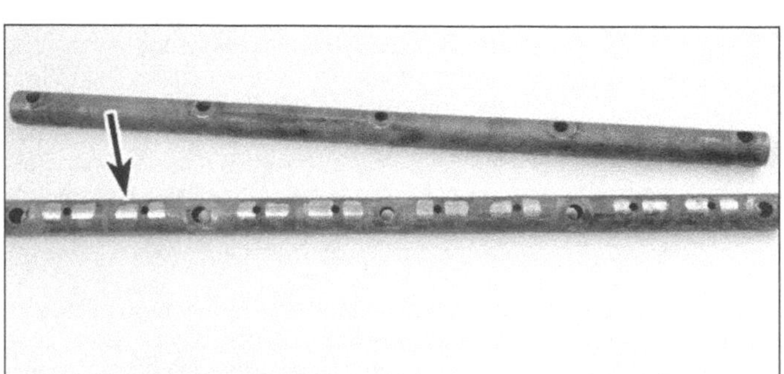

4.6 Inspect all the surfaces of the rocker arm shaft, looking carefully at metal-to-metal contact areas (arrow)

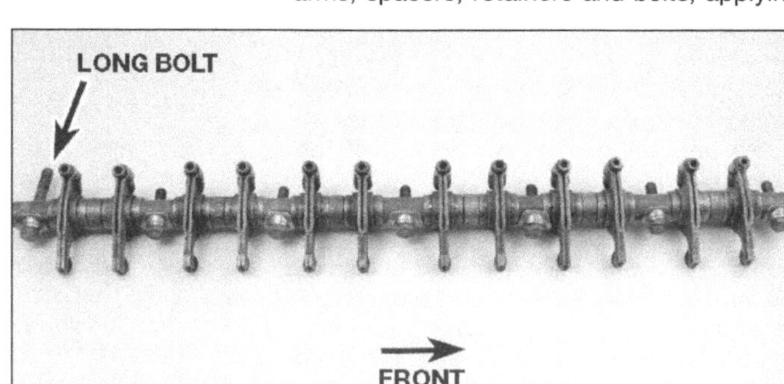

4.8 The one long bolt goes in the rear position

5.8 **Once the spring is depressed, the keepers can be removed with a small magnet or a needle-nose pliers (a magnet is preferred to prevent dropping the keepers)**

5.17 Be sure to install the seals on the correct valves

1 Exhaust valve seal *2 Intake valve seal*

moly-base grease or engine assembly lube to the wear areas on the rocker arm shaft **(see illustration). Caution:** *Make sure the oil holes in the rocker shaft are located correctly so the rocker assemblies are being oiled properly* **(see illustration 4.5).** *Also, make sure there is a spacer between intake and exhaust rocker arms for the same cylinder.*

9 Apply moly-base grease or engine assembly lube to the tops of the valve stems.

11 Install the pushrods in the correct positions, applying moly-base grease or engine assembly lube to the rounded lower end and the cupped upper end.

12 Install the rocker arm shaft with the correct end facing the front of the engine.

13 Install the rocker shaft bolts (the long bolt is installed at the rear).

14 Tighten each bolt a little at a time, working from bolt to bolt, until the torque listed in this Chapter's Specifications is obtained. Be careful not to bend the pushrods.

15 Attach a new valve cover gasket to the cover. Notice that there are tabs provided in the cover to retain the gasket. It may be necessary to apply RTV-type sealant to the corners of the cover to retain the gasket there.

16 The remainder of the installation procedure is the reverse of removal.

17 Start the engine and run it until it reaches normal operating temperature, then make sure that there are no leaks.

5 Valve springs, retainers and seals - replacement

Refer to illustrations 5.8, 5.17, 5.19 and 5.21

Note: *Broken valve springs and defective valve stem seals can be replaced without removing the cylinder head. Two special tools and a compressed air source are normally required to perform this operation, so read through this Section carefully and rent or buy the tools before beginning the job. If compressed air isn't available, a length of nylon rope can be used to keep the valves from falling into the cylinder during this procedure.*

1 Remove the valve cover(s) (See Section 3).

2 Remove the spark plugs (see Chapter 1).

3 Turn the crankshaft until the number one piston is at top dead center on the compression stroke (see Part 2C, Section 4).

4 Remove the rocker arm shaft (see Section 4). If you intend to use a lever-type spring compressor, remove the rocker arms and reinstall the shaft.

5 Thread an adapter, available at most auto parts stores, into the spark plug hole and connect an air hose from a compressed air source to it. Most auto parts stores can supply the air hose adapter. **Note:** *Many cylinder compression gauges utilize a screw-in fitting that may work with your air hose quick-disconnect fitting.*

6 Apply compressed air to the cylinder. The valves should be held in place by the air pressure. If the valve faces or seats are in poor condition, leaks may prevent the air pressure from retaining the valves - valve reconditioning is indicated.

7 If you don't have access to compressed air, an alternative method can be used. Position the piston at a point just before TDC on the compression stroke, then feed a long piece of nylon rope through the spark plug hole until it fills the combustion chamber. Be sure to leave the end of the rope hanging out of the engine so it can be removed easily. Use a large breaker bar and socket to rotate the crankshaft in the normal direction of rotation (clockwise) until slight resistance is felt.

8 Stuff shop rags into the cylinder head holes around the valves to prevent parts and tools from falling into the engine, then use a valve spring compressor to compress the spring. Remove the keepers with small needle-nose pliers or a magnet **(see illustration). Note:** *Several different types of tools are available for compressing the valve springs with the head in place. One type, shown here, grips the lower spring coils and presses on the retainer as the knob is turned, while the lever-type (Chrysler No. C-3906, or equivalent) utilizes the rocker arm shaft for leverage. Both types work very well, although the lever type is usually less expensive.*

9 Remove the valve spring and retainer.

10 Remove the old valve stem seals.

11 If there are any valve spring shims that have been used, to get the correct assembled height, note where and how many shims are used at each valve and then remove the shims. **Caution:** *The shims AND valve springs must be reinstalled in their original locations for correct reassembly.*

12 Wrap a rubber band or tape around the top of the valve stem so the valve won't fall into the combustion chamber, then release the air pressure. **Note:** *If a rope was used instead of air pressure, turn the crankshaft slightly in the direction opposite normal rotation.*

13 Inspect the valve stem for damage. Rotate the valve in the guide and check the end for eccentric movement, which would indicate that the valve is bent.

14 Move the valve up-and-down in the guide and make sure it doesn't bind. If the valve stem binds, either the valve is bent or the guide is damaged. In either case, the head will have to be removed for repair.

15 Reapply air pressure to the cylinder to retain the valve in the closed position, then remove the tape or rubber band from the valve stem. If a rope was used instead of air pressure, rotate the crankshaft in the normal direction of rotation until slight resistance is felt.

16 If shims were used on any of the valves, place them back in their original positions. This will give the same valve spring installed height as prior to disassembly.

17 If you're working on an exhaust valve, install the new shield on the valve stem and push it down to the top of the valve guide **(see illustration)**.

18 If you're working on an intake valve, push a new valve stem seal down over the valve guide, using the valve stem as a guide, but don't force the seal against the top of the guide.

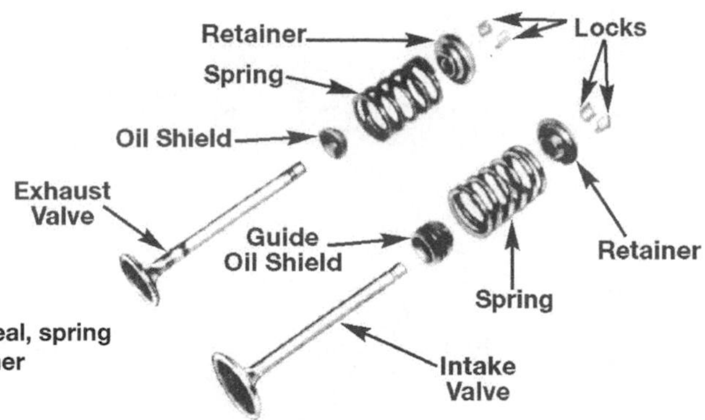

5.19 Install the seal, spring and retainer

5.21 Apply a small dab of grease to each keeper as shown here before installation - it'll hold them in place on the valve stem as the spring is released

19 Install the spring and retainer or rotator in position over the valve **(see illustration)**. **Caution:** *Different-length springs are used on the intake and exhaust valves - accidentally installing an intake spring on an exhaust valve can cause major engine damage!*

20 Compress the valve spring assembly only enough to install the keepers in the valve stem.

21 Position the keepers in the valve stem groove. Apply a small dab of grease to the inside of each keeper to hold it in place if necessary **(see illustration)**. Remove the pressure from the spring tool and make sure the keepers are seated.

22 Disconnect the air hose and remove the adapter from the spark plug hole. If a rope was used in place of air pressure, pull it out of the cylinder.

23 Repeat the above procedure on the remaining cylinders, following the firing order sequence (see Chapter 1). Bring each piston to top dead center on the compression stroke before applying air pressure or inserting the rope into the cylinder.

24 Reinstall the rocker arm assemblies and the valve covers.

25 Start the engine, then check for oil leaks and unusual sounds coming from the valve cover area.

6 Intake and exhaust manifolds - removal and installation

Removal

Refer to illustrations 6.5, 6.6a and 6.6b

1 Disconnect the cable from the negative battery terminal.

2 Remove the air cleaner, vacuum lines and heat stove flexible connector (see Chapter 4).

3 Disconnect the throttle linkage (see Chapter 4).

4 If the manifold is going to be replaced, remove the carburetor, EGR and divorced choke assembly (see Chapter 4) so that it can be installed on the new manifold.

5 Disconnect the exhaust pipe at the exhaust manifold **(see illustration)**.

6 Remove the nuts attaching the intake/exhaust manifold assembly to the cylinder head **(see illustration)** and detach the manifold assembly **(see illustration)**.

7 Remove the three vertical bolts (or stud nuts) to separate the intake manifold from the exhaust manifold. The bolts are in the center of the intake manifold, around the carburetor mounting point.

8 Use a gasket scraper to remove all traces of sealant and old gasket material, then wipe the mating surfaces with a cloth saturated with lacquer thinner or acetone. If there is old sealant or oil on the mating surfaces when the manifolds are installed, vacuum or exhaust leaks may develop. Don't forget to clean the intake-to-exhaust manifold mating surfaces. **Note:** *Aerosol gasket removal solvents are available from auto parts stores and may prove helpful.*

9 Check the gasket surfaces for warpage with a straightedge and feeler gauges. They must not be warped more than 0.009 inch per foot of manifold length. Check the manifolds carefully for cracks and distortion - it's common to find cracks in these long exhaust manifolds. Replace the manifolds if any cracks or distortion are found.

Installation

10 Use new gaskets when rejoining and installing the manifold assembly.

11 Install, but DO NOT TIGHTEN, the bolts attaching the intake manifold to the exhaust manifold.

12 Install the manifold assembly and tighten

6.5 Remove the two bolts and nuts to disconnect the exhaust pipe at the exhaust manifold (arrow); the second nut is on the opposite side of the manifold

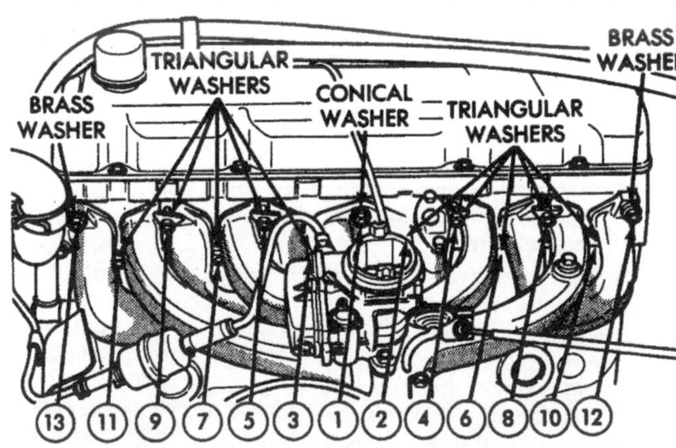

6.6a Intake/exhaust manifold nut TIGHTENING sequence - reverse this sequence when loosening the bolts. Be sure to note the locations of the special conical and triangular washers when removing them

6.6b The intake/exhaust manifolds come off the engine as a single unit

7.5 Remove the nuts (1) and separate and lower the center link, then remove the engine-to-transmission brace (2) and torque converter inspection cover (3)

7.9a Remove the oil pan bolts (arrows) . . .

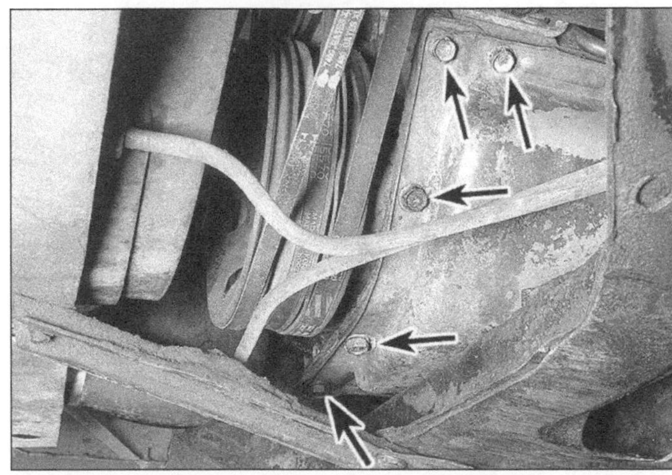

7.9b . . . and don't miss the ones at the front, just forward of the body K-member (arrows)

the fasteners finger tight only! Washers spanning the intake and exhaust manifold flanges must be flat - replace any that are distorted by previous over-tightening. Install the steel conical washers with the cup side facing the nut. Install the brass washers with the flat sides facing the manifold. Install the nuts with the cone-shaped side facing the washer.

13 Carefully tighten all fasteners to only 10 in-lbs; do not over-tighten them!

14 Tighten the inboard intake-to-exhaust manifold bolts or nuts to 10 ft-lbs.

15 Tighten the outboard intake-to-exhaust manifold bolts or nuts to 10 ft-lbs.

16 Repeat the procedure, tightening the intake-to-exhaust manifold bolts or nuts to the torque listed in this Chapter's Specifications.

17 Tighten the manifold assembly nuts to the torque listed in this Chapter's Specifications, following the proper sequence **(see illustration 6.6a)**. Work up to the final torque in three steps and DO NOT OVER TIGHTEN THEM!

18 The rest of the installation procedure is the reverse of removal.

7 Oil pan - removal and installation

Removal

Refer to illustrations 7.5, 7.9a, 7.9b and 7.10

1 Disconnect the battery cable from the negative battery terminal.

3 Raise the vehicle and support it securely on jackstands (see Chapter 1).

4 Drain the engine oil (see Chapter 1).

5 Using a puller, separate the center link from both the pitman arm and the idler **(see illustration)** - refer to Chapter 10, if necessary. Leave the inner tie-rod ends connected to the center-link. No re-alignment should be necessary after reassembly. Let it hang down as far as possible. **Note:** *It may also be necessary to disconnect the exhaust pipe from the exhaust manifold* **(see illustration 6.5)**, *on some models.*

6 Remove the engine-to-transmission brace **(see illustration 7.5).**

7 On vehicles equipped with an automatic transmission, remove the torque converter inspection cover **(see illustration 7.5)**.

8 Support the engine at the right front corner of the engine block. **Warning:** *DON'T support the engine on the crankshaft pulleys or on the vibration damper.* Remove the engine mount bolts and raise the vehicle approximately 1-1/2 to 2 inches. On some models it may be necessary to separate the engine fan shroud from the radiator and set it back over the fan (see Chapter 3) in order to raise the engine enough to remove the oil pan. **Note:** *Watch the radiator hoses closely; you should be able to raise the engine the small amount needed without draining the cooling system and disconnecting the radiator hoses, but, since the upper radiator hose is so short, it may need to be disconnected.*

9 Remove the oil pan mounting bolts **(see illustrations)**, separate the oil pan from the engine and remove the oil pan. If the pan is stuck to the block, strike it with a rubber mallet to break the gasket seal. Do not pry between the pan and block or damage to the gasket surfaces may occur and oil leaks will develop. **Note:** *It may be necessary to rotate the crankshaft when removing the oil pan to get the crankshaft counterweights in a posi-*

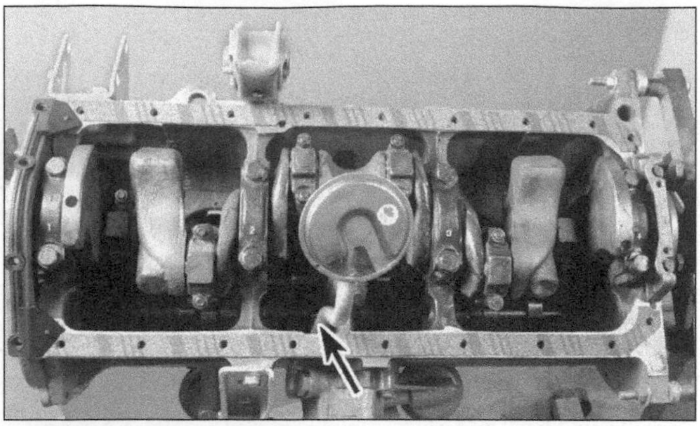

7.10 Unscrew the oil pick-up tube/screen assembly from the block and either clean or replace it at this time

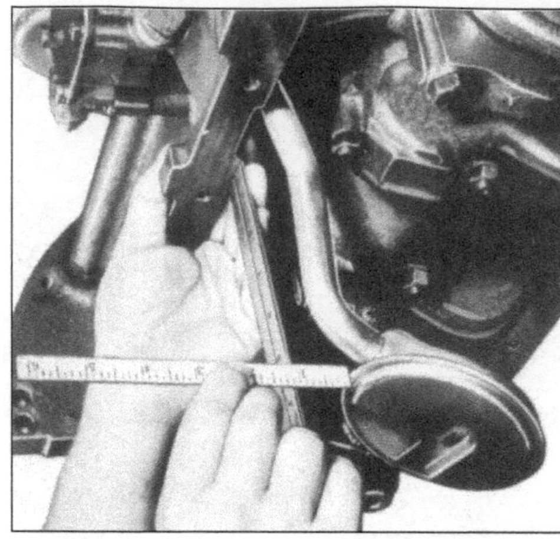

7.15 Checking the alignment of the oil pump pick-up tube (it must be the specified distance from the inside edge of the block)

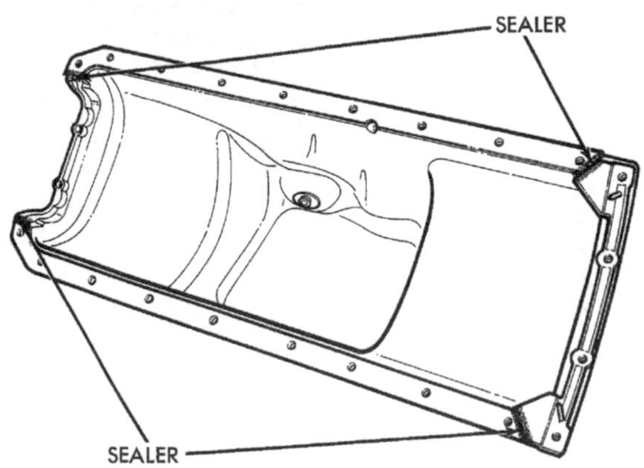

7.16 RTV-type sealant should be applied to the gasket-to-seal junctions before installing the oil pan (attach the side rail gaskets to the oil pan with contact-cement-type gasket adhesive)

8.5 Use a bolt-type puller to remove the vibration damper

tion so the pan can be moved past it.

10 The oil pick-up tube and screen should be cleaned at this time. Unscrew the tube/screen assembly **(see illustration)** from the block and clean it with solvent. Check the condition of the screen and replace it if necessary.

11 Clean the oil pan with solvent and wipe it dry with a clean cloth.

Installation

Refer to illustration 7.15 and 7.16

12 Scrape away all traces of gasket material from the mounting surfaces of the oil pan and the block. Gasket removal solvents are available from auto parts stores and may prove helpful. After all gasket material is removed, wipe the surfaces clean with a rag soaked in lacquer thinner or acetone.

13 Check the oil pan mounting flange for distortion and straighten it, if necessary.

14 Thread the oil pick-up tube and screen into the block, applying thread sealant to the threads, and tighten it until it is it tight, then proceed to the next Step to position it correctly. **Caution:** *Be absolutely certain that the pick-up screen is properly tightened so that no air can be sucked into the oiling system at this connection; otherwise, the oil pump could lose its prime and serious engine damage could result.*

15 Check the position of the tube/screen assembly with steel rules **(see illustration)**. The screen should be about 1-1/8 inch from the inside edge of the block. Turn the screen as necessary so the dimension is correct. Also, the screen must touch the bottom of the oil pan when the oil pan is installed. But also be certain that the pick-up tube is tightly threaded into the pump body so there is no possibility that any air can sucked into the oiling system at that connection.

16 Install the oil pan, using new gaskets. Attach the side-rail gaskets to the pan with contact-cement-type gasket adhesive such as Fel-Pro's High-Tack. Apply a bead of RTV sealant to the four points where the gaskets join **(see illustration)**.

17 Install the bolts and tighten them to the torque listed in this Chapter's Specifications, starting from the center and working out in each direction. Work up to the final torque in three steps.

18 The rest of the installation procedure is the reverse of removal. Fill the crankcase with the proper amount of oil and, after starting the engine, check carefully for leaks at the oil pan gasket sealing surfaces.

8 Crankshaft oil seals - replacement

Front seal

Removal

Refer to illustrations 8.5, 8.6a, 8.6b and 8.8

Note: *The front crankshaft oil seal on all models can be replaced once the timing chain cover is removed. But, on early models (1971 though 1974), the seal MUST be replaced from inside the cover, while on later models (1975 through 1983), the seal can be replaced by carefully prying it out of the timing chain cover while it is still bolted to the engine block.*

1 Drain the cooling system (see Chapter 1).

2 Unbolt the fan assembly and remove the fan and shroud (see Chapter 3).

3 Remove the radiator (see Chapter 3)

4 Unbolt and remove the pulley that is bolted to the vibration damper.

5 Remove the vibration damper assembly from the front end of the crankshaft. A special puller, which can be found at your local parts store, will be required for this operation **(see illustration)**.

8.6a Remove all the timing cover bolts (arrows) - don't miss the oil pan-to-cover bolts that are removed from below

8.6b Separate the front cover from the block by prying it enough to break the gasket's hold, but be careful not gouge or scratch the gasket surface on either part or a leak could develop - also note that there are two dowel pins that align the cover with the block, so the cover should be removed as straight as possible to prevent distorting the dowel holes in the cover

1971 through 1974 models

6 Remove the timing chain cover **(see illustrations)**. **Note:** *This procedure can be done by removing the front oil pan bolts, that connect the pan and front cover, and loosening the remaining oil pan bolts so the pan can be slightly lowered. It can be extremely difficult on some models, however, to do this without removal of the oil pan. If necessary, remove the oil pan (see Section 7).*

7 After the front cover is removed, remove all traces of gasket sealant from the cover-to-block mating surfaces with a scraper, then wipe the surfaces clean with a rag soaked in lacquer thinner or acetone. Remove the old rubber end seal from the front cover. Support the cover on top of two blocks of wood and drive the seal out from the front with a hammer and punch. **Caution:** *Be careful not to scratch, gouge or distort the area that the seal fits into or a leak will develop.*

1975 and later models

8 If the seal is being removed while the cover is still attached to the engine block, carefully pry the seal out of the cover with a seal puller or a large screwdriver **(see illustration)**. **Caution:** *Be careful not to scratch, gouge or distort the area that the seal fits into or a leak will develop.*

9 Clean the bore to remove any old seal material and corrosion. Wipe the surface clean with a rag soaked in lacquer thinner or acetone.

Installation

Refer to illustration 8.10

10 After either method of seal removal is complete and the new seal is ready to be installed, coat the metal outside diameter of the seal with RTV sealant. Position the seal on the cover so the seal lip (end with the spring) will face IN, toward the engine. Drive the seal into the bore with a large socket and hammer until it's completely seated

8.8 On 1975 and later models, pry the old seal out with a seal puller or screwdriver - the front cover can remain attached to the block

(see illustration). Select a socket that's the same outside diameter as the seal and make sure the new seal is pressed into place until it bottoms against the cover flange. **Note:** *The seal on 1974 and earlier models is driven in from the back side of the cover and, on 1975 and later models, the seal is driven in from the front.*

11 If the front cover was removed (1975 and later models), put a new cover-to-block gasket and rubber end seal on the front cover, attaching the gasket to the cover with RTV sealant. Install the front cover, tightening the cover-to-block bolts to the torque listed in this Chapter's Specifications. Apply beads of RTV sealant just to the rear of the rubber timing cover end seal, to make sure there's a good seal at this problem sealing area, then install and tighten the oil pan bolts to the torque listed in this Chapter's Specifications.

12 Lubricate the seal lips with engine oil and reinstall the vibration damper. A special

8.10 Drive the new seal into place with a socket and hammer

threaded tool, available at most auto parts stores, is normally used to install the vibration damper. However, if care is taken, a large wooden block and a hammer can be used to seat the damper. Take special care to ensure the Woodruff key and keyway in the damper line up before pressing the damper into place. The remainder of installation is the reverse of removal.

Rear seal

Note: *If you're installing a new seal during overhaul, ignore the steps in this procedure that concern removal of external parts. Also, since the crankshaft is already removed, it's not necessary to use any special tools to remove the upper seal half; remove and install the upper seal half the same way as the lower seal half.*

Removal

Refer to illustrations 8.15, 8.16a and 8.16b

13 To replace the crankshaft rear oil seal, split rubber seal halves should be used in

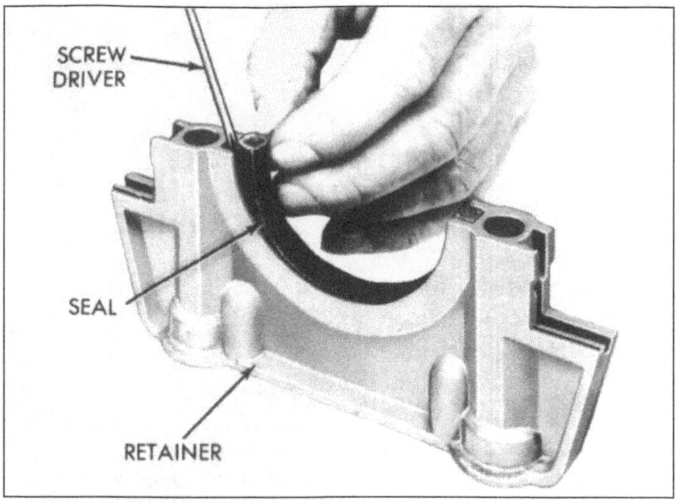

8.15 Carefully pry the old rear seal half out of the retainer groove

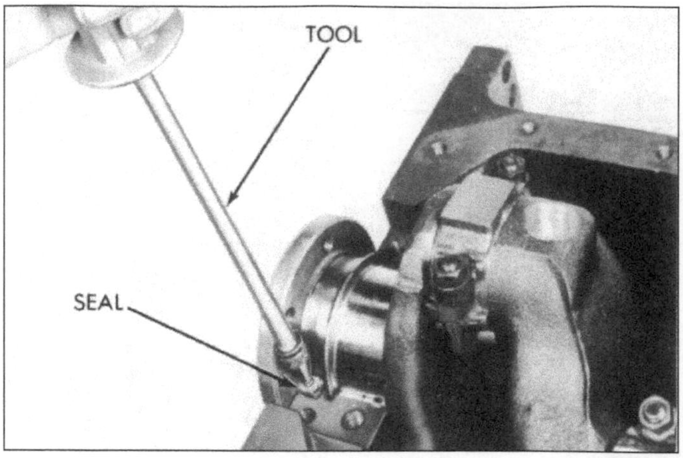

8.16a If a rope-type seal is installed, a special seal puller is recommended for removal of the upper half of the rear main seal (a sheet-metal or wood screw and a pair of pliers will work just as well)

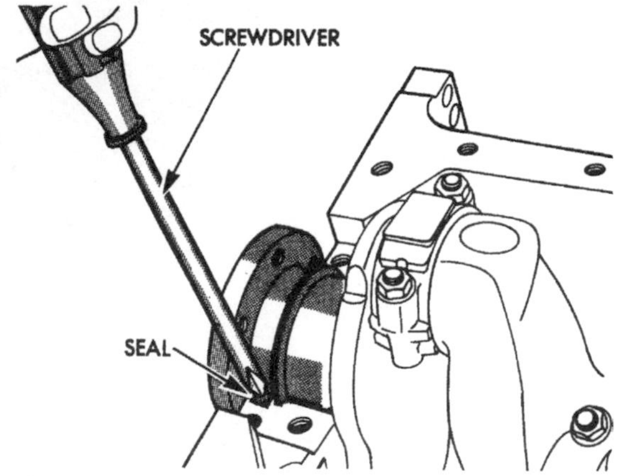

8.16b if a rubber seal is installed, push on one end with a screwdriver until the other end protrudes enough to be grasped and removed with a pair of pliers

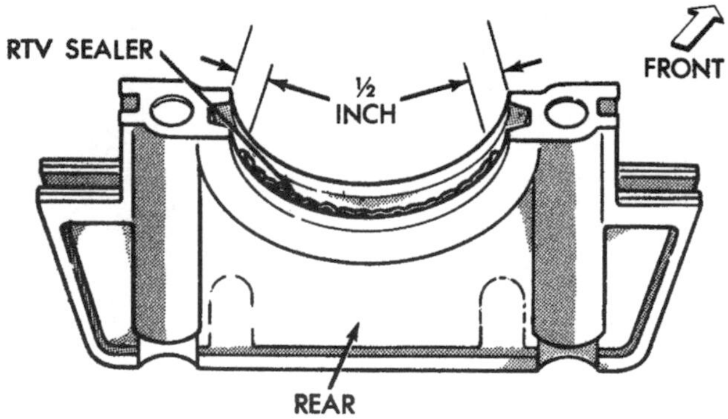

8.20 A 1/8-inch bead of RTV-type sealant should be applied to the retainer groove, to within 1/2- inch of each end

place of any rope-type seals installed during original engine assembly.

14 Remove the oil pan (see Section 7), then remove the bolts and detach the rear seal retainer from the rear main bearing cap.

15 Remove lower rope-type oil seal by prying from the side **(see illustration)** with a small screwdriver. **Note:** *Most replacement seals are of split-rubber type composition. The seals make it possible to replace the upper rear seal without removing the crankshaft. The seal must be used as an upper and lower set and cannot be combined with the original "rope" style oil seal.*

16 If a rope-type seal is installed, remove the upper rope piece by screwing a puller type tool (KD-492, or equivalent) into the end of the seal **(see illustration)**, being careful not to damage the crankshaft. Pull the seal out with the tool while an assistant is rotating the crankshaft. **Caution:** *Be very careful not to nick or gouge the crankshaft seal contact surface as this is done.* If a rubber seal is installed, push one end **(see illustration)** until the other end moves out far enough to be grasped with a pair of pliers.

Installation

Refer to illustrations 8.20, 8.22 and 8.24

17 Wipe the crankshaft surface clean, then oil lightly before installing a new seal.

18 Lubricate the lip of one new seal half with engine oil.

19 Insert the new seal half into the engine block while holding the seal (with paint stripe to the rear - lip facing forward) tightly against crankshaft to make sure that the sharp edge of the groove in the block does not shave or nick the back of the seal. Install seal in the block groove. Rotate the crankshaft, if necessary, while sliding the seal into the groove. Care must be exercised not to damage the sealing lip.

20 Apply a 1/8-inch bead of RTV sealant to the bottom of the retainer groove, starting and finishing 1/2-inch from the ends of the groove **(see illustration)**, then install the other half of the seal in the retainer.

21 Install the other half of the seal into the lower seal retainer with the paint stripe to rear (lip facing forward).

22 Install the two side seals into grooves in seal retainer **(see illustration)** with contact cement-type gasket adhesive.

23 Lightly grease the retainer side seals for ease of assembly into the block.

24 Install the lower seal retainer after apply-

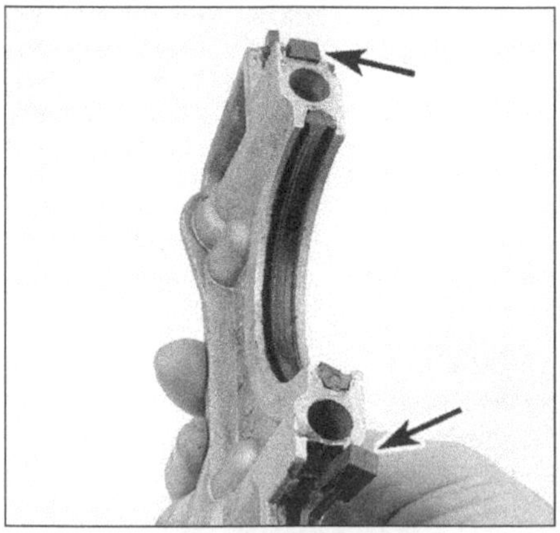

8.22 When installing the two side seals (arrows), be sure to use a contact-cement-type gasket sealer to hold them in place during installation

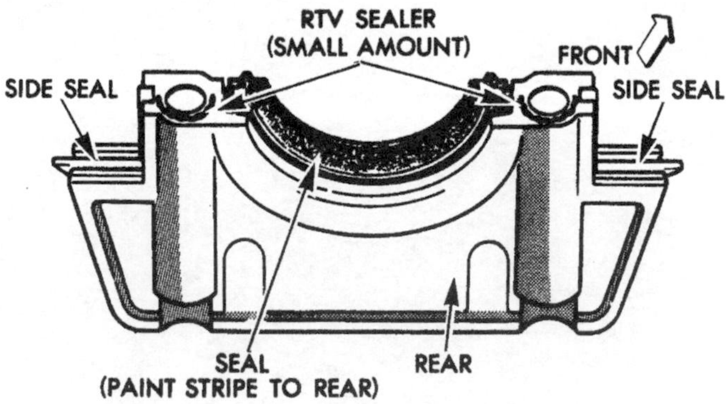

8.24 A small amount of RTV-type sealant should be applied to the seal retainer mating surfaces - DO NOT get any sealant on the lips or ends of the seal!

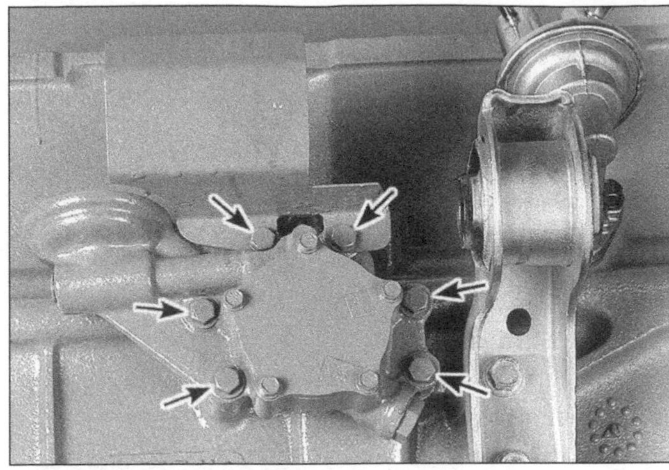

9.3a Remove the mounting bolts (arrows) and detach the pump assembly from the block

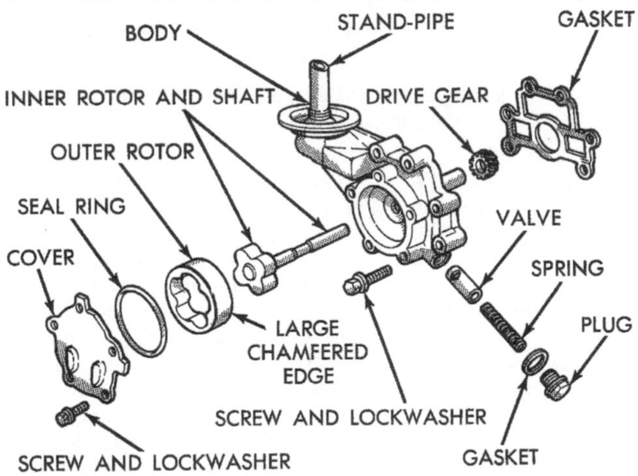

9.3b Oil pump - exploded view

ing a small amount of RTV sealant around each bolt hole **(see illustration)**. **Caution:** *DON'T allow sealant to get on the seal ends or lip.* Tighten the bolts to the torque listed in this Chapter's Specifications.

25 The remainder of installation is the reverse of removal. Add the correct type and quantity of oil to the engine.

26 Start the engine and check for oil leaks.

9 Oil pump- removal and installation

Removal

Refer to illustrations 9.3a and 9.3b

1 Disconnect the battery cable from the negative battery terminal.

2 The oil pump is externally mounted to the block and is located on the right side (passenger's side) of the cylinder block near the engine mount. Drain the oil and remove the oil filter.

3 Remove the mounting bolts and detach the pump assembly from the block **(see illustrations)**. **Note:** *The oil pump can be disassembled* **(see illustration)** *and rebuilt, but the cost of a new or remanufactured oil pump is probably less than the individual parts would be. We recommend replacing the oil pump if there's any doubt at all about its condition.*

Installation

4 Clean the gasket mating surfaces on the block and pump. Position the pump on the engine block with a new gasket. Install the mounting bolts and tighten them to the torque listed in this Chapter's Specifications.

5 Run the engine and check for oil pressure and leaks.

10 Timing chain, cover and sprockets- removal, inspection and installation

Timing chain slack check (cover on engine)

Note: *This procedure will allow you to check the amount of slack in the timing chain without removing the timing chain cover from the engine.*

1 Disconnect the negative battery cable from the battery.

2 Place the engine at TDC (Top Dead Center) (see Chapter 2C).

3 Keeping the piston on the compression stroke, place the number one piston about 30-degrees before TDC (BTDC).

4 Remove the distributor cap (see Chapter 1).

5 You must rotate the crankshaft very slowly by hand for accuracy on this check, so get a 1/2-inch drive breaker bar, extension and correct-size socket to use on the large bolt in the center of the crankshaft pulley. **Note:** *Not all models are equipped with a bolt in the center of the pulley, but all crankshafts are threaded so a bolt of the correct size and pitch could be installed.*

6 Turn the crankshaft clockwise until the number one piston is at TDC. This will take up the slack on the left side of the timing chain.

7 Mark the position of the distributor rotor on the distributor housing.

8 Slowly turn the crankshaft counterclockwise until the slightest movement is seen at the distributor rotor. Stop and note how far the number one piston has moved away from the TDC mark by looking at the ignition timing marks.

9 If the mark has moved more than 10 degrees, the timing chain is probably worn excessively. Remove the timing chain cover for a more accurate check.

Cover removal

10 Remove the vibration damper (see Section 8).

11 Loosen the front oil pan bolts, then remove the timing cover bolts (see Section 8). **Note:** *This procedure can be done by removing the front oil pan bolts and loosening the remaining oil pan bolts so the oil pan can be slightly lowered. It can be extremely difficult on some models, however, to do this without removing the oil pan to properly reseal the gasket areas. If necessary, remove the oil pan.*

12 Detach the timing cover and gasket (see Section 8).

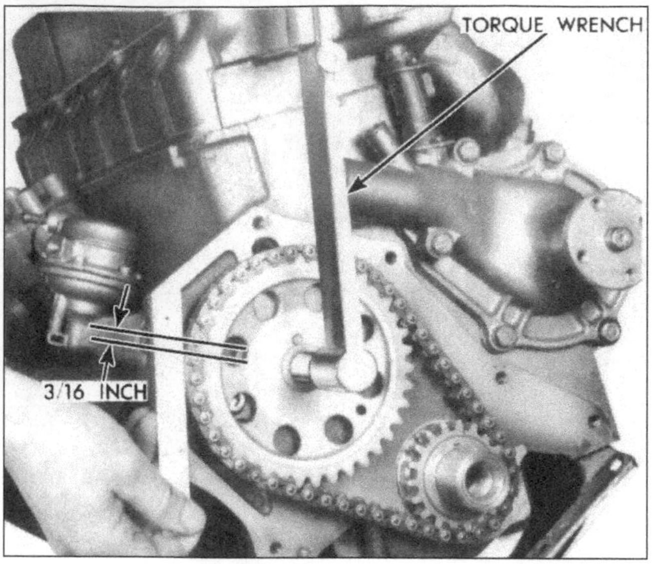

10.14 Attach a torque wrench, as shown, and measure the amount of movement to determine the need for timing chain replacement

10.15 Inspect the timing chain and both gears for wear or damage

1 *Camshaft gear (original-style aluminum gear with Nylon teeth)*
2 *Crankshaft gear*

13 Remove the oil slinger from the end of the crankshaft, if equipped.

Timing chain and gear inspection (cover removed from engine)

Refer to illustrations 10.14 and 10.15

14 Attach a socket and torque wrench to the camshaft sprocket bolt **(see illustration)** and apply force in the normal direction of crankshaft rotation (30 ft-lbs if the cylinder head is still in position, complete with rocker arms, or 15 ft-lbs if the cylinder head has been removed). Don't allow the crankshaft to rotate. If necessary, jam the crankshaft sprocket so that it can't move. Using a ruler, note the amount of movement of the chain. If it exceeds 3/16-inch a new timing chain will be required. **Note:** *Whenever a new timing chain is required, the entire set (chain, camshaft and crankshaft sprockets) must be replaced as an assembly.*

10.17 The timing marks must be aligned as shown here when installing the timing chain and sprockets

15 Inspect the camshaft gear for damage or wear. The camshaft gear on early models may have been steel, but most cam gears will be an aluminum gear with a nylon coating on the teeth **(see illustration)**. This nylon coating may be cracked or be breaking off in small pieces. These pieces tend to end up in the oil pan and may eventually plug the oil pump pick-up screen. If the pieces have come off the camshaft gear, the oil pan should be removed to properly clean or replace the oil pump pick-up screen (see Section 7).

16 Inspect the crankshaft gear for damage or wear. The crankshaft gear is a steel gear, but the teeth can be grooved or worn enough to cause a poor meshing of the gear and the chain.

Chain removal

Refer to illustration 10.17

17 Be sure the timing marks are aligned **(see illustration)**. Hold a pry bar through one of the camshaft sprocket holes to prevent the camshaft from turning. Remove the bolt from the camshaft sprocket and withdraw the camshaft sprocket and chain. The chain can be disengaged from the teeth of the crankshaft sprocket.

18 The sprocket on the crankshaft can be removed with a two or three jaw puller, but be careful not to damage the threads in the end of the crankshaft. **Note:** *If the timing chain cover oil seal has been leaking, refer to Section 8 and install a new one.*

Installation

19 Installation is the reverse of removal. Make sure the timing marks on the gears are properly aligned **(see illustration 10.17)**. Note the metal camshaft gear that replaces the original aluminum/plastic combination (this is preferred for long life). Refer to Section 8 to reinstall the front cover. Be sure to apply a 1/8-inch bead of RTV-type sealant just to the rear of the timing cover rubber seal (always use a new timing cover lower seal). Also use a new timing cover gasket.

20 A special threaded tool (Chrysler No. C-3732A or equivalent) is normally used to install the vibration damper. However, if care is taken, a large wooden block and a hammer can be used to seat the damper. Take special care to ensure the Woodruff key and keyway in the damper line up before pressing the damper into place.

11 Cylinder head - removal and installation

Removal

1 Drain the cooling system (see Chapter 1).

2 Remove the air cleaner, disconnect the accelerator linkage and unplug the vacuum line from the distributor and the carburetor (see Chapter 4).

3 Disconnect the spark plug wires from the spark plugs.

4 Disconnect the heater hoses from the engine.

5 Disconnect the electrical connector from the coolant temperature sending unit (see Chapter 3).

6 Disconnect the exhaust pipe from the exhaust manifold (see Section 6).

7 Where applicable, disconnect the diverter valve vacuum line and remove the air tube assembly.

8 Remove the manifolds (see Section 6).

9 Remove the rocker arm cover, shaft and rocker arms (see Sections 3 and 4).

10 Remove the pushrods in sequence and label them so they can be installed in their

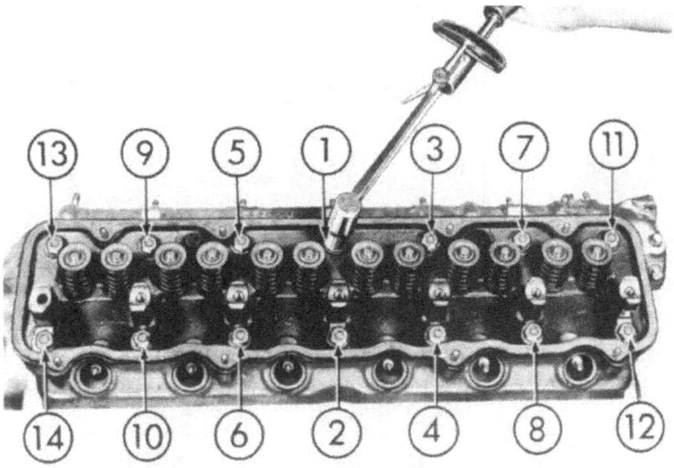

11.15 Cylinder head bolt TIGHTENING sequence - when loosening the bolts, reverse this sequence

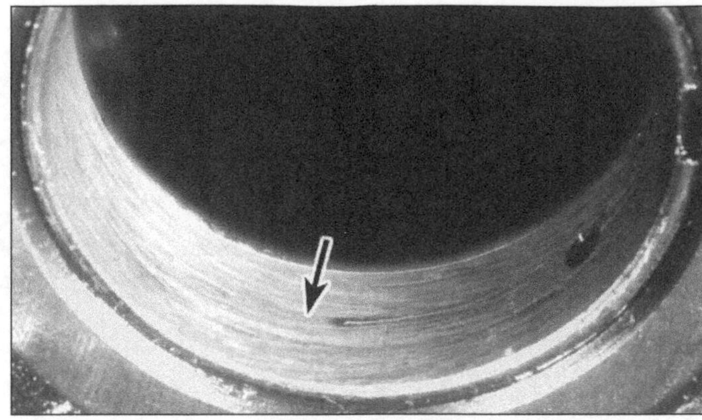

12.17 If the camshaft bearing surfaces are excessively worn or damaged, like the one shown here, the block should be taken to an automotive machine shop to have new bearings installed - this job can be done at home, but it is very involved and should be performed by an automotive machine shop

original locations. A numbered box or rack will keep them properly organized.

11 Remove the cylinder head bolts, loosening the bolts in the reverse order of the tightening sequence **(see illustration 11.15)** and lift the cylinder head from the engine block. It may be necessary to break the gaskets seal by prying up the cylinder head. Try to find a casting protrusion to pry against. **Caution:** *Do not wedge any tools between the cylinder head and block gasket mating surfaces.*

Installation

Refer to illustrations 11.15

Note: *Clean all the head bolts and run a bottoming tap through the bolt holes in the block and a die down each head bolt (clamp the bolt head in a vise while doing this). Lightly lubricate the bolts before assembly.*

12 Scrape away all traces of gasket material from the mounting surfaces of the cylinder head and the block. Also clean the cylinder head-to-manifold mating surfaces. While scraping, stuff rags into the cylinder bores and pushrod holes in the block to keep gasket material from falling in. If any material does fall in, use a vacuum cleaner to remove it.

Gasket removal solvents are available from auto parts stores and may prove helpful. After all gasket material is removed, wipe the surfaces clean with a rag soaked in lacquer thinner or acetone.

13 Position the head gasket over the dowel pins on the block, making sure that it is facing the right direction and that the correct surface is exposed. Gaskets are often marked "front" and "this side up" to ensure correct installation.

14 Carefully lower the cylinder head onto the block.

15 Coat the cylinder head retaining bolts with light engine oil and thread the bolts into the block. Tighten the bolts in the proper sequence **(see illustration)**. Work up to the final recommended torque listed in this Chapter's Specifications in three steps.

16 The remainder of the installation procedure is the reverse of removal.

12 Camshaft and lifters - removal, inspection and installation

Camshaft lobe wear check

1 In order to determine the extent of cam lobe wear, the valve lift should be checked prior to camshaft removal. Refer to Section 5 and remove the rocker arm cover.

2 Position the number one piston at TDC on the compression stroke (see Chapter 2C).

3 Beginning with the number one cylinder valves, mount a dial indicator on the engine and position the plunger against the top surface of the first rocker arm. The plunger should be directly above and in line with the valve, not the pushrod.

4 Zero the dial indicator, then very slowly turn the crankshaft in the normal direction of rotation until the indicator needle stops and begins to move in the opposite direction. The point at which it stops indicates maximum valve lift.

5 Record this figure for future reference, then reposition the piston at TDC on the compression stroke.

6 Move the dial indicator to the remaining number one cylinder rocker arm and repeat the check. Be sure to record the results for each valve.

7 Repeat the check for the remaining valves. Since each piston must be at TDC on the compression stroke for this procedure, work from cylinder-to-cylinder following the firing order sequence (see Chapter 1).

8 After the check is complete, compare the result to the Specifications. If valve lift is less than specified, cam lobe wear has probably occurred and a new camshaft should be Installed.

Removal

9 Refer to the appropriate Sections and remove the rocker arms and shafts, the pushrods, the cylinder head and the timing chain and camshaft sprocket. The radiator should be removed as well (see Chapter 3).

Note: *It may also be necessary to cut a hole in the front bulkhead, immediately in front of the camshaft, to allow the camshaft to be withdrawn from the engine. If this is done, pop rivet a piece of sheet metal over the hole after the camshaft is installed.*

10 There are several ways to extract the lifters from the bores. A special tool designed to grip and remove lifters is manufactured by many tool companies and is widely available, but it may not be required in every case. On newer engines without a lot of varnish buildup, the lifters can often be removed with a small magnet. A pair of locking pliers can also be used to grip the top of each lifter, but do this only if the lifters are to be replaced with new ones. **Caution:** *The top lip of the lifter is easily broken if to much pressure is applied by the locking pliers.*

11 Before removing the lifters, arrange to store them in a clearly labeled box to ensure that they are reinstalled in their original locations. Remove the lifters and store them where they will not get dirty. Do not attempt to withdraw the camshaft with the lifters in place.

12 Remove the distributor (see Chapter 5).

13 Remove the oil pump (see Section 9).

14 Remove the fuel pump (see chapter 4).

15 Thread a long bolt into the camshaft sprocket bolt hole to use as a handle and give you more leverage when removing the camshaft from the block.

16 Carefully pull the camshaft out. Support the cam near the block so the lobes do not nick or gouge the bearings as it is withdrawn. If you feel resistance, don't pull harder or you'll damage the bearings. Lift and rotate the camshaft until it aligns better.

Inspection

Camshaft and bearings

Refer to illustrations 12.17 and 12.18

17 After the camshaft has been removed from the engine, cleaned with solvent and dried, inspect the bearing journals for uneven wear, pitting and evidence of seizure. If the

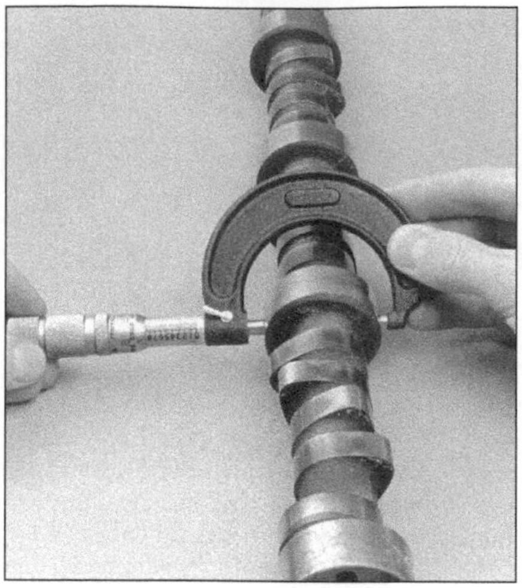

12.18 The camshaft bearing journal diameters are checked to pinpoint excessive wear and out-of-round conditions

journals are damaged, the bearing inserts in the block are probably damaged as well **(see illustration)**. Both the camshaft and bearings will have to be replaced with new ones.

18 Measure the bearing journals with a micrometer **(see illustration)** to determine if they are excessively worn or out-of-round.

19 Check the camshaft lobes for heat discoloration, score marks. chipped areas, pitting and uneven wear. If the lobes are in good condition and the valve lift measurements are as specified, the camshaft can be reused.

Lifters

Refer to illustration 12.21

20 Clean the lifters with solvent and dry them thoroughly without mixing them up.

21 Check each lifter wall, pushrod seat and foot for scuffing, score marks and uneven wear. Each lifter foot (the surface that rides on the cam lobe) must be slightly convex, although this can be difficult to determine by eye (holding the foot of the lifter against the wall (side) of another lifter and rocking the lifter back-and-forth will usually determine if the lifter foot is convex). If the base of the lifter is concave or damaged **(see illustration),** the lifters and camshaft must be replaced. If the lifter walls are damaged or worn (which is not very likely), inspect the lifter bores in the engine block as well. If the pushrod seats are worn, check the pushrod ends.

22 If new lifters are being installed, a new camshaft must also be installed. If a new camshaft is installed, then use new lifters as well. Never install used lifters unless the original camshaft is used and the lifters can be installed in their original locations.

Bearing replacement

23 Camshaft bearing replacement requires special tools and expertise that place it beyond a job you want to get into. The tool for bearing removal/installation is available at

12.21 If the bottom of any lifter is worn concave, scratched or galled, REPLACE THE ENTIRE SET AND THE CAMSHAFT WITH NEW PARTS

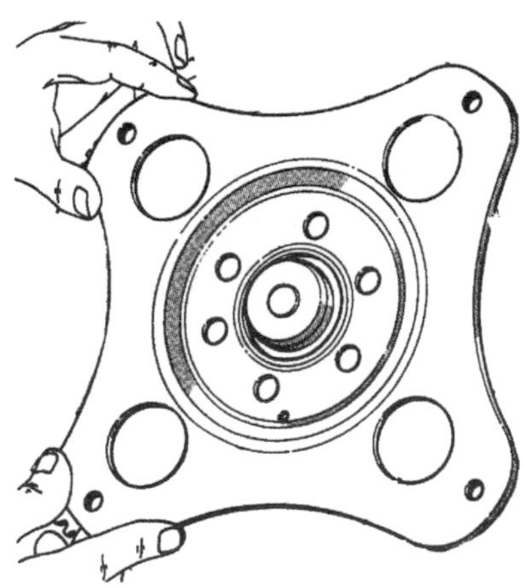

13.3a The bolt holes on most driveplates are offset to ensure correct installation

stores that carry automotive tools, possibly even found at a tool rental business. It is advisable though, if the bearings are bad, that the engine should be removed and the block taken to an automotive machine shop to ensure that the job is done correctly.

Installation

Refer to illustration 12.24

24 Lubricate the camshaft bearing journals and cam lobes with moly-base grease or engine assembly lube **(see illustration).**

25 Slide the camshaft into the engine. Support the cam near the block and be careful not to scrape or nick the bearings.

26 Lubricate the lifters with clean engine oil and install them in the block. If the original lifters are being reinstalled, be sure to return them to their original locations. If a new camshaft was installed, be sure to install new lifters as well.

27 The remaining installation steps are the reverse of removal.

28 Before starting and running the engine, change the oil and install a new oil filter (see Chapter 1).

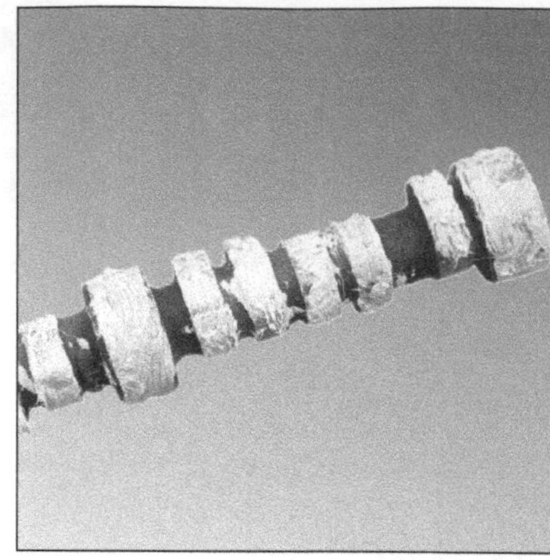

12.24 Be sure to apply moly-base grease or engine assembly lube to the cam lobes and bearing journals before installing the camshaft

13.3b Here's a flywheel with a staggered bolt pattern - note the two bolts (arrows) are spaced much farther apart than the others

13 Flywheel/driveplate - removal and installation

Removal

Refer to illustration 13.3a and 13.3b

1 Raise the vehicle and support it securely on jackstands, then refer to Chapter 7 and remove the transmission.

2 If you're working on a model with a manual transmission, remove the pressure plate and clutch disc **(see Chapter 8)**. Now is a good time to check/replace the clutch components and pilot bearing.

3 Most models have a flywheel/driveplate with offset bolt holes mounting it to the crankshaft to ensure correct alignment during installation **(see illustrations)**. If the vehicle being worked on doesn't have offset bolts, or if there is a doubt about realignment later, make a paint mark or a mark of some type before removal, in order to re-install the flywheel/driveplate, in the correct position.

4 Remove the bolts that secure the fly-

14.7 Detach the mount bolts from the engine block and bracket (arrows) . . .

14.8 . . . then remove the through-bolt and nut (arrows) and lift the mount from the engine compartment (driver's [left] side mount shown, passenger's [right] side similar)

wheel/driveplate to the crankshaft. If the crankshaft turns, wedge a screwdriver in the ring gear teeth to jam the flywheel.

5 Remove the flywheel/driveplate from the crankshaft. If you're removing a flywheel, be careful because it's fairly heavy. Be sure to support it while removing the last bolt.

Inspection

6 If inspecting a flywheel, clean it to remove grease and oil. Inspect the surface for cracks, rivet grooves, burned areas and score marks. Light scoring can be removed with emery cloth. Check for cracked and broken ring gear teeth. **Note:** *On automatic transmission-equipped vehicles, the ring gear teeth are part of the torque converter. Check the condition of the teeth before reassembly.* Lay the flywheel on a flat surface and use a straightedge to check for warpage. If necessary, take the flywheel to an automotive machine shop to have it resurfaced.

Installation

7 Position the flywheel/driveplate against the crankshaft. Align the staggered bolt holes; this will ensure correct installation. Before installing the bolts, apply thread locking compound to the threads.

8 Wedge a screwdriver in the ring gear teeth to keep the flywheel/driveplate from turning as you tighten the bolts to the torque listed in this Chapter's Specifications. Follow a criss-cross pattern and work up to the final torque listed in this Chapter's Specifications in three or four steps.

9 The remainder of installation is the reverse of the removal procedure.

14 Engine mounts - check and replacement

Refer to illustrations 14.7 and 14.8

1 Engine mounts seldom require attention, but broken or deteriorated mounts should be replaced immediately or the added strain placed on the driveline components may cause damage.

Check

2 During the check, the engine must be raised slightly to remove the weight from the mounts.

3 Raise the vehicle and support it securely on jackstands, then position the jack under the engine oil pan. Place a large block of wood between the jack head and the oil pan, then carefully raise the engine just enough to take the weight off the mounts.

4 Check the mounts to see if the rubber is cracked, hardened or separated from the metal plates. Sometimes the rubber will split right down the center. Rubber preservative may be applied to the mounts to slow deterioration.

5 Check for relative movement between the mount brackets and the engine or (use a large screwdriver or pry bar to attempt to move the mounts). If movement is noted, lower the engine and tighten the mount fasteners.

Replacement

6 Disconnect the cable from the negative battery terminal, then, if you're replacing a right-side mount, raise the vehicle and support it securely on jackstands. **Note:** *The left (driver's) side mount can usually be removed from above.*

7 Raise the engine slightly, then remove the mount bracket-to-engine bolts **(see illustration)**.

8 Remove the nut and withdraw the mount through-bolt **(see illustration)**.

9 Remove the mount from the engine compartment.

10 Installation is the reverse of removal. Use thread locking compound on the mount bolts and be sure to tighten them securely.

Notes

Chapter 2 Part B
V8 engines

Contents

Specifications

Bore and stroke

318 cubic inch	3.91 x 3.31 inches
340 cubic inch	4.040 x 3.31 inches
360 cubic inch	4.00 x 3.58 inches
383 cubic inch	4.25 x 3.375 inches
400 cubic inch	4.342 x 3.375 inches
440 cubic inch	4.320 x 3.750 inches

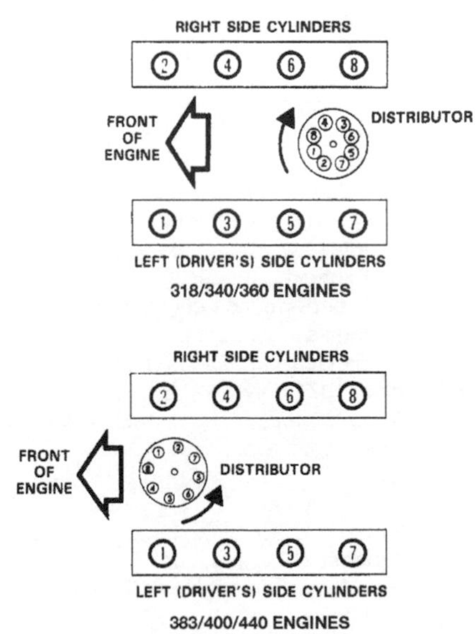

Cylinder location and distributor rotation

General

Cylinder numbers (front-to-rear)	
Left (driver's) side	1-3-5-7
Right side	2-4-6-8
Firing order	1-8-4-3-6-5-7-2
Distributor rotation (viewed from above)	
Small-block	Clockwise
Big-block	Counterclockwise
Minimum compression	
Engine warm, spark plugs removed and a wide-open-throttle	100 psi
Maximum variation between cylinders	40 psi

Camshaft

Journal diameters (all engines)	
No. 1 (front of engine)	1.998 to 1.999 inches
No. 2	1.982 to 1.983 inches
No. 3	1.967 to 1.968 inches
No. 4	1.951 to 1.952 inches

Camshaft (continued)

No. 5 (rear of engine)	
Small-block	1.5605 to 1.5615 inches
Big-block	1.748 to 1.749 inches
Bearing inside diameter	
No. 1	2.000 to 2.001 inches
No. 2	1.984 to 1.985 inches
No. 3	1.969 to 1.970 inches
No. 4	1.953 to 1.954 inches
No. 5	
Small-block	1.5625 to 1.5635 inches
Big-block	1.750 to 1.751 inches
Bearing oil clearance	0.001 to 0.003 inch
End play	
Standard	0.002 to 0.010 inch
Service limit	0.010 inch

Oil pump

Minimum pressure at curb idle	8 psi
Operating pressure	30-to-80 psi at 2000 rpm
Pressure drop from a clogged filter	7-to-9 psi
Oil pump cover surface variation limit.	0.0015 inch maximum
Outer rotor thickness limit	
Small-block	0.825 inch minimum
Big-block	0.943 inch minimum
Outer rotor diameter limit	2.469 inches minimum
Inner rotor thickness limit	
Small-block	0.825 inch minimum
Big-block	0.943 inch minimum
Clearance over rotors	0.004 inch maximum
Outer rotor clearance limit	0.014 inch maximum
Rotor tip clearance limit	0.010 inch maximum

Valve lift

318 engine	
1971 through 1974	
Intake	0.373 inch
Exhaust	0.399 inch
1975 through 1989	
Intake	0.373 inch
Exhaust	0.400 inch
340 engine (1971 through 1973)	
Intake	0.429 inch
Exhaust	0.444 inch
360 engine	
1971 through 1974	
Intake	0.410 inch
Exhaust	0.410 inch
1975 through 1980	
Standard engine (two-barrel carburetor)	
Intake	0.410 inch
Exhaust	0.410 inch
High-performance engine (four-barrel carburetor)	
Intake	0.429 inch
Exhaust	0.444 inch
383 engine (1971 only)	
Intake	0.425 inch
Exhaust	0.435 inch
400 engine	
Standard engine	
Intake	0.434 inch
Exhaust	0.430 inch
High-performance engine (1971 through 1975 only)	
Intake	0.450 inch
Exhaust	0.464 inch
440 engine	
Standard engine	
Intake	0.434 inch
Exhaust	0.430 inch

High-performance engine	
1971	
Intake	0.450 inch
Exhaust	0.458 inch
1972 through 1975	
Intake	0.450 inch
Exhaust	0.464 inch
1976 through 1978	
Intake	0.449 inch
Exhaust	0.464 inch

Torque specifications	**Ft-lbs** *(unless otherwise indicated)
Camshaft sprocket bolt	
Small-block	50
Big-block	35
Camshaft thrust plate bolts (Small-block only)	18
Cylinder head bolts	
Small-block	
1971 through 1977	95
1978 through 1980	105
1981 through 1989	95
Big-block	70
Exhaust manifold bolts (small-block)	20
Exhaust manifold nuts	
Small-block	15
Big-block	30
Exhaust pipe flange nuts	24
Flywheel/driveplate bolt	55
Intake manifold bolts	45
Intake manifold/valley pan end bolts (Big-block)	10
Oil pan bolts	
Small-block	20
Big-block	16
Oil pump mounting bolts	30
Rocker arm shaft bracket bolts	
Small-block	16
Big-block	25
Timing chain cover bolts	
Small-block	35
Big-block	17
Main bearing cap bolts	85
Oil pan bolts	200 in-lbs
Oil pump mounting bolts	30
Pulley-to-damper bolts	
Small-block	200 in lbs
Big-block	9
Rear oil seal retainer bolts (big-block engines)	25
Timing cover bolts	
Small-block engines	35
Big-block engines	17
Rocker arm shaft bolts	
Small-block engines	200 in-lbs
Big-block engines	25
Valve cover	
Nuts	80 in-lbs
Studs	115 in-lbs
Vibration damper/pulley-to-crankshaft bolt	
Small-block	100
Big-block	135
Water pump housing bolts (big-block engines)	30

***Note:** *Refer to Part C for additional specifications*

1 General information

This part of Chapter 2 is devoted to in-vehicle repair procedures for V8 engines. All information concerning engine removal and installation and engine block and cylinder head overhaul can be found in Part C of this Chapter.

Since the repair procedures included in this Part are based on the assumption that the engine is still installed in the vehicle, if they are being used during a complete engine overhaul (with the engine already out of the vehicle and on a stand) many of the steps included here will not apply.

The specifications included in this Part of Chapter 2 apply only to the procedures found here. The specifications necessary for rebuilding the block and cylinder heads are included in Part C.

3.4 Remove the PCV valve or breather tube and move the spark plug wires aside

1 PCV valve *2 Spark plug wires*

3.6 Valve cover mounting nuts (arrows) - driver's side shown

Three basic engine designs are covered in this Chapter.

The "A" series (sometimes called "LA") is a small-block design in displacements of 318 cubic inches (5.2 liters), 340 cubic inches (5.6 liters) and 360 cubic inches (5.9 liters).

The "B" series (sometimes called "LB") is a big-block design displacing 383 cubic inches (6.5 liters) or 400 cubic inches (6.6 liters).

The "RB" engine is a big-block design based on the "B" series design. The primary difference between the "B" and "RB" engines is the "deck height" (height of the cylinder head gasket surface). The 440 cubic inch (7.3 liter) "RB" engine has a deck height about 1/2-inch higher than on the "B" series of engines so it can achieve its larger displacement through a longer stroke. Also, the "RB" engine has slightly larger crankshaft main bearing journals and longer crankshaft "throws" to achieve its bigger displacement through a longer stroke. Although not covered in this manual, the 426 "Hemi" engine, available in 1971, is also an engine of the "RB" design.

2 Repair operations possible with the engine in the vehicle

Many major repair operations can be accomplished without removing the engine from the vehicle.

Clean the engine compartment and the exterior of the engine with some type of pressure washer before any work is done. A clean engine will make the job easier and will help keep dirt out of the internal areas of the engine.

Depending on the components involved, it may be a good idea to remove the hood to improve access to the engine as repairs are performed (refer to Chapter 11 if necessary).

If oil or coolant leaks develop, indicating a need for gasket or seal replacement, the repairs can generally be made with the engine in the vehicle. The oil pan gasket, the cylinder head gaskets, intake and exhaust manifold gaskets, timing cover gaskets and the crankshaft oil seals are all accessible with the engine in place.

Exterior engine components, such as the water pump, the starter motor, the alternator, the distributor and the carburetor, as well as the intake and exhaust manifolds, can be removed for repair with the engine in place.

Since the cylinder heads can be removed without pulling the engine, valve component servicing can also be accomplished with the engine in the vehicle.

Replacement of, repairs to or inspection of the timing chain and sprockets and the oil pump are all possible with the engine in place.

In extreme cases caused by a lack of necessary equipment, repair or replacement of piston rings, pistons, connecting rods and rod bearings is possible with the engine in the vehicle. However, this practice is not recommended because of the cleaning and preparation work that must be done to the components involved.

3 Valve covers - removal and installation

Removal

Refer to illustrations 3.4, 3.6 and 3.7

1 Disconnect the battery cable from the negative battery terminal.

2 Remove the air cleaner assembly (see Chapter 4).

3 Detach any air injection (AIR) system components that are in the way, if equipped (see Chapter 6).

4 Remove the breather tube or PCV valve and hose **(see illustration)**.

5 Label the spark plug wires and position the brackets/wires out of the way.

6 Remove the valve cover mounting bolts **(see illustration)**.

7 Remove the valve cover **(see illustration)**. **Note:** *If the cover is stuck to the head, bump the cover with a block of wood and a hammer to release it. If it still will not come loose, try to slip a flexible putty knife between the head and cover to break the seal. Don't pry at the cover-to-head joint, as damage to the sealing surface and cover flange will result and oil leaks will develop.*

3.7 Move the wires and hoses aside and guide the valve cover out

Installation

Refer to illustration 3.8

8 The mating surfaces of each cylinder head and valve cover must be perfectly clean when the covers are installed. Use a gasket scraper to remove all traces of sealant or old gasket **(see illustration)**, then wipe the mating surfaces with a cloth saturated with lacquer thinner or acetone. If there is sealant or oil on the mating surfaces when the cover is installed, oil leaks may develop.

9 Make sure any threaded holes are clean. Run a tap into them to remove corrosion and restore damaged threads.

10 Mate the new gaskets to the covers before the covers are installed. Apply a thin coat of RTV sealant to the cover flange, then position the gasket inside the cover lip and

3.8 Remove all traces of old gasket material

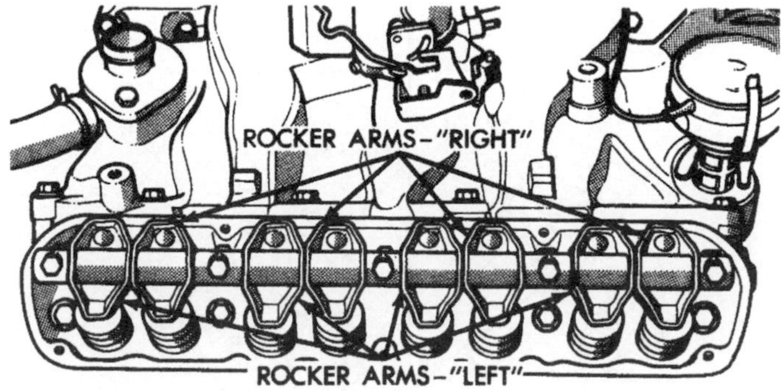

4.2a The correct rocker arm locations (left and right) on the rocker shaft - note the locations of the five bolts along the shaft that secure the shaft to the cylinder head

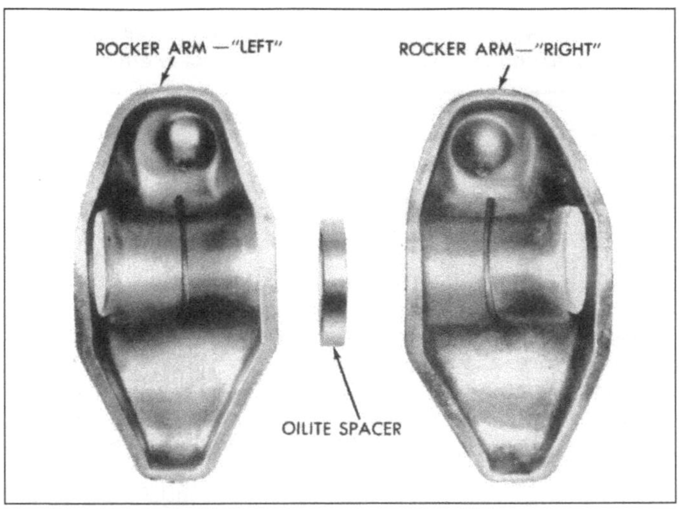

4.2b Big-block engines use "Oilite" spacers between rocker arms for the same cylinders

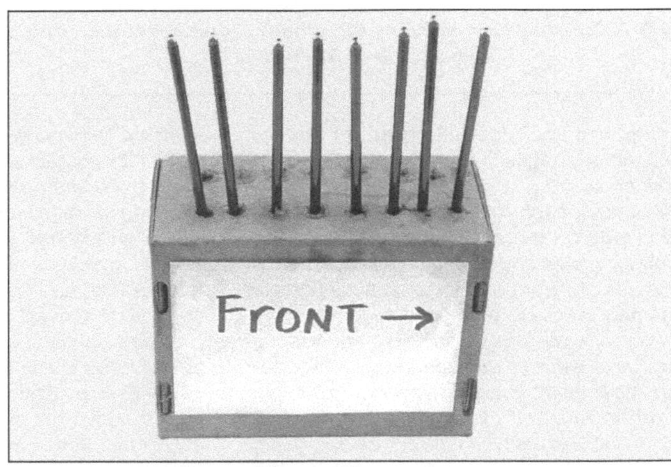

4.3 A perforated cardboard box can be used to store the pushrods to ensure that they're reinstalled in their original locations - note the label indicating the front of the engine

allow the sealant to set up so the gasket adheres to the cover (if the sealant is not allowed to set, the gasket may fall out of the cover as it is installed on the engine).

11 Carefully position the cover on the head and install the nuts/bolts.

12 Tighten the bolts in three steps to the torque listed in this Chapter's Specifications. **Caution:** *DON'T over-tighten the valve cover bolts.*

13 The remaining installation steps are the reverse of removal.

14 Start the engine and check carefully for oil leaks as the engine warms up.

4 Rocker arms, shafts and pushrods - removal, inspection and installation

Removal

Refer to illustrations 4.2a, 4.2b and 4.3

1 Remove the valve covers from the cylinder heads (see Section 3).

2 Loosen the rocker arm shaft bolts **(see illustration)** a little at a time, working back and forth until they're all loose. Lift off the rocker arm shaft with the bolts in place and, if they are removed from the shafts, keep the rocker arms in order (the rocker arms and shafts must be reinstalled in their original positions). **Note 1:** *We recommend removing the rocker arms from the shafts for cleaning and inspection. The best way to keep the rocker arms in order once they're removed is to tie a piece of heavy wire (such as an unbent coat-hanger wire) around a short dowel or similar object, then slide each rocker arm, starting from the engine-front end, onto the end of the wire in the same orientation as it was on the shaft. When finished, you'll have the rocker arms stored in the correct order in a way that will allow them to be cleaned and inspected easily. Just remember that the end of the wire with the dowel represents the front of the engine. Mark each dowel (as well as the rocker shaft) to indicate which side of the engine it is from. Also, mark the rocker shafts to indicate which end faces front.* **Note 2:** *On big-block (383, 400 and 440) engines, there is an "Oilite" spacer used between rocker arms that are on the same cylinder* **(see illustration).** *On the 318, 340 and 360, there are no "Oilite" spacers, but the spacers under the bolts are not the same size and must be reinstalled in their original locations, so note where they go and, if necessary, make a drawing showing the locations of the long and short spacers under the bolts.*

3 Remove the pushrods and store them separately to make sure they don't get mixed up during installation **(see illustration). Caution:** *All valve train components must go back in their original position. Organize and store the parts in a way that they won't get mixed up.*

Inspection

Refer to illustration 4.5

4 Check each rocker arm for wear, cracks and other damage, especially where the pushrods and valve stems contact the rocker arm faces.

5 Make sure the oil-feed holes in each

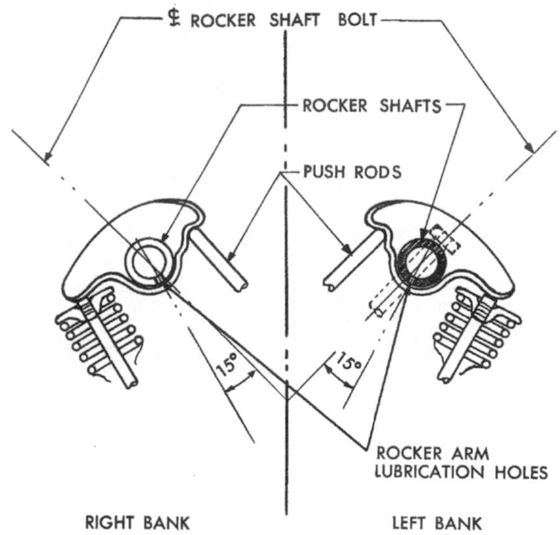

4.5 A Cutaway view showing the proper alignment of the oiling holes in the rocker arm shaft

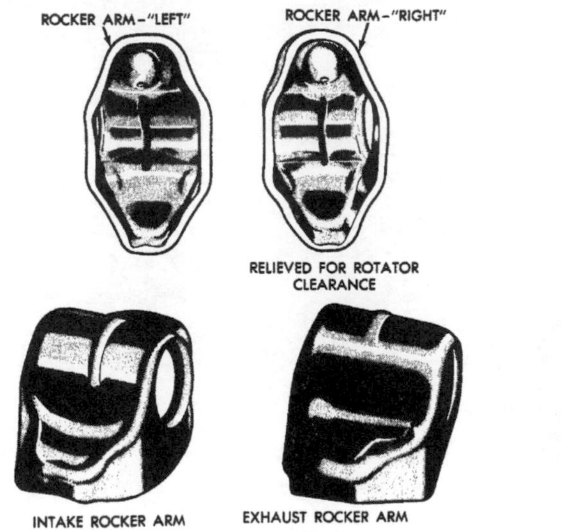

4.9 Rocker arm identification details showing the offset of the pushrod on both right and left side rocker arms

rocker arm shaft **(see illustration)** are not plugged so that each rocker arm gets proper lubrication.

6 Check each rocker arm pivot area and wear area on the shaft for excessive wear, scoring, cracks and galling. If the rocker arms or shafts are worn or damaged, replace them with new ones. **Note:** *Keep in mind that there is no valve adjustment on these engines, so excessive wear or damage in the valve train can easily result in excessive valve clearance, which in turn will cause valve "tapping" or "clattering" noises when the engine is running.*

7 Inspect the pushrods for cracks and excessive wear at the ends. Roll each pushrod across a piece of plate glass to see if it is bent (if it wobbles, it is bent). Replace the pushrods if any of these conditions are present.

Installation

Refer to illustrations 4.9 and 4.11

8 Assemble the rocker arms onto the rocker arm shafts in the same orientation as they were originally.

9 If replacing rocker arms, note that, on engines with exhaust valve rotators, exhaust rockers must have clearance reliefs **(see illustration)**.

10 Lubricate the lower end of each pushrod with clean engine oil or moly-base grease and install them in their original locations. Make sure each pushrod seats completely in the lifter socket.

11 Apply moly-base grease to the ends of the valve stems and the upper ends of the pushrods before positioning the rocker arm shafts onto the pedestals on the engine **(see illustration)**. Install the shafts so the notch on the end of each shaft is pointing toward the centerline of the engine, and toward the front of the engine on the driver's (left) side or toward the rear on the passenger's (right) side. **Caution:** *Be sure the wear areas on the shaft are facing down, as they originally were. DON'T try to gain some extra service life by turning the shaft over, since the oil holes in the shaft must face down.*

12 Tighten the rocker-arm shaft bolts evenly, in several stages, working back and forth. Apply moly-base grease to the contact surfaces to prevent damage before engine oil pressure builds up. Tighten the bolts to the torque listed in this Chapter's Specifications.

13 Refer to Section 3 and install the valve covers. Start the engine, listen for unusual valve train noses and check for oil leaks at the valve cover joints.

5 Valve springs, retainers and seals - replacement

Refer to illustrations 5.5, 5.8, 5.15, 5.17 and 5.19

Note: *Broken valve springs and defective valve stem seals can be replaced without removing the cylinder head. Two special tools and a compressed air source are normally required to perform this operation, so read through this Section carefully and rent or buy the tools before beginning the job. If compressed air isn't available, a length of nylon rope can be used to keep the valves from falling into the cylinder during this procedure.*

1 Remove the valve covers (See Section 3).

2 Remove the spark plugs (see Chapter 1).

3 Turn the crankshaft until the number one piston is at top dead center on the compression stroke (see Chapter 2, Part C).

4 Remove the rocker arm shaft (see Section 4). If you intend to use a lever-type spring compressor, remove the rocker arms from the shaft and re-install the shaft without the rocker arms.

4.11 Lubricate the pushrod ends and the valve stems with a moly-based grease before installing the rocker arms

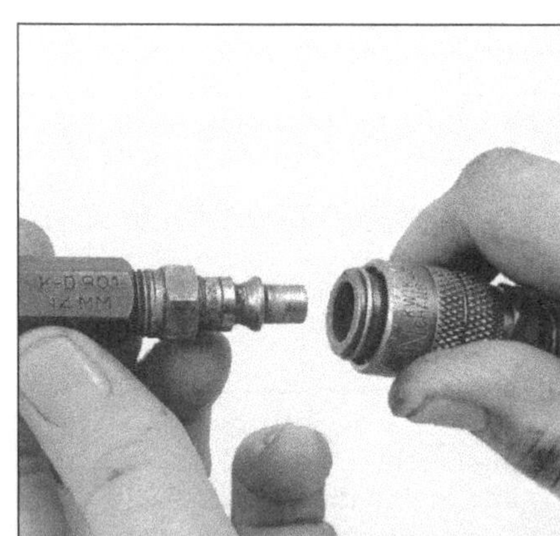

5.5 This is what the air hose adapter that fits into the spark plug hole looks like - they're commonly available from auto parts stores

5.8 Once the spring is depressed, the keepers can be removed with a small magnet or needle-nodes pliers (a magnet is preferred to prevent dropping the keepers)

5.15 Be sure to install the seals on the correct valve stems

1 *Exhaust seal (shield)* 2 *Intake seal*

5 Thread an adapter into the spark plug hole and connect an air hose from a compressed air source to it **(see illustration)**. Most auto parts stores can supply the air hose adapter. **Note:** *Many cylinder compression gauges utilize a screw-in fitting that may work with your air hose quick-disconnect fitting.*

6 Apply compressed air to the cylinder. The valves should be held in place by the air pressure. If the valve faces or seats are in poor condition, leaks may prevent the air pressure from retaining the valves - valve reconditioning is indicated.

7 If you don't have access to compressed air, an alternative method can be used. Position the piston at a point just before TDC on the compression stroke, then feed a long piece of nylon rope through the spark plug hole until it fills the combustion chamber. Be sure to leave the end of the rope hanging out of the engine so it can be removed easily. Use a large breaker bar and socket to rotate the crankshaft in the normal direction of rotation (clockwise) until slight resistance is felt.

8 Stuff shop rags into the cylinder head holes around the valves to prevent parts and tools from falling into the engine, then use a valve spring compressor to compress the spring. Remove the keepers with small needle-nose pliers or a magnet **(see illustration)**. **Note:** *Several different types of tools are available for compressing the valve springs with the head in place. One type, shown here, grips the lower spring coils and presses on the retainer as the knob is turned, while the lever-type utilizes the rocker arm shaft for leverage. Both types work very well, although the lever type is usually less expensive.*

9 Remove the valve spring and retainer (intake valve) or rotator (exhaust valve). **Caution:** *Different-length springs are used on the intake and exhaust valves - accidentally installing an intake spring on an exhaust valve can cause major engine damage.*

10 Remove the old valve stem seals. **Caution:** *The exhaust valve oil seal is referred to by the manufacturer as an oil shield. The intake valve seal fits over the valve guide,*

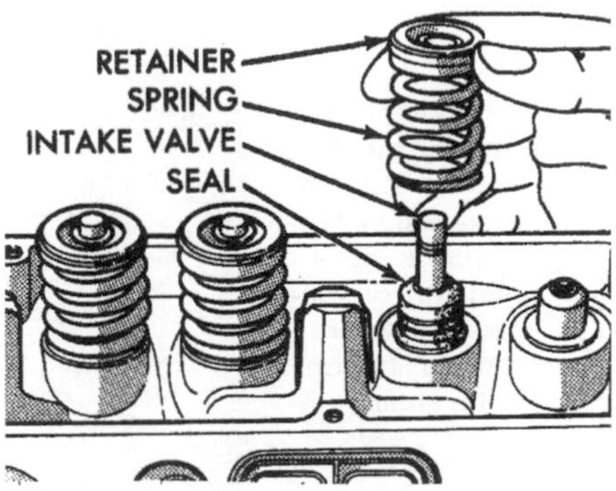

5.17 Install the seal, spring and retainer or rotator (exhaust valve only)

while the exhaust valve "shield" is what's commonly referred to in the industry as an "umbrella-type" seal. If you install an exhaust valve shield on an intake valve, high oil consumption will result.

11 Wrap a rubber band or tape around the top of the valve stem so the valve won't fall into the combustion chamber, then release the air pressure. **Note:** *If a rope was used instead of air pressure, turn the crankshaft slightly in the direction opposite normal rotation.*

12 Inspect the valve stem for damage. Rotate the valve in the guide and check the end for eccentric movement, which would indicate that the valve is bent.

13 Move the valve up-and-down in the guide and make sure it doesn't bind. If the valve stem binds, either the valve is bent or the guide is damaged. In either case, the head will have to be removed for repair.

14 Reapply air pressure to the cylinder to retain the valve in the closed position, then remove the tape or rubber band from the valve stem. If a rope was used instead of air pressure, rotate the crankshaft in the normal direction of rotation until slight resistance is felt. Remove the old valve seals.

5.19 Apply small dab of grease to each keeper as shown here before installation - it'll hold them in place on the valve stem as the spring is released

15 If you're working on an exhaust valve, install the new shield on the valve stem and push it down to the top of the valve guide **(see illustration)**.

16 If you're working on an intake valve, push a new valve stem seal down over the valve guide, using the valve stem as a guide, but don't force the seal against the top of the guide.

17 Install the spring and retainer or rotator in position over the valve **(see illustration)**. **Caution:** *Different length springs are used on the intake and exhaust valves - accidentally installing an intake spring on an exhaust valve can cause major engine damage!*

18 Compress the valve spring assembly only enough to install the keepers in the valve stem.

19 Position the keepers in the valve stem groove. Apply a small dab of grease to the inside of each keeper to hold it in place if necessary **(see illustration)**. Remove the pressure from the spring tool and make sure the keepers are seated.

20 Disconnect the air hose and remove the

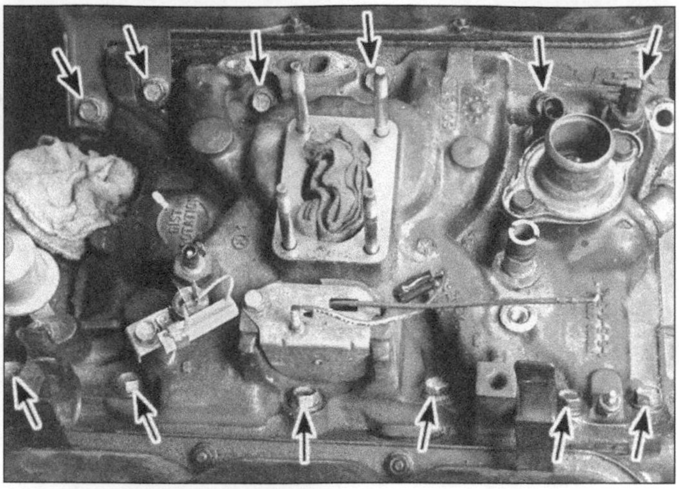

6.8a Intake manifold bolt locations (typical)

6.8b Pry up on the corner of the manifold, if necessary

adapter from the spark plug hole. If a rope was used in place of air pressure, pull it out of the cylinder.

21 Repeat the above procedure on the remaining cylinders, following the firing order sequence (see the Specifications). Bring each piston to top dead center on the compression stroke before applying air pressure or inserting the rope into the cylinder.

22 Reinstall the rocker arm assemblies and the valve covers.

23 Start the engine, then check for oil leaks and unusual sounds coming from the valve cover area.

6.9 The manifold is heavy and must be removed from an awkward position; be sure to get a good hold on the manifold before trying to lift it off the engine

6 Intake manifold - removal and installation

Removal

Refer to illustrations 6.8a, 6.8b and 6.9

1 Disconnect the battery cable from the negative battery terminal.

2 On small-block (318/340/360) engines, drain the cooling system (see Chapter 1). **Note:** *On big-block (383/400/440) engines, no coolant is routed through the intake manifold, so it's not necessary to drain the cooling system.*

3 Remove the air cleaner assembly. **Note:** *If you intend to resurface, clean or replace the manifold, remove the carburetor* (see Chapter 4).

4 Remove the alternator (see Chapter 5). On air-conditioned models, you'll probably also need to unbolt the air conditioning compressor and set it aside, without disconnecting the refrigerant lines.

5 Label and then disconnect fuel lines, wires and vacuum hoses that run from the vehicle to the intake manifold connections and carburetor.

6 On small-block (318/340/360) engines, disconnect the heater hose and water pump bypass hose from the intake manifold (see Chapter 3).

7 Disconnect the accelerator linkage (see Chapter 4) and, if equipped, the cruise control linkage. Then, on models with automatic transmissions, disconnect the throttle rod at the carburetor (see Chapter 7B). **Caution:** *On 1971 through 1973 models with a one-piece throttle rod, be sure to mark around the bolt so you can return the throttle rod to its original adjustment on reassembly; incorrect adjustment can lead to early transmission failure. On models with three-piece throttle rods, simply pry the intermediate rod off the upper bellcrank (it's a ball-and-socket joint). On 1974 and later models with one-piece rods, disassemble the rod from the linkage at the carburetor.*

8 Loosen the intake manifold mounting bolts **(see illustration)** in 1/4-turn increments until they can be removed by hand. **Note:** *Small-block (318/340/360) engines use six bolts on each side of the manifold (12 total); big-block engines (383/400/440) use only four bolts per side (8 total).* The manifold will probably be stuck to the cylinder heads and force may be required to break the gasket seal. A pry bar can be positioned to pry up the front of the manifold **(see illustration)** to break the bond made by the gasket. **Caution:** *Do not pry between the block and manifold or the heads and manifold or damage to the gasket sealing surfaces may result and vacuum leaks could develop.*

9 Remove the intake manifold **(see illustration).** As the manifold is lifted from the engine, be sure to check for and disconnect anything still attached or wrapped over the manifold. On big-block engines, remove the end bolts and lift off the engine valley cover, which is integral with the intake manifold gasket.

Installation

Refer to illustrations 6.13, 6.14, 6.17a and 6.17b

All engines

Note: *The mating surfaces of the cylinder heads, block and manifold must be perfectly clean when the manifold is installed. Gasket removal solvents in aerosol cans are available at most auto parts stores and may be helpful when removing old gasket material that is stuck to the heads and manifold. Be sure to follow the directions printed on the container.*

10 Remove carbon deposits from the exhaust crossover passages. Use a gasket scraper to remove all traces of sealant and old gasket material, then wipe the mating

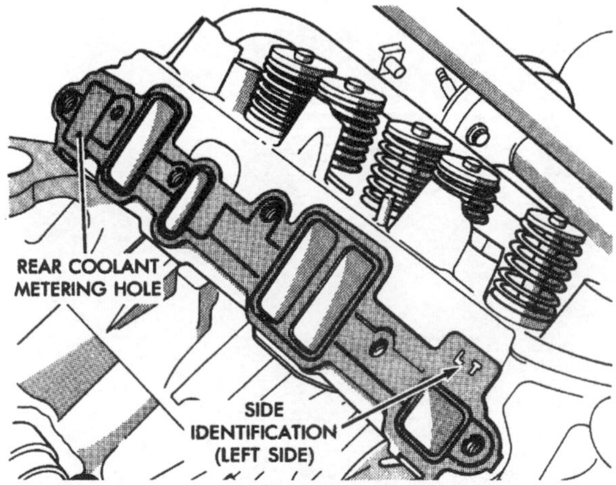

6.13 The gaskets must be installed on the proper side; some may be marked, as shown, but you'll often need to line up the holes with the ports on the head to be sure they're installed correctly

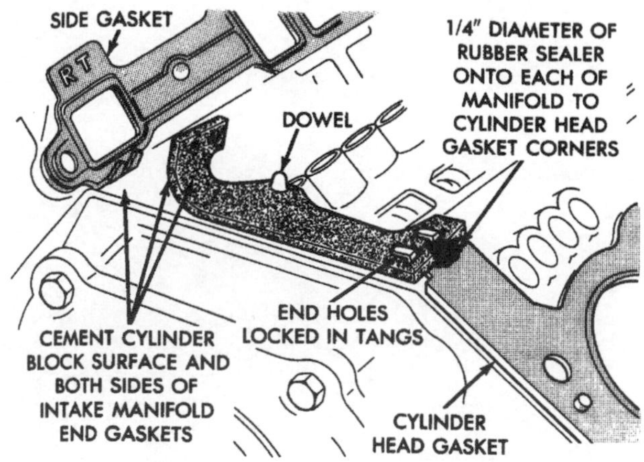

6.14 Fit the end seals over the dowels and engage the end tangs with the holes - apply a bead of RTV sealant to each of the four corners where the manifold gaskets and end seals meet

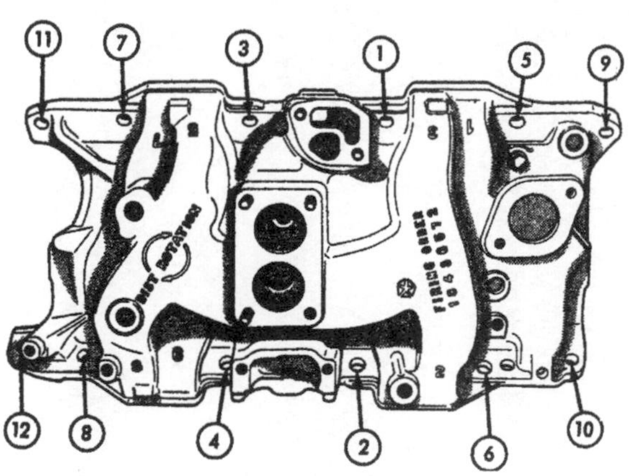

6.17a Intake manifold bolt tightening sequence (small-block engine)

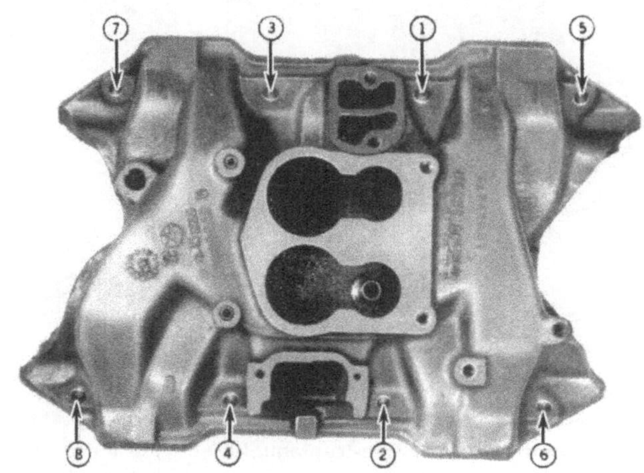

6.17b Intake manifold bolt tightening sequence (big-block engine)

surfaces with a cloth saturated with lacquer thinner or acetone. If there is old sealant or oil on the mating surfaces when the manifold is installed, oil or vacuum leaks may develop. Cover the lifter valley with shop rags to keep debris out of the engine. Use a vacuum cleaner to remove any gasket material that falls into the intake ports in the heads.

11 Use a tap of the correct size to chase the threads in the bolt holes, then use compressed air (if available) to remove the debris from the holes. **Warning:** *Wear safety glasses or a face shield to protect your eyes when using compressed air.*

Small-block engines only

12 Apply a thin coat of RTV sealant to the cylinder-head side of the new intake manifold gaskets.

13 Position the side gaskets on the cylinder heads. Look to see if the gaskets are marked LT for left or RT for right **(see illustration).** If they are, this will ensure proper installation. Make sure they are installed on the correct side and all intake port openings, coolant passage holes and bolt holes are aligned correctly.

14 Install the front and rear end seals on the block. Apply a thin, uniform coating of quick-dry cement to the intake manifold end seals and the cylinder block contact surfaces. Engage the dowels and the end tangs **(see illustration)**. Refer to the instructions with the gasket set for further information.

Big-block engines only

15 Apply a bead of RTV sealant to each of the four gasket-surface corners where the cylinder head meets the engine block, then apply a thin coat of RTV to the gasket surfaces on the block, between the cylinder heads; coat both sides of the intake manifold sealing surfaces with a sealer such as Fel-Pro Brush Tack, then install the gasket/valley pan, tightening the end bolts to the torque listed in this Chapter's Specifications.

All engines

16 Carefully set the manifold in place. **Caution:** *Do not disturb the gaskets and DO NOT move the manifold fore-and-aft after it contacts the front and rear seals or the gaskets will be pushed out of place and you may not notice the problem until you see the oil leak.*

17 Install the intake manifold bolts and hand tighten them. Then tighten the bolts, following the tightening sequence **(see illustrations),** to the torque listed in this Chapter's Specifications.

18 The remaining installation steps are the reverse of removal. Change the engine oil and filter. On small-block engines, add coolant. Start the engine and check carefully for oil, vacuum and coolant leaks at the intake manifold joints.

7 Exhaust manifolds - removal and installation

Removal

Warning: *Allow the engine to cool completely before performing this procedure.*

1 Disconnect the battery cable from the

7.4 Remove the nuts (arrows) and detach the air pre-heater stove (if equipped)

7.6 Unbolt the air injection tube (arrow) (if equipped)

7.8 Remove the exhaust pipe-to-manifold nuts (arrows)

7.9 If new studs are being installed, use sealant on the studs (arrow) where they screw into the heads

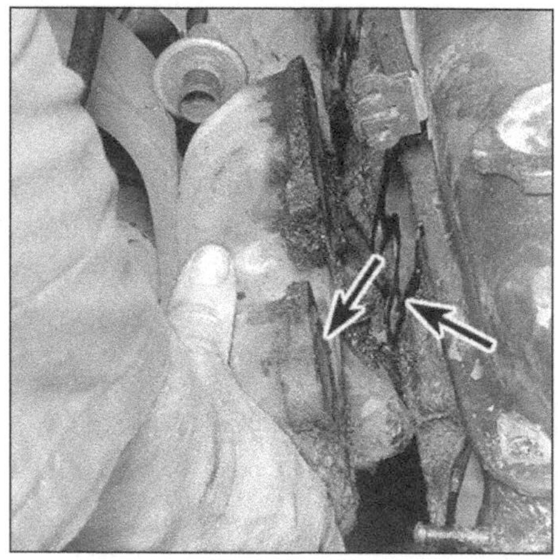

7.10 Clean the manifold and head gasket surfaces of all traces of old gasket material

negative battery terminal.

2 Disconnect the spark plug wires and remove the spark plugs (see Chapter 1). If there is any danger of mixing the plug wires up, it's recommended that they be labeled with pieces of numbered tape.

Left (driver's) side

Refer to illustrations 7.4 and 7.6

3 Remove the air cleaner assembly and duct for access to the manifold.

4 Detach the air injection hose/tube and remove the air pre-heater stove, if equipped **(see illustration)**.

5 On 1980 and later models equipped with an oxygen sensor, unplug the connector to the sensor.

6 Unbolt the air injection tube, if equipped, from the exhaust manifold **(see illustration)**.

Both sides

Refer to illustrations 7.8 and 7.9

7 Set the parking brake and block the rear wheels. Raise the front of the vehicle and support it securely on jackstands.

8 Disconnect the exhaust pipe from the manifold outlets **(see illustration)**. **Note:** *Often a short period of soaking with penetrating oil is necessary to remove frozen exhaust attaching nuts/bolts.*

9 Loosen the outer fasteners first, then the center ones to separate the manifold from the head. **Note:** *If the studs come out with the nuts, install new studs* **(see illustration)**. *Apply sealant to the coarse thread ends or coolant leaks may develop.*

Installation

Refer to illustration 7.10

10 Installation is basically the reverse of the removal procedure. Clean the manifold and head gasket surfaces **(see illustration)** and check the manifold heat control valve for ease of operation. If the valve is stuck, free it with penetrating oil.

11 Install the fasteners. Tighten them to the torque listed in this Chapter's Specifications. Work from the center to the ends and approach the final torque in three steps.

12 Apply anti-seize compound to the exhaust manifold-to-exhaust pipe bolts and tighten them securely.

8 Cylinder heads - removal and installation

Removal

Refer to illustrations 8.6, 8.7a and 8.7b

1 Drain the cooling system (see Chapter 1).

2 Remove the valve covers (see Section 3).

3 Remove the intake manifold (see Section 6).

4 Detach both exhaust manifolds from the cylinder heads (see Section 7) - you can usually leave them attached at the exhaust pipe.

5 Remove the rocker arm shaft assembly and pushrods (see Section 4). **Caution:** *Again, as mentioned in Section 4, keep all the parts in order so they are reinstalled in the same location.*

6 Loosen the head bolts in 1/4-turn increments until they can be removed by hand. Reverse the sequence shown in **illustration 8.16a or 8.16b**. **Note:** *On all engines, there will be different-length head bolts for different*

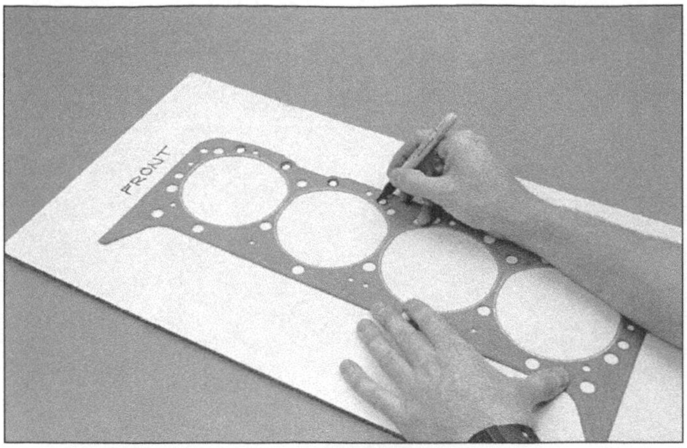

8.6 To avoid mixing up the head bolts, use a new gasket to transfer the bolt pattern to a piece of cardboard, then punch holes to accept the bolts (typical gasket shown)

8.7a pry on a casting protrusion to break the head loose

8.7b Once loose, get a good hold on the cylinder head and remove it from the engine block

8.10 Remove all traces of old gasket material

locations, so store the bolts in a cardboard holder **(see illustration)** *or some type of container as they are removed. This will ensure that the bolts are reinstalled in their original holes.*

7 Lift the heads off the engine. If resistance is felt, do not pry between the head and block as damage to the mating surfaces will result. To dislodge the head, place a block of wood against the end of it and strike the wood block with a hammer, or lift on a casting protrusion **(see illustrations)**. Store the heads on blocks of wood to prevent damage to the gasket sealing surfaces.

8 Cylinder head disassembly and inspection procedures are covered in detail in Chapter 2, Part C.

Installation

Refer to illustrations 8.10, 8.12, 8.13, 8.16a and16b

9 The mating surfaces of the cylinder heads and block must be perfectly clean when the heads are installed. Gasket removal solvents are available at auto parts stores and may prove helpful.

10 Use a gasket scraper to remove all traces of carbon and old gasket material **(see illustration)**, then wipe the mating surfaces with a cloth saturated with lacquer thinner or acetone. If there is oil on the mating surfaces when the heads are installed, the gaskets may not seal correctly and leaks may develop. When working on the block, cover the lifter valley with shop rags to keep debris out of the engine. Use a vacuum cleaner to remove any debris that falls into the cylinders.

11 Check the block and head mating surfaces for nicks, deep scratches and other damage. If damage is slight, it can be removed with emery cloth. If it is excessive, machining may be the only alternative.

12 Use a tap of the correct size to chase the threads in the head bolt holes in the block. Mount each bolt in a vise and run a die down the threads to remove corrosion and restore the threads **(see illustration)**. Dirt, corrosion, sealant and damaged threads will affect torque readings.

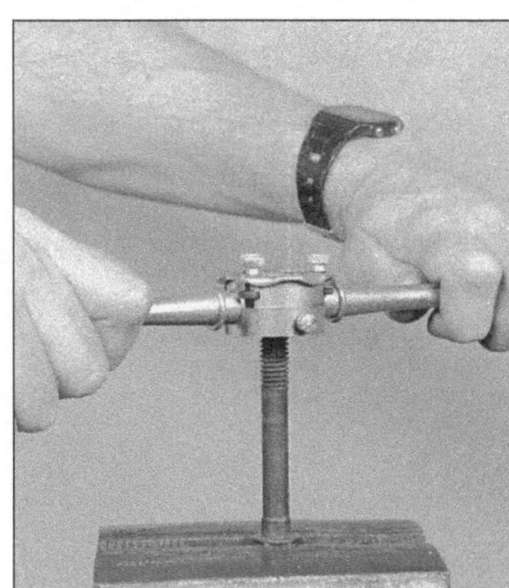

8.12 A die should be used to remove sealant and corrosion from the bolt threads prior to installation

8.13 Install the new head gasket over the dowels (arrow); there is a dowel at each end of the cylinder head

8.16a Cylinder head tightening sequence for "small-block" engines

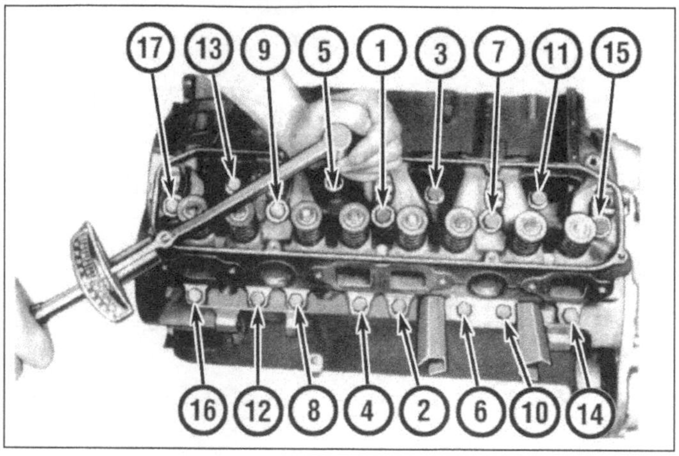

8.16b Cylinder head tightening sequence for "big-block" engines

9.12 Remove the crankshaft pulley bolts (arrows) and the large vibration damper bolt in the center (viewed from below)

13 Position the new gaskets over the dowels in the block **(see illustration)**.
14 Carefully position the heads on the block without disturbing the gaskets.
15 Before installing the head bolts, coat the threads with a non-hardening sealant such as Permatex No. 2.
16 Install the bolts in their original locations and tighten them finger tight. Following the recommended sequence **(see illustrations)**, tighten the bolts in several steps to the torque listed in this Chapter's Specifications.
17 The remaining installation steps are the reverse of removal.
18 Add coolant and change the oil and filter (see Chapter 1). Start the engine, set the ignition timing and check the engine for proper operation and coolant or oil leaks.

9 Timing cover, chain and sprockets - removal, inspection and installation

Timing chain slack check (cover on engine)

Note: *This procedure will allow you to check the amount of slack in the timing chain without removing the timing chain cover from the engine.*
1 Disconnect the negative battery cable from the battery.
2 Place the engine at TDC (Top Dead Center) (see Chapter 2C).
3 Keeping the piston on the compression stroke, place the number one piston about 30-degrees before TDC (BTDC).
4 Remove the distributor cap (see Chapter 1).
5 You must rotate the crankshaft very slowly by hand for accuracy on this check, so get a 1/2-inch drive breaker bar, extension and correct-size socket to use on the large bolt in the center of the crankshaft pulley.
6 Turn the crankshaft clockwise until the number one piston is at TDC. This will take up the slack on the left side of the timing chain.
7 Mark the position of the distributor rotor on the distributor housing.
8 Slowly turn the crankshaft counterclockwise until the slightest movement is seen at the distributor rotor. Stop and note how far the number one piston has moved away from the TDC mark by looking at the ignition timing marks.
9 If the mark has moved more than 10 degrees, the timing chain is probably worn excessively. Remove the timing chain cover for a more accurate check.

Cover removal

Refer to illustrations 9.12, 9.15, 19.18a and 19.18b

10 Detach any accessories such as the power steering pump, alternator and air conditioning compressor that block access to the timing chain cover. Leave the hoses connected to the A/C compressor and power steering pump and tie the units aside. Refer to Chapters 3, 5 and 10 for additional information. Unbolt the accessory brackets from the front of the engine.
11 On small-block engines, remove the water pump (see Chapter 3). On big-block engines, unbolt and remove the water-pump housing from the engine block - the water pump can remain bolted to the water pump housing, although, if the water pump is in questionable condition, now is an excellent time to replace it (see Chapter 3). **Note:** *On some big-block engines, one of the water-pump housing bolts cannot be removed from its hole because of interference with a heater-*

9.15 Use a bolt-type puller to remove the vibration damper

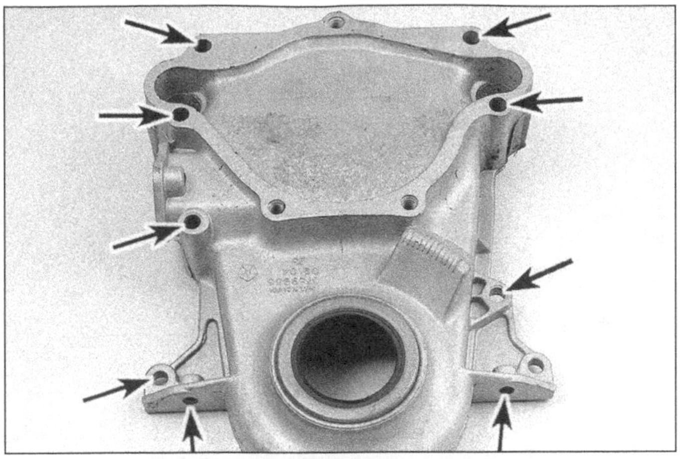

9.18a Timing chain cover bolt hole locations on small-block engines (removed from engine for clarity)

9.18b Remove the cover from the engine to get at the chain and gears

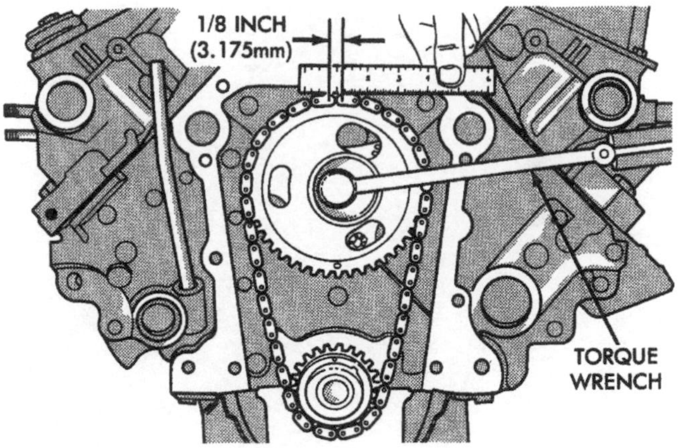

9.20 Position a ruler over the chain to measure slack

hose fitting. Remove all the other bolts first, then loosen the trapped bolt a little at a time, while moving the housing away from the engine block. If the bolt must be removed (for example, to replace the housing), unscrew the heater-hose fitting from the housing.

12 Remove the bolts **(see illustration)** and separate the crankshaft pulley(s) from the vibration damper.

13 Position the number six piston at TDC on the compression stroke (see Chapter 2C). **Caution:** *Once this has been done, do not turn the crankshaft until the timing chain and sprockets have been reinstalled.*

14 Remove the large vibration damper bolt **(see illustration 9.12)**. To keep the crankshaft from turning, remove the starter (see Chapter 5) and have an assistant jam a large screwdriver against the ring gear teeth.

15 Use a vibration damper puller (commonly available from auto parts stores) to detach the vibration damper **(see illustration)**. **Caution:** *Do not use a puller with jaws that grip the outer edge of the damper. The puller must be the type that utilizes bolts to apply force to the damper hub only.*

16 On small-block engines, remove the fuel pump and cap the lines (see Chapter 4).

17 Remove the front two oil pan bolts, that thread into the timing chain cover - they're most easily accessed from below. **Note:** *Even though this procedure can be done without the removal of the oil pan, it is extremely difficult on some models and oil pan removal may actually simplify the job (see Section 11).* Use an X-acto knife to cut the cover free from the front of the oil pan gasket (on small-block engines, only cut along the top edge - do not cut the rounded area). If the pan has been in place for an extended period of time, it is likely the pan gasket will break when the cover is removed.

18 Remove the remaining mounting bolts (there are eight bolts on big-block engines) **(see illustration)** and separate the timing chain cover from the block and oil pan **(see illustration)**. The cover may be stuck; if so, use a putty knife to break the gasket seal. The cover is easily damaged, so DO NOT attempt to pry it off. **Note:** *Big-block engines have a bolt-on timing tab. Note the location and be sure to reinstall it in the same way.*

19 Remove the oil slinger from the end of the crankshaft, if equipped. Note how it's installed so you can reinstall it the same way on reassembly. It's a good idea to replace the crankshaft front seal at this time (see Section 13).

Timing chain and gear inspection (cover removed from engine)

Refer to illustration 9.20

20 Attach a socket and torque wrench to the camshaft sprocket bolt **(see illustration)** and apply force in the normal direction of crankshaft rotation (30 ft-lbs if the cylinder head is still in position complete with rocker arms, or 15 ft-lbs if the cylinder head has been removed). Don't allow the crankshaft to rotate. If necessary, jam the crankshaft sprocket so that it can't move. Using a ruler, note the amount of movement of the chain. If it exceeds 1/8-inch a new timing chain will be required. **Note:** *Whenever a new timing chain is required, the entire set (chain, camshaft and crankshaft sprockets) must be replaced as an assembly.*

21 Inspect the camshaft gear for damage or wear. The camshaft gear on some models

9.23 With the timing marks aligned, remove the camshaft sprocket bolt

1 *Timing marks* 2 *Camshaft sprocket bolt*

9.24a The sprocket on the camshaft and crankshaft can be removed with a two or three-jaw puller. . .

9.24b . . . or two screwdrivers

9.28a Slip the chain and camshaft sprocket in place over the crankshaft sprocket with the timing mark (arrow) at the bottom

is steel, but most original-equipment cam gears will be an aluminum gear with a nylon coating on the teeth. This nylon coating may be cracked or be breaking off in small pieces. These pieces tend to end up in the oil pan and may eventually plug the oil pump pickup screen. If the pieces have come off the camshaft gear, the oil pan should be removed to properly clean or replace the oil pump pickup screen.

22 Inspect the crankshaft gear for damage or wear. The crankshaft gear is a steel gear, but the teeth can be grooved or worn enough to cause a poor meshing of the gear and the chain.

Chain removal

Refer to illustrations 9.23, 9.24a and 9.24b

23 Be sure the timing marks are aligned. Hold a prybar through one of the camshaft sprocket holes to prevent the camshaft from turning. Remove the bolt from the camshaft sprocket **(see illustration)**, then detach the fuel pump eccentric (small-block engines only).

24 The sprockets on the camshaft and crankshaft can be removed with a two or three-jaw puller or by using two screwdrivers **(see illustrations)**, but be careful not to damage the threads in the end of the crankshaft. **Note:** *If the timing chain cover oil seal has been leaking, refer to Section 13 and install a new one.*

Installation

Refer to illustrations 9.28a, 9.28b, 9.28c and 9.32

25 Use a gasket scraper to remove all traces of old gasket material and sealant from the cover and engine block (and the water pump housing and engine block on big-block engines). Stuff a shop rag into the opening at the front of the oil pan to keep debris out of the engine. On small-block engines, remove the rubber semi-circular seal from the oil pan. Wipe the cover and block sealing surfaces with a cloth saturated with lacquer thinner or acetone.

26 On big-block engines, check the cover flange for distortion, particularly around the bolt holes. Straighten any distortion by placing the sealing flange on a block of wood and tapping it with the rounded end of a ball-peen hammer.

27 If you're installing a new crankshaft sprocket, be sure to align the keyway in the crankshaft sprocket with the Woodruff key in the end of the crankshaft. **Note:** *Timing chains must be replaced as a set with the camshaft and crankshaft gears. Never put a new chain on old gears.* Align the sprocket with the Woodruff key and press the sprocket onto the crankshaft with the vibration damper bolt, a large socket and some washers or tap it gently into place until it is completely seated. **Caution:** *If resistance is encountered, do not hammer the sprocket onto the crankshaft. It may eventually move onto the shaft, but it may be cracked in the process and fail later, causing extensive engine damage.*

28 Loop the new chain over the camshaft sprocket, then turn the sprocket until the timing mark is at the bottom **(see illustration)**. Mesh the chain with the crankshaft sprocket and position the camshaft sprocket on the end of the cam. If necessary, turn the camshaft so the key fits into the sprocket keyway with the timing mark in the 6 o'clock

9.28b The timing marks MUST align as shown here

9.28c The keyway must align with the key (arrows)

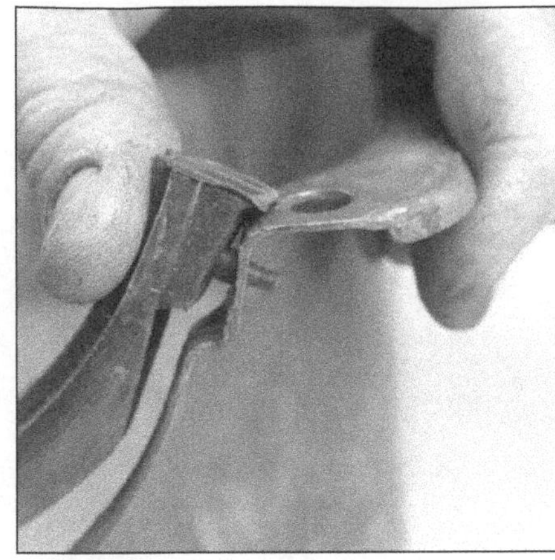

9.32 Be sure to engage the oil seal tang in the oil pan hole

position **(see illustration)**. When the chain is installed, the timing marks MUST align as shown **(see illustration)**. **Note:** *The number six piston MUST be at TDC on the compression stroke as the chain and sprockets are installed.*

29 Apply a thread locking compound to the camshaft sprocket bolt threads, then install the fuel pump eccentric (small-block engines) and tighten the bolt to the torque listed in this Chapter's Specifications.

30 Lubricate the chain with clean engine oil.

31 Check for cracks and deformation of the oil pan gasket before installing the timing cover. If the gasket has deteriorated or is damaged, it must be replaced before reinstalling the timing cover. **Note 1:** *On small-block engines, the rubber semi-circular seal must be replaced whenever the oil pan is removed; additionally, if the exposed upper short pieces of the oil pan rail gaskets are in bad shape, they will need to be replaced also.* **Note 2:** *On all models, if the portion of oil pan gasket that extends beyond the block is damaged, you may be able to repair the area without replacing the entire oil pan gasket. First, using an X-acto knife, trim off the front portion of the gasket at the point where it meets the engine block, then trim off the front portion of the new oil pan rail gasket so it's exactly the same size. Cover the exposed inside area of the oil pan with a rag, then clean all traces of gasket material off the area where the gasket was removed. Attach the new gasket piece to the front of the oil pan with contact-cement-type gasket adhesive.*

32 On small-block engines, clean all traces of old gasket material off the semi-circular area at the front of the oil pan, finishing off with lacquer thinner or acetone to de-grease the area. Install a new semi-circular rubber seal to the oil pan; be sure to pull the oil seal "nubs" securely through the holes at the front of the oil pan **(see illustration)**.

33 Apply a thin layer of RTV sealant to both sides of the new cover gasket and the corners of the pan and block, then position the new cover gasket on the engine. The sealant will hold it in place. **Note:** *On small-block engines, be sure to apply a dab of RTV to the two points where the semi-circular rubber seal meets the oil pan side-rail gaskets as well as the two points where the oil pan side-rail gaskets meet the block.*

33 Install the timing chain cover on the block. On big-block engines, tighten the bolts, a little at a time, until you reach the torque listed in this Chapter's Specifications. On small-block engines, install all front cover bolts, except the ones that also secure the water pump, tightening them finger-tight.

34 On small-block engines, install the water pump and tighten the timing chain cover bolts to the torque listed in this Chapter's Specifications. On big-block engines, install the water pump housing, using new gaskets and tightening the bolts to the torque listed in this Chapter's Specifications.

35 Install the oil pan bolts, bringing the oil pan up against the timing chain cover.

36 Lubricate the oil seal contact surface of the vibration damper hub with moly-base grease or clean engine oil, then install the damper on the end of the crankshaft. The keyway in the damper must be aligned with the Woodruff key in the crankshaft nose. If the damper cannot be seated by hand, slip the large washer over the bolt, install the bolt and tighten it to pull the damper into place. Tighten the bolt to the torque listed in this Chapter's Specifications.

38 The remaining installation steps are the reverse of removal.

39 Add coolant and check the oil level. Run the engine and check for oil and coolant leaks.

10 Camshaft and lifters - removal, inspection and installation

Camshaft lobe lift check

Refer to illustration 10.3

1 In order to determine the extent of cam lobe wear, the lobe lift should be checked

10.3 When checking the camshaft lobe lift, the dial indicator plunger must be positioned directly above and in-line with the pushrod

prior to camshaft removal.

2 Remove the valve covers (see Section 3).

3 Beginning with the number one cylinder, mount a dial indicator on the engine and position the plunger against the top surface of the first rocker arm. Set the number one cylinder at TDC on the compression stroke (see Chapter 2C). The plunger should be directly above and in line with the pushrod **(see illustration)**.

4 Zero the dial indicator, then very slowly turn the crankshaft in the normal direction of rotation (clockwise) until the indicator needle stops and begins to move in the opposite direction. The point at which it stops indicates maximum cam lobe lift.

5 Record this figure for future reference, then reposition the piston at TDC on the compression stroke.

6 Move the dial indicator to the other number one cylinder rocker arm and repeat the check. Be sure to record the results for each valve.

7 Repeat the check for the remaining valves. Since each piston must be at TDC on the compression stroke for this procedure, work from cylinder-to-cylinder following the

10.10 Pull up on the distributor driveshaft gear to remove it - upon installation, with the number one cylinder at TDC, be sure the slot is aligned as shown (small-block) or parallel to the crankshaft (big-block)

10.12a The lifters in an engine that has accumulated many miles may have to be removed with a special tool (arrow) - store them in an organized manner to make sure they're reinstalled in their original locations

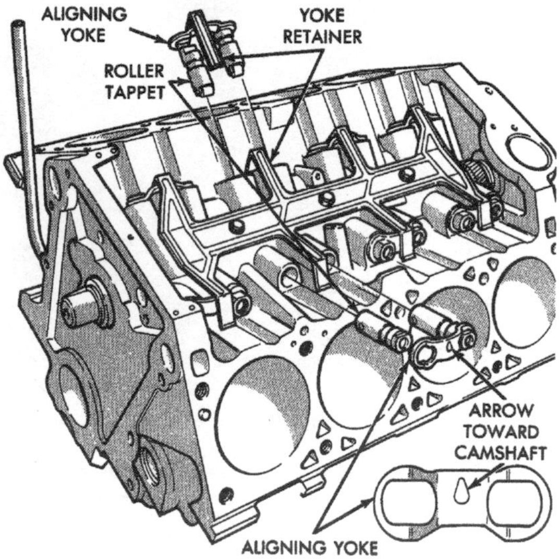

10.12b Vehicles with roller lifters have alignment yokes with directional marks; be sure they get installed in the direction shown

10.13 Remove the camshaft thrust plate and oil tab

1 Oil tab bolt
2 Thrust plate bolts

firing order sequence.

8 After the check is complete, multiply all the recorded figures by 1.5 to obtain the total valve lift for each valve (the rocker arms have a 1.5:1 ratio). Compare the results to the specifications in this Chapter. If the valve lift is 0.003 inch less than specified, cam lobe wear has occurred and a new camshaft should be installed.

Removal

Refer to illustrations 10.10, 10.12a, 10.12b, 10.13, 10.14 and 10.15

9 Refer to the appropriate Sections and remove the intake manifold (small-block engines only), rocker arms, pushrods and the timing chain and camshaft sprockets. **Note:** *On big-block engines (383, 400 and 440 cubic inch), the lifters can be removed through the pushrod openings in the cylinder head with the use of a special removal tool that's available from most auto parts stores; therefore, intake manifold removal may not be necessary.*

10 Remove the distributor drive **(see illustration)**. Also remove the radiator and air conditioning condenser (see Chapter 3). **Note:** *To provide enough clearance for camshaft removal on some models, it may be necessary to unbolt and remove the vertical hood latch support bar and maybe even the grille (see Chapter 11).*

11 There are several ways to extract the lifters from the bores. A special tool designed to grip and remove lifters is manufactured by many tool companies and is widely available, but it may not be required in every case. On newer engines without a lot of varnish buildup, the lifters can often be removed with a small magnet or even with your fingers. A machinist's scribe with a bent end can be used to pull the lifters out by positioning the point under the retainer ring inside the top of each lifter. **Caution:** *Do not use pliers to remove the lifters unless you intend to replace them with new ones (along with the camshaft). The pliers will damage the precision machined and hardened lifters, rendering them useless.*

12 Before removing the lifters, arrange to store them in a clearly labeled box to ensure that they are reinstalled in their original locations **(see illustration)**. **Note:** *On engines equipped with roller lifters, the yoke retainer must be removed before the lifters are withdrawn* **(see illustration)**. Remove the lifters and store them where they will not get dirty. **Caution:** *Do not attempt to withdraw the camshaft with the lifters in place.*

13 On small-block engines, unbolt and remove the camshaft thrust plate and oil tab **(see illustration).** Note how the oil tab is installed so you can return it to its original location on reassembly. On big-block engines, remove the oil pump (see Section 12).

10.14 Thread a long bolt into the camshaft sprocket bolt hole to use as a handle

10.15 As the camshaft is being removed, support it near the block so the lobes do not nick or gouge the bearings

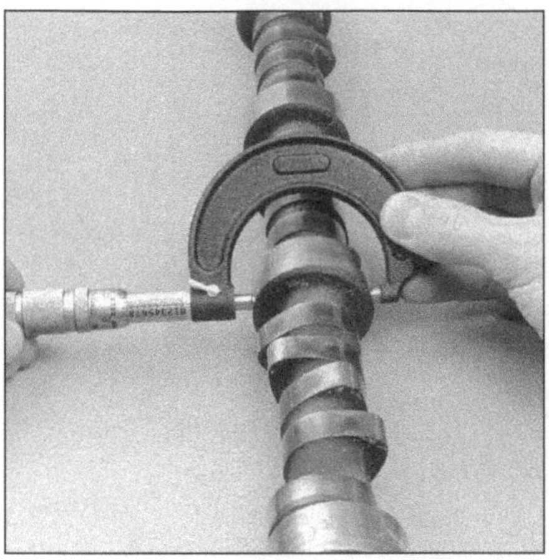

10.17 Check the diameter of each camshaft bearing journal to pinpoint excessive wear and out-of-round conditions

14 On all engines, thread a long bolt into the camshaft sprocket bolt hole to use as a handle when removing the camshaft from the block **(see illustration)**.

15 Carefully pull the camshaft out. Support the cam near the block so the lobes do not nick or gouge the bearings as it is withdrawn **(see illustration)**.

Inspection

Refer to illustration 10.17

Camshaft and bearings

16 After the camshaft has been removed from the engine, cleaned with solvent and dried, inspect the bearing journals for uneven wear, pitting and evidence of seizure. If the journals are damaged, the bearing inserts in the block are probably damaged as well. Both the camshaft and bearings will have to be replaced. **Note:** *Camshaft bearing replacement requires special tools and expertise that probably place it beyond a job you want to get into at home. The tool for bearing removal/installation is available at stores that carry automotive tools, possibly even found*

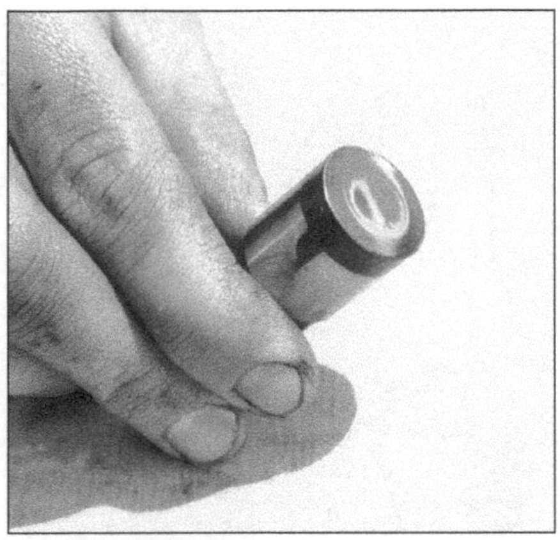

10.20a if the bottom of any lifters is worn concave, scratched or galled, replace the entire set with new lifters

at a tool rental business. It is advisable though, if the bearings are bad, that the engine should be removed and the block taken to an automotive machine shop to ensure that the job is done correctly.

17 Measure the bearing journals with a micrometer to determine if they are excessively worn or out-of-round **(see illustration)**.

18 Check the camshaft lobes for heat discoloration, score marks, chipped areas, pitting and uneven wear. If the lobes are in good condition and if the lobe lift measurements are as specified in this Chapter, the camshaft can be reused.

Conventional lifters

Refer to illustration 10.20a, 10.20b, 10.20c and 10.20d

Note: *Some engines are fitted with 0.008 inch oversize lifters at the factory. These may be identified by a 3/8 inch diamond-shaped stamp on the top pad at the front of the engine and a flat ground on the outside surface of each oversize lifter bore.*

19 Clean the lifters with solvent and dry them thoroughly without mixing them up.

20 Check each lifter wall, pushrod seat and foot for scuffing, score marks and uneven

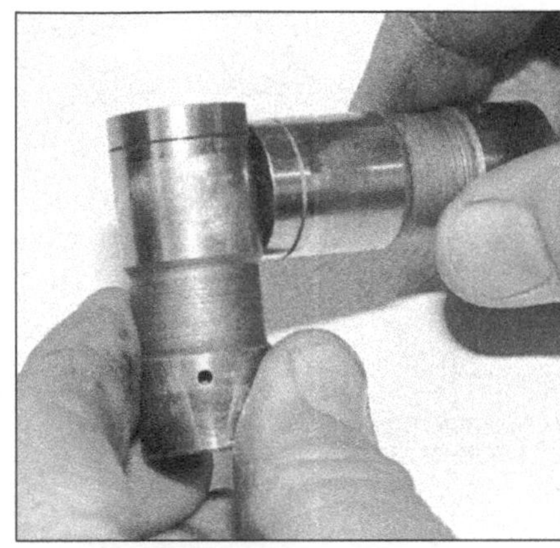

10.20b The foot of each lifter should be slightly convex - the side of another lifter can be used as a straightedge to check it; if it appears flat, it is worn and must not be reused

10.20c If the lifters are pitted or rough, they shouldn't be reused

wear. Each lifter foot (the surface that rides on the cam lobe) must be slightly convex, although this can be difficult to determine by eye. If the base of the lifter is concave **(see illustrations)**, the lifters and camshaft must be replaced. If the lifter walls are damaged or

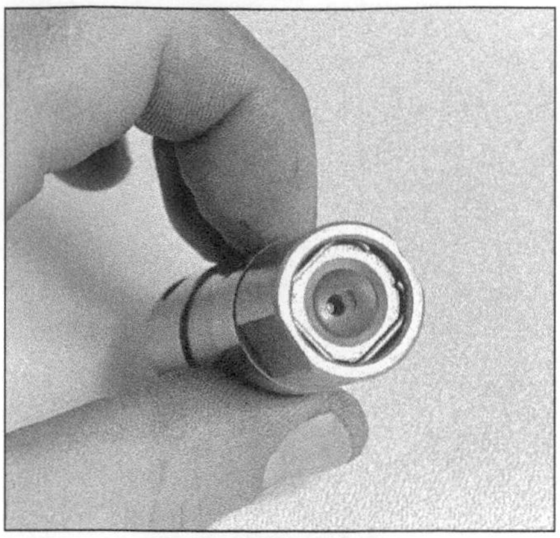

10.20d Check the pushrod seat in the top of each lifter for wear

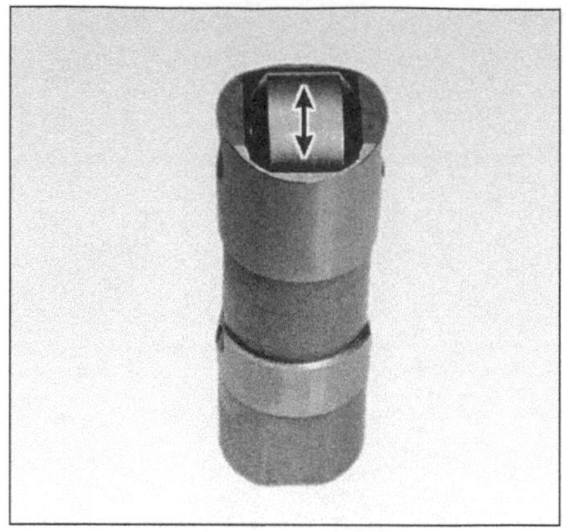

10.22 The roller on the roller lifters must turn freely - check for wear and excessive play as well

10.24 Be sure to apply moly—based grease or engine assembly lube to the cam lobes and bearing journals before installing the camshaft

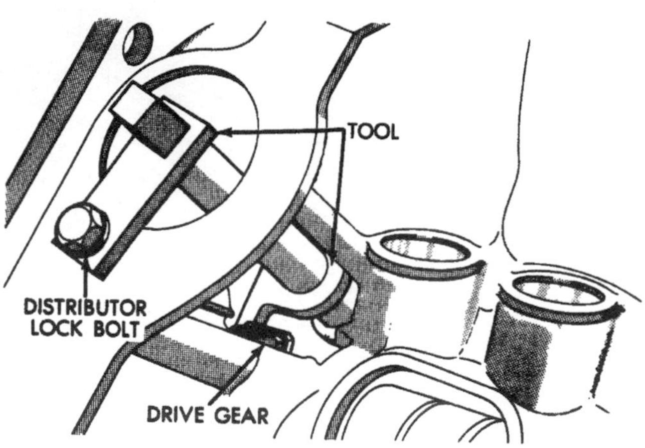

10.25 Camshaft holding tool C-3509 (installed position)

11.4 Separate the Pitman and idler arm from the steering center link (arrows) (see Chapter 10) and let the steering linkage hang down

worn (which is not very likely), inspect the lifter bores in the engine block as well. If the pushrod seats are worn, check the pushrod ends.

21 If new lifters are being installed, a new camshaft must also be installed. If a new camshaft is installed, then use new lifters as well. Never install used lifters unless the original camshaft is used and the lifters can be installed in their original locations.

Roller lifters

Refer to illustration 10.22

22 Check the rollers carefully for wear and damage and make sure they turn freely without excessive play **(see illustration)**. The inspection procedure for conventional lifters also applies to roller lifters.

23 Used roller lifters can be reinstalled with a new camshaft and the original camshaft can be used if new lifters are installed.

Installation

Refer to illustration 10.24, 10.25 and 10.27

24 Lubricate the camshaft bearing journals and cam lobes with moly-base grease or engine assembly lube **(see illustration)**.

25 Position a camshaft holding tool (C-3509, or equivalent) through the distributor hole **(see illustration)**. This tool will prevent the camshaft from being pushed in too far and prevents the welch plug in the back of the block from being knocked out. The tool should remain in place until the timing chain has been installed. **Caution:** *If the tool is unavailable, work carefully to avoid knocking the plug out. If the welch plug seal is damaged, a massive oil leak will occur.*

26 Slide the camshaft into the engine. Support the cam near the block and be careful not to scrape or nick the bearings. On small-block engines, install the camshaft thrust plate and oil tab.

27 On all engines, install the timing chain and sprockets (see Section 9). Align the timing marks on the crankshaft and camshaft sprockets.

28 Install the distributor drive gear **(see illustration 10.10)**. On big-block engines, install the oil pump.

29 On all engines, lubricate the lifters with clean engine oil and install them in the block. If the original lifters are being reinstalled, be sure to return them to their original locations. If a new camshaft was installed, be sure to install new lifters as well.

30 The remaining installation steps are the reverse of removal.

31 Change the oil, add Mopar Crankcase Conditioner part no. 3419130, or equivalent, and install a new oil filter (see Chapter 1).

32 Start the engine, check for oil pressure and leaks and set the ignition timing.

11 Oil pan - removal and installation

Removal (small-block engines)

All models

Refer to illustrations 11.4 and 11.6

1 Disconnect the battery cable from the negative battery terminal. Raise the vehicle and support it securely on jackstands (see Chapter 1).

2 Drain the engine oil and replace the oil filter (see Chapter 1).

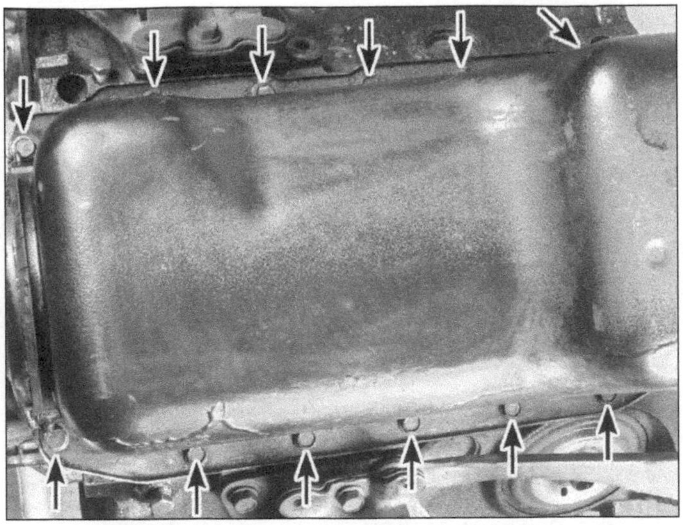

11.6 Remove the bolts (arrows) around the perimeter of the oil pan (typical)

11.9 Here's the oil pan lowered from the engine - note how the lowered steering center link provides clearance

3 Unbolt and remove the lower engine-to-transmission support struts and the torque converter inspection cover (see Chapter 7, Part B).

4 Separate the Pitman and idler arm from the steering center link **(see illustration)** and lower the steering linkage as far a possible to make room for oil pan removal.

5 On single-exhaust systems only, that use a cross-over pipe under the oil pan, disconnect the exhaust pipe from the exhaust manifolds (see Section 7) and lower the pipe as far as possible. **Note:** *High-performance models use dual exhaust pipes and moving the exhaust isn't required for oil pan removal.*

6 Remove all the oil pan bolts **(see illustration)**, then lower the pan from the engine. The pan will probably stick to the engine, so strike the pan with a rubber mallet until it breaks the gasket seal with the engine. **Caution:** *Before using force on the oil pan, be sure all the bolts have been removed.* At this point, the oil pan is free from the engine; you can very carefully try to slide it out the rear, but, on most models, it probably will still hang up on the transmission (models through 1981) or starter (1982 and later models). Follow the procedure below that applies to your model.

Models through 1981

Refer to illustration 11.9

7 Place a jack under the transmission and remove the rear transmission mount through-bolt (see Chapter 7, Part B).

8 Raise the transmission with the jack.

9 On all except high-performance models, you should be able to remove the oil pan now by lowering it and sliding it to the rear **(see illustration)**. On high-performance models, you'll find you can slide the pan to the rear far enough to clear the transmission bellhousing, but the pan will still hang up on the windage tray - leave the pan in this position and proceed to Step 10.

High-performance models only

10 On high-performance models through 1981, disconnect the radiator fan shroud and set it back over the fan (see Chapter 3).

11 Lower the transmission and reinstall the mount through-bolt.

12 Remove the engine mount through bolts (see Section 14).

13 Place a jack at the front of the engine on the vibration damper, using a block of wood on the jack pad, and raise the engine about 1 inch. **Warning:** *Make sure the engine is stable on the jack before attempting to remove the oil pan, and never, under any circumstances, place any part of your body between the engine or oil pan and any other part when it's supported only by a jack - assume the engine could slip off the jack at any moment!*

14 Remove the oil pan very carefully by sliding it out towards the rear.

1982 and later models

15 Remove the starter and upper starter mounting stud **(see Chapter 5).**

16 Remove the oil pan by lowering it and sliding it out to the rear.

Removal (big-block engines)

17 Disconnect the exhaust pipes from the exhaust manifolds (see Section 7).

18 Remove the clamp attaching the exhaust pipe at the front and lower the exhaust pipe.

19 Remove all the oil pan bolts, then lower the pan from the engine. The pan will probably stick to the engine, so strike the pan with a rubber mallet until it breaks the gasket seal with the engine. **Caution:** *Before using force on the oil pan, be sure all the bolts have been removed.*

20 On all except 1971 through 1976 B-body models, you should be able to remove the oil pan at this point. Rotate the crankshaft until the front throw is all the way up, then remove the pan to the rear. Turn the pan counterclockwise to clear oil pick-up tube as the pan is lowered. On 1971 through 1976 B-body models, you'll probably find you can't slide the pan to the rear far enough to clear the transmission bellhousing; proceed to Step 21.

1971 through 1976 B-body models only

21 On 1971 through 1976 B-body models with big-block engines, disconnect the radiator fan shroud and set it back over the fan (see Chapter 3).

22 Remove the engine mount through bolts (see Section 14).

23 Place a jack at the front of the engine on the vibration damper, using a block of wood on the jack pad, and raise the engine about 1-1/2 to 2 inches. **Warning:** *Make sure the engine is stable on the jack before attempting to remove the oil pan, and never, under any circumstances, place any part of your body between the engine or oil pan and any other part when it's supported only by a jack - assume the engine could slip off the jack at any moment!*

24 Remove the oil pan very carefully by sliding it out towards the rear.

Installation (all models)

Refer to illustration 11.27

25 Wash out the oil pan with solvent.

26 Thoroughly clean the mounting surfaces of the oil pan and engine block of old gasket material and sealer. If the oil pan is distorted at the bolt-hole areas, straighten the flange by supporting it from below on a 1x4 wood block and tapping the bolt holes with the rounded end of a ball-peen hammer. Wipe the gasket surfaces clean with a rag soaked in lacquer thinner or acetone.

27 On small-block engines, install the side-rail gaskets on the engine. Attach them with contact-cement-type adhesive such as Fel-Pro's High-tack. Attach the rubber end seals

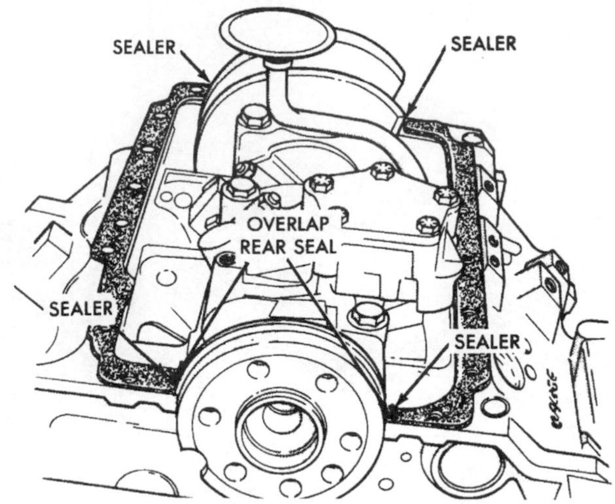

11.27 Apply RTV sealer to the four corners, as shown

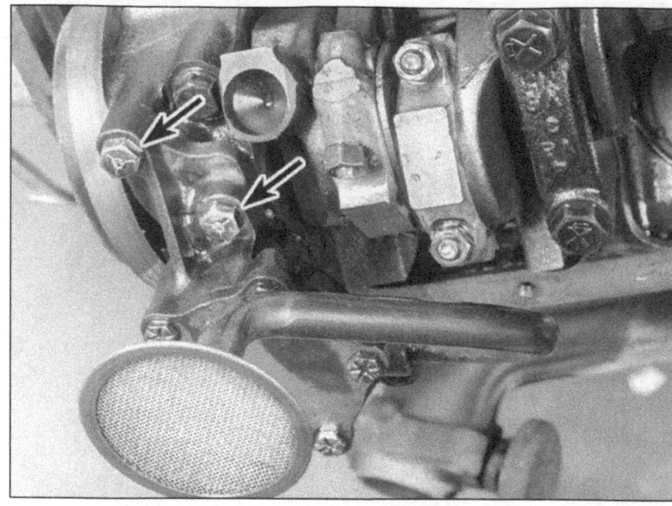
12.2 Remove the bolts (arrows) and lower the oil pump

12.6a Inspect the end of the driveshaft for excessive wear and replace it, if necessary

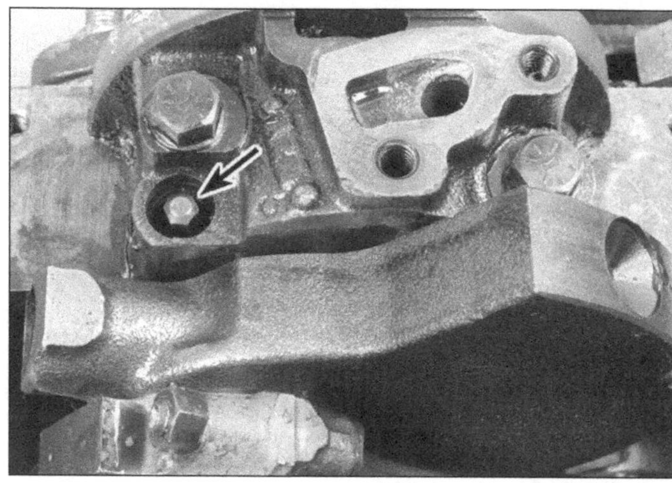
12.6b The oil pump driveshaft (arrow) must align with the pump

to the oil pan, pulling the "nubs" through the holes in the pan. Apply a bead of RTV sealant in the four corners where the side-rail gaskets meet the front and rear rubber gaskets **(see illustration)**.

28 On big-block engines, apply a small dab of RTV sealant to the points where the front cover meets the engine block and where the rear seal retainer meets the engine block. Then attach the oil pan gasket to the engine block with contact-cement-type adhesive such as Fel-Pro's High-tack.

28 On all models, lift the pan into position, being careful not to disturb the gasket, and install the bolts finger tight.

29 Starting at the ends and alternating from side-to-side towards the center, tighten the bolts to the torque listed in this Chapter's Specifications.

30 The remainder of the installation procedure is the reverse of removal.

31 Add the proper type and quantity of oil (see Chapter 1), start the engine and check for leaks before placing the vehicle back in service.

12 Oil pump - removal and installation

Small-block engines

Refer to illustration 12.2

Removal

1 Unbolt and lower the oil pan (see Section 11). **Note:** *The pump assembly can usually be removed by lowering the pan* **(see illustration 11.9b).**

2 While supporting the oil pump, remove the pump-to-rear main bearing cap bolts **(see illustration)**.

3 Lower the pump and pickup screen assembly from the vehicle and clean the gasket surfaces.

Installation

Refer to illustrations 12.6a and 12.6b

4 Thread the oil pickup tube and screen into the oil pump and tighten it. **Caution:** *Be absolutely certain that the pickup screen is properly tightened so that no air can be sucked into the oiling system at this connection.*

5 Prime the pump by pouring clean motor oil into the pickup tube.

6 Position the pump on the engine with a new gasket and make sure the pump driveshaft is aligned with the oil pump **(see illustrations)**.

7 Install the mounting bolts and tighten them to the torque listed in this Chapter's Specifications.

8 Install the oil pan and add oil.

9 Run the engine and check for oil pressure and leaks.

Big-block engines

Refer to illustration 12.11

10 Remove the oil filter (see Chapter 1).

11 Remove the bolts **(see illustration)** and pull the oil pump assembly off the engine. The drive-gear extension may bind a bit on the way out, so rock the assembly back and forth to permit easier removal.

12 Using a scraper, remove all traces of old gasket material from the gasket surfaces of the pump and the engine block. Gasket removal solvents are available from

12.11 The oil pump on big-block engines is located on the lower right side of the block - remove the bolts (arrows) to remove it

13.3 Carefully drive out the seal with a hammer and punch

13.4 On 1978 and later small-block engines, you can pry the old seal out with a seal puller or screwdriver - there's no need to remove the timing cover

13.6a On 1978 and later small-block engines, drive the new seal into place with a socket and a hammer (timing cover installed)

13.6b Tapping the new seal into place on a small-block model (timing cover removed)

13.7 If the sealing surface of the damper has a wear groove from contact with the seal, repair sleeves, like the one shown here, are available from many auto parts stores - when pressed into place, these sleeves provide a flat surface for the new seal and will help prevent an oil leak

auto parts stores and may prove helpful. After the gasket is removed, wipe the gasket surfaces clean with a rag soaked in lacquer thinner or acetone.

13 Coat the new oil pump gasket with RTV sealant, then place it on the oil pump. Install the bolts through the pump to hold the gasket in place, then install the oil pump on the block, tightening the bolts to the torque listed in this Chapter's Specifications.

14 Replace the oil filter and add the proper type and quantity of oil (see Chapter 1).

13 Crankshaft oil seals - replacement

Front seal

Removal

Refer to illustrations 13.3 and 13.4

1 Remove the vibration damper (see Section 9).

2 On all big-block engines and 1977 and earlier small-block engines, remove the timing chain cover **(see Section 9).** On 1978 and later small-block engines, the seal can be replaced with the cover installed, although replacement is usually easier with the cover removed.

3 If the seal is being replaced with the timing chain cover removed (which is the only way on big-block engines and earlier small-block engines), support the cover on top of two blocks of wood and drive the seal out from the back (later small-block engines) or front (big-block engines and earlier small-block engines) with a hammer and punch **(see illustration)**. **Caution:** *Be careful not to scratch, gouge or distort the area that the seal fits into or a leak will develop.*

4 If the seal is being removed while the cover is still attached to the engine block (later small-block engines only), carefully pry the seal out of the cover with a seal puller or a large screwdriver **(see illustration)**. **Caution:** *Be careful not to scratch, gouge or distort the area that the seal fits into or a leak will develop.*

Installation

Refer to illustrations 13.6a, 13.6b and 13.7

5 Clean the bore to remove any old seal material and corrosion. Position the new seal in the bore with the seal lip (usually the side with the spring) facing IN (toward th inside of the engine). A small amount of oil applied to the outer edge of the new seal will make installation easier - don't overdo it!

6 After either method of seal removal is complete and the new seal is ready to be installed, drive the seal into the bore with a large socket and hammer until it's completely seated **(see illustrations)**. Select a socket that's the same outside diameter as the seal and make sure the new seal is pressed into place until it bottoms against the cover flange.

7 Check the surface of the damper that the oil seal rides on **(see illustration)**. If the

13.12 **Remove the bolts (arrows) and detach the rear main bearing cap from the engine**

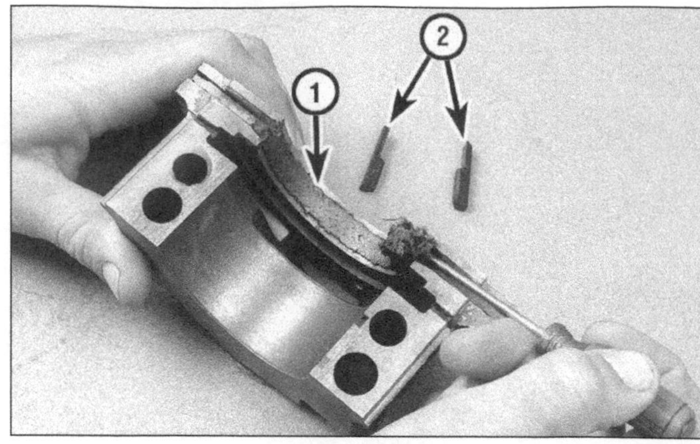

13.13 Lift the old seal sections out of the bearing cap with a small screwdriver

1 Rear main seal (rope or rubber) *2 Side seals*

surface has been grooved from long-time contact with the seal, there is a press-on sleeve available to put a new surface on the damper for the seal to ride on. This sleeve is pressed into place with a hammer and a block of wood. These sleeves are commonly available from auto parts stores.

8 Lubricate the seal lips with engine oil and reinstall the vibration damper (see Section 9).

Rear seal

All engines

Refer to illustrations 13.12. 13.13 and 13.14

Note: *If you're installing a new seal during overhaul, ignore the steps in this procedure that concern removal of external parts. Also, since the crankshaft is already removed, it's not necessary to use any special tools to remove the upper seal half; remove and install the upper seal half the same way as the lower seal half.*

9 Remove the oil pan (see Section 11).

10 Remove the oil pump on small-block engines (see Section 12).

11 The rear main seal can be replaced with the engine in the vehicle. There are two different types of material used by the manufacturer. The original material used on most models is the "rope" style, two-piece seal. The newer version of the rear main seal is also a two-piece seal, but made from neoprene rubber. Most aftermarket part suppliers have substituted the rubber for the older "rope" style, because it works better than the original design, but the seal must be used as a complete upper and lower set. The "rope" and "rubber" styles of seals CANNOT be combined.

12 On small-block engines, remove the bolts and detach the rear main bearing cap from the engine **(see illustration)**. On big-block engines, unbolt and remove the rear oil seal retainer that's located behind the rear main bearing cap. **Note:** *A twelve-point socket is necessary to remove the two seal retainer bolts on big-block engines.*

13 Remove the lower half of the oil seal from the bearing cap or retainer. On small-block engines equipped with side seals, also remove the side seals from the cap by prying them from the side with a small screwdriver **(see illustration)** and grasping the seals with a needle-nose pliers, if necessary.

14 Remove upper half of the rope oil seal (in the engine block) by screwing a tool (KD-492, or equivalent) into the end of the seal **(see illustration)**, being careful not to damage the crankshaft. Pull the seal out with the tool while rotating the crankshaft. If a neoprene rubber-type seal is installed on your engine, use a small brass punch to carefully press on one end of the upper half of the seal until the other side protrudes, then pull the seal out from the other side with needle-nose pliers. Clean the bearing cap and engine block surfaces carefully to degrease them and remove any sealant.

15 On small-block engines equipped with cap side seals, insert the seals into the slots in bearing cap and press them carefully into place. **Note:** *On many aftermarket seals, the side seals are integral with the lower seal half.* Generally, the side seal with the yellow paint goes into right side of the cap (when installed on the engine) and the narrow sealing edge should face up. Make sure the seals are fully seated, that they match the contours of the cap and they protrude slightly from the cap; then, using a new, single-edged razor blade, cut the ends flush with side edges of the cap. On small-block engines not equipped with side seals, apply a dab of RTV sealant to each side of the cap, adjacent to the seal. **Caution:** *Do not apply RTV to the seal itself.*

16 On all engines, lightly oil the lips of the new (neoprene rubber) crankshaft seals. **Caution:** *Always wipe the crankshaft surface clean. then oil it lightly before installing a new seal.*

17 Rotate a new seal half into cylinder block with the paint stripe toward the rear (the seal lip must face toward the front of the engine). **Caution:** *Hold your thumb firmly against the outside diameter of the seal as you're rotating it into place. This will prevent the seal outside diameter from being shaved from contact with the sharp edge of the engine block. If the seal gets shaved, you may wind up with an oil leak.*

18 Place the other seal half in the bearing cap or seal retainer with the paint stripe toward the rear.

Small-block engines

19 Install the rear main bearing cap and tighten the bolts to the torque listed in this Chapter's Specifications.

20 Install the oil pump and oil pan.

21 The remainder of installation is the reverse of removal. Fill the pan with oil, run the engine and check for leaks.

Big-block engines

22 Clean the seal retainer side grooves in preparation for installing the new side seals. **Caution:** *Perform the following operations as rapidly as possible. These side seals are made from a material that expands quickly when oiled.*

13.14 Screw a removal tool into one end of the old upper seal section and pull it out (engine removed for a clearer view of the procedure)

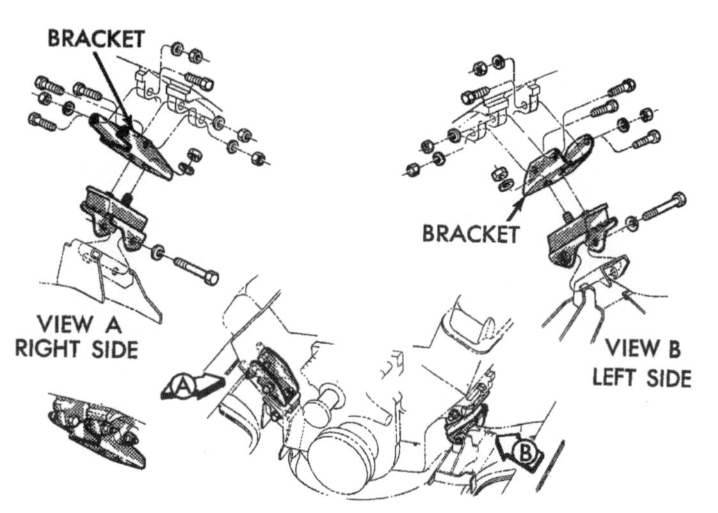

14.1a An exploded view of typical early-style engine mounts (small-block shown, big-block similar)

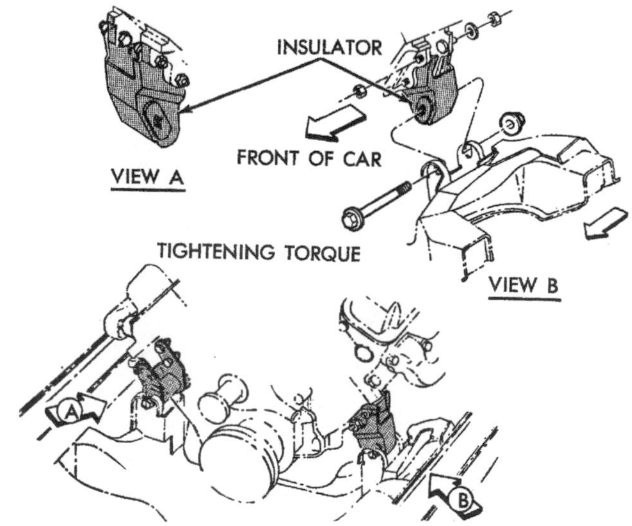

14.1b An exploded view of typical later-style small-block engine mounts

23 Apply mineral spirits or diesel fuel to the side seals. Do not apply any sealer to side seals, as this will prevent the oil from swelling the seals. **Caution:** *Failure to pre-oil the seals will result in an oil leak.*

24 Install the side seals immediately in the seal retainer grooves Apply RTV sealant to the bottom of the retainer on both sides.

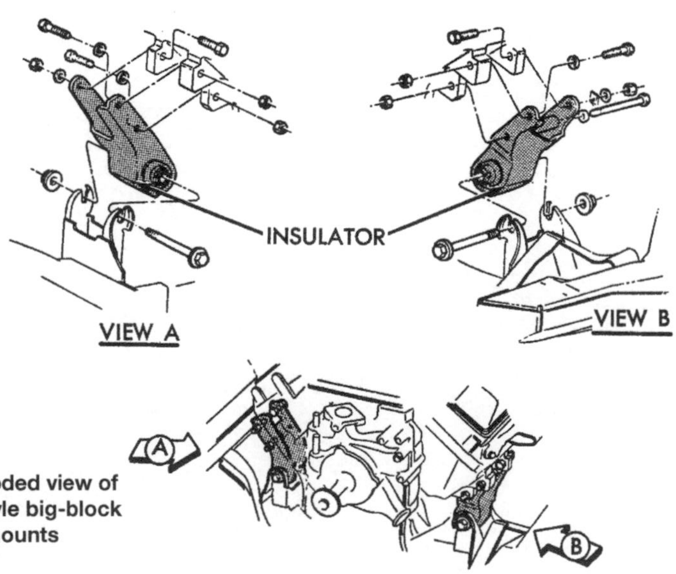

14.1c An exploded view of typical later-style big-block engine mounts

25 Install seal retainer and tighten screws to the torque listed in this Chapter's Specifications.

26 The remainder of installation is the reverse of removal. Fill the pan with oil, run the engine and check for leaks.

14 Engine mounts - check and replacement

Refer to illustrations 14.1a, 14.1b and 14.1c

Refer to Chapter 2, Part A for this procedure, but use the accompanying illustrations for V8 engine mount locations and descriptions.

15 Flywheel/driveplate - removal and installation

Refer to Chapter 2, Part A for this procedure, but be sure to use the torque value listed in this Chapter's Specifications.

Notes

Chapter 2 Part C
General engine overhaul procedures

Contents

Specifications

Six-cylinder engine

General

Oil pressure	30 to 70 psi above 2000 rpm
Cylinder compression	100 psi minimum; no more than 40 psi variation among cylinders

Cylinder block

Cylinder bore diameter (standard)	3.400 to 3.402 inches
Taper limit	0.010 inch
Out-of-round limit	0.005 inch
Deck (head gasket surface) warpage limit	0.00075 inch times the span length of the deck (maximum)

Pistons and rings

Piston-to-cylinder bore clearance	0.0005 to 0.0015 inch
Piston ring side clearance	
Compression rings	0.0015 to 0.0030 inch
Oil ring (steel rails)	0.0002 to 0.005 inch
Piston ring end gap	
Compression rings	0.010 to 0.020 inch
Oil ring (steel rails)	0.015 to 0.055 inch

Crankshaft and connecting rods

Crankshaft main journal	
Diameter (standard)	2.7495 to 2.7505 inches
Out-of-round/taper limit	0.001 inch
Main bearing oil clearance	
Standard	
1971 through 1976	0.0005 to 0.0015 inch
1977 through 1980	0.0002 to 0.0022 inch
1981 through 1986	0.0010 to 0.0025 inch
Service limit	0.0025 inch

Six-cylinder engine (continued)

Crankshaft connecting rod journal	
Diameter	2.1865 to 2.1875 inches
Taper limit	0.001 inch
Out-of-round limit	0.001 inch
Standard bearing oil clearance	
1971 through 1976	0.0005 to 0.002 inch
1977	0.0005 to 0.0025 inch
1978 and later	0.0010 to 0.0022 inch
Oil clearance service limit	0.0025 inch
Connecting rod side clearance	
1971 through 1976	0.006 to 0.012 inch
1977 and 1978	0.006 to 0.025 inch
1979	0.002 to 0.022 inch
1980	0.002 to 0.009 inch
1981 and later	0.007 to 0.013 inch
Crankshaft endplay	
1971 through 1976	0.002 to 0.007 inch
1977 through 1980	0.002 to 0.009 inch
1981 and later	0.0035 to 0.0095 inch

Cylinder head and valve train

Head gasket surface warpage limit	0.00075 inch times the span length of the head (maximum)
Valve stem diameter (standard)	
Intake	0.372 to 0.373 inch
Exhaust	0.371 to 0.372 inch
Valve stem-to-guide clearance	
Intake	
Standard	0.001 to 0.003 inch
Service limit	0.008 inch
Exhaust	
Standard	0.002 to 0.004 inch
Service limit	0.008 inch
Valve spring free length	1-59/64 inches
Valve spring installed height	1-5/8 to 1-11/16 inches
Valve seat width	
Intake	0.070 to 0.090 inch
Exhaust	0.040 to 0.060 inch

318, 340 and 360 cubic inch (small-block) V8 engines

General

Oil pressure	30 to 80 psi at 3000 rpm
Cylinder compression	100 psi minimum; no more than 40 psi variation among cylinders

Cylinder block

Cylinder bore diameter	
318 cu in	3.910 to 3.912 inches
340 cu in	4.040 to 4.042 inches
360 cu in	4.000 to 4.002 inches
Taper limit	0.010 inch
Out-of-round limit	0.005 inch
Deck (head gasket surface) warpage limit	0.00075 inch times the span length of the deck (maximum)

Pistons and rings

Piston-to-cylinder bore clearance	0.0005 to 0.0015 inch
Piston ring side clearance	
Compression rings	0.0015 to 0.0030 inch
Oil ring (steel rails)	0.002 to 0.005 inch
Piston ring end gap	
Compression rings	0.010 to 0.020 inch
Oil ring (steel rails)	
318 cu in	0.010 to 0.062 inch
340 cu in	0.015 to 0.055 inch
360 cu in	0.015 to 0.062 inch

Crankshaft and connecting rods

Main journal	
Diameter	
318 and 340 cu in	2.4995 to 2.5005 inches
360 cu in	2.8095 to 2.8105 inches
Taper/out-of-round limits	0.001 inch
Main bearing oil clearance	
No. 1 journal (1981 and later only)	
Standard	0.0005 to 0.0015 inch
Service limit	0.0015 inch
All other journals (and No. 1 journal on 1980 and earlier models)	
Standard	0.0005 to 0.0020 inch
Service limit	0.0025 inch
Connecting rod journal	
Diameter	2.124 to 2.125 inches
Taper/out-of-round limits	0.001 inch
Connecting rod bearing oil clearance	
1971 through 1978	
318 and 340 cu in	
Standard	0.0002 to 0.0022 inch
Service limit	0.003 inch
360 cu in	
Standard	0.0005 to 0.0025 inch
Service limit	0.0025 inch
1979 and later (318 and 360 cu in)	
Standard	0.0005 to 0.0025 inch
Service limit	0.0025 inch
Connecting rod side clearance	0.006 to 0.014 inch
Crankshaft endplay	
318 and 340 cu in	0.002 to 0.007 inch
360 cu in	0.002 to 0.009 inch

Cylinder head and valve train

Head gasket surface warpage limit	0.00075 inch times the span length of the head (maximum)
Valve stem diameter (standard)	
Intake	0.372 to 0.373 inch
Exhaust	0.371 to 0.372 inch
Valve stem-to-guide clearance	
Intake	
Standard	0.001 to 0.003 inch
Service limit	0.008 inch
Exhaust	
Standard	0.002 to 0.004 inch
Service limit	0.008 inch
Valve spring free length	
Intake	1-13/16 inches
Exhaust	
Without rotators	2.00 inches
With rotators	1-13/16 inches
Valve spring installed height	
Intake	1-5/8 to 1-11/16 inches
Exhaust	
Without rotators	1-5/8 to 1-11/16 inches
With rotators	1-29/64 to 1-33/64 inches
Valve seat width	
Intake	0.065 to 0.085 inch
Exhaust	0.080 to 0.100 inch

383, 400 and 440 cubic inch V8 engines

General

Oil pressure	30 to 80 psi at 2000 rpm
Cylinder compression pressure	100 psi minimum; no more than 40 psi variation among cylinders

Cylinder block

Cylinder bore diameter	
383 cu in	4.2495 to 4.2515 inches
400 cu in	4.342 to 4.344 inches
440 cu in	4.320 to 4.322 inches

383, 400 and 440 cubic inch V8 engines (continued)

Taper limit	0.010 inch
Out-of-round	0.005 inch
Deck (cylinder head gasket surface) warpage limit	0.00075 inch times the span length of the deck (maximum)

Pistons and rings

Piston-to-cylinder clearance	0.0003 to 0.0013 inch
Piston ring side clearance	
Compression rings	
Through 1976	0.0025 to 0.004 inch
1977 and later	0.0015 to 0.003 inch
Oil ring (steel rails)	
Through 1976	0.001 to 0.003 inch
1977 and later	0.0002 to 0.005 inch
Piston ring end gap	
Compression rings	0.013 to 0.023 inch
Oil ring (steel rails)	0.015 to 0.055 inch

Crankshaft and connecting rods

Main journal diameter	
383 and 400 cu in	2.4625 to 2.6255 inch
440 cu in	2.7495 to 2.7505 inches
Main journal taper/out-of-round limits	0.001 inch
Main bearing oil clearance	
Standard	0.0005 to 0.002 inch
Service limit	0.0025 inch
Connecting rod journal	
Diameter	2.375 to 2.376 inches
Taper/out-of-round limits	0.001 inch
Connecting rod bearing oil clearance	
Standard	
Through 1976	0.001 to 0.0025 inch
1977 and later	0.0005 to 0.0030 inch
Service limit	0.0030 inch
Connecting rod side clearance	0.009 to 0.017 inch
Crankshaft endplay	
Through 1976	0.002 to 0.007 inch
1977 and later	0.002 to 0.009 inch

Cylinder head and valve train

Head gasket surface warpage limit	0.00075 inch times the span length of the head (maximum)
Valve stem diameter	
Standard engines (2 bbl carburetor)	
Intake	0.3723 to 0.3730 inch
Exhaust	0.3713 to 0.3730 inch
Hi-performance engines (4 bbl carburetor)	
Intake	0.3718 to 0.3725 inch
Exhaust	0.3708 to 0.3725 inch
Valve stem-to-guide clearance	
Intake	
Through 1976	
Standard	0.0016 to 0.0033 inch
Service limit	0.008 inch
1977 and later	
Standard	0.0011 to 0.0028 inch
Service limit	0.008 inch
Exhaust	
Standard	
Through 1976	0.0015 to 0.0032 inch
1977 and later	0.0011 to 0.0038 inch
Service limit (all models)	0.008 inch
Valve spring free length	2-37/64 inches
Valve spring installed height	1-53/64 to 1-57/64 inches
Valve seat width (intake and exhaust)	0.040 to 0.060 inch

Torque specifications*

	Ft-lbs
Main bearing cap bolts (all engines)	85
Connecting rod cap nuts (all engines)	45
Crankshaft rear main seal retainer bolts	
Six-cylinder engines	30
383/400/440 cu in (big block) V8 engines	25

***Note:** *Refer to either Part A or Part B for additional specifications.*

1 General information

Included in this portion of Chapter 2 are the general overhaul procedures for cylinder heads and internal engine components. The information ranges from advice concerning preparation for an overhaul and the purchase of replacement parts to detailed, step-by-step procedures covering removal and installation of internal engine components and the inspection of parts.

The following Sections have been written based on the assumption that the engine has been removed from the vehicle. For information concerning in-vehicle engine repair, as well as removal and installation of the external components necessary for the overhaul, see Parts A and B of this Chapter and Section 9 of this Part.

The Specifications included here in Part C are only those necessary for the inspection and overhaul procedures which follow. Refer to Parts A and B for additional specifications.

2 Engine removal - methods and precautions

If you have decided that an engine must be removed for overhaul or major repair work, several preliminary steps should be taken.

Locating a suitable work area is extremely important. A shop is, of course, the most desirable place to work. Adequate work space, along with storage space for the vehicle, will be needed. If a shop or garage is not available, at the very least a flat, level, clean work surface made of concrete or asphalt is required.

Cleaning the engine compartment and engine before beginning the removal procedure will help keep tools clean and organized.

An engine hoist or A-frame will be needed. Make sure that the equipment is rated in excess of the combined weight of the engine and its accessories. Safety is of primary importance, considering the potential hazards involved in lifting the engine out of the vehicle.

If the engine is being removed by a novice, a helper should be available. Advice and aid from someone more experienced would also be helpful. There are many instances when one person cannot simultaneously perform all of the operations required when lifting the engine out of the vehicle.

3.4a The oil pressure sending unit on V8 engines is located at the rear of the engine, just behind the intake manifold (adjacent to the distributor on small-block engines)

Plan the operation ahead of time. Arrange for or obtain all of the tools and equipment you will need prior to beginning the job. Some of the equipment necessary to perform engine removal and installation safely and with relative ease are (in addition to an engine hoist) a heavy duty floor jack, complete sets of wrenches and sockets as described in the front of this manual, wooden blocks and plenty of rags and cleaning solvent for mopping up spilled oil, coolant and gasoline. If the hoist is to be rented, make sure that you arrange for it in advance and perform beforehand all of the operations possible without it. This will save you money and time.

Plan for the vehicle to be out of use for a considerable amount of time. A machine shop will be required to perform some of the work which the do-it-yourselfer cannot accomplish due to a lack of special equipment. These shops often have a busy schedule, so it would be wise to consult them before removing the engine in order to accurately estimate the amount of time required to rebuild or repair components that may need work.

Always use extreme caution when removing and installing the engine. Serious injury can result from careless actions. Plan ahead, take your time and a job of this nature, although major, can be accomplished successfully.

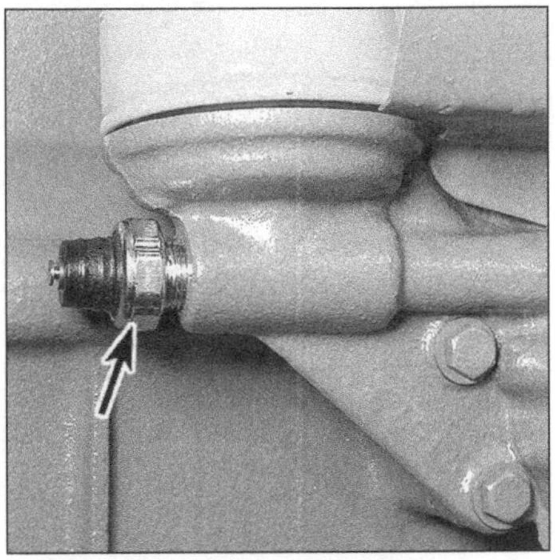

3.4b On six-cylinder engines, the sending unit is screwed into the oil pump housing, just below the rear of the oil filter. The type of sending unit shown here is for a system that uses an indicator light on the dash, while the canister-type shown for the V8 engines in the previous illustration is for a system that uses a gauge.

3 Engine overhaul - general information

Refer to illustrations 3.4a, 3.4b and 3.4c

It is not always easy to determine when, or if, an engine should be completely overhauled, as a number of factors must be considered.

High mileage is not necessarily an indication that an overhaul is needed, while low mileage does not preclude the need for an overhaul. Frequency of servicing is probably the most important consideration. An engine that has had regular and frequent oil and filter changes, as well as other required maintenance, will most likely give many thousands of miles of reliable service. Conversely, a neglected engine may require an overhaul very early in its life.

Excessive oil consumption is an indication that piston rings and/or valve guides are in need of attention. Make sure that oil leaks are not responsible before deciding that the rings and/or guides are bad. Test the cylinder compression (see Section 5) or have a leak-down test performed by an experienced tune-up mechanic to determine the extent of the work required.

If the engine is making obvious knocking or rumbling noises, the connecting rod

3.4c Unscrew the oil pressure sending unit and install a pressure gauge in its place. On gauge-type sending units, there are flats below the canister to unscrew it with a wrench. For sending units on systems with an indicator light, it's best to purchase a special socket designed for removing sending units - this will prevent damaging the sending unit during removal.

4.4 Mark the cap and base of the distributor, just below the number one spark plug terminal (arrows)

4.6a On small-block V8 engines, align the groove on the crankshaft damper with the 0-degree mark cast into the driver's side of the timing chain cover

and/or main bearings are probably at fault. To accurately test oil pressure, temporarily connect a mechanical oil pressure gauge in place of the oil pressure sending unit **(see illustrations)**. Compare the reading to the pressure listed in this Chapter's Specifications. If the pressure is extremely low, the bearings and/or oil pump are probably worn out.

Loss of power, rough running, excessive valve train noise and high fuel consumption rates may also point to the need for an overhaul, especially if they are all present at the same time. If a complete tune-up does not remedy the situation, major mechanical work is the only solution.

An engine overhaul involves restoring the internal parts to the specifications of a new engine. During an overhaul, the piston rings are replaced and the cylinder walls are reconditioned (rebored and/or honed). If a rebore is done, new pistons are required. The main bearings, connecting rod bearings and camshaft bearings are generally replaced with new ones and, if necessary, the crankshaft may be reground to restore the journals. Generally, the valves are serviced as well, since they are usually in less-than-perfect condition at this point. While the engine is being overhauled, other components, such as the distributor, starter and alternator, can be rebuilt as well. The end result should be a like-new engine that will give many thousands of trouble-free miles. **Note:** *Critical cooling system components such as the hoses, the drivebelts, the thermostat and the water pump MUST be replaced with new parts when an engine is overhauled. The radiator should be checked carefully to ensure that it isn't clogged or leaking. If in doubt, replace it with a new one. Also, we do not recommend overhauling the oil pump - always install a new one when an engine is rebuilt.*

Before beginning the engine overhaul, read through the entire procedure to familiarize yourself with the scope and requirements of the job. Overhauling an engine is not difficult if you take your time and follow all procedures carefully. But it is time consuming. Plan on the vehicle being tied up for a minimum of two weeks, especially if parts must be taken to an automotive machine shop for repair or reconditioning. Check on availability of parts and make sure that any necessary special tools and equipment are obtained in advance. Most work can be done with typical hand tools, although a number of precision measuring tools are required for inspecting parts to determine if they must be replaced. Often an automotive machine shop will handle the inspection of parts and offer advice concerning reconditioning and replacement. **Note:** *Always wait until the engine has been completely disassembled and all components, especially the engine block, have been inspected before deciding what service and repair operations must be performed by an automotive machine shop.* Since the block's condition will be the major factor to consider when determining whether to overhaul the original engine or buy a rebuilt one, never purchase parts or have machine work done on other components until the block has been thoroughly inspected. As a general rule, time is the primary cost of an overhaul, so it does not pay to install worn or sub-standard parts.

As a final note, to ensure maximum life and minimum trouble from a rebuilt engine, everything must be assembled with care in a spotlessly clean environment.

4 Top Dead Center (TDC) for number one piston - locating

Refer to illustrations 4.4, 4.6a, 4.6b and 4.7

1 Top Dead Center (TDC) is the highest point in the cylinder that each piston reaches as it travels up-and-down when the crankshaft turns. Each piston reaches TDC on the compression stroke and again on the exhaust stroke, but TDC generally refers to piston position on the compression stroke. The timing marks are referenced to the number one piston at TDC on the compression stroke.

2 Positioning the pistons at TDC is an essential part of many procedures such as camshaft removal, timing chain replacement and distributor removal.

3 In order to bring any piston to TDC, the crankshaft must be turned using one of the methods outlined below. When looking at the front of the engine, normal crankshaft rotation is clockwise. **Warning:** *Before beginning this procedure, be sure to place the transmission in Neutral and disable the ignition system by disconnecting the coil wire from the distributor cap and grounding it on the engine block.*

a) *The preferred method is to turn the crankshaft with a large socket and breaker bar attached to the large bolt that is threaded into the front of the crankshaft.* **Note:** *There is no bolt in the front of the crankshaft on many six-cylinder engines, so you will have to get 3/4 inch x 16 threads-per-inch bolt that is about 1-1/2 or 2-inches long.*

b) *A remote starter switch, which may save some time, can also be used. Attach the switch leads to the S (switch) and B (battery) terminals on the starter solenoid. Once the piston is close to TDC, use a socket and breaker bar as described in the previous paragraph.*

c) *If an assistant is available to turn the ignition switch to the Start position in short bursts, you can get the piston close to TDC without a remote starter switch. Use a socket and breaker bar as described in Paragraph (a) to complete the procedure.*

4 Scribe or paint a small mark on the dis-

4.6b The timing plate on six-cylinder and big-block V8 engines is on the passenger's side of the engine

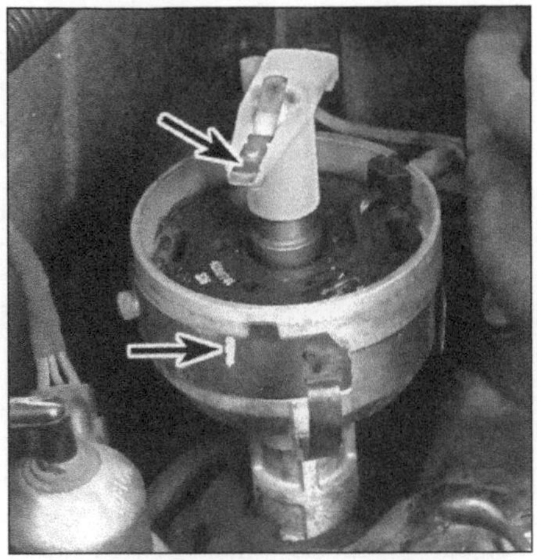

4.7 The rotor should align with the mark on the distributor base (arrows) below the number one terminal on the cap

tributor body directly below the number one spark plug wire terminal in the distributor cap **(see illustration)**.

5 Remove the distributor cap as described in Chapter 1.

6 Turn the crankshaft (see Step 3 above) until the line on the vibration damper is aligned with the zero or "TDC" mark on the timing indicator **(see illustrations)**.

7 The rotor should now be pointing directly at the mark on the distributor base **(see illustration)**. If it is 180-degrees off, the piston is at TDC on the exhaust stroke.

8 If the rotor is 180-degrees off, turn the crankshaft one complete turn (360-degrees) clockwise. The rotor should now be pointing at the mark. When the rotor is pointing at the number-one spark plug wire terminal in the distributor cap (which is indicated by the mark on the distributor body) and the timing marks are aligned, the number one piston is at TDC on the compression stroke.

9 After the number one piston has been positioned at TDC on the compression stroke, TDC for any of the remaining cylinders is located by turning the engine over in the normal direction of rotation until the rotor is pointing to the spark plug wire terminal in the distributor cap (you will have to install and remove the distributor cap as you are turning the engine over) for the piston you wish to bring to TDC.

5 Compression check

Refer to illustration 5.4

1 A compression check will tell you what mechanical condition the pistons, rings, valves and head gaskets of your engine are in. Specifically, it can tell you if the compression is down due to leakage caused by worn piston rings, defective valves and seats or a blown head gasket. **Note:** *The engine must be at normal operating temperature for this check and the battery must be fully charged.*

2 Begin by cleaning the area around the spark plugs before you remove them (compressed air works best for this). This will prevent dirt from getting into the cylinders as the compression check is being done. Remove all of the spark plugs from the engine.

5.4 A compression gauge with a threaded fitting for the spark plug hole is preferred over the type that requires hand pressure to maintain the seal - be sure to open the throttle valve as far as possible during the compression check

3 Block the throttle wide open and disconnect the primary wires from the coil.

4 With the compression gauge in the number one spark plug hole, crank the engine over at least four compression strokes and watch the gauge **(see illustration)**. The compression should build up quickly in a healthy engine. Low compression on the first stroke, followed by gradually increasing pressure on successive strokes, indicates worn piston rings. A low compression reading on the first stroke, which does not build up during successive strokes, indicates leaking valves or a blown head gasket (a cracked head could also be the cause). Record the highest gauge reading obtained.

5 Repeat the procedure for the remaining cylinders and compare the results to the Specifications.

6 Add some engine oil (about three squirts from a plunger-type oil can) to each cylinder, through the spark plug hole, and repeat the test.

7 If the compression increases after the oil is added, the piston rings are definitely worn. If the compression does not increase significantly, the leakage is occurring at the valves or head gasket. Leakage past the valves may be caused by burned valve seats and/or faces or warped, cracked or bent valves.

8 If two adjacent cylinders have equally low compression, there is a strong possibility that the head gasket between them is blown. The appearance of coolant in the combustion chambers or the oil would verify this condition. Generally, coolant will cause the oil on the dipstick to look light-colored (milky).

9 If the compression is unusually high, the combustion chambers are probably coated with carbon deposits. If that is the case, the cylinder heads should be removed and decarbonized.

10 If compression is way down or varies greatly between cylinders, it would be a good idea to have a leak-down test performed by an automotive repair shop. This test will pinpoint exactly where the leakage is occurring and how severe it is.

6 Vacuum gauge diagnostic checks

A vacuum gauge provides valuable information about what is going on in the engine at a low cost. You can check for worn rings or cylinder walls, leaking head or intake manifold gaskets, incorrect carburetor adjustments, restricted exhaust, stuck or burned valves, weak valve springs, improper ignition or valve timing and ignition problems.

Unfortunately, vacuum gauge readings are easy to misinterpret, so they should be used in conjunction with other tests to confirm the diagnosis.

Both the gauge readings and the rate of needle movement are important for accurate interpretation. Most gauges measure vacuum in inches of mercury (in-Hg). As vacuum increases (or atmospheric pressure decreases), the reading will increase. Also, for every 1,000-foot increase in elevation above sea level, the gauge readings will decrease about one inch of mercury.

Connect the vacuum gauge directly to intake manifold vacuum, not to ported (carburetor) vacuum. Be sure no hoses are left disconnected during the test or false readings will result.

Before you begin the test, allow the engine to warm up completely. Block the wheels and set the parking brake. With the transmission in Park, start the engine and allow it to run at normal idle speed. **Warning:** *Carefully inspect the fan blades for cracks or damage before starting the engine. Keep your hands and the vacuum tester clear of the fan and do not stand in front of the vehicle or in line with the fan when the engine is running.*

Read the vacuum gauge; an average, healthy engine should normally produce about 17 to 22 inches of vacuum with a fairly steady needle. Refer to the following vacuum gauge readings and what they indicate about the engines condition:

1 A low, steady reading usually indicates a leaking gasket between the intake manifold and carburetor or throttle body, a leaky vacuum hose, late ignition timing or incorrect camshaft timing. Check ignition timing with a timing light and eliminate all other possible causes, utilizing the tests provided in this Chapter before you remove the timing chain cover to check the timing marks.

2 If the reading is three to eight inches below normal and it fluctuates at that low reading, suspect an intake manifold gasket leak at an intake port.

3 If the needle has regular drops of about two to four inches at a steady rate, the valves are probably leaking. Perform a compression or leak-down test to confirm this.

4 An irregular drop or down-flick of the needle can be caused by a sticking valve or an ignition misfire. Perform a compression or leak-down test and read the spark plugs.

5 A rapid vibration of about four inches-Hg vibration at idle combined with exhaust smoke indicates worn valve guides. Perform a leak-down test to confirm this. If the rapid vibration occurs with an increase in engine speed, check for a leaking intake manifold gasket or head gasket, weak valve springs, burned valves or ignition misfire.

6 A slight fluctuation, say one inch up and down, may mean ignition problems. Check all the usual tune-up items and, if necessary, run the engine on an ignition analyzer.

7 If there is a large fluctuation, perform a compression or leak-down test to look for a weak or dead cylinder or a blown head gasket.

8 If the needle moves slowly through a wide range, check for a clogged PCV system, incorrect idle fuel mixture, carburetor/throttle body or intake manifold gasket leaks.

9 Check for a slow return after revving the engine by quickly snapping the throttle open until the engine reaches about 2,500 rpm and let it shut. Normally the reading should drop to near zero, rise above normal idle reading (about 5 in-Hg over) and then return to the previous idle reading. If the vacuum returns slowly and doesn't peak when the throttle is snapped shut, the rings may be worn. If there is a long delay, look for a restricted exhaust system (often the muffler or catalytic converter). An easy way to check this is to temporarily disconnect the exhaust ahead of the suspected part and re-test.

7 Engine rebuilding alternatives

The do-it-yourselfer is faced with a number of options when performing an engine overhaul. The decision to replace the engine block, piston/connecting rod assemblies and crankshaft depends on a number of factors, with the number one consideration being the condition of the block. Other considerations are cost, access to machine shop facilities, parts availability, time required to complete the project and the extent of prior mechanical experience on the part of the do-it-yourselfer.

Some of the rebuilding alternatives include:

Individual parts - If the inspection procedures reveal that the engine block and most engine components are in reusable condition, purchasing individual parts may be the most economical alternative. The block, crankshaft and piston/connecting rod assemblies should all be inspected carefully. Even if the block shows little wear, the cylinder bores should be surface honed.

Crankshaft kit - This rebuild package consists of a reground crankshaft and a matched set of pistons and connecting rods. The pistons will already be installed on the connecting rods. Piston rings and the necessary bearings will be included in the kit. These kits are commonly available for standard cylinder bores, as well as for engine blocks which have been bored to a standard oversize.

Short block - A short block consists of an engine block with a camshaft, timing chain and gears, crankshaft and piston/connecting rod assemblies already installed. All new bearings are incorporated and all clearances will be correct. The existing cylinder head(s), valve train components and external parts can be bolted to the short block with little or no machine shop work necessary.

Long block - A long block consists of a short block plus an oil pump, cylinder head(s) and valve train components. All components are installed with new bearings, seals and gaskets incorporated throughout. The installation of manifolds and external parts is all that is necessary.

Give careful thought to which alternative is best for you and discuss the situation with local automotive machine shops, auto parts dealers or parts store counterpeople before ordering or purchasing replacement parts.

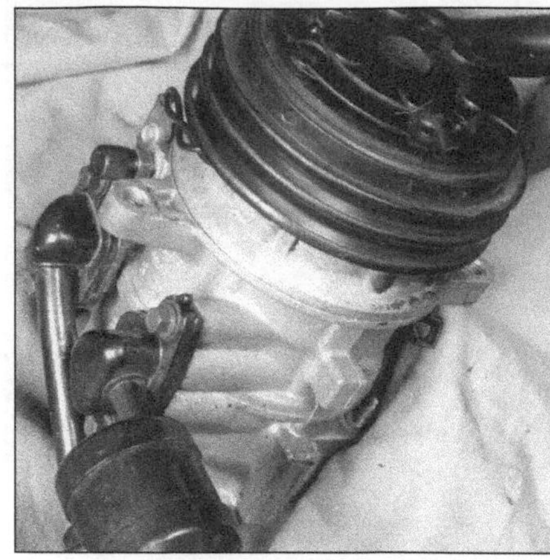

8.1 Unbolt the air conditioning compressor and set it aside without disconnecting the hoses, if possible

8 Engine - removal and installation

Refer to illustrations 8.1, 8.10, 8.16, 8.17, 8.19, 8.20, 8.25a and 8.25b

Warning: *Gasoline is extremely flammable, so take extra precautions when disconnecting any part of the fuel system. Don't smoke or allow open flames or bare light bulbs in or near the work area and don't work in a garage where a natural gas appliance (such as a clothes dryer or water heater) is installed. If you spill gasoline on your skin, rinse it off immediately. Have a fire extinguisher rated for gasoline fires handy and know how to use it! Also, the air conditioning system is under high pressure - have a dealer service department or service station discharge the system before disconnecting any of the hoses or fittings.*

Note: *Read through the following steps carefully and familiarize yourself with the procedure before beginning work.*

Removal

1 On air-conditioned models, inspect the compressor mounting to determine if the compressor can be unbolted and moved aside without disconnecting the refrigerant lines **(see illustration)**. If this is not possible, have the system discharged by a dealer or service station (see **Warning** above).

2 Remove the hood (see Chapter 11).

3 Remove the air cleaner assembly and all hoses connected to it (see Chapter 4).

4 Drain the cooling system and remove the drivebelts (see Chapter 1). **Note:** *Mark all the accessory drivebelts with markers to indicate which components they drive, i.e. alternator, air conditioning compressor, power steering pump, and also number the belts to indicate which order they go onto the pulleys. This will eliminate confusion on reassembly, which might be days later.*

5 Remove the radiator, shroud and fan

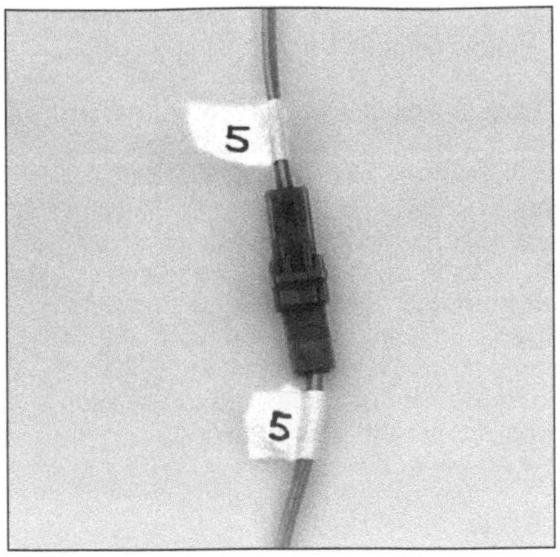

8.10 Label each wire before unplugging the connector

(see Chapter 3).

6 Detach the radiator and heater hoses from the engine.

7 Disconnect the accelerator cable (see Chapter 4) and throttle rod (automatic transmission only - see Chapter 7B) from the carburetor.

8 Remove the power steering pump and brackets (if equipped) without disconnecting the hoses and tie it out of the way (see Chapter 10).

9 Remove the alternator (see Chapter 5).

10 Label and disconnect all wires from the engine **(see illustration)**. Masking tape and/or a touch-up paint applicator work well for marking items. **Note:** *Take instant photos or sketch the locations of components and brackets to help with reassembly.*

11 Disconnect the fuel lines at the engine (see Chapter 4) and plug the lines to prevent fuel loss.

12 Label and remove all vacuum lines from the intake manifold.

13 Raise the vehicle and support it securely on jackstands.

14 Drain the engine oil (see Chapter 1).

15 Disconnect the exhaust pipe(s) from the exhaust manifold(s).

16 Support the transmission with a floor jack or use a length of chain to keep it in place when the engine is separated for removal **(see illustration)**.

17 Remove the engine-to-transmission brace, if so equipped **(see illustration)**.

18 Disconnect the wires from the starter solenoid and remove the starter (see Chapter 5).

19 Remove the flywheel or torque converter inspection cover **(see illustration)**.

20 If equipped with an automatic transmission, remove the torque converter-to-driveplate bolts **(see illustration)**. In order to turn the engine to reach all the bolts, reinstall the bolt in the front end of the crankshaft and use it to turn the driveplate to the needed positions. **Note:** *There is no bolt in the front of the crankshaft on six-cylinder engines; you will have to get 3/4-inch x 16 threads-per-inch bolt that is about 1-1/2 or 2-inches long.*

21 Remove the transmission-to-engine bolts.

22 Attach an engine hoist to the engine and take the weight off the engine mounts.

8.16 Support the transmission with a length of chain (as shown) or a floor jack to keep it in place when the engine is separated for removal

8.17 On six-cylinder and small-block V8 engines, unbolt and remove the engine-to-transmission brace (arrow) from each side of the engine

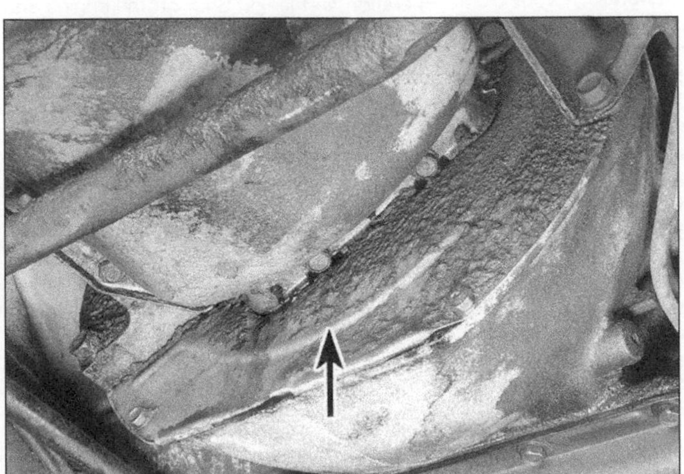

8.19 Remove the bolts and take off the torque converter inspection cover (arrows)

8.20 On automatic-transmission models, remove the four bolts that attach the driveplate to the torque converter - you'll have to rotate the engine for access to each one

8.25a Lifting a V8 engine from the engine compartment. The engine tilter shown here is very helpful for angling the engine as it's being lifted out. To attach the tilter, we used long bolts threaded into the holes at the front and rear of the cylinder heads. On V8 engines, if you have only a chain, try installing longer bolts and washers in the left front and right rear intake manifold bolt holes to hook up the chain, or try looping the chain around the exhaust manifolds (the rear of the right manifold and the front of the left one) and use bolts, nuts and washers to secure the chain to itself.

8.25b Lifting out a six-cylinder engine - if a tilter isn't available, try hooking up a chain to the right rear cylinder head bolt and looping the chain around the front runner of the intake manifold, securing the front of the chain to itself with a bolt, nut and washer

9.3a Before removing the engine accessories and their attaching brackets, note how they're attached (front view of a small-block V8 engine shown)

23 Remove the engine mount through bolts or stud nuts (see Chapter 2A or 2B).

24 Lift the engine slightly and separate the engine from the transmission. Carefully work it forward to separate it from the transmission. If you're working on a vehicle with an automatic transmission, be sure the torque converter stays in the transmission (clamp a pair of locking pliers to the housing to keep the converter from sliding out). If you're working on a vehicle with a manual transmission, the input shaft must be completely disengaged from the clutch.

25 Check to make sure everything is disconnected, then lift the engine out of the vehicle **(see illustrations)**. The engine will probably need to be tilted and/or maneuvered as it's lifted out, so have an assistant handy. **Warning:** *Do not place any part of your body under the engine when it is supported only by a hoist or other lifting device.*

26 Remove the flywheel/driveplate and mount the engine on an engine stand or set the engine on the floor and support it so it doesn't tip over. Then disconnect the engine hoist.

Installation

27 Check the engine mounts. If they're worn or damaged, replace them.

28 On manual transmission models, inspect the clutch components (see Chapter 8). On automatic models, inspect the converter seal and bushing.

29 Apply a dab of grease to the pilot bushing on manual transmission models.

30 Attach the hoist to the engine and carefully lower the engine into the engine compartment.

31 Carefully guide the engine into place. Follow the procedure outlined in Chapter 7 for transmission attachment. **Caution:** *Do not use the bolts to force the engine and transmission into alignment. It may crack or damage major components.*

32 Align the engine to the engine mounts. Install the through bolts and tighten them securely.

33 Reinstall the remaining components in the reverse order of removal.

34 Add coolant, oil, power steering and transmission fluid as needed (see Chapter 1).

35 Run the engine and check for proper operation and leaks. Shut off the engine and recheck fluid levels.

9 Engine overhaul - disassembly sequence

Refer to illustrations 9.3a, 9.3b and 9.3c

1 It is much easier to disassemble and work on the engine if it is mounted on a portable engine stand. These stands can often be rented quite cheaply from an equipment rental yard. Before the engine is mounted on a stand, the flywheel/driveplate should be removed from the engine (refer to Chapter 2A or 2B).

2 If a stand is not available, it is possible to disassemble the engine with it blocked up on a sturdy workbench or on the floor. Be extra careful not to tip or drop the engine when working without a stand.

3 If you are going to obtain a rebuilt engine, all external components must come

9.3b Small-block V8 engine - left side view

9.3c Small-block V8 engine - right side view

off first, to be transferred to the replacement engine **(see illustrations)**, just as they will if you are doing a complete engine overhaul yourself. These include:

Alternator and brackets
Emissions control components
Distributor, spark plug wires and spark plugs
Thermostat and housing cover
Water pump
Carburetor
Intake/exhaust manifolds
Oil pump assembly
Engine mounts
Clutch and flywheel/driveplate

Note: *When removing the external components from the engine, pay close attention to details that may be helpful or important during installation. Note the installed position of gaskets, seals, spacers, pins, washers, bolts and other small items.*

4 If you are obtaining a short block, which consists of the engine block, crankshaft, pistons and connecting rods all assembled, then the cylinder heads, oil pan and oil pump will have to be removed as well. See *Engine rebuilding alternatives* for additional information regarding the different possibilities to be considered.

5 If you are planning a complete overhaul, the engine must be disassembled and the internal components removed in the general following order:

Valve cover(s)
Intake and exhaust manifolds
Rocker arms and pushrods
Valve lifters
Cylinder head(s)
Oil pan
Timing chain cover
Timing chain and sprockets
Oil pump
Camshaft
Piston/connecting rod assemblies
Crankshaft and main bearings

6 Critical cooling system components such as the hoses, the drivebelts, the thermostat and the water pump MUST be replaced with new parts when an engine is overhauled. Also, we do not recommend overhauling the oil pump - always install a new one when an engine is rebuilt.

7 Before beginning the disassembly and overhaul procedures, make sure the following items are available:

Common hand tools
Small cardboard boxes or plastic bags for storing parts
Gasket scraper
Ridge reamer
Vibration damper puller
Micrometers
Telescoping gauges
Dial indicator set
Valve spring compressor
Cylinder surfacing hone
Piston ring groove cleaning tool
Electric drill motor
Tap and die set
Wire brushes
Oil gallery brushes
Cleaning solvent

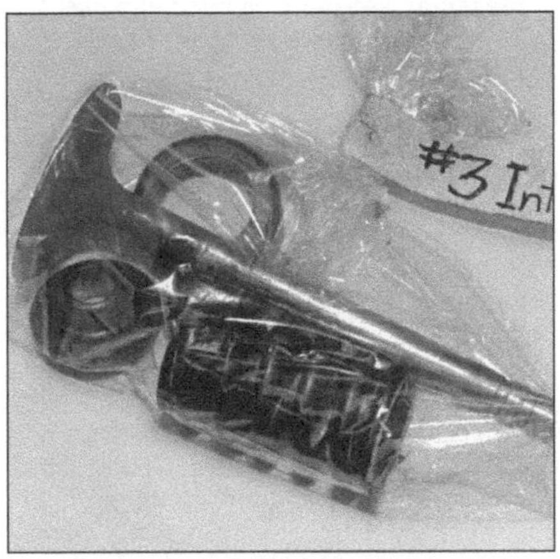

10.2 A small plastic bag, with an appropriate label, can be used to store the valve train components so they can be kept together and reinstalled in the original position

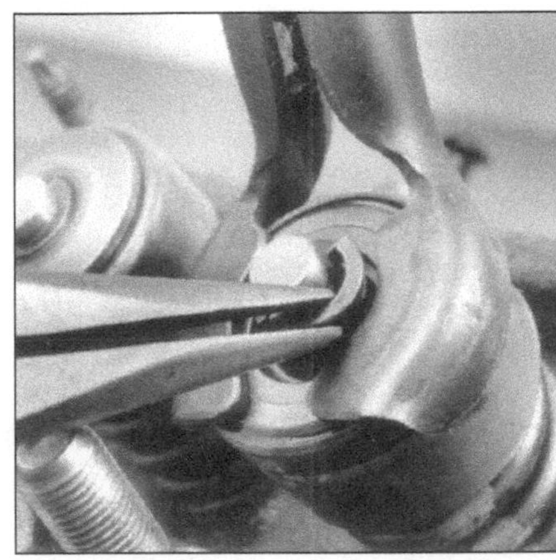

10.3a Use a valve spring compressor to compress the spring, then remove the keepers from the valve stem with a magnet or needle-nose pliers

10 Cylinder head - disassembly

Refer to illustrations 10.2, 10.3a and 10.3b

Note: *New and rebuilt cylinder heads are commonly available for most engines at dealerships and auto parts stores. Due to the fact that some specialized tools are necessary for the disassembly and inspection procedures, and replacement parts may not be readily available, it may be more practical and economical for the home mechanic to purchase replacement heads rather than taking the time to disassemble, inspect and recondition the originals.*

1 Cylinder head disassembly involves removal of the intake and exhaust valves and related components. If they are still in place, remove the rocker arms. Label the parts or store them separately so they can be reinstalled in their original locations.

2 Before the valves are removed, arrange to label and store them, along with their related components, so they can be kept separate and reinstalled in the same valve guides they are removed from **(see illustration)**.

3 Compress the springs on the first valve

10.3b If the valve won't pull through the guide, deburr the edge of the stem end and the area around the top of the keeper groove with a file or whetstone

with a spring compressor and remove the keepers **(see illustration)**. Carefully release the valve spring compressor and remove the retainer and (if used) rotators, the shield, the springs and the spring seat or shims (if used). Remove the oil seal(s) from the valve stem and the umbrella-type seal from over the guide boss (if used), then pull the valve from the head. If the valve binds in the guide (won't pull through), push it back into the head and deburr the area around the keeper groove with a fine file or whetstone **(see illustration)**.

4 Repeat the procedure for the remaining valves. Remember to keep all the parts for each valve together so they can be reinstalled in the same locations.

5 Once the valves and related components have been removed and stored in an organized manner, the head should be thoroughly cleaned and inspected. If a complete engine overhaul is being done, finish the engine disassembly procedures before beginning the cylinder head cleaning and inspection process.

11 Cylinder head - cleaning and inspection

1 Thorough cleaning of the cylinder heads and related valve train components, followed by a detailed inspection, will enable you to decide how much valve service work must be done during the engine overhaul.

Cleaning

2 Scrape away all traces of old gasket material and sealing compound from the head gasket, intake manifold and exhaust manifold sealing surfaces. Be very careful not to gouge the cylinder head. Special gasket removal solvents, which soften gaskets and make removal much easier, are available at auto parts stores.

3 Remove any built up scale from the coolant passages.

4 Run a stiff wire brush through the various holes to remove any deposits that may have formed in them.

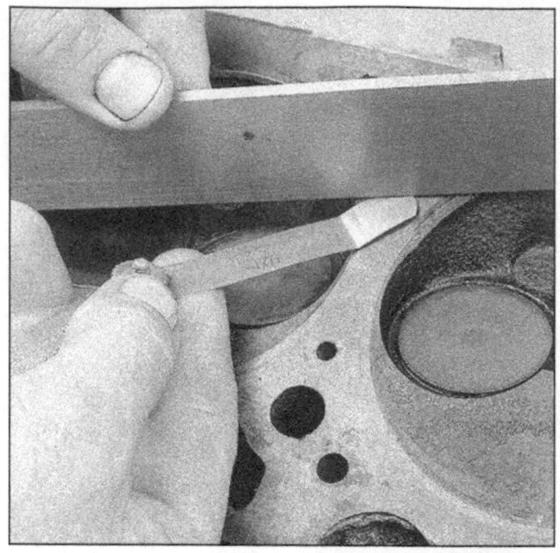

11.12 Check the cylinder head gasket surface for warpage by trying to slip a feeler gauge under the straightedge (see this Chapter's Specifications for the maximum warpage allowed and use a feeler gauge of that thickness)

5 Run an appropriate size tap into each of the threaded holes to remove any corrosion and thread sealant that may be present. If compressed air is available, use it to clear the holes of debris produced by this operation.

6 Check the condition of the spark plug threads.

7 Clean the cylinder head with solvent and dry it thoroughly. Compressed air will speed the drying process and ensure that all holes and recessed areas are clean. **Note:** *Decarbonizing chemicals are available and may prove very useful when cleaning cylinder heads and valve train components. They are very caustic and should be used with caution. Be sure to follow the instructions on the container.*

8 On six-cylinder models, clean the rocker arms, shafts and pushrods with solvent and dry them thoroughly (don't mix them up during the cleaning process). Compressed air will speed the drying process and can be used to clean out the oil passages.

9 Clean all the valve springs, shields, keepers and retainers (or rotators) with solvent and dry them thoroughly. Do the components from one valve at a time to avoid mixing up the parts.

10 Scrape off any heavy deposits that may have formed on the valves, then use a wire brush mounted in a drill motor to remove deposits from the valve heads and stems. Again, make sure the valves do not get mixed up.

Inspection

Cylinder head

Refer to illustrations 11.12 and 11.14

11 Inspect the head very carefully for cracks, evidence of coolant leakage and other damage. If cracks are found, a new cylinder head should be obtained.

11.14 A dial indicator can be used to determine the valve stem-to-guide clearance - move the valve stem as indicated by the arrows

12 Using a straightedge and feeler gauge, check the head gasket mating surface for warpage **(see illustration)**. If the warpage exceeds the limit specified in this Chapter, it can be resurfaced at an automotive machine shop.

13 Examine the valve seats in each of the combustion chambers. If they are pitted, cracked or burned, the head will require valve service with special equipment that is beyond the scope of most home mechanics, especially from an equipment stand point.

14 Check the valve stem-to-guide clearance by measuring the lateral movement of the valve stem with a dial indicator attached securely to the head **(see illustration)**. The valve must be in the guide and approximately 1/16-inch off the seat. The total valve stem movement indicated by the gauge needle must be divided by two to obtain the actual

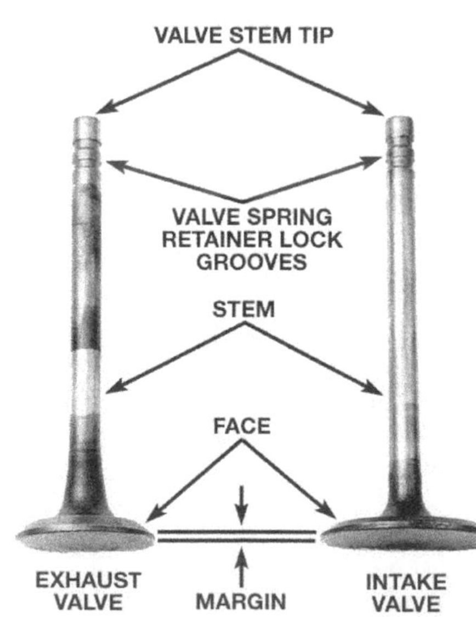

11.15 Check for valve wear at the points shown here

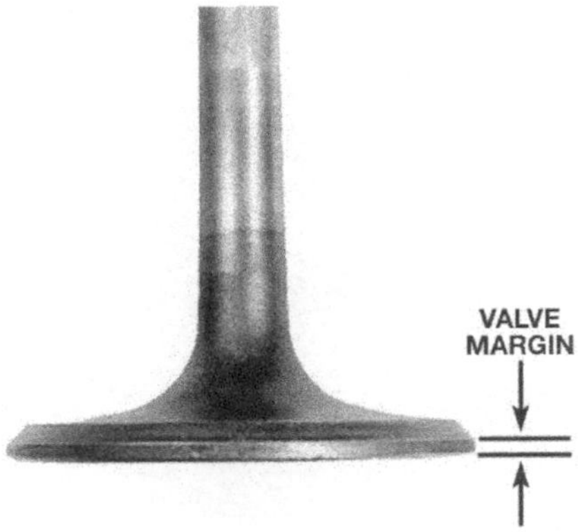

11.16 The margin width on each valve must be as specified (if no margin exists, the valve cannot be reused)

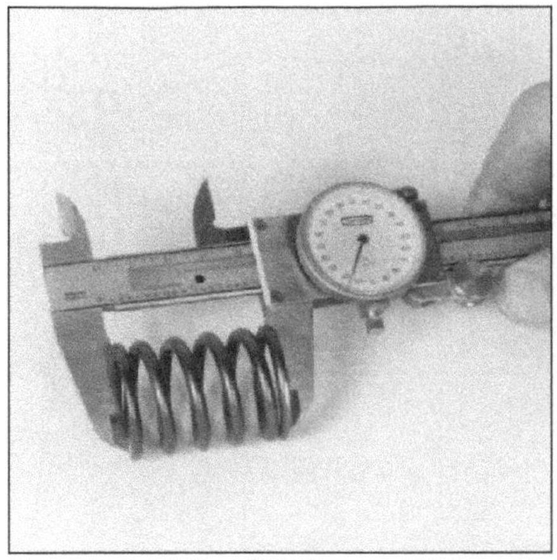

11.17 Measure the free length of each valve spring with a dial or vernier caliper

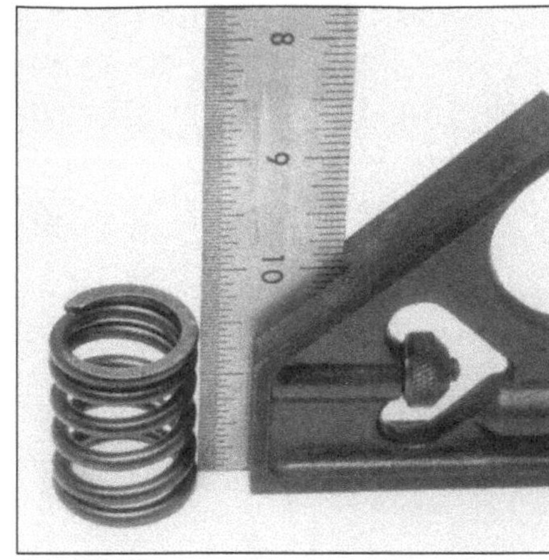

11.18 Check each valve spring for squareness - if it is bent, it should be replaced

clearance. After this is done, if there is still some doubt regarding the condition of the valve guides they should be checked by an automotive machine shop (the cost should be minimal).

Valves

Refer to illustrations 11.15 and 11.16

15 Carefully inspect each valve face for uneven wear **(see illustration)**, deformation, cracks, pits and burned spots. Check the valve stem for scuffing and galling and the neck for cracks. Rotate the valve and check for any obvious indication that it is bent. Look for pits and excessive wear on the end of the stem. The presence of any of these conditions indicates the need for valve service by an automotive machine shop.

16 Measure the margin width on each valve. Any valve with a margin that is narrower than 0.047 (3/64) inch should be discarded and replaced with a new one **(see illustration)**.

Valve components

Refer to illustrations 11.17, 11.18 and 11.19

17 Check each valve spring for wear (on the ends) and pits. Measure the free length and compare it to the Specifications in this Chapter **(see illustration)**. Any springs that are shorter than specified have sagged and should not be reused. The tension of all springs should be checked with a special fixture before deciding that they are suitable for use in a rebuilt engine (take the springs to an automotive machine shop for this check).

18 Stand each spring on a flat surface and check it for squareness **(see illustration)**. If any of the springs are distorted or sagged, replace all of them with new parts.

19 Check the spring retainers (or rotators) and keepers for obvious wear **(see illustration)** and cracks. Any questionable parts should be replaced with new ones, as extensive damage will occur if they fail during engine operation.

Rocker arm components

20 Check the rocker arm faces (the areas that contact the pushrod ends and valve stems) for pits, wear, galling, score marks and rough spots. Check the rocker arm pivot contact areas and shafts. Look for cracks in each rocker arm.

21 Inspect the pushrod ends for scuffing and excessive wear. Roll each pushrod on a flat surface, such as plate glass, to determine if it is bent.

22 Check the rocker arm bolt holes in the cylinder heads for damaged threads.

23 Any damaged or excessively worn parts must be replaced with new ones.

24 If the inspection process indicates that the valve components are in generally poor condition and worn beyond the limits specified in this Chapter, which is usually the case in an engine that is being overhauled, reassemble the valves in the cylinder head and refer to Section 12 for valve servicing recommendations.

12 Valves - servicing

1 Because of the complex nature or the job and the special tools and equipment needed, servicing of the valves, the valve seats and the valve guides (commonly known as a "valve job") is best left to a professional.

2 The home mechanic can remove and disassemble the heads, do the initial cleaning and inspection, then reassemble and deliver the heads to a dealer service department or an automotive machine shop for the actual valve servicing.

3 The dealer service department, or automotive machine shop, will remove the valves and springs, recondition or replace the valves and valve seats, recondition the valve guides, check and replace the valve springs, spring retainers or rotators and keepers (as necessary), replace the valve seals with new ones, reassemble the valve components and make sure the installed spring height is correct. The cylinder head gasket surface will also be resurfaced if it is warped.

4 After the valve job has been performed by a professional, the head will be in like-new condition. When the head is returned, be sure to clean it again before installation on the engine to remove any metal particles and abrasive grit that may still be present from the valve service or head resurfacing operations. Use compressed air, if available, to blow out all the oil holes and passages.

11.19 The exhaust valve rotators can be checked by turning the inner and outer sections in opposite directions to feel for smooth movement and excessive play

13 Cylinder head - reassembly

Refer to illustrations 13.5, 13.6, and 13.8

1 Regardless of whether or not the heads were sent to an automotive machine shop for valve servicing, make sure they are clean before beginning reassembly.

2 If the heads were sent out for valve servicing, the valves and related components will already be in place. Begin the reassembly procedure with Step 8.

3 Beginning at one end of the head, lubricate and install the first valve. Apply moly-base grease or clean engine oil to the valve stem.

4 Different types of valve stem oil seals

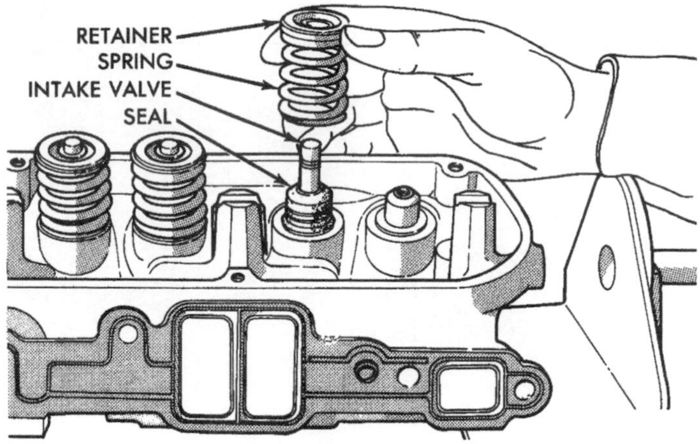

13.5 The intake valves have retainers; the exhaust valves use rotators

13.6 Apply a small dab of grease to each keeper as shown here before installation - it'll hold them in place on the valve stem as the spring is released

13.8 Valve spring installed height is the distance from the spring seat on the head to the top of the spring

14.1 A ridge reamer is required to remove the ridge from the top of each cylinder - do this before removing the pistons!

14.3 Check the connecting rod side clearance (endplay) with a feeler gauge as shown here

are used on the intake and exhaust. On some applications, an umbrella type seal which extends down over the valve guide boss is used over the valve stem.

5 Drop the spring seat or shim(s), if used, over the valve guide and set the valve spring and retainer (or rotator) in place **(see illustration)**.

6 Compress the springs with a valve spring compressor. Position the keepers in the upper groove, then slowly release the compressor and make sure the keepers seat properly. Apply a small dab of grease to each keeper to hold it in place if necessary **(see illustration)**.

7 Repeat the procedure for the remaining valves. Be sure to return the components to their original locations - do not mix them up!

8 Check the installed valve spring height with a ruler graduated in 1/32-inch increments **(see illustration)** or a dial caliper. If the heads were sent out for service work, the installed height should be correct (but don't automatically assume that it is). The measurement is taken from the top of each spring seat or shim(s) to the bottom of the retainer/rotator. If the height is greater than listed in this Chapter's specifications, shims can be added under the springs to correct it. **Caution:** *Do not, under any circumstances, shim the springs to the point where the installed height is less than specified. A condition called "coil bind" will be caused when the spring is compressed.*

14 Piston/connecting rod assembly - removal

Refer to illustrations 14.1, 14.3, 14.4 and 14.5

Note: *Prior to removing the piston/connecting rod assemblies, remove the cylinder head(s), the oil pan and the oil pump, if it is mounted in the crankcase, and the oil pump pickup screen assembly by referring to the appropriate Sections in Chapter 2, Part A or Part B.*

1 Completely remove the ridge at the top of each cylinder with a ridge reaming tool **(see illustration)**. Follow the manufacturer's instructions provided with the tool. Failure to remove the ridge before attempting to remove the piston/connecting rod assemblies will result in piston breakage.

2 After the cylinder ridges have been removed, turn the engine upside-down so the crankshaft is facing up.

3 Before the connecting rods are removed, check the endplay (side clearance) with feeler gauges. Slide them between the first connecting rod and the crankshaft throw until the play is removed **(see illustration)**. The endplay is equal to the thickness of the feeler gauge(s). If the endplay exceeds the service limit listed in this Chapter's Specifications, new connecting rods will be required. If new rods (or a new crankshaft) are installed, the endplay may fall under the specified minimum. If it does, the rods will have to be machined to restore it - consult an automotive machine shop for advice if necessary. Repeat the procedure for the remaining connecting rods.

4 Check the connecting rods and caps for identification marks **(see illustration)**. If they are not plainly marked, use a small center-punch to make the appropriate number of indentations on each rod and cap.

5 Loosen each of the connecting rod cap nuts 1/2-turn at a time until they can be removed by hand. Remove the number one connecting rod cap and bearing insert. Do not drop the bearing insert out of the cap.

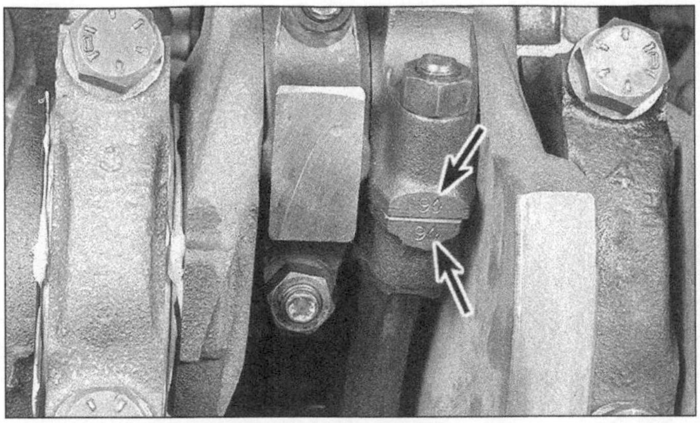

14.4 The connecting rods should be marked with numbers at the parting line of each rod and cap - the number corresponds to the engine cylinder number - on reassembly, be sure the numbers are on the same side, facing each other, as shown, and the rod/piston goes back into the correct cylinder

14.5 To prevent damage to the crankshaft journals and cylinder walls, slip sections of hose over the rod bolts before removing the piston/rod assemblies

Slip a short length of plastic or rubber hose over each connecting rod cap bolt to protect the crankshaft journal and cylinder wall when the piston is removed **(see illustration)**. Push the connecting rod/piston assembly out through the top of the engine. Use a wooden hammer handle to push on the upper bearing insert in the connecting rod. If resistance is felt, double-check to make sure that all of the ridge was removed from the cylinder.

6 Repeat the procedure for the remaining cylinders. After removal, reassemble the connecting rod caps and bearing inserts in their respective connecting rods and install the cap nuts finger tight. Leaving the old bearing inserts in place until reassembly will help prevent the connecting rod bearing surfaces from being accidentally nicked or gouged.

15 Crankshaft - removal

Refer to illustrations 15.2, 15.3, 15.4a and 15.4b

Note: *The crankshaft can be removed only after the engine has been removed from the vehicle. It is assumed that the flywheel or driveplate, vibration damper or crankshaft pulley, timing chain, oil pan, oil pump and piston/connecting rod assemblies have already been removed.*

1 Before the crankshaft is removed, check the endplay. Mount a dial indicator with the stem in line with the crankshaft and touching one of the crank throws.

2 Push the crankshaft all the way to the rear and zero the dial indicator. Next, pry the crankshaft to the front as far as possible and check the reading on the dial indicator **(see illustration)**. The distance that it moves is the endplay. If it is greater than the maximum listed in this Chapter's Specifications, check the crankshaft thrust surfaces for wear. If no wear is evident, new main bearings should correct the endplay.

3 If a dial indicator is not available, feeler gauges can be used. Gently pry or push the crankshaft all the way to the front of the engine. Slip feeler gauges between the crankshaft and the front face of the thrust main bearing to determine the clearance **(see illustration)**.

15.2 Checking crankshaft endplay with a dial indicator

4 Check the main bearing caps to see if they are marked to indicate their locations. They should be numbered consecutively from the front of the engine to the rear **(see illustration)**. If they aren't, mark them with

15.3 Checking crankshaft endplay with a feeler gauge

15.4a The main bearing caps are marked to indicate their locations (arrows) - they should be numbered consecutively from the front of the engine to the rear

15.4b If the main cap marks aren't clear, mark the caps with number stamping dies or a center-punch

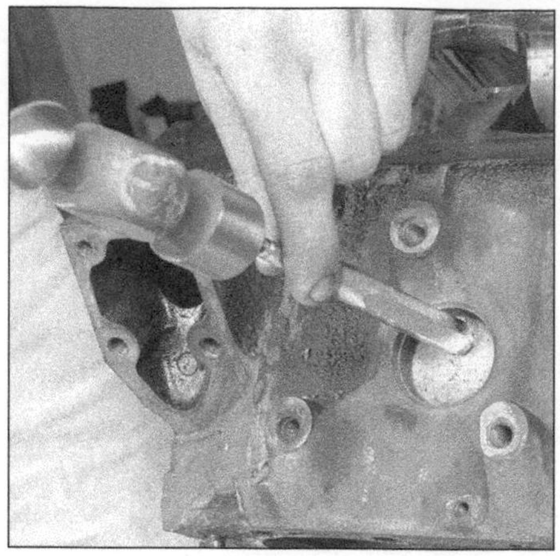

16.1a A hammer and a large punch can be used to knock the core plugs sideways in their bores

16.1b Pull the core plugs from the block with pliers

16.4 Remove all the oil galley plugs/bolts (arrows) to more thoroughly clean all debris from the oiling system (typical galley plugs shown)

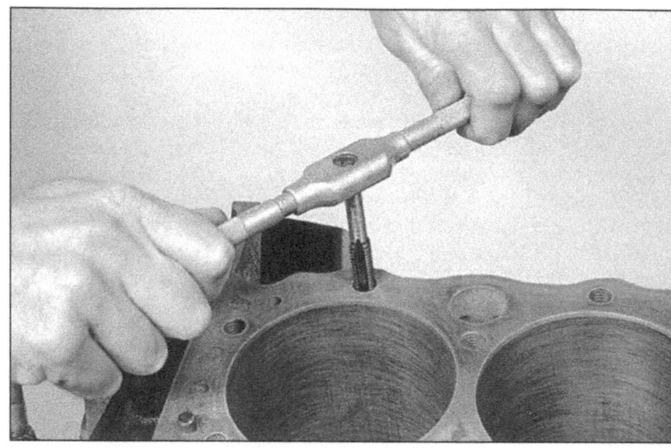

16.8 All bolt holes in the block - particularly the main bearing cap and head bolt holes - should be cleaned and restored with a tap (be sure to remove debris from the holes after this is done)

number stamping dies or a center-punch **(see illustration)**.

5 Loosen each of the main bearing cap bolts 1/4-turn at a time each, until they can be removed by hand.

6 Gently tap the caps with a soft-face hammer, then separate them from the engine block. If necessary, use the bolts as levers to remove the caps. Try not to drop the bearing inserts if they come out with the caps.

7 Carefully lift the crankshaft out of the engine. It is a good idea to have an assistant available, since the crankshaft is quite heavy. With the bearing inserts in place in the engine block and main bearing caps, return the caps to their respective locations on the engine block and tighten the bolts finger tight.

16 Engine block - cleaning

Refer to illustrations 16.1a, 16.1b, 16.4, 16.8 and 16.10

1 Using the wide end of a punch **(see illustration)** tap in on the outer edge of the core plug to turn the plug sideways in the bore. Then, using a pair of pliers, pull the core plug from the engine block **(see illustration)**. Don't worry about the condition of the old core plugs as they are being removed because they will be replaced on reassembly with new plugs.

2 Using a gasket scraper, remove all traces of gasket material from the engine block.

3 Remove the main bearing caps or cap assembly and separate the bearing inserts from the caps and the engine block. Tag the bearings, indicating which cylinder they were removed from and whether they were in the cap or the block, then set them aside.

4 Remove all of the threaded oil gallery plugs from the block **(see illustration)**. The plugs are usually very tight - they may have to be drilled out and the holes retapped. If drilled out or damaged during removal, use new plugs when the engine is reassembled.

5 If the engine is extremely dirty it should be taken to an automotive machine shop to be steam cleaned or hot tanked.

6 After the block is returned, clean all oil holes and oil galleries one more time. Brushes specifically designed for this purpose are available at most auto parts stores. Flush the passages with warm water until the water runs clear, dry the block thoroughly and wipe all machined surfaces with a light, rust preventive oil. If you have access to compressed air, use it to speed the drying process and to blow out all the oil holes and galleries. **Warning:** *Wear eye protection when using compressed air!*

7 If the block isn't extremely dirty or sludged up, you can do an adequate cleaning job with hot soapy water and a stiff brush. Take plenty of time and do a thorough job. Regardless of the cleaning method used, be sure to clean all oil holes and galleries very thoroughly, dry the block completely and coat all machined surfaces with light oil.

8 The threaded holes in the block must be clean to ensure accurate torque readings during reassembly. Run the proper size tap into each of the holes to remove rust, corrosion, thread sealant or sludge and restore damaged threads **(see illustration)**. If possible, use compressed air to clear the holes of debris produced by this operation. Now is a good time to clean the threads on the head bolts and the main bearing cap bolts as well.

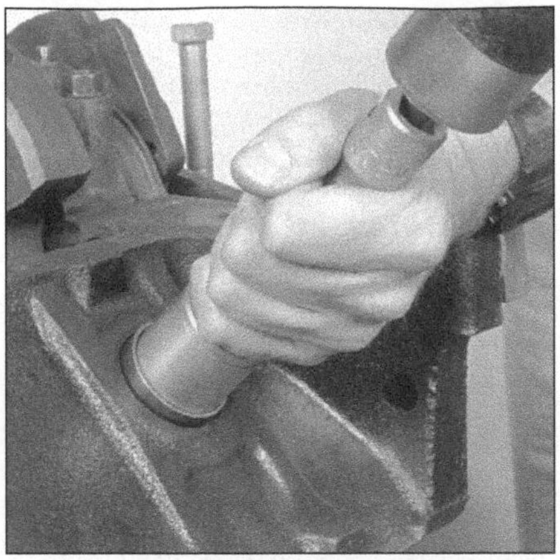

16.10 A large socket on an extension can be used to drive the new core plugs into the bores

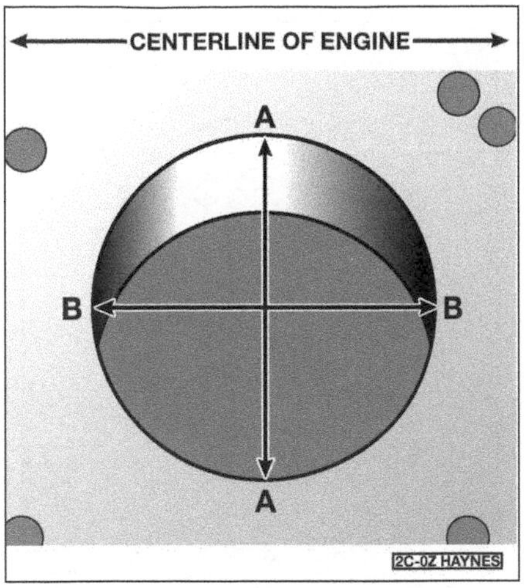

17.4a Measure the diameter of each cylinder just under the wear ridge (A), at the center (B) and at the bottom

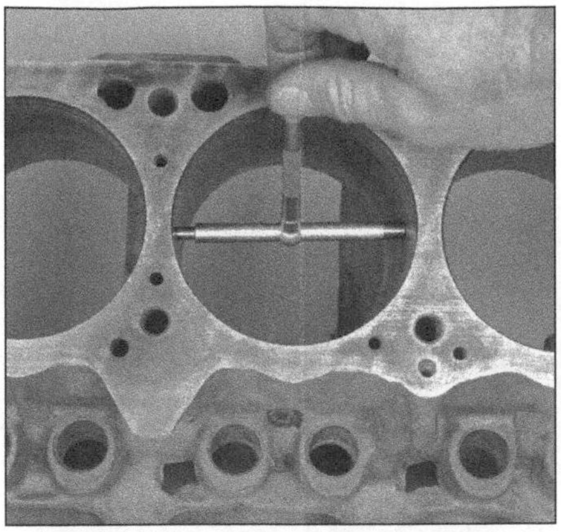

17.4b The ability to "feel" when the telescoping gauge is at the correct point will be developed over time, so work slowly and repeat the check until you're satisfied the bore measurement is accurate

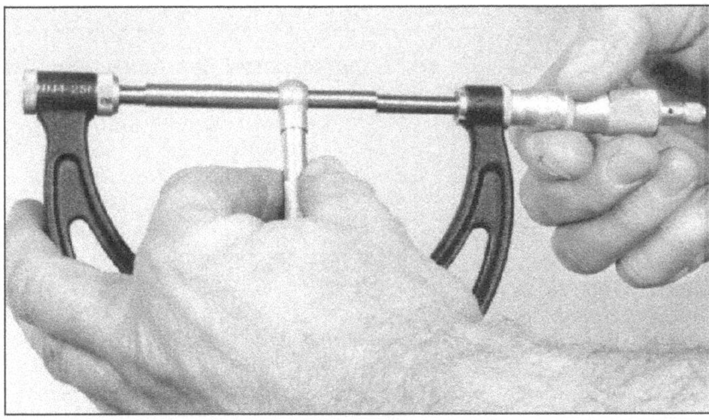

17.4c The gauge is then measured with a micrometer to determine the bore size

9 Reinstall the main bearing caps and tighten the bolts finger tight.

10 After coating the sealing surfaces of the new core plugs with Permatex no. 2 sealant, install them in the engine block **(see illustration)**. Make sure they're driven in straight or leakage could result. Special tools are available for this purpose, but a large socket, with an outside diameter that will just slip into the core plug, a 1/2-inch drive extension and a hammer will work just as well.

11 Apply non-hardening sealant (such as Permatex no. 2 or Teflon tape) to the new oil gallery plugs and thread them into the holes in the block. Make sure they're tightened securely.

12 If the engine isn't going to be reassembled right away, cover it with a large plastic trash bag to keep it clean.

17 Engine block - inspection

Refer to illustrations 17.4a, 17.4b and 17.4c

1 Before the block is inspected, it should be cleaned (see Section 14).

2 Visually check the block for cracks, rust and corrosion. Look for stripped threads in the threaded holes. It's also a good idea to have the block checked for hidden cracks by an automotive machine shop that has the special equipment to do this type of work. If defects are found, have the block repaired, if possible, or replaced.

3 Check the cylinder bores for scuffing and scoring.

4 Measure the diameter of each cylinder at the top (just under the ridge area), center and bottom of the cylinder bore, parallel to the crankshaft axis **(see illustrations)**.

5 Next, measure each cylinder's diameter at the same three locations across the crankshaft axis. Compare the results to the Specifications.

6 If the required precision measuring tools aren't available, the piston-to-cylinder clearances can be obtained, though not quite as accurately, using feeler gauge stock. Feeler gauge stock comes in 12-inch lengths and various thicknesses and is generally available at auto parts stores.

7 To check the clearance, select a feeler gauge and slip it into the cylinder along with the matching piston. The piston must be positioned exactly as it normally would be. The feeler gauge must be between the piston and cylinder on one of the thrust faces (90-degrees to the piston pin bore).

8 The piston should slip through the cylinder (with the feeler gauge in place) with moderate pressure.

9 If it falls through or slides through easily, the clearance is excessive and a new piston will be required. If the piston binds at the lower end of the cylinder and is loose toward the top, the cylinder is tapered. If tight spots are encountered as the piston/feeler gauge is rotated in the cylinder, the cylinder is out-of-round.

10 Repeat the procedure for the remaining pistons and cylinders.

11 If the cylinder walls are badly scuffed or scored, or if they're out-of-round or tapered beyond the limits given in the Specifications, have the engine block rebored and honed at an automotive machine shop. If a rebore is done, oversize pistons and rings will be required.

12 Using a precision straightedge and a feeler gauge, check the block deck (the surface that mates with the cylinder head) the same way the cylinder head was checked for distortion (see Section 11). If it's distorted beyond the specified limit, it can be resurfaced by an automotive machine shop.

13 If the cylinders are in reasonably good condition and not worn to the outside of the limits, and if the piston-to-cylinder clearances can be maintained properly, then they don't have to be rebored. Honing is all that's necessary (see Section 18).

18 Cylinder honing

Refer to illustrations 18.3a and 18.3b

1 Prior to engine reassembly, the cylinder bores must be honed so the new piston rings will seat correctly and provide the best possible combustion chamber seal. **Note:** *If you do not have the tools or do not want to tackle the honing operation, most automotive machine shops will do it for a reasonable fee.*

2 Before honing the cylinders, install the main bearing caps and tighten the bolts to the specified torque.

3 Two types of cylinder hones are com-

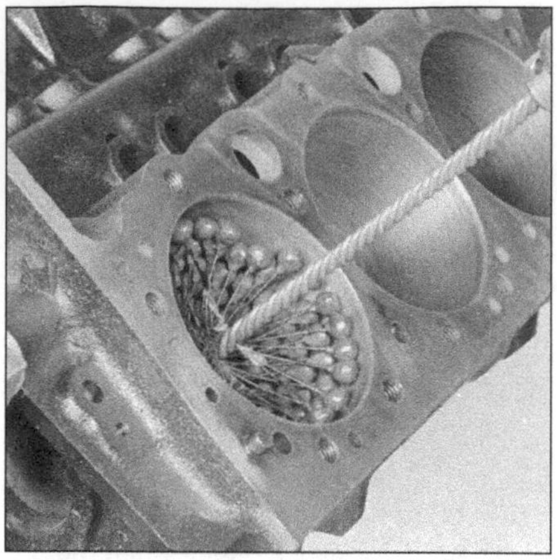

18.3a A "bottle brush" hone will produce better results if you've never honed cylinders before

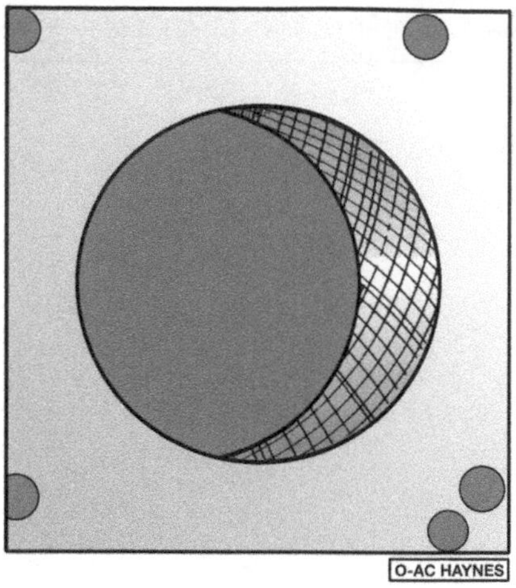

18.3b The cylinder hone should leave a smooth, crosshatch pattern with the lines intersecting at approximately a 60-degree angle

monly available - the flex hone or "bottle brush" type and the more traditional surfacing hone with spring-loaded stones. Both will do the job, but for the less experienced mechanic the "bottle brush" hone will probably be easier to use. You will also need plenty of light oil or honing oil, some rags and an electric drill motor. Proceed as follows:

a) *Mount the hone in the drill motor, compress the stones and slip it into the first cylinder* **(see illustration)**.
b) *Lubricate the cylinder with plenty of oil, turn on the drill and move the hone up-and-down in the cylinder at a pace which will produce a fine crosshatch pattern on the cylinder walls. Ideally, the crosshatch lines should intersect at approximately a 60-degree angle* **(see illustrations)**. *Be sure to use plenty of lubricant and do not take off any more material than is absolutely necessary to produce the desired finish.* **Note:** *Piston ring manufacturers may specify a smaller crosshatch angle than the traditional 60-degrees, typically 45-degrees, so read and follow any instructions printed on the piston ring packages.*
c) *Do not withdraw the hone from the cylinder while it is running. Instead, shut off the drill and continue moving the hone up-and-down in the cylinder until it comes to a complete stop, then compress the stones and withdraw the hone. If you are using a - "bottle brush" type hone, stop the drill motor, then turn the chuck in the normal direction of rotation while withdrawing the hone from the cylinder.*
d) *Wipe the oil out of the cylinder and repeat the procedure for the remaining cylinders.*

4 After the honing job is complete, chamfer the top edges of the cylinder bores with a small file so the rings will not catch when the pistons are installed. Be very careful not to nick the cylinder walls with the end of the file.

5 The entire engine block must be washed again very thoroughly with warm, soapy water to remove all traces of the abrasive grit produced during the honing operation. **Note:** *The bores can be considered clean when a white cloth - dampened with clean engine oil - used to wipe down the bores does not pick up any more honing residue, which will show up as gray areas on the cloth. Be sure to run a brush through all oil holes and galleries and flush them with running water.*

6 After rinsing, dry the block and apply a coat of light rust preventive oil to all machined surfaces. Wrap the block in a plastic trash bag to keep it clean and set it aside until reassembly.

19 Piston/connecting rod assembly - inspection

Refer to illustrations 19.4a, 19.4b and 19.10

1 Before the inspection process can be carried out, the piston/connecting rod assemblies must be cleaned and the original piston rings removed from the pistons. **Note:** *Always use new piston rings when the engine is reassembled.*

2 Using a piston ring installation tool, carefully remove the rings from the pistons. Be careful not to nick or gouge the pistons in the process.

3 Scrape all traces of carbon from the top of the piston. A hand-held wire brush or a piece of fine emery cloth can be used once the majority of the deposits have been scraped away. Do not, under any circumstances, use a wire brush mounted in a drill motor to remove deposits from the pistons. The piston material is soft and may be eroded away by the wire brush.

4 Use a piston ring groove cleaning tool to remove carbon deposits from the ring grooves. If a tool isn't available, a piece broken off the old ring will do the job. Be very careful to remove only the carbon deposits - don't remove any metal and do not nick or scratch the sides of the ring grooves **(see illustrations)**.

5 Once the deposits have been removed, clean the piston/rod assemblies with solvent and dry them with compressed air (if available). **Warning:** *Wear eye protection when using compressed air!* Make sure the oil return

19.4a The piston ring grooves can be cleaned with a special tool, as shown here . . .

19.4b . . . or a section of a broken ring

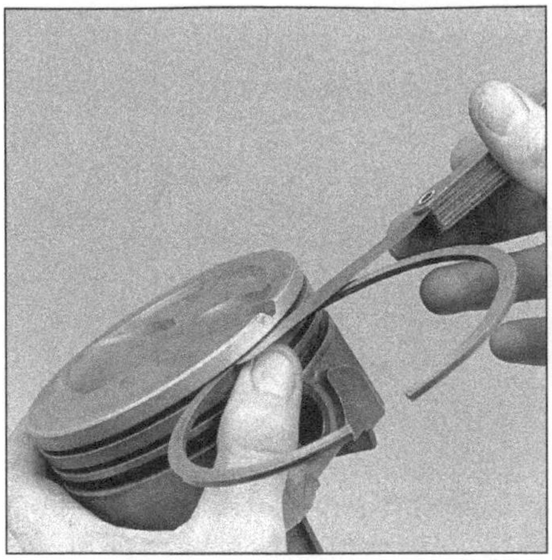

19.10 Check the ring side clearance with a feeler gauge at several points around the groove

20.1 The oil holes should be chamfered so sharp edges don't gouge or scratch the new bearings

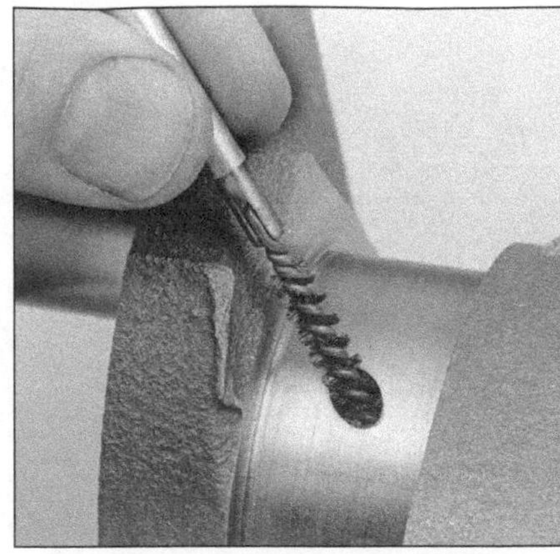

20.2 Use a wire or stiff plastic bristle brush to clean the oil passages in the crankshaft - flush the passages out with solvent

holes in the back sides of the ring grooves and the oil hole in the lower end of each rod are clear.

6 If the pistons and cylinder walls aren't damaged or worn excessively, and if the engine block is not rebored, new pistons won't be necessary. Normal piston wear appears as even vertical wear on the piston thrust surfaces and slight looseness of the top ring in its groove. New piston rings, however, should always be used when an engine is rebuilt.

7 Carefully inspect each piston for cracks around the skirt, at the pin bosses and at the ring lands.

8 Look for scoring and scuffing on the thrust faces of the skirt, holes in the piston crown and burned areas at the edge of the crown. If the skirt is scored or scuffed, the engine may have been suffering from overheating and/or abnormal combustion, which caused excessively high operating temperatures. The cooling and lubrication systems should be checked thoroughly. A hole in the piston crown is an indication that abnormal combustion (preignition) was occurring. Burned areas at the edge of the piston crown are usually evidence of spark knock (detonation). If any of the above problems exist, the causes must be corrected or the damage will occur again. The causes may include intake air leaks, incorrect fuel/air mixture, incorrect ignition timing and EGR system malfunctions.

9 Corrosion of the piston, in the form of small pits, indicates that coolant is leaking into the combustion chamber and/or the crankcase. Again, the cause must be corrected or the problem may persist in the rebuilt engine.

10 Measure the piston ring side clearance by laying a new piston ring in each ring groove and slipping a feeler gauge in beside it **(see illustration)**. Check the clearance at three or four locations around each groove. Be sure to use the correct ring for each groove - they are different. If the side clearance is greater than specified, new pistons will have to be used.

11 Check the piston-to-bore clearance by measuring the bore (see Section 17) and the piston diameter. Make sure the pistons and bores are correctly matched. Measure the piston across the skirt, at a 90-degree angle to the piston pin, the specified distance down from the top of the piston or the lower edge of the oil ring groove. Subtract the piston diameter from the bore diameter to obtain the clearance. If it's greater than specified, the block will have to be rebored and new pistons and rings installed.

12 Check the piston-to-rod clearance by twisting the piston and rod in opposite directions. Any noticeable play indicates excessive wear, which must be corrected. The piston/connecting rod assemblies should be taken to an automotive machine shop to have the pistons and rods re-sized and new pins installed.

13 If the pistons must be removed from the connecting rods for any reason, they should be taken to an automotive machine shop. While they are there have the connecting rods checked for bend and twist, since automotive machine shops have special equipment for this purpose. **Note:** *Unless new pistons and/or connecting rods must be installed, do not disassemble the pistons and connecting rods.*

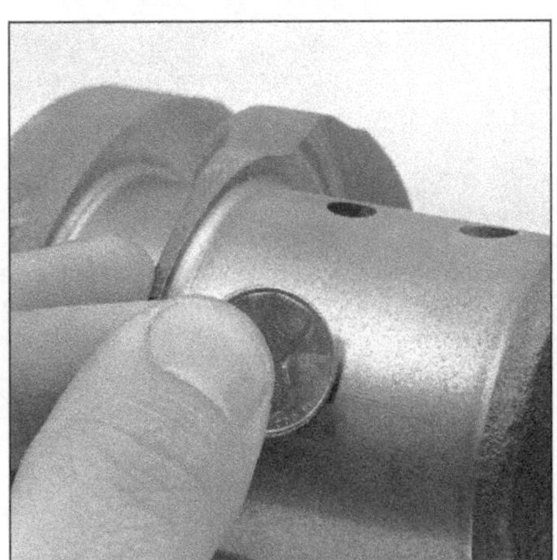

20.4 Rubbing a penny lengthwise on each journal will reveal its condition - if copper rubs off and is embedded in the crankshaft, the journals should be reground

14 Check the connecting rods for cracks and other damage. Temporarily remove the rod caps, lift out the old bearing inserts, wipe the rod and cap bearing surfaces clean and inspect them for nicks, gouges and scratches. After checking the rods, replace the old bearings, slip the caps into place and tighten the nuts finger tight. **Note:** *If the engine is being rebuilt because of a connecting rod knock, be sure to check with an automotive machine shop on the possibility of re-sizing either the small or large end of the rod. If this isn't possible install new rods.*

20 Crankshaft - inspection

Refer to illustrations 20.1, 20.2, 20.4, 20.6a, 20.6b and 20.8

1 Remove all burrs from the crankshaft oil holes with a stone, file or scraper **(see illustration)**.

2 Clean the crankshaft with solvent and dry it with compressed air (if available). Be sure to clean the oil holes with a stiff brush and flush them with solvent **(see illustration)**.

3 Check the main and connecting rod bearing journals for uneven wear, scoring, pits and cracks.

4 Rub a penny across each journal several times **(see illustration)**. If a journal picks up copper from the penny, it's too rough and must be reground.

5 Check the rest of the crankshaft for cracks and other damage. It should be magnafluxed to reveal hidden cracks - an automotive machine shop will handle the procedure.

6 Using a micrometer, measure the diameter of the main and connecting rod journals and compare the results to the Specifications

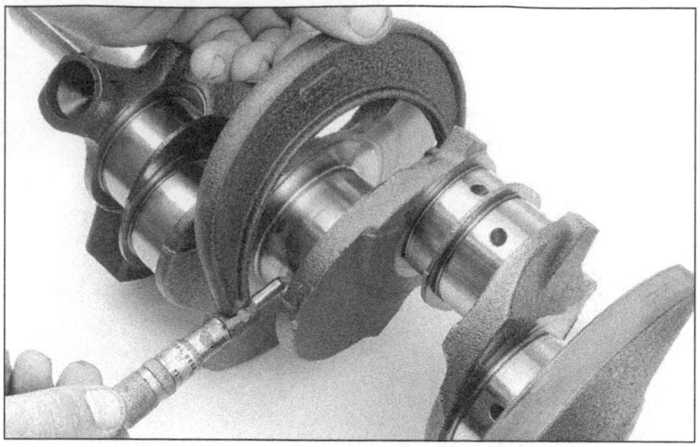

20.6a Measure the diameter of each crankshaft journal at several points to detect taper and out-of-round conditions

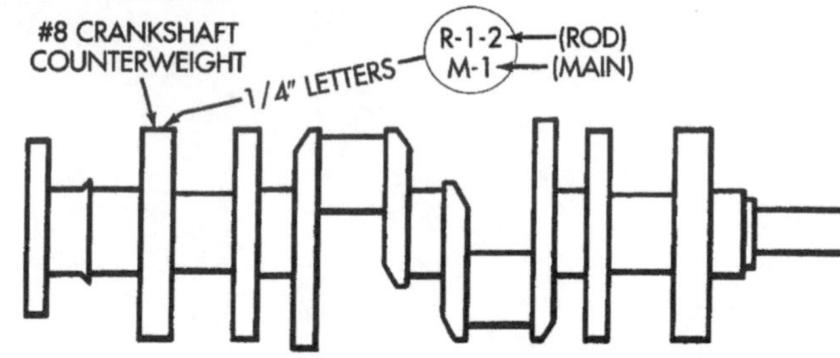

20.6b The V8 engine crankshaft may have a letter/number stamped into the counterweight of the number eight journal - this marking will reveal the location (journal number) of the undersize rod (R) or (M) main journal bearing(s)

20.8 If the seals have worn grooves in the crankshaft journals, or if the seal contact surfaces are nicked or scratched, the new seals will leak

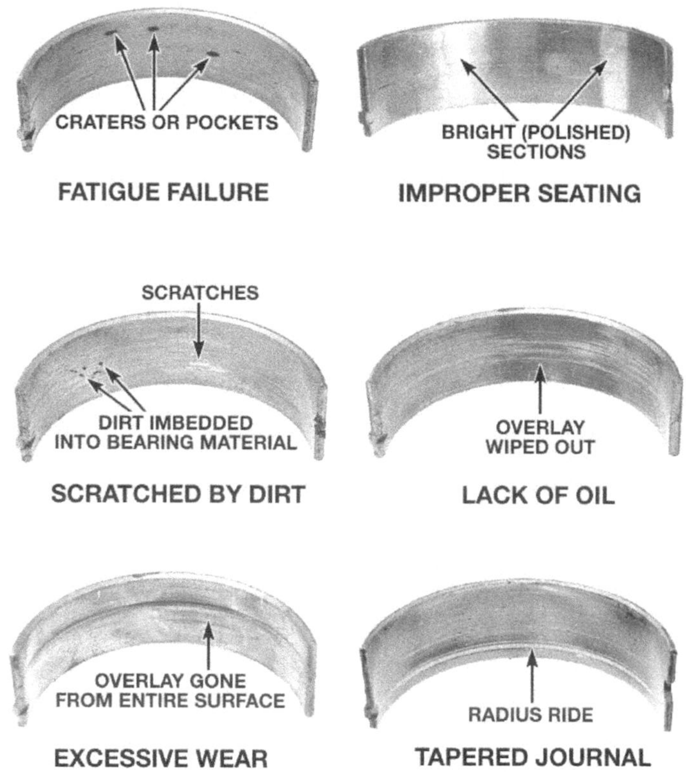

21.1 Before discarding the used bearings, examine them for indications of any possible problems with the crankshaft, noting the bearing location it came from. Here are some typical bearing failures

(see illustrations). By measuring the diameter at a number of points around each journal's circumference, you'll be able to determine whether or not the journal is out-of-round. Take the measurement at each end of the journal, near the crank throws, to determine if the journal is tapered. Crankshaft runout should be checked also, but large V-blocks and a dial indicator are needed to do it correctly. If you don't have the equipment, have a machine shop check the runout.

7 If the crankshaft journals are damaged, tapered, out-of-round or worn beyond the limits given in the Specifications, have the crankshaft reground by an automotive machine shop. Be sure to use the correct size bearing inserts if the crankshaft is reconditioned.

8 Check the oil seal journals at each end of the crankshaft for wear and damage **(see illustration)**. If the seal has worn a groove in the journal, or if it's nicked or scratched, the new seal may leak when the engine is reassembled. In some cases, an automotive machine shop may be able to repair the journal by pressing on a thin sleeve. If repair isn't feasible, a new or different crankshaft should be installed.

9 Examine the main and rod bearing inserts (see Section 21).

21 Main and connecting rod bearings - inspection and selection

Inspection

Refer to illustration 21.1

1 Even though the main and connecting rod bearings should be replaced with new ones during the engine overhaul, the old bearings should be retained for close examination, as they may reveal valuable information about the condition of the engine **(see illustrations)**.

2 Bearing failure occurs because of lack of lubrication, the presence of dirt or other foreign particles, overloading the engine and corrosion. Regardless of the cause of bearing failure, it must be corrected before the engine is reassembled to prevent it from happening again.

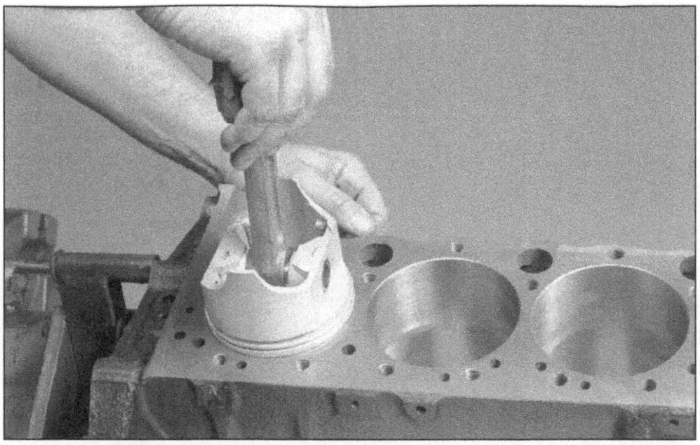

23.3 When checking piston ring end gap, the ring must be square in the cylinder bore (this is done by pushing the ring down with the top of a piston as shown)

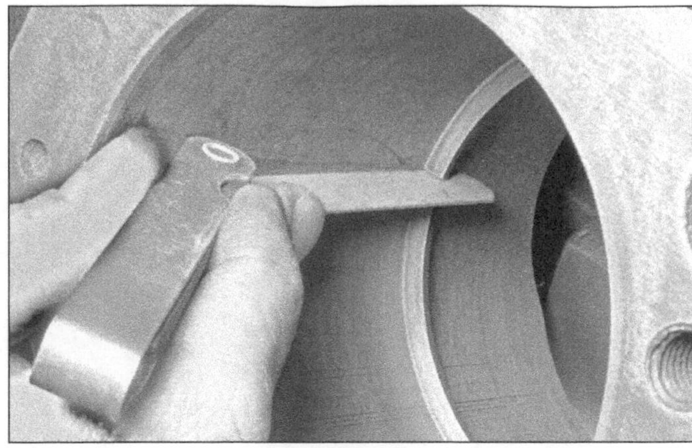

23.4 With the ring square in the cylinder, measure the end gap with a feeler gauge

3 When examining the bearings, remove them from the engine block, the main bearing caps, the connecting rods and the rod caps and lay them out on a clean surface in the same general position as their location in the engine. This will enable you to match any bearing problems with the corresponding crankshaft journal.

4 Dirt and other foreign particles get into the engine in a variety of ways. It may be left in the engine during assembly, or it may pass through filters or the PCV system. It may get into the oil, and from there into the bearings. Metal chips from machining operations and normal engine wear are often present. Abrasives are sometimes left in engine components after reconditioning, especially when parts are not thoroughly cleaned using the proper cleaning methods. Whatever the source, these foreign objects often end up embedded in the soft bearing material and are easily recognized. Large particles will not embed in the bearing and will score or gouge the bearing and journal. The best prevention for this cause of bearing failure is to clean all parts thoroughly and keep everything spotlessly clean during engine assembly. Frequent and regular engine oil and filter changes are also recommended.

5 Lack of lubrication (or lubrication breakdown) has a number of interrelated causes. Excessive heat (which thins the oil), overloading (which squeezes the oil from the bearing face) and oil leakage or throw-off (from excessive bearing clearances, worn oil pump or high engine speeds) all contribute to lubrication breakdown. Blocked oil passages, which usually are the result of misaligned oil holes in a bearing shell, will also oil starve a bearing and destroy it. When lack of lubrication is the cause of bearing failure, the bearing material is wiped or extruded from the steel backing of the bearing. Temperatures may increase to the point where the steel backing turns blue from overheating.

6 Driving habits can have a definite effect on bearing life. Full throttle, low speed operation (lugging the engine) puts very high loads on bearings, which tends to squeeze out the oil film. These loads cause the bearings to flex, which produces fine cracks in the bearing face (fatigue failure). Eventually the bearing material will loosen in pieces and tear away from the steel backing. Short trip driving leads to corrosion of bearings because insufficient engine heat is produced to drive off the condensed water and corrosive gases. These products collect in the engine oil, forming acid and sludge. As the oil is carried to the engine bearings, the acid attacks and corrodes the bearing material.

7 Incorrect bearing installation during engine assembly will lead to bearing failure as well. Tight fitting bearings leave insufficient bearing oil clearance and will result in oil starvation. Dirt or foreign particles trapped behind a bearing insert result in high spots on the bearing which lead to failure.

8 If you need to use a STANDARD size main bearing, install one that has the same number as the original bearing.

9 If you need to use a STANDARD size rod bearing, install one that has the same number as the number stamped into the connecting rod cap.

10 Remember, the oil clearance is the final judge when selecting new bearing sizes for either main or rod bearings. If you have any questions or are unsure which bearings to use, get help from a dealer parts or service department or other parts supplier.

22 Engine overhaul - reassembly sequence

1 Before beginning engine reassembly, make sure you have all the necessary new parts, gaskets and seals as well as the following items on hand:

Common hand tools
A torque wrench
Piston ring installation tool
Piston ring compressor
Short lengths of rubber or plastic hose to fit over connecting rod bolts,
Plastigage
Feeler gauges
A fine-tooth file
New engine oil
Engine assembly lube or moly-base grease
Gasket sealant
Thread locking compound

2 In order to save time and avoid problems, engine reassembly must be done in the following general order:

Piston rings
Crankshaft and main bearings
Piston/connecting rod assemblies
Rear main oil seal
Cylinder head(s) and rocker arms or lifters
Camshaft
Timing chain and gears and cover
Oil pump
Oil pick-up tube
Oil pan
Valve cover(s)
Intake and exhaust manifolds
Flywheel/driveplate

23 Piston rings - installation

Refer to illustrations 23.3, 23.4, 23.5, 23.10a, 23.10b and 23.13

1 Before installing the new piston rings, the ring end gaps must be checked. It's assumed that the piston ring side clearance has been checked and verified correct (see Section 19).

2 Lay out the piston/connecting rod assemblies and the new ring sets so the ring sets will be matched with the same piston and cylinder during the end gap measurement and engine assembly.

3 Insert the top (number one) ring into the first cylinder and square it up with the cylinder walls by pushing it in with the top of the piston **(see illustration)**. The ring should be near the bottom of the cylinder, at the lower limit of ring travel.

4 To measure the end gap, slip feeler gauges between the ends of the ring until a gauge equal to the gap width is found **(see**

23.5 If the end gap is too small, clamp a file in a vise and file the ring ends (pulling from the vise towards you only) to enlarge the gap slightly

23.10a Installing the spacer/expander in the oil control ring groove . .

illustration). The feeler gauge should slide between the ring ends with a slight amount of drag. Compare the measurement to the Specifications. If the gap is larger or smaller than specified, double-check to make sure you have the correct rings before proceeding.

5 If the gap is too small, it must be enlarged or the ring ends may come in contact with each other during engine operation, which can cause serious damage to the engine. The end gap can be increased by filing the ring ends very carefully with a fine file. Mount the file in a vise equipped with soft jaws, slip the ring over the file with the ends contacting the file face and slowly move the ring to remove material from the ends **(see illustration)**. **Caution:** *When performing this operation, file only from the outside in.*

6 Excess end gap is not critical unless it is greater than 0.040-inch. Again, double-check to make sure you have the correct rings for your engine.

7 Repeat the procedure for the rest of the rings. Remember to keep rings, pistons and cylinders matched up.

8 Repeat the procedure for each ring that will be installed in the first cylinder and for each ring in the remaining cylinders. Remember to keep rings, pistons and cylinders matched up.

9 Once the ring end gaps have been checked/corrected, the rings can be installed on the pistons.

10 The oil control ring (lowest one on the piston) is usually installed first. It's composed of three separate components. Slip the spacer/expander into the groove **(see illustration)**. If an anti-rotation tang is used, make sure it's inserted into the drilled hole in the ring groove. Next, install the lower side rail. Don't use a piston ring installation tool on the oil ring side rails, as they may be damaged. Instead, place one end of the side rail into the groove between the spacer/expander and the ring land, hold it firmly in place and slide a finger around the piston while pushing the rail into the groove **(see illustration)**. Next, install the upper side rail in the same manner.

23.10b . . . followed by the side rails - DO NOT use a piston ring installation tool when installing the oil ring side rails

11 After the three oil ring components have been installed, check to make sure that both the upper and lower side rails can be turned smoothly in the ring groove.

12 The number two (middle) ring is installed next. It's usually stamped with a mark which must face up, toward the top of the piston. **Note:** *Always follow the instructions printed on the ring package or box - different manufacturers may require different approaches. Do not mix up the top and middle rings, as they have different cross sections.*

13 Use a piston ring installation tool and make sure the identification mark is facing the top of the piston, then slip the ring into the middle groove on the piston. Don't expand the ring any more than necessary to slide it over the piston **(see illustration)**.

14 Install the number one (top) ring in the same manner. Make sure the mark is facing up. Be careful not to confuse the number one and number two rings.

15 Repeat the procedure for the remaining pistons and rings.

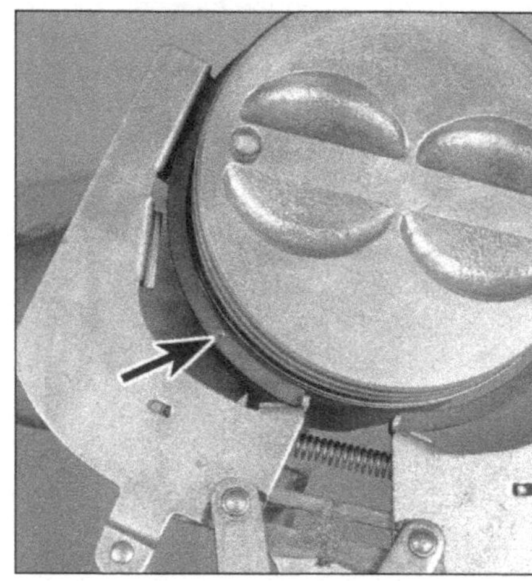

23.13 Installing the compression rings with a ring expander - the mark (arrow) must face up

24 Crankshaft - installation and main bearing oil clearance check

1 Crankshaft installation is the first major step in engine reassembly. It's assumed at this point that the engine block and crankshaft have been cleaned, inspected and repaired or reconditioned.

2 Position the engine with the bottom facing up.

3 Remove the main bearing cap bolts and lift out the caps. Lay the caps out in the proper order to ensure correct installation.

4 If they're still in place, remove the old bearing inserts from the block and the main bearing caps. Wipe the main bearing surfaces of the block and caps with a clean, lint-free cloth. They must be kept spotlessly clean!

Main bearing oil clearance check

Refer to illustration 24.10 and 24.14

5 Clean the back sides of the new main

24.10 Lay the Plastigage strips (arrow) on the main bearing journals, parallel to the crankshaft centerline

24.14 Compare the width of the crushed Plastigage to the scale on the envelope to determine the main bearing oil clearance (always take the measurement at the widest point of the Plastigage); be sure to use the correct scale - standard and metric ones are included

bearing inserts and lay the bearing half with the oil groove and hole in each main bearing saddle in the block (on all engines covered by this manual, the bearings with oil holes go in the block and those without oil holes go in the caps). Lay the other bearing half from each bearing set in the corresponding main bearing cap. Make sure the tab on each bearing insert fits into the recess in the block or cap. Also, the oil holes in the block must line up with the oil holes in the bearing insert. **Caution:** *Do not hammer the bearings into place and don't nick or gouge the bearing faces. No lubrication should be used at this time.*

6 The thrust bearings must be installed in the number three main bearing saddle.

7 Clean the faces of the bearings in the block and the crankshaft main bearing journals with a clean, lint-free cloth. Check or clean the oil holes in the crankshaft, as any dirt here can go only one way - straight through the new bearings.

8 Once you're certain the crankshaft is clean, carefully lay it in position in the main bearings. No lubricant should be used at this time.

9 Before the crankshaft can be permanently installed, the main bearing oil clearance must be checked.

10 Trim several pieces of the appropriate size Plastigage (they must be slightly shorter than the width of the main bearings) and place one piece on each crankshaft main bearing journal, parallel with the journal axis **(see illustration)**.

11 Clean the faces of the bearings in the caps and install the caps in their respective positions (don't mix them up) with the arrows pointing toward the front of the engine. Don't disturb the Plastigage. Apply a light coat of oil to the bolt threads and the under sides of the bolt heads, then install them.

12 Working from the center out, tighten the main bearing cap bolts, in three steps, to the torque listed in this Chapter's Specifications. Don't rotate the crankshaft at any time during this operation!

13 Remove the bolts and carefully lift off the main bearing caps. Keep them in order. Don't disturb the Plastigage or rotate the crankshaft. If any of the main bearing caps are difficult to remove, tap them gently from side-to-side with a soft-face hammer to loosen them.

14 Compare the width of the crushed Plastigage on each journal to the scale printed on the Plastigage envelope to obtain the main bearing oil clearance **(see illustration)**. Check the Specifications to make sure it's correct.

15 If the clearance is not as specified, the bearing inserts may be the wrong size which means different ones will be required (see Section 21). Before deciding that different inserts are needed, make sure that no dirt or oil was between the bearing inserts and the caps or block when the clearance was measured. If the Plastigage is noticeably wider at one end than the other, the journal may be tapered (see Section 20).

16 Carefully scrape all traces of the Plastigage material off the main bearing journals and/or the bearing faces. Don't nick or scratch the bearing faces.

Final crankshaft installation

17 Carefully lift the crankshaft out of the engine. Clean the bearing faces in the block, then apply a thin, uniform layer of clean moly-base grease or engine assembly lube to each of the bearing surfaces. Coat the thrust washers as well.

18 Install the rear main oil seal halves into the engine block and rear main bearing cap (small-block V8 engines) or seal retainer (six-cylinder and big-block V8 engines) (refer to Chapter 2A or 2B for details). It's not necessary to "shoehorn" the upper seal half into the engine - simply press it into place while the crankshaft is removed. Lubricate the crankshaft surfaces that contact the oil seals with moly-base grease, engine assembly lube or clean engine oil.

19 Make sure the crankshaft journals are clean, then lay the crankshaft back in place in the block. Clean the faces of the bearings in the caps or cap assembly, then apply the same lubricant to them. Install the caps in their respective positions with the arrows pointing toward the front of the engine. **Note:** *If you haven't already done so, be sure to install the thrust washers in the number three main bearing saddle.*

20 Apply a light coat of oil to the bolt threads and the under sides of the bolt heads, then install them. Tighten all except the number three cap bolts (the one with the thrust washers on the block side) to the torque listed in this Chapter's Specifications (work from the center out and approach the final torque in three steps). Tighten the number three cap bolts to 10-to-12 ft-lbs. Tap the ends of the crankshaft forward and backward with a lead or brass hammer to line up the thrust washer and crankshaft surfaces. Retighten all main bearing cap bolts to the specified torque, following the recommended sequence.

21 Rotate the crankshaft a number of times by hand to check for any obvious binding.

22 Check the crankshaft endplay with a feeler gauge or a dial indicator (see Section 15). The endplay should be correct if the crankshaft thrust faces aren't worn or damaged and new thrust washers have been installed.

23 Install a new rear main oil seal, then bolt the retainer to the block (see Chapter 2A or 2B).

25 Pistons/connecting rods - installation and rod bearing oil clearance check

Refer to illustrations 25.3, 25.5, 25.9, 25.11, 25.13 and 25.17

1 Before installing the piston/connecting

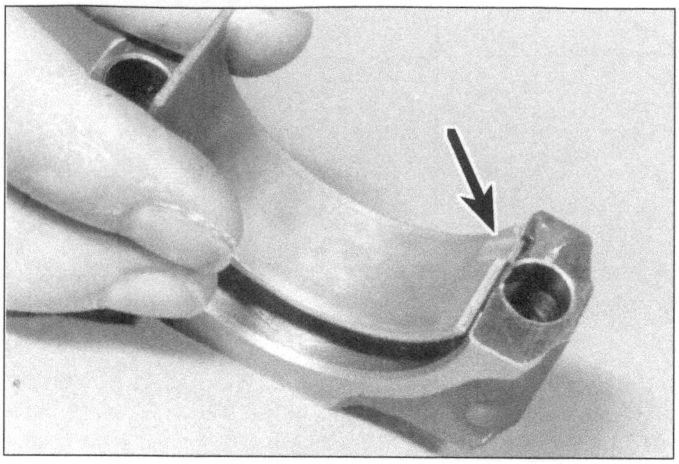

25.3 Make sure the bearing tang fits securely into the notch in the rod cap

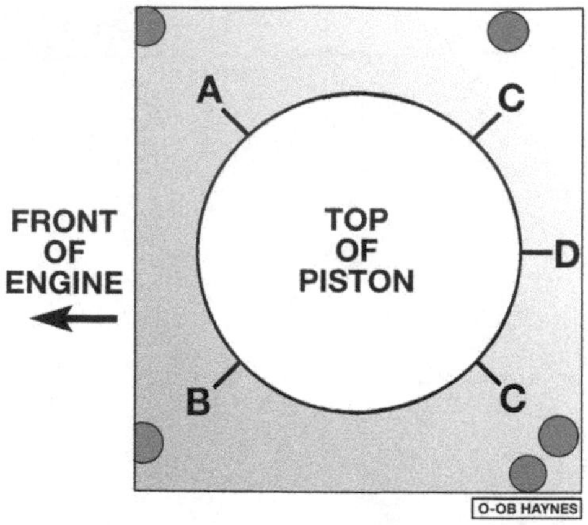

25.5 Ring end gap positions - align the oil spacer gap at A, the oil ring side rails at D, one inch either side of the pin centerline), and the compression rings at B and C, one inch either side of the pin centerline

25.9 When installed, the mark/notch (arrow) in the piston must face the front of the engine

25.11 The piston can be driven gently into the cylinder bore with the end of a wooden or plastic hammer handle

rod assemblies, the cylinder walls must be perfectly clean, the top edge of each cylinder must be chamfered, and the crankshaft must be in place.

2 Remove the cap from the end of the number one connecting rod (refer to the marks made during removal). Remove the original bearing inserts and wipe the bearing surfaces of the connecting rod and cap with a clean, lint-free cloth. They must be kept spotlessly clean.

Connecting rod bearing oil clearance check

3 Clean the back side of the new upper bearing insert, then lay it in place in the connecting rod. Make sure the tab on the bearing fits into the recess in the rod **(see illustration)** so the oil holes line up. Don't hammer the bearing insert into place and be very careful not to nick or gouge the bearing face. Don't lubricate the bearing at this time.

4 Clean the back side of the other bearing insert and install it in the rod cap. Again, make sure the tab on the bearing fits into the recess in the cap, and don't apply any lubricant. It's critically important that the mating surfaces of the bearing and connecting rod are perfectly clean and oil free when they're assembled.

5 Position the piston ring gaps at staggered intervals around the piston **(see illustration)**.

6 Slip a section of plastic or rubber hose over each connecting rod cap bolt **(see illustration 14.5)**.

7 Lubricate the piston and rings with clean engine oil and attach a piston ring compressor to the piston. Leave the skirt protruding about 1/4-inch to guide the piston into the cylinder. The rings must be compressed until they're flush with the piston.

8 Rotate the crankshaft until the number one connecting rod journal is at BDC (bottom dead center) and apply a coat of engine oil to the cylinder walls.

9 With the mark on top of the piston **(see illustration)** facing the front (timing chain end) of the engine, gently insert the piston/connecting rod assembly into the number one cylinder bore and rest the bottom edge of the ring compressor on the engine block.

10 Tap the top edge of the ring compressor to make sure it's contacting the block around its entire circumference.

11 Gently tap on the top of the piston with the end of a wooden or plastic hammer handle **(see illustration)** while guiding the end of the connecting rod into place on the crankshaft journal. The piston rings may try to pop out of the ring compressor just before entering the cylinder bore, so keep some pressure on the ring compressor. Work slowly, and if any resistance is felt as the piston enters the cylinder, stop immediately. Find out what's hanging up and fix it before proceeding. Do not, for any reason, force the piston into the cylinder - you might break a ring and/or the piston.

12 Once the piston/connecting rod assembly is installed, the connecting rod bearing oil clearance must be checked before the rod cap is permanently bolted in place.

13 Cut a piece of the appropriate size Plastigage slightly shorter than the width of the connecting rod bearing and lay it in place on

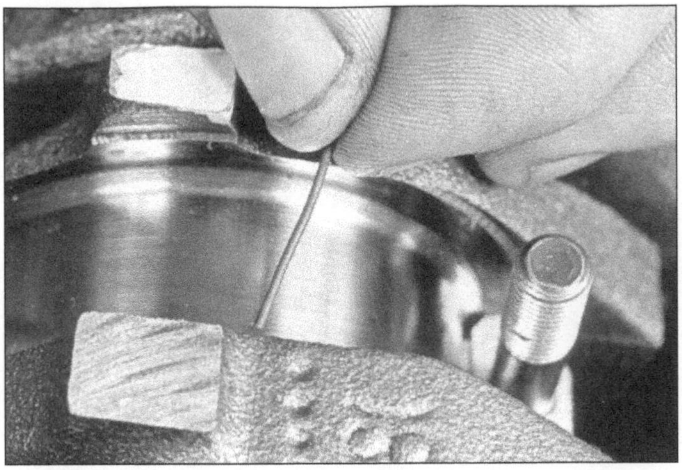

25.13 Lay the Plastigage strips on each rod bearing journal, parallel to the crankshaft centerline

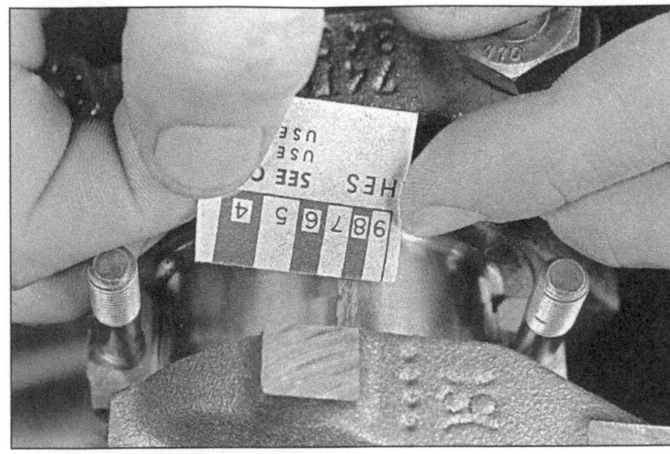

25.17 Measuring the width of the crushed Plastigage to determine the rod bearing oil clearance (be sure to use the correct scale - standard and metric ones are included)

the number one connecting rod journal, parallel with the journal axis **(see illustration)**.

14 Clean the connecting rod cap bearing face, remove the protective hoses from the connecting rod bolts and install the rod cap. Make sure the mating mark on the cap is on the same side as the mark on the connecting rod. Check the cap to make sure the front mark is facing the timing chain end of the engine.

15 Apply a light coat of oil to the undersides of the nuts, then install and tighten them to the torque listed in this Chapter's Specifications. Use a thin-wall socket to avoid erroneous torque readings that can result if the socket is wedged between the rod cap and nut. If the socket tends to wedge itself between the nut and the cap, lift up on it slightly until it no longer contacts the cap. Do not rotate the crankshaft at any time during this operation.

16 Remove the nuts and detach the rod cap, being very careful not to disturb the Plastigage.

17 Compare the width of the crushed Plastigage to the scale printed on the Plastigage envelope to obtain the oil clearance **(see illustration)**. Compare it to the Specifications to make sure the clearance is correct.

18 If the clearance is not as specified, the bearing inserts may be the wrong size (which means different ones will be required). Before deciding that different inserts are needed, make sure that no dirt or oil was between the bearing inserts and the connecting rod or cap when the clearance was measured. Also, recheck the journal diameter. If the Plastigage was wider at one end than the other, the journal may be tapered (see Section 20).

Final connecting rod installation

19 Carefully scrape all traces of the Plastigage material off the rod journal and/or bearing face. Be very careful not to scratch the bearing - use your fingernail or the edge of a credit card.

20 Make sure the bearing faces are perfectly clean, then apply a uniform layer of clean moly-base grease or engine assembly lube to both of them. You'll have to push the piston into the cylinder to expose the face of the bearing insert in the connecting rod - be sure to slip the protective hoses over the rod bolts first.

21 Slide the connecting rod back into place on the journal, remove the protective hoses from the rod cap bolts, install the rod cap and tighten the nuts to the torque listed in this Chapter's Specifications.

22 Repeat the entire procedure for the remaining pistons/connecting rods.

23 The important points to remember are:

a) *Keep the back sides of the bearing inserts and the insides of the connecting rods and caps perfectly clean when assembling them.*
b) *Make sure you have the correct piston/rod assembly for each cylinder.*
c) *The dimple on the piston must face the front (timing chain end) of the engine.*
d) *Lubricate the cylinder walls with clean oil.*
e) *Lubricate the bearing faces when installing the rod caps after the oil clearance has been checked.*

24 After all the piston/connecting rod assemblies have been properly installed, rotate the crankshaft a number of times by hand to check for any obvious binding.

25 As a final step, the connecting rod side clearance (endplay) must be checked (see Section 14).

26 Compare the measured side clearance to the Specifications to make sure it's correct. If it was correct before disassembly and the original crankshaft and rods were reinstalled, it should still be right. If new rods or a new crankshaft were installed, the side clearance may be inadequate. If so, the rods will have to be removed and taken to an automotive machine shop for re-sizing.

27 Adjust the oil pickup tube and screen the the correct distance above the bottom of the oil pan (see Chapter 2A or 2B) and reinstall the oil pan.

28 The rest of the assembly procedure is the reverse of the removal procedure.

26 Initial start-up and break-in after overhaul

Warning: *Have a fire extinguisher handy when starting the engine for the first time.*

1 Once the engine has been installed in the vehicle, double-check the engine oil and coolant levels. Add one pint of Chrysler Crankcase Conditioner (part no. 4318002) or equivalent.

2 Start the engine. It may take a few moments for the gasoline to reach the carburetor but the engine should start without a great deal of effort.

3 After the engine starts, it should be allowed to warm up to normal operating temperature. While the engine is warming up, make a thorough check for oil and coolant leaks.

4 Shut the engine off and recheck the engine oil and coolant levels.

5 Drive the vehicle to an area with minimum traffic, accelerate at full throttle from 30 to 50 mph, then allow the vehicle to slow to 30 mph with the throttle closed. Repeat the procedure 10 or 12 times. This will load the piston rings and cause them to seat properly against the cylinder walls. Check again for oil and coolant leaks.

6 Drive the vehicle gently for the first 500 miles (no sustained high speeds) and keep a constant check on the oil level. It is not unusual for an engine to use oil during the break-in period.

7 At approximately 500 to 600 miles, change the oil and filter.

8 For the next few hundred miles, drive the vehicle normally. Do not pamper it or abuse it.

9 After 2000 miles, change the oil and filter again and consider the engine fully broken in.

Notes

Chapter 3
Cooling, heating and air conditioning systems

Contents

Specifications

General

Coolant capacity	See Chapter 1
Drivebelt tension	See Chapter 1
Radiator pressure cap rating	14 to 16 psi
Thermostat opening temperature	185 to 195-degrees F

Torque specifications

	Ft-lbs (unless otherwise indicated)
Fan-to-water pump attaching bolts	
1971	15 to 18
1972 through 1989	200 in-lbs
Viscous fan drive-to-fan blade assembly attaching bolts (all)	20
Thermostat housing bolts	
1971 through 1979 (all)	30
1980 through 1983	
Six cylinder	30
V8	200 in-lbs
1984 through 1989 (all)	200 in-lbs
Water pump attaching bolts	30

1 General information

The cooling system consists of a radiator and coolant reserve system, a radiator pressure cap, a thermostat (of varying temperatures depending on year and model), a four, five or seven blade fan, and a crankshaft pulley-driven water pump.

The radiator cooling fan is mounted on the front of the water pump. There are two types of fans used; one incorporates a fluid drive fan clutch (viscous fan clutch) which saves horsepower and reduces noise, the second is a basic fan with a solid hub attaching to the water pump. On either type a fan shroud is mounted on the rear of the radiator to direct air flow through the radiator.

The system is pressurized by a spring-loaded radiator cap, which, by maintaining pressure, increases the boiling point of the coolant. If the coolant temperature goes above this increased boiling point, the extra pressure in the system forces the radiator cap valve off its seat and exposes the overflow pipe or hose. The overflow pipe/hose leads to a coolant recovery system. This consists of a plastic reservoir into which the coolant that normally escapes due to expansion is retained. When the engine cools, the excess coolant is drawn back into the radiator by the vacuum created as the system cools, maintaining the system at full capacity. This is a continuous process and provided the level in the reservoir is correctly maintained, it is not necessary to add coolant to the radiator.

Coolant in the right side of the radiator circulates up the lower radiator hose to the water pump, where it is forced through the water passages in the cylinder block. The coolant then travels up into the cylinder head, circulates around the combustion chambers and valve seats, travels out of the cylinder head past the open thermostat into the upper radiator hose and back into the radiator.

When the engine is cold, the thermostat restricts the circulation of coolant to the engine. When the minimum operating temperature is reached, the thermostat begins to open, allowing coolant to return to the radiator.

Automatic transmission-equipped models have a cooler element incorporated into the radiator to cool the transmission fluid.

The heating system works by directing air through the heater core mounted in the dash and then to the interior of the vehicle by a system of ducts. Temperature is controlled by mixing heated air with fresh air, using a system of flapper doors in the ducts, and a heater motor.

Air conditioning is an optional accessory, consisting of an evaporator core located under the dash, a condenser in front of the radiator, a receiver-drier in the engine compartment and a belt-driven compressor mounted at the front of the engine.

3.7 Loosen the clamp (arrow) and disconnect the upper radiator hose from the thermostat cover (V8 shown)

2 Antifreeze - general information

Warning: *Do not allow antifreeze to come in contact with your skin or painted surfaces of the vehicle. Rinse off spills immediately with plenty of water. Antifreeze is highly toxic if ingested. Never leave antifreeze lying around in an open container or in puddles on the floor; children and pets are attracted by it's sweet smell and may drink it. Check with local authorities about disposing of used antifreeze. Many communities have collection centers which will see that antifreeze is disposed of safely.*

The cooling system should be filled with a water/ethylene glycol based antifreeze solution which will prevent freezing down to at least -20-degrees F (even lower in cold climates). It also provides protection against corrosion and increases the coolant boiling point.

The cooling system should be drained, flushed and refilled at least every other year (see Chapter 1). The use of antifreeze solutions for periods of longer than two years is likely to cause damage and encourage the formation of rust and scale in the system.

Before adding antifreeze to the system, check all hose connections. Antifreeze can leak through very minute openings.

The exact mixture of antifreeze to water which you should use depends on the relative weather conditions. The mixture should contain at least 50-percent antifreeze, but should never contain more than 70-percent anti-freeze.

3 Thermostat - check and replacement

Refer to illustrations 3.7, 3.8, 3.9 and 3.10

Warning: *The engine must be completely cool when this procedure is performed.*

Note: *Don't drive the vehicle without a thermostat! The computer (if so equipped) may stay in open loop and emissions and fuel economy will suffer.*

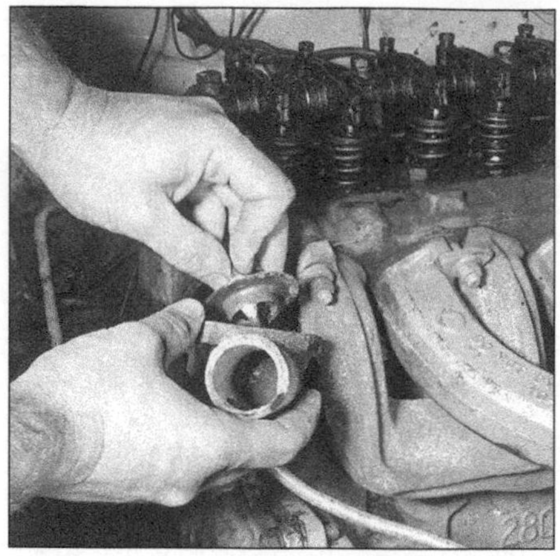

3.8 Remove the bolts and lift the thermostat cover off (Six cylinder shown)

Check

1 Before condemning the thermostat, check the coolant level, drivebelt tension and temperature gauge (or light) operation.

2 If the engine takes a long time to warm up, the thermostat is probably stuck open. Replace the thermostat.

3 If the engine runs hot, check the temperature of the upper radiator hose. If the hose isn't hot, the thermostat is probably stuck shut. Replace the thermostat.

4 If the upper radiator hose is hot, it means the coolant is circulating and the thermostat is open. Refer to the *Troubleshooting* section for the cause of overheating.

5 If an engine has been overheated, you may find damage such as leaking head gaskets, scuffed pistons and warped or cracked cylinder heads.

Replacement

6 Drain coolant (about 1 gallon) from the radiator, until the coolant level is below the thermostat housing.

7 Disconnect the upper radiator hose from the thermostat cover **(see illustration)**.

8 Remove the bolts and lift the cover off **(see illustration)**. It may be necessary to tap the cover with a soft-face hammer to break the gasket seal.

9 Note how it's installed, then remove the thermostat **(see illustration)**. Be sure to use a replacement thermostat with the correct opening temperature (see this Chapter's Specifications).

10 Use a scraper or putty knife to remove all traces of old gasket material and sealant from the mating surfaces **(see illustration)**. **Caution:** *Be careful not to gouge or damage the gasket surfaces, because a leak could develop after assembly. Make sure no gasket material falls into the coolant passages; it is a good idea to stuff a rag in the passage. Wipe the mating surfaces with a rag saturated with lacquer thinner or acetone.*

11 Install the thermostat and make sure the correct end faces out - the spring is directed toward the engine **(see illustration 3.9)**.

12 Apply a thin coat of RTV sealant to both

3.9 Note how it's installed, then remove the thermostat

3.10 Remove all traces of old gasket material and sealant, but avoid gouging the mating surfaces

sides of the new gasket and position it on the engine side, over the thermostat, and make sure the gasket holes line up with the bolt holes in the housing.

13 Carefully position the cover and install the bolts. Tighten them to the torque listed in this Chapter's Specifications - do not overtighten the bolts or the cover may crack or become distorted.

14 Reattach the radiator hose to the cover and tighten the clamp - now may be a good time to check and replace the hoses and clamps (see Chapter 1).

15 Refer to Chapter 1 and refill the system, then run the engine and check carefully for leaks.

16 Repeat steps 1 through 5 to be sure the repairs corrected the previous problem(s).

4 Engine cooling fan and clutch - check, removal and installation

Warning. *Keep hands, tools and clothing away from the fan. To avoid injury or damage DO NOT operate the engine with a damaged fan. Do not attempt to repair fan blades - replace a damaged fan with a new one.*

4.6 Loosen the water pump pulley bolts (arrows) - two additional bolts cannot be seen on the other side of the pulley

4.9 If the engine has a non-viscous fan, a spacer will be used as shown

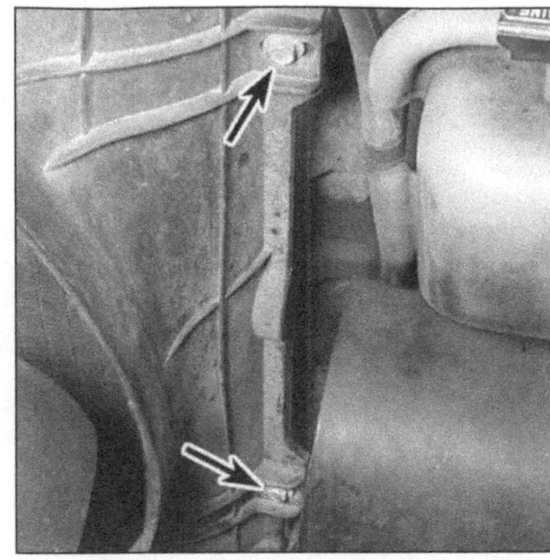
4.10a Remove the fan shroud bolts (arrows) - there are two on either side of the radiator

Check

Viscous clutch models only

Warning: *In order to check the fan clutch the engine will need to be at operating temperature, so while going through checks prior to Step 6 be careful that the engine IS NOT STARTED while the checks are being completed. Severe personal injury can result!*

1 Symptoms of failure of the fan clutch are continuous noisy operation, looseness leading to vibration and evidence of silicone fluid leaks.

2 Rock the fan back and forth by hand to check for excessive bearing play.

3 With the engine cold, turn the blades by hand. The fan should turn freely.

4 Visually inspect for substantial fluid leakage from the clutch assembly, a deformed bi-metal spring or grease leakage from the cooling fan bearing. If any of these conditions exist, replace the fan clutch.

5 When the engine is warmed up, turn off the ignition switch and disconnect the cable from the negative battery terminal. Turn the fan by hand. Some resistance should be felt. If the fan turns easily, replace the fan clutch.

Removal and installation

All models

Refer to illustrations 4.6, 4.9, 4.10a, 4.10b and 4.11

6 Loosen the water pump pulley bolts **(see illustration)**.

7 Remove the drivebelts from the water pump pulley (see Chapter 1).

8 It may be necessary to drain about two quarts of coolant (see Chapter 1) and detach the upper hose from the radiator. **Note:** *Some engine and vehicle model combinations may have enough room to remove the fan assembly from the water pump without disconnection of any radiator hoses.*

9 Unbolt the fan assembly and detach it from the water pump. **Note:** *If the engine has a non-viscous fan there will be a spacer used. Be careful not to loose it once removed* **(see illustration).**

4.10b On some models it may be necessary to remove the fan and shroud together

10 Remove the fan shroud bolts and lift the fan and shroud from the vehicle **(see illustrations)**. **Note:** *Here again it may be possible, with some vehicle and engine combinations, to remove the fan assembly without removal of the shroud from the radiator but removal of the shroud usually makes fan removal much easier.*

11 The fan clutch (if equipped) can be unbolted from the fan blade assembly for replacement **(see illustration)**. **Caution:** *To prevent silicone fluid from draining from the clutch assembly into the fan drive bearing and ruining the lubricant, DON'T place the drive unit in a position with the rear of the shaft pointing down.*

12 Installation is the reverse of removal.

5 Radiator and coolant reservoir tank - removal and installation

Refer to illustrations 5.2a, 5.2b, 5.2c, 5.3, 5.4, 5.5 and 5.6

Warning: *The engine must be completely cool when this procedure is performed.*

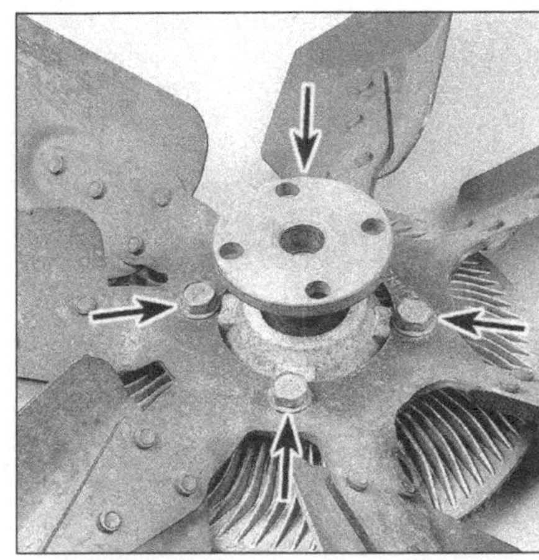
4.11 Remove the fan-to-clutch mounting bolts (arrows)

Radiator

Removal

1 Disconnect the battery cable at the negative battery terminal.

2 Drain the cooling system as described in Chapter 1, then disconnect the overflow hose and the upper and lower radiator hoses from the radiator **(see illustrations)**.

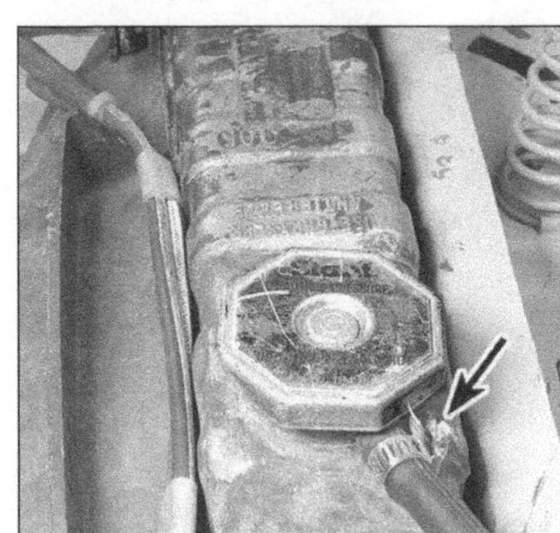
5.2a Disconnect the overflow hose from the radiator filler neck

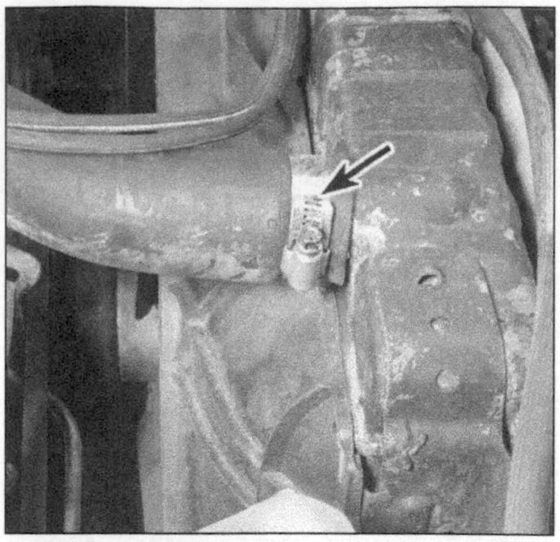

5.2b Loosen the clamp (arrow) on the upper radiator hose . . .

3 If equipped with an automatic transmission or engine oil cooler, remove the cooler lines from the radiator **(see illustration)** - be careful not to damage the lines or fittings. Plug the ends of the disconnected lines to prevent leakage and stop dirt from entering the system. Have a drip pan ready to catch any spills.

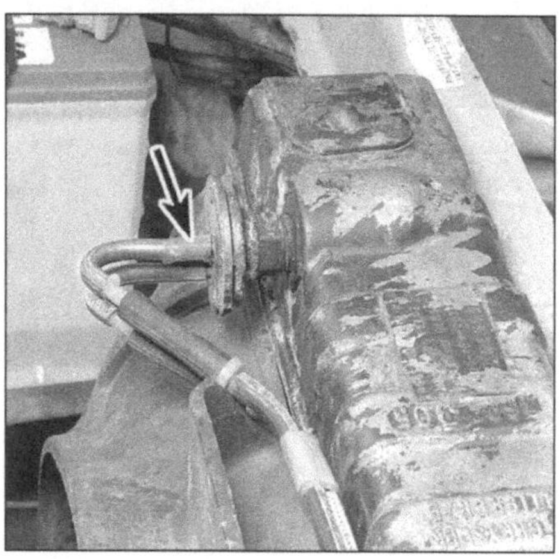

5.4 Disconnect the vacuum lines at the valve located at the top of the radiator (arrow)

5.6 Unbolt the fan shroud and slip it back over the fan and out of the way for radiator removal

5.2c . . . and the clamp (arrow) on the lower radiator hose and separate the hoses from the radiator

4 Disconnect the vacuum lines at the valve located at the top of the radiator **(see illustration)**, if equipped.

5 Remove the radiator mounting bolts **(see illustration).**

6 Unbolt the fan shroud **(see illustration)** and slip it back over the fan.

5.5 Remove the radiator mounting bolts (arrows) - there are two more in the same location on the other side of the radiator

5.11 Disconnect the overflow hose (arrow) at the reservoir

5.3 If equipped with an automatic transmission or engine oil cooler, remove the cooler lines (arrow) from the radiator (only one of the lines shown here) and always use a backup wrench to avoid twisting off a fitting or line

7 Lift the radiator from the engine compartment. Take care not to contact the fan blades.

8 Prior to installation of the radiator, replace any damaged hose clamps and radiator hoses.

Installation

9 Radiator installation is the reverse of removal.

10 After installation, fill the system with the proper mixture of antifreeze, and also check the automatic transmission fluid level, where applicable (see Chapter 1).

Coolant reservoir tank

Removal and installation

Refer to illustrations 5.11 and 5.12

11 Remove the coolant overflow hose from the reservoir **(see illustration)**.

12 Remove the screws and detach the reservoir **(see illustration)**.

5.12 The coolant reservoir tank shown here is secured to the inner fender by two screws (arrows) - some models have the tank mounted at the core support at the right of the radiator

6.2 Check the weep hole (arrow) for leakage (pump removed for clarity)

6.4 Check the pump for loose or rough bearings (this can be done with the fan and pulley in place)

7.5 Detach the coolant hoses from the water pump (arrows) - V8 engine shown

13 Prior to installation make sure the reservoir is clean and free of debris which could be drawn into the radiator (wash it with soap and water if necessary).

14 Installation is the reverse of removal.

6 Water pump - check

Refer to illustrations 6.2 and 6.4

1 Water pump failure can cause overheating and serious damage to the engine. There are three ways to check the operation of the water pump while it is installed on the engine. If any one of the following quick checks indicates water pump problems, it should be replaced immediately.

2 A seal protects the water pump impeller shaft bearing from contamination by engine coolant. If this seal fails, a weep hole in the water pump snout will leak coolant **(see illustration)** (an inspection mirror can be used to look at the underside of the pump if the hole isn't on top). If the weep hole is leaking, shaft bearing failure will follow. Replace the water pump immediately.

3 Besides contamination by coolant after a seal failure, the water pump impeller shaft bearing can also be prematurely worn out by an improperly tensioned drivebelt. When the bearing wears out, it emits a high pitched squealing sound. If such a noise is coming from the water pump during engine operation, the shaft bearing has failed - replace the water pump immediately. **Note:** *Do not confuse belt noise with bearing noise.*

4 To identify excessive bearing wear before the bearing actually fails, grasp the water pump pulley and try to force it up-and-down or from side-to-side **(see illustration)**. If the pulley can be moved either horizontally or vertically, the bearing is nearing the end of its service life. Replace the water pump.

7 Water pump - removal and installation

Refer to illustrations 7.5, 7.7a, 7.7b and 7.8

Removal

1 Disconnect the battery cable from the negative battery terminal.

2 Drain the coolant (see Chapter 1).

3 Loosen the water pump pulley bolts or fan mounting bolts, then remove the drivebelts (see Chapter 1).

4 Remove the fan and shroud assembly (see Section 5) and water pump pulley.

5 Detach the coolant hoses from the water pump **(see illustration)**.

6 On many models the compressor, alternator, power steering and air pump brackets will be attached to the water pump. Where necessary, loosen or remove the components so the brackets can be moved aside and the water pump bolts removed. **Note:** *Do not disconnect the power steering or air conditioning compressor hoses; unbolt the units and tie them aside.*

7 Remove the water pump mounting bolts and detach it. It may be necessary to tap the pump with a soft-face hammer to break the gasket seal **(see illustrations)**.

Installation

8 Clean the sealing surfaces of all gasket material on both the water pump and block **(see illustration)**. Wipe the mating surfaces with a rag saturated with lacquer thinner or acetone.

7.7a Remove the water pump mounting bolts and detach the pump (if it is difficult to remove use a soft face hammer to break the seal made by the gasket)

7.7b Water pump mounting bolt locations (arrows) - big block V8 shown

7.8 Remove all traces of old gasket material - use care to avoid gouging the soft aluminum

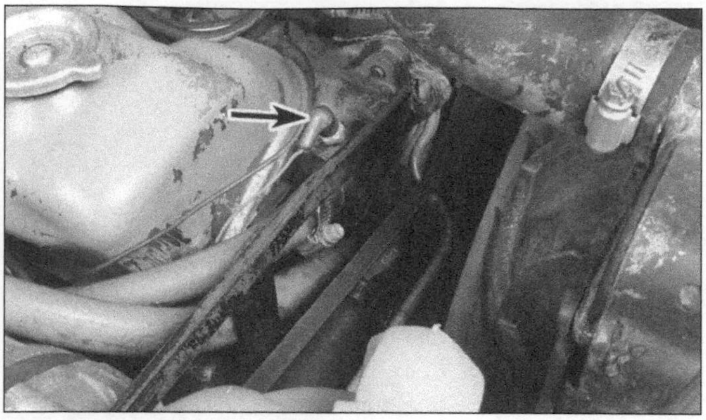

8.1 The coolant temperature sending unit (arrow) is located near the front of the intake manifold (six cylinder shown)

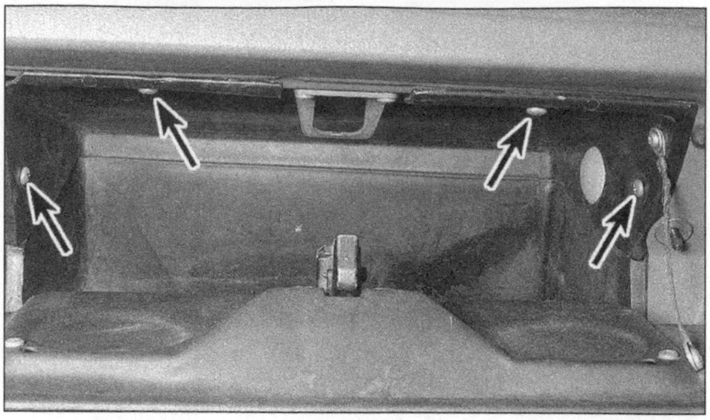

9.7 Remove the screws (arrows) and pull the glove box insert from the dash

9 Apply a thin layer of RTV sealant to both sides of the new gasket.

10 Install the gasket on the water pump.

11 Place the water pump in position and install the bolts finger tight. Use caution to ensure that the gasket doesn't slip out of position. Remember to replace any mounting brackets secured by the water pump mounting bolts. Tighten the bolts to the torque listed in this Chapter's Specifications.

12 Install the water pump pulley, fan assembly and pulley bolts.

13 Install the coolant hoses and hose clamps. Tighten the hose clamps securely.

14 Install the drivebelts (see Chapter 1) and tighten the fan/pulley mounting bolts securely.

15 Add coolant to the specified level (see Chapter 1).

16 Reconnect the battery cable to the negative battery terminal.

17 Start the engine and check for the proper coolant level and the water pump and hoses for leaks.

8 Coolant temperature sending unit - check and replacement

Refer to illustration 8.1

Warning: *Wait until the engine is completely cool before beginning this procedure.*

Check

1 The coolant temperature indicator system is composed of a light or temperature gauge mounted in the dash and a coolant temperature sending unit mounted on the engine **(see illustration)**. Some vehicles have more than one sending unit, but only one is used for the indicator system and the other is used to send temperature information to the computer.

2 If an overheating indication occurs, check the coolant level in the system and then make sure the wiring between the light or gauge and the sending unit is secure and all fuses are intact.

3 When the ignition switch is turned on and the starter motor is turning, the indicator light (if equipped) should be on (overheated engine indication).

4 If the light is not on, the bulb may be burned out, the ignition switch may be faulty or the circuit may be open. Test the circuit by grounding the wire to the sending unit while the ignition is on (engine NOT running). If the gauge deflects full scale or the light comes on, replace the sending unit.

5 As soon as the engine starts, the light should go out and remain out unless the engine overheats. Failure of the light to go out may be due to a grounded wire between the light and the sending unit, a defective sending unit or a faulty ignition switch. Check the coolant to make sure it's the proper type. **Note:** *Plain water may have too low a boiling point to activate the sending unit.*

Replacement

6 Make sure the engine is cool before removing the defective sending unit. There will be some coolant loss as the unit is removed, so be prepared to catch it.

7 Prepare the new sending unit by coating the threads with sealant or wrapping them with Teflon tape. Disconnect the electrical connector and simply unscrew the sensor from the engine and install the replacement.

8 Check the coolant level after the replacement unit has been installed and top up the system, if necessary (see Chapter 1).

9 Heater and air conditioning blower motor - check, removal and installation

Check

Refer to illustrations 9.7 and 9.8

1 Check the fuse and all connections in the circuit for looseness and corrosion. Make sure the battery is fully charged.

2 With the transmission in Park, the parking brake securely set, turn the ignition switch to the Run position. It isn't necessary to start the vehicle.

3 Disconnect the blower motor electrical connector and connect a voltmeter to the blower motor connector and ground.

4 Move the blower switch through each of its positions and note the voltage readings. Changes in voltage indicates that the motor speeds will also vary as the switch is moved to the different positions.

9.8 Unplug the connections at the blower motor resistor (arrow) located in the heater case behind the glove box

5 If there is voltage present, but the blower motor does not operate, the blower motor is probably faulty. If there is no voltage present, check the blower resistor as follows:

6 Disconnect the voltmeter from the blower motor and connect the electrical connector back to the motor.

7 Remove the screws (arrows) and the glove box insert from the dash **(see illustration)**.

8 Disconnect the electrical connector from the blower motor resistor **(see illustration)**. With the ignition on, check for voltage at each of the terminals in the connector as the switch is moved to the different positions. If the voltmeter responds correctly to the switch then the resistor is probably faulty. If there is no voltage present from the switch then the switch, control panel or related wiring is probably faulty.

Blower motor removal

Note: *There are several different blower motor locations and removal procedures, depending on year and body type (see table).*

9 Disconnect the battery cable at the negative battery terminal.

Heater and air conditioning blower motor removal procedures (by body type*)

Body designation	A	B	C	E	F	J	M	R
Models where heater housing must be removed	1971 thru 1976 (no a/c)	1971 thru 1979 (no a/c)		1971 thru 1974 (no a/c)	1976 thru 1980 (no a/c)		1977 thru 1981 (no a/c)	
Models where right inner fender panel must be removed			1971 thru 1978					
Models where blower motor can be removed from engine compartment	1971 thru 1976 (with a/c)	1971 thru 1979 (with a/c)		1971 thru 1976 (with a/c)				
Models where blower motor is removed from inside vehicle					1976 thru 1980 (with a/c)	1979 thru 1983	All with a/c (US) and all 1979 thru 1981 (Canada)	1979 thru 1981

**For more information or clarification of models, years and body identification refer to the model chart at the front of this manual.*

Models where heater housing must be removed

Refer to illustrations 9.11, 9.15, 9.21 and 9.23

10 Drain the cooling system (see Chapter 1).

11 Disconnect the heater hoses at the heater core inlet and outlet **(see illustration)** and plug the open fittings.

12 Remove the mounting nuts from studs around the blower motor and remove the flange and air seal from the heater core tubes.

13 Remove the lower dash trim, instrument trim panels and center console, if equipped (and on 1979 through 1981 R body models also remove the glove compartment).

14 Label and disconnect the wiring harness electrical connectors.

15 Remove the housing to plenum support rod (brace), which is on the right side of the housing above the fresh air opening **(see illustration)**.

16 Detach the antenna cable from the retaining clip on the right end of the heater housing.

17 Remove the defroster duct from the fitting at the top of the heater case assembly.

18 Disconnect the heater control and the vacuum feed line from the check valve. Label and disconnect any remaining vacuum hoses.

19 Disconnect the vent, temperature and mode control cables.

20 Remove the blower motor vent tube.

21 On later models, remove the heater housing assembly mounting nuts from the firewall **(see illustration)**.

22 Disconnect the wires from the blower motor resistor.

23 Move the heater housing to the rear until the studs clear the holes. Make sure the heater core inlet and outlet pipes are securely plugged to prevent coolant from spilling into the passenger compartment, then turn the unit so the studs point down and remove the assembly from under the passenger side of

9.11 Loosen the hose clamps (arrows) and disconnect the heater hoses from the heater core tubes at the firewall

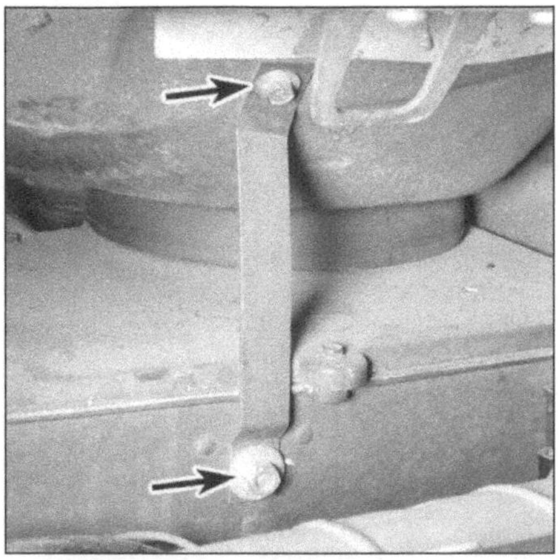

9.15 Remove the heater case-to-dash support brace (arrows) (glove box removed for clarity)

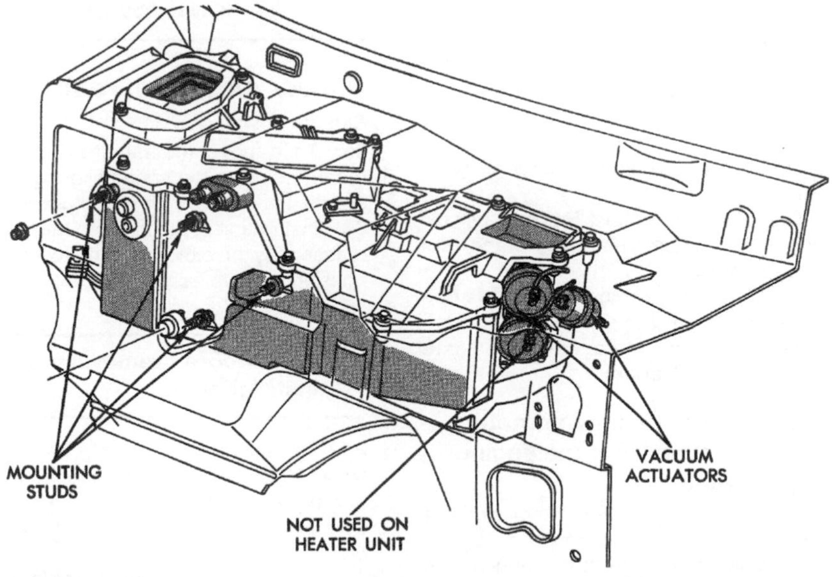

9.21 Remove the heater housing assembly mounting nuts from the firewall (arrows)

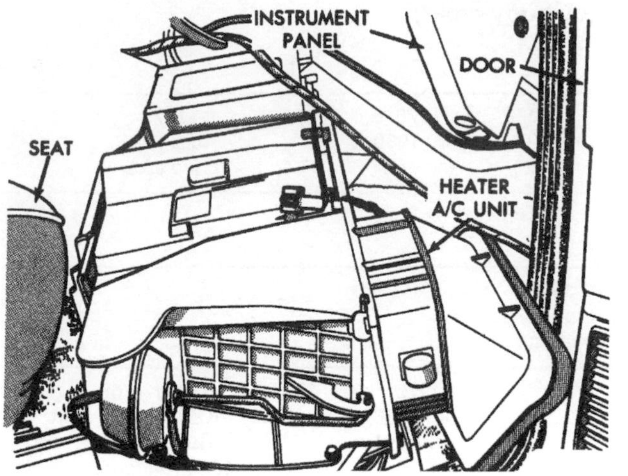

9.23 Pull the heater housing to the rear (towards the seat) until the studs clear the firewall, twist the unit so the studs turn towards the floor and slide the assembly from the car through the passenger door

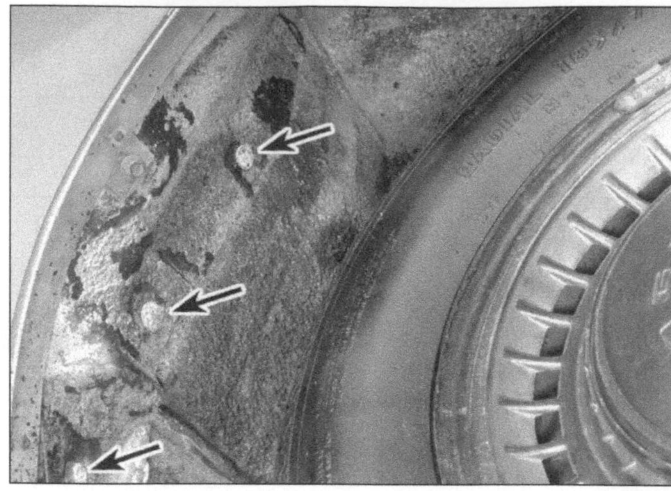

9.26 Remove the bolts (arrows) and take off the inner fender shield to gain access to the blower motor

the dash **(see illustration)**.

24 Unbolt and remove the blower motor from the heater case.

Models where the right inner fender panel must be removed

Refer to illustration 9.26

25 Raise the front end of the car and place it on jack stands and turn the wheels to the right. **Note:** *This is only to make the job as easy as possible but the this procedure can be done with the car on the ground and by turning the wheels all the way to the right.*

26 Unbolt and remove the inner fender panel **(see illustration)**. This will allow you to access the blower motor for removal.

27 Unbolt and remove the blower motor.

Models where the blower motor is removed from inside vehicle

Refer to illustrations 9.31 and 9.35

28 Remove the lower dash or remove the glove compartment, depending if the blower motor drops straight down for removal or if the blower motor is removed by taking it out the side of the heater case.

29 Disconnect the blower motor electrical connector.

30 On 1979 through 1981 R body cars only, remove the plenum mounting brace **(see illustration 9.15).**

31 On all vehicles, remove the screws from the perimeter of the blower motor and remove the blower motor from the heater housing **(see illustration)**.

Models where the blower motor can be removed from the engine compartment

32 Disconnect the blower motor electrical connector and remove the cooling tube.

33 Remove the nuts from the studs retaining the blower motor to the firewall.

34 Remove the blower motor.

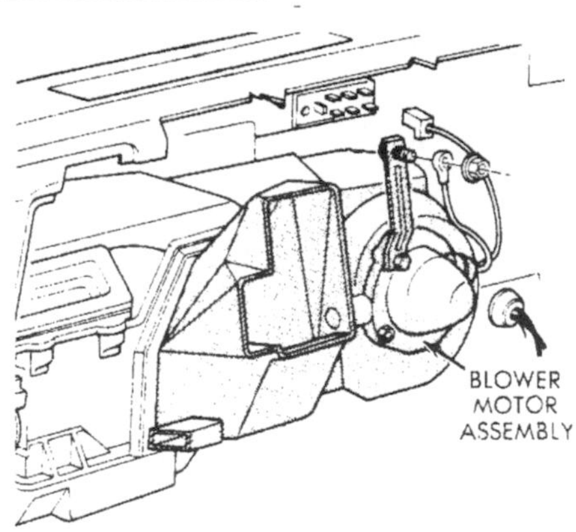

9.31 Remove the screws from the perimeter of the blower motor (arrow) and remove the blower motor from the heater housing (some models have the blower motor mounted up from beneath the blower housing)

Blower motor fan replacement

Refer to illustration 9.35

35 To remove the fan from the blower motor, squeeze the spring clip together and slip the fan off the shaft **(see illustration)**.

36 Installation of the fan onto the motor and reassembly of blower motor to the heater housing is the reverse of removal.

10 Heater core - removal and installation

Refer to illustrations 10.3 and 10.4

Note: *This procedure applies to models without air conditioning. To remove the heater core from models equipped with air conditioning see Section 16.*

1 Disconnect the battery cable from the negative battery terminal. Drain the cooling system (see Chapter 1).

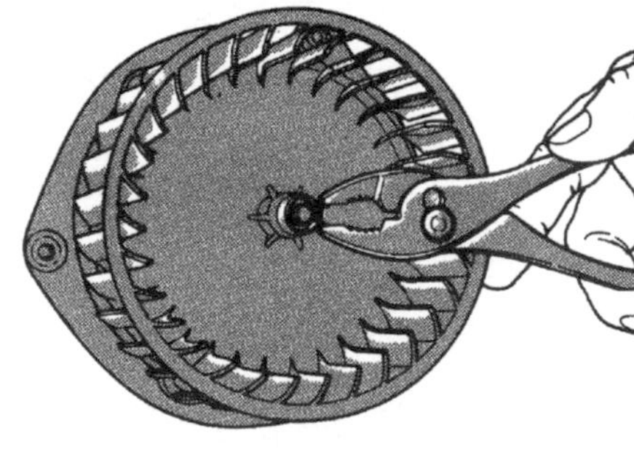

9.35 Remove the fan by squeezing the spring clip together and slipping the fan off the shaft

2 Remove the heater housing assembly (refer to Section 9, steps 10 through 23) on all models except C-body (see the front on this manual for model clarification).

3 On C-body, remove the passenger side housing only **(see illustration)**. The procedure is essentially the same as other models, except do not remove the bolts around the blower motor. Instead, remove the nuts from the studs around the engine-side housing.

4 After removing the heater housing, it will be necessary to either separate the halves by removing the retaining clips or screws or remove the heater core cover from the housing (depending on model type) **(see illustration)**.

5 Remove any seal retainers and seals.

6 Remove the mounting screw from the heater core retaining bracket and lift the core from the housing.

7 Installation is the reverse of removal.

8 Refill the cooling system (see Chapter 1).

9 Start the engine and check for proper operation.

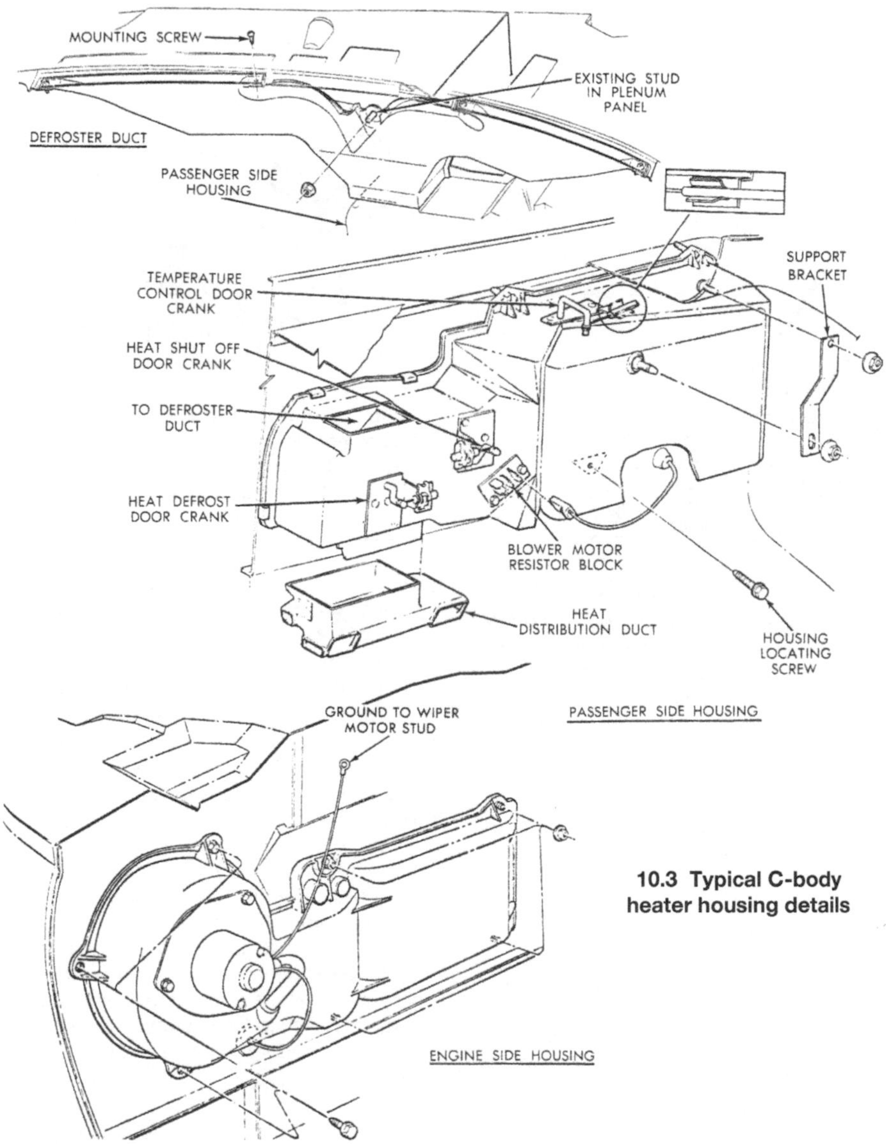

10.3 Typical C-body heater housing details

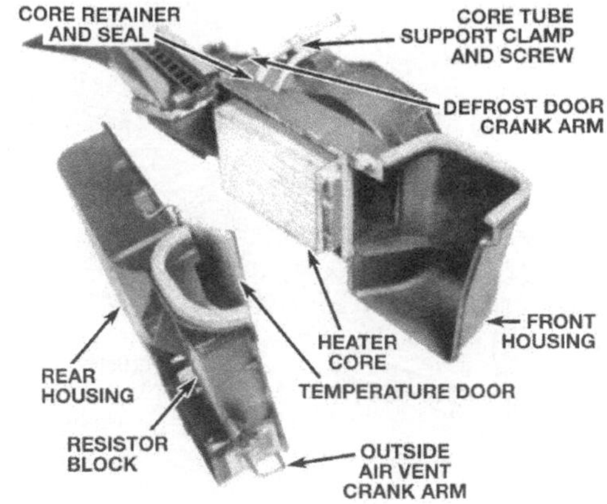

10.4 Separate the heater housing to gain access to the heater core

11 Heater and air conditioning control assembly - removal and installation

Refer to illustrations 11.7a, 11.7b, 11.7c, 11.8a and 11.8b

Note: *For the many years and models covered there are several locations of heater and air conditioning control assemblies. Please refer to the illustration that applies to the vehicle being worked on.*

1 Disconnect the battery cable from the negative battery terminal.

2 Remove the trim bezel that surrounds the heater/air conditioning control assembly (see Chapter 11). **Note:** *On E body models it will not be an entire bezel but only a piece of trim molding that surrounds the control assembly.*

3 Some earlier models with a high, center mounted controls will require radio removal (see Chapter 11).

4 On E body vehicles, remove the left spot cooler air duct.

5 On 1979 and earlier C body vehicles remove the instrument upper cover and trim pad (see Chapter 12).

6 If the control knobs must be removed simply pull them off. Some earlier models may require loosening the set screws in the bottom of the control lever knobs to remove the knobs.

7 Remove the control assembly retaining screws, nuts or bolts **(see illustrations)**. Depending on the model and year of the vehicle, the mounting of the control assembly may be attached from the front, usually with screws, which are seen after the trim has been removed, or from the rear, behind the dash (usually with nuts) and can only be removed by reaching up from under the dash.

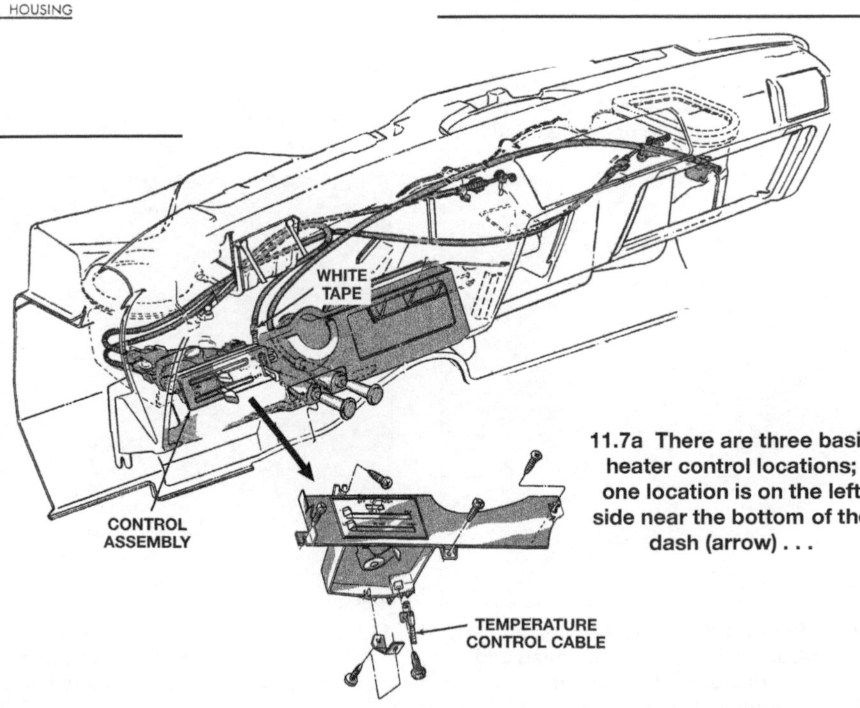

11.7a There are three basic heater control locations; one location is on the left side near the bottom of the dash (arrow) . . .

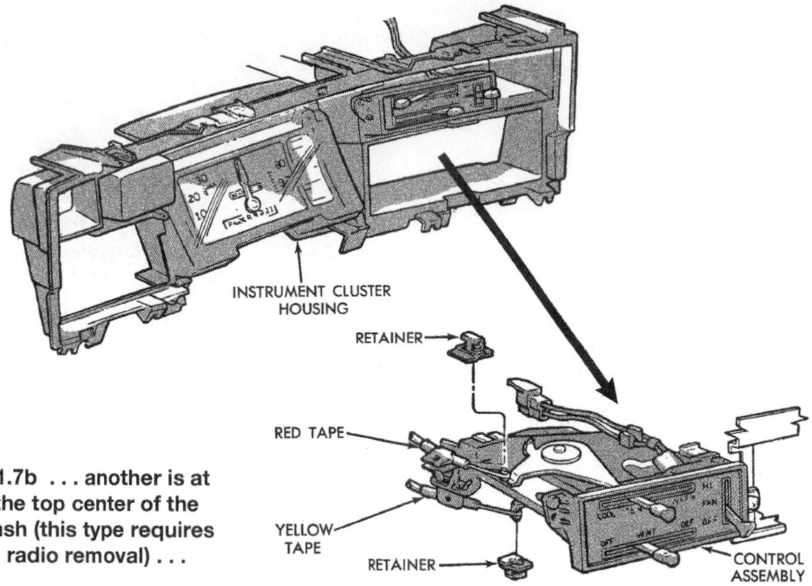

11.7b . . . another is at the top center of the dash (this type requires radio removal) . . .

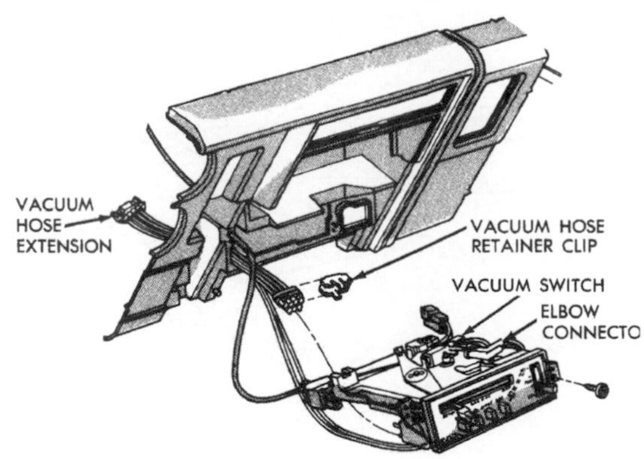

11.7c . . . and later models have a lower center heater control mounting position and can be removed after taking off the trim bezel

8 Remove the heater/air conditioning control assembly from the dash and disconnect the control cables, electrical connections and vacuum lines from the control head **(see illustrations)**. **Note:** *When disconnecting the vacuum lines, they may be in one large cluster or attached individually.* **Caution:** *Be careful when removing the vacuum lines to avoid cracking the plastic connectors and causing a vacuum leak (possibly internal within the control head).*

9 To install the control, reverse the removal procedure.

12 Air conditioning and heating system - check and maintenance

Refer to illustration 12.6

Warning: *The air conditioning system is under high pressure. DO NOT loosen any fittings or remove any components until after the system has been discharged. Air conditioning refrigerant should be properly discharged into an EPA-approved container at a dealer service department or an automotive air conditioning facility. Always wear eye protection when disconnecting air conditioning system fittings.*

1 The following maintenance steps should be performed on a regular basis to ensure that the air conditioner continues to operate at peak efficiency.

a) Check the tension of the drivebelt and adjust if necessary (see Chapter 1).

b) Check the condition of the hoses. Look for cracks, hardening and deterioration. **Warning:** *Do not replace air conditioning hoses until the system has been discharged by a dealer or air conditioning shop.*

c) Check the fins of the condenser for leaves, bugs and other foreign material. A soft brush and compressed air can be used to remove them.

d) Check the wire harness for correct routing, broken wires, damaged insulation, etc. Make sure the electrical connectors are clean and tight.

e) Maintain the correct refrigerant charge.

2 The system should be run for about 10 minutes at least once a month. This is particularly important during the winter months because long-term non-use can cause hardening of the internal seals.

3 Because of the complexity of the air conditioning system and the special equipment required to effectively work on it, accurate troubleshooting of the system should be left to a professional technician. One probable cause for poor cooling that can be determined by the home mechanic is a low refrigerant charge. Should the system lose its cooling ability, the following procedure will help you pinpoint the cause.

Check

4 Warm the engine up to normal operating temperature.

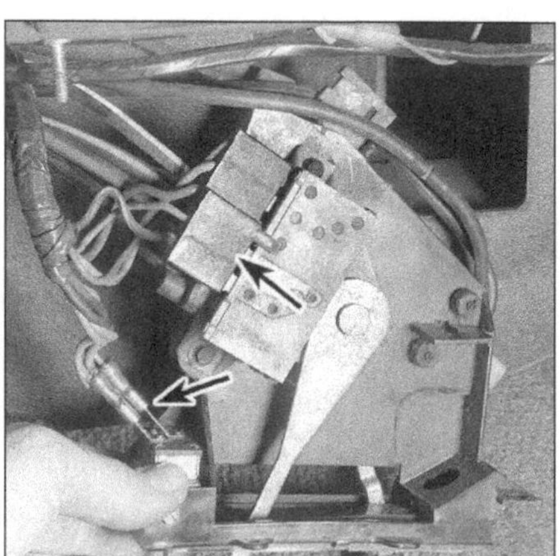

11.8a Remove the heater/air conditioning control assembly from the dash and disconnect electrical connector (arrows) . . .

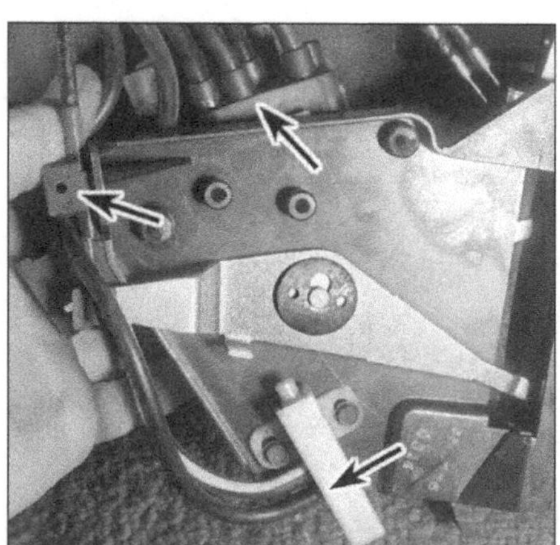

11.8b . . . vacuum lines and control cables (arrows) from the control head (typical control head assembly shown)

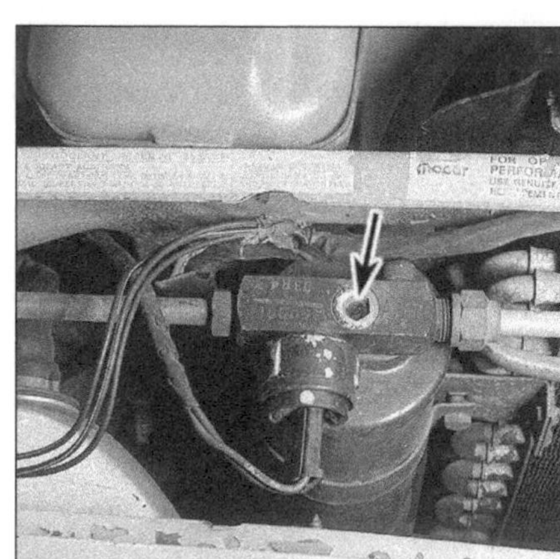

12.6 Inspect the sight glass (arrow) - if the refrigerant looks foamy, it's low (but if the refrigerant appears clear, the system is properly charged)

13.2a Disconnect the electrical connector (arrow) from the air conditioning compressor . .

13.2b . . .and remove the refrigerant line fitting bolts (arrows) from the top of the compressor

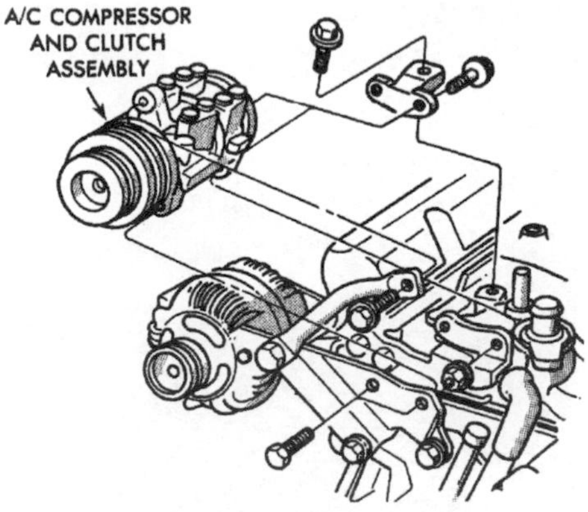

13.4 Air conditioning compressor mounting location (typical)

5 Place the air conditioning temperature selector at the coldest setting and put the blower at the highest setting. Open the doors (to make sure the air conditioning system doesn't cycle off as soon as it cools the passenger compartment).

6 With the compressor engaged - the clutch will make an audible click and the center of the clutch will rotate - inspect the sight glass **(see illustration)**. If the refrigerant looks foamy, it's low. Charge the system as described later in this Section. If the refrigerant appears clear, the system is properly charged.

7 Place a thermometer in the dashboard vent nearest the evaporator and add refrigerant until the indicated temperature is around 40 to 45-degrees F.

Adding refrigerant

Note: *Because of recent Federal regulations enacted by the Environmental Protection Agency, 14-ounce cans of refrigerant may not be available in your area. If this is the case, it will be necessary to take the vehicle to a licensed air conditioning technician for charging. If you decide to add refrigerant from one of the large 30 lb. cans available, you will need a set of manifold gauges, all the necessary fittings, adapters and hoses to hook everything up and a copy of the* Haynes Automotive Heating and Air Conditioning Manual *before tackling this job.*

13 Air conditioning compressor - removal and installation

Refer to illustrations 13.2a, 13.2b and 13.4

Removal

Warning: *The air conditioning system is under high pressure. DO NOT loosen any fittings or remove any components until after the system has been discharged. Air conditioning refrigerant should be properly discharged into an EPA-approved container at a dealer service department or an automotive air conditioning facility. Always wear eye protection when disconnecting air conditioning system fittings.*

Note: *The receiver drier (see Section 15) should be replaced whenever the compressor is replaced.*

1 Disconnect the battery cable at the negative battery terminal.

2 Disconnect the electrical connector **(see illustration)** from the air conditioning compressor and remove the line fitting bolts from the top of the compressor **(see illustration)**. Plug the open fittings to prevent the entry of dirt and moisture.

3 Remove the drivebelt(s) (refer to Chapter 1).

4 Remove the compressor-to-bracket bolts and nuts and lift the compressor from the engine compartment **(see illustration)**.

Installation

5 If a new compressor is being installed, pour the oil from the old compressor into a graduated container and add that exact amount of new refrigerant oil to the new compressor. Also follow any directions included with the new compressor.

6 If only the compressor is being replaced, the clutch may be transferred from the original to the new compressor.

7 Place the compressor in position on the bracket and install the nuts and bolts finger tight. Once all the compressor mounting nuts and bolts are installed, tighten them securely.

8 Install the drivebelt(s) (see Chapter 1).

9 Reconnect the electrical connector to the compressor. Install the line fitting bolt(s) to the compressor, using new O-rings lubricated with clean refrigeration oil, and tighten them securely.

10 Reconnect the battery cable to the negative battery terminal.

11 Take the vehicle to a dealer service department or an air conditioning shop and have the system evacuated, recharged and leak tested.

14 Air conditioning condenser - removal and installation

Refer to illustrations 14.3, 14.4a, 14.4b, 14.5a, 14.5b, 14.6 and 14.7,

Warning: *The air conditioning system is under high pressure. DO NOT loosen any fittings or remove any components until after the system has been discharged. Air conditioning refrigerant should be properly discharged into an EPA-approved container at a dealer service department or an automotive air conditioning facility. Always wear eye protection when disconnecting air conditioning system fittings.*

1 Disconnect the battery cable at the negative battery terminal.

2 Unbolt the radiator from the core support and raise the radiator up about one inch and block it in that position to gain access to the two lower vertical core support bolts that are threaded in from the engine side of the core support. **Note:** *It won't be necessary to drain the cooling system or disconnect the hoses if the radiator is only being raised enough to remove the lower two bolts for the core support.*

3 Remove the upper vertical support brace bolts at the center of the core support **(see illustration).**

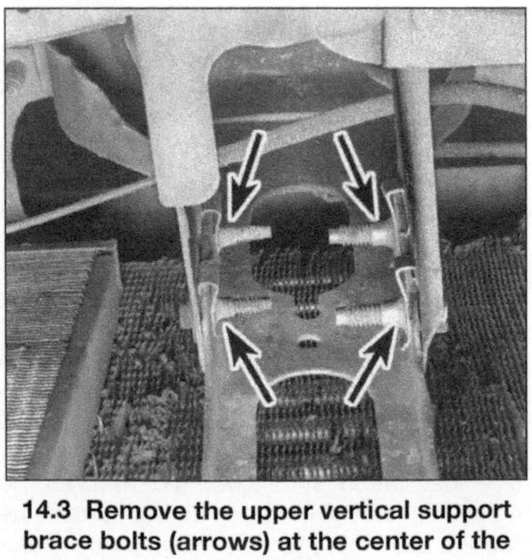

14.3 Remove the upper vertical support brace bolts (arrows) at the center of the core support

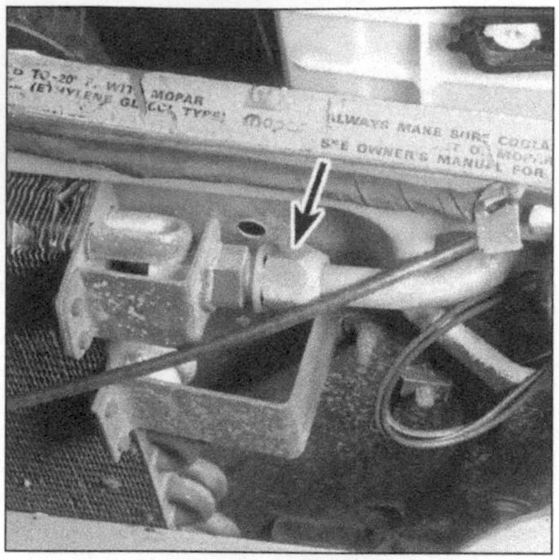

14.4a Disconnect the refrigerant line fittings (arrow) - depending on the model the connection may be of the type shown here . . .

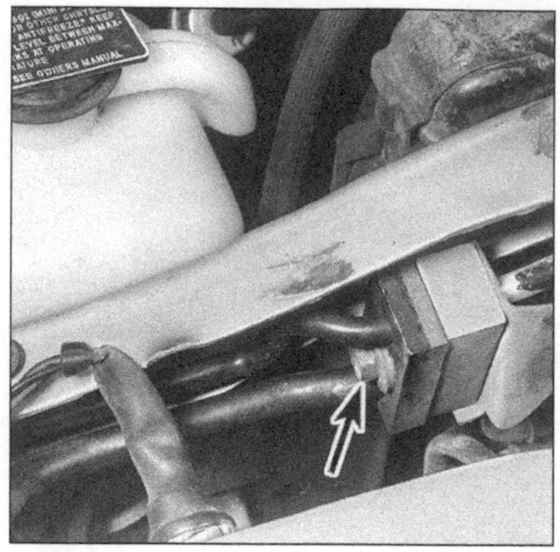

14.4b . . . or the refrigerant lines may be connected to the condenser (arrow) as shown here

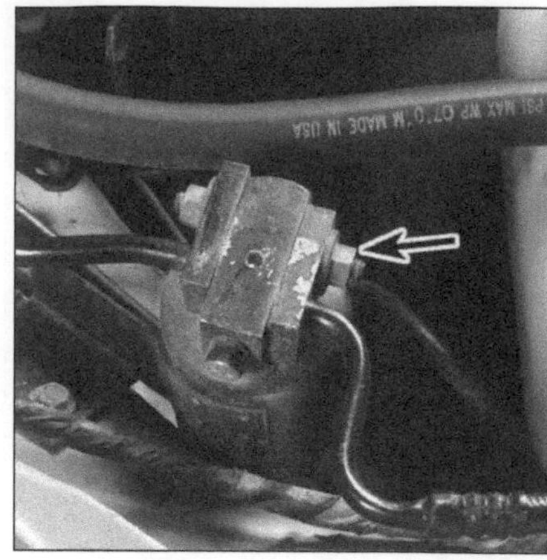

14.5a Disconnect the receiver-drier from the condenser high pressure hose (arrow)

14.5b If the receiver-drier is the style attached to the condenser it can be removed by loosening the clamp bolt, but it does not need to be removed; it can come out while still attached to it's mounting bracket on the condenser.

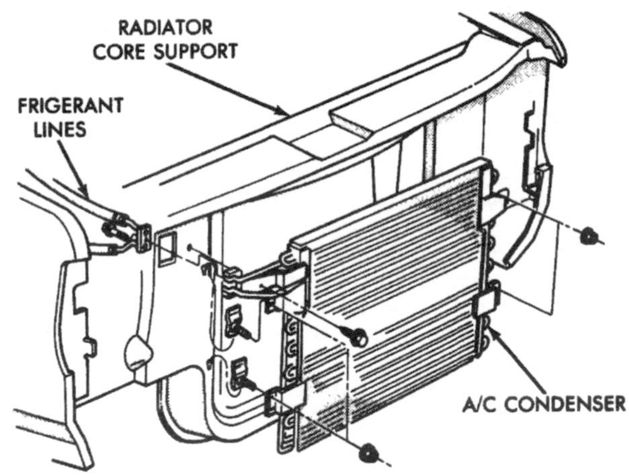

14.6 Unbolt the condenser from the core support by removing the bolts and nuts

4 Disconnect the refrigerant line fittings **(see illustrations)** and cap the open fittings to prevent the entry of dirt and moisture. **Caution:** *Always use a back-up wrench to avoid twisting off any refrigerant line connections.*

5 Disconnect the receiver-drier **(see illustrations)**. **Note:** *If the receiver-drier is the style attached to the condenser, it does not need to be removed - it can come out while still attached to it's mounting bracket on the condenser.*

6 Unbolt the condenser from the core support **(see illustration)**.

7 Remove the hood latch assembly bracket that is bolted to the core support **(see illustration)**.

8 Tilt the condenser forward and lift the condenser from the vehicle. **Caution:** *The condenser is made of aluminum - be careful not to damage it during removal.*

9 Installation is the reverse of removal. Be sure to use new O-rings on the refrigerant line fittings (lubricate the O-rings with clean refrigerant oil).

10 Have the system evacuated, recharged and leak tested by the shop that discharged it. If a new condenser is installed, add 1 oz of new refrigerant oil to the system.

14.7 Remove the hood latch assembly bracket (arrow)

15 Air conditioning receiver/drier - removal and installation

Refer to illustration 15.2

Removal

Warning: *The air conditioning system is under high pressure. DO NOT loosen any fittings or remove any components until after the system has been discharged. Air conditioning refrigerant should be properly discharged into an EPA-approved container at a dealer service department or an automotive air conditioning facility. Always wear eye protection when disconnecting air conditioning system fittings.*

1 Disconnect the battery cable at the negative battery terminal.

2 Disconnect the refrigerant inlet and outlet lines **(see illustration)**. Cap or plug the open lines immediately.

3 Remove the mounting bolt and slide the receiver-drier assembly up and out of the mounting bracket.

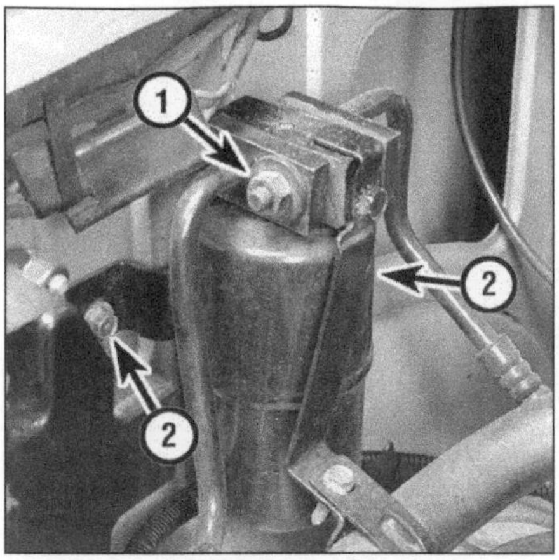

15.2 The refrigerant line connections may be the kind shown here, or a large tube nut on the refrigerant line that screws onto the receiver/drier

1 *Nut*
2 *Mounting bolts*

Installation

4 If you are replacing the receiver-drier, add one ounce of fresh refrigerant oil to the new unit.

5 Place the new receiver-drier into position, install the mounting bolt and tighten it securely.

6 Install the inlet and outlet lines, using clean refrigerant oil on the new O-rings. Tighten the through-bolt securely.

7 Connect the cable to the negative terminal of the battery.

8 Have the system evacuated, recharged and leak tested by a dealer service department or an air conditioning shop.

16 Air conditioning evaporator/heater core - removal and installation

Warning: *The air conditioning system is under high pressure. DO NOT loosen any fittings or remove any components until after the system has been discharged. Air conditioning refrigerant should be properly discharged into an EPA-approved container at a dealer service department or an automotive air conditioning facility. Always wear eye protection when disconnecting air conditioning system fittings.*

Removal

Refer to illustrations 16.6a, 16.6b, 16.6c, 16.12a and 16.12b

1 All models, except early C-body, require complete removal of the heater/evaporator case for either heater or evaporator core removal. Removal of the heater core *only* on some early C-body's can be accomplished without discharging the air conditioning system. Refer to Section 10 for removal of the heater core only on an early model C-body, otherwise follow the steps below.

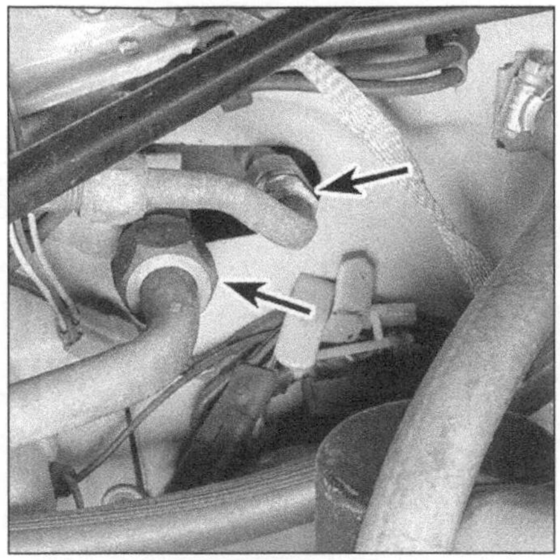

16.6a Disconnect the refrigerant lines (arrows) at the firewall

2 Have the air conditioning system discharged (see Warning above).

3 Disconnect the cable from the negative battery terminal.

4 Drain the cooling system (see Chapter 1).

5 Remove the heater hoses from the heater inlet and outlet tubes and plug the tubes. Remove the retainer and air seal, if equipped.

6 Locate the refrigerant lines to the evaporator, under the hood at the firewall area, and disconnect them **(see illustrations)**. Use a back-up wrench when removing the refrigerant lines to prevent twisting the tubing. **Note:** *Plug all refrigerant lines to keep moisture out of the system.*

7 Remove the grommet(s) that insulate the refrigerant lines from the firewall. Remove the drain tube.

8 Disconnect the thermoswitch electrical connector, if equipped. Label and remove the vacuum lines from the engine and push the lines and grommet through the firewall.

9 Remove the nuts from the studs retaining the heater/evaporator assembly to the firewall.

10 In the passenger compartment, remove any instrument panel trim, glove box, ash tray, center console, or other items necessary to provide adequate clearance for removal of the unit. Disconnect the control cables, vacuum harness, electrical connections and air distribution ducts from the heater/evaporator case **(see Sections 9 and 10)**.

16.6b On models with an expansion valve (arrow), separate the refrigerant lines from the valve . . .

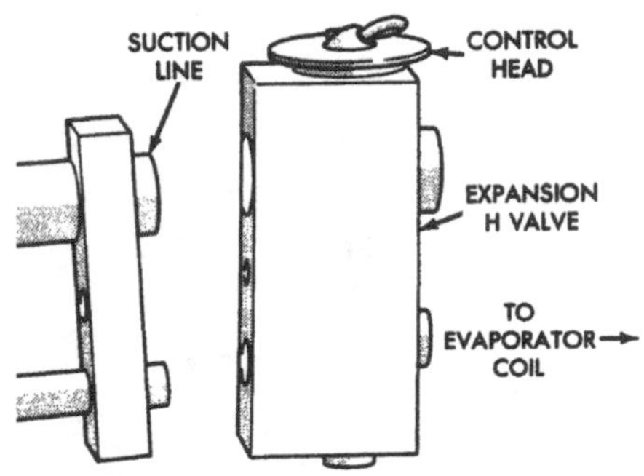

16.6c . . . then separate the valve from the evaporator coil at the firewall

11 Remove the mounting brace and remove the heater/evaporator case from under the instrument panel.

12 Disassemble the heater/evaporator case, depending on the year and model of the vehicle, by removing the screws and taking off the cover or separate the two halves

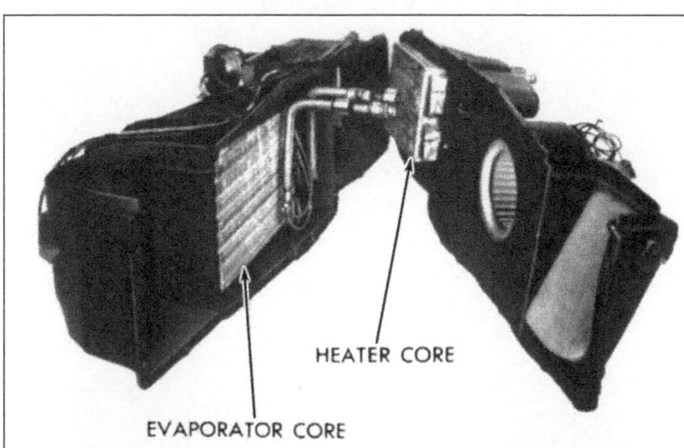

16.12a Separate the housing to gain access to the heater or evaporator core

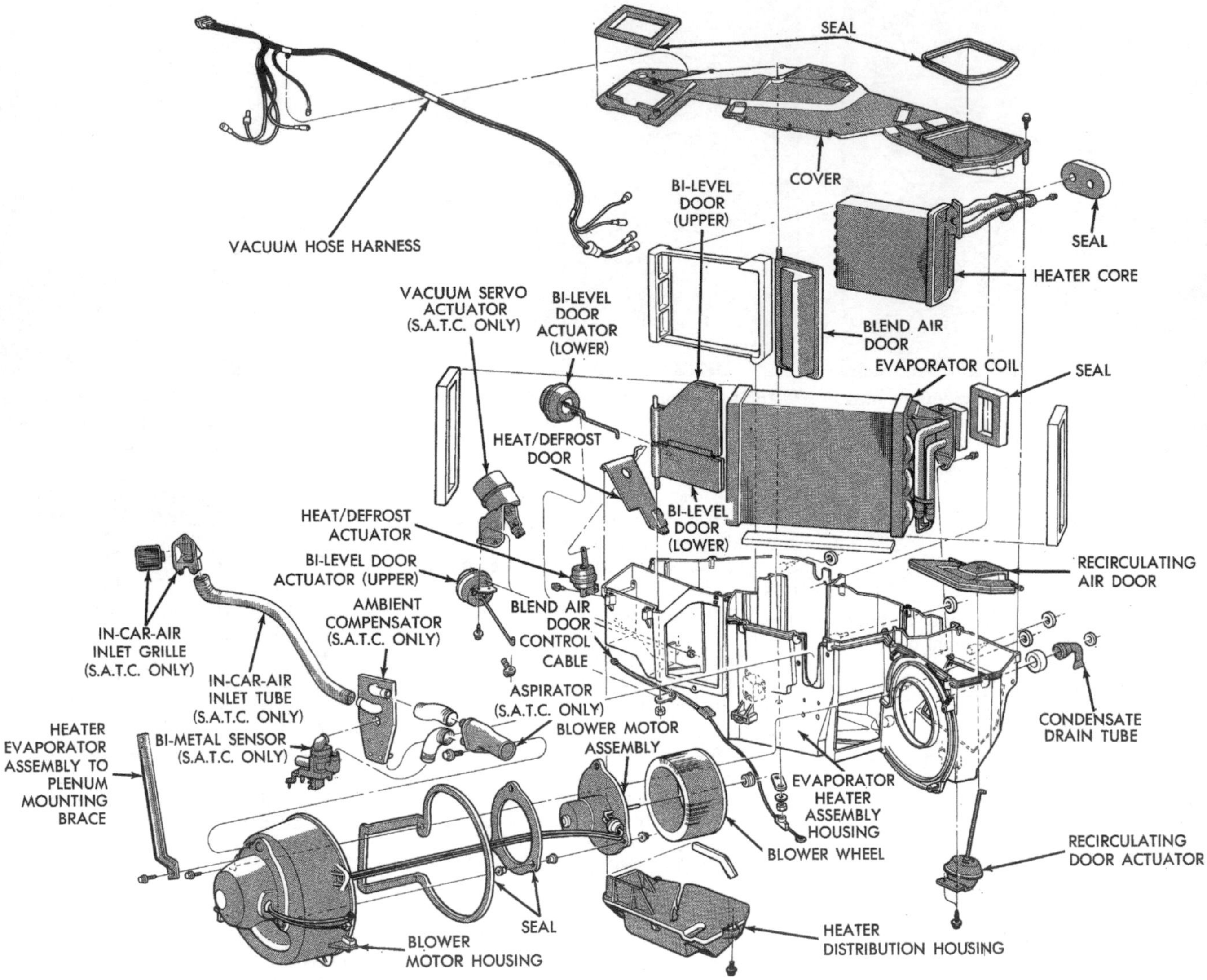

16.12b Exploded view of the heater/evaporator housing assembly (typical later models)

(see illustrations). Remove the hold-down clamps or brackets, then remove the evaporator and/or heater core.

Installation

10 Installation is the reverse of the removal procedure. Be sure to re-fill the cooling system (see Chapter 1). Have the air conditioning system evacuated, recharged and leak tested by the shop that discharged it.

Chapter 4
Fuel and exhaust systems

Contents

1 General information

The fuel system on all models consists of a fuel tank mounted in various locations on the chassis, a mechanically operated fuel pump and a carburetor. A combination of metal and rubber fuel hoses is used to connect these components.

The carburetor is either a one-, two- or four-barrel type, depending on engine displacement and year of production. Later models are equipped with electronic feedback carburetors. These carburetors are linked with a variety of sensors and output actuators to help control the amount of emissions.

Here is a list of the various carburetors used on these models:

One-barrel carburetors

Holley Models 1945 and 1920 (installed on early six-cylinder engines)

Holley Model 6145 (this is a feedback version of the Model 1945, used on later six-cylinder engines)

Carter BBS (installed on early 198 six-cylinder engines)

Two-barrel carburetors

Carter BBD (installed on early 225 six-cylinder and 318 V8 engines

Holley Model 2245 (installed on 360 and 400 V8 engines)

Holley Model 2210 (installed on 360 V8 engines)

Holley Model 2280 (installed on 318 V8 engines)

Holley Model 6280 (feedback version of Model 2280, installed on later 318 V8 engines)

Carter AVS (installed on early few early 383 and 440 V8 engines)

Rochester 2GV (installed on a few 1971 models only)

Four-barrel carburetors

Holley Model 4160 (installed on a few 1971 383 engines)

Carter Thermo-quad (installed on 360, 400 and 440 V8 engines - this is the most common four-barrel)

Rochester Quadrajet (installed onto some 1985 and later V8 engines)

The fuel system (especially the carburetor) is interrelated with the emissions control systems on all vehicles produced for sale in the United States. Components of the emissions control systems are described in Chapter 6.

2 Fuel pump/fuel pressure check

Refer to illustration 2.1

Warning: *Gasoline is extremely flammable, so take extra precautions when you work on any part of the fuel system. Don't smoke or allow open flames or bare light bulbs near the work area, and don't work in a garage where a natural gas-type appliance (such as a water heater or a clothes dryer) with a pilot light is present. Since gasoline is carcinogenic, wear latex gloves when there's a possibility of being exposed to fuel, and, if you spill any fuel on your skin, rinse it off immediately with soap and water. Mop up any spills immediately and do not store fuel-soaked rags where they could ignite. When you perform any kind of work on the fuel system, wear safety glasses and have a Class B type fire extinguisher on hand.*

Note: *It is a good idea to check the fuel pump and lines for any obvious damage or fuel leakage. Also check all hoses from the tank to the pump, particularly the suction hoses at the fuel tank and pump which, if they have are cracked, may not allow fuel to the fuel pump. It is also possible for the fuel pump diaphragm to rupture internally and leak fuel into the crankcase. If you have excess fuel consumption or fuel smell, and you can't find any external leaks, check the condition of the engine oil for any signs of fuel mixing with the engine oil; the oil level will usually be abnormally high and the oil will be thinned out and have a fuel smell.*

2.1 Install a fuel pressure gauge onto a T-fitting and connect the assembly between the fuel inlet and the carburetor bowl inlet valve at the carburetor

1 Disconnect the fuel line from the carburetor and install a T-fitting. Connect a fuel pressure gauge to the T-fitting with a section of fuel hose that is no longer than six inches **(see illustration)**.

2 Start the engine and allow it to idle. The pressure on the gauge should be 2-1/2 to 8 psi. It should remain constant and return to zero slowly when the engine is shut off.

Note: *If the engine will not start, crank the engine until you get a reading on the gauge.*

3 An instant pressure drop indicates a faulty outlet valve and the fuel pump must be replaced.

4 If the pressure is too high, check the air vent to see if it is plugged before replacing the pump.

5 If the pressure is too low, be sure the fuel hoses and lines are in good shape and not plugged, then replace the pump.

3 Fuel lines and fittings - repair and replacement

Warning: *Gasoline is extremely flammable, so take extra precautions when you work on any part of the fuel system. Don't smoke or allow open flames or bare light bulbs near the work area, and don't work in a garage where a natural gas-type appliance (such as a water heater or a clothes dryer) with a pilot light is present. Since gasoline is carcinogenic, wear latex gloves when there's a possibility of being exposed to fuel, and, if you spill any fuel on your skin, rinse it off immediately with soap and water. Mop up any spills immediately and do not store fuel-soaked rags where they could ignite. When you perform any kind of work on the fuel system, wear safety glasses and have a Class B type fire extinguisher on hand.*

Inspection

1 Once in a while, you will have to raise the vehicle to service or replace some component (an exhaust pipe hanger, for example). Whenever you work under the vehicle, always inspect the fuel lines and fittings for possible damage or deterioration.

2 Check all hoses and pipes for cracks, kinks, deformation or obstructions.

3 Make sure all hose and pipe clips attach their associated hoses or pipes securely to the underside of the vehicle.

4 Verify all hose clamps attaching rubber hoses to metal fuel lines or pipes are snug enough to assure a tight fit between the hoses and pipes.

Replacement

5 If you must replace any damaged sections, use hoses approved for use in fuel systems or pipes made from steel only (it's best to use an original-type pipe from a dealer that's already flared and pre-bent). Do not install substitutes constructed from inferior or inappropriate material, as this could result in a fuel leak and a fire.

6 Always, before detaching or disassembling any part of the fuel line system, note the routing of all hoses and pipes and the orientation of all clamps and clips to assure that replacement sections are installed in exactly the same manner.

7 Before detaching any part of the fuel system, be sure to relieve the fuel tank pressure by removing the fuel filler cap.

8 Always use new hose clamps after loosening or removing them.

9 While you're under the vehicle, it's a good idea to check the following related components:

a) *Check the condition of the fuel filter - make sure that it's not clogged or damaged (see Chapter 1).*

b) *Inspect the evaporative emission control (EVAP) system. Verify that all hoses are attached and in good condition (see Chapter 6).*

4 Fuel tank - removal and installation

Refer to illustrations 4.5, 4.6a, 4.6b, 4.8, 4.10 and 4.11

Warning: *Gasoline is extremely flammable, so take extra precautions when you work on any part of the fuel system. Don't smoke or allow open flames or bare light bulbs near the work area, and don't work in a garage where a natural gas-type appliance (such as a water heater or a clothes dryer) with a pilot light is present. Since gasoline is carcinogenic, wear latex gloves when there's a possibility of being exposed to fuel, and, if you spill any fuel on your skin, rinse it off immediately with soap and water. Mop up any spills immediately and do not store fuel-soaked rags where they could ignite. When you perform any kind of work on the fuel system, wear safety glasses and have a Class B type fire extinguisher on hand.*

Note: *The following procedure is much easier to perform if the fuel tank is empty. Some tanks have a drain plug for this purpose. If the tank does not have a drain plug, drain the fuel into an approved fuel container using a commercially available siphoning kit (NEVER start a siphoning action by mouth) or wait until the fuel tank is nearly empty, if possible.*

1 Remove the fuel tank filler cap to relieve fuel tank pressure.

2 Detach the cable from the negative terminal of the battery.

3 If the tank still has fuel in it, you can drain it at the fuel feed line after raising the vehicle. If the tank has a drain plug, remove it and allow the fuel to collect in an approved gasoline container.

4 Raise the vehicle and place it securely on jackstands.

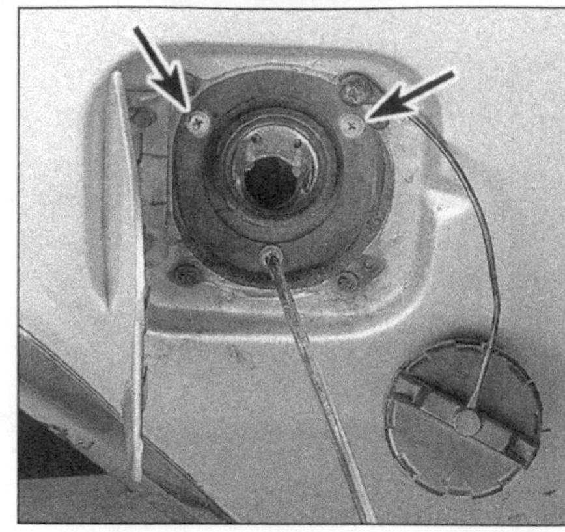

4.5 Remove the screws (arrows) and separate the fuel filler tube from the body of the vehicle (M-body shown)

5 Disconnect the fuel filler neck from the body of the vehicle **(see illustration)**.

6 Disconnect the fuel lines and the vapor return line **(see illustrations)**. **Note:** *The fuel feed and return lines and the vapor return line are three different diameters, so reattachment is simplified. If you have any doubts, however, clearly label the three lines and the fittings. Be sure to plug the hoses to prevent leakage and contamination of the fuel system.*

7 If there is still fuel in the tank, siphon it out from the fuel feed port. Remember - NEVER start the siphoning action by mouth! Use a siphoning kit, which can be purchased at most auto parts stores.

8 If the fuel tank is equipped with a protective shield, remove the nuts and detach it from the underside of the chassis **(see illustration)**.

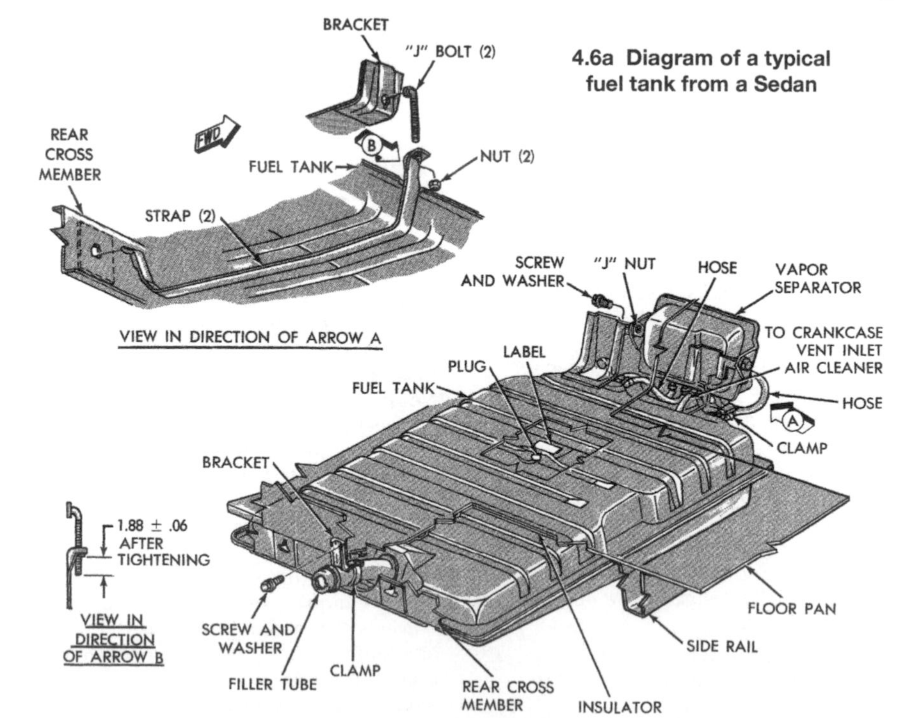

4.6a Diagram of a typical fuel tank from a Sedan

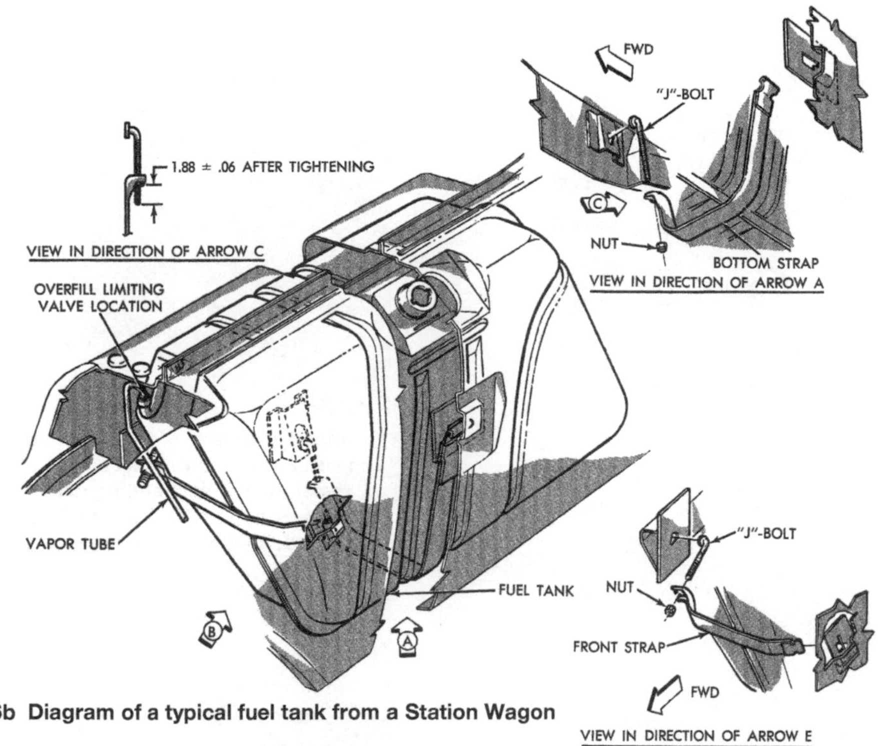

4.6b Diagram of a typical fuel tank from a Station Wagon

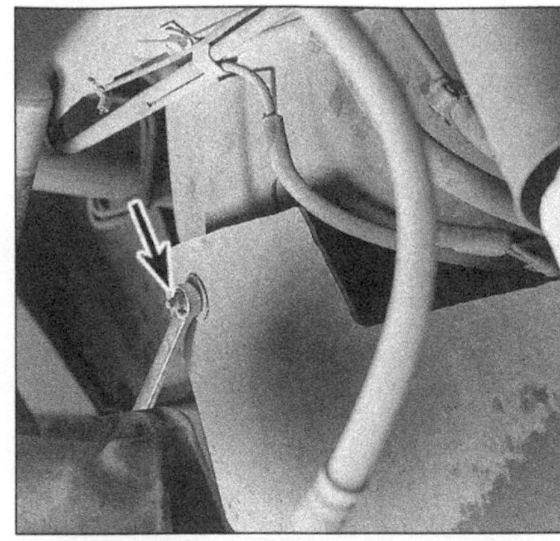

4.8 Remove the nuts (arrows) from the fuel tank protective shield and detach it from the underside of the vehicle

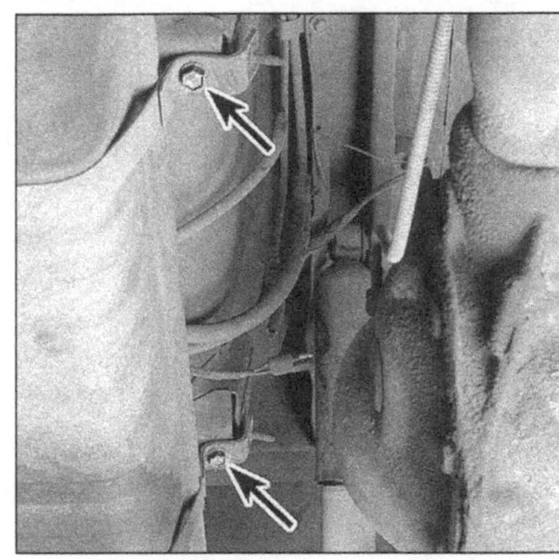

4.10 Unscrew the fuel tank strap bolts (arrows) from the fuel tank mounts

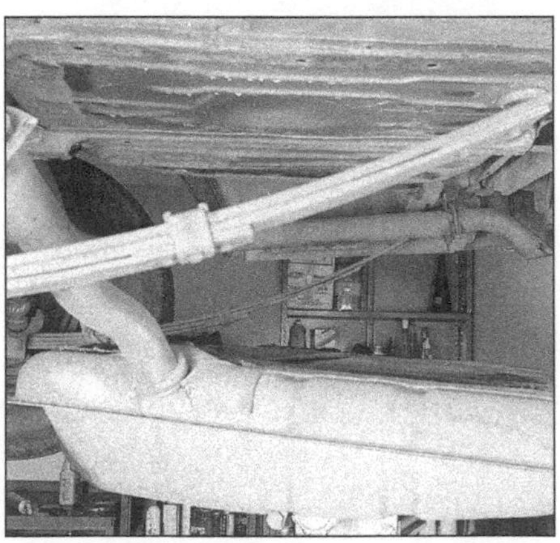

4.11 Carefully lower the fuel tank without damaging the fuel filler tube

9 Support the fuel tank with a floor jack. Position a piece of wood between the jack head and the fuel tank to protect the tank.

10 Disconnect both fuel tank retaining straps and pivot them down until they are hanging out of the way **(see illustration)**.

11 Lower the tank enough to disconnect the electrical wires and ground strap from the fuel pump/fuel gauge sending unit, if you have not already done so **(see illustration)**.

12 Remove the tank from the vehicle.

13 Installation is the reverse of removal.

5 Fuel tank cleaning and repair - general information

1 All repairs to the fuel tank or filler neck should be carried out by a professional who has experience in this critical and potentially dangerous work. Even after cleaning and flushing of the fuel system, explosive fumes can remain and ignite during repair of the tank.

2 If the fuel tank is removed from the vehicle, it should not be placed in an area where sparks or open flames could ignite the fumes coming out of the tank. Be especially careful inside garages where a natural gas-type appliance is located, because the pilot light could cause an explosion.

6 Fuel pump - removal and installation

Refer to illustrations 6.3 and 6.4

Warning: *Gasoline is extremely flammable, so take extra precautions when you work on any part of the fuel system. Don't smoke or allow open flames or bare light bulbs near the work area, and don't work in a garage where a natural gas-type appliance (such as a water heater or a clothes dryer) with a pilot light is present. Since gasoline is carcinogenic, wear latex gloves when there's a possibility of being exposed to fuel, and, if you spill any fuel on your skin, rinse it off immediately with soap and water. Mop up any spills immediately and do not store fuel-soaked rags where they could ignite. When you perform any kind of work on the fuel system, wear safety glasses and have a Class B type fire extinguisher on hand.*

1 On V8 engines, the fuel pump is located on the right (passenger's) side of the engine, towards the front of the engine, either on or near the timing chain cover. On six-cylinder engines, the pump is also located on the right side of the engine, but it's closer to the center of the engine.

2 On all models, place rags under the fuel pump to catch any gasoline which is spilled during removal.

3 Carefully unscrew the fuel line fittings and detach the lines from the pump. A flare-nut wrench along with a backup wrench

6.3 Remove the two bolts (arrows) from the fuel pump and lift the assembly from the engine compartment

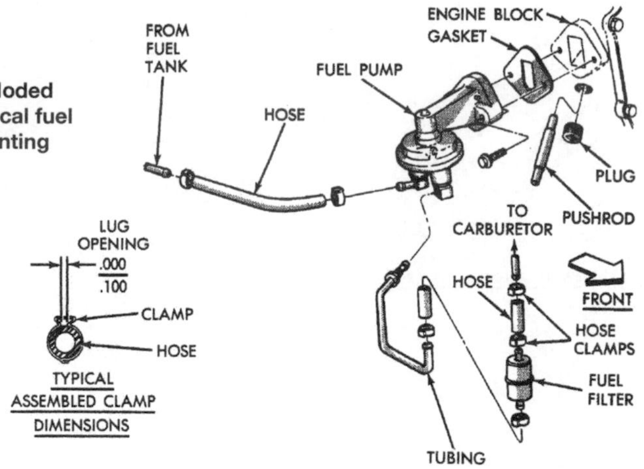

6.4 An exploded view of a typical fuel pump mounting

7.4 Install the leads of the ohmmeter onto the sending unit terminals and check the resistance of the sending unit. This particular fuel tank is almost empty, as indicated by the high resistance value

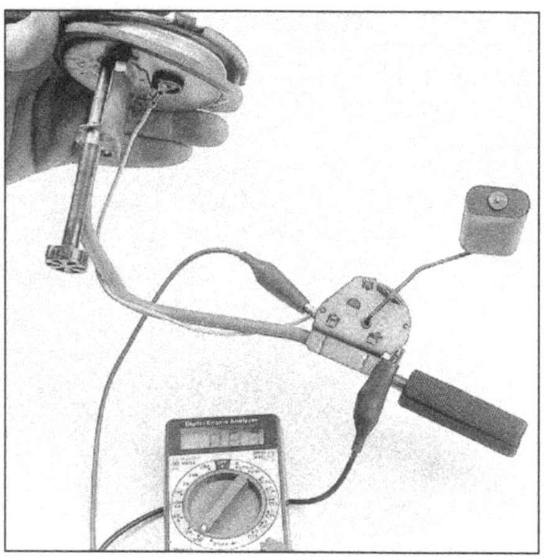

7.7 Install the probes of the ohmmeter onto the terminals of the sending unit, then move the float through the complete length of travel and observe the resistance. The sending unit should indicate low resistance when it reaches the top of its travel distance (full tank) and then high resistance when it reaches the bottom of its travel (empty tank)

7.11 Use a brass punch to tap the sending unit assembly cover until the tabs align with the indentations on the fuel tank ring

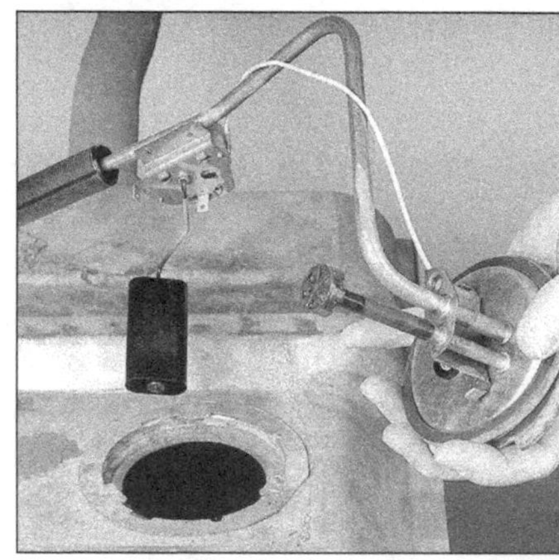

7.12 Carefully lift the sending unit from the fuel tank without damaging the float unit

should be used on the pressure-side fitting to prevent damage to the line fittings **(see illustration)**.

4 Unbolt and remove the fuel pump **(see illustration)**.

5 Before installation, coat both sides of the gasket surface with RTV sealant, position the gasket and fuel pump against the block and install the bolts, tightening them securely.

6 Attach the lines to the pump and tighten the pressure fitting securely (use a flare-nut wrench, if one is available, to prevent damage to the fittings).

7 Run the engine and check for leaks.

7 Fuel level sending unit - check and replacement

Refer to illustrations 7.4, 7.7, 7.11 and 7.12

Warning: *Gasoline is extremely flammable, so take extra precautions when you work on any part of the fuel system. Don't smoke or allow open flames or bare light bulbs near the work area, and don't work in a garage where a natural gas-type appliance (such as a water heater or a clothes dryer) with a pilot light is present. Since gasoline is carcinogenic, wear latex gloves when there's a possibility of being exposed to fuel, and, if you spill any fuel on your skin, rinse it off immediately with soap and water. Mop up any spills immediately and do not store fuel-soaked rags where they could ignite. When you perform any kind of work on the fuel system, wear safety glasses and have a Class B type fire extinguisher on hand.*

Check

1 Before performing any tests on the fuel level sending unit, completely fill the tank with fuel.

2 Raise the vehicle and secure it with jackstands.

3 Disconnect the fuel level sending unit electrical connector located on the fuel tank, usually at the rear.

4 Position the ohmmeter probes onto the electrical connector of the sending unit and check for resistance **(see illustration)**. Use the 200-ohm scale on the ohmmeter.

5 With the fuel tank completely full, the resistance should be about 5.0 to 8.0 ohms.

6 Reconnect the electrical connector and lower the vehicle and drive it until the tank is nearly empty.

7 Check the resistance. The resistance of the sending unit should be about 60 to 75 ohms. **Note:** *If there is any doubt concerning the resistance of the sending unit, remove the sending unit (see Steps 9 through 14) and check it by attaching the ohmmeter leads directly to the electrical terminals* **(see illustration)**.

8 If the readings are incorrect, replace the sending unit.

Replacement

9 Remove the fuel tank from the vehicle (see Section 4).

10 If necessary, carefully peel back the insulation material from the top of the fuel tank without damaging it.

11 Using a brass punch and hammer, tap the inside sending unit cover **(see illustration)** until the tabs align with the indents on the fuel tank retaining ring.

8.2 Using a small screwdriver, pry the tabs on the electrical connectors and separate them from the computer (models with a Spark Control Computer)

8.3 Lift the air cleaner assembly from the engine compartment

9.1a Remove the retainer clip from the throttle shaft and slide the accelerator cable off the shaft

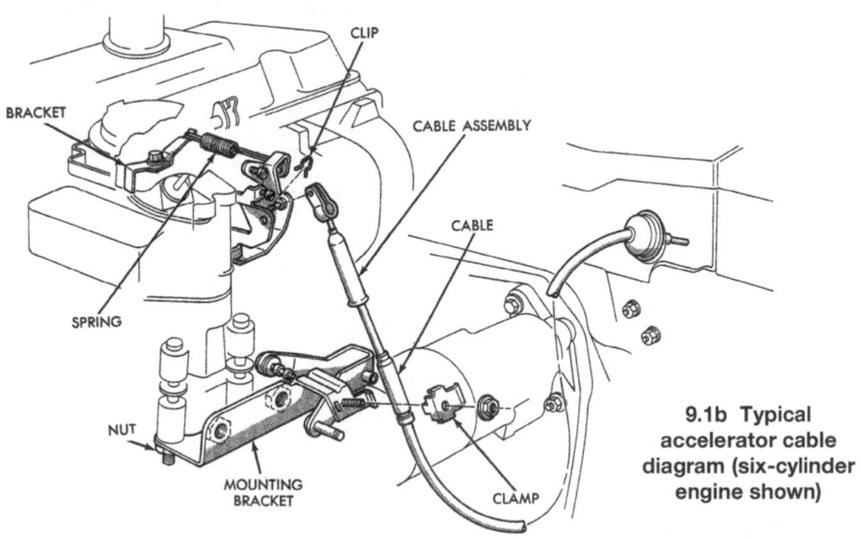

9.1b Typical accelerator cable diagram (six-cylinder engine shown)

12 Lift the sending unit from the tank. Carefully angle the sending unit out of the opening without damaging the fuel level float located at the bottom of the assembly **(see illustration)**.
13 Disconnect the electrical connector from the sending unit.
14 Installation is the reverse of removal.

8 Air cleaner housing - removal and installation

Refer to illustrations 8.2 and 8.3

1 Remove the air cleaner from the air cleaner housing (see Chapter 1).
2 Disconnect any vacuum hose or electrical connectors that are attached to the air cleaner **(see illustration)** and mark them with paint or marked pieces of tape for reassembly purposes.
3 Lift the air cleaner housing from the engine compartment **(see illustration)**.
4 Installation is the reverse of removal.

9 Accelerator cable - removal, installation and adjustment

Refer to illustrations 9.1a, 9.1b, 9.2 and 9.4

Removal

1 Remove the accelerator cable from the throttle linkage by prying the retainer clip off the shaft using a small screwdriver **(see illustrations)**.
2 Loosen the accelerator cable locknut and remove the cable from the bracket assembly **(see illustration)**.
3 Detach the screws and the clips retaining the lower instrument trim panel on the driver's side and remove the trim piece (see Chapter 11).
4 Pull the cable end out and then up from the accelerator pedal recess **(see illustration)**.

9.2 Loosen the clamp nut (arrow) and remove the cable from the clamp

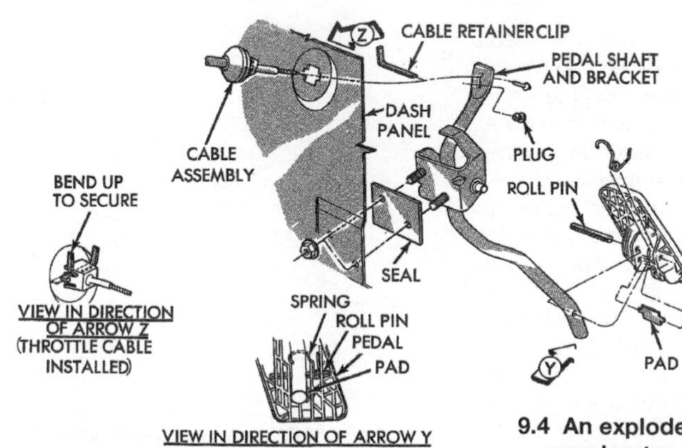

9.4 An exploded view of a typical accelerator cable and pedal

5 Push the cable through the firewall into the engine compartment. To disconnect the cable at the firewall, remove the cable retainer clip, if equipped, then use pliers to squeeze the tabs on the cable retainer from inside the vehicle **(see illustration 9.4, View Z)**.

Installation

6 Installation is the reverse of removal. **Note:** *To prevent possible interference, flexible components (hoses, wires, etc.) must not be routed within two inches of moving parts, unless routing is controlled.*

7 Operate the accelerator pedal and check for any binding condition by completely opening and closing the throttle.

8 If necessary, at the engine compartment side of the firewall, apply sealant around the accelerator cable to prevent water from entering the passenger compartment.

Adjustment

9 Warm the engine until normal operating temperature is attained and allow the engine to idle. Set the specified idle speed, if necessary.

10 Loosen the cable clamp **(see illustration 9.2)** and slide the accelerator cable away from the carburetor until all the cable slack has been removed, but don't move the throttle lever on the carburetor.

11 Slide the cable back toward the carburetor to allow 1/4-inch play in the cable tension. Tighten the clamp assembly.

12 Raise the engine rpms and then allow the engine to settle back down to idle. Check for proper adjustment and carburetor throttle actuation and re-adjust the cable if necessary.

13 Shut off the engine and remove the air cleaner assembly. Have an assistant press the accelerator all the way to the floor while you observe the movement of the throttle plate(s). The throttle plate(s) must be fully open (completely vertical). If not, readjustment is necessary.

10 Carburetor - diagnosis and overhaul

Warning: *Gasoline is extremely flammable, so take extra precautions when you work on any part of the fuel system. Don't smoke or allow open flames or bare light bulbs near the work area, and don't work in a garage where a natural gas-type appliance (such as a water heater or a clothes dryer) with a pilot light is present. Since gasoline is carcinogenic, wear latex gloves when there's a possibility of being exposed to fuel, and, if you spill any fuel on your skin, rinse it off immediately with soap and water. Mop up any spills immediately and do not store fuel-soaked rags where they could ignite. When you perform any kind of work on the fuel system, wear safety glasses and have a Class B type fire extinguisher on hand.*

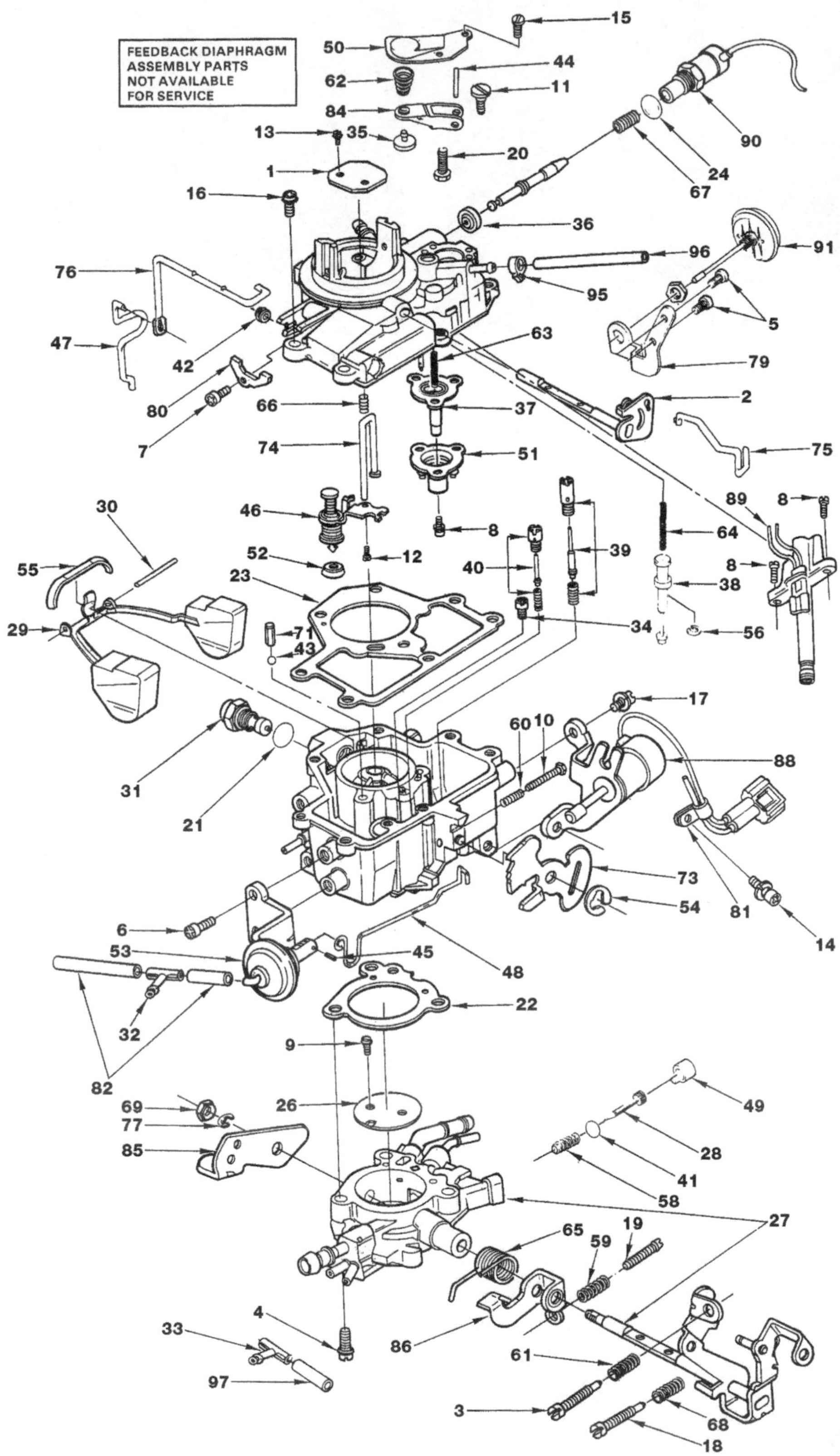

10.7a An exploded view of the Holley Models 1945 and 6145 carburetors

Index Number	Part Name	Index Number	Part Name
1	Choke Plate	50	ECS Vent Cover
2	Choke Shaft & Lever Assembly	51	Power Valve Diaphragm Body
3	Fast Idle Adjusting Screw	52	Accelerator Pump Cup
4	Throttle Body Screw & L.W.	53	Choke Diaphragm Assy.
5	Dashpot Bracket Screw	54	Fast Idle Cam Retainer
6	Choke Diaphragm Bracket Screw	55	Float Shaft Retainer
7	Pump Rod Clamp Screw	56	Power Valve Piston Retainer
8	Power Valve Diaphragm Body Screw & L.W. (Duty Cycle Solenoid)	57	Wire Tie-Down Strap*
		58	Idle Adj. Needle Spring
9	Throttle Plate Screw	59	Power Valve Lever Adj. Screw Spring
10	Curb Idle Adjusting Screw		
11	ECS Hinge Retaining Screw	60	Curb Idle Adj. Screw Spring
12	Retaining Plate Screw	61	Fast Idle Adj. Screw Spring
13	Choke Plate Screw	62	ECS Vent Spring
14	Wire Hold-Down Clamp Screw	63	Power Valve Diaphragm Spring
15	ECS Vent Cover Screw	64	Power Valve Piston Spring
16	Air Horn Screw & L.W.	65	Power Valve Lever Return Spring
17	Idle Solenoid Bracket Screw & L.W	66	Power Valve Rod Drive Spring
18	Solenoid Adjusting Screw	67	Switch Vent Solenoid Spring
19	Power Valve Lever Adjusting Screw	68	Idle Solenoid Adj. Screw Spring
20	ECS Vent Valve Adjusting Screw	69	Pump Operating Lever Nut
21	Fuel Inlet Fitting (Needle & Seat Assy.) Gasket	70	Dashpot Lock Nut
		71	Pump Discharge Valve Weight
22	Throttle Body Gasket	72	Power Valve Diaphragm Spacer*
23	Main Body Gasket	73	Fast Idle Cam
24	Switch Vent Solenoid Gasket	74	Power Valve Rod
25	Power Valve Spacer Gasket	75	Fast Idle Rod
26	Throttle Plate	76	Pump Operating Rod
27	Throttle Body & Shaft Assy.	77	Pump Operating Lever Nut L.W.
28	Idle Adjusting Needle	78	Insulator Washer*
29	Float & Hinge Assy.	79	Dashpot Bracket
30	Float Hinge Shaft	80	Pump Rod Clamp
31	Fuel Inlet Needle & Seat Assy.	81	Wire Hold-Down Clamp
32	Choke Diaphragm Hose Connector	82	Choke Diaphragm Vacuum Hose
33	Spark Tube Hose Connector	83	Idle Screw Insulator*
34	Main Jet	84	ECS Vent Lever
35	ECS Vent Valve	85	Pump Operating Lever
36	Switch Vent Seal	86	Power Valve Lever
37	Power Valve Diaphragm & Stem Assy.	87	Idle Stop Switch*
		88	Idle Stop Solenoid
38	Power Valve Piston	89	Duty Cycle Solenoid
39	Power Valve Assembly (Vacuum)	90	Switch Vent Solenoid & Armature Assy.
40	Power Valve Assembly (Mech.)		
41	Idle Adj. Needle "0" Ring	91	Dashpot Assy.
42	Pump-Rod Seal	92	A/C Speed-Up Solenoid*
43	Pump Discharge Valve Checkball	93	T.P.T. Assy.*
44	ECS Lever Hinge Pin	94	T.P.T. Bracket*
45	Choke Diaphragm Link Pin	95	Power Valve Vacuum Hose Clamp
46	Pump Stem & Retainer Plate Assy.	96	Power Valve Vacuum Hose
47	Pump Operating Link	97	Spark Tube Vacuum Hose
48	Choke Diaphragm Link		
49	Idle Needle Limiter Cap or Plug		

*Parts not shown on T.V. 44-2.

Parts list for the Holley Models 1945 and 6145 carburetors

Diagnosis

1 A thorough road test and check of carburetor adjustments should be done before any major carburetor service. Specifications for some adjustments are listed on the *Vehicle Emissions Control Information* (VECI) label found in the engine compartment.

2 Carburetor problems usually show up as flooding, hard starting, stalling, severe backfiring and poor acceleration. A carburetor that's leaking fuel and/or covered with wet looking deposits definitely needs attention.

3 Some performance complaints directed at the carburetor are actually a result of loose, out-of-adjustment or malfunctioning engine or electrical components. Others develop when vacuum hoses leak, are disconnected or are incorrectly routed. The proper approach to analyzing carburetor problems should include the following items:

a) Inspect all vacuum hoses and actuators for leaks and correct installation (see Chapters 1 and 6).
b) Tighten the intake manifold and carburetor mounting nuts/bolts evenly and securely.
c) Perform a compression test and vacuum test (see Chapter 2C).
d) Clean or replace the spark plugs as necessary (see Chapter 1).
e) Check the spark plug wires (see Chapter 1).
f) Inspect the ignition primary wires.
g) Check the ignition timing (follow the instructions printed on the Emissions Control Information label).
h) Check the fuel pump pressure/volume (see Section 2).
i) Check the heat control valve in the air cleaner for proper operation (see Chapter 1).
j) Check/replace the air filter element (see Chapter 1).
k) Check the PCV system (see Chapter 6).
l) Check/replace the fuel filter (see Chapter 1). Also, the strainer in the tank could be restricted.
m) Check for a plugged exhaust system.
n) Check EGR valve operation (see Chapter 6).
o) Check the choke - it should be completely open at normal engine operating temperature (see Chapter 1).
p) Check for fuel leaks and kinked or dented fuel lines (see Chapters 1 and 4)
q) Check accelerator pump operation with the engine off (remove the air cleaner cover and operate the throttle as you look into the carburetor throat - you should see a stream of gasoline enter the carburetor).
r) Check for incorrect fuel or bad gasoline.
s) Check the valve clearances (if applicable) and camshaft lobe lift (see Chapters 1 and 2)
t) Have a dealer service department or repair shop check the electronic engine and carburetor controls.

4 Diagnosing carburetor problems may require that the engine be started and run with the air cleaner off. While running the engine without the air cleaner, backfires are possible. This situation is likely to occur if the carburetor is malfunctioning, but just the removal of the air cleaner can lean the fuel/air mixture enough to produce an engine backfire. **Warning:** *Do not position any part of your body, especially your face, directly over the carburetor during inspection and servicing procedures. Wear eye protection!*

Overhaul

Refer to illustrations 10.7a through 10.7j

5 Once it's determined that the carburetor needs an overhaul, several options are available. If you're going to attempt to overhaul the carburetor yourself, first obtain a good-quality carburetor rebuild kit (which will include all necessary gaskets, internal parts, instructions and a parts list). You'll also need some special solvent and a means of blowing out the internal passages of the carburetor with air.

6 An alternative is to obtain a new or rebuilt carburetor. They are readily available from dealers and auto parts stores. Make absolutely sure the exchange carburetor is identical to the original. A tag is usually attached to the top of the carburetor or a number is stamped on the float bowl. It will help determine the exact type of carburetor you have. When obtaining a rebuilt carburetor or a rebuild kit, make sure the kit or carburetor matches your application exactly. Seemingly insignificant differences can make a large difference in engine performance.

7 If you choose to overhaul your own carburetor, allow enough time to disassemble it carefully, soak the necessary parts in the cleaning solvent (usually for at least one-half

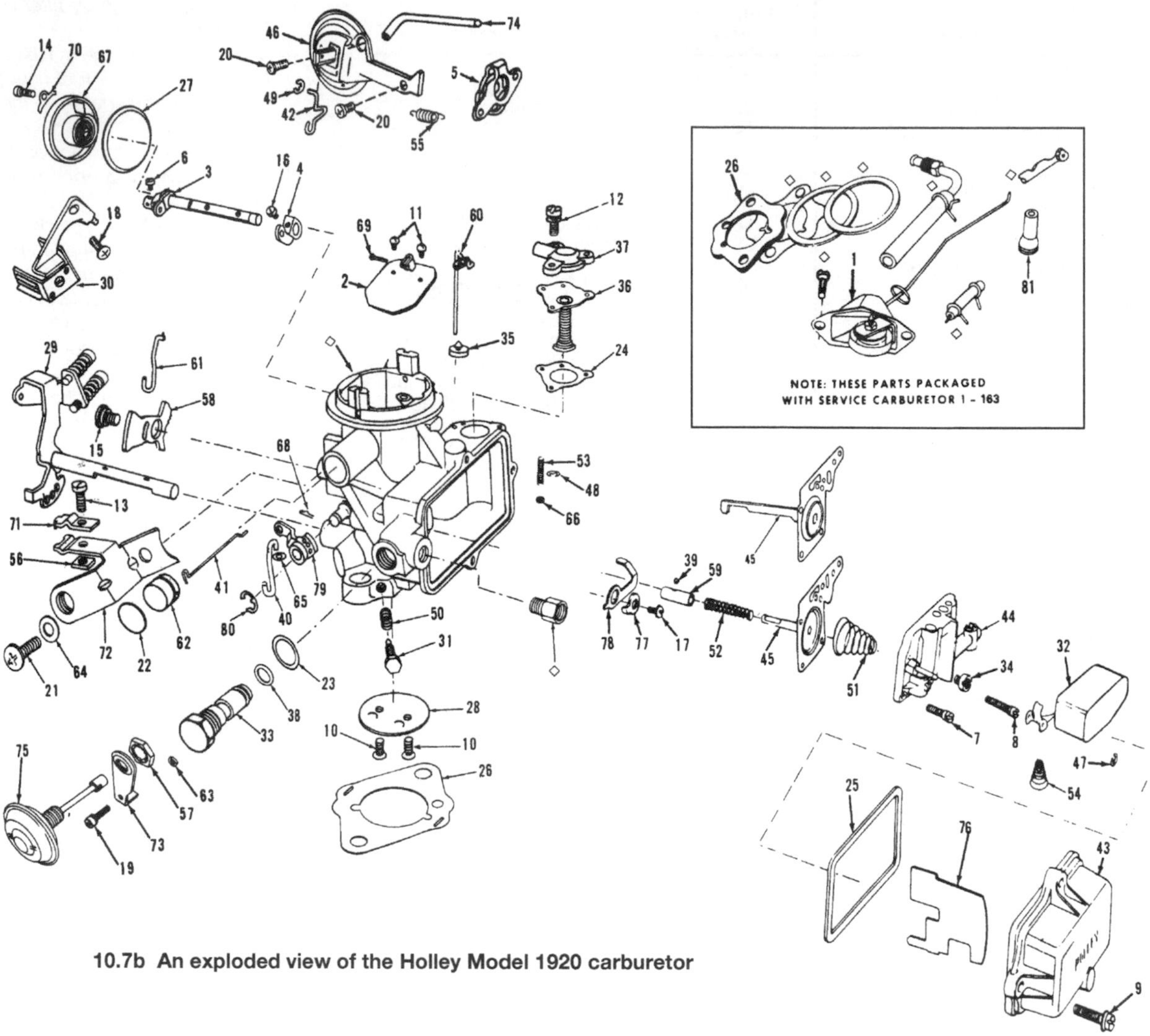

10.7b An exploded view of the Holley Model 1920 carburetor

Index Number	Part Name
1	Automatic Choke Assembly
2	Choke Plate Assembly
3	Choke Shaft & Lever Assy.
4	Choke Lever
5	Choke Control Lever
6	Choke Control Lev. Swivel Scr.
7	Main Well & Economizer Body Scr. & L.W. (Short)
8	Main Well & Economizer Body Scr. & L.W. (Long)
9	Fuel Bowl Scr. & L.W.
10	Throt. Plate Scr. & L.W.
11	Choke Plate Scr. & L.W.
12	Econ. Body Scr. & L.W.
13	Wire Clamp Scr. & L.W.
14	Therm. Cover Clamp Scr. & L.W.
15	Fast Idle Cam Screw
16	Choke Lever Screw
17	Mechanical Vent Scr. & L.W.
18	Throt. Adapter Scr. & L.W.
19	Dashpot Bracket Screw
20	Choke Brkt. Retainer Screw
21	Wire Bracket Screw
22	Choke Piston Plug
23	Fuel Inlet Seat Gasket
24	Economizer Body Gasket
25	Float Bowl Gasket
26	Flange Gasket

Index Number	Part Name
27	Therm. Housing Gasket
28	Throttle Plate
29	Throt. Lever & Shaft Assy.
30	Throt. Lever Drive Assy.
31	Idle Adjusting Needle
32	Float & Lever Assy.
33	Fuel Inlet Needle & Seat Assy.
34	Main Jet
35	Mechanical Vent Valve
36	Economizer Stem Assembly
37	Economizer Body Cover Assy.
38	Fuel Inlet Seat "O" Ring Seal
39	Diaphragm Push Rod Sleeve Ball
40	Pump Operating Link
41	Choke Piston Link
42	Choke Diaphragm Link
43	Float Bowl
44	Main Well & Econ. Body & Plugs Assy.
45	Pump Diaphragm & Rod Assy.
46	Choke Diaphragm Assy. Complete
47	Float Spring Retainer
48	Mechanical Vent Spring Retainer
49	Choke Diaphragm Link Retainer
50	Idle Adjusting Needle Spring
51	Pump Return Spring
52	Pump Operating Spring
53	Mechanical Vent Spring
54	Float Spring

Index Number	Part Name
55	Choke Lever Spring
56	Wire Clamp Screw Nut
57	Dashpot Nut
58	Fast Idle Cam
59	Pump Push Rod Sleeve
60	Mechanical Vent Rod
61	Fast Idle Rod
62	Choke Piston Assy.
63	Dashpot Bracket Lock Washer
64	Wire Brkt. Scr. L.W.
65	Pump Operating Link Washer
66	Mechanical Vent Spring Washer
67	Choke Therm. & Cap Assy.
68	Pump Operating Link Retainer
69	Choke Piston Link Retainer Ping
70	Choke Therm. Cover Clamp
71	Wire Clamp
72	Wire Bracket
73	Dashpot Bracket
74	Choke Vacuum Hose
75	Dashpot Assembly
76	Fuel Inlet Baffle Assy.
77	Mechanical Vent Control Lever
78	Mechanical Vent Operating Lever
79	Pump Operating Lever
80	Pump Operating Lever Retainer
81	Throttle Lever Bushing

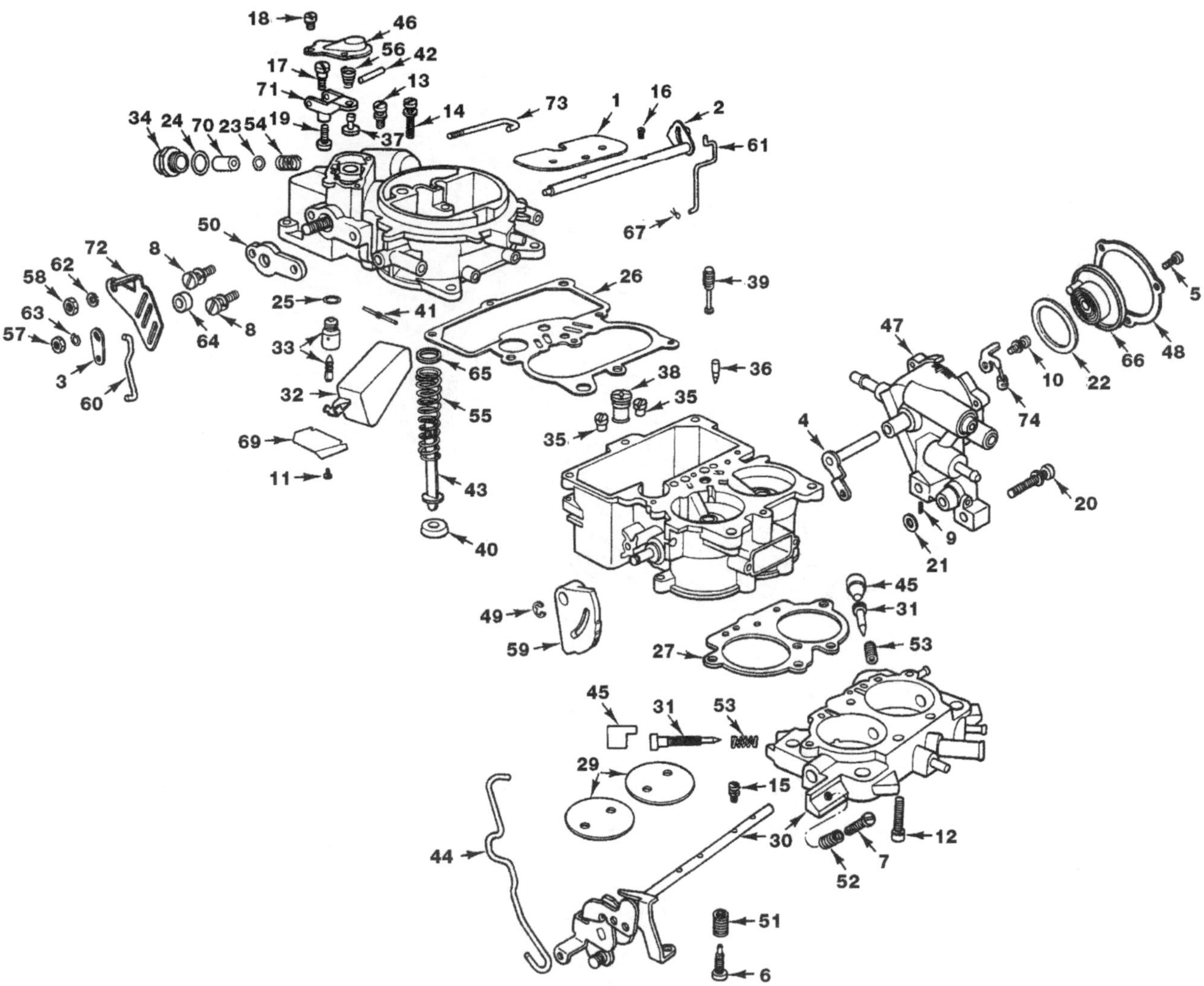

10.7c An exploded view of the Holley Models 2210 and 2245 carburetors

Index Number	Part Name
1	Choke Plate
2	Choke Shaft & Lever Assembly
3	Fast Idle Lever
4	Choke Housing Lever & Shaft Assembly
5	Thermostat Cap Clamp Screw
6	Fast Idle Adjusting Screw
7	Curb Idle Adjusting Screw
8	Pump Shaft Lever Screw
9	Choke Piston Adjusting Screw
10	Choke Lever screw & L.W.
11	Fuel Bowl Baffle Screw & L.W.
12	Throttle Body Screw & L.W.
13	Air Horn Screw & L.W. (Short)
14	Air Horn Screw & L.W. (Long)
15	Throttle Plate Screw & L.W.
16	Choke Plate Screw
17	E.C.S. Hinge Retainer Screw
18	E.C.S. Vent Cover Screw
19	Vent Valve Adjusting Screw
20	Choke Housing Screw & L.W.
21	Choke Housing Gasket
22	Thermostat Housing Gasket
23	Fuel Inlet Filter Gasket
24	Fuel Inlet Fitting Gasket
25	Fuel Inlet Seat Gasket
26	Main Body Gasket
27	Throttle Body Gasket
28	Flange Gasket
29	Throttle Plate
30	Throttle Body & Shaft Assembly
31	Idle Adjusting Needle
32	Float & Hinge Assembly
33	Fuel Inlet Needle & Seat Assembly
34	Fuel Inlet Fitting
35	Main Jet
36	Pump Discharge Needle Valve
37	Air Vent Valve
38	Power Valve Assembly
39	Power Valve Piston Assembly
40	Pump Piston Cup
41	Float Hinge Pin
42	E.C.S. Lever Hinge Pin
43	Pump Stem Assembly
44	Pump Link
45	Idle Needle Limiter Cap
46	E.C.S. Vent Cover
47	Choke Housing Assembly
48	Thermostat Cap Retainer Ring
49	"E" Ring Retainer
50	Pump Shaft Retainer
51	Fast Idle Adjusting Screw Spring
52	Curb Idle Adjusting Screw Spring
53	Idle Needle Adjusting Screw Spring
54	Fuel Inlet Filter Spring
55	Pump Drive Spring
56	E.C.S. Vent Spring
57	Fast Idle Lever Nut
58	Pump Lever Nut
59	Fast Idle Cam
60	Fast Idle Rod
61	Choke Rod
62	Pump Lever L.W.
63	Fast Idle Lever L.W.
64	Washer
65	Pump Drive Spring Washer
66	Thermostat & Cap Assembly
67	Rod Retainer
68	Choke Vacuum Hose
69	Fuel Bowl Baffle
70	Fuel Inlet Filter
71	Vent Lever
72	Pump Lever
73	Pump Lever Shaft
74	Choke Thermostat Lever & Piston Assembly

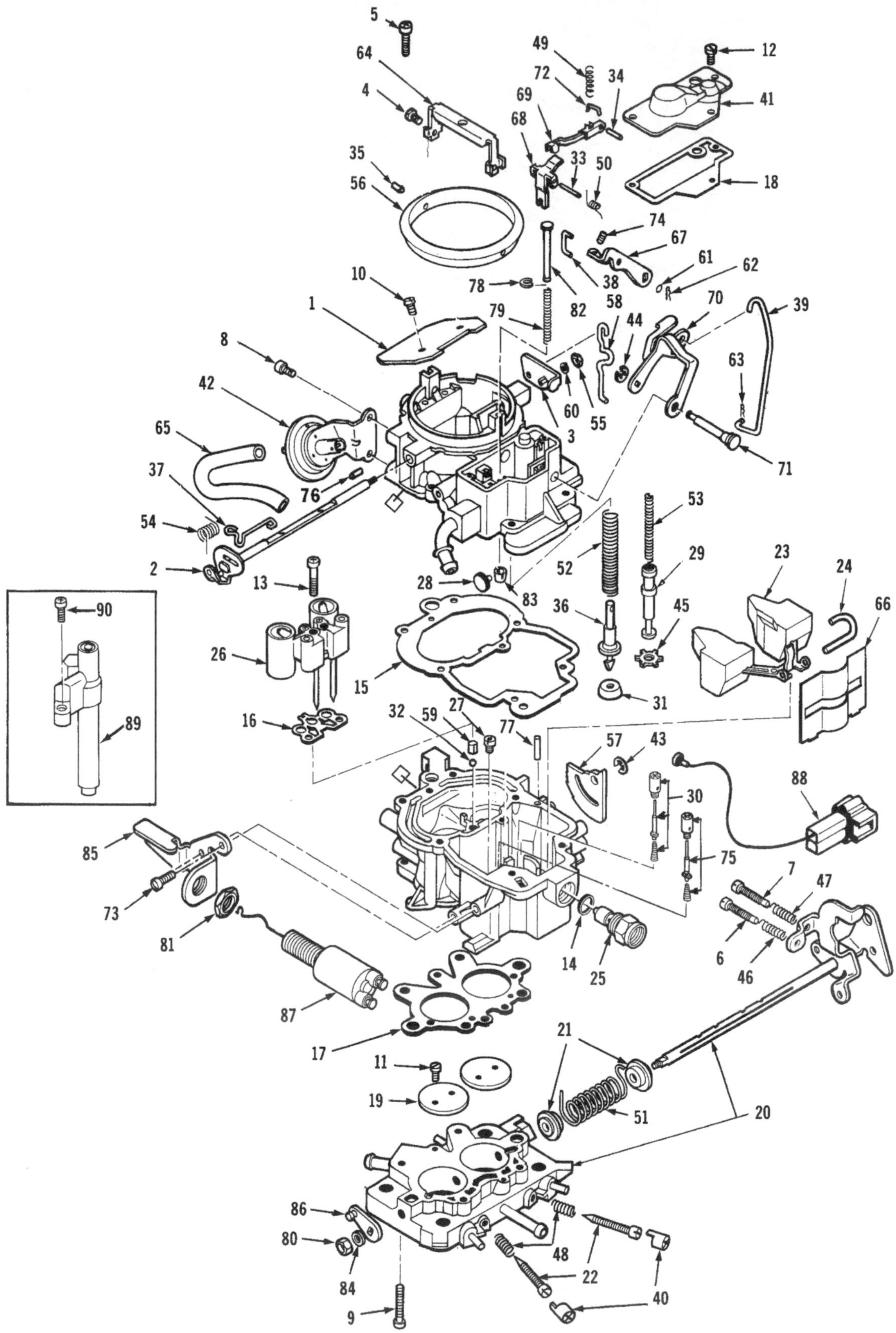

10.7d An exploded view of the Holley Models 2280 and 6280 carburetors

Index Number	Part Name	Index Number	Part Name
1	Choke Plate	43	Fast Idle Cam Retainer
2	Choke Shaft & Lever Assembly	44	Pump Operating Lever Retainer
3	Dechoke Lever	45	Power Valve Piston Retainer
4	Air Cleaner Bracket Screw	46	Throttle Stop Screw Spring
5	Air Horn to Main Body Screw & L.W.	47	Fast Idle Adjusting Screw Spring
6	Throttle Stop Screw	48	Idle Adjusting Needle Spring
7	Fast Idle Adjusting Screw	49	Vent Valve Operating Lever Spring
8	Bracket Retaining	50	Vent Valve Spring
9	Throttle Body to Main Body Screw & L.W.	51	Throttle Return Spring
10	Choke Plate Screw	52	Pump Drive Spring
11	Throttle Plate Screw	53	Power Valve Piston Spring
12	Cover Screw	54	Choke Shaft Spring
13	Nozzle Bar Screw	55	Dechoke Lever Nut
14	Fuel Inlet Fitting Gasket	56	Air Cleaner Ring
15	Main Body Gasket	57	Fast Idle Cam
16	Nozzle Bar Gasket	58	Fast Idle Rod
17	Throttle Body Gasket	59	Pump Discharge Weight
18	E.C.S. Cover Gasket	60	Dechoke Lever Nut L.W.
19	Throttle Plate	63	Pump Link Retainer
20	Throttle Body & Shaft Assembly	64	Air Cleaner Bracket
		65	Vacuum Hose
21	Throttle Return Spring Bushings	66	Float Baffle
22	Idle Adjusting Needle	67	Pump Lever
23	Float & Hinge Assembly	68	Vent Valve Lever
24	Float Hinge Shaft & Retainer	69	Vent Valve Operating Lever
25	Fuel Inlet Needle & Seat Assembly	70	Pump Operating Lever
		71	Pump Operating Lever Shaft
26	Nozzle Bar Assembly	72	Vent Valve Retainer
27	Main Jet	73	T.P.T. Bracket Screw
28	Vent Valve	74	Power Valve Adjusting Screw
29	Power Valve Piston	75	Power Valve Assembly (Staged)
30	Power Valve Assembly		
31	Pump Cup	76	Choke Diaphragm Link Pin
32	Pump Discharge Ball	77	Roll-Pin
33	Vent Valve Lever Pin	78	Power Valve Stem Retainer
34	Vent Valve Operating Lever Pin	79	Power Valve Spring
35	Air Cleaner Ring Retaining Pin	80	T.P.T. Lever Nut
36	Pump Steam & Head Assembly	81	T.P.T. Assembly Locknut
37	Choke Diaphragm Link	82	Mechanical Power Valve Stem
38	Pump Stem Link	83	Mechanical Power Valve Stem Cap
39	Pump Link		
40	Idle Needle Limiter	84	T.P.T. Lever Lockwasher
41	Accelerator Pump Housing Cover	85	T.P.T. Bracket
		86	T.P.T. Lever
42	Choke Diaphragm Assembly Complete	87	T.P.T. Assembly
		88	T.P.T. Wire Assembly
		89	Duty-Cycle Solenoid

Parts list for Holley Models 2280 and 6280 carburetors

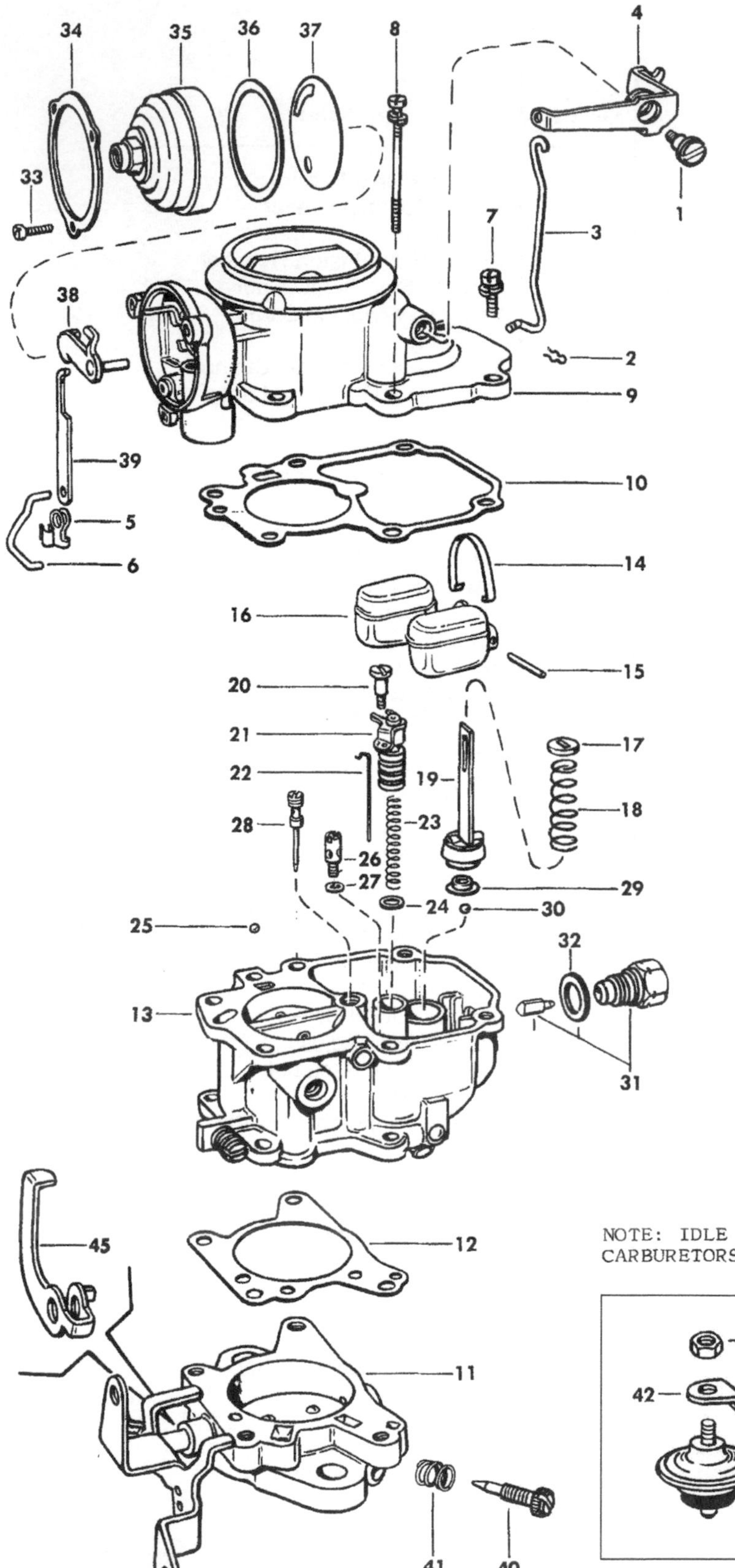

1	Screw - Pump arm
2	Spring - Pin
3	Rod - Throttle connector
4	Arm - Pump
5	Retainer
6	Rod - Choke connector
7	Screw and Lock Washer Assembly
8	Screw and Lock Washer Assembly
9	Air Horn Assembly
10	Gasket - Body
11	Flange Assembly - Body
12	Gasket - Body flange
13	Main Body
14	Retainer - Float pin
15	Pin - Float
16	Float Assembly
17	Washer - Pump spring retainer
18	Spring - Pump
19	Piston Assembly - Pump
20	Screw - Step-up piston guide
21	Piston Assembly - Step-up
22	Rod - Step-up
23	Spring - Step-up piston
24	Gasket - Step-up piston
25	Ball - Pump discharge
26	Jet Assembly - Main metering
27	Gasket - Main metering jet
28	Jet - Low speed
29	Retainer - Pump intake ball
30	Ball - Pump intake
31	Needle and Seat Assembly
32	Gasket - Needle seat
33	Screw
34	Retainer - Coil housing
35	Coil and Housing Assembly - Thermostatic
36	Gasket - Coil housing
37	Plate - Choke baffle
38	Lever Assembly - Choke trip
39	Link - Fast idle
40	Screw - Idle adjusting
41	Spring - Idle adjusting screw
42	Bracket - Dashpot
43	Nut
44	Dashpot Assembly
45	Lever - Dashpot (with Powerflyte)
46	Bracket - Kickdown switch
47	Nut
48	Switch Assembly - Kickdown

NOTE: IDLE ADJUSTING SCREW (40) IS NOT REMOVABLE FROM CARBURETORS USED ON ENGINES WITH CLEAN AIR PACKAGE.

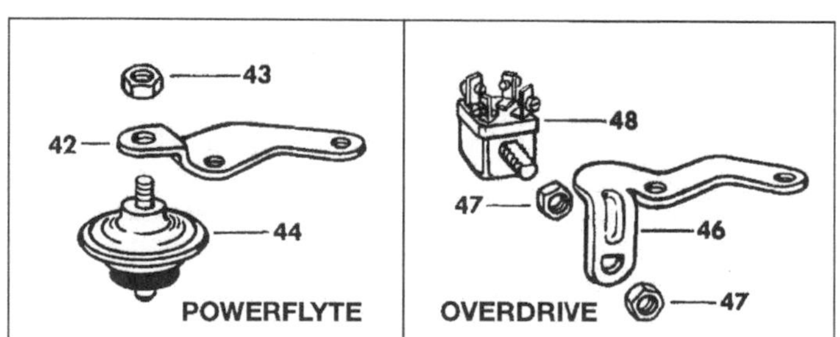

10.7e An exploded view of the Carter BBS carburetor

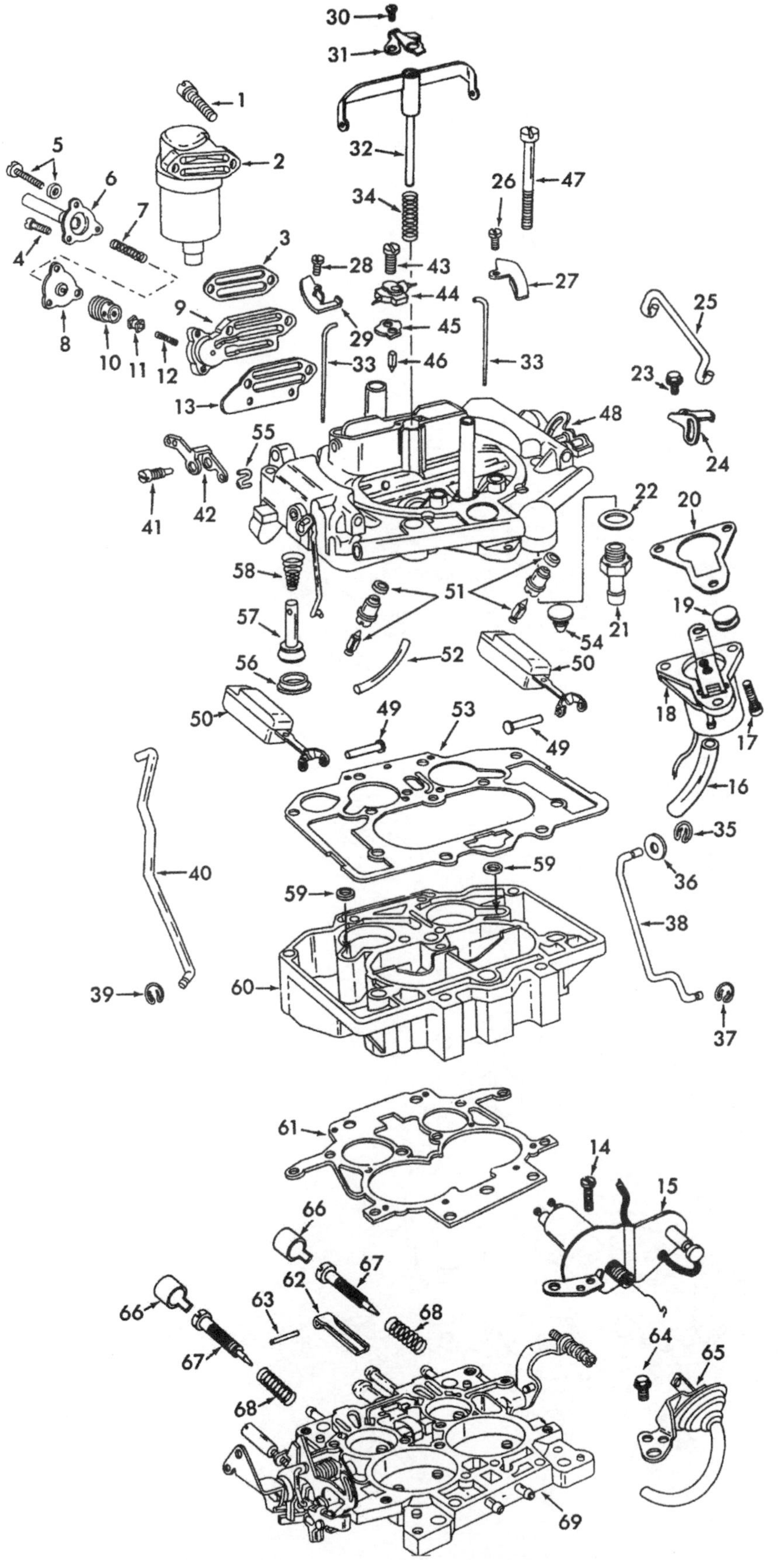

1. Screw & Lockwasher (2)—Altitude Compensator
2. Altitude Compensator Assy. (Some Models)
3. Gasket—Altitude Compensator
4. Screw (2)—Diaphragm Cover
5. Screw & Seal Washer (1)—Cover
6. Cover—Diaphragm
7. Spring—Diaphragm
8. Diaphragm Assy.—Enrichment Valve
9. Housing—Idle Enrichment Valve
10. Valve Seat—Enrichment
11. Valve—Enrichment Seat
12. Spring—Enrichment Valve
13. Gasket—Enrichment Valve Housing
14. Screw—Transducer Bracket
15. Transducer, Idle Stop Switch & Bracket Assy.
16. Hose—Bowl Vent Solenoid
17. Screw & Lockwasher (3)—Bowl Vent Solenoid
18. Bowl Vent Solenoid Assy.
19. Valve—Bowl Vent
20. Gasket—Bowl Vent Solenoid Assy.
21. Fitting—Fuel Inlet
22. Gasket—Inlet Fitting
23. Screw—Rod Retainer
24. Retainer—Choke Connector Rod
25. Rod—Choke Connector
26. Screw—Metering Rod Cover Plate
27. Plate—Cover
28. Screw—Metering Rod Cover Plate
29. Plate—Cover
30. Screw—Plate
31. Plate—Step Up Piston Cover
32. Step Up Piston & Rod Assy.
33. Rod (2)—Step Up
34. Spring—Step Up Piston
35. Retainer—Choke Pull Off Rod
36. Washer—Choke Pull Off Rod
37. Retainer—Choke Pull Off Rod
38. Rod—Choke Pull Off
39. Retainer—Pump Rod
40. Rod—Pump Arm Connector
41. Screw—Pump Arm
42. Arm—Pump
43. Screw—Pump Jet Housing
44. Housing—Pump Jet
45. Gasket—Housing
46. Needle—Pump Disc. Check
47. Screw (10)—Bowl Cover
48. Bowl Cover Assy.
49. Pin (2)—Float Lever
50. Float Assy. (2)
51. Rotary Disc—Fuel Inlet Valve & Gasket (2)
52. Tube—Pump Passage
53. Gasket—Bowl Cover
54. Valve—Bowl Vent - (Mechanical Model)
55. S Link—Pump
56. Check Valve Assy.—Pump Intake
57. Pump Assy.
58. Spring—Pump
59. O Ring (2)—Main Well Seal
60. Bowl Assy.—Fuel
61. Gasket—Throttle Body
62. Lever—Step Up Piston
63. Pin—Lever
64. Screw—Choke Pull Off Bracket
65. Choke Pull Off Assy.
66. Cap (2)—Idle Limiter
67. Needle (2)—Idle Adjusting
68. Spring (2)—Idle Adjusting Needle
69. Throttle Body Assy.

10.7f An exploded view of the Carter ThermoQuad carburetor

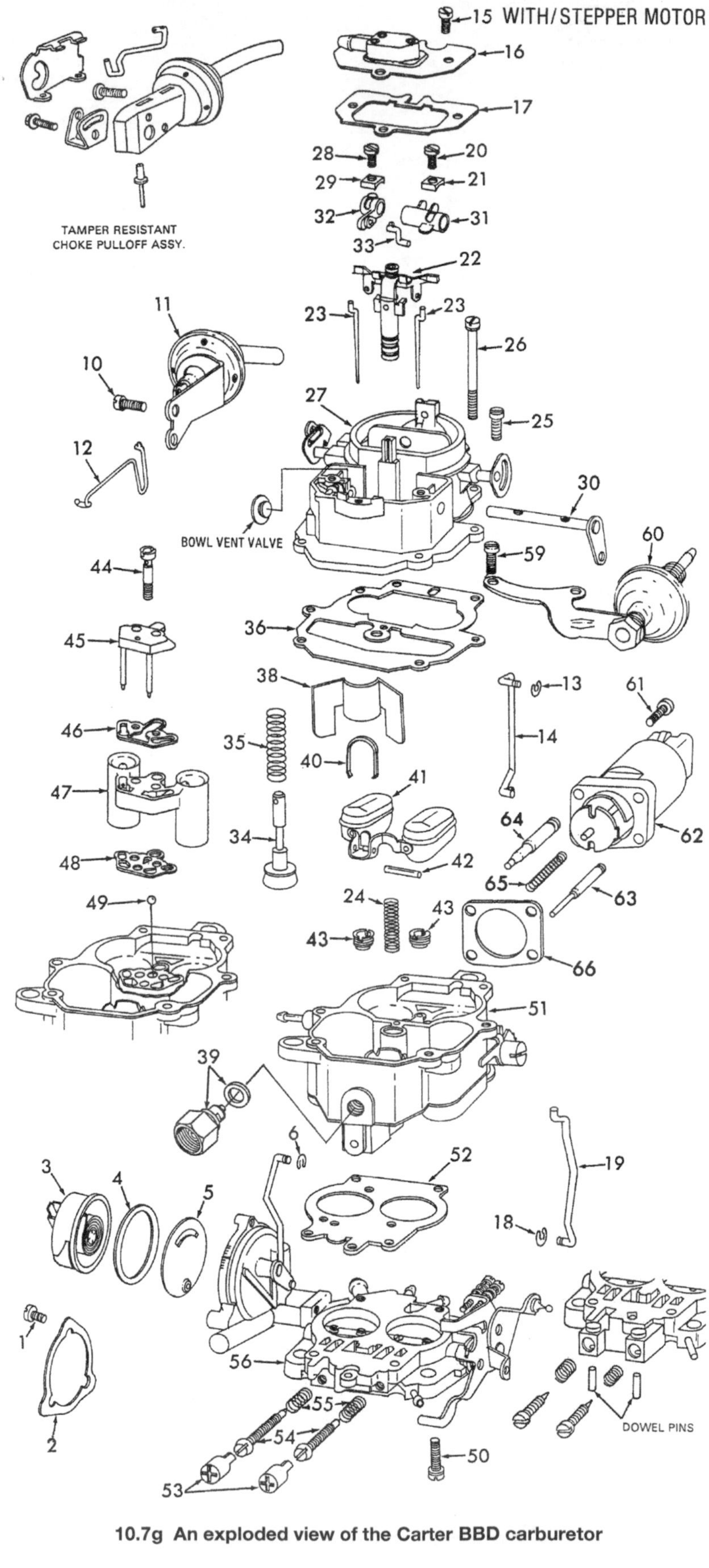

10.7g An exploded view of the Carter BBD carburetor

1. SCREW (3) - CHOKE COVER RETAINER
2. RETAINER (3) - CHOKE COVER
3. CHOKE COVER ASSY.
4. GASKET - CHOKE COVER
5. BAFFLE - CHOKE HOUSING
6. RETAINER - CHOKE ROD
7. SCREW (2) - CHOKE HOUSING
8. HOUSING ASSY. - CHOKE
9. SEAL - VAC. PASSAGE
10. SCREW (2) - CHOKE PULL OFF
11. CHOKE PULL OFF ASSY.
12. LINK - CHOKE PULL OFF
13. RETAINER - FAST IDLE ROD
14. ROD - FAST IDLE
15. SCREW (3) - DUST COVER
16. COVER ASSY. - DUST
17. GASKET - COVER
18. RETAINER - PUMP ROD
19. ROD - PUMP
20. SCREW - METERING ROD ARM WASHER
21. WASHER - METERING ROD ARM
22. PISTON ASSY. - VACUUM
23. ROD (2) - METERING
24. SPRING - VACUUM PISTON
25. SCREW & LOCKWASHER (4) - BOWL COVER
26. SCREW & LOCKWASHER (2) - BOWL COVER
27. BOWL COVER ASSY.
28. SCREW - PUMP ARM WASHER
29. WASHER - PUMP ARM
30. SHAFT - PUMP
31. ARM - METERING ROD
32. ARM - PUMP LINK
33. LINK - PUMP
34. PUMP ASSY.
35. SPRING - PUMP
36. GASKET - BOWL COVER
37. BALL - PUMP INTAKE CHECK (LARGE)
38. BAFFLE - FUEL BOWL
39. ROTORY VALVE & GASKET ASSY.
40. RETAINER - FLOAT PIN
41. FLOAT ASSY.
42. PIN - FLOAT
43. JET (2) - MAIN
44. SCREW (2) - VENTURI CLUSTER
45. COVER - VENTURI
46. GASKET - VENTURI COVER
47. CLUSTER ASSY. - VENTURI
48. GASKET - VENTURI CLUSTER
49. BALL - PUMP DISC. CHECK (SMALL)
50. SCREW & LOCKWASHER (2) -THROTTLE BODY
51. BOWL ASSY.
52. GASKET - THROTTLE BODY
53. CAP (2) - IDLE LIMITER
54. NEEDLE (2) - IDLE ADJUSTING
55. SPRING (2) - IDLE ADJ. NEEDLE
56. THROTTLE BODY ASSY.
57. SCREW - FAST IDLE
58. CAM - FAST IDLE
59. SCREW & LOCKWASHER (2) - DASH-POT
60. DASH-POT & BRACKET
61. SCREW (4) - STEPPER MOTOR
62. STEPPER MOTOR ASSY.
63. PIN - METERING (MAIN SYSTEM)
64. PIN - METERING (LOW SPEED)
65. SPRING - STEPPER MOTOR
66. GASKET STEPPER MOTOR

Parts list for the the Carter BBD carburetor

day or according to the instructions listed on the carburetor cleaner) and reassemble it, which will usually take much longer than disassembly **(see illustrations)**. When disassembling the carburetor, match each part with the illustration in the carburetor kit and lay the parts out in order on a clean work surface. Overhauls by inexperienced mechanics can result in an engine which runs poorly or not at all. To avoid this, use care and patience when disassembling the carburetor so you can reassemble it correctly.

8 Because carburetor designs are constantly modified by the manufacturer in order to meet increasingly more stringent emissions regulations, it isn't feasible to include a step-by-step overhaul of each type. You'll receive a detailed, well illustrated set of instructions with the carburetor overhaul kit.

11 Electric choke heater - testing

Refer to illustration 11.1

Caution: *If there is any loss of electrical current to the choke heater, operation of any type, including idling, should be avoided. Loss of power to the choke will cause the choke to remain partly closed during engine operation. A very rich air-to-fuel mixture will be created and result in abnormally high exhaust system temperatures, which may cause damage to the catalytic converter or other underbody parts of the vehicle.*

Note: *This system is used on most 1973 and later models, except some mid-1980's and later four-barrel carburetors (Rochester Quadrajet). For more information and detailed test procedures for the entire system, refer to Chapter 6.*

1 With the engine cold, connect a jumper wire from the choke heater wire to the posi-

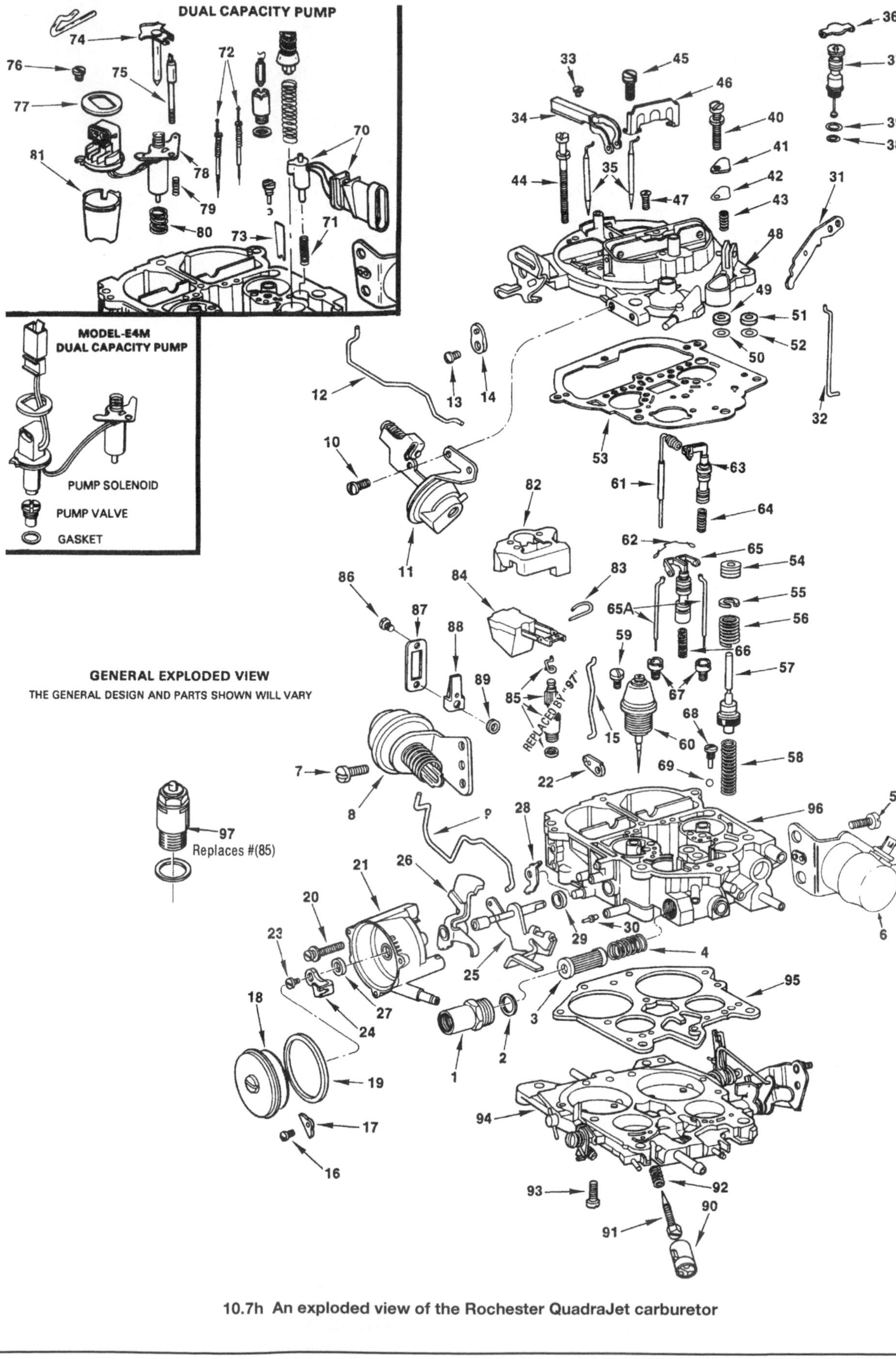

10.7h An exploded view of the Rochester QuadraJet carburetor

1. Fuel Filter Nut
2. Fuel Filter Nut Gasket
3. Fuel Filter
4. Fuel Filter Spring
5. Solenoid Screw
6. Idle Speed Solenoid
7. Rear Vac. Break Screw
8. Rear Vac. Break Assembly
9. Rear Vac. Break Rod
10. Front Vac. Break Screw
11. Front Vac. Break Assy.
12. Front Vac. Break Rod
13. Choke Lever Screw
14. Choke Lever
15. Choke Rod
16. Choke Cover Retainer Rivet
17. Choke Cover Retainer
18. Choke Cover Assembly
19. Choke Cover Gasket
20. Choke Housing Screw
21. Choke Housing Assy.
22. Intermediate Choke Lever
23. Choke Coil Lever Screw
24. Choke Coil Lever
25. Intermediate Choke Shaft
26. Fast Idle Cam
27. Choke Housing Seal
28. Sec. Lockout Cam
29. Inter. Choke Shaft Seal
30. Vacuum Passage Tube
31. Pump Lever
32. Pump Rod
33. Sec. Metering Rod Holder Screw
34. Sec. Metering Rod Holder
35. Sec. Metering Rods
36. Air Bleed Valve Cover
37. Idle Air Bleed Valve
38. Lower Bleed Valve O-Ring
39. Upper Bleed Valve O-Ring
40. Vent Cover Screw
41. Vent Cover
42. Vent Cover Gasket
43. Vent Valve Spring
44. Airhorn Screw-Long
45. Airhorn Screw-Short
46. Air Baffle
47. Airhorn Screw-Special
48. Airhorn Assy.
49. Pump Stem Seal
50. Pump Stem Seal Retainer
51. T.P.S. Plunger Seal
52. T.P.S. Plunger Seal Retainer
53. Airhorn Gasket
54. Pump Stem Spacer
55. Pump Spring Retainer
56. Pump Spring
57. Pump Stem Assy.
58. Pump Return Spring
59. Aneroid Assy. Screw
60. Aneroid Assy.
61. Auxiliary Metering Rod
62. Metering Rod Spring
63. Auxiliary Power Piston
64. Auxiliary Power Piston Spring
65. Power Piston Assy.
65A. Main Metering Rods
66. Power Piston Assy. Spring
67. Main Jets
68. Pump Discharge Ball Screw
69. Pump Discharge Ball
70. Throttle Position Sensor Assy.
71. T.P.S. Spring
72. Main Metering Rod & Spring
73. Pump Well Baffle
74. Solenoid Plunger
75. Sol. Lean Mixture Adj. Screw, See Special Instruction.
76. ECM Connector Screw
77. ECM Connector Gasket
78. ECM Connector & Solenoid Assy.
79. Lean Mixture Screw
80. Solenoid Spring
81. Well Insert
82. Fuel Bowl Insert
83. Float Hinge Pin
84. Float & Lever Assy.
85. Needle & Seat Assy.
86. Hot Idle Compensator Cover Screw
87. Hot Idle Compensator Cover
88. Hot Idle Compensator Assy.
89. Hot Idle Compensator Gasket
90. Idle Limiter Cap (2)
91. Idle Needle (2)
92. Idle Needle Spring (2)
93. Throttle Body Screw
94. Throttle Body
95. Throttle Body Gasket
96. Main Body
97. Rotory Inlet Valve

10.7h An exploded view of the Rochester QuadraJet carburetor

tive battery terminal. The choke heater housing should begin becoming hot and the choke plate in the carburetor should slowly open after some time. Be sure it opens completely in about five minutes or replacement is necessary **(see illustration)**. **Note:** *As the heater is heating the choke coil, occasionally tap the accelerator to allow the choke to open.*

2 If the choke heater tests okay, but the choke valve does not open in normal operation, check for voltage to the choke heater when the engine is cold and the ignition switch is in the ON position.

3 If no voltage is present in Step 2, repair the open in the power circuit to the choke heater (be sure to check the fuse first!).

12 Carburetor adjustments

Note 1: *These carburetor adjustments are strictly on-vehicle adjustments. During overhaul, refer to the instructions included in the overhaul kit for complete procedures and any additional adjustments that are required.*

Note 2: *For additional information concerning overhaul and adjustments on the Holley and Rochester carburetors, refer to the* Haynes Holley Carburetor Manual *and the* Haynes Rochester Carburetor Manual.

Choke plate opening

Note: *For more information on the choke system, refer to Chapter 1 and the previous Section on the electric choke heater system.*

1 The choke plate opening is a critical carburetor adjustment directly involved with cold-running conditions and choke enrichment during cranking. The choke unloader is a mechanical device that partially opens the choke valve at wide open throttle to eliminate choke enrichment during hard acceleration.

2 Engines which have been stalled or flooded by excessive choke enrichment can be cleared by the use of the choke unloader. With the throttle valve at wide open throttle, the choke plate should be slightly open to allow a sufficient amount of intake air into the carburetor venturi.

11.1 Apply 12 volts to the choke heater connector. The choke should open in about five minutes

Holley carburetors

3 These models are equipped with a choke thermostat coil mounted in the intake manifold, which actuates the choke via a rod. With the carburetor installed and the engine cold, the coil should hold the choke lightly closed (possibly open slightly, but no more than about 1/16-inch).

4 If the choke is closed tightly or open too far, remove the coil from the intake manifold; the center mount for the thermostat coil should be slotted so you can adjust the coil's tension with a screwdriver. If there's no adjustment at the choke coil's center mount, bend the rod to adjust the choke. If the coil is broken, corroded or deformed, replace it.

Carter carburetors

5 These models are equipped with a choke thermostat coil mounted in the intake manifold, which actuates the choke via a rod. With the carburetor installed and the engine cold, the coil should hold the choke lightly closed (possibly open slightly, but no more than about 1/16-inch).

6 If the choke is closed tightly or open too far, remove the coil from the intake manifold; the center mount for the thermostat coil should be slotted so you can adjust the coil's tension with a screwdriver. If there's no adjustment at the choke coil's center mount, bend the rod to adjust the choke. If the coil is broken, corroded or deformed, replace it.

Rochester carburetors

2GV models

7 These models are equipped with a choke thermostat coil mounted in the intake manifold, which actuates the choke via a rod. With the carburetor installed and the engine cold, the coil should hold the choke almost, but not quite, completely closed.

8 If the choke is closed tightly or open too far, remove the coil from the intake manifold; the center mount for the thermostat coil should be slotted so you can adjust the coil's tension with a screwdriver. If there's no

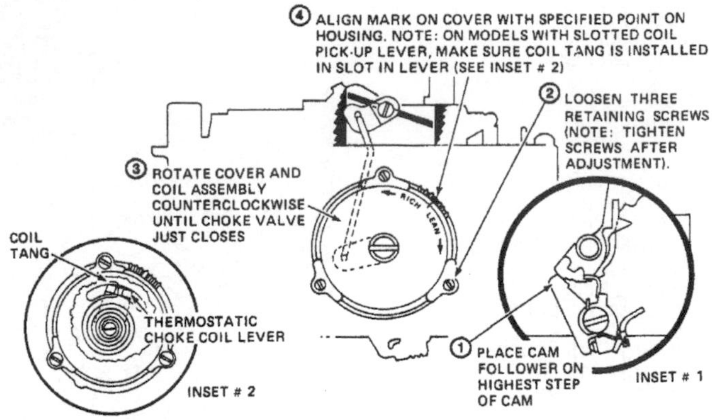

12.9 Choke coil adjustment details on the Rochester Quadrajet

12.12a Adjusting the idle mixture screw on the Holley Model 1945

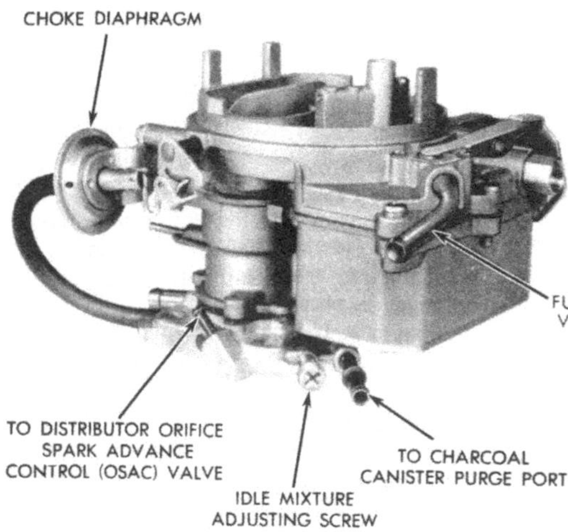

12.12b Component locations on the Holley Model 2245 (note that this idle mixture screw has a limiter cap)

12.12c Adjusting the idle mixture screw on the Holley Model 6280

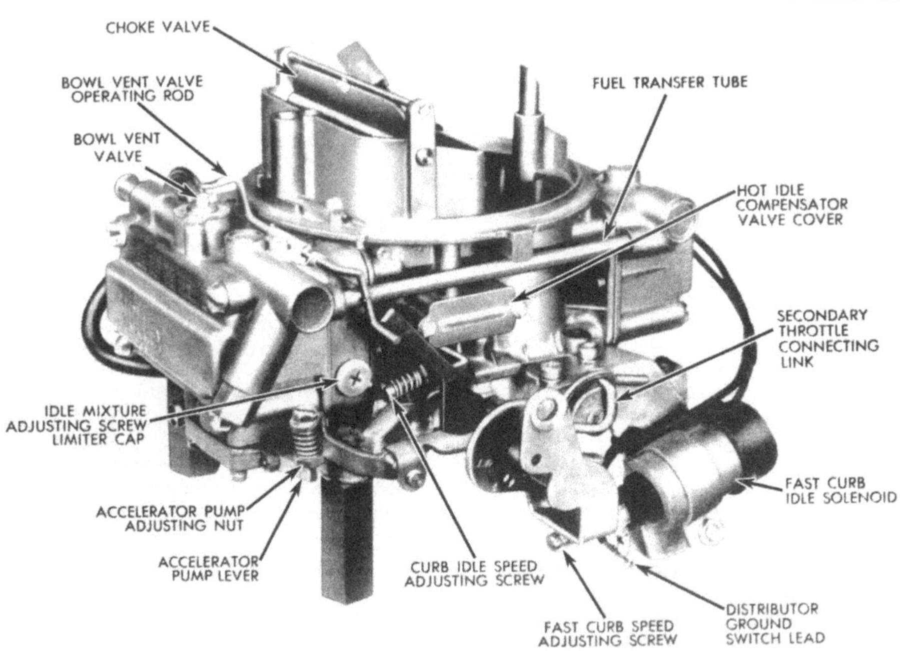

12.12d Component locations on the Holley Model 4160

adjustment at the choke coil's center mount, bend the rod to adjust the choke. If the coil is broken, corroded or deformed, replace it.

Quadrajet models

Refer to illustration 12.9

9 All the Quadrajet carburetors installed on to the models covered in this manual are equipped with an integral choke system. Since choke coils lose their tension over time, you may find, on older coils, that lining up the marks does not provide enough choke. Readjust the coil by rotating the cover counterclockwise until the choke plate closes completely, then rotate it clockwise until the plate just moves away from the fully closed position (about 1/16-inch, measured at the rear of the air horn) **(see illustration)**.

10 Re-tighten the three cover screws. **Note:** *The metal clips beneath the screws must be positioned so they exert their maximum spring force against the cover. If they're installed backwards, the choke cover will have a tendency to rotate, changing the choke setting.*

Idle mixture

Refer to illustrations 12.12a through 12.12g

Note: *Be sure to check the ignition timing,*

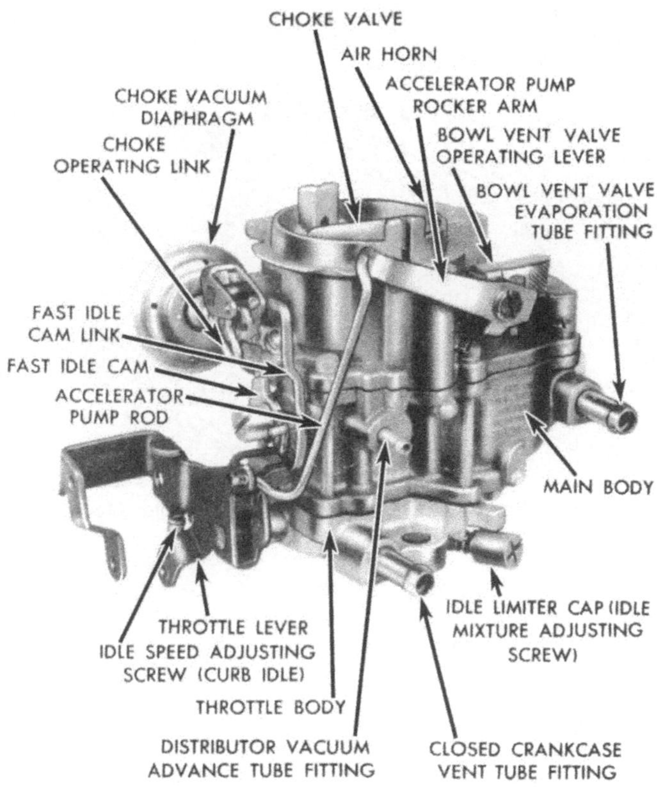

12.12e Component locations on the Carter BBS

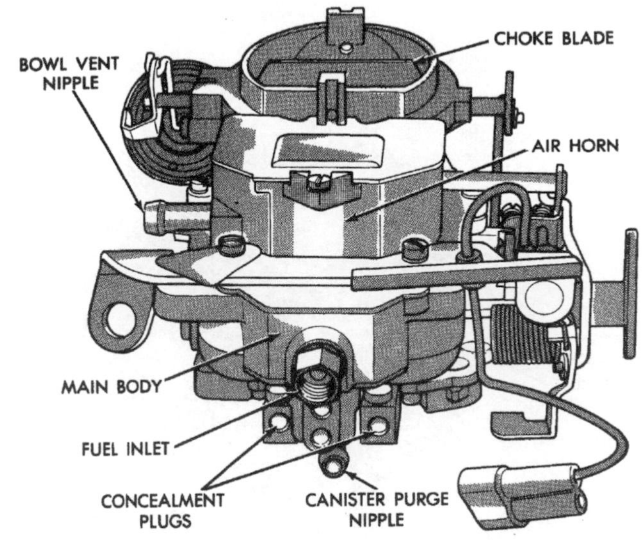

12.12f Component locations on the Carter BBD. Note that these mixture screws are covered by concealment plugs

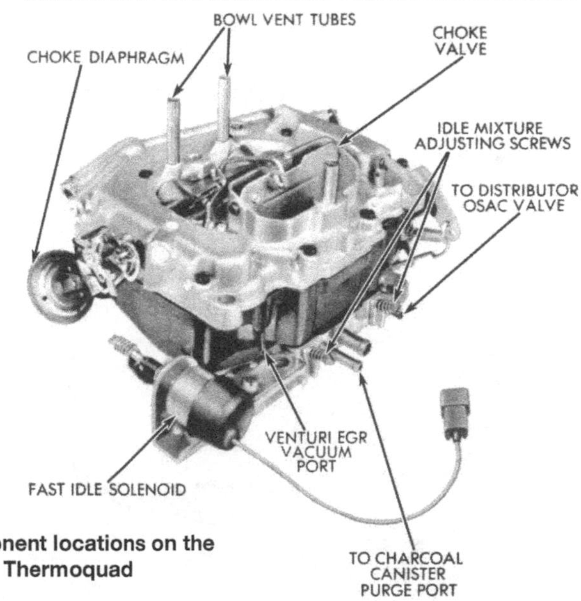

12.12g Component locations on the Carter Thermoquad

the dwell on the ignition points (1971 and 1972 models), the condition of the spark plugs and ignition wires and, if necessary, perform a complete tune-up before attempting to make the following carburetor adjustment. Often, problems with the mixture are due to a vacuum leak (see Troubleshooting *at the front of this manual and the vacuum gauge checks in Chapter 2C).*

Note: *On carburetors that have the factory concealment plugs installed over the idle mixture screws, do not adjust the idle mixture, as it has been factory pre-set. Also, on models with limiter caps on the mixture screws, do not adjust the mixture any richer than the caps will allow (do not break off the caps). If the engine runs too rich at the factory settings, there's something wrong with the choke or the carburetor is faulty. If the engine is running too lean at the factory settings, there's a vacuum leak or the carburetor is faulty. If necessary, have the system diagnosed by a qualified shop.*

11 After the engine has reached normal operating temperature, set the parking brake and block the wheels, then remove the air cleaner and turn the idle speed adjusting screw out as far as possible without the engine running rough, since the throttle plate(s) must be as nearly closed as possible when adjusting the idle mixture.

12 Working with one idle mixture screw at a time (one-barrel models have only one screw, but all others have two screws), turn the screw clockwise until the idle speed drops a noticeable amount **(see illustrations)**. Now slowly turn the screw out until the maximum idle speed is reached, but no further.

13 On two- and four-barrel models, repeat the procedure for the other idle mixture screw. Keep track of the number of turns each screw is off the seat to be sure they are each out approximately the same number of turns. This will verify that the idle circuit is operating properly and balance is maintained. **Note:** *If it was necessary to turn the idle speed screw in (to keep the engine running) before setting the idle mixture, adjust the idle speed, then perform the mixture adjustment procedure again. This will ensure that the idle mixture is set with the throttle plates fully closed, so the engine is drawing the fuel mixture only from the idle circuit.*

Idle speed

Refer to illustrations 12.17a through 12.17g

Note: *In order to adjust the idle speed (curb idle speed) on engines equipped with feedback carburetor systems (Models 6280, 6145, BBD, Thermoquad, Quadrajet etc.), it will be necessary to prevent certain systems (EVAP, EGR, Oxygen sensor feedback, idle stop solenoid, etc.) from operating. Refer to the VECI label in the engine compartment of your vehicle for a list of the steps necessary to make this adjustment.*

14 Engine idle speed is the speed at which the engine operates when no accelerator pedal pressure is applied, as when stopped at a traffic light. This speed is critical to the performance of the engine itself, as well as many engine subsystems.

15 A hand-held tachometer must be used

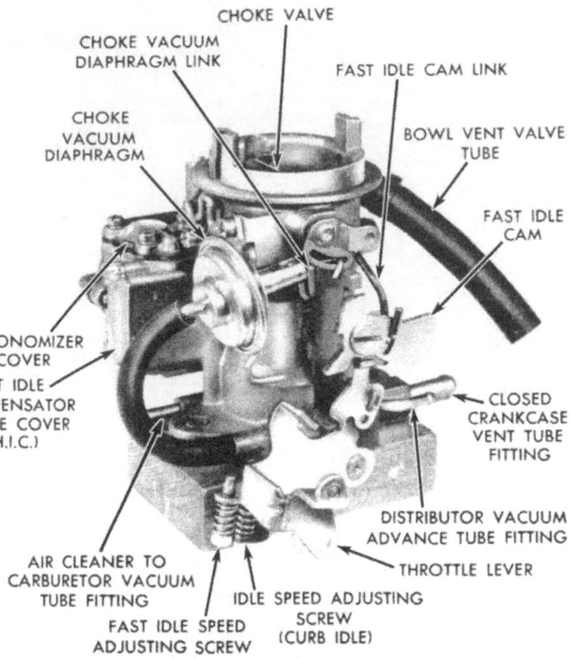

12.17a Component locations on the Holley Model 1920

12.17b Adjusting the idle speed screw on the Holley Model 1945

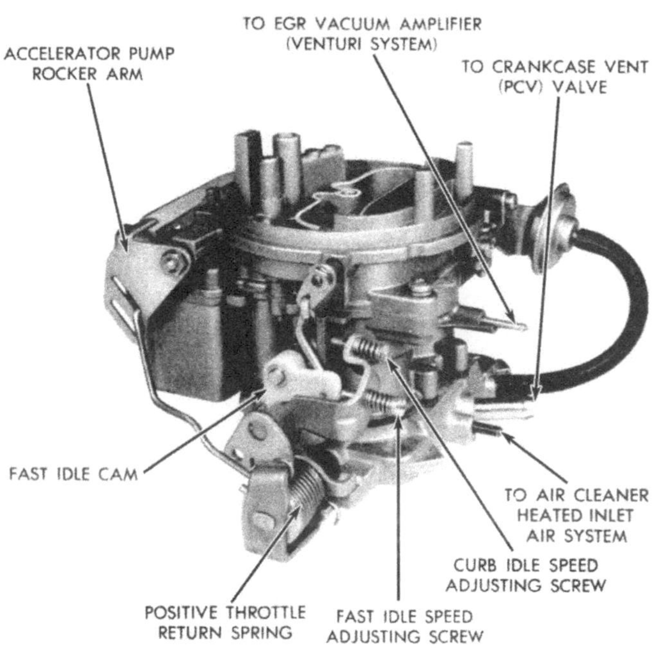

12.17c Component locations on the Holley Model 2245

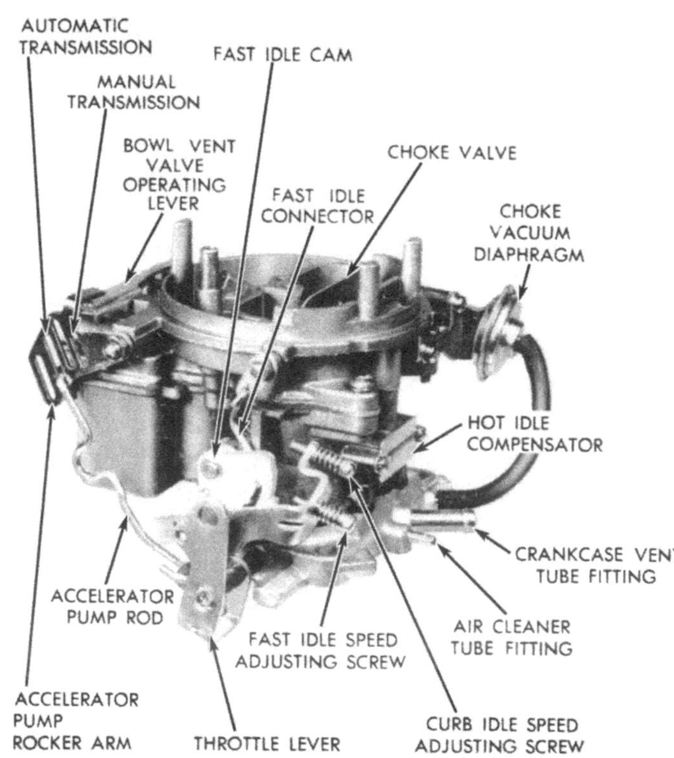

12.17d Component locations on the Holley Model 2210

when adjusting idle speed to get an accurate reading. The exact hook-up for these meters varies with the manufacturer, so follow the particular directions included.

16 All vehicles covered in this manual should have a tune-up decal or *Emissions Control Information* label located on the inside of the hood. The printed instructions for setting idle speed can be found on this decal, and should be followed since they are for your particular engine.

17 After adjusting the idle mixture, adjust the idle speed. Turn the idle speed adjusting screw **(see Illustrations)** clockwise to increase the idle speed and counterclockwise to decrease it. Set the idle speed to the specifications on the VECI label located in the engine compartment. If the label is missing, set the idle speed to approximately 750 rpm with the transmission in Park or Neutral.

Fast-idle speed

Refer to illustrations 12.18a through 12.18e

Note: *In order to adjust the fast idle speed on engines equipped with feedback carburetor systems (Models 6280, 6145, BBD, Thermo-quad, Quadrajet etc.), it may be necessary to prevent certain systems (EVAP, EGR, Oxygen sensor feedback, idle stop solenoid, etc.) from operating. Refer to the VECI label for*

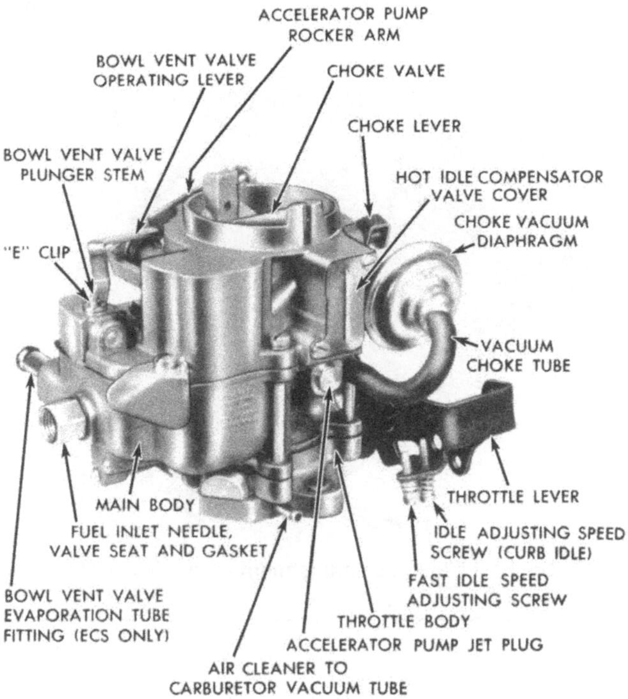

12.17e Component locations on the Carter BBS

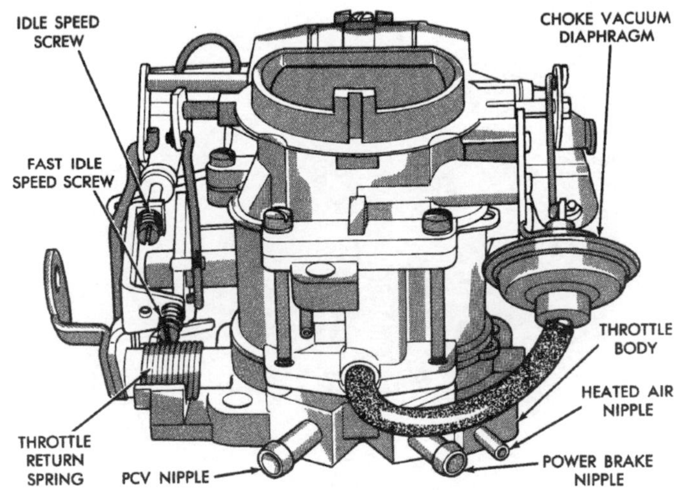

12.17f Component locations on the Carter BBD

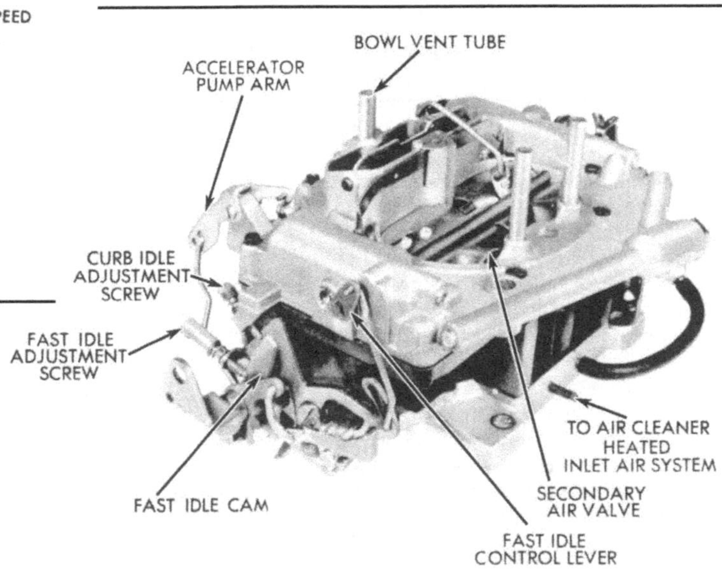

12.17g Component locations on the Carter Thermoquad

your particular model for a list of the steps necessary to make this adjustment.

Note: *Many of the locations of the fast idle adjusting screw for the various carburetors are pictured in the preceding illustrations.*

18 Set the parking brake and block the wheels. Remove the air cleaner and place the fast-idle speed adjusting screw on the second highest step on the fast-idle cam **(see illustrations)**.

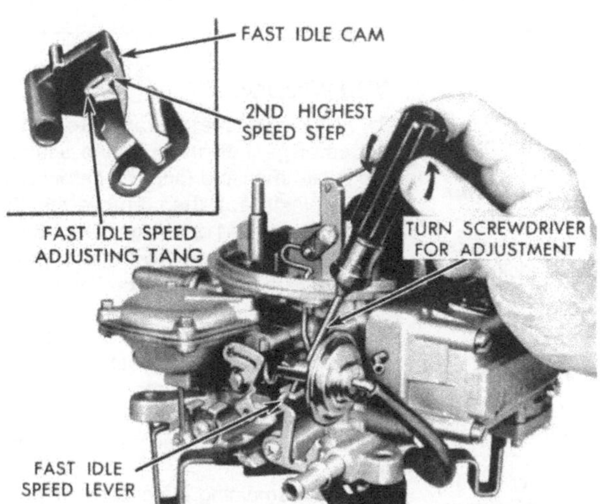

12.18a Fast idle speed adjustment on the Holley Model 4160 carburetor

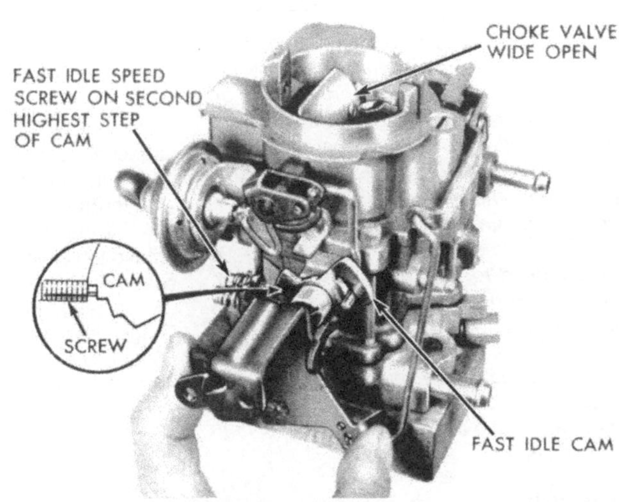

12.18b Fast idle speed adjustment on the Carter BBS

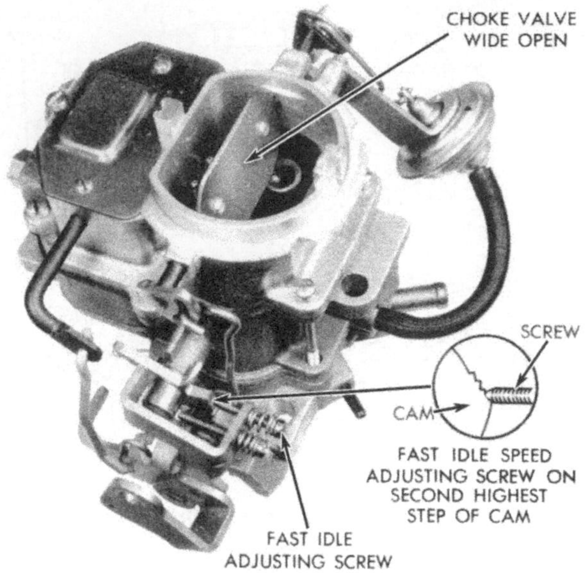

12.18c Fast idle speed adjustment on the Carter BBD

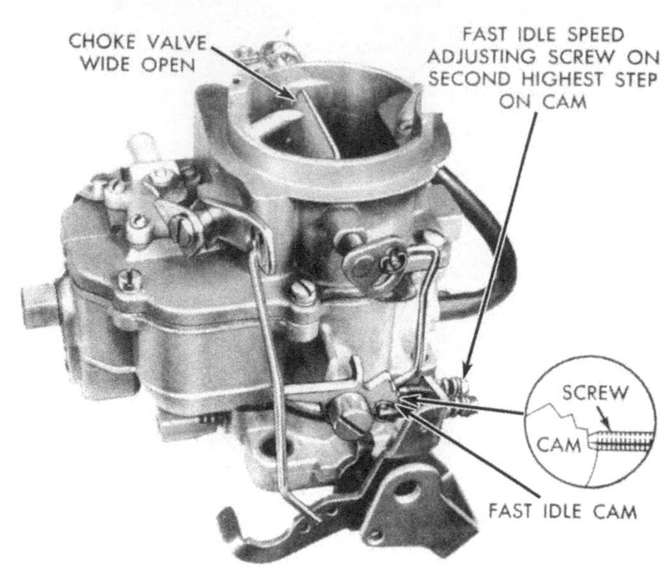

12.18d Fast idle speed adjustment on the Rochester 2GV

19 Connect a tachometer according to the tool manufacturer's instructions.
20 Start and run the engine to normal operating temperature.
21 With the choke fully open, turn the fast-idle adjusting screw until the correct fast-idle speed is obtained. Double-check to make sure the fast idle adjusting screw remains on the second highest step of the cam. Refer to the VECI label in the engine compartment. If the label is not available, set the fast-idle speed to approximately 1,700 rpm.
22 To verify the fast-idle speed is correct, return the engine to idle, then reposition the adjusting screw on the second highest step of the fast-idle cam. Readjust the fast-idle speed, if necessary.

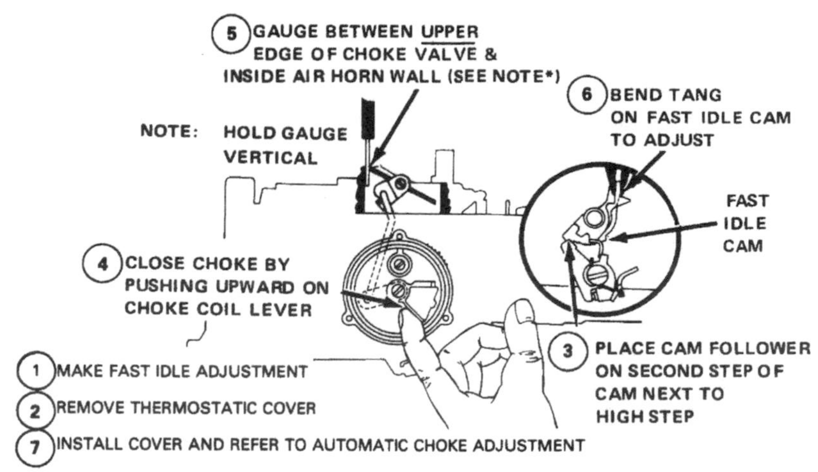

12.18e Fast idle speed adjustment on the Rochester Quadrajet

13 Carburetor - removal and installation

Warning: *Gasoline is extremely flammable, so take extra precautions when you work on any part of the fuel system. Don't smoke or allow open flames or bare light bulbs near the work area, and don't work in a garage where a natural gas-type appliance (such as a water heater or a clothes dryer) with a pilot light is present. Since gasoline is carcinogenic, wear latex gloves when there's a possibility of being exposed to fuel, and, if you spill any fuel on your skin, rinse it off immediately with soap and water. Mop up any spills immediately and do not store fuel-soaked rags where they could ignite. When you perform any kind of work on the fuel system, wear safety glasses and have a Class B type fire extinguisher on hand.*

Removal

Refer to illustrations 13.6 and 13.8

1 Remove the fuel filler cap to relieve fuel tank pressure.
2 Remove the air cleaner from the carburetor. Be sure to label all vacuum hoses attached to the air cleaner housing.
3 Disconnect the accelerator cable from the throttle lever (see Section 9).
4 If the vehicle is equipped with an automatic transmission, disconnect the kickdown cable from the throttle lever.
5 Clearly label all vacuum hoses and fittings, then disconnect the hoses.
6 Disconnect the fuel line from the carburetor **(see illustration)**.
7 Label the wires and terminals, then unplug all the electrical connectors.
8 Remove the mounting fasteners **(see illustration)** and detach the carburetor from the intake manifold. Remove the carburetor mounting gasket. Stuff a rag into the intake manifold openings.

Installation

9 Use a gasket scraper to remove all traces of gasket material and sealant from the intake manifold (and the carburetor, if it's being reinstalled), then remove the shop rag from the manifold openings. Clean the mating surfaces with lacquer thinner or acetone.
10 Place a new gasket on the intake manifold.
11 Position the carburetor on the gasket and install the mounting fasteners.
12 To prevent carburetor distortion or damage, tighten the fasteners to approximately 16 ft-lbs in a criss-cross pattern, 1/4-turn at a time.
13 The remaining installation steps are the reverse of removal.
14 Check and, if necessary, adjust the idle speed (see Section 12).
15 If the vehicle is equipped with an auto-

13.6 Use a flare-nut wrench and a back-up wrench when detaching the fuel line from the carburetor

13.8 Remove the carburetor nuts (arrows) from the base of the carburetor body (other two nuts not visible)

matic transmission, refer to Chapter 7B for the kickdown cable adjustment procedure.

16 Start the engine and check carefully for fuel leaks.

14 Exhaust system - servicing and general information

Warning: *Inspection and repair of exhaust system components should be done only after enough time has elapsed after driving the vehicle to allow the system components to cool completely. Also, when working under the vehicle, make sure it is securely supported on jackstands.*

1 The exhaust system consists of the exhaust manifold(s), the catalytic converter(s), the muffler, the tailpipe and all connecting pipes, brackets, hangers and clamps. The exhaust system is attached to the body with mounting brackets and rubber hangers. If any of the parts are improperly installed, excessive noise and vibration will be transmitted to the body.

2 Conduct regular inspections of the exhaust system to keep it safe and quiet. Look for any damaged or bent parts, open seams, holes, loose connections, excessive corrosion or other defects which could allow exhaust fumes to enter the vehicle. Deteriorated exhaust system components should not be repaired; they should be replaced with new parts.

3 If the exhaust system components are extremely corroded or rusted together, welding equipment will probably be required to remove them. The convenient way to accomplish this is to have a muffler repair shop remove the corroded sections with a cutting torch. If, however, you want to save money by doing it yourself (and you don't have a welding outfit with a cutting torch), simply cut off the old components with a hacksaw. If you have compressed air, special pneumatic cutting chisels can also be used. If you do decide to tackle the job at home, be sure to wear safety goggles to protect your eyes from metal chips and work gloves to protect your hands.

4 Here are some simple guidelines to follow when repairing the exhaust system:

- a) *Work from the back to the front when removing exhaust system components.*
- b) *Apply penetrating oil to the exhaust system component fasteners to make them easier to remove.*
- c) *Use new gaskets, hangers and clamps when installing exhaust system components.*
- d) *Apply anti-seize compound to the threads of all exhaust system fasteners during reassembly.*
- e) *Be sure to allow sufficient clearance between newly installed parts and all points on the underbody to avoid overheating the floor pan and possibly damaging the interior carpet and insulation. Pay particularly close attention to the catalytic converter and heat shield.*

Notes

Chapter 5
Engine electrical systems

Contents

Specifications

Ignition coil resistance

Primary	
1971 through 1979	
Chrysler/Essex	1.41 to 1.55 ohms
Chrysler/Prestolite	1.65 to 1.79 ohms
1980 through 1989	1.34 to 1.55 ohms
Secondary	
1971 through 1979	
Chrysler/Essex	8,000 to 10,700 ohms
Chrysler/Prestolite	9,400 to 11,700 ohms
All others	8, 000 to 12,000 ohms

Pick-up coil resistance

Pick-up coil resistance	150 to 900 ohms

Ballast resistor

Single type	
Breaker point ignition	0.5 to 0.6 ohms
Electronic ignition	1.12 to 1.38 ohms
Dual type (electronic ignition)	
Primary resistor (thermal)	0.5 to 0.6 ohms
Primary resistor (non-thermal)	1.12 to 1.38 ohms
Auxiliary resistor	4.75 to 5.75 ohms

Air gap

Single pick-up coil	
1972 through 1976	0.008 inches
1977 and later	0.006 inches
Dual pick-up coils	
"START"	0.008 inches
"RUN"	
1977 through 1980	0.010 inches
1981 through 1985	0.012 inches

1 General information and precautions

The engine electrical systems include all ignition, charging and starting components. Because of their engine-related functions, these components are discussed separately from chassis electrical devices such as the lights, the instruments, etc. (which are included in Chapter 12).

Always observe the following precautions when working on the electrical systems:

a) *Be extremely careful when servicing engine electrical components. They are easily damaged if checked, connected or handled improperly.*
b) *The alternator is driven by an engine drivebelt which could cause serious injury if your hands, hair or clothes become entangled in it with the engine running.*
c) *Both the alternator and the starter are connected directly to the battery and could arc or even cause a fire if mishandled, overloaded or shorted out.*
d) *Never leave the ignition switch on for long periods of time with the engine off.*
e) *Don't disconnect the battery cables while the engine is running.*
f) *Maintain correct polarity when connecting a battery cable from another source, such as a vehicle, during jump starting.*
g) *Always disconnect the negative cable first and hook it up last or the battery may be shorted by the tool being used to loosen the cable clamps.*

It's also a good idea to review the safety-related information regarding the engine electrical systems located in the *Safety First* section near the front of this manual before beginning any operation included in this Chapter.

2 Battery - removal and installation

Warning: *Hydrogen gas is produced by the battery, so keep open flames, bare light bulbs and lighted tobacco away from it at all times. Always wear eye protection when working around a battery. Rinse off spilled electrolyte immediately with large amounts of water.*

1 Disconnect both cables from the battery terminals. **Caution:** *Always disconnect the negative cable first and hook it up last or the battery may be shorted by the tool being used to loosen the cable clamps.*

2 Remove the battery hold-down clamp.

3 Lift out the battery. Use the proper lifting technique - the battery is heavy.

4 While the battery is out, inspect the battery carrier (tray) for corrosion (see Chapter 1).

5 If you are replacing the battery, make sure that you get an identical battery, with the same dimensions, amperage rating, "cold cranking" rating, etc.

6 Installation is the reverse of removal.

3 Battery - emergency jump starting

Refer to the *Booster battery (jump) starting* procedure at the front of this manual.

4 Battery cables - check and replacement

1 Periodically inspect the entire length of each battery cable for damage, cracked or burned insulation and corrosion. Poor battery cable connections can cause starting problems and decreased engine performance.

2 Check the cable-to-terminal connections at the ends of the cables for cracks, loose wire strands and corrosion. The presence of white, fluffy deposits under the insulation at the cable terminal connection is a sign that the cable is corroded and should be replaced. Check the terminals for distortion, missing mounting bolts and corrosion.

3 When removing the cables, always disconnect the negative cable first and hook it up last or the battery may be shorted by the tool used to loosen the cable clamps. Even if only the positive cable is being replaced, be sure to disconnect the negative cable from the battery first (see Chapter 1 for further information regarding battery cable removal).

4 Disconnect the old cables from the battery, then trace each of them to their opposite ends and detach them from the starter solenoid and ground. Note the routing of each cable to insure correct installation.

5 If you are replacing either or both of the old cables, take them with you when buying new cables. It is vitally important that you replace the cables with identical parts. Cables have characteristics that make them easy to identify: positive cables are usually red, larger in cross-section and have a larger diameter battery post clamp; ground cables are usually black, smaller in cross-section and have a slightly smaller diameter clamp for the negative post.

6 Clean the threads of the solenoid or ground connection with a wire brush to remove rust and corrosion. Apply a light coat of battery terminal corrosion inhibitor, or petroleum jelly, to the threads to prevent future corrosion.

7 Attach the cable to the solenoid or ground connection and tighten the mounting nut/bolt securely.

8 Before connecting a new cable to the battery, make sure that it reaches the battery post without having to be stretched.

9 Connect the positive cable first, followed by the negative cable.

5 Ignition system - general information

Refer to illustrations 5.4, 5.6 and 5.7

1 The ignition system is designed to ignite the fuel/air charge entering each cylinder at just the right moment. It does this by producing a high voltage spark between the electrodes of each spark plug.

2 The types of systems installed on these vehicles evolved through the years to accommodate more strict emissions standards adapted by major automotive manufacturers. Some of the changes in the ignition systems overlapped in certain years and others changed with certain models while some did not. Here is a general listing of the various systems:

3 ***Breaker point ignition systems:*** (points and condenser type). This ignition

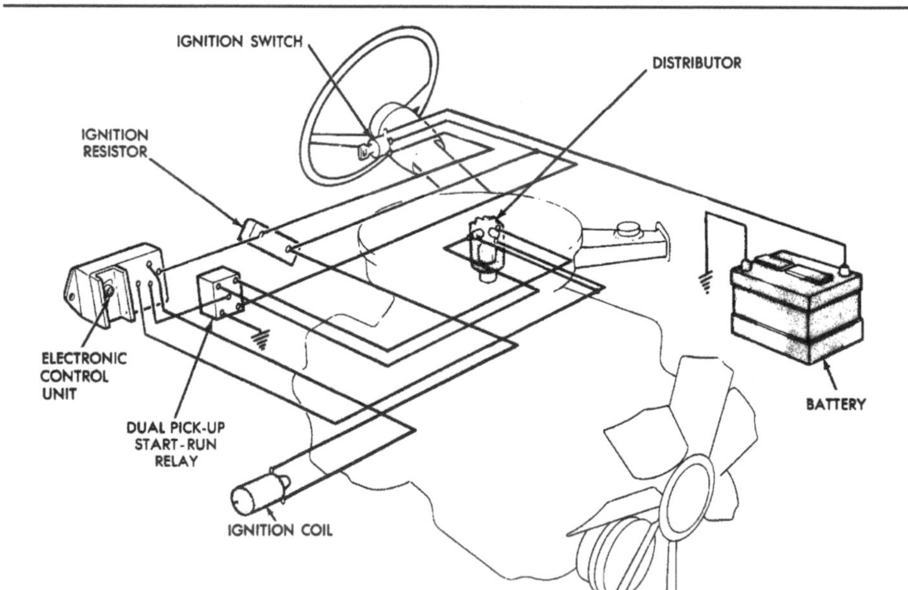

5.4 Schematic of the electronic ignition system (dual pick-up coil system shown)

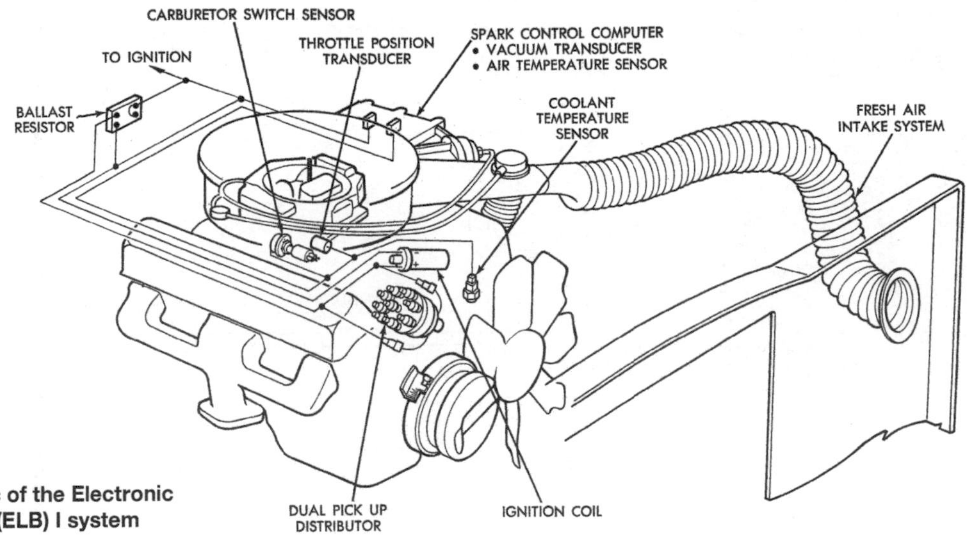

5.6 Schematic of the Electronic Lean Burn (ELB) I system

system controls the ignition spark using the conventional points and condenser arrangement and timing is controlled by mechanical advance components. This system was originally installed on 1971 and 1972 models only.

4 ***Electronic ignition systems* (see illustration)**: This breakerless ignition system basically controls the ignition spark without any electrical (computer-controlled) timing advance or retard capabilities. These systems were installed on certain models ranging from 1972 (select models only) until 1989. The components in these systems include a module (electronic control unit) usually mounted on the firewall, a coil and one or two magnetic pick-up coils mounted in the distributor. The dual pick-up coil systems were installed on models in 1981 through 1989. These systems also included a dual pick-up START-RUN relay that is used to transfer power to the operating pick-up coil. On 1972 through 1979 models, a five-pin connector is used on the module as well as a dual ballast resistor. On 1980 through 1989 models, a four-pin connector and a single ballast resistor was standard equipment. Timing is advanced using centrifugal or vacuum advance components depending on the year and model.

5 ***Spark Control Computer (SCC) systems***: On these systems, the spark timing is constantly adjusted by the computer in response to input from the various sensors located on the engine. This system was divided into two different systems:

6 ***Electronic Lean Burn (ELB) I system* (see illustration)**: This system was installed on most models in 1976 and 1977 (except 1977 LeBaron and Diplomat). This ELB system consists of a spark control computer, a magnetic pulse distributor, five engine sensors and a dual ballast resistor. The spark control computer is made of two modules. The computer is housed in a plastic box mounted on the air cleaner consisting of the program schedule unit and the ignition control unit. One eight-wire electrical connector and one four-wire connector link the computer with the other sensors (coolant temperature sensor, throttle position transducer and carburetor switch sensor). The air temperature sensor and the vacuum transducer are mounted in the computer housing.

7 ***Electronic Lean Burn (ELB) II system* (see illustration):** This system was installed on most models from 1978 through 1989. This system is a redesigned version of the ELB I system. The system in 1979 was renamed the ESC or Electronic Spark Control system. In 1980 it was renamed again to ESA or Electronic Spark Advance. All models except 1980 California models use a spark control computer. California models in 1980

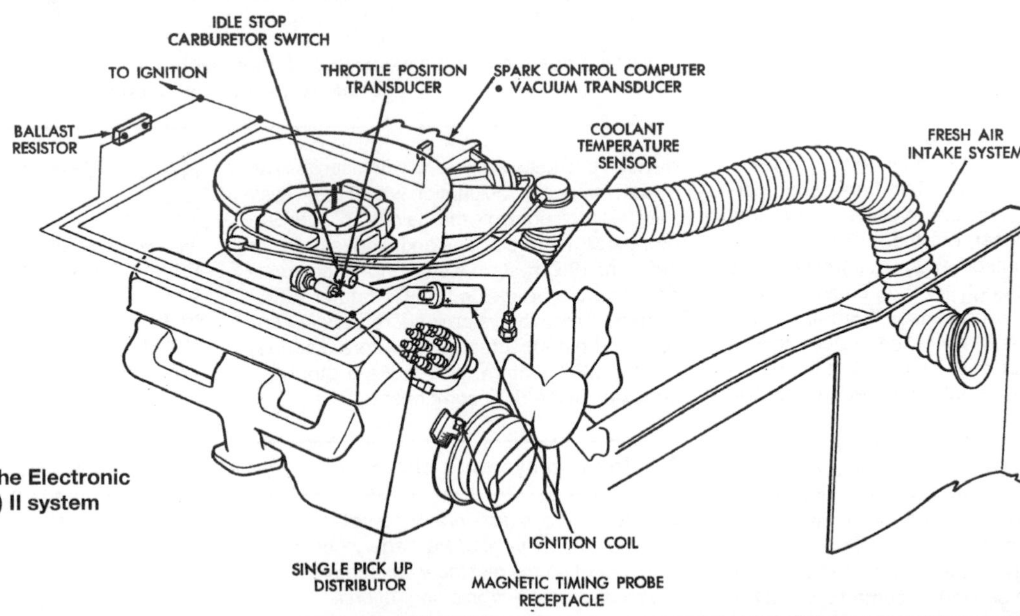

5.7 Schematic of the Electronic Lean Burn (ELB) II system

6.1 To use a calibrated ignition tester, simply disconnect a spark plug wire, clip the tester to a convenient ground (like a valve cover bolt) and operate the starter - if there is enough power to fire the plug, sparks will be visible between the electrode tip and the tester body

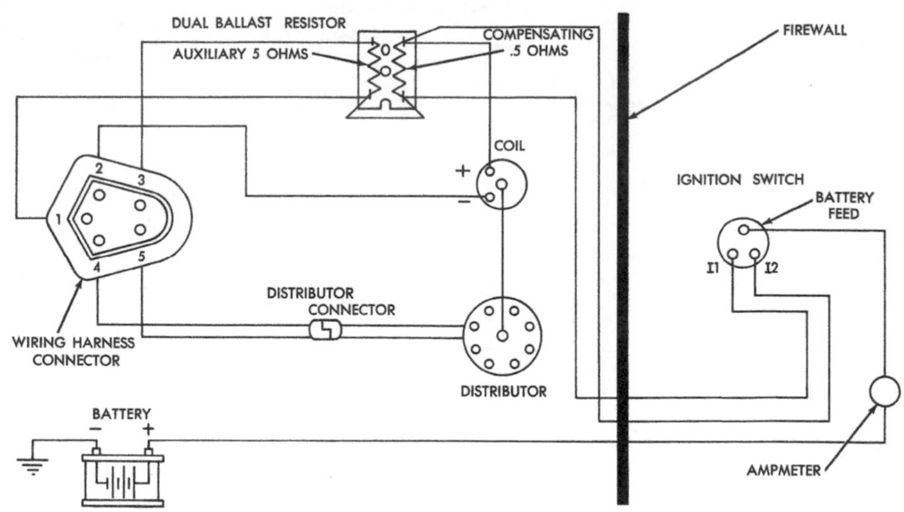

6.11b Schematic of the electronic ignition system on 1972 through 1979 models (5-pin connector, dual ballast resistor)

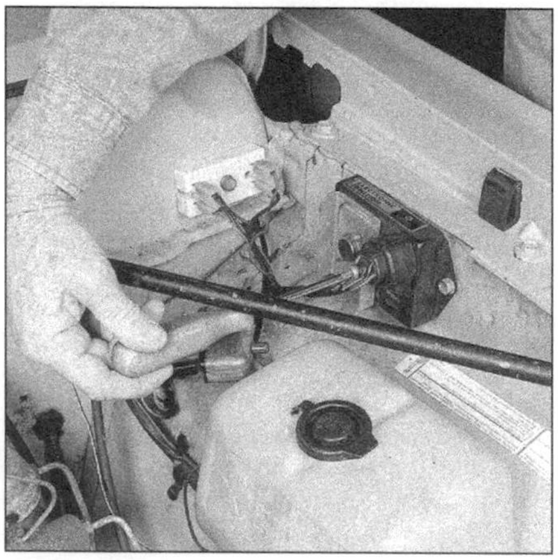

6.11a Remove the retaining screw that locks the electrical connector onto the ignition module

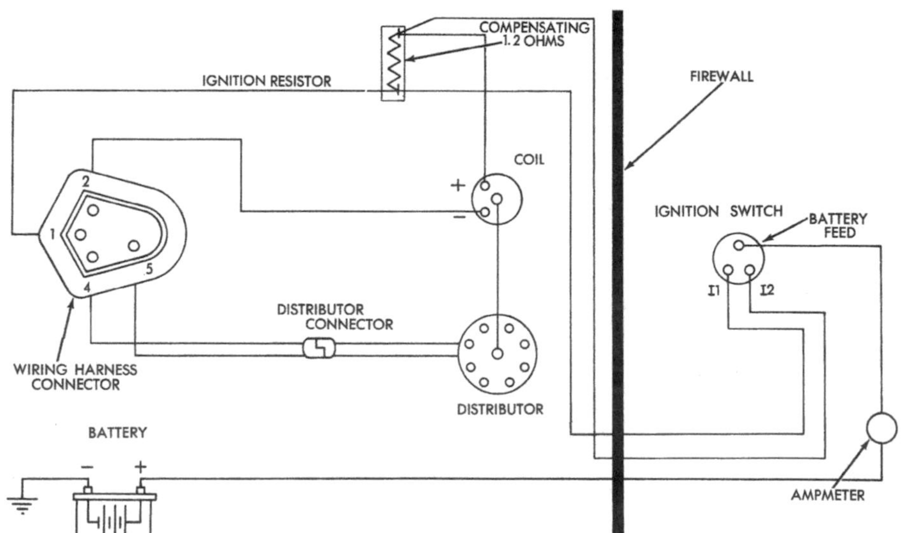

6.11c Schematic of the electronic ignition system on the 1980 and later models (4-pin connector, single ballast resistor)

use a special spark combustion computer. ELB II is different in the programming changes in the computer as well as the additional and different type of sensors installed on the engine. Most systems include the same sensors as the ELB I (coolant temperature sensor, throttle position transducer, carburetor switch, air temperature sensor and vacuum transducer) but the ELB II also uses a detonation sensor (1980 California models and all 1981 through 1989 V8 engines) and a charge temperature sensor (both are in the intake manifold). Many of the system checks will be similar for both systems but the ELB II will require additional tests because of the added sensors.

8 ***Fuel Control Computer systems* (see Chapter 6, Section 2)**: The electronic fuel control (EFC) or more commonly called the electronic feedback carburetor (also EFC) emission system relies on an electronic signal, which is generated by an exhaust gas sensor (oxygen sensor) to control a variety of devices and keep emissions within limits. The system works in conjunction with a three-way catalyst to control the levels of carbon monoxide, hydrocarbons and oxides of nitrogen. The EFC system also works in conjunction with the Spark Control Computer (SCC) system. The two systems share certain sensors and output actuators therefore diagnosing the EFC system will require a thorough check of all the SCC components.

6 Ignition system - check

Caution: *Always disconnect the cable from the negative terminal of the battery before disconnecting the electrical connectors from the module or electronic control unit.*

All ignition systems

Refer to illustration 6.1

1 If the engine will not start even though it cranks over, check for spark at the spark plug by installing a calibrated ignition system tester to the end of the plug wire **(see illustration)**. The tool is available at most auto parts stores. Be sure to order the correct tool for your particular ignition system (breaker type [points] or breakerless type [electronic]).

2 Connect the clip on the tester to a ground such as a metal bracket or valve cover bolt, crank the engine and watch the end of the tester for a bright blue, well defined spark.

3 If sparks occur, sufficient voltage is reaching the plugs. However, the plugs them-

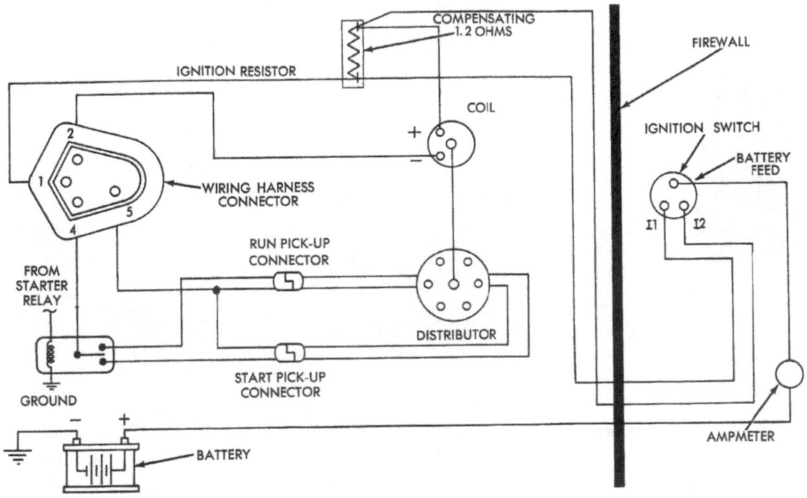

6.11d Schematic of the electronic ignition system on 1981 and later models (4-pin connector, single ballast resistor with dual pick-up coils)

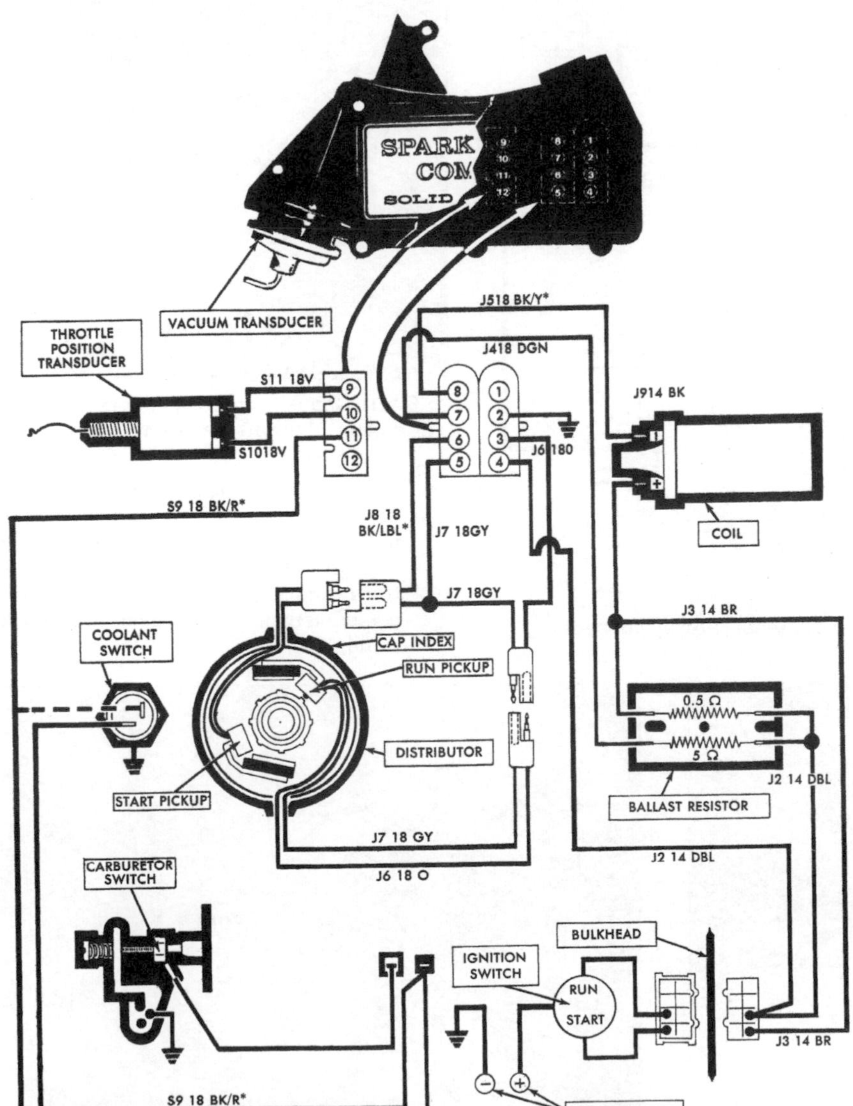

6.11e Schematic of the SCC system (ELB I systems) (8-way and 4-way connectors)

selves may be fouled, so remove and check them as described in Chapter 1 or replace them with new ones.

4 If no spark occurs, remove the distributor cap and check the cap and rotor as described in Chapter 1. If moisture is present, use WD-40 or something similar to dry out the cap and rotor, then reinstall the cap and repeat the spark test.

5 If there is still no spark, the tester should be attached to the wire from the coil and the test repeated again.

6 If no spark occurs, check the primary wire connections at the coil to make sure they are clean and tight. Make any necessary repairs, then repeat the check again.

7 If sparks now occur, the distributor cap, rotor, plug wires or spark plugs may be defective. If there is still no spark, the coil-to-cap wire may be bad. Remove the wire from the distributor and have an assistant hold it about 1/4-inch away from the engine block with insulated pliers. Crank the engine over (be careful of moving engine parts) and check for spark at the gap between the secondary wire terminal and the engine block.

Breaker type ignition systems (points)

Note: *Some of the following check(s) may require a voltmeter, ohmmeter and/or a jumper cable.*

8 If there is still no spark, check the ignition points (see Chapter 1). If the points appear to be in good condition (no pits or burned spots on the point surface) and the primary wires are hooked up correctly, adjust the points as described in Chapter 1.

9 Measure the voltage at the points with a voltmeter. With the ignition on, the points should produce a voltmeter reading of at least 10.5 volts. If not, the battery must be recharged or replaced. If the reading is 10.5 volts or more, record the reading for future reference.

10 If there is still no spark at the plugs, check for a grounded or open circuit in the distributor/points circuit. There may be a damaged resistor block or points terminal causing the ignition voltage to become shorted or diminished.

Breakerless type ignition systems (electronic)

Refer to illustrations 6.11a, 6.11b, 6.11c, 6.11d, 6.11e, 6.11f, 6.12a, 6.12b, 6.14, 6.16a, 6.16b, 6.20, 6.22, 6.23 and 6.28

Note: *These tests include checks for the early electronic ignition systems as well as for the Spark Computer Control (SCC) systems installed on later models*

11 If the battery voltage is satisfactory, turn the ignition off and disconnect the electronic control unit (ECU) at the wiring harness connector **(see illustrations)**. On SCC equipped ignition systems, disconnect the 8-way (ELB I systems) or 10-way connector (ELB II and later SCC systems) **(see illustrations)**.

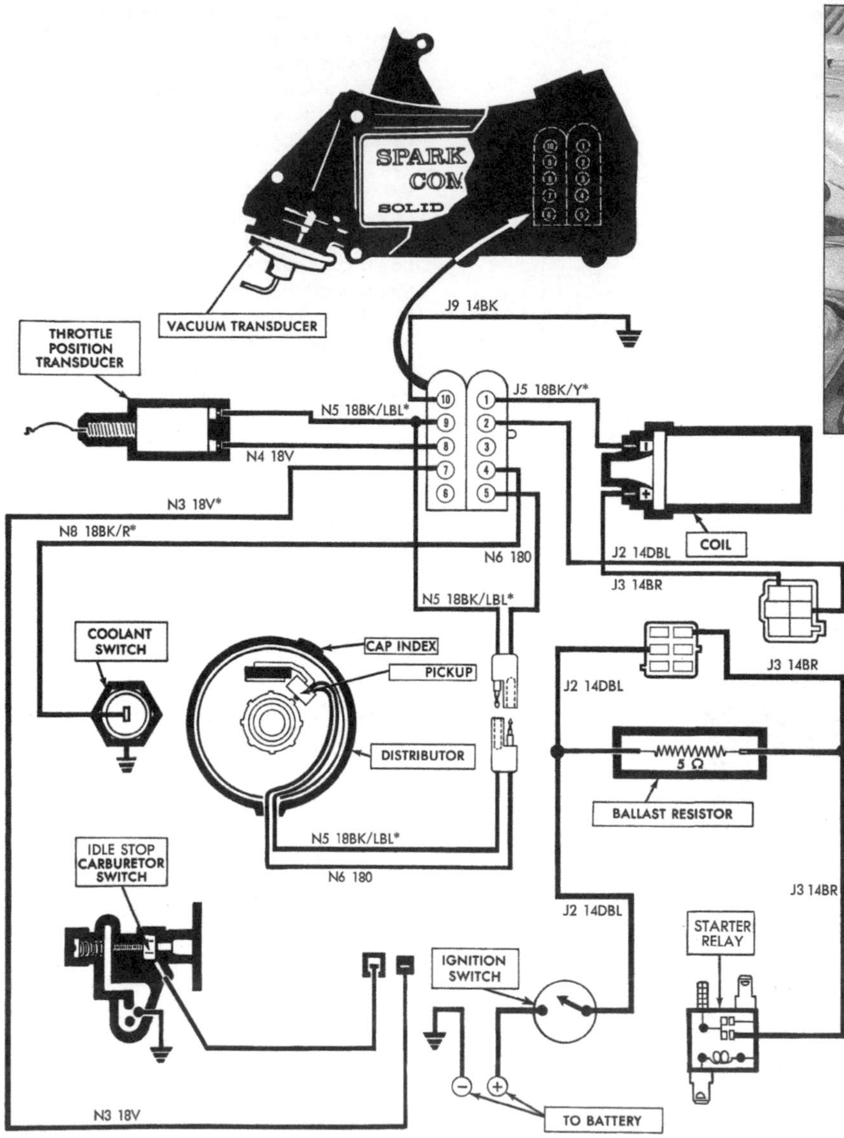

6.11f Schematic of the SCC system (ELB II systems) (10-way connector)

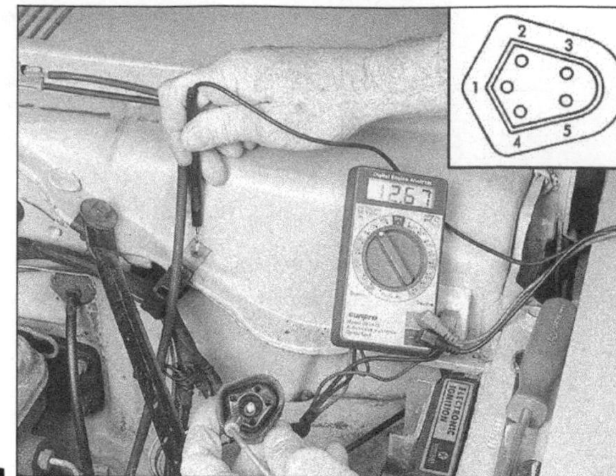

6.12a Check for battery voltage on cavity number 2 of the ignition module electrical (ECU) connector (1974 model shown)

6.12b Check for battery voltage on cavity number 1 of the 10-way electrical connector

12 Check the voltage at cavity 2 of the ECU connector or cavity 1 of the 10-way connector. It should be within one volt of battery voltage **(see illustrations)**. On ELB I systems, check for battery voltage at cavity number 8 on the 8-way connector **(see illustration 6.11e)**.

6.14 Check the voltage of the carburetor switch - it should be greater than five volts

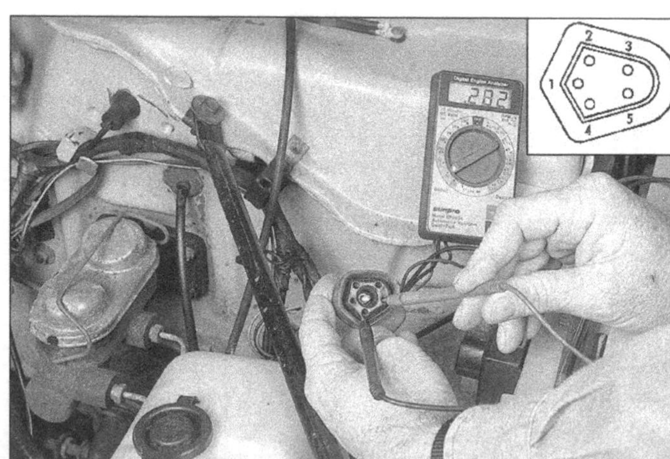

6.16a Check the resistance between cavities number 4 and 5 - it should be between 150 to 900 ohms

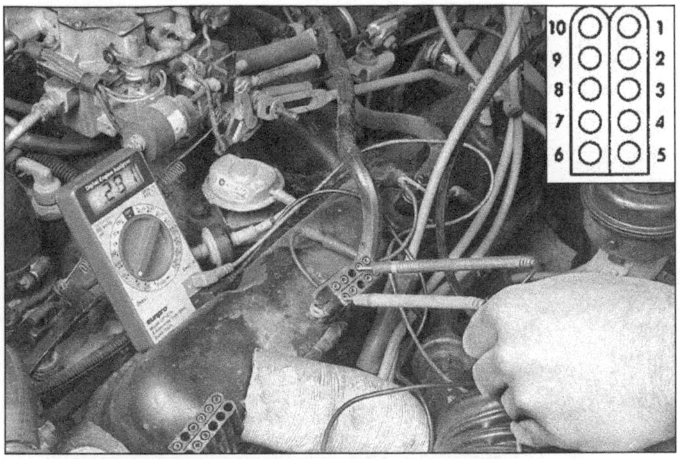

6.16b Check the resistance between cavities number 5 and number 9 on the 10-way connector - it should be between 150 to 900 ohms

6.20 Check for continuity on the ground circuit (pin number 5) and ground on the ignition module (1974 model shown)

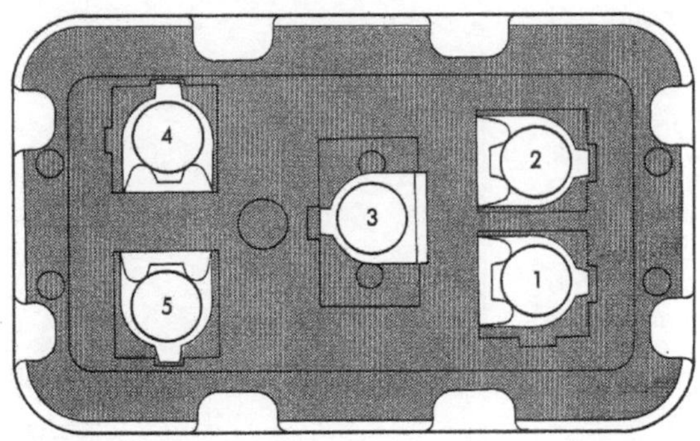

6.22 Check the resistance across terminal numbers 4 and 5 on the dual pick-up start relay - it should be between 20 and 30 ohms

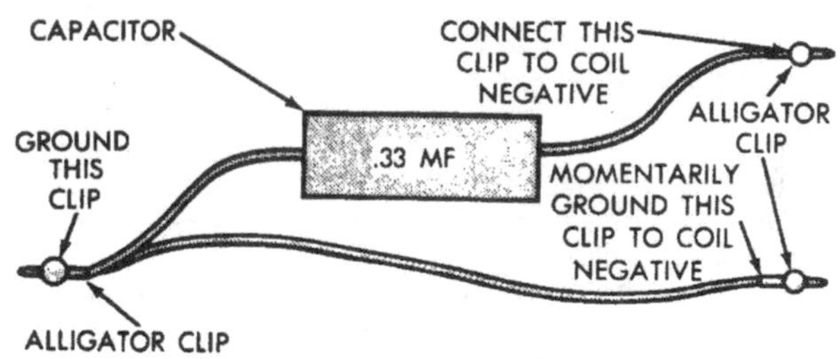

6.23 A homemade jumper cable used to test the primary ignition system

13 If the reading is not within one volt of battery voltage, there is probably a short circuit between the ECU or 10-way connector and the coil negative terminal. Use a continuity tester or ohmmeter to locate the short and repeat Step 4.

14 If the voltage reading is satisfactory, check the voltage at the carburetor switch on SCC equipped ignitions **(see illustration)**. Place a thin insulator between the idle adjusting screw and carburetor switch. The reading should be at least five volts. Measure the voltage at cavity 2 of the 10-way connector, cavity 4 on the 8-way connector or cavity 1 of the ECU connector. It should be within one volt of battery voltage.

15 If the reading is not within one volt of battery voltage, there is probably a short circuit between the ECU connector or 10 or 8-way connector and the ignition switch. Use a continuity tester or ohmmeter to locate the short and repeat Step 14.

16 If the voltage at cavities 1, 4 or 2 is as specified, check the resistance between the cavities with an ohmmeter. Measure between cavities 4 and 5 on the ECU connector **(see illustration)** or cavities 3 and 9 and 5 and 9 of the 10-way connector **(see illustration)** or between cavities 6 and 5 and 3 and 5 of the 8-way connector. The resistance should be 150 to 900 ohms. This test checks the resistance of the pick-up coils and wiring harness at the module/computer. Refer to Section 8 for additional checks.

17 If the resistance is within the specified range, proceed to Step 18. If the reading does not fall into the specified range, disconnect the pick-up leads from the distributor and check the resistance (ignition off) of the pick-up coils (see Section 8). The ohmmeter reading should be between 150 and 900 ohms. This test checks the resistance of the pick-up coils only.

18 If resistance at the pick-up leads is within the specified range, the wire between the ECU or 10 or 8-way connector and the pick-up coil(s) is open or shorted and must be repaired.

19 If the resistance at the pick-up lead is not within the specified range, ground one of the ohmmeter leads and check for a ground or short circuit at each pick-up coil lead. A zero reading should be indicated.

20 If a zero reading is not indicated, check the pick-up leads for a ground or short circuit. If the pick-up leads are in good condition, test the ignition module. Check the ground circuit on ECU equipped ignitions **(see illustration)**. Make sure there is a good ground at pin 5 of the ECU. Make sure the ECU connections are clean and the mounting screws are tight. Refer to Section 9 for additional information on the ignition module.

21 If a zero reading was indicated and the resistance reading at the pick-up leads did not fall within the specified range, one of the pick-up coils is defective and must be replaced.

22 On dual pick-up coil systems, check the START-RUN relay. Disconnect the electrical connectors from pin numbers 4 and 5 and using an ohmmeter, measure the resistance of the relay between the same terminals **(see illustration)**. Resistance should be between 20 and 30 ohms.

23 If there was no spark at the secondary coil wire while cranking the engine, a jumper cable will have to be fashioned to carry out further testing. Use two lengths of wire and three alligator clips to fabricate a three connector jumper cable. Splice a 0.33 MF capacitor into one of the wires **(see illustration)**.

24 Disconnect the ECU or 10 or 8-way connector and connect one of the double leads to the coil negative terminal. Connect the opposite end (single lead) to a good ground.

25 Using the remaining wire lead, momentarily make contact with the coil negative terminal while holding the secondary coil wire about 1/4-inch away from the engine block.

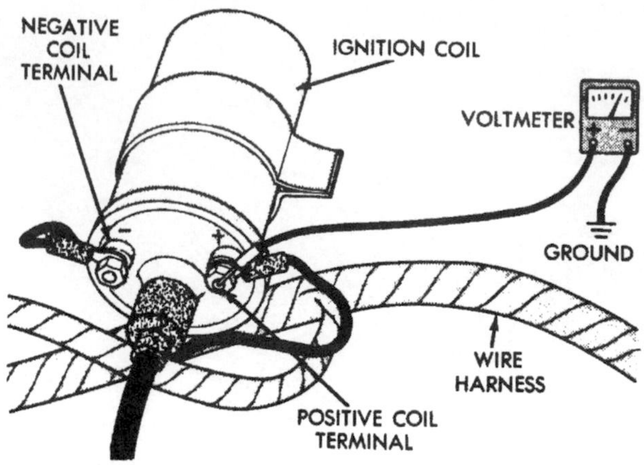

6.28 Checking the voltage at the coil positive terminal

7.4 Checking the coil primary resistance

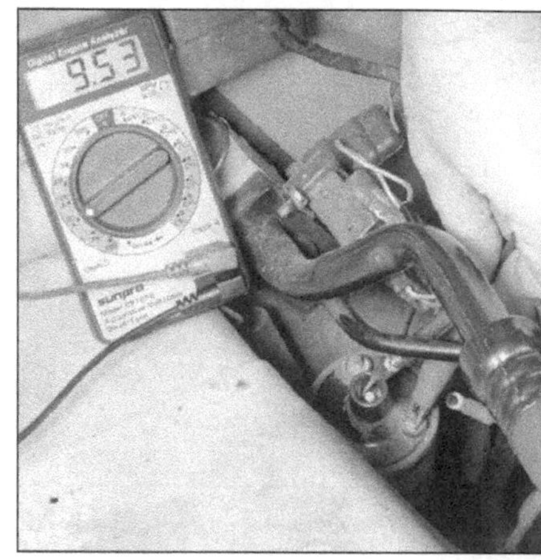

7.5 Checking the coil secondary resistance

26 If there is now spark at the secondary coil wire, the SCC or ECU is faulty and should be replaced.

27 If there is still no spark at the secondary coil wire, measure the voltage at the battery terminals with a voltmeter. Record the measurement.

28 Measure the voltage at the coil positive terminal. It should be within one volt of the battery voltage **(see illustration)**. If the correct voltage is indicated, proceed to Step 29 for electronic ignition systems. Proceed to Step 34 for SCC equipped ignition systems. Also, check the resistance of the ballast resistor (see Section 10).

29 If sufficient voltage is not measured at the coil positive terminal, check the wiring between the battery and the coil terminal for an open or short circuit. Replace or repair any faulty wiring. On SCC equipped systems, proceed to Step 34.

30 If the wiring is not the problem, the next step is to replace the starter relay.

31 Repeat Step 29. If there is still insufficient voltage, the ballast resistor is defective and must be replaced (see Section 10).

32 If sufficient voltage was measured at the coil positive terminal, check the voltage at the coil negative terminal. It should be within one volt of the battery voltage.

33 If there is battery voltage at the coil negative terminal, but there is no spark when shorting the terminal (see Step 29), the coil is defective and must be replaced.

34 If the specified voltage was measured at the coil positive terminal on SCC equipped systems, check for voltage at the coil negative terminal. It should be within one volt of battery voltage.

35 If the voltage at the negative coil terminal is within this range, but no spark was produced when shorting the secondary coil lead (Step 29), the ignition coil is faulty and should be replaced. Check the ignition coil (see Section 7).

36 If the voltage at the negative coil terminal is not within this range, the coil is faulty and should be replaced.

37 If voltage at the negative coil terminal is within the specified range and a spark was produced when shorting the secondary coil lead, the wiring should be checked further.

38 With the ignition off, use an ohmmeter to check for continuity between the carburetor switch and cavity 10 of the 4-way connector on the ELB I systems or cavity 8 of the 10-way connector (ELB II and later SCC systems).

39 If the circuit is not continuous, isolate the open circuit with the ohmmeter and repair the wiring.

40 If the circuit is continuous, the spark control computer is probably faulty. It might be a good idea to have the computer diagnosed by a dealer service department before purchasing a replacement.

7 Ignition coil - check and replacement

Check

Refer to illustrations 7.4 and 7.5

1 Mark the wires and terminals with pieces of numbered tape, then remove the primary wires and the high-tension lead from the coil.

2 Clean the outer case and check it for cracks and other damage.

3 Clean the coil primary terminals and check the coil tower terminal for corrosion. Clean it with a wire brush if any corrosion is found.

4 Check the coil primary resistance by attaching the leads of an ohmmeter to the positive and negative terminals. Compare the measured resistance to the value listed in this Chapter's Specifications **(see illustration)**.

5 Check the coil secondary resistance by hooking one of the ohmmeter leads to one of the primary terminals and the other ohmmeter lead to the large center terminal **(see illustration)**. Compare the measured resistance to the value listed in this Chapter's Specifications.

6 If the measured resistances are not as specified, the coil is defective and should be replaced with a new one.

7 It is essential for proper operation of the ignition system that all coil terminals and wires be kept clean and dry.

Removal and installation

8 To remove the coil, loosen the screw on the mounting clamp and slide the coil out of the clamp.

9 Installation is the reverse of removal.

8 Pick-up coil(s) - check and replacement

Refer to illustration 8.2a and 8.2b

Check

1 Before replacing the pick-up coil(s), check the ignition system thoroughly (see Section 6).

2 Disconnect the electrical connector(s) from the pick-up coil(s) at the distributor and using an ohmmeter, check for resistance **(see illustrations)**. There should be 150 to 900 ohms resistance.

3 If the readings are incorrect, replace the pick-up coil(s).

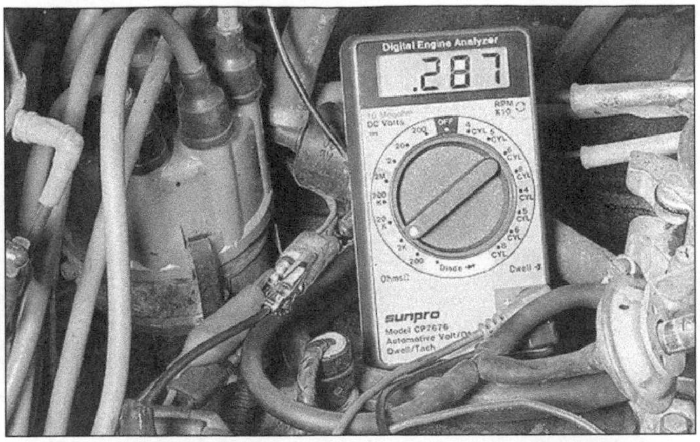

8.2a Disconnect the electrical connector and check the resistance of the pick-up coil at the distributor

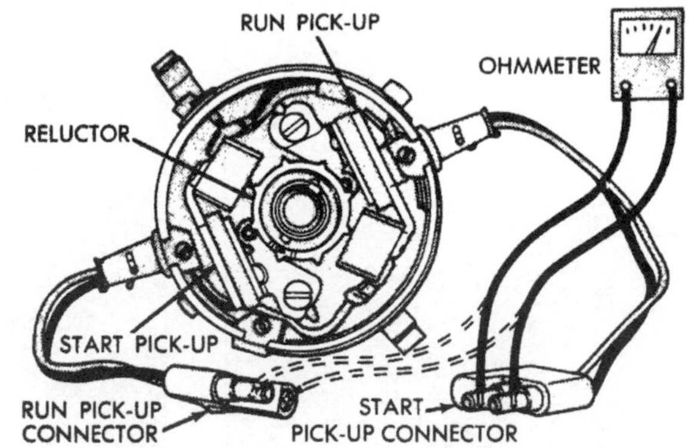

8.2b On dual pick-up coil systems, the resistance should be the same for the START pick-up coil and the RUN pick-up coil

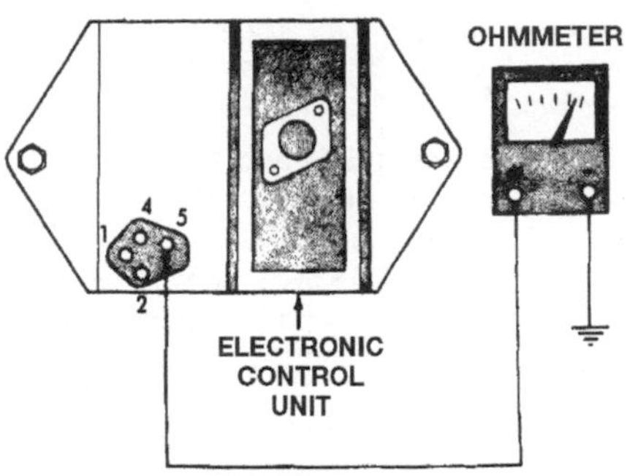

9.1 Checking for ground at the ECU (ignition module) connector (4 pin electronic ignition system shown)

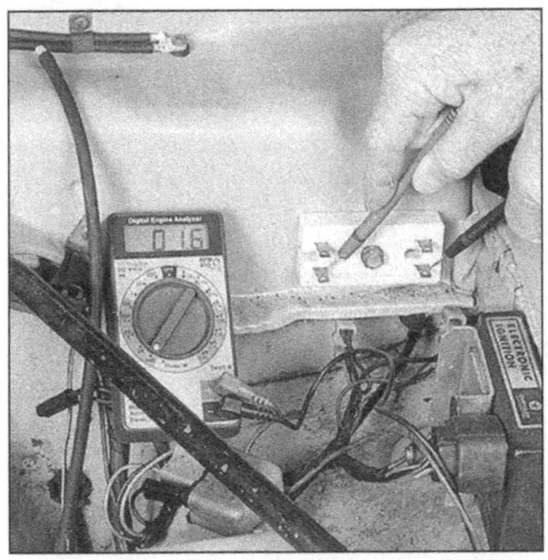
10.3a Checking the primary side of the ballast resistor

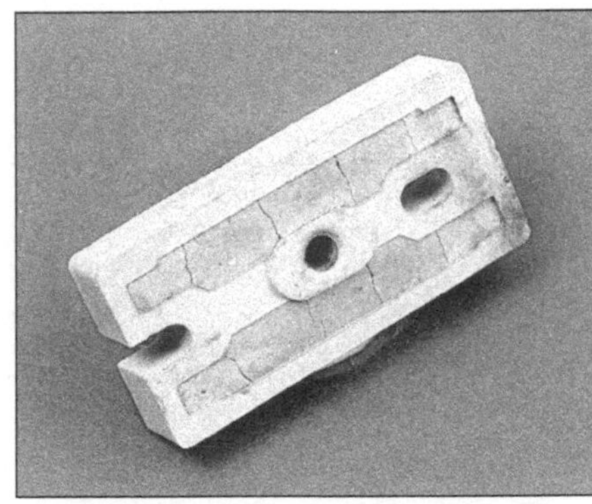
10.3b Remove the ballast resistor from the firewall and check the backside; non-thermal types will have the resistor coils sealed from the atmosphere (shown) while thermal types will have the resistor coils exposed

Replacement

4 Remove the distributor as described in Section 12 and carefully clamp it in a vise equipped with soft jaws (don't apply excessive pressure). **Note:** *On some early models, the pick-up coil can be detached after removing the mounting screw and separating the wires from the retainers on the upper plate and distributor housing.*

5 Using two screwdrivers not more than 7/16-inch wide, remove the reluctor by prying it up **(see illustration 12.5b)**. Be careful not to damage the reluctor teeth.

6 Remove the two mounting screws from the vacuum advance unit, if equipped. Disconnect the control arm from the pick-up plate.

7 Remove the pick-up coil leads from the distributor housing.

8 Remove the two mounting screws from the pick-up plate.

9 Lift out (do not completely remove) the pick-up coil and plate as an assembly. Depress the retainer clip on the underside of the plate assembly and detach the assembly from the distributor. On most models, the pick-up coil and plate are replaced as an assembly and are not separable.

10 Installation is the reverse of removal, but be sure to place a small amount of distributor cam lubricant on the plate support pins before installing the plate assembly.

11 Set the air gap as described in Section 13.

9 Electronic ignition system module - check and replacement

Refer to illustrations 9.1

Check

1 Connect one ohmmeter lead to a good ground and the other lead to the control unit pin number 5 **(see illustration)**.

2 The ohmmeter should show continuity between ground and the control unit pin. If continuity does not exist, try removing the control unit (module), cleaning the back side of the housing and tightening the bolts a little tighter to make a better connection on the firewall.

3 Repeat step 1. If continuity still does not exist, replace the control unit.

Replacement

4 Remove the electrical connector from the control unit (module), remove the bolts and lift the assembly from the engine compartment.

10 Ballast resistor - check and replacement

Check

Refer to illustrations 10.3a, 10.3b and 10.3c

1 The ballast resistor limits voltage to the coil during low speed operation but allows it to increase as the engine speed increases. While the engine is cranking, the ballast resistor is bypassed to ensure adequate voltage through the coil.

2 On models equipped with a dual ballast resistor, voltage is limited by a pair of resistors; the primary resistor and the auxiliary resistor. Voltage to the electronic control unit is limited by the auxiliary side of the ballast resistor. Some earlier models were equipped with speed limiter circuitry to prevent engine damage from excessive rpm's.

3 Disconnect the electrical leads from the ballast resistor and using an ohmmeter, check the resistance of the ballast resistor. On single ballast resistors, the resistance

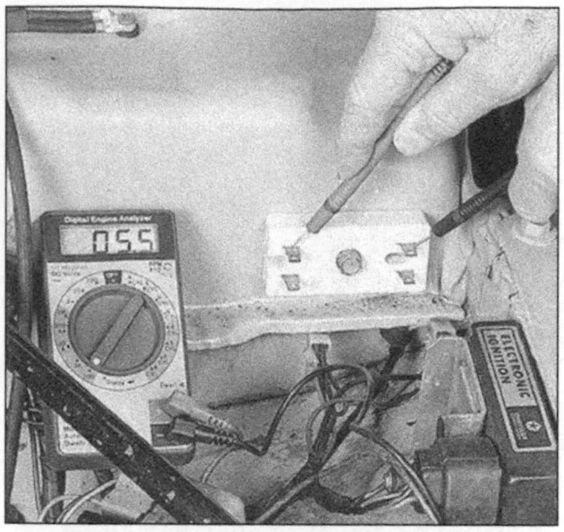

10.3c Check the resistance of the auxiliary side of the ballast resistor - it should be 4.75 to 5.75 ohms

should be 1.12 to 1.38 ohms (electronic ignition systems) or 0.5 to 0.6 ohms (points type ignition systems). On dual ballast resistors, first check the primary side **(see illustrations)**. The primary side is either a thermal type (wire resistor is openly exposed) (it should be 0.5 ohms), or the non-thermal type (wire resistor hidden in case) (it should be 1.12 to 1.38 ohms). Next check the auxiliary side of the dual ballast resistor **(see illustration)**. It should be 4.75 to 5.75 ohms.

Replacement

4 Disconnect the electrical leads from the ballast resistor.
5 Remove the screws from the ballast resistor. and lift it from the engine compartment.
6 Installation is the reverse of removal.

11 Spark Control Computer (SCC) system - check and component replacement

Note: *Refer to Chapter 6, Section 3 for additional checks and replacement procedures for sensors and output actuators incorporated into the SCC system that are not covered in this section*

General description

1 Some of the later model vehicles covered in this manual are equipped with an SCC ignition system **(see illustrations 5.6 and 5.7)**. This system delivers an infinitely variable spark advance. Some of these ignition advance modes occur only during cold engine operation, thereby reducing HC and nitrous oxide emissions.
2 The computer unit, mounted on the side of the air cleaner housing, constantly sends and receives signals to provide optimum ignition timing for current driving conditions.
3 Depending on the engine temperature and throttle position, vacuum is applied to the diaphragm in the distributor vacuum unit and the ignition timing is changed to reduce emissions and improve cold engine driveability. Here is a list of the various components, sensors and output actuators involved with these SCC systems:

SCC computer
Throttle position transducer (1976 through 1979 models)
Throttle control system (1980 through 1989 models)
Carburetor switch
Coolant temperature sensor
Charge air temperature sensor
Detonation (knock) sensor
Oxygen sensor
Mixture control solenoid (also called duty cycle solenoid, oxygen feedback solenoid)

11.8 Check the ignition timing while applying vacuum to the vacuum transducer on the SC computer - the ignition timing should advance 38 to 50-degrees with the engine running at 2000 rpm

Check

SCC computer

Refer to Illustration 11.8

4 Visually check all vacuum hoses for cracks and kinks as well as correct routing. Make sure the engine is warmed up to operating temperature and all the sensors are working correctly (see Chapter 6, Section 3).
5 Make sure the ignition timing is adjusted properly (see Chapter 1).
6 Make sure the curb idle adjusting screw is not touching the carburetor switch (see Chapter 4). If necessary, insert a piece of paper between them to prevent grounding.
7 Remove the vacuum line from the vacuum transducer and plug it with a suitable tool.
8 Connect a hand-held vacuum pump to the vacuum transducer **(see illustration)** and set the vacuum at 10 inches Hg for six cylinder engines or 16 inches Hg for eight cylinder engines.
9 Increase the engine speed to 2,000 rpm, wait one minute and observe that the timing is advanced 38 to 50-degrees depending upon the year and model.
10 If the timing does not advance into the specified range, have the computer checked by a dealer service department.

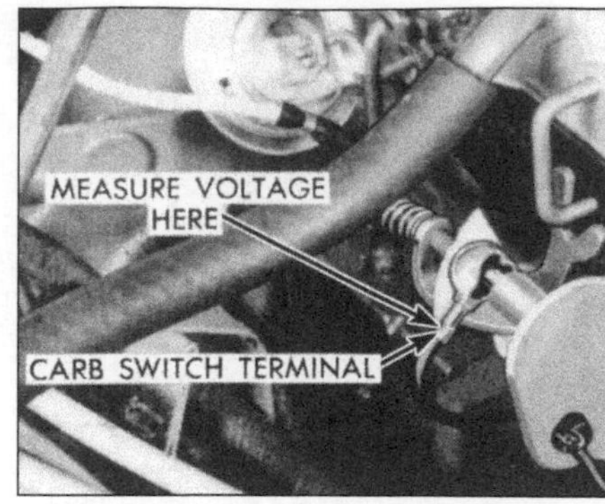

11.11 Install a piece of paper between the idle adjustment screw and the carburetor switch, then check for voltage at the switch terminal (ELB I systems)

Carburetor switch

Note: *Grounding the carburetor switch eliminates all spark advance on most of the SCC systems.*

ELB I systems (1976 and 1977)

Refer to illustrations 11.11, 11.13 and 11.15

11 With the ignition key OFF, disconnect the 8-way connector from the SCC computer and install a piece of paper between the idle adjustment screw and the carburetor switch **(see illustration)**.
12 With the ignition key ON, connect the positive probe of the voltmeter to the carburetor switch and the negative lead to ground.

a) *If the voltage was more than 5 volts but less than 10 volts, proceed to step number 15.*
b) *If the voltage was less than 5 volts, proceed to step number 14.*
c) *If the voltage is more than 10 volts, proceed to step number 13.*

13 Turn the ignition switch OFF and disconnect the 8-way connector on the SCC computer **(see illustration)**. Connect the probes of an ohmmeter to terminal number 2 and ground. There should be no continuity. If there is continuity, replace the SCC computer.

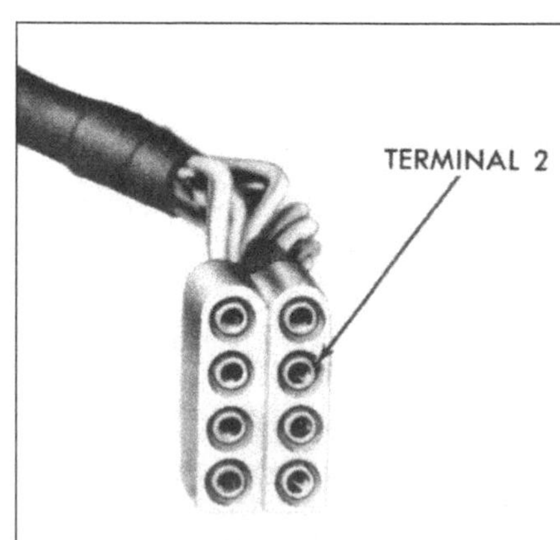

11.13 Check for continuity between terminal number 2 and ground (ELB I systems)

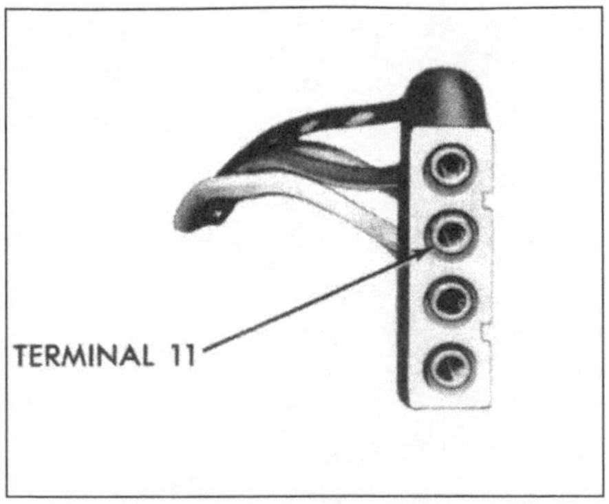

11.15 Check for continuity between terminal number 11 and the carburetor switch (ELB I systems)

14 If voltage is less than 5 volts, turn the ignition switch to OFF, disconnect the 8-way terminal from the SCC computer and check for voltage (key ON) at terminal number 4. There should be battery voltage (+/- 1 volt).
15 If the voltage is correct, disconnect the single connector from the SCC computer and check for continuity to the carburetor switch on terminal number 11 **(see illustration)**. There should be continuity. If there is no continuity, check the wire harness for shorts or damage to the harness.
16 If the continuity is correct, check for continuity between terminal number 2 and ground on the 8-way connector (see Step 13).
17 Next, check for battery voltage on the 8-way connector from the ignition coil and from the ballast resistor (terminal numbers 7 and 8). Refer to the wiring schematic for the ELB I system in **illustration 6.11e**.
18 If the ohmmeter indicates continuity, battery voltage is reaching the SCC control unit and all the circuits check out, replace the SC computer.

ELB II systems

Refer to illustration 11.22

19 With the ignition key OFF, disconnect the 10-way connector from the SCC computer and install a piece of paper between the idle adjustment screw and the carburetor sensor switch **(see illustration 11.11)**. **Note:** *On 1977 through 1979 models, disconnect the coolant temperature sensor wire connector before performing the following tests.*
20 With the ignition key ON, connect the positive probe of the voltmeter to the carburetor switch and the negative lead to ground.

a) *If the voltage was more than 5 volts but less than 10 volts, proceed to step number 24*
b) *If the voltage was less than 5 volts, proceed to step number 22.*
c) *If the voltage is more than 10 volts, proceed to step number 21*

21 Turn the ignition switch OFF and disconnect the 10-way connector on the SCC computer **(see illustration 6.12b)**. Connect the probes of a voltmeter to terminal number 2 and ground. With the ignition key ON (engine not running), the voltmeter should

11.22 Continuity should exist between terminal number 7 and the carburetor switch (ELB II systems)

indicate battery voltage.
22 If voltage is less than 5 volts, first check for voltage on terminal number 2 (see Step 21) and then with the ignition key OFF, check for continuity between terminal number 7 and the carburetor switch **(see illustration)**. Continuity should exist. **Note:** *On 1981 through 1989 models continuity should NOT exist with the throttle closed. Open the throttle and the circuit should indicate continuity.*
23 Next, check for continuity to the carburetor switch on terminal number 10 **(see illustration 11.22)**. There should be NO continuity. If there is continuity, check the wire harness for shorts or damage to the harness.
24 If the voltage is between 5 and 10 volts, check the pick-up coil resistance (see Section 8).
25 If the ohmmeter indicates continuity, battery voltage is reaching the SCC control unit, pick-up coil resistance is correct and all the circuits check out, replace the SC computer.

Throttle position transducer (ELB I and ELB II systems - 1976 through 1979 models only)

Refer to illustrations 11.29, 11.30, 11.32 and 11.33

26 Before performing the following tests, adjust the throttle position transducer. The air temperature inside the computer must be below 135-degrees F. If necessary, remove the air cleaner and allow the computer to cool.
27 With the engine cooled down, start it back up and wait 90 seconds and connect a jumper wire from the carburetor switch to ground. Disconnect the electrical connector on the transducer and observe the timing (see Chapter 1).
28 Reconnect the transducer electrical connector and recheck the timing. It should remain the same. If the timing advances, loosen the transducer locknut and turn the transducer clockwise until it reaches specifications. Turn the locknut an additional 1/2 turn and tighten the locknut. Conversely, if

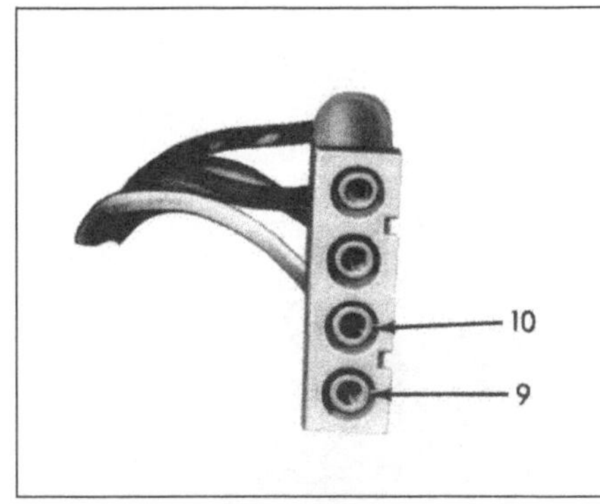

11.29 Check the resistance of the throttle position transducer (ELB I system) between terminals 9 and 10 of the four-way electrical connector at the SC computer

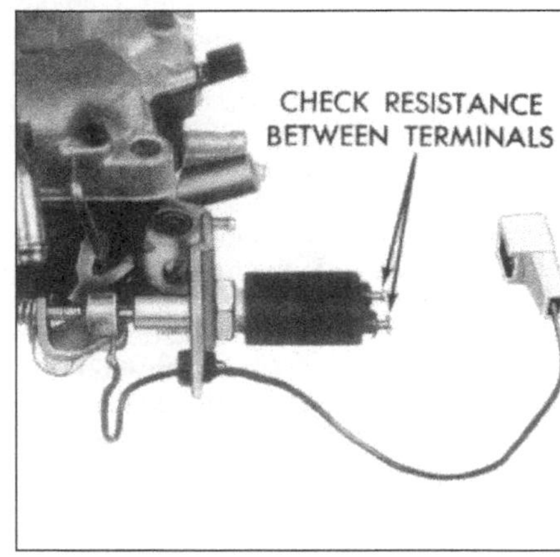

11.30 Unplug the connector and check the resistance across the terminals of the throttle position transducer

the ignition timing is retarded, turn the transducer counter-clockwise until the base setting is achieved, add an extra 1/2 turn and tighten the locknut.
29 Turn the ignition key OFF and disconnect the single 4-way connector (ELB I systems) or the 10-way connector (ELB II systems) on the SC computer. Check the resistance between terminals number 9 and number 10 (ELB I system) **(see illustration)** or terminals 8 and 9 **(ELB I system) (see illustration 6.11f)**. It should be between 50 and 90 ohms. This test checks the resistance of the throttle position transducer and its circuit. **Note:** *Refer to illustrations 6.11e and 6.11f for the exact pin designations on the computer electrical connectors for the ELB I and ELB II systems.*
30 If the readings are incorrect, disconnect the electrical connectors at the transducer and measure the resistance **(see illustration)**. It should be between 50 and 90 ohms. If the resistance is incorrect, replace the transducer.
31 Reconnect the wires to the transducer and with the ignition key ON (engine not running), check for voltage at the transducer terminals.

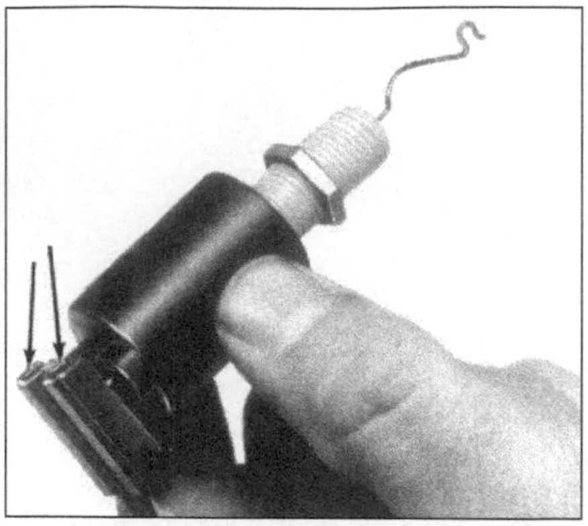

11.32 Check the voltage at each terminal while opening and closing the throttle

32 Open and close the throttle (wide open) and observe that the voltage changes approximately 2 volts **(see illustration)**. **Note:** *If the voltage readings are incorrect, do not fault the transducer and replace it yet. It still may be a computer problem.*

33 Position the throttle on the fast idle cam and ground the carburetor switch using a jumper wire. To perform the test, it will be necessary to remove the transducer, install the electrical connector onto the transducer terminals and move the transducer core (wire) all the way inside **(see illustration)**.

34 Start the engine and wait approximately 90 seconds and pull the transducer core out approximately one inch. Confirm that the ignition timing advances 4 to 12-degrees. Now move the core back into the transducer and confirm that the timing returns to the initial setting.

35 If there was no advance or return, replace the SC computer. It may be necessary to replace the throttle position transducer also. Repeat the test and check the results.

36 Repeat the test again and this time have an assistant move the core in and out of the transducer rapidly about 5 or 6 times. The timing should advance 7 to 14-degrees for approximately one second and then return to base setting. If the advance is not indicated replace the computer.

Throttle control system

Refer to illustration 11.42

37 The SCC incorporates into the computer an integral system called the throttle control system. A solenoid mounted on the carburetor is energized when the air conditioning switch, electronic backlite (EBL) or electronic timers are activated.

38 Connect a tachometer to the engine (see Chapter 1).

39 Start the engine and warm it up to operating temperature.

40 Depress the accelerator pedal and release it. The idle speed should remain slightly higher than the curb idle speed momentarily during deceleration.

41 On vehicles equipped with air conditioning or EBL, activate the system and confirm

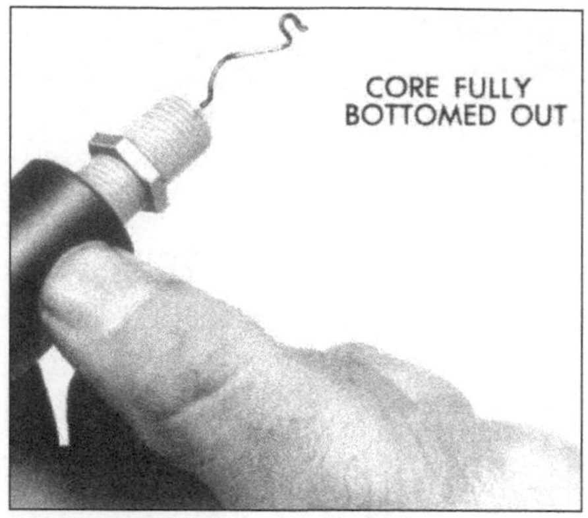

11.33 With the core pulled out, the timing should advance 4 to 12-degrees

that the idle speed increases to compensate for the additional load. **Note:** *The air conditioning clutch will cycle ON and OFF during the test so do not be confused with the actual idle speed adjustments.*

42 If the idle speed does not increase, turn off the ignition, disconnect the electrical connector from the solenoid **(see illustration)** and check the resistance from the black wire to ground. It should be between 15 and 35 ohms.

43 Start the vehicle and before the timer delay has elapsed, measure the voltage from the black wire of the 3-way connector. The voltmeter should read charging voltage.

44 If it does not, have the SCC computer checked by a dealer service department or other repair shop.

Replacement

Spark Control Computer

Refer to illustration 11.46

45 Remove the air cleaner from the carburetor (see Chapter 4).

46 Remove the bolts from inside the air cleaner housing **(see illustration)** and separate the SC computer from the housing.

47 Installation is the reverse of removal.

11.46 Remove the bolts (arrows) and separate the computer from the air cleaner housing

11.42 Disconnect the carburetor solenoid and check the resistance between the black wire and ground - it should read 15 to 35 ohms

12 Distributor - removal and installation

Refer to illustrations 12.3, 12.5a and 12.5b

Removal

1 Unplug the primary electrical connector from the distributor.

2 Look for a raised "1" on the distributor cap. This marks the location for the number one cylinder spark plug wire terminal. If the cap doesn't have a mark for the number one terminal, locate the number one spark plug and trace the wire back to the terminal on the cap.

3 Mark the relationship of the distributor base to the engine block to ensure the distributor is installed correctly **(see illustration)**. Remove the distributor cap (see Chapter 1) and turn the engine over until the rotor is pointing toward the number one spark plug terminal (see locating TDC procedure in Chapter 2).

12.3 Be sure to mark the position of the distributor to the engine block (arrow) as well as the position of the rotor to the distributor body (arrow) before removing the assembly from the block

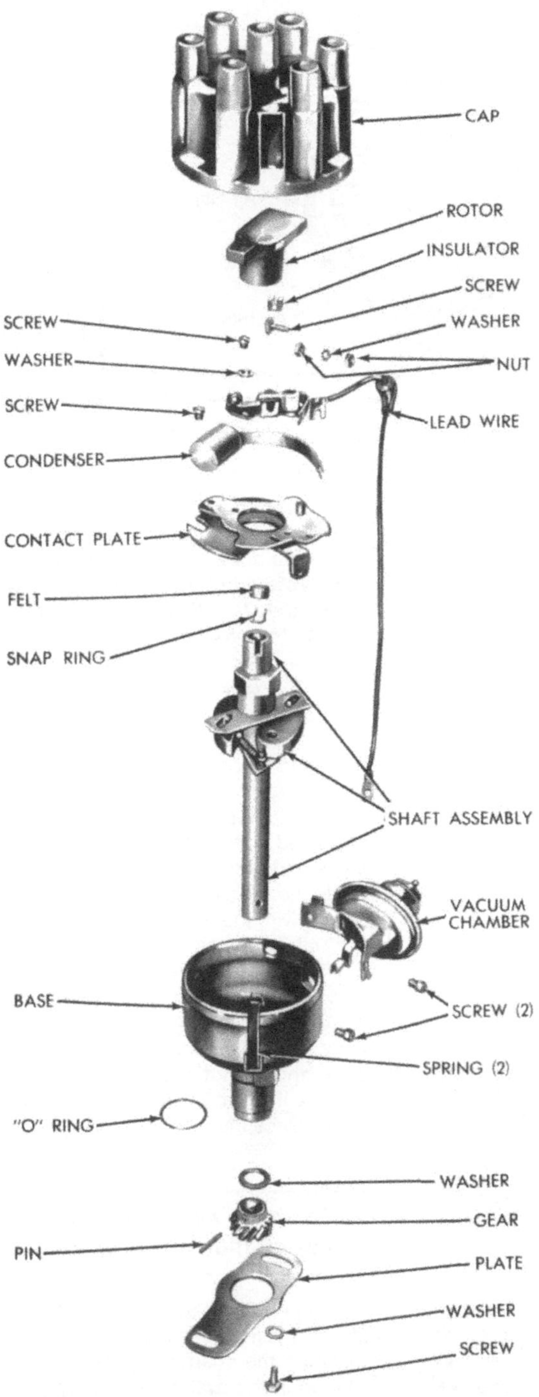

12.5a Exploded view of a breaker-point type distributor

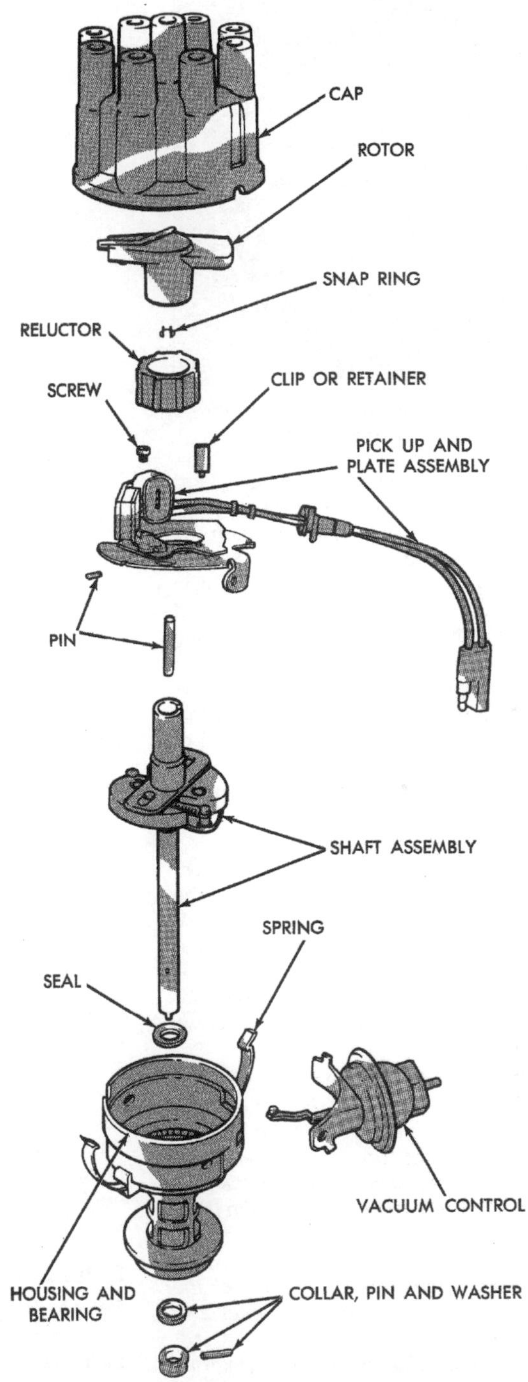

12.5b Exploded view of an electronic ignition distributor (this distributor is exactly the same for SCC systems except that the distributor weights are absent)

4 Make a mark on the edge of the distributor base directly below the rotor tip and in line with it.

5 Remove the distributor hold-down bolt and clamp, then pull the distributor straight up to remove it **(see illustrations)**. **Caution:** *DO NOT turn the crankshaft while the distributor is out of the engine or the alignment marks will be useless.*

Installation

Note: *If the crankshaft has been turned while the distributor is out, the number one piston must be repositioned at TDC. This can be done by feeling for compression pressure at the number one plug hole as the crankshaft is turned. Once compression is felt, align the crankshaft timing mark with the zero on the engine timing scale.*

Note: *Be sure to inspect the distributor gear before installing it back into the engine block. Six-cylinder models are equipped with a nylon drive gear that may need to be replaced because of wear and damage.*

6 Insert the distributor into the engine in exactly the same relationship to the block that it was in when removed.

7 On six-cylinder models, to mesh the

13.4 Checking the air gap on a single pick-up coil distributor

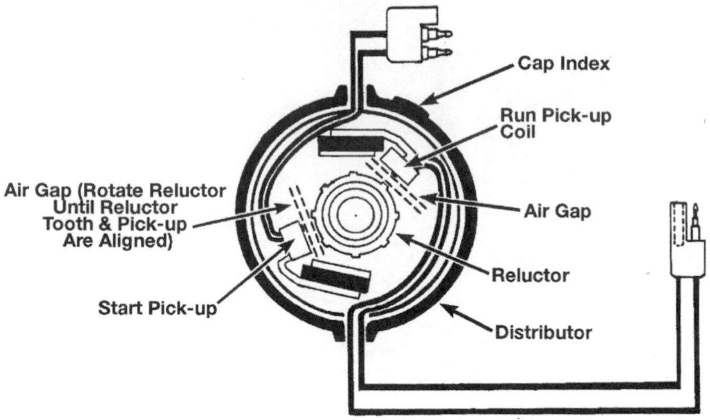

13.12 Pick-up coil details (dual pick-up distributor)

helical gears on the camshaft and distributor, you may have to turn the rotor slightly. Recheck the alignment marks between the distributor base and the block to verify the distributor is in the same position it was before removal. Also check the rotor to see if it's aligned with the mark you made on the edge of the distributor base.

8 On V8 models, engage the shaft with the slot in the driveshaft. Note that it's possible to install the distributor 180-degrees out of time.

9 Place the hold-down clamp (if equipped) in position and loosely install the bolt.

10 Install the distributor cap.

11 Plug in the primary electrical connector.

12 Reattach the spark plug wires to the plugs (if removed).

13 Connect the cable to the negative terminal of the battery.

14 Check the ignition timing (refer to Chapter 1) and tighten the distributor hold-down bolt securely.

15 If the engine was rotated with the distributor removed, perform the following:

a) *Rotate the crankshaft until the number one piston is at top dead center (TDC) on the compression stroke. The mark on the damper should be aligned with the 0 degree (TDC) mark on the timing chain case. Do not confuse this setting with the companion cylinder TDC. Check the TDC setting procedure in Chapter 2 to verify valve position.*

b *Install the distributor rotor onto the distributor shaft. Rotate the rotor approximately to the "CYL NO. 1" alignment mark on the distributor cap.*

c) *Install the distributor into the engine (to its original position). Engage the distributor shaft with the slot in the oil pump drive gear (V8) or engage the gears (six-cylinder - see step 7).*

d) *Install the distributor hold-down clamp and bolt. Do not tighten it at this time.*

e) *Slowly rotate the distributor until the rotor is aligned with the "CYL. NO. 1" alignment mark on the distributor cap.*

f) *Tighten the hold-down clamp bolt securely.*

g) *Install the distributor cap.*

13 Air gap - check and adjustment

Single pick-up coil type

Refer to illustrations 13.4

1 Release the spring clips or hold-down screws and remove the distributor cap. Secure the cap out of way.

2 Align one of the reluctor teeth attached to the distributor shaft with the pick-up coil tooth.

3 Loosen the pick-up coil adjusting screw.

4 Insert a non-magnetic type feeler gauge of the correct thickness between the reluctor and the pick-up coil **(see illustration)**. Refer to the Specifications listed in this Chapter for the correct gap and use a feeler gauge of that thickness.

5 Move the pick-up coil until light contact is made with the feeler gauge.

6 Tighten the pick-up coil adjusting screw.

7 Remove the feeler gauge. No force should be required to remove the gauge.

8 Using a feeler gauge 0.002-inch larger than specified, check the air gap. The feeler gauge should not fit between the teeth (do not force it). Reset the air gap if necessary.

9 Replace the distributor cap.

Dual pick-up coil type

Refer to illustrations 13.12

10 Some models equipped with Electronic Ignition systems from 1976 through 1989 are equipped with dual pick-up coils. These systems include ELB I, Combustion Computer, SC computer and Electronic Fuel Control computer systems. The ignition module or computer is separated into two integral units contained within the computer housing; the program schedule unit and the ignition control unit. The program schedule regulates the RUN pick-up coil (running condition) while the ignition control unit controls the START pick-up coil (starting condition) within the distributor. The air gap for each pick-up coil is set to a different specification designed for the different functions.

11 Release the spring clips or hold-down screws and remove the distributor cap. Secure the cap out of way.

12 Align one of the reluctor teeth attached to the distributor shaft with the pick-up coil tooth. **Note:** *First adjust the START pick-up coil and then the RUN pick-up coil* **(see illustration)**. *The procedure is the same for both but the air gap specifications are different. The START pick-up coil electrical connector is the larger of the two connectors.*

13 Loosen the pick-up coil adjusting screw.

14 Insert a non-magnetic type feeler gauge of the correct thickness between the reluctor and the pick-up coil. Refer to the Specifications listed in this Chapter for the correct gap and use a feeler gauge of that thickness.

15 Move the pick-up coil until light contact is made with the feeler gauge.

16 Tighten the pick-up coil adjusting screw.

17 Remove the feeler gauge. No force should be required to remove the gauge.

18 Using a feeler gauge 0.002-inch larger than specified, check the air gap. The feeler gauge should not fit between the teeth (do not force it). Reset the air gap if necessary.

19 Replace the distributor cap.

14 Charging system - general information and precautions

The charging system includes the alternator with either an external voltage regulator regulating the charging system output. Also included in the system is a charge indicator, the battery, a fusible link and the wiring between all the components. The charging system supplies electrical power for the ignition system, the lights, the radio, etc. The alternator is driven by a drivebelt at the front of the engine.

The fusible link is a short length of insulated wire integral with the engine compartment wiring harness. The link is four wire gauges smaller in diameter than the circuit it protects. Production fusible links and their identification flags are identified by the flag color. See Chapter 12 for additional information regarding fusible links.

The charging system doesn't ordinarily

require periodic maintenance. However, the drivebelt, battery and wires and connections should be inspected at the intervals outlined in Chapter 1.

The dashboard charge warning light should come on when the ignition key is turned to Start, then go off immediately. If it remains on, there is a malfunction in the charging system (see Section 15). Some vehicles are also equipped with a voltmeter. If the voltmeter indicates abnormally high or low voltage, check the charging system

Be very careful when making electrical circuit connections to a vehicle equipped with an alternator and note the following:

- a) *When reconnecting the wires to the alternator from the battery, be sure to note the polarity.*
- b) *Before using arc welding equipment to repair any part of the vehicle, disconnect the wires from the alternator and the battery terminals.*
- c) *Never start the engine with a battery charger connected.*
- d) *Always disconnect both battery leads before using a battery charger.*
- e) *The alternator is turned by an engine drivebelt which could cause serious injury if your hands, hair or clothes become entangled in it with the engine running.*
- f) *Because the alternator is connected directly to the battery, it could arc or cause a fire if overloaded or shorted out.*
- g) *Wrap a plastic bag over the alternator and secure it with rubber bands before steam cleaning the engine.*

15 Charging system - check

1 If a malfunction occurs in the charging circuit, don't automatically assume that the alternator is causing the problem.
First check the following items:

- a) *Check the drivebelt tension and condition (see Chapter 1). Replace it if it's worn or deteriorated.*
- b) *Make sure the alternator mounting and adjustment bolts are tight.*
- c) *Inspect the alternator wiring harness and the electrical connectors at the alternator and voltage regulator (if equipped). They must be in good condition and tight.*
- d) *Check the fusible link (if equipped) located between the starter solenoid and the alternator. If it's burned, determine the cause, repair the circuit and replace the link (the vehicle won't start and/or the accessories won't work if the fusible link blows). Sometimes a fusible link may look good, but still be bad. If in doubt, remove it and check for continuity.*
- e) *Start the engine and check the alternator for abnormal noises (a shrieking or squealing sound indicates a bad bearing).*

16.2 Loosen the pivot bolt (arrow) and the adjustment bolt on the alternator brackets

- f) *Check the specific gravity of the battery electrolyte. If it's low, charge the battery (doesn't apply to maintenance free batteries).*
- g) *Make sure the battery is fully charged (one bad cell in a battery can cause overcharging by the alternator).*
- h) *Disconnect the battery cables (negative first, then positive). Inspect the battery posts and the cable clamps for corrosion. Clean them thoroughly if necessary (see Chapter 1). Reconnect the cable to the positive terminal.*
- i) *With the key off, connect a test light between the negative battery post and the disconnected negative cable clamp.*
  - 1) *If the test light does not come on, reattach the clamp and proceed to the next Step.*
  - 2) *If the test light comes on, there is a short (drain) in the electrical system of the vehicle. The short must be repaired before the charging system can be checked.*
  - 3) *Disconnect the alternator wiring harness.*
    - (a) *If the light goes out, the alternator is bad.*
    - (b) *If the light stays on, pull each fuse until the light goes out (this will tell you which component is shorted).*

2 Using a voltmeter, check the battery voltage with the engine off. If should be approximately 12-volts.

3 Start the engine and check the battery voltage again. It should now be approximately 14 to 15 volts.

4 Turn on the headlights. The voltage should drop, and then come back up, if the charging system is working properly.

5 If the voltage reading is more than the specified charging voltage, replace the voltage regulator (refer to Section 17). If the voltage is less, the alternator diode(s), stator or rectifier may be bad or the voltage regulator may be malfunctioning.

16 Alternator - removal and installation

Refer to illustration 16.2

1 Detach the cable from the negative terminal of the battery.

2 Loosen the alternator adjustment and pivot bolts and detach the drivebelt **(see illustration)**.

3 Detach the wires from the alternator.

4 Remove the adjustment and pivot bolts and separate the alternator from the engine.

5 If you are replacing the alternator, take the old alternator with you when purchasing a replacement unit. Make sure the new/rebuilt unit is identical to the old alternator. Look at the terminals - they should be the same in number, size and location as the terminals on the old alternator. Finally, look at the identification markings - they will be stamped in the housing or printed on a tag or plaque affixed to the housing. Make sure these numbers are the same on both alternators.

6 Many new/rebuilt alternators do not have a pulley installed, so you may have to switch the pulley from the old unit to the new/rebuilt one. When buying an alternator, find out the shop's policy regarding pulleys - some shops will perform this service free of charge.

7 Installation is the reverse of removal.

8 After the alternator is installed, adjust the drivebelt tension (see Chapter 1).

9 Check the charging voltage to verify proper operation of the alternator (see Section 16).

17 Voltage regulator - replacement

Refer to illustration 17.1

1 Disconnect the electrical connector from the voltage regulator **(see illustration)**.

2 Remove the mounting screws and regulator from the firewall.

3 Installation is the reverse of removal.

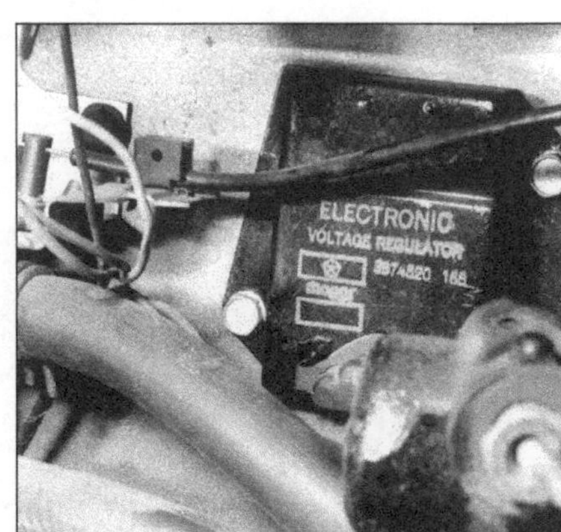

17.1 The voltage regulator is mounted in the engine compartment and can be detached after unplugging the electrical connectors and removing the mounting bolts

18.2a Typical mounting locations for the alternator brushes on early models

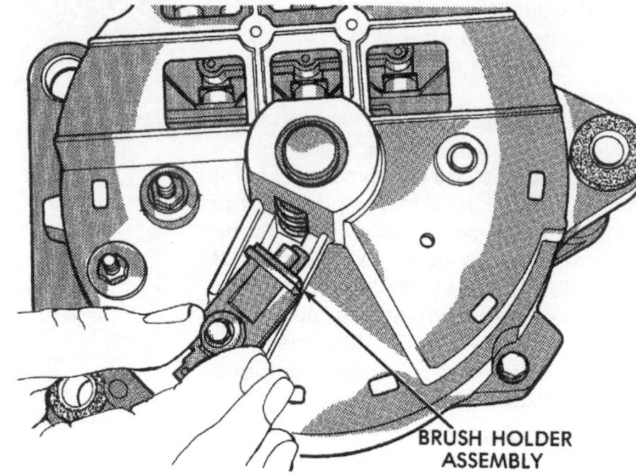

18.2b Typical mounting locations for the alternator brushes on later models

Before mounting the regulator to the firewall, make sure the regulator and the firewall are clean (bare metal) so a good ground will be achieved.

18 Alternator brushes - replacement

Refer to illustrations 18.2a, 18.2b, 18.2c and 18.3

1 Remove the alternator from the vehicle (if necessary).

2 Remove the brush holder mounting screws a little at a time to prevent distortion of the holder **(see illustrations)**.

3 Rotate the brush holder and separate it from the rear of the alternator **(see illustration)**.

4 If the brushes are worn badly, or if they do not move smoothly in the brush holder, replace the brush holder assembly with a new one.

5 Insert the holder into position, making sure the brushes seat correctly.

6 Hold the brush holder securely in place and install the screws. Tighten them evenly, a little at a time, so the holder is not distorted. Once the screws are snug, tighten them securely.

7 Connect the negative battery cable.

19 Starting system - general information and precautions

The sole function of the starting system is to turn over the engine quickly enough to allow it to start.

The starting system consists of the battery, the starter motor, the starter relay, the starter solenoid and the wires connecting them. The solenoid is mounted directly on the starter motor or is a separate component located in the engine compartment.

The solenoid/starter motor assembly is installed on the lower part of the engine, next to the transmission bellhousing.

When the ignition key is turned to the Start position, the starter solenoid is actuated through the starter relay and control circuit. The starter solenoid then connects the battery to the starter. The battery supplies the electrical energy to the starter motor, which does the actual work of cranking the engine.

The starter motor on a vehicle equipped with a manual transmission can only be operated when the clutch pedal is depressed; the starter on a vehicle equipped with an automatic transmission can only be operated when the transmission selector lever is in Park or Neutral.

Always observe the following precautions when working on the starting system:

a) *Excessive cranking of the starter motor can overheat it and cause serious damage. Never operate the starter motor for more than 15 seconds at a time without pausing to allow it to cool for at least two minutes.*
b) *The starter is connected directly to the battery and could arc or cause a fire if mishandled, overloaded or shorted out.*
c) *Always detach the cable from the negative terminal of the battery before working on the starting system.*

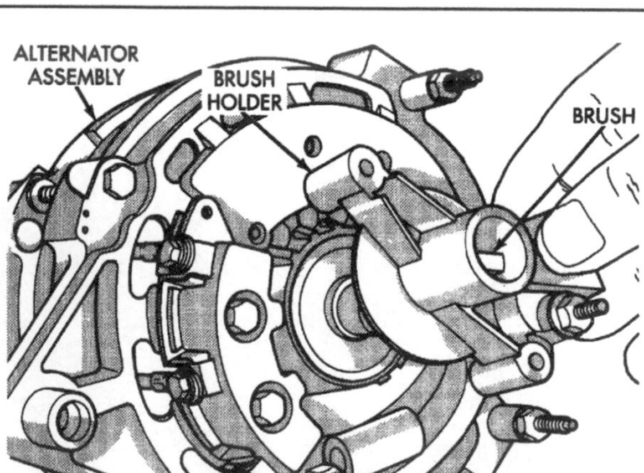

18.2c Other late models have brush holders that fit over the end of the shaft - remove the dust cover to gain access to the brush holder

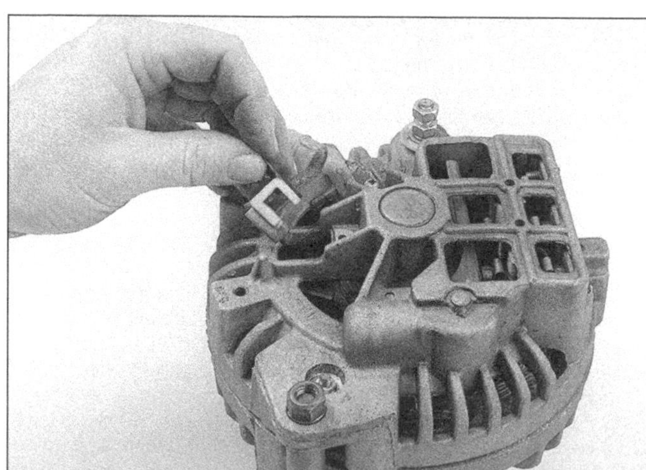

18.3 The brushes on early model alternators slip out from under the rear cover

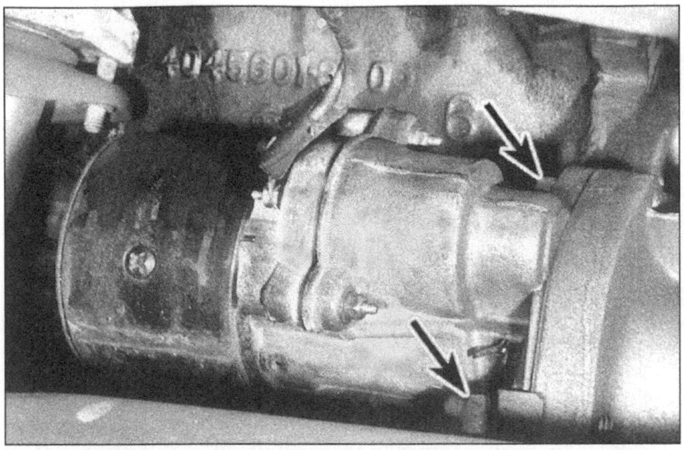
21.5a Starter motor assembly on a V8 engine - it's retained by two bolts (arrows)

21.5b Starter motor assembly on a six-cylinder engine

20 Starter motor - testing in vehicle

Note: *Before diagnosing starter problems, make sure the battery is fully charged.*

1 If the starter motor does not turn at all when the ignition switch is operated, make sure the shift lever is in Neutral or Park.

2 Make sure that the battery is charged and that all cables, both at the battery and starter solenoid terminals, are clean and secure.

3 If the starter motor spins but the engine is not cranking, the overrunning clutch in the starter motor is slipping and the starter motor must be replaced.

4 If, when the switch is actuated, the starter motor does not operate at all but the solenoid clicks, then the problem lies with either the battery, the main solenoid contacts or the starter motor itself, or the engine is seized.

5 If the solenoid plunger cannot be heard when the switch is actuated, the battery is bad, the fusible link is burned (the circuit is open), the starter relay is malfunctioning or the solenoid itself is defective.

6 To check the solenoid, connect a jumper lead between the battery (positive terminal) and the ignition switch terminal (the small terminal) on the solenoid. If the starter motor now operates, the solenoid is OK and the problem is in the ignition switch, neutral start switch or in the wiring.

7 If the starter motor still does not operate, remove the starter/solenoid unit for disassembly, testing and repair.

8 If the starter motor cranks the engine at an abnormally slow speed, first make sure that the battery is charged and that all terminal connections are tight. If the engine is partially seized, or has the wrong viscosity oil in it, it will crank slowly.

9 Run the engine until normal operating temperature is reached, then disconnect the coil wire from the distributor cap and ground it on the engine.

10 Connect a voltmeter positive lead to the battery positive post and then connect the negative lead to the negative post.

11 Crank the engine and take the voltmeter readings as soon as a steady figure is indicated. Do not allow the starter motor to turn for more than 15 seconds at a time. A reading of nine volts or more, with the starter motor turning at normal cranking speed, is normal. If the reading is nine volts or more but the cranking speed is slow, the motor is faulty. If the reading is less than nine volts and the cranking speed is slow, the solenoid contacts are probably burned, the starter motor is bad, the battery is discharged or there is a bad connection.

21 Starter motor - removal and installation

Refer to illustrations 21.5a and 21.5b

1 Disconnect the cable from the negative terminal of the battery.

2 Remove the heat shield (if equipped).

3 Detach the battery cable from the starter.

4 Disconnect the solenoid wire from the solenoid.

5 Remove the starter mounting nuts/bolts **(see illustrations)**.

6 Remove the starter assembly.

7 Installation is the reverse of removal.

22 Starter motor brushes - replacement

Refer to illustrations 22.5 and 22.19

1 Remove the starter motor assembly (see Section 21).

2 Remove the through bolts and separate the end head assembly from the field frame assembly.

3 Carefully pull the armature out of the gear housing and field frame assembly.

4 Pull the field frame assembly away from the gear housing just far enough to expose the terminal screw.

5 Support the brush terminal screw with a finger and remove the screw **(see illustration)**.

6 Remove the field frame assembly.

7 Remove the mounting nuts from the solenoid and brush holder plate assembly and separate the assembly from the gear housing.

8 Remove the nut and washers from the solenoid terminal.

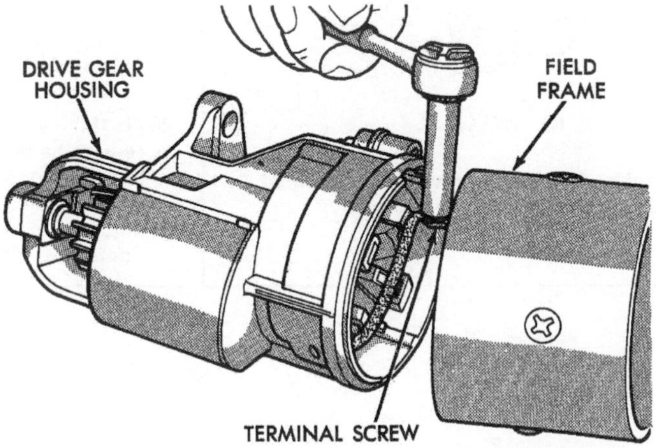

22.5 When removing the terminal screw, it's a good idea to position your finger under the terminal so it doesn't bend

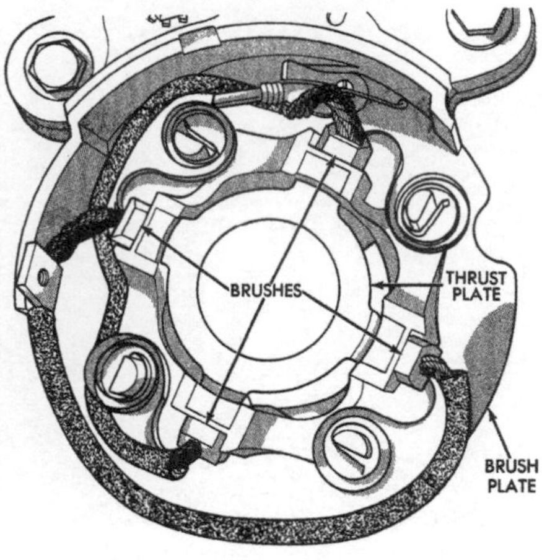

22.19 Correct positioning of the starter brushes

9 Unwind the solenoid wire from the starter brush terminal.
10 Separate the solenoid wire and the solenoid from the brush plate.
11 The brushes can now be removed from the brush holders and new ones installed.
12 Check the condition of the contacts in the brush holder plate. If they are badly burned, replace the plate and brushes as an assembly.
13 Use a new gasket when attaching the brush holder plate to the solenoid. If a new gasket is not available, apply a thin bead of RTV sealant to the outside edge of the gasket.
14 Insert the solenoid wire through the hole in the brush holder.
15 Insert the brush holder plate mounting bolts and attach the plate to the solenoid.
16 Install the nut and washers on the solenoid terminal.
17 Wind the solenoid wire tightly around the brush terminal and solder it in place with a high temperature resin core solder and resin flux.
18 Carefully install the solenoid and brush plate assembly in the gear housing and install the bolts.
19 Position the brushes against the thrust plate **(see illustration)**.
20 Install the brush terminal screw.
21 Carefully install the field frame assembly and armature. Rotate the armature shaft slightly to engage it with the reduction gear.
22 Install the thrust washer onto the armature shaft.
23 Install the end head assembly and through bolts. Tighten the through bolts securely.

23 Starter relay - check and replacement

Refer to illustrations 23.4a and 23.4b

Check

1 If the starter doesn't crank when the ignition key is turned to Start, but the battery and ignition switch have checked out OK, check for power to the starter solenoid as an assistant turns the key. If there is no power present at the starter solenoid when the key is turned to start, check the starter relay.
2 Place the gear selector lever in the Neutral or Park position (automatic transmission) or the Neutral position (manual transmission). Set the parking brake.
3 Using a test light, check for battery voltage between the battery positive terminal of the starter relay (terminal no. 1) and ground (**see illustration 23.4** for terminal locations). If there is no power present, repair the wire between the battery and the relay.
4 If there is power available on terminal no. 1, connect a jumper wire between terminals 1 and 4 (ignition) **(see illustration)**. If the engine cranks, the starter relay is good. If the engine doesn't crank, connect another jumper wire between the ground terminal (no. 5) and a good ground **(see illustration)**.
5 Bridge terminals 1 and 4 again - if the engine now cranks, the relay is OK. Check the ground wire from the relay and the neutral safety switch or shift linkage adjustment. If these items check out OK, replace the starter relay.

Replacement

6 Disconnect the cable from the negative terminal of the battery.
7 The starter relay is located on the left side of the firewall on some models. On others, it's located on the left inner fender panel. If equipped, remove the cover from the relay.
8 Remove the relay mounting screw, remove the relay and unplug the electrical connector.
9 Installation is the reverse of removal.

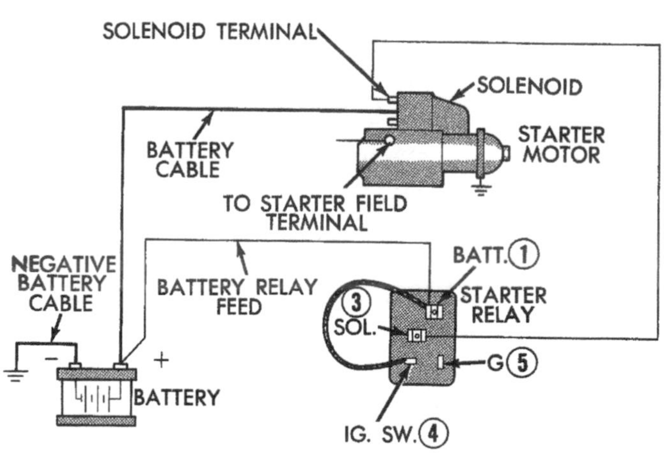

23.4a If the engine cranks when a jumper wire is connected between terminals 1 and 4, the relay is good

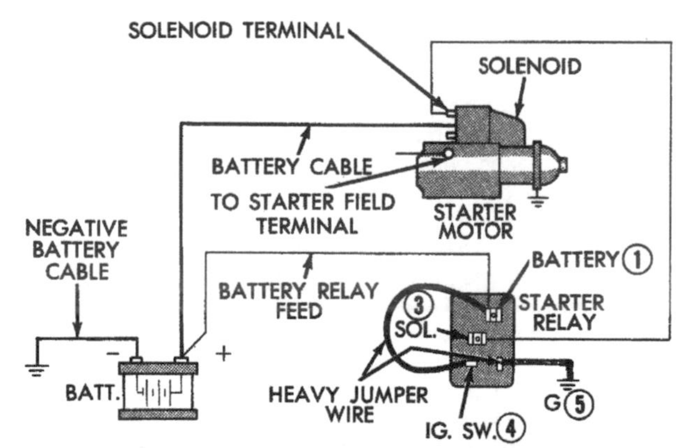

23.4b If the engine doesn't crank when bridging terminals 1 and 4, repeat the test with another jumper wire connected between terminal no. 5 and ground - if the engine cranks, the circuit between terminal no. 5 and ground is bad, the shift linkage is out of adjustment or the neutral start switch is out of adjustment or defective - if the engine doesn't crank, replace the relay

Chapter 6
Emissions control systems

Contents

1 General information

Refer to illustration 1.1 and 1.6a through 1.6av

1 To prevent pollution of the atmosphere from incompletely burned and evaporating gases, and to maintain good driveability and fuel economy, a number of emission control systems are incorporated. The combination of systems used depends on the year in which the vehicle was manufactured, the locality to which it was originally delivered (referred to in this Chapter as California, Canadian or Federal (which is any state except California) and the engine type. The systems incorporated on the vehicles with which this manual is concerned include the **(see illustration):**

Air Injection (AIR) system
Air aspirator system

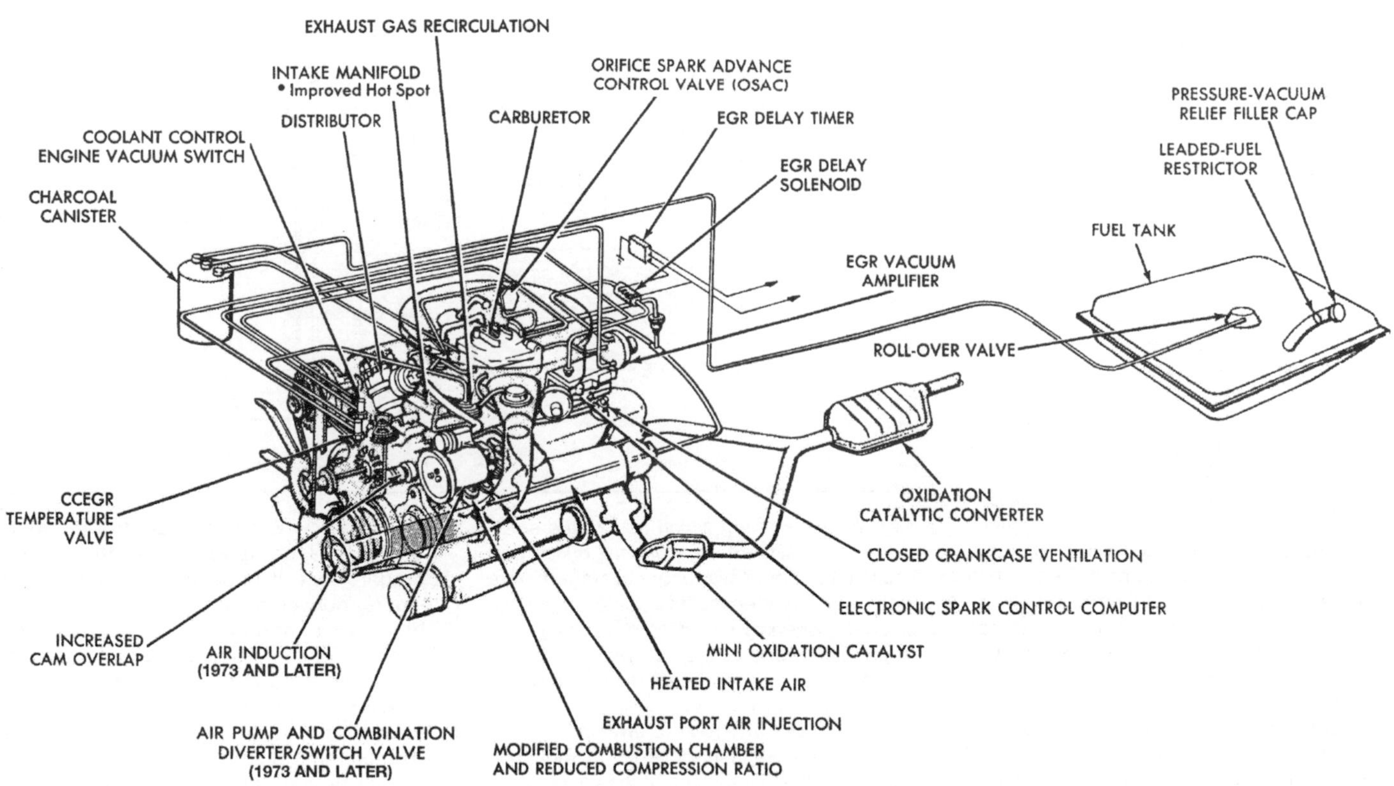

1.1 A typical emissions control system - the system shown here is for a later model with a full complement of emissions control devices - your model may not use all the devices shown here

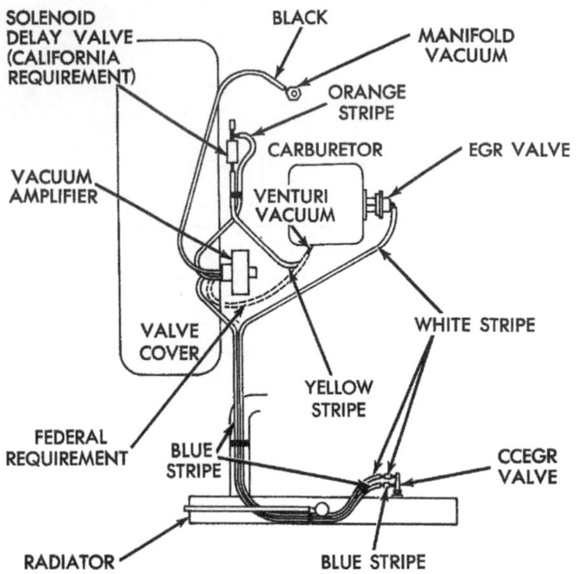

1.6a 1973 and 1974 EGR vacuum hose routing for the 198 and 225 CID six-cylinder engines with Venturi EGR system

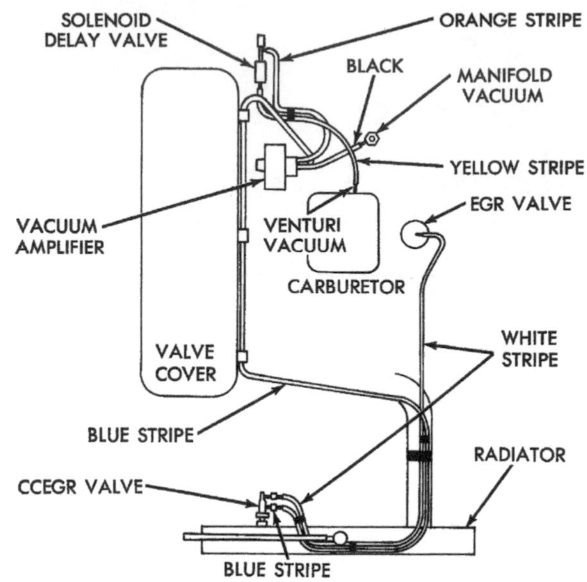

1.6b 1973 and 1974 EGR vacuum hose routing for the 318 CID V8 engines with Venturi EGR system (California models)

Catalytic converter (CAT)
Electric assist choke system
Exhaust Gas Recirculation (EGR) system
Feedback carburetor system
Fuel evaporative emission control (EVAP) system
Heated Inlet Air (HIA) system
Idle enrichment system
Orifice Spark Advance Control (OSAC)
Positive Crankcase Ventilation (PCV) system
Spark Control Computer (SCC) system
Vacuum throttle positioner

All of these systems are linked, directly or indirectly, to the emissions control system.

2 The Sections in this Chapter include general descriptions, checking procedures within the scope of the home mechanic and component replacement procedures (when possible).

3 Before assuming that an emissions control system is malfunctioning, check the fuel and ignition systems carefully. The diagnosis of some emission control devices requires specialized tools, equipment and training. If checking and servicing become too difficult or if a procedure is beyond your ability, consult a dealer service department. Remember, the most frequent cause of emissions problems is simply a loose or broken vacuum hose or wire, so always check the hose and wiring connections first.

4 This doesn't mean, however, that emission control systems are particularly difficult to maintain and repair. You can quickly and easily perform many checks and do most of the regular maintenance at home with common tune-up and hand tools. **Note:** *Because of a Federally mandated extended warranty (five years or 50,000 miles at the time this manual was written) which covers the emission control system components, check with your dealer about warranty coverage before working on any emissions-related systems. Once the warranty has expired, you may wish to perform some of the component checks and/or replacement procedures in this Chapter to save money.*

5 Pay close attention to any special precautions outlined in this Chapter. It should be noted that the illustrations of the various systems may not exactly match the system installed on your vehicle because of changes made by the manufacturer during production or from year-to-year.

6 A Vehicle Emissions Control Information (VECI) label is located in the engine compartment. It is usually located on the inner fenderwell or on the radiator support. This label contains important emissions specifications and adjustment information, as well as a vacuum hose schematic with emissions components identified **(see illustrations)**. When servicing the engine or emissions systems, the VECI label in your particular vehicle should always be checked for up-to-date information. **Note:** *In the event your VECI label is damaged or missing, contact a dealer parts department and order a replacement.*

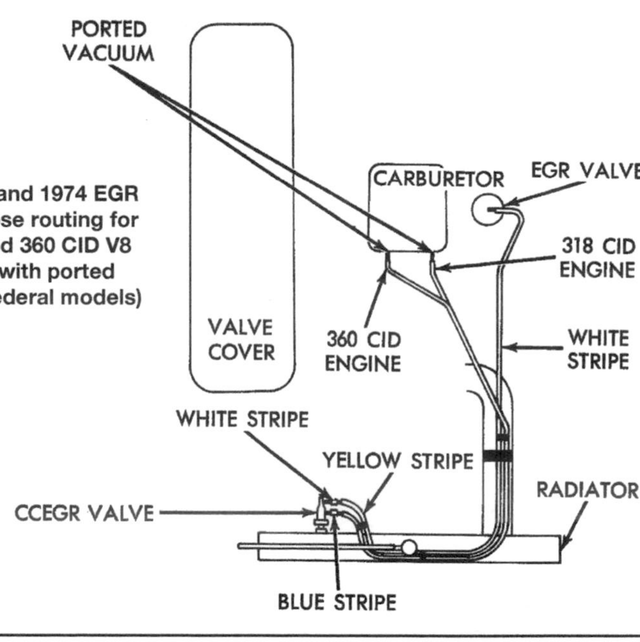

1.6c 1973 and 1974 EGR vacuum hose routing for the 318 and 360 CID V8 engines with ported vacuum (Federal models)

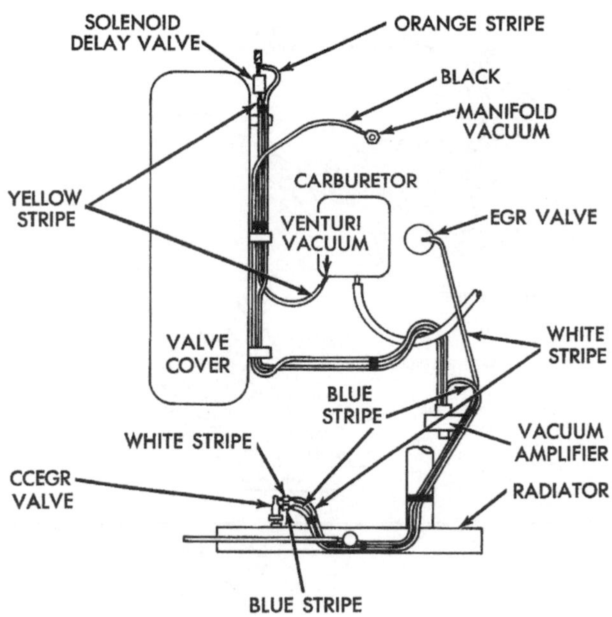

1.6d 1973 and 1974 EGR vacuum hose routing for the 360 CID V8 engines with Venturi EGR system (California models). Note: *Federal models are not equipped with a solenoid delay valve.*

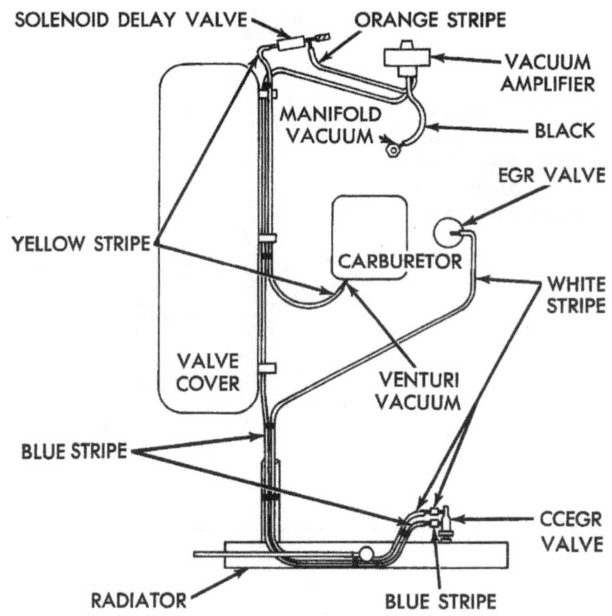

1.6e 1973 and 1974 EGR vacuum hose routing for the 400 and 440 CID V8 engines with Venturi EGR system (California models). Note: *Federal models are not equipped with a solenoid delay valve.*

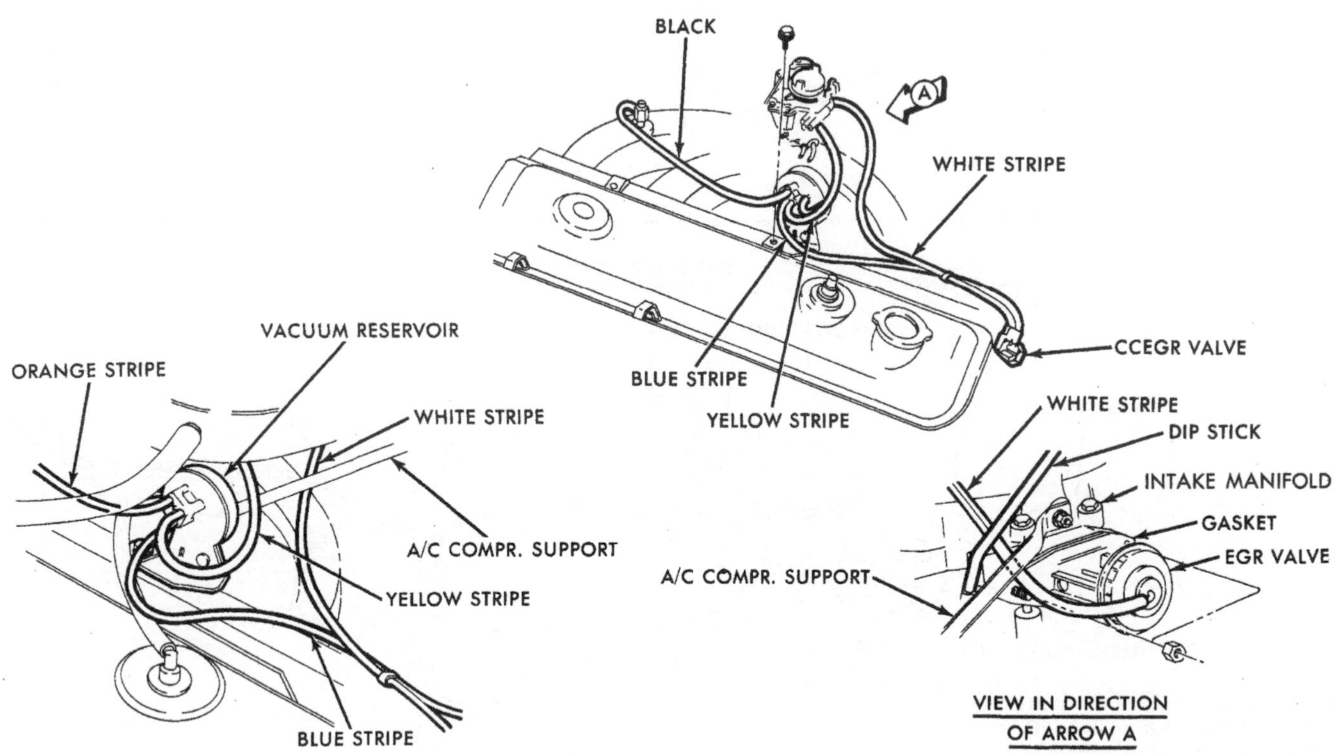

1.6f 1975 and 1976 EGR vacuum hose routing for the 225 CID six-cylinder engines (Federal and Canadian models)

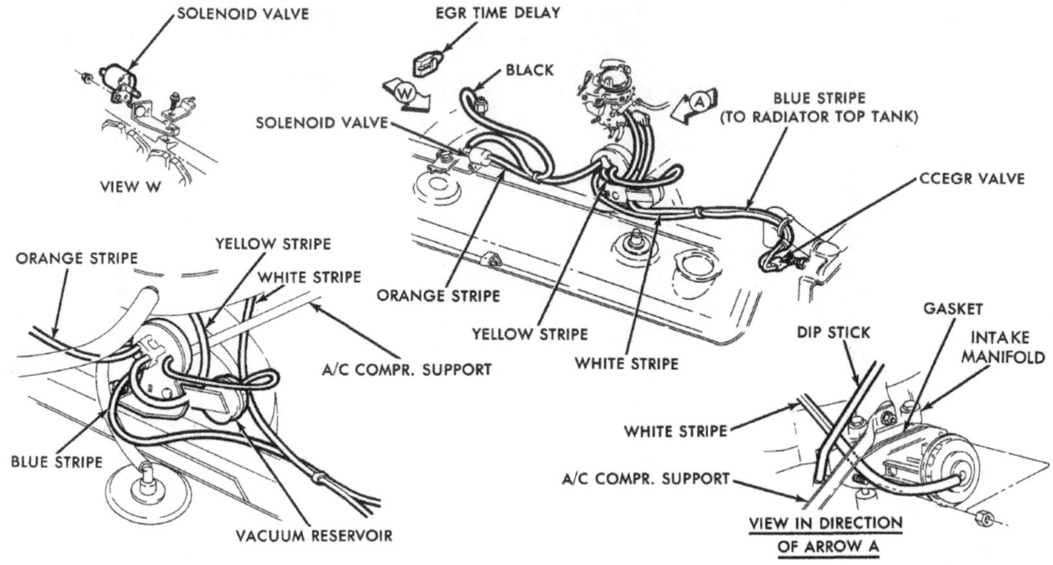

1.6g 1975 and 1976 EGR vacuum hose routing for the 225 CID six-cylinder engines (California models)

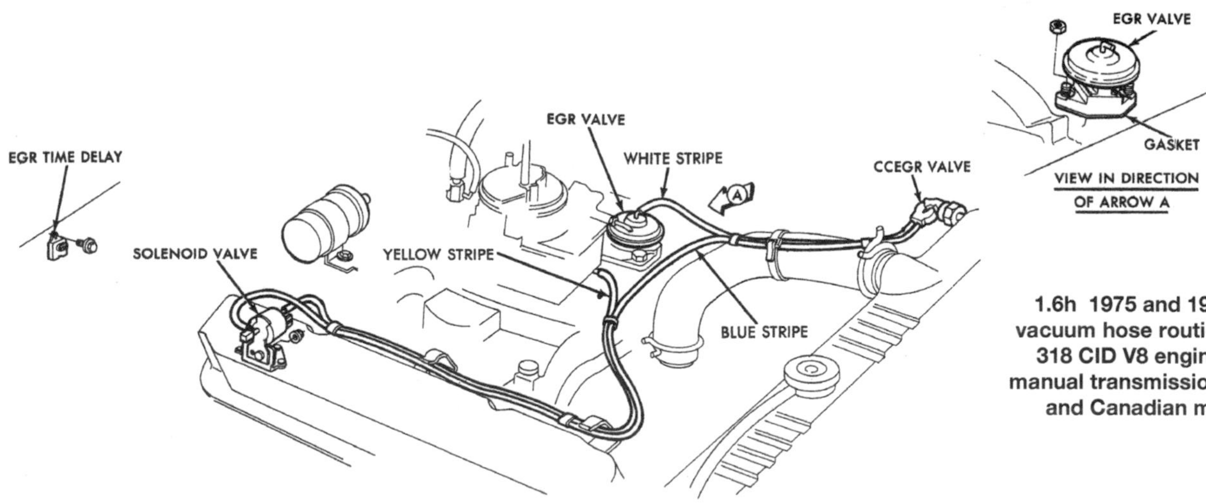

1.6h 1975 and 1976 EGR vacuum hose routing for the 318 CID V8 engines with manual transmission (Federal and Canadian models)

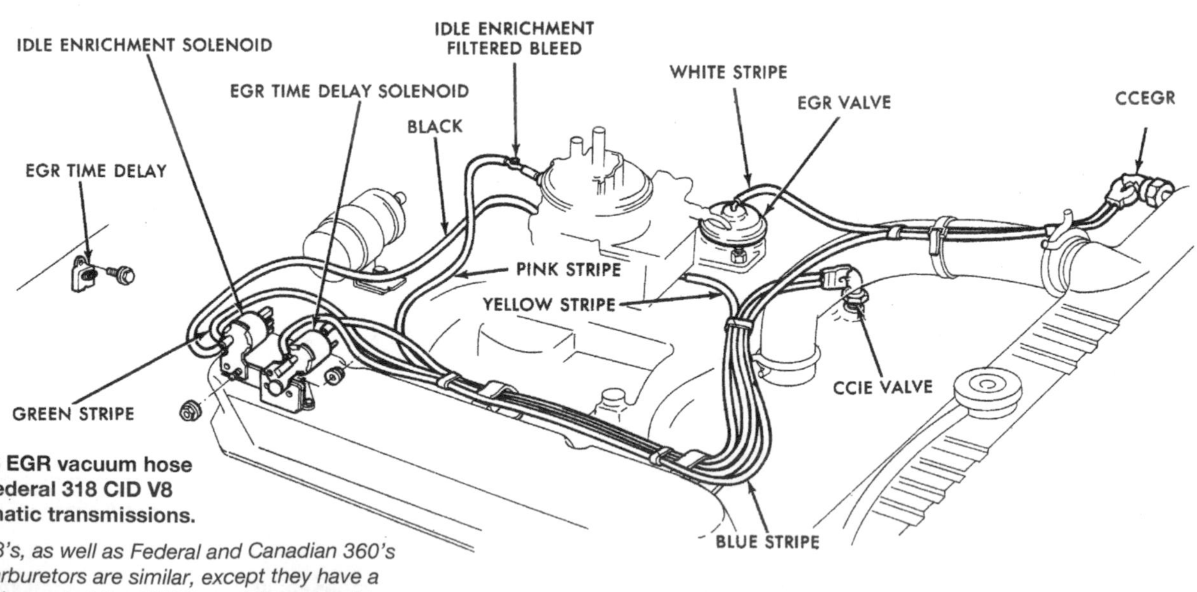

1.6i 1975 and 1976 EGR vacuum hose routing for the Federal 318 CID V8 engines with automatic transmissions.

Note: *California 318's, as well as Federal and Canadian 360's with two-barrel carburetors are similar, except they have a vacuum amplifier instead of the EGR time-delay solenoid.*

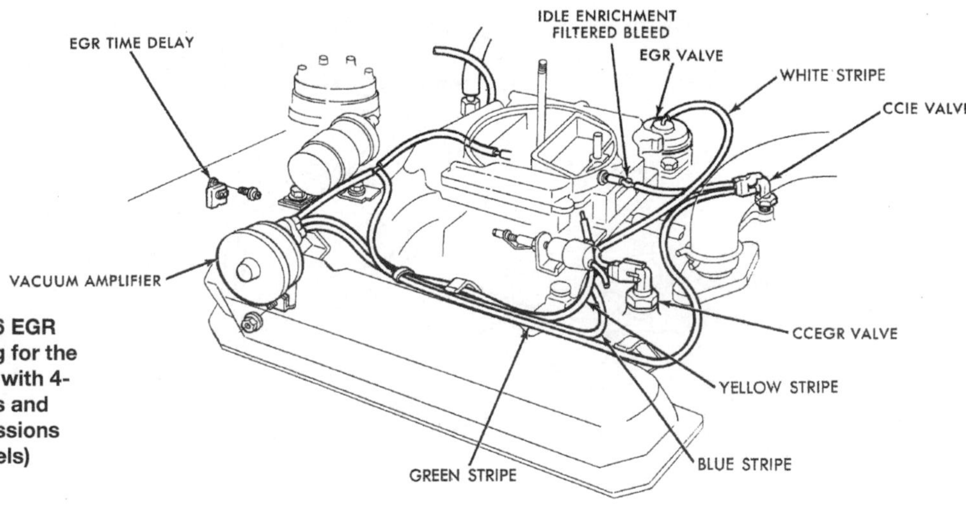

1.6j 1975 and 1976 EGR vacuum hose routing for the 360 CID V8 engines with 4-barrel carburetors and automatic transmissions (California models)

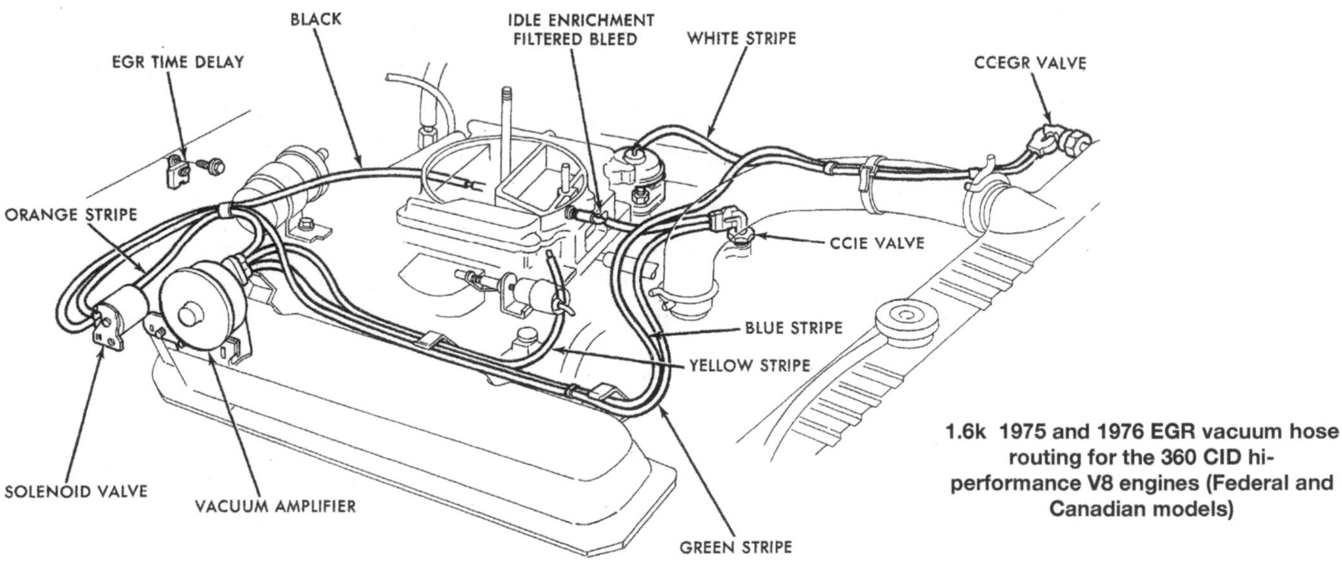

1.6k 1975 and 1976 EGR vacuum hose routing for the 360 CID hi-performance V8 engines (Federal and Canadian models)

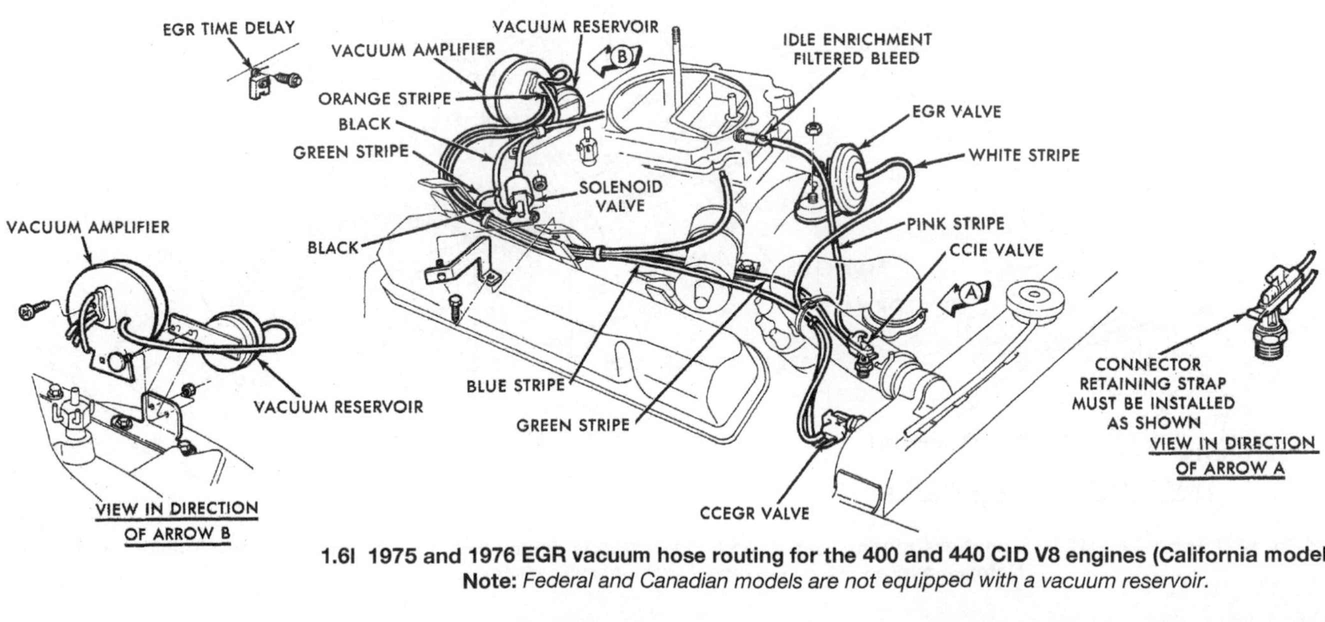

1.6l 1975 and 1976 EGR vacuum hose routing for the 400 and 440 CID V8 engines (California models).
Note: *Federal and Canadian models are not equipped with a vacuum reservoir.*

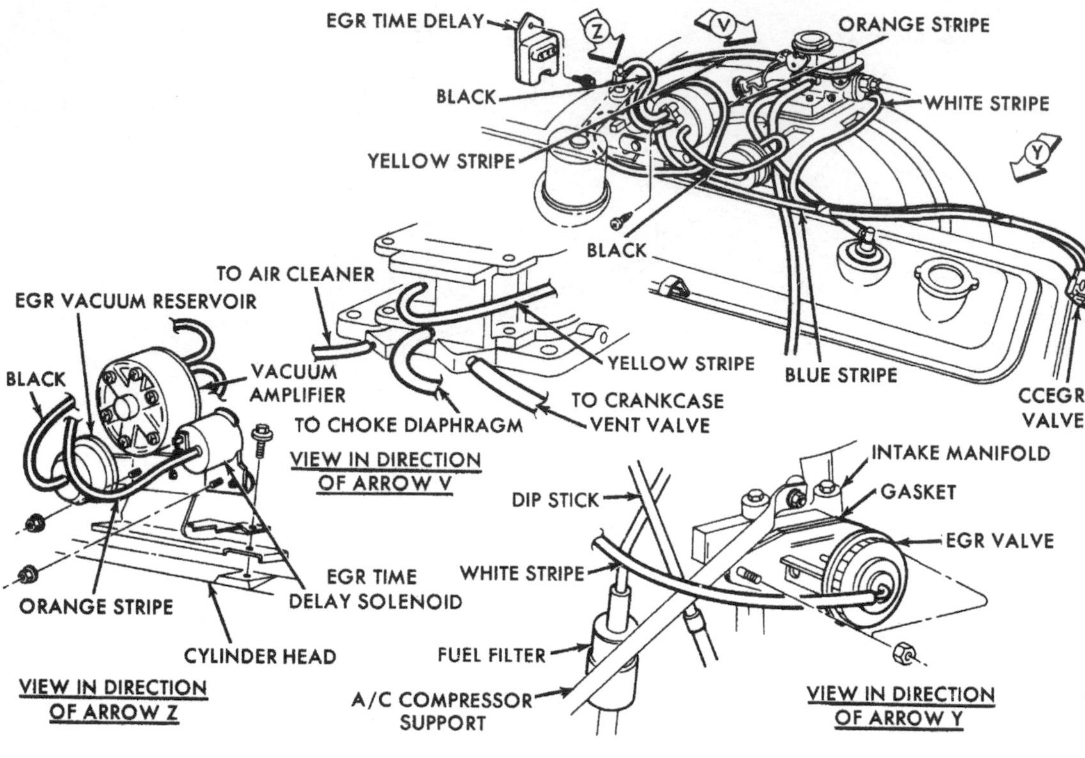

1.6m 1977 and 1978 EGR vacuum hose routing for the 225 CID six-cylinder engines with 2 barrel carburetors (Federal and Canadian models)

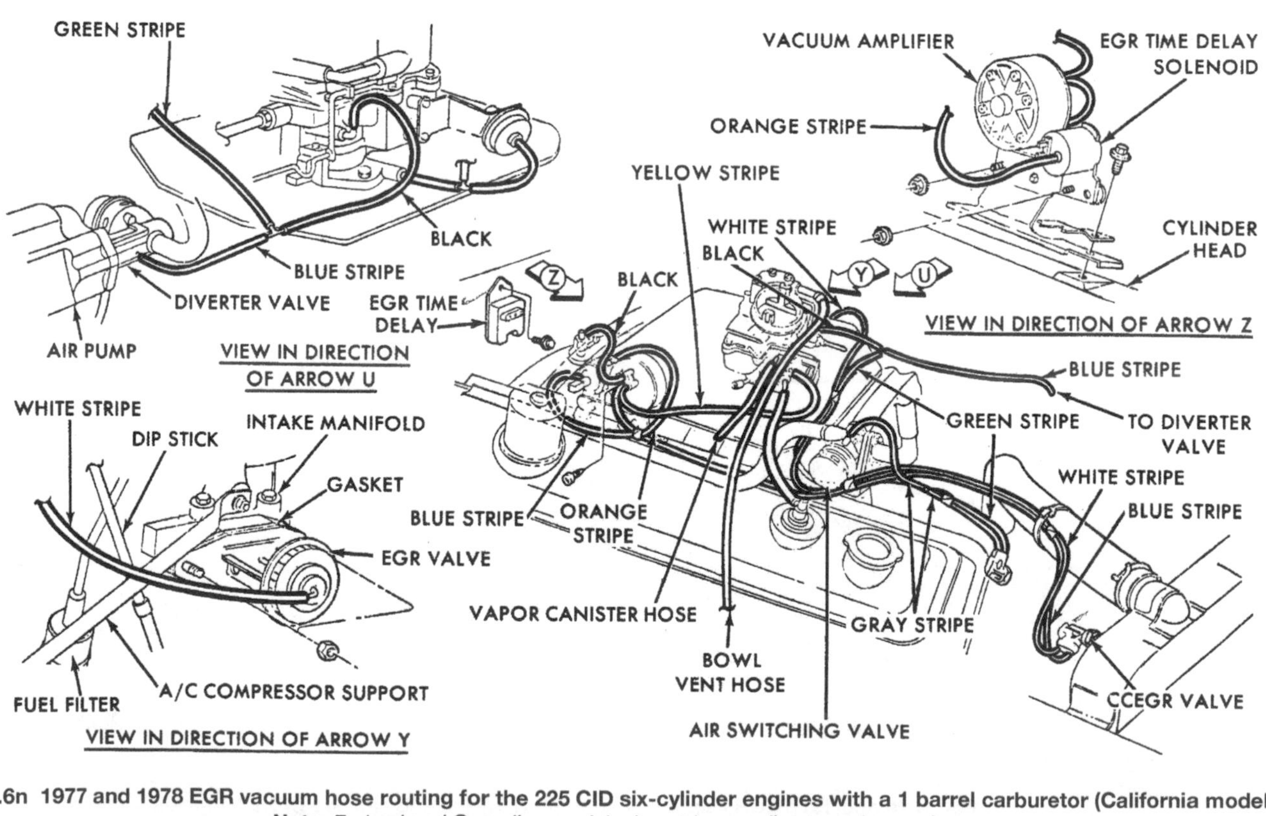

1.6n 1977 and 1978 EGR vacuum hose routing for the 225 CID six-cylinder engines with a 1 barrel carburetor (California models). Note: *Federal and Canadian models do not have a diverter valve or air pump*

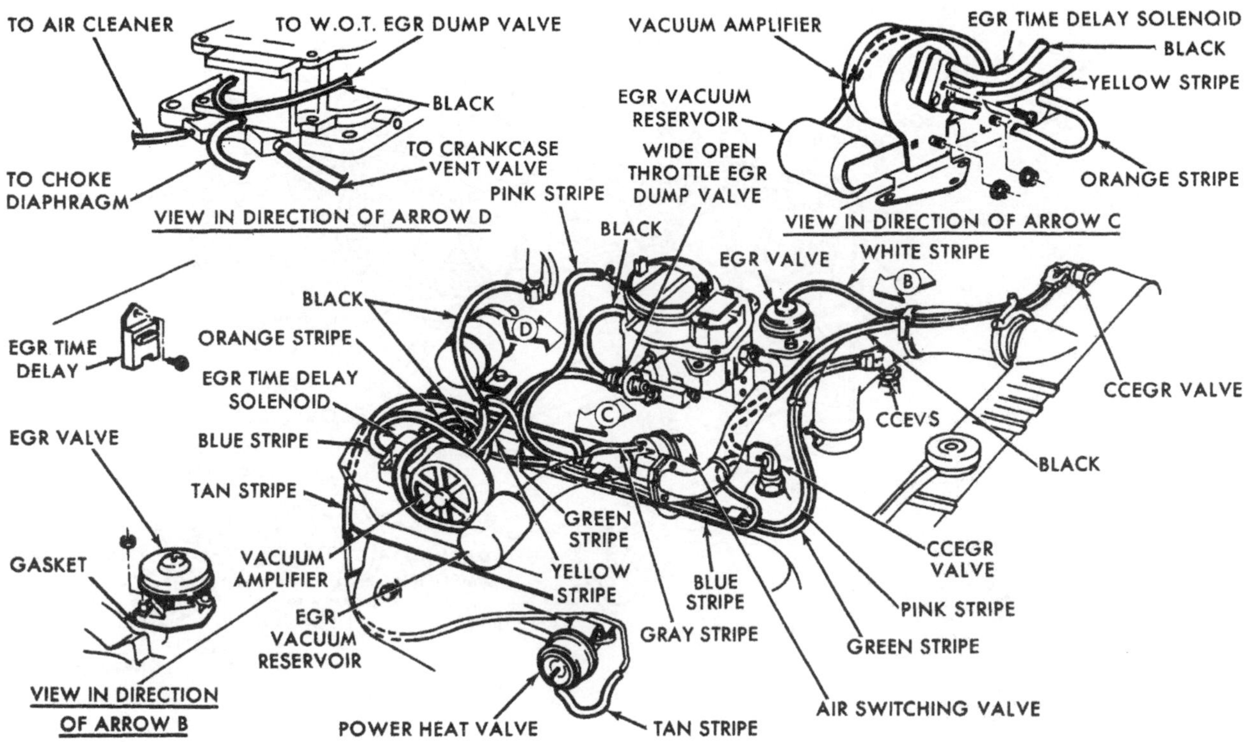

1.6o 1977 and 1978 EGR vacuum hose routing for the 318 CID V8 engines with automatic transmissions (California models).
Note: *Federal and Canadian models do not have an EGR vacuum reservoir or power heat valve.*

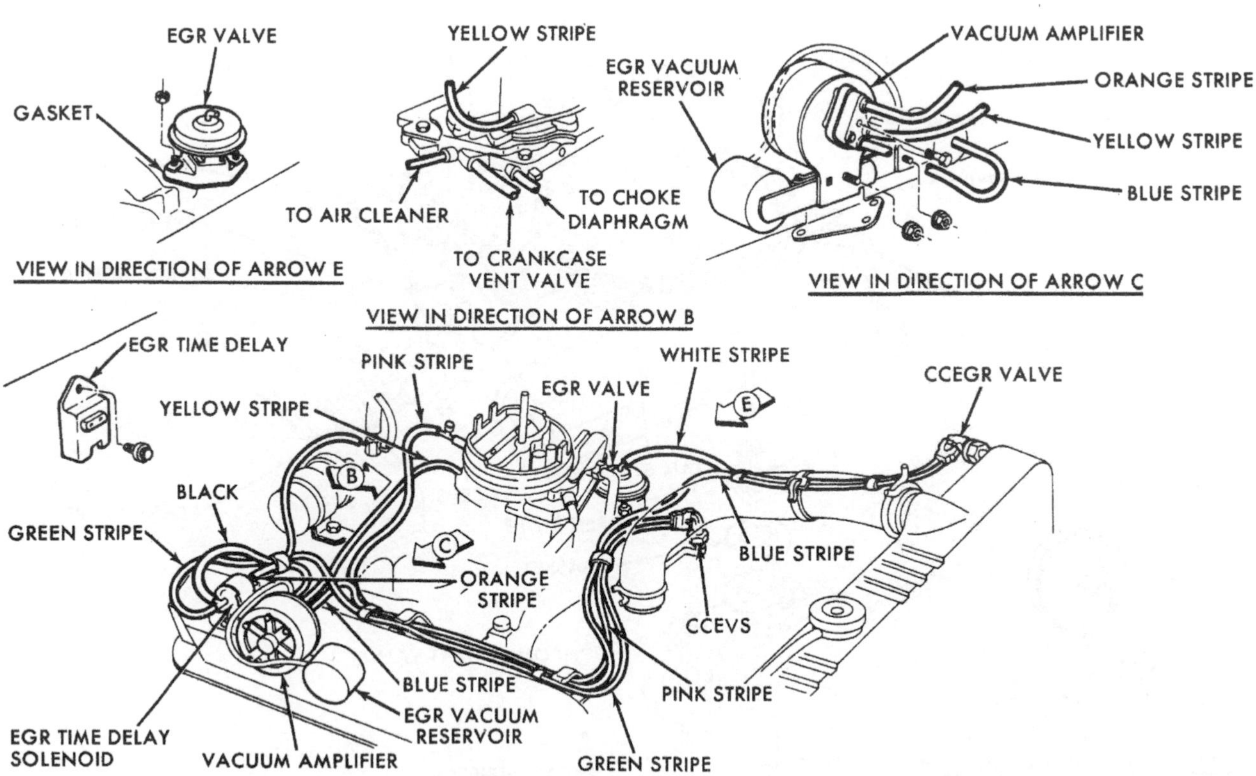

1.6p 1977 and 1978 EGR vacuum hose routing for the 360 CID V8 engines with 2 barrel carburetors and automatic transmissions (Federal and California models)

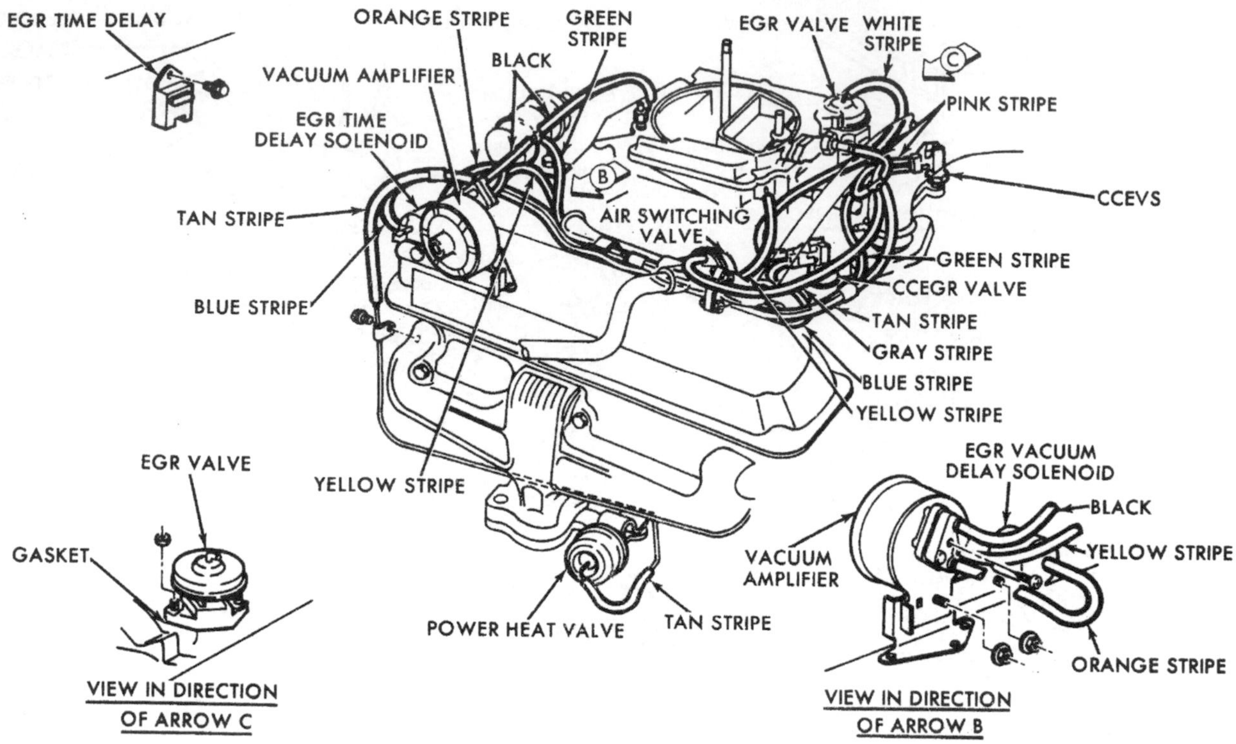

1.6q 1977 and 1978 EGR vacuum hose routing for the 360 CID V8 engines with 4 barrel carburetor and automatic transmissions (California models)

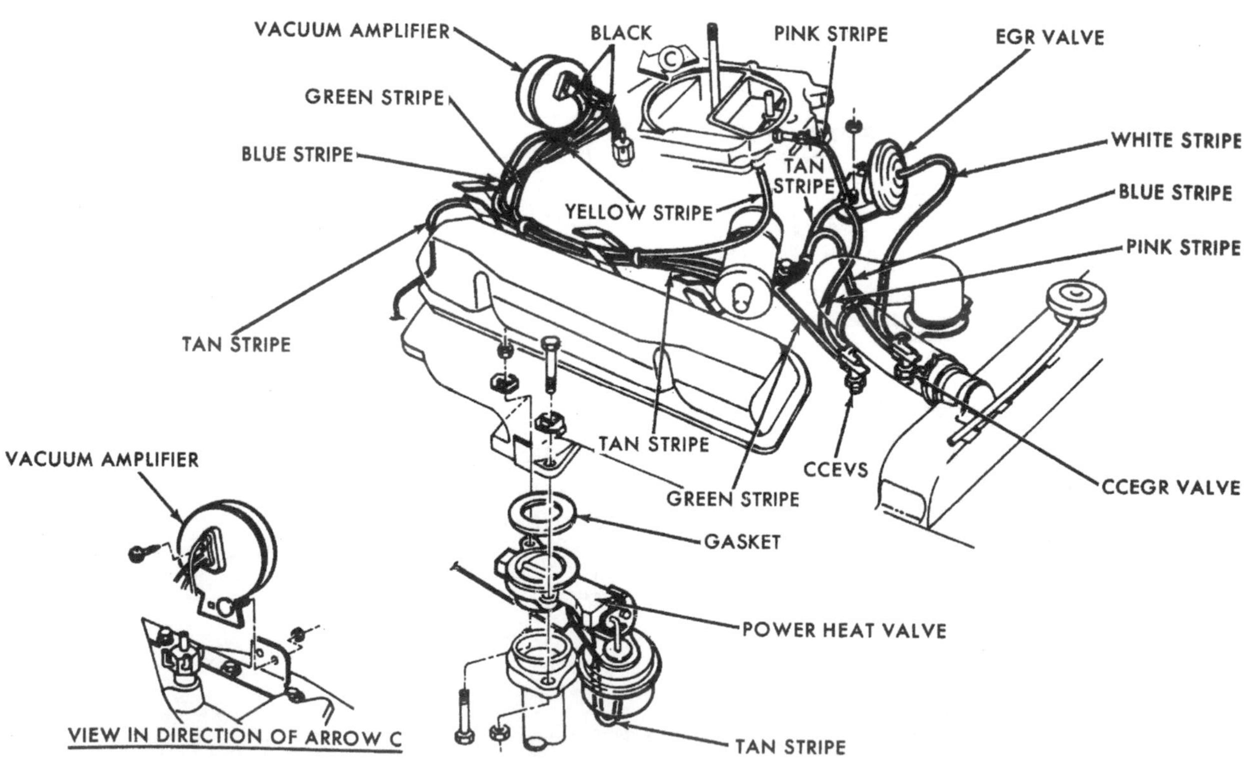

1.6r 1977 and 1978 EGR vacuum hose routing for the 440 CID V8 engines with 4 barrel carburetors (California models) (B-body only)

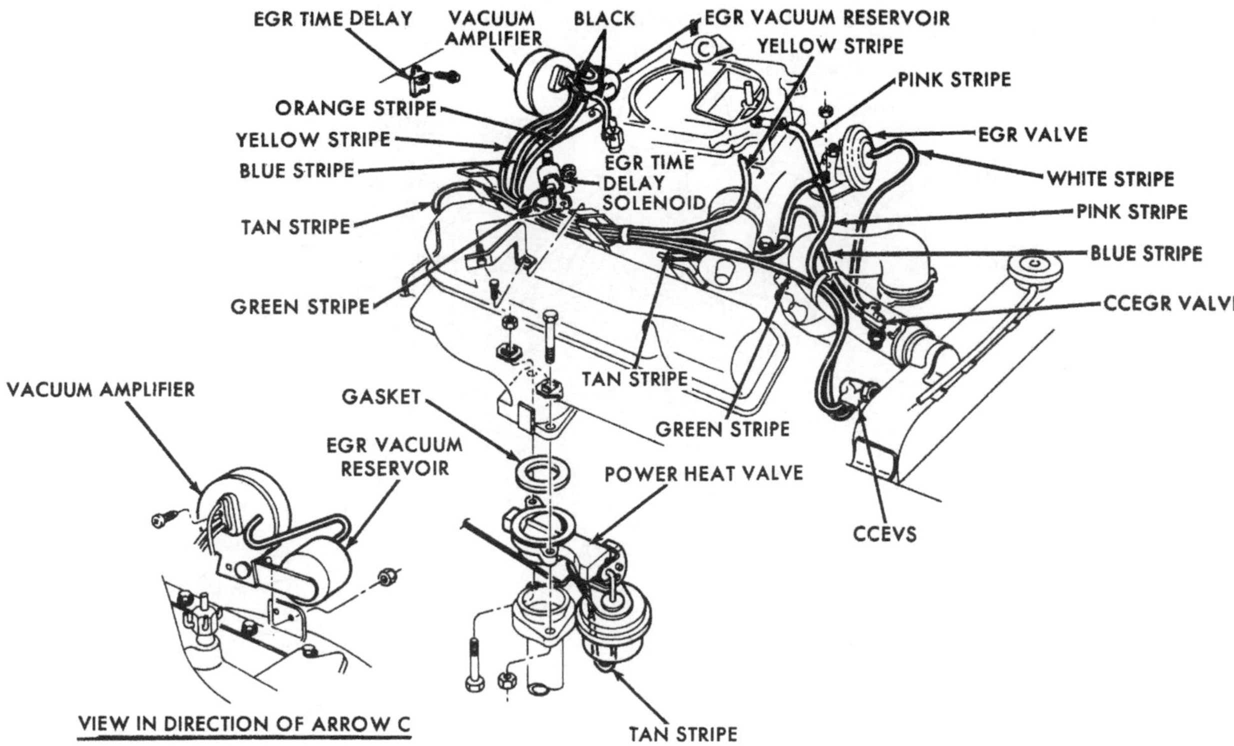

1.6s 1977 and 1978 EGR vacuum hose routing for the 440 CID engines with 4 barrel carburetors (California models) (C body only)

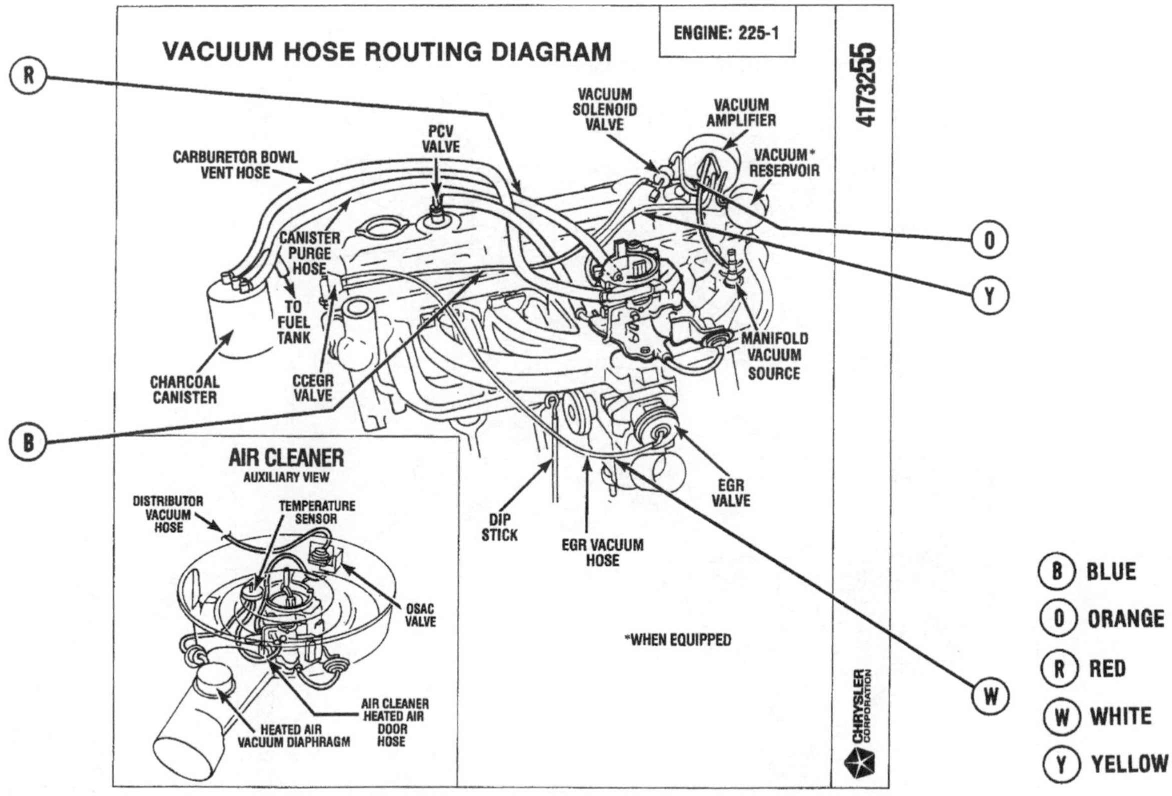

1.6t 1979 vacuum hose routing for the 225 CID six-cylinder engines with 1 barrel carburetors (Federal and Canadian models). **Note:** *Canadian models do not have a CCEGR.*

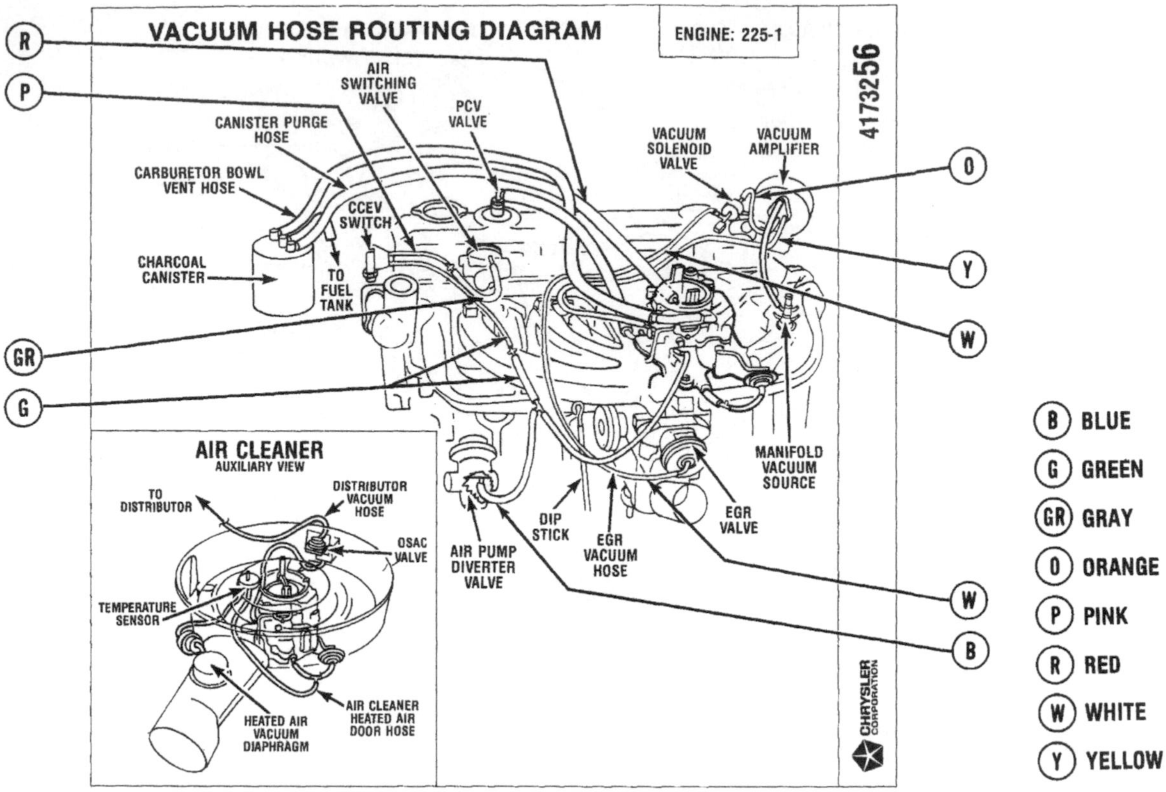

1.6u 1979 vacuum hose routing for the 225 CID six-cylinder engines with 1 barrel carburetors and automatic transmissions (California models)

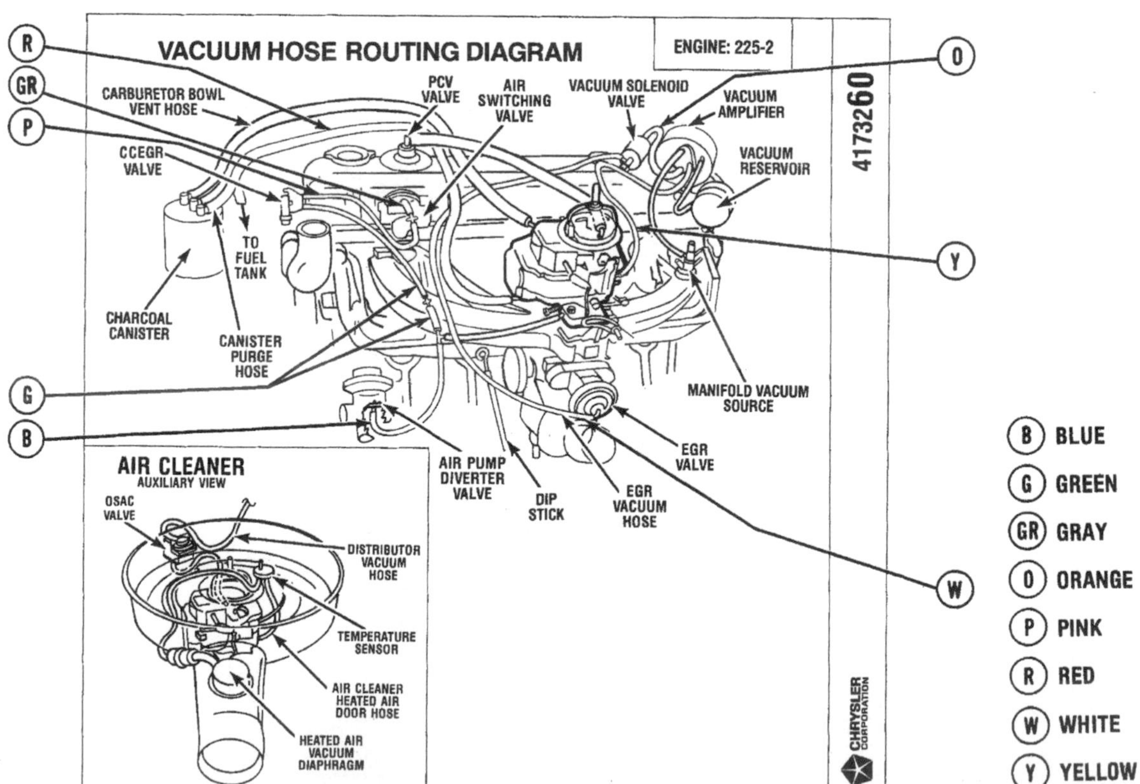

1.6v 1979 vacuum hose routing for the 225 CID six-cylinder engines with 2 barrel carburetors and automatic transmissions (Federal models)

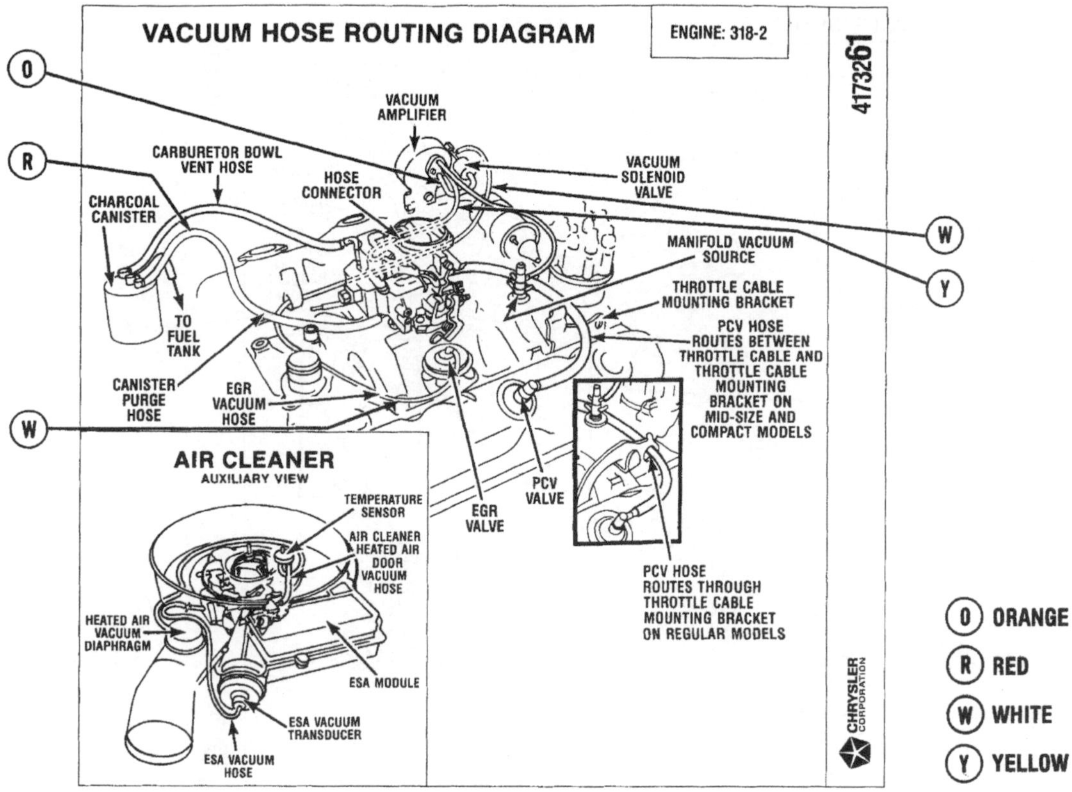

1.6w 1979 vacuum hose routing for the 318 CID V8 engines with 2 barrel carburetors and automatic transmissions (Federal and Canadian models)

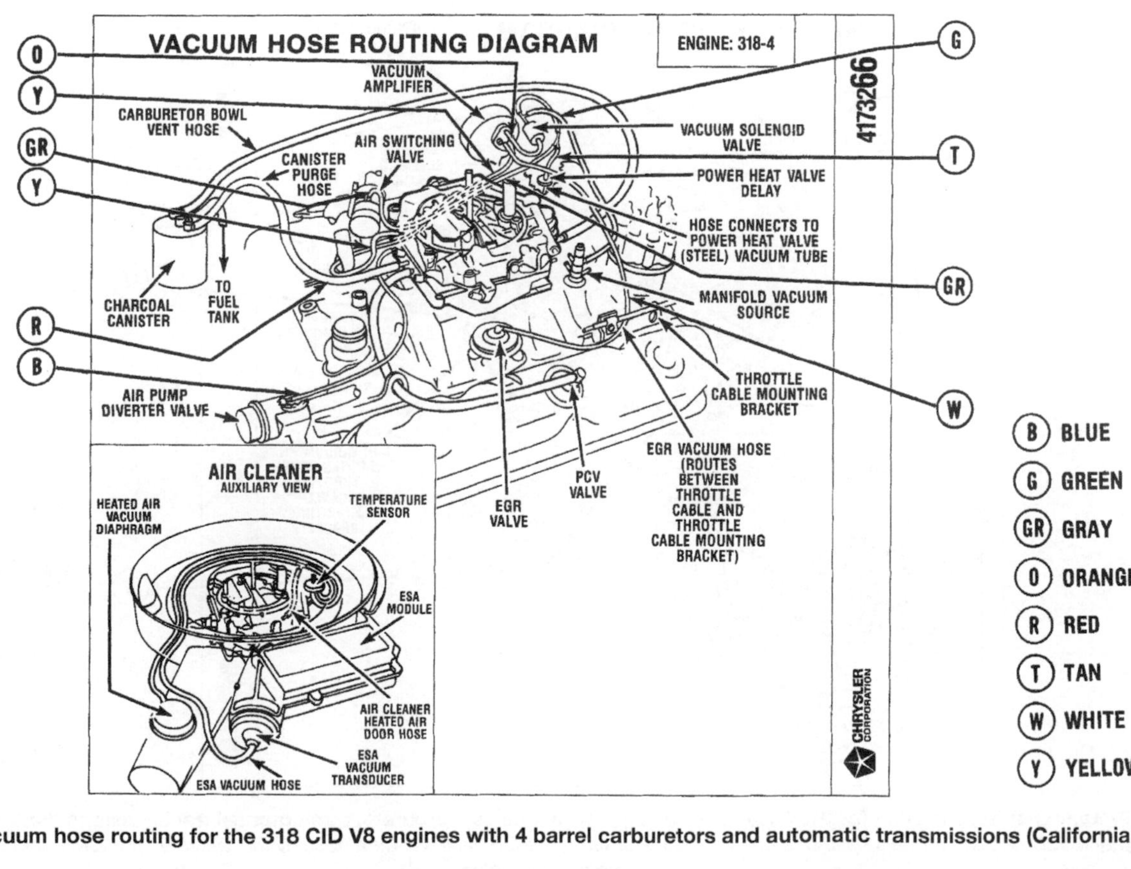

1.6x 1979 vacuum hose routing for the 318 CID V8 engines with 4 barrel carburetors and automatic transmissions (California models)

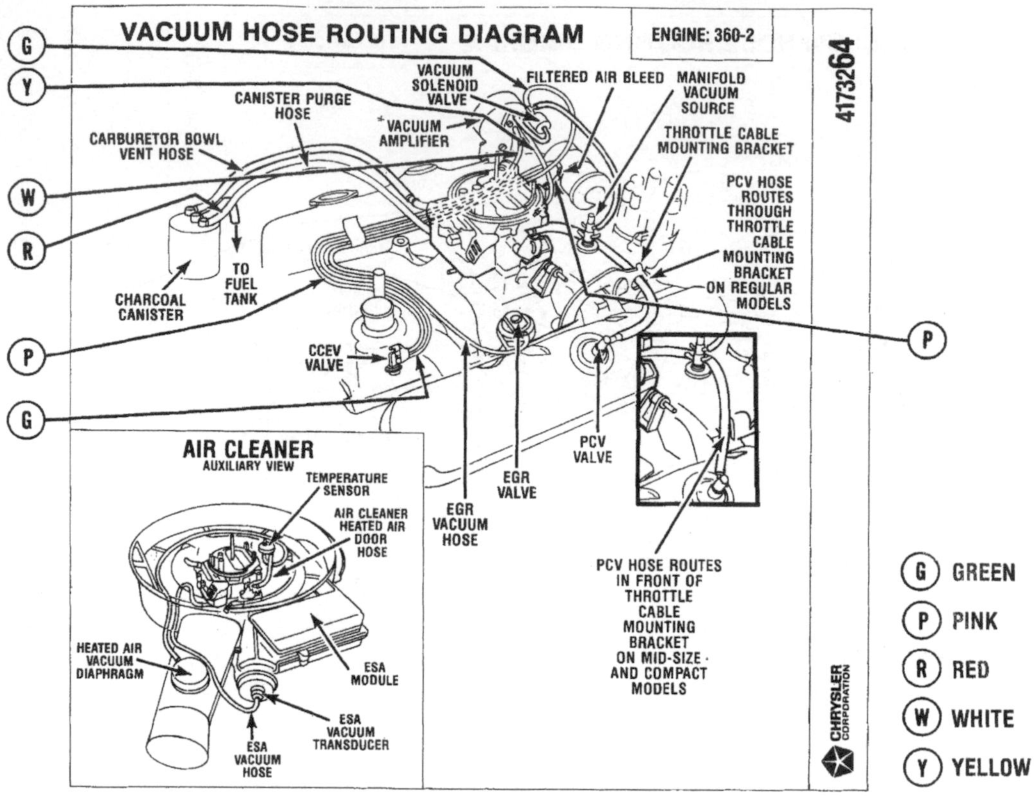

1.6y 1979 vacuum hose routing for the 360 CID V8 engines with 2 barrel carburetors and automatic transmissions (Canadian models with ESA and Catalyst). Note: *Canadian models without catalyst do not have a vacuum amplifier; Federal models do.*

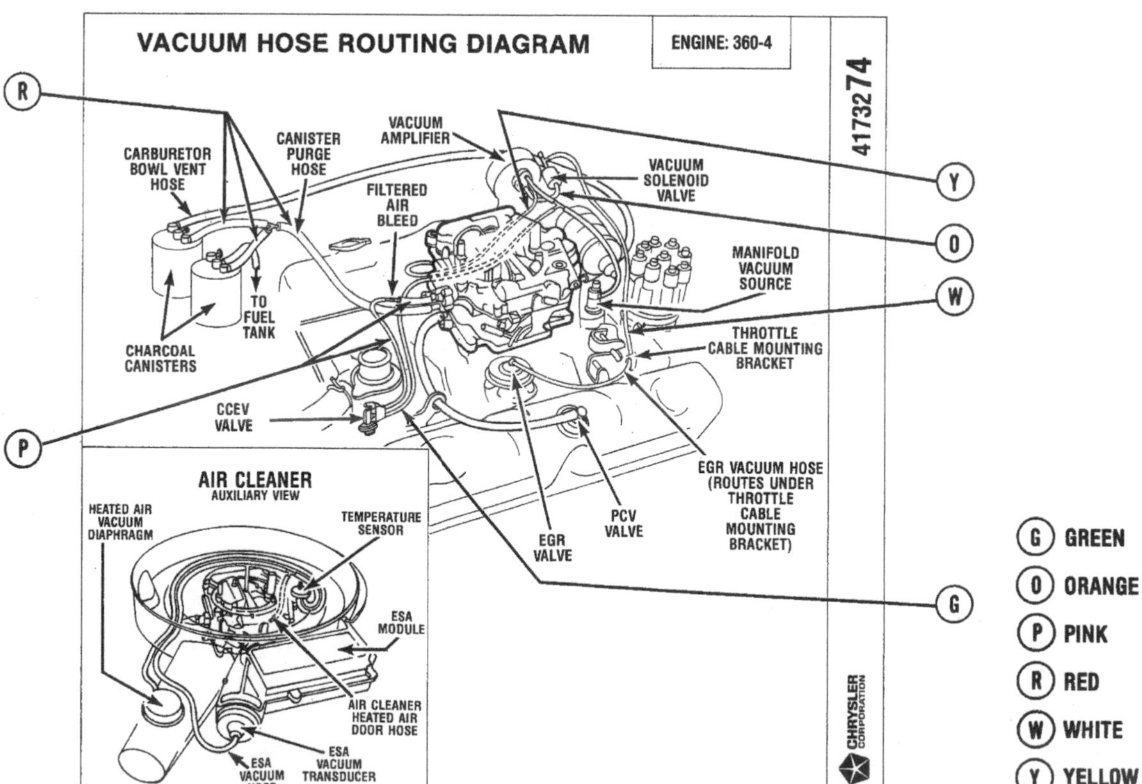

1.6z 1979 vacuum hose routing for the 360 CID V8 engines with 4 barrel carburetors and manual transmissions (Federal models). Note: *Automatic transmission models have one charcoal canister and the EGR valve vacuum hose routes between the cable and the throttle cable mounting bracket.*

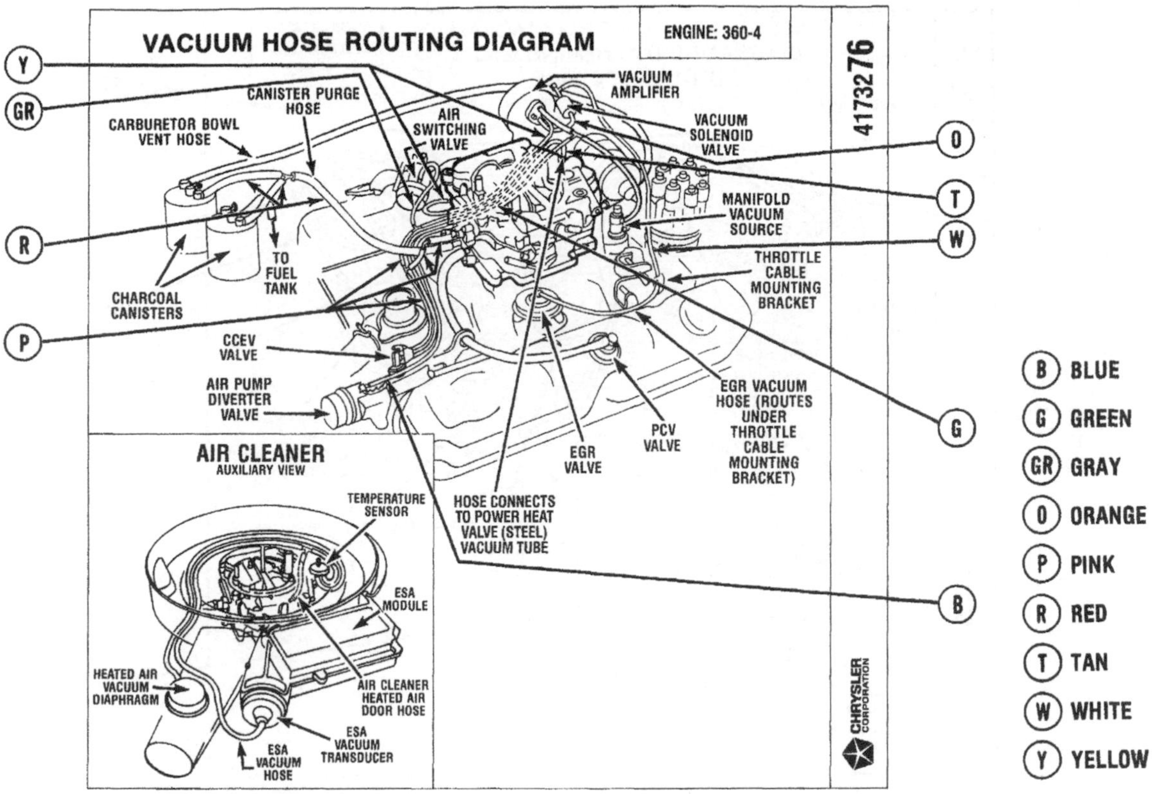

1.6aa 1979 vacuum hose routing for the 360 CID V8 engines with 4 barrel carburetors and automatic transmissions (California models with ESA and without Catalyst). Note: *Canadian models do not have an air switching valve.*

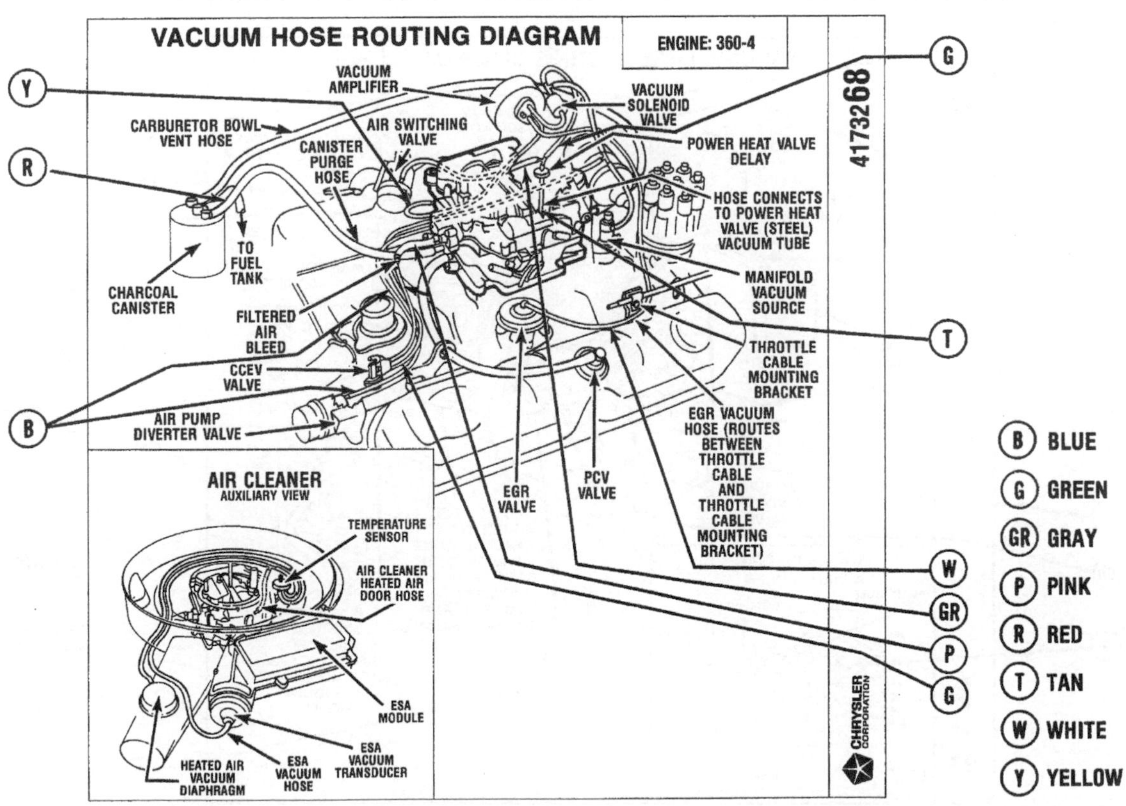

1.6ab 1979 vacuum hose routing for the 360 CID V8 engines with 4 barrel carburetors and automatic transmissions (California models) Note: *Manual transmission models have 2 charcoal canisters and the EGR valve vacuum hose routes under the throttle cable mounting bracket.*

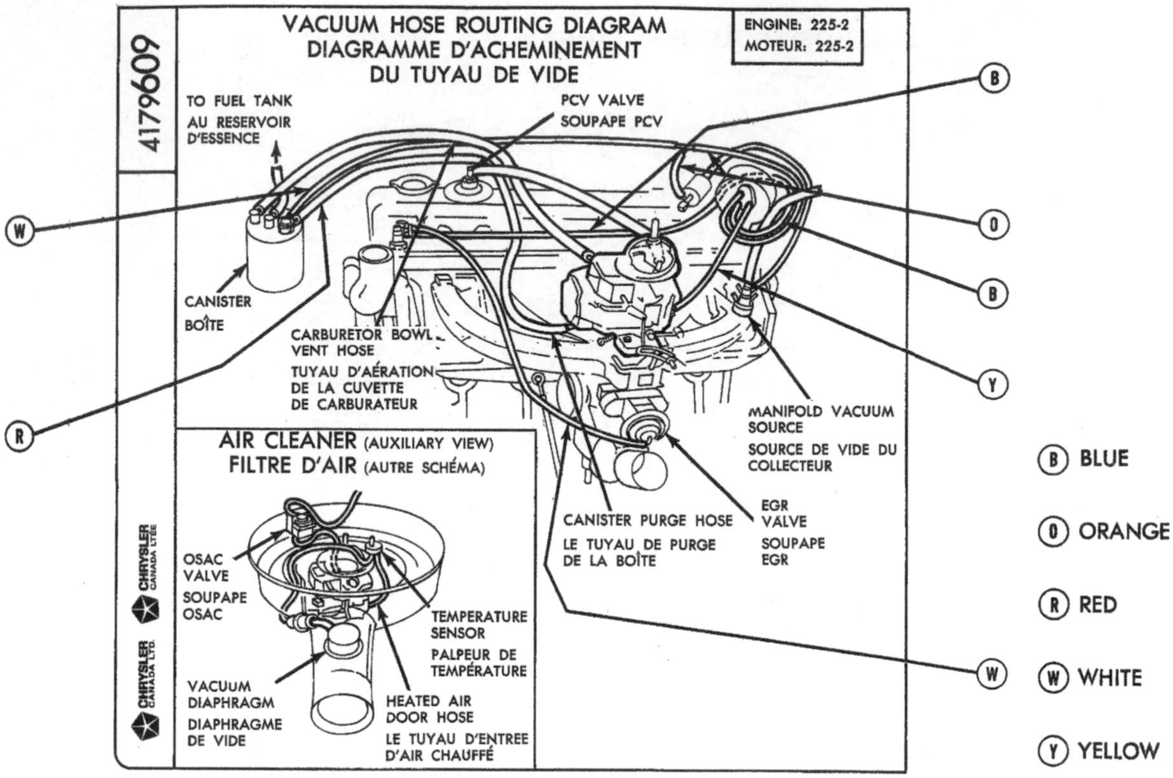

1.6ac 1980 vacuum hose routing for the 225 CID six-cylinder engines with 2-barrel carburetors (Canada models)

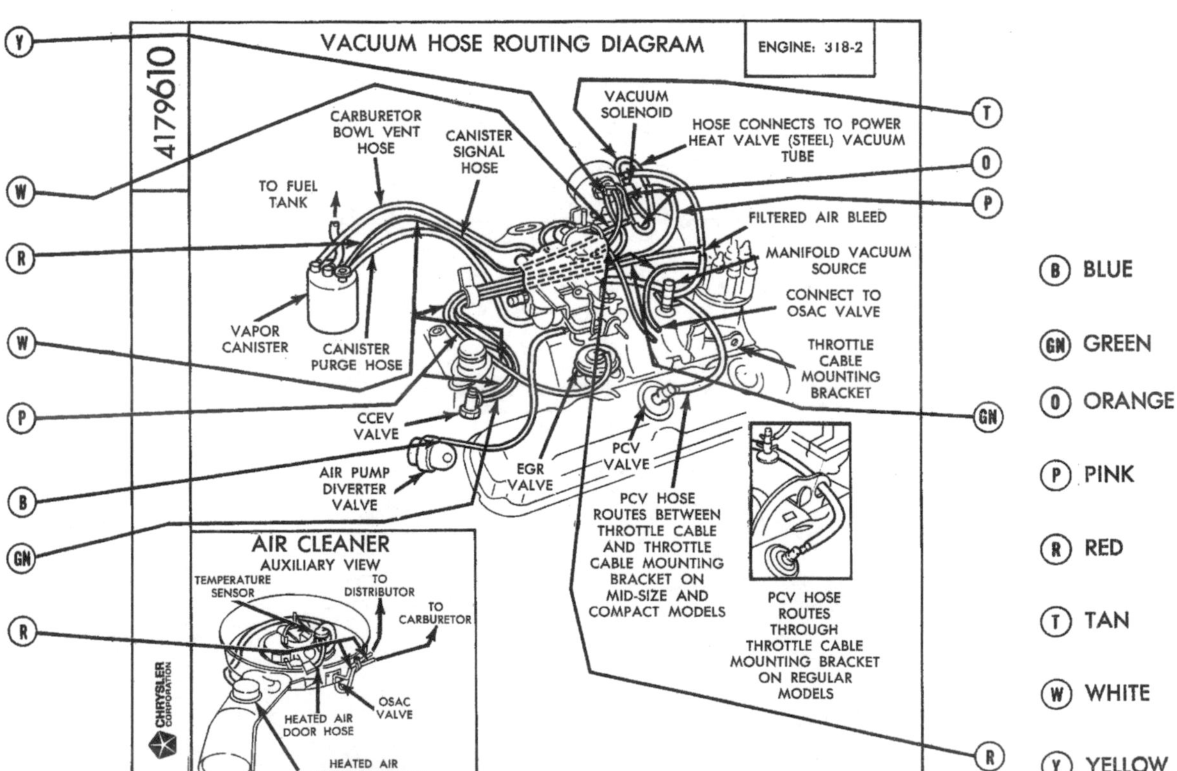

1.6ad 1980 vacuum hose routing for the 318 CID V8 engines with 2 barrel carburetors (Federal models). **Note:** *Canadian models do not have an air pump diverter valve and no CCEV valve.*

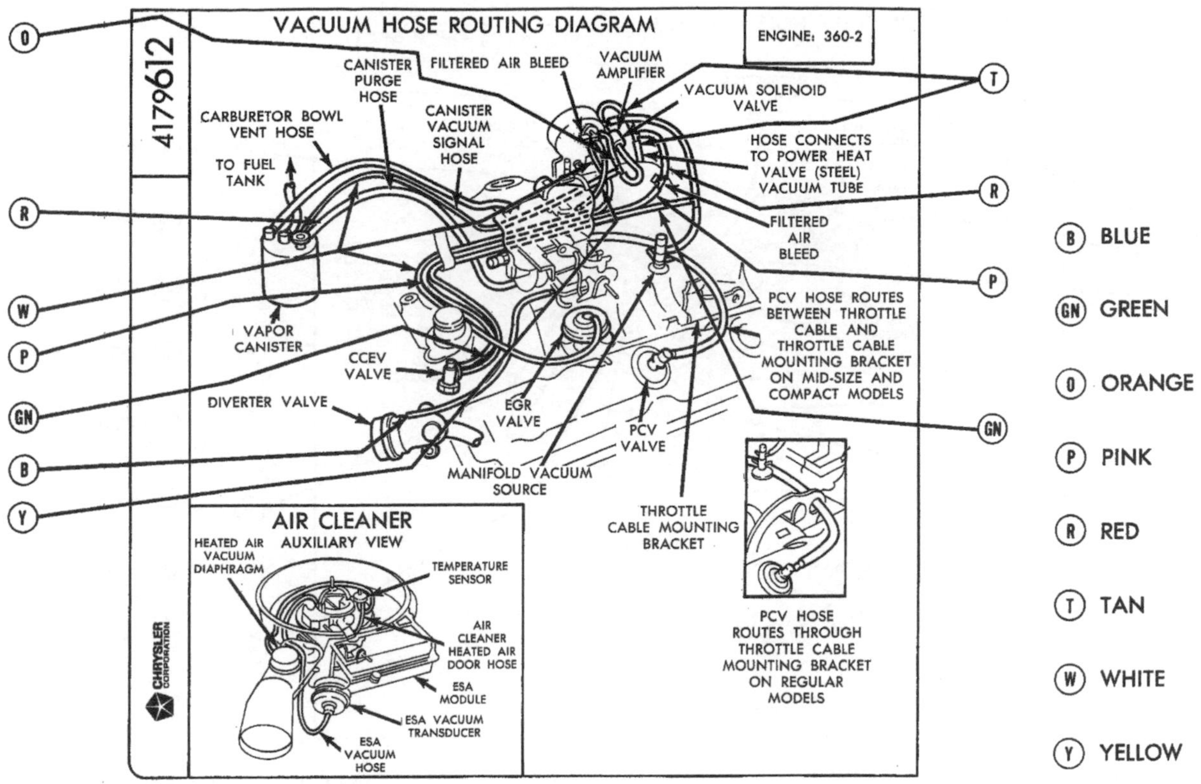

1.6ae 1980 vacuum hose routing for the 360 CID V8 engines with 2 barrel carburetors (Federal models). Note: *Canadian models do not have a diverter valve or CCEV valve.*

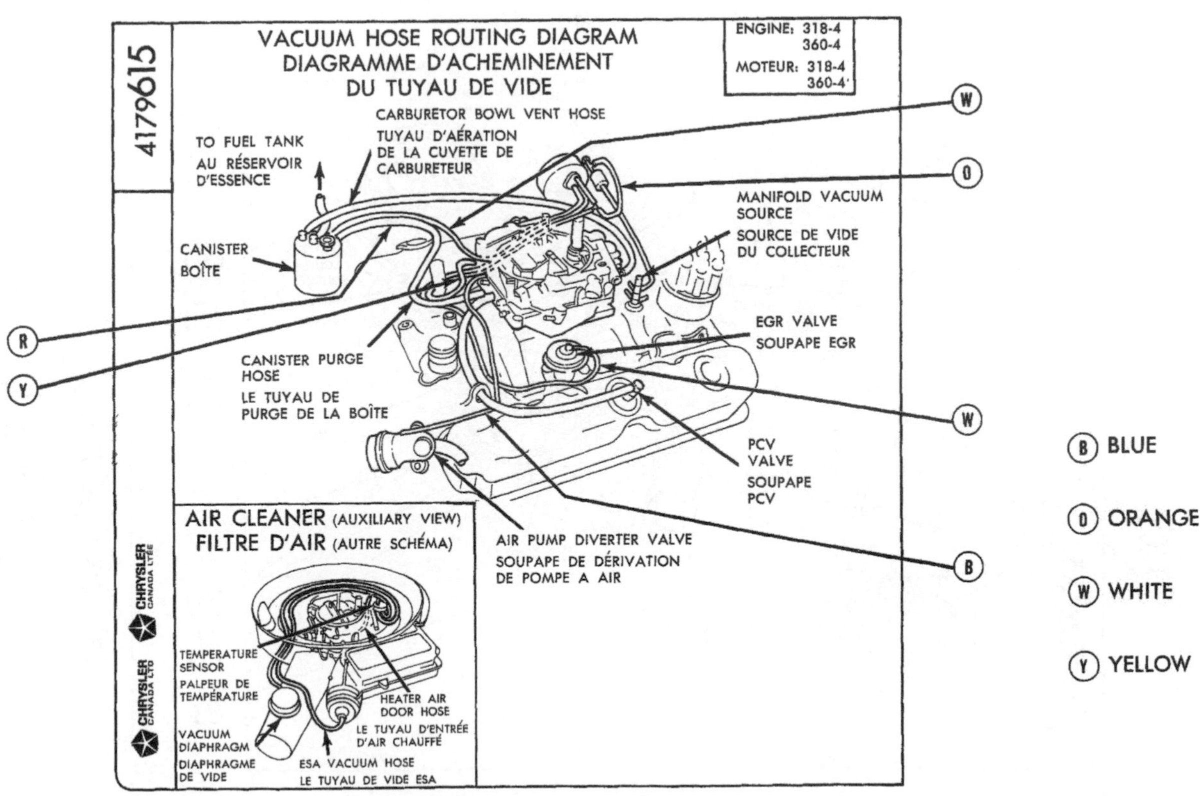

1.6af 1980 vacuum hose routing for the 318 and 360 CID V8 engines with 4 barrel carburetors (Canadian models) (M body). Note: *Canadian model R body does not have a CCEV valve and has two charcoal canisters.*

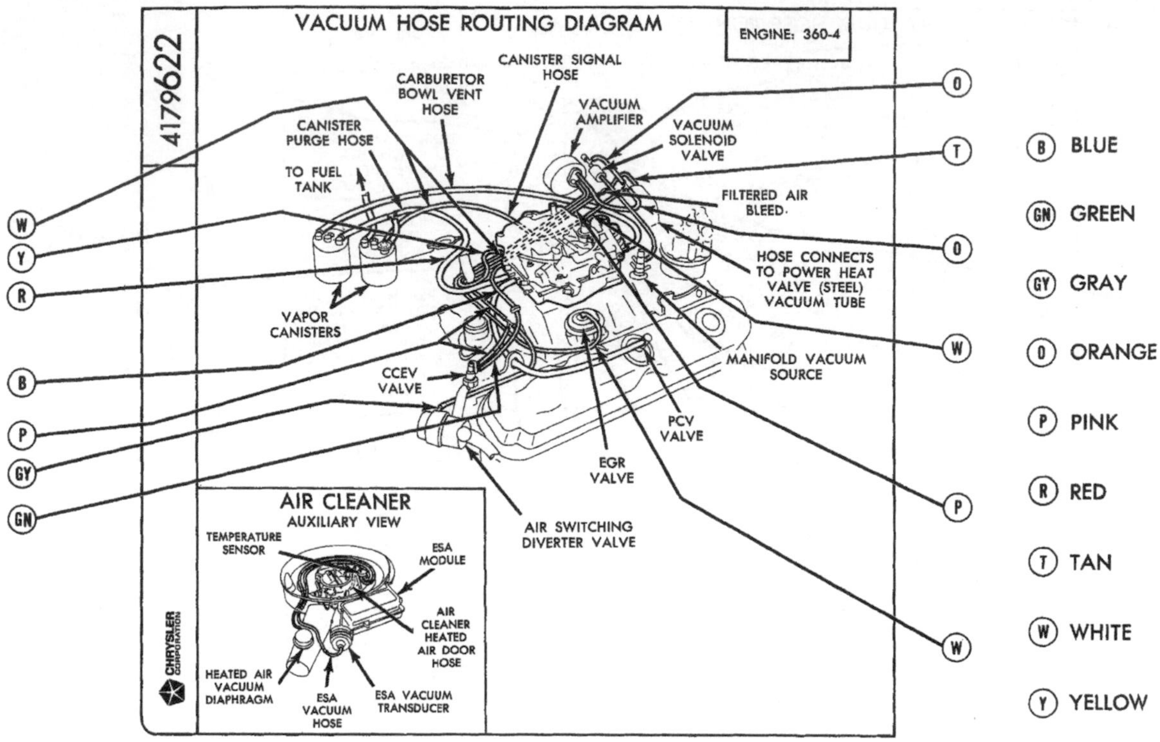

1.6ag 1980 vacuum hose routing for the 360 CID V8 engines with 4 barrel carburetors (Federal models) (R body). Note: *Federal M body models have only one vapor canister.*

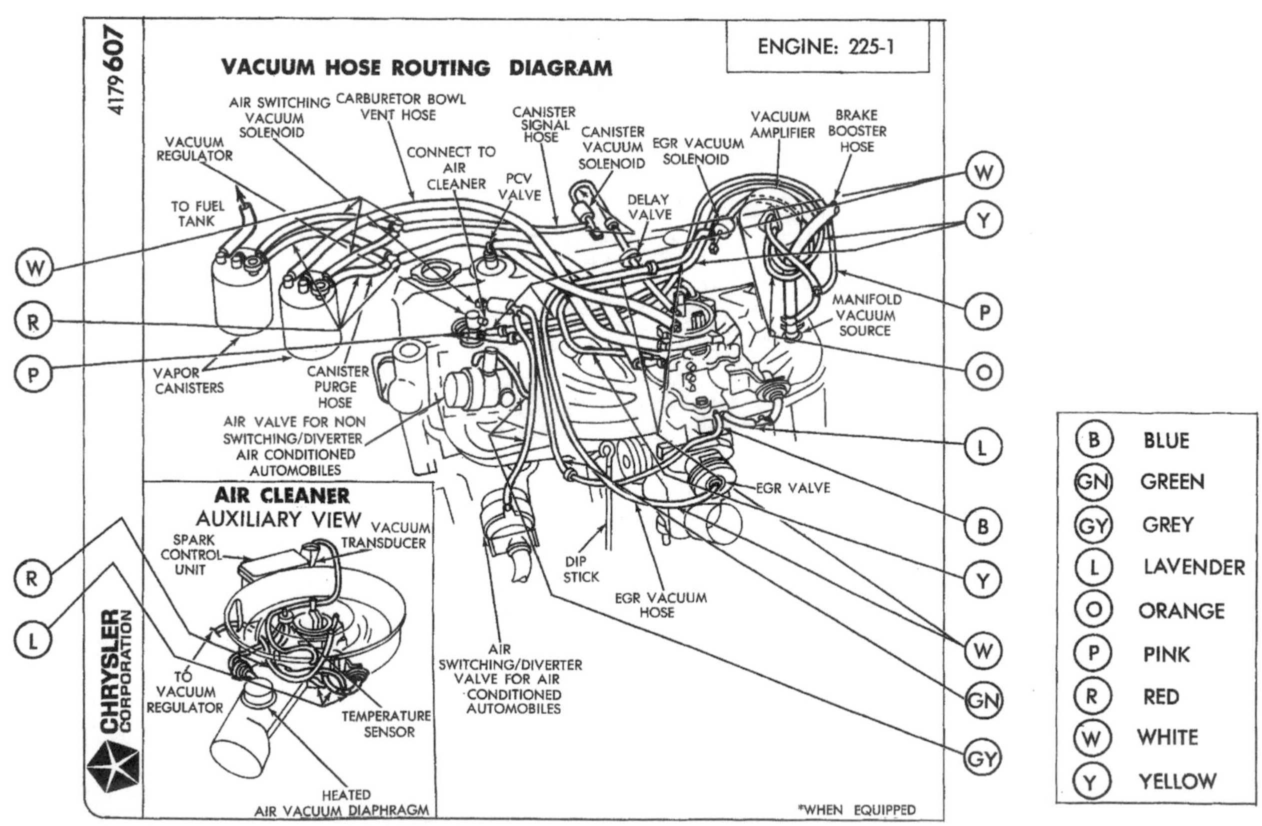

1.6ah 1980 vacuum hose routing for the 225 CID six-cylinder engines with 1 barrel carburetors (California models) (R body).
Note: *Canadian models do not have an air switching/valve, Canadian and Federal models do not have canister solenoid, air switching vacuum solenoid and only one vacuum canister.*

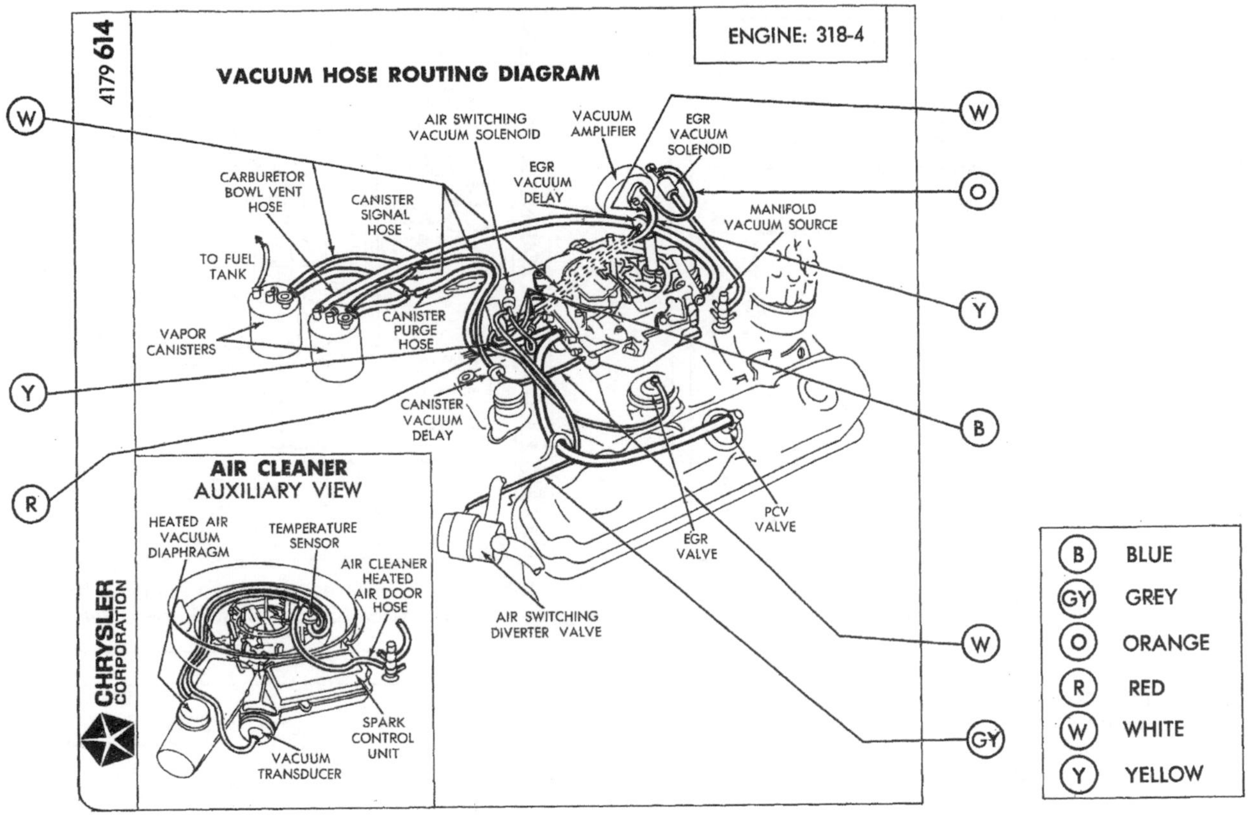

1.6ai 1980 vacuum hose routing for the 318 CID V8 engines with 4 barrel carburetors (California models) (M and R body).
Note: *On California R body models, the canister hoses to the fuel tank and manifold vacuum have been reversed.*

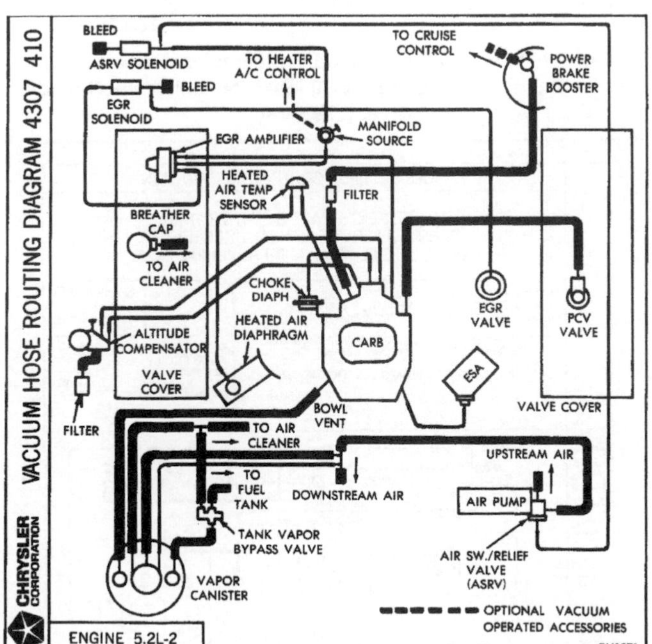

1.6aj 1984 vacuum hose routing for the 318 CID V8 engines with 2 barrel carburetor (Federal and California). Note: *Canadian models do not have an air pump, altitude compensator or EGR solenoid.*

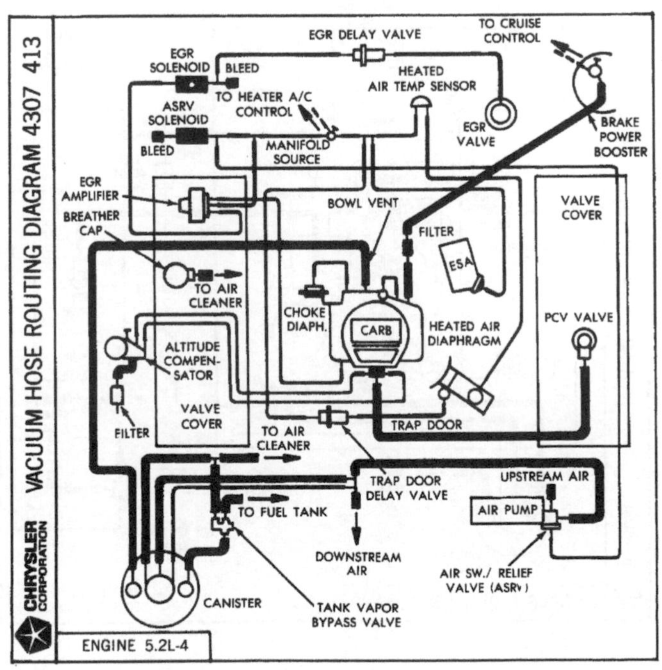

1.6ak 1984 vacuum hose routing for the 318 CID V8 engines with 4 barrel carburetor (Federal and Canada, four-hose routing)

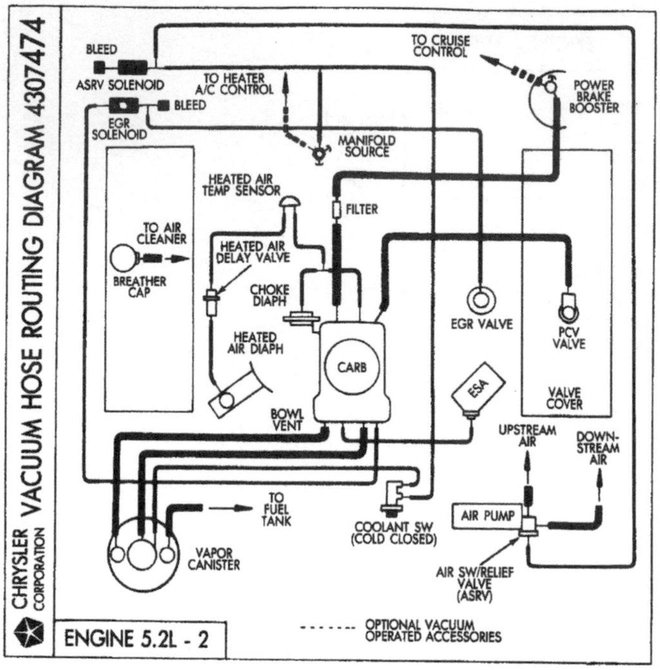

1.6al 1985 through 1989 vacuum hose routing for the 318 CID V8 engines with 2 barrel carburetor (Federal and California).
Note: *Canadian models do not have an EGR solenoid, ASRV solenoid or heated air delay valve.*

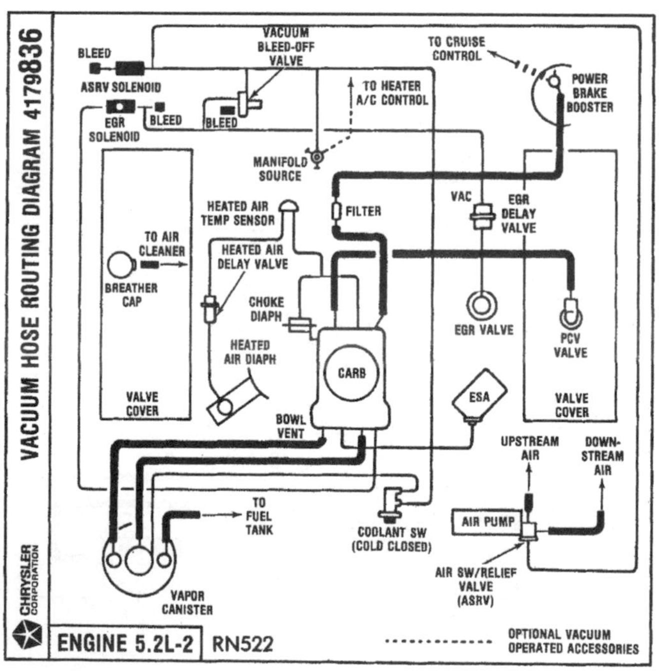

1.6an 1986 through 1989 vacuum hose routing for the 318 CID V8 engines with 2 barrel carburetor (Federal 2-hose routing).
Note: *California models do not have an EGR delay valve.*

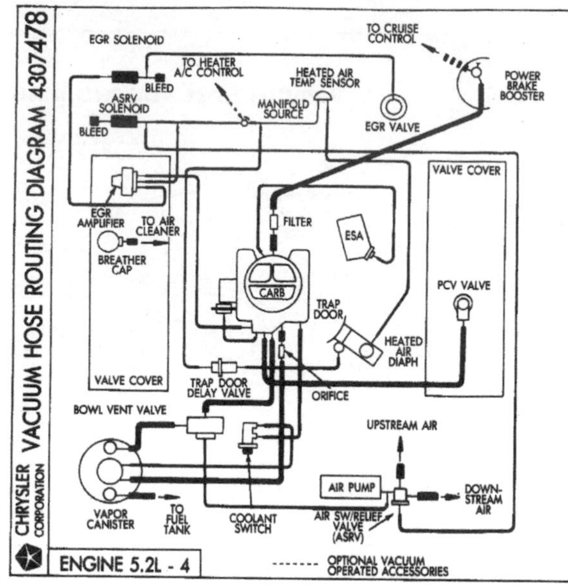

1.6am 1985 through 1989 vacuum hose routing for the 318 CID V8 engines with 4 barrel carburetor (Canadian and Federal models)

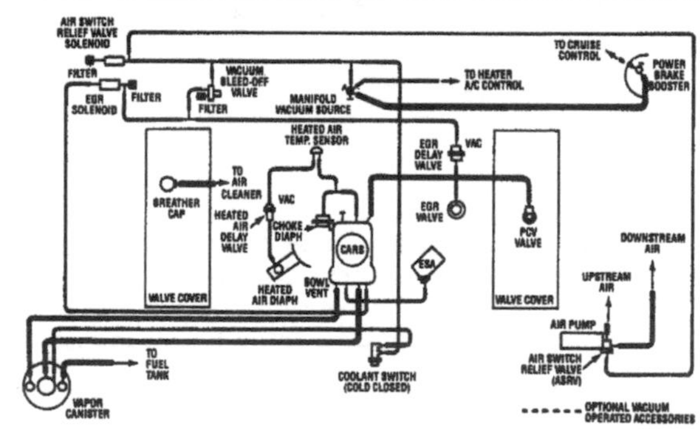

1.6ao 1988 and 1989 combined vehicle emission, adjustment and vacuum hose routing (VECI) label for California, Canadian and Federal

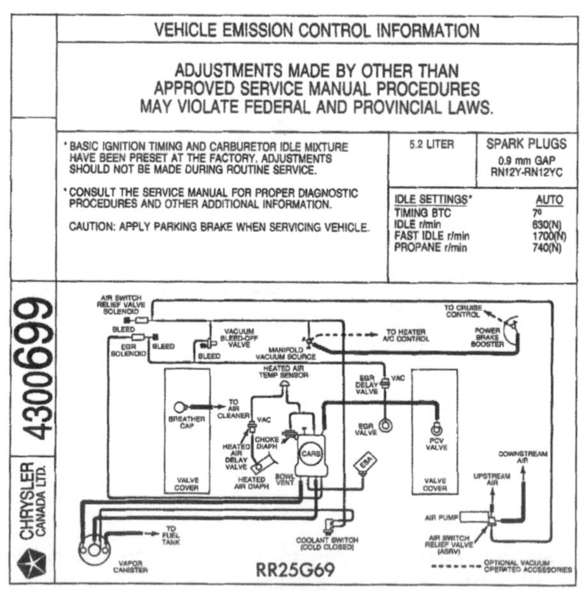

1.6ap Typical 1988 and 1989 Canadian VECI label

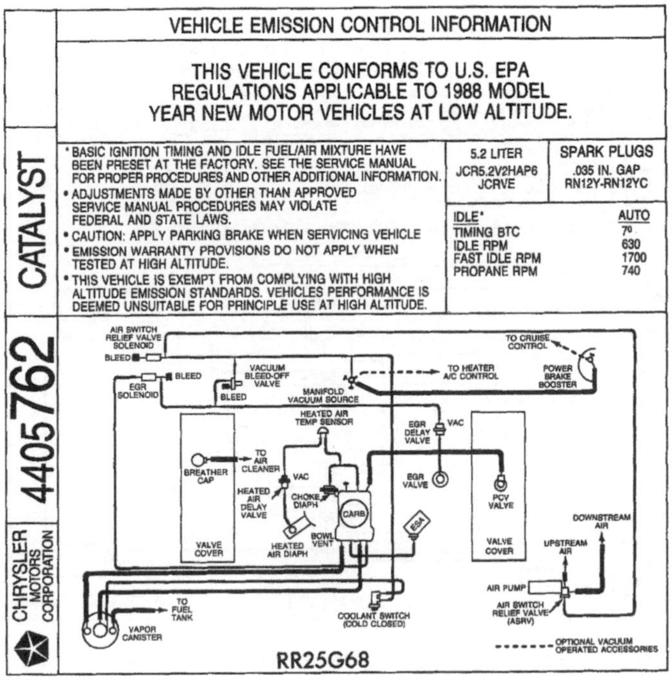

1.6aq Typical 1988 and 1989 Federal 5.2L (318 CID) V8 engine with 2 barrel carburetor VECI label

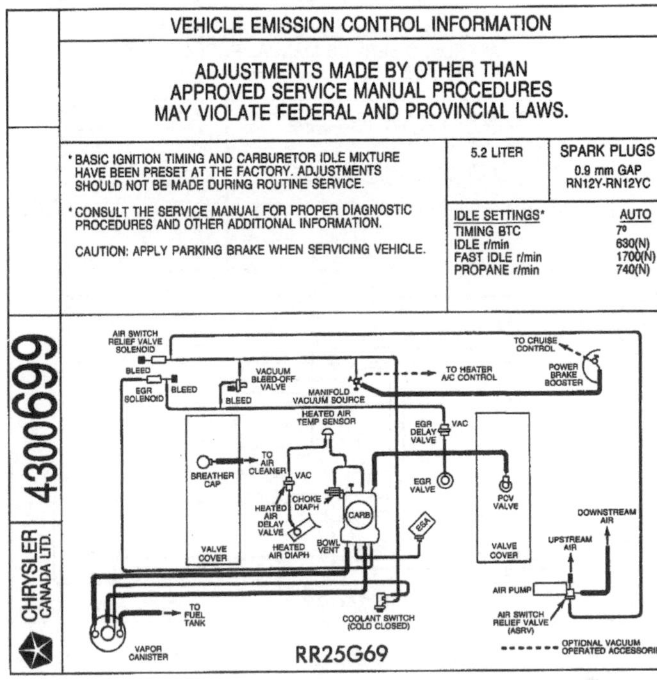

1.6ar Typical 1988 and 1989 Canadian 5.2L (318 CID) V8 engine with 2 barrel carburetor VECI label

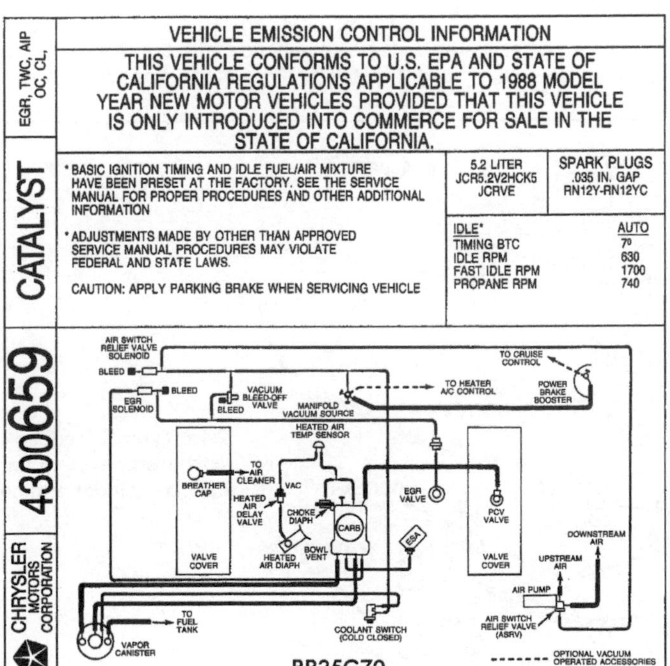

1.6as Typical 1988 and 1989 California 5.2L (318 CID) V8 engine with 2 barrel carburetor VECI label

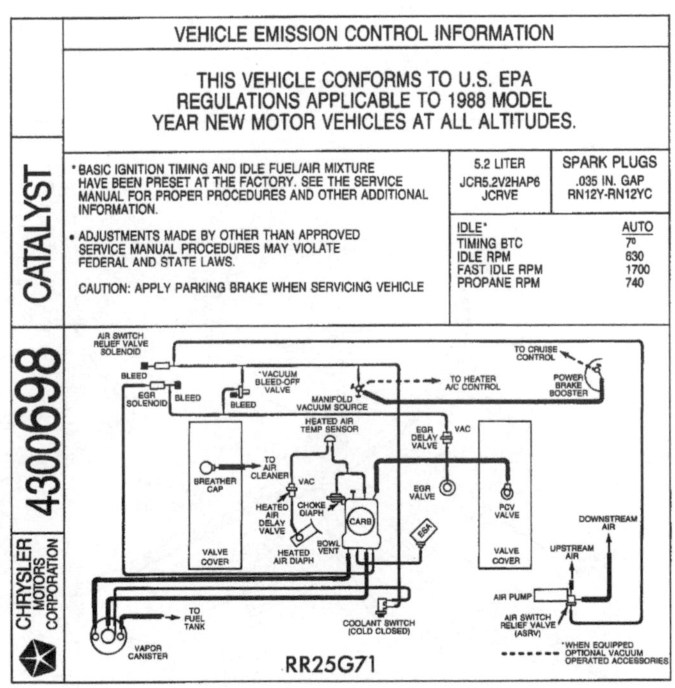

1.6at Typical 1988 and 1989 Federal 5.2L (318 CID) V8 engine with 2 barrel carburetor VECI label

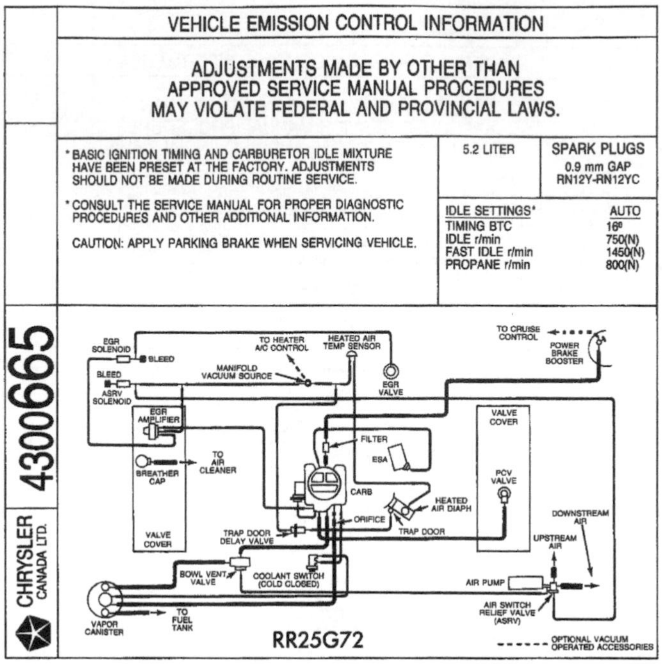

1.6au Typical 1988 and 1989 Canadian 5.2L (318 CID) V8 engine with 4 barrel carburetor VECI label

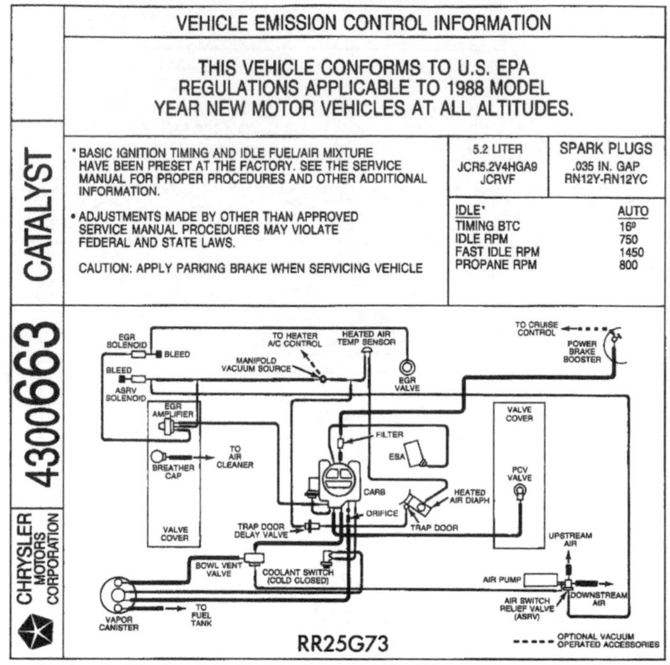

1.6av Typical 1988 and 1989 Federal, High-altitude 5.2L (318 CID) V8 engine with 4 barrel carburetor VECI label

2 Electronic feedback carburetor (EFC) systems - general information

Refer to illustrations 2.1a and 2.1b

Note 1: *Refer to Chapter 5, Section 11 for additional checks and replacement procedures for components and output actuators incorporated into the SCC system that are not covered in this section.*

Note 2: *The Electronic Feedback Carburetor systems were installed on some models starting in 1980. The computer. which is mounted on the air cleaner, will usually read* ***Electronic Fuel Control system****.*

1 The electronic feedback carburetor (EFC) emission system relies on an electronic signal, which is generated by an exhaust gas oxygen sensor, to control a variety of devices and keep emissions within limits **(see illustrations)**. The system works in conjunction with a three-way catalyst to control the levels of carbon monoxide, hydrocarbons and oxides of nitrogen. The EFC system also works in conjunction with the Spark Control Computer (SCC) system. The two systems

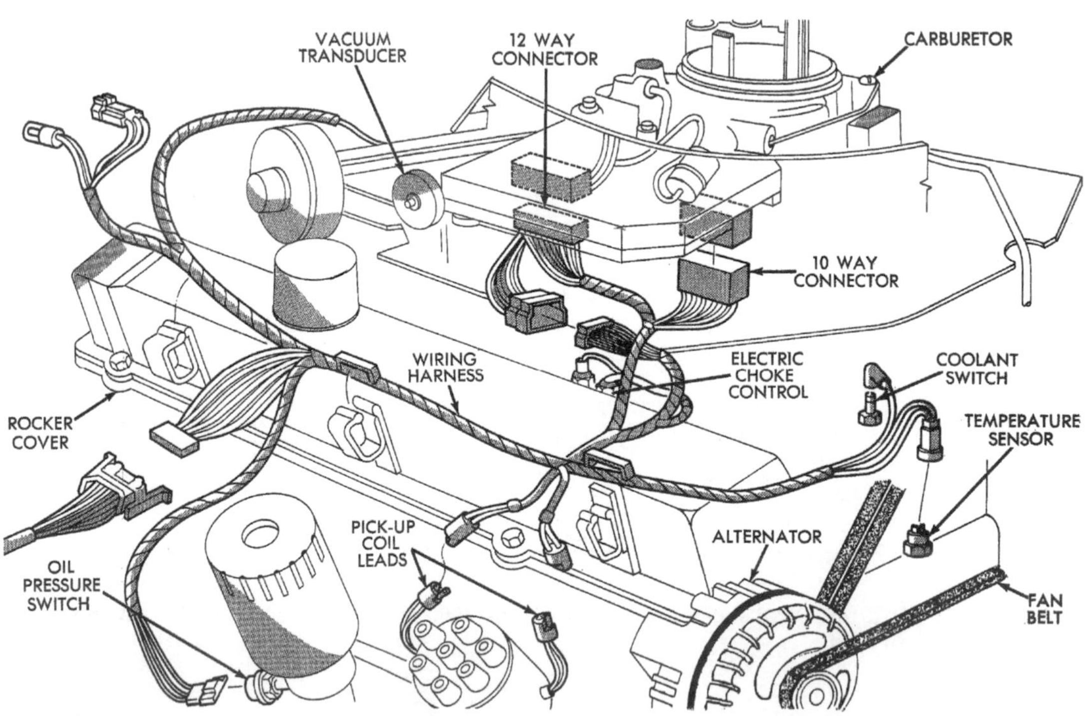

2.1a Typical feedback carburetor system for the six-cylinder engine

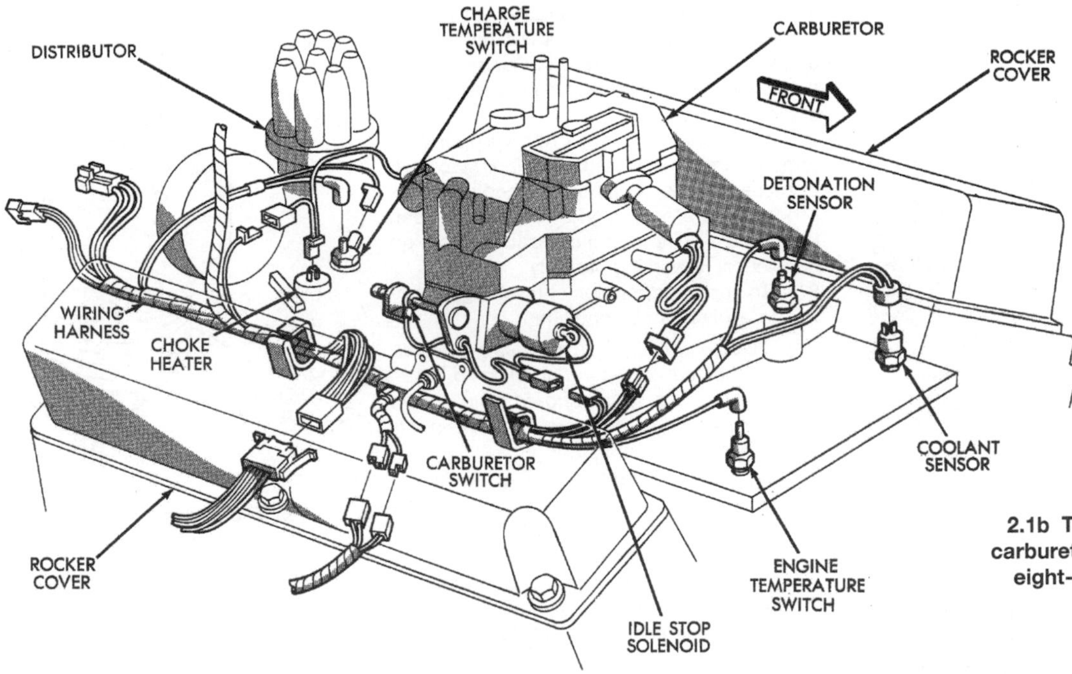

2.1b Typical feedback carburetor system for the eight-cylinder engine

share certain sensors and output actuators; therefore, diagnosing the EFC system will require a thorough check of all the SCC components. Refer to Chapter 5, Section 6 and Section 11 for complete diagnostic procedures for the SCC system.

2 The system operates in two modes: open loop and closed loop. When the engine is cold, the air/fuel mixture is controlled by the computer in accordance with a program designed in at the time of production. The air/fuel mixture during this time will be richer to allow for proper engine warm-up. When the engine is at operating temperature, the system operates in closed loop and the air/fuel mixture is varied depending on the information supplied by the exhaust gas oxygen sensor.

3 The various types of feedback carburetor systems are separated by the different models of feedback carburetors installed on the engine. Here is a list of the feedback carburetors;

Carter BBD
Carter Thermoquad
Holley Model 6145
Holley Model 6280
Rochester Quadrajet

4 Here is a list of the various sensors and output actuators involved with these feedback carburetor systems:

Coolant temperature sensor
Charge air temperature sensor
Detonation (knock) sensor
Oxygen sensor
Magnetic pick-up assembly
Vacuum transducer
Mixture control solenoid (also called duty-cycle solenoid, oxygen feedback solenoid)
SCC computer
Throttle position transducer (1976 through 1979 models)
Throttle control system (1980 through 1989 models)
Carburetor switch

5 Refer to Section 3 for the diagnostic checks for the feedback carburetor system components.

3 Information sensors and output actuators - check and replacement

Engine coolant temperature sensor

Refer to illustration 3.2

Note: *The coolant temperature sensor is used only on 1977 and later models that are equipped with a Spark Control Computer system or feedback carburetor system.*

General description

1 The coolant sensor on 1981 through 1989 models is a thermistor (a resistor which varies the value of its resistance in accordance with temperature changes). The change in the resistance values will directly affect the voltage signal from the computer. As the sensor temperature DECREASES, the resistance values will DECREASE. As the sensor temperature INCREASES, the resistance values will INCREASE. The coolant sensor on 1977 through 1980 models is a switch. When the engine is cold, the ohmmeter should indicate continuity and when the engine is warmed-up to operating temperature, the ohmmeter should indicate NO continuity (open circuit).

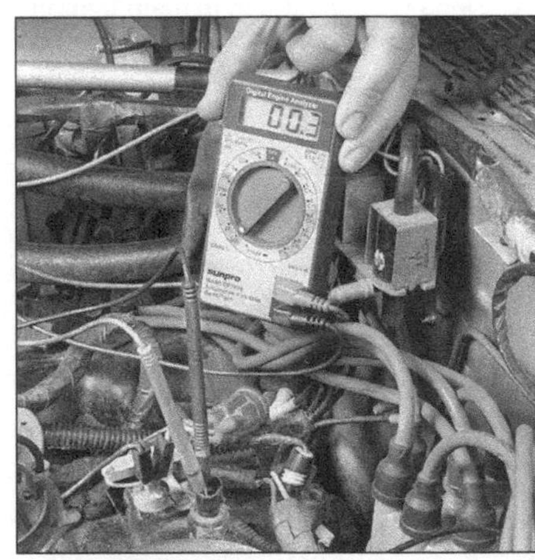

3.2 The coolant temperature sensor resistance will increase as the engine temperature increases

Check

2 To check the sensor on 1981 through 1989 models, install the probes of the ohmmeter onto the terminals of the coolant temperature sensor while it is completely cold (50 to 80-degrees F = 500 to 1,000 ohms). Next, start the engine and warm it up until it reaches operating temperature **(see illustration)**. The resistance should be higher (180 to 200-degrees F = 1,200 to 1,400 ohms). **Note:** *Access to the coolant temperature sensor makes it difficult to position electrical probes on the terminals. If necessary, remove the sensor and perform the tests in a pan of heated water to simulate the conditions.*

Replacement

3 To remove the sensor, release the locking tab, unplug the electrical connector, then carefully unscrew the sensor. **Caution:** *Han-*

3.7 With the engine warmed to operating temperature, disconnect the oxygen sensor connector, hook up the digital voltmeter positive lead to the oxygen sensor, the negative lead to ground and check for a milli-volt signal from the oxygen sensor (which generates its own voltage) - the voltage should fluctuate between 0.1 and 0.9 volts if the sensor is working properly. Keep in mind that an overly rich mixture can also cause a high signal voltage

dle the coolant sensor with care. Damage to this sensor will affect the operation of the entire feedback carburetor system.

4 Before installing the new sensor, wrap the threads with Teflon sealing tape to prevent leakage and thread corrosion.

Installation is the reverse of removal.

Oxygen sensor

Refer to illustration 3.7

Note 1: *The oxygen sensor is used only on 1980 and later models that are equipped with a feedback carburetor.*

Note 2: *A faulty oxygen sensor is one of the most common causes of high fuel consumption and poor driveability on a vehicle equipped with a feedback carburetor.*

General description and check

5 The oxygen sensor, which is located in the exhaust manifold, monitors the oxygen content of the exhaust gas stream. The oxygen content in the exhaust reacts with the oxygen sensor to produce a voltage output which varies from 0.1-volt (high oxygen, lean mixture) to 0.9-volts (low oxygen, rich mixture). The electronic control unit constantly monitors this variable voltage output to determine the ratio of oxygen to fuel in the mixture. The electronic control unit alters the air/fuel mixture ratio by controlling the pulse width (open time) of the mixture control solenoid. A mixture ratio of 14.7 parts air to 1 part fuel is the ideal mixture ratio for minimizing exhaust emissions, thus allowing the catalytic converter to operate at maximum efficiency. It is this ratio of 14.7 to 1 which the electronic control unit and the oxygen sensor attempt to maintain at all times.

6 The oxygen sensor produces no voltage when it is below its normal operating temperature of about 600-degrees F. During this initial period before warm-up, the system operates in open loop mode.

7 If the engine reaches normal operating temperature and/or has been running for two or more minutes, and if the oxygen sensor is producing a steady signal voltage between 0.1 and 0.9-volts **(see illustration)**, the oxygen sensor is working properly.

8 The proper operation of the oxygen sensor depends on four conditions:

a) ***Electrical*** *- The low voltages generated by the sensor depend upon good, clean connections which should be checked whenever a malfunction of the sensor is suspected or indicated.*
b) ***Outside air supply*** *- The sensor is designed to allow air circulation to the internal portion of the sensor. Whenever the sensor is removed and installed or replaced, make sure the air passages are not restricted.*
c) ***Proper operating temperature*** *- The electronic control unit will not react to the sensor signal until the sensor reaches approximately 600-degrees F. This factor must be taken into consideration when evaluating the performance of the sensor.*
d) ***Unleaded fuel*** *- The use of unleaded fuel is essential for proper operation of the sensor. Make sure the fuel you are using is of this type.*

9 In addition to observing the above conditions, special care must be taken whenever the sensor is serviced.

a) *The oxygen sensor has a permanently attached pigtail and electrical connector which should not be removed from the sensor. Damage or removal of the pigtail or electrical connector can adversely affect operation of the sensor.*
b) *Grease, dirt and other contaminants should be kept away from the electrical connector and the louvered end of the sensor.*
c) *Do not use cleaning solvents of any kind on the oxygen sensor.*
d) *Do not drop or roughly handle the sensor.*
e) *The silicone boot must be installed in the correct position to prevent the boot from being melted and to allow the sensor to operate properly.*

Replacement

Note: *Because it is installed in the exhaust manifold or pipe, which contracts when cool, the oxygen sensor may be very difficult to loosen when the engine is cold. Rather than risk damage to the sensor (assuming you are planning to reuse it in another manifold or pipe), start and run the engine for a minute or two, then shut it off. Be careful not to burn yourself during the following procedure.*

10 Disconnect the cable from the negative terminal of the battery.

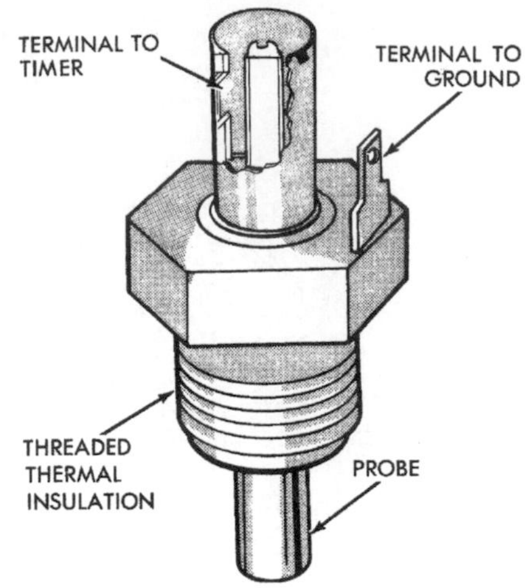

3.19a A diagram of an early-type charge temperature sensor

11 Raise the vehicle and place it securely on jackstands.

12 Carefully disconnect the electrical connector from the sensor.

13 Carefully unscrew the sensor from the exhaust manifold. **Caution:** *Excessive force may damage the threads.*

14 Anti-seize compound must be used on the threads of the sensor to facilitate future removal. The threads of new sensors will already be coated with this compound, but if an old sensor is removed and reinstalled, recoat the threads.

15 Install the sensor and tighten it securely.

16 Reconnect the electrical connector of the pigtail lead to the main engine wiring harness.

17 Lower the vehicle and reconnect the cable to the negative terminal of the battery.

Charge temperature switch (sensor)

Check

Refer to illustrations 3.19a and 3.19b

18 The charge temperature switch or sensor, used on 1980 and later models with a feedback carburetor, monitors the air temperature of the air intake charge. It is located in the intake manifold. To test the charge temperature switch, turn the ignition switch off and disconnect the lead wire from the charge temperature switch. **Note:** *The charge temperature sensor on 1988 and 1989 models is a thermistor (a resistor which varies the value of its resistance in accordance with temperature changes). The change in the resistance values will directly affect the voltage signal from the computer. As the sensor temperature DECREASES, the resistance values will INCREASE. As the sensor temperature INCREASES, the resistance values will DECREASE.*

19 On models with a switch (mostly 1980 through 1987 models), connect one ohmmeter lead to a good ground or the terminal

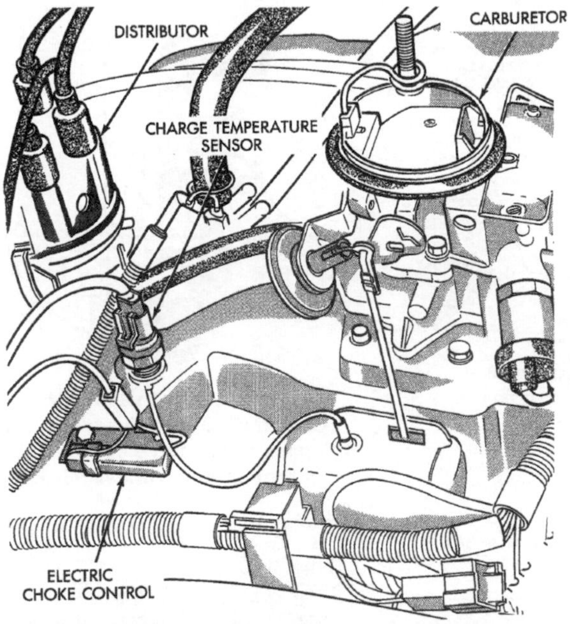

3.19b The charge temperature sensor on 1988 and 1989 models is a thermistor type (resistance varies with temperature)

ground **(see illustration)** and the other lead to the center terminal of the switch. On later models, with a sensor, install the probes onto the two terminals **(see illustration)**.

20 **Working on early models (switch type):** On a cold engine, continuity should be indicated. If the charge temperature switch is cooler than 60-degrees F, resistance should be less than 100 ohms. If there is no continuity, the switch is defective and should be replaced. On a hot engine (normal operating temperature), the ohmmeter should indicate no continuity. If continuity exists, the charge temperature switch is defective and should be replaced.

21 **Working on later models (sensor type):** On a cold engine (approximately 70 degrees F), the resistance of the air charge temperature sensor should be approximately 2,500 ohms. On a warm engine (approximately 200 degrees F), the resistance of the sensor should be approximately 6,000 ohms.

Component replacement

22 The charge temperature switch/sensor can be replaced by simply unthreading the component and installing the replacement. The charge temperature switch/sensor is located on the intake manifold, near the carburetor. To prevent a possible vacuum leak, put thread sealing compound on the threads of the new sensor before screwing it in.

Mixture control solenoid

Check

Refer to illustration 3.24

23 The function of the mixture control solenoid is to provide limited regulation of the fuel-air ratio of a feedback carburetor in response to the electronic signals sent by the computer. This is accomplished by metering the main fuel jets in the carburetor with the use of a plunger that extends into the fuel passage and shuts the fuel off (lean) and on (rich), allowing the fuel/air mixture ratio to change. By controlling the duration of this voltage signal, the ratio of power ON-time versus the power OFF-time is called the duty cycle. The mixture control solenoid is of a slightly different design and mounted in different locations on the carburetor, depending upon the type (Quadrajet, Carter BBD, ThermoQuad, etc.).

24 Maintain an engine speed of 1500 rpm. Disconnect the mixture control solenoid connector from the solenoid **(see illustration)**. Average engine speed should increase a minimum of 50 rpm.

25 Reconnect the feedback solenoid connector. The engine speed should slowly return to 1500 rpm. If rpm does not change as specified, there's a problem in the feedback carburetor system, possibly the mixture-control solenoid, although the oxygen sensor is usually the cause. First check all electrical connections carefully, then check the oxygen sensor. If the oxygen sensor is OK and all connections are OK, the mixture-control solenoid is probably bad or out of adjustment. If you have a Rochester Quadrajet, the Lean and Rich stops may be out of adjustment; see the *Haynes Rochester Carburetor Manual* for the adjustment procedure.

Component replacement

26 The location and the removal procedure differs according to the type of carburetor on the particular feedback carburetor system. Removal and installation will require partial dismantling of the carburetor: If you have a Holley 6280 or 6145, refer to the *Haynes Holley Carburetor Manual* for the procedure. If you have a Rochester Quadrajet, refer to the *Haynes Rochester Carburetor Manual* for the procedure.

Detonation sensor

Check

Refer to illustration 3.29

27 Install a timing light to the engine, start the engine and run it on the second highest step of the fast-idle cam (at least 1200 rpm).

28 Connect a hand-held vacuum pump to the vacuum transducer (attached to the computer on the air cleaner) and apply 16 inches of vacuum.

29 Tap lightly on the manifold near the detonation sensor with the end of a wrench or

3.24 The mixture-control solenoid connector location on the Rochester Quadrajet - the connector on all models is located on the carburetor air horn

3.29 Tap on the intake manifold near the detonation sensor and observe the timing retard

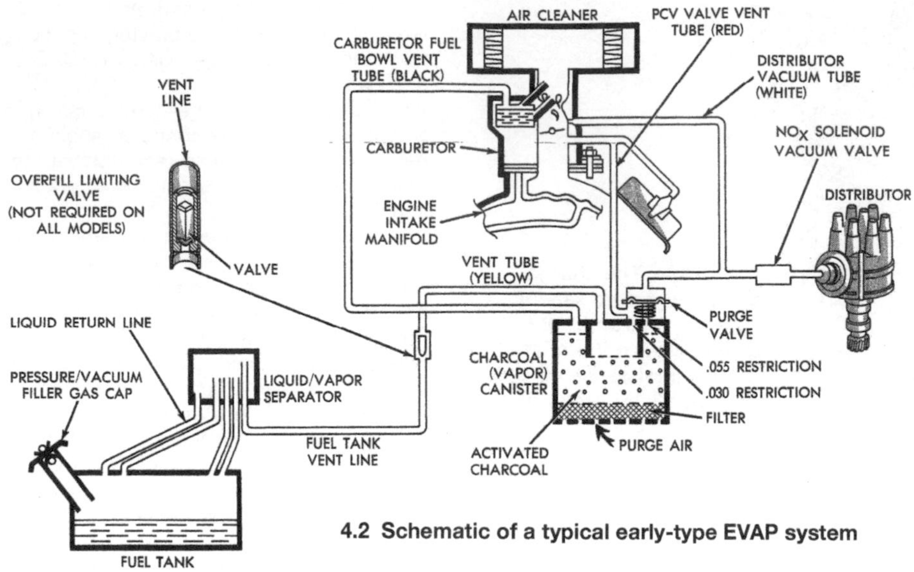

4.2 Schematic of a typical early-type EVAP system

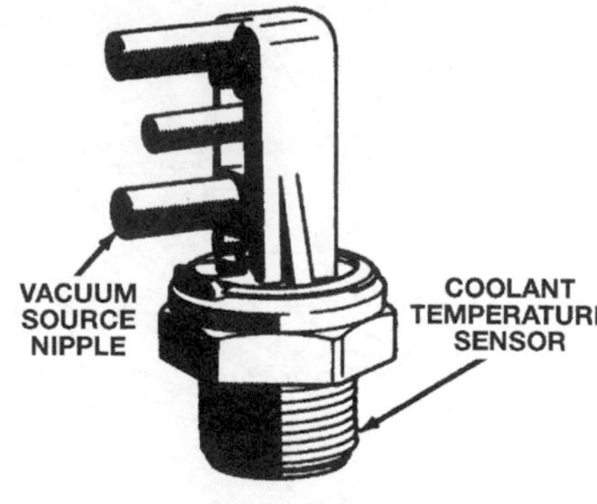

4.6 The Coolant Vacuum Switch Cold Closed (CVSCC) valve allows vacuum to pass to the purge control solenoid at temperatures above 98-degrees F

screwdriver and observe that the ignition retards not more than 11 degrees **(see illustration)**.

30 If there is not any noticeable change in the timing, replace the sensor.

Replacement

31 Disconnect the electrical connector from the sensor and unscrew the sensor from the intake manifold.

32 Installation is the reverse of removal.

4 Evaporative emissions control (EVAP) system

General description

Refer to illustrations 4.2, 4.6, 4.10 and 4.13

1 This system is designed to trap and store fuel that evaporates from the fuel system that would normally enter the atmosphere in the form of hydrocarbon (HC) emissions.

2 The system consists of a charcoal-filled vapor canister **(see illustration)**, a purge control valve (early systems), a purge control solenoid (later systems), a rollover valve (later systems), overfill-limiting valve (early systems), pressure vacuum filler cap (on tank inlet) and connecting lines and hoses (see the VECI label attached to the inside of the engine compartment). **Note:** *Not all the models covered in this manual are equipped with the same components. Early systems are more simplified. Refer to the VECI label for the exact location and component breakdown for the particular model. Also refer to the vacuum diagrams in Section 1. Note that many early models are not equipped with a purge control valve or solenoid.*

3 When the engine is off and pressure begins to build up in the fuel tank (caused by fuel expansion), the charcoal in the canister absorbs the fuel vapor. When the engine is started (cold), the charcoal continues to absorb and store fuel vapor. As the engine warms up, the stored fuel vapors are routed to the intake manifold or air cleaner and combustion chambers where they are burned during normal vehicle operation.

4 On early models the charcoal canister is purged using the purge control valve. When vacuum reaches a specified level, the control diaphragm opens and allow vapor to pass from the canister into the carburetor venturi where they are burned during the normal combustion process. This is a mechanically operated system.

5 On later models, the charcoal canister is purged using engine vacuum via the purge control solenoid which is controlled by the Spark Control Computer (SCC). This type is an electrically operated system.

6 The purge control valve or solenoid operates in conjunction with the Coolant Vacuum Switch Cold Closed (CVSCC) valve **(see illustration)**. This valve is usually mounted in the thermostat housing. The CVSCC and the purge control valves remain shut at low engine temperatures. When the engine is operating at normal temperatures, the CVSCC valve opens, allowing vacuum to be applied to the purge control valve so fuel vapors can be vented into the intake manifold.

7 On all models equipped with Spark Control Computer or Electronic Fuel Control system, before the SCC will energize the purge control solenoid for canister purging, the following conditions must be met:

a) *The coolant temperature must be above 90-degrees F.*
b) *The engine must have been running for approximately 1.5 minutes or more.*
c) *The vehicle speed must be above 5 to 15 mph.*
d) *The engine speed must be above 800 to 1200 rpm.*
e) *The engine vacuum must be below 30 inches.*
f) *The carburetor switch must be open.*

4.10 With the ignition key on, check for voltage to the purge-control solenoid

8 Some models are also equipped with an external bowl vent valve which is controlled by air pump operation. Some Canadian and Federal models are equipped with a thermal bowl vent valve (located in the bowl vent line to the canister) which is controlled by engine temperature to aid in cold starting.

9 All models are equipped with a relief valve, mounted in the fuel tank filler cap, which is calibrated to open when the fuel tank when vacuum or pressure reaches a certain level **(see illustration)**. This vents the fuel tank and relieves extreme tank vacuum or pressure.

Checks

Purge control solenoid

10 Check the power to the solenoid. With the ignition key on (engine not running), check for voltage at the electrical connector **(see illustration)**. There should be voltage (12 volts) present. If not, check the wiring harness (see Chapter 12). If the solenoid is suspect, use jumper wires and apply battery voltage to the solenoid and check for solenoid activation.

11 Also check for vacuum to the solenoid. With the engine running, install a vacuum

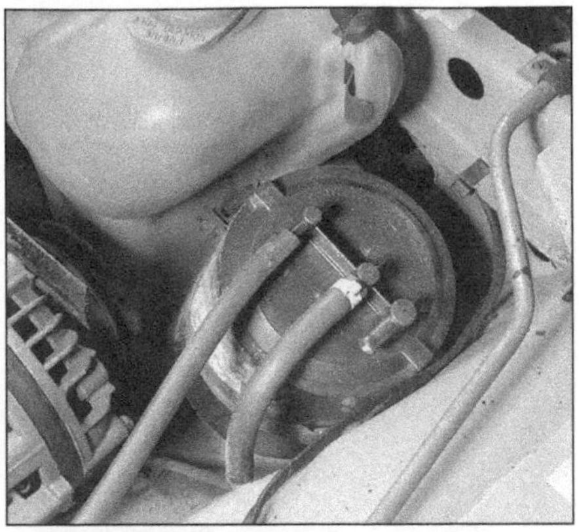

4.13 Check all the hoses on the charcoal canister for leaks, breaks or misrouting - note that the right hose in this photo is disconnected

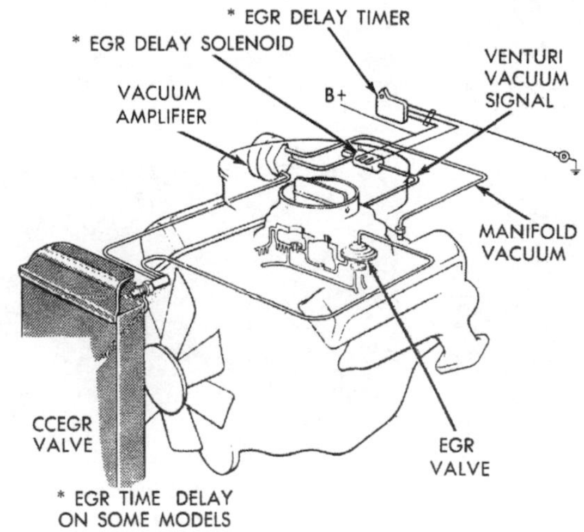

5.1a A typical EGR system with Venturi Vacuum EGR System

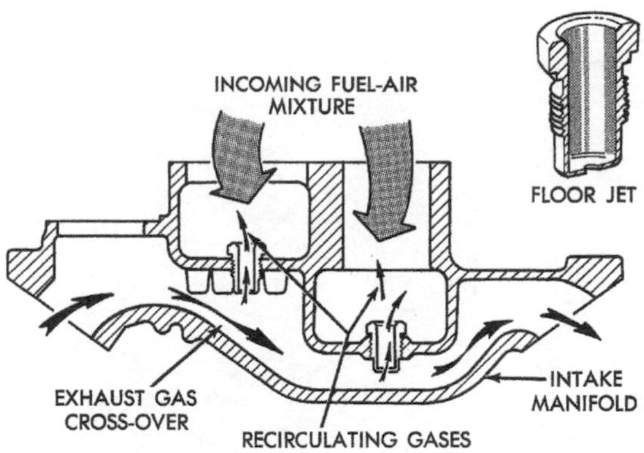

5.1b The very early floor-jet-type EGR valves (orifices) often require servicing because of the carbon deposits that plug the openings - you can frequently clean them with a thin wire while holding the throttle plates open

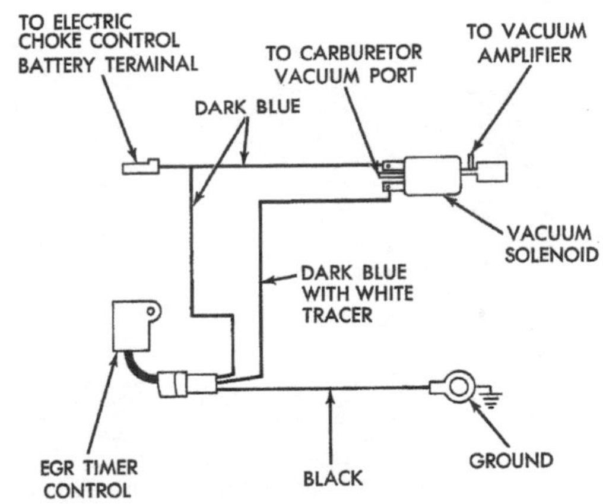

5.4 Schematic of the EGR time delay system

gauge and check for a vacuum signal. If vacuum is not present, check all the lines for damage or improper vacuum hose routing.

Charcoal canister

12 Remove the filter at the bottom of the canister and inspect it for damage and saturation. If necessary, replace it with a new one.

13 Check all the hoses, valves and connectors for damage or cracked hoses **(see illustration).**

14 Replace any component that is damaged.

Purge control valve

15 Disconnect the upper (smaller) vacuum hose from the purge control valve and attach a hand vacuum pump.

16 Disconnect the lower (larger) hose from the purge control valve and attach a short length of clean hose to the fitting on the valve.

17 Try to blow through the hose connected to the lower hose fitting; air should not pass through the valve.

18 Apply 15 inches of vacuum to the upper hose fitting with the vacuum pump, then try again to blow through the lower hose fitting; air should not pass through the valve. If the valve does not operate as specified, the valve is faulty and should be replaced.

5 Exhaust Gas Recirculation (EGR) system

General description

Refer to illustration 5.1a, 5.1b and 5.4

1 This system recirculates a portion of the exhaust gases into the intake manifold or carburetor in order to reduce the combustion temperatures and decrease the amount of nitrogen oxide (NOx) produced **(see illustration)**. The main component in the system is the EGR valve. **Note:** *Some very early models built for California are equipped with jets in the floor of the intake manifold* ***(see illustration) and no EGR valve.*** *This simple EGR system uses an orifice in each jet that allows a fixed amount of exhaust gas to be drawn through by engine vacuum to dilute incoming fuel/air mixture. These jets often get plugged with carbon and will require servicing.*

2 Early models are equipped with either a Ported Vacuum Control system or a Venturi Vacuum Control system. The Venturi Vacuum system utilizes a vacuum tap at the throat of the carburetor venturi to provide the control signal. This signal is a low amplitude signal that requires a vacuum amplifier to increase the level of vacuum needed to operate the EGR valve. The ported vacuum control system uses a slot in the carburetor body which is exposed to an increasing percentage of manifold vacuum as the throttle valve is opened up during acceleration. Ported systems do not require a vacuum amplifier.

3 The early models are also equipped with a Coolant Control Exhaust Gas Recirculation (CCEGR) valve mounted on the radiator. When the coolant temperature in the top radiator tank reaches 65 degrees F, the valve opens so that vacuum is applied to the EGR valve controlling exhaust gas recirculation.

4 Some models are equipped with an EGR delay system that uses an electrical timer mounted on the firewall in the engine compartment which controls an engine-mounted solenoid **(see illustration)**. The timer prevents any exhaust gas recirculation for at least 35 seconds after the engine has started.

5.9a Apply vacuum to the EGR valve and put your finger on the underside of the valve, feeling for movement - the valve diaphragm should move up (small-block V8 engine shown)

5.9b This particular EGR valve does not hold vacuum because the diaphragm is ruptured (225 CID six-cylinder engine shown)

5.15 Remove the rubber junction block along with the vacuum hoses from the vacuum amplifier

5.16 Check to make sure engine vacuum is reaching the vacuum amplifier by using a vacuum gauge to check for vacuum on the corresponding line

5.17 Also check to make sure vacuum is reaching the CCEGR valve and the EGR valve from the amplifier. Replace any damaged or broken hoses.

5.18 Watch carefully for a change in vacuum supply as battery voltage is applied to the solenoid - when voltage is applied, vacuum should drop

5 Engines equipped with SCC (see Chapter 5) and/or EFC (see Section 2) systems are equipped with a charge temperature switch that is activated above 60 degrees F, thus allowing EGR timer to control exhaust gas recirculation. The CTS is located in the number 6 intake manifold runner on six cylinder engines or the number 8 intake manifold runner on eight cylinder engines. Refer to Section 3 for the charge temperature switch diagnostic procedures.

6 These later models are also equipped with an EGR solenoid. This solenoid is electrically operated by the computer and it is designed to control vacuum to the EGR valve.

Checking

Refer to illustrations 5.9a, 5.9b, 5.15, 5.16, 5.17, 5.18 and 5.19

7 Check all hoses for cracks, kinks, broken sections and proper connection. Inspect all system connections for damage, cracks and leaks.

8 To check the EGR system operation, bring the engine up to operating temperature and, with the transmission in Neutral (parking brake set and tires blocked to prevent movement), allow it to idle for 70 seconds. Open the throttle abruptly so the engine speed is between 2000 and 3000 rpm and then allow it to close. The EGR valve stem should move if the control system is working properly. The test should be repeated several times. Movement of the stem indicates the control system is functioning correctly.

9 If the EGR valve stem does not move, check all of the hose connections to make sure they are not leaking or clogged. Disconnect the vacuum hose and apply ten inches of vacuum with a hand pump **(see illustrations)**. If the stem still does not move, replace the EGR valve with a new one. If the valve does open, measure the valve travel to make sure it is approximately 1/8-inch. Also, the engine should run roughly when the valve is open. If it doesn't, the passages are probably clogged.

10 Apply vacuum with the pump and then clamp the hose shut. The valve should stay open for 30 seconds or longer. If it does not, the diaphragm is leaking and the valve should be replaced with a new one.

11 If the EGR valve does not operate with the CCEGR or CVSCC bypassed, the carburetor must be removed to check and clean the slotted port in the throttle bore and the vacuum passages and orifices in the throttle body. Use solvent to remove deposits and check for flow with light air pressure.

12 If the engine idles roughly and it is suspected the EGR valve is not closing, remove the EGR valve and inspect the poppet and seat area for deposits.

13 If the deposits are more than a thin film of carbon, the valve should be cleaned. To clean the valve, apply solvent and allow it to penetrate and soften the deposits, making sure that none gets on the valve diaphragm, as it could be damaged.

14 Use a vacuum pump to hold the valve open and carefully scrape the deposits from the seat and poppet area with a tool. Inspect the poppet and stem for wear and replace the valve with a new one if wear is found.

5.19 Also check to make sure vacuum is reaching the EGR valve from the EGR solenoid (1986 5.2L shown). Replace any damaged or broken hoses

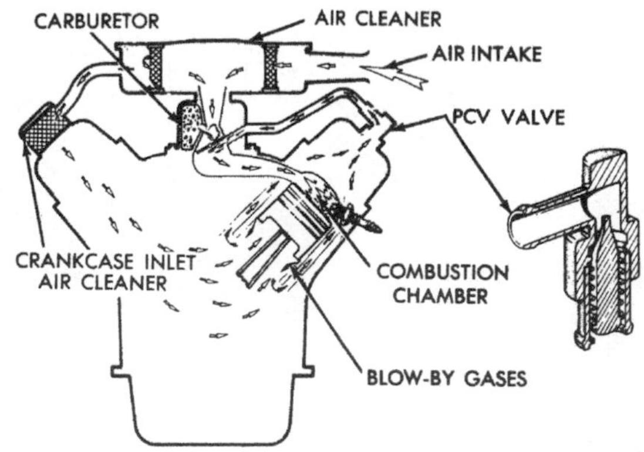

6.2 A typical PCV system (V8 engine shown)

15 Locate the vacuum amplifier (if equipped) and remove the rubber junction block along with the vacuum hoses **(see illustration)**.
16 With the engine running, check for a vacuum signal from the intake manifold **(see illustration)**. This test will determine if manifold vacuum is reaching the vacuum amplifier from the intake manifold.
17 Apply vacuum to the port (vacuum hose) that is routed to the EGR valve or the CCEGR valve **(see illustration)**. The gauge should hold vacuum. If not, check for broken hoses or ruptured diaphragms in the EGR valve or CCEGR valve.
18 Check the EGR solenoid (if equipped). Using a hand-held vacuum pump, apply vacuum and then apply battery voltage. The solenoid should activate and allow vacuum to pass through the solenoid **(see illustration)**.
19 Also remove the vacuum hose(s) from the solenoid and check for vacuum to the EGR valve **(see illustration)**.

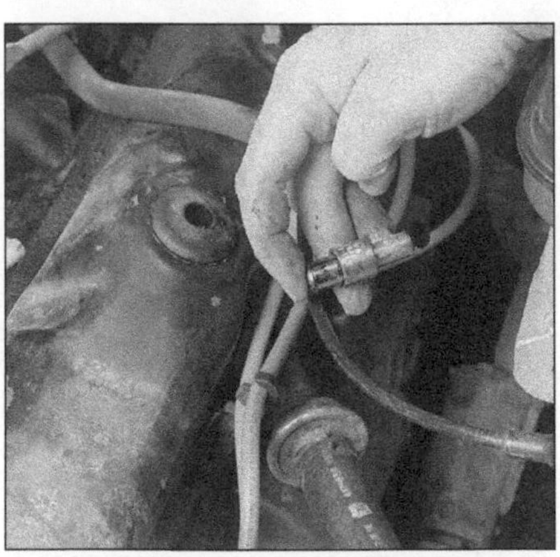

6.3 Place the tip of your finger over the end of the PCV valve and check for engine vacuum. There should be a strong vacuum signal. This indicates that the system is drawing crankcase vapors from the engine

EGR valve replacement

20 Unscrew the fitting attaching the EGR valve tube to the intake manifold (if equipped). Remove the two nuts retaining the valve and remove the assembly.
21 Check the tube for signs of leaking or cracks.
22 Installation is the reverse of removal.

6 Positive Crankcase Ventilation (PCV) system

General description

Refer to illustrations 6.2 and 6.3

1 This system is designed to reduce hydrocarbon emissions (HC) by routing blow-by gases (fuel/air mixture that escapes from the combustion chamber past the piston rings into the crankcase) from the crankcase to the intake manifold and combustion chambers, where they are burned during engine operation.
2 The system is very simple and consists of rubber hoses and a small, replaceable metering valve (PCV valve) **(see illustration)**.

Checking and component replacement

3 With the engine running at idle, pull the PCV valve out of the valve cover and place your finger over the valve inlet **(see illustration)**. A strong vacuum will be felt if the valve is operating properly. Replace the valve with a new one if it is not functioning as described. Do not attempt to clean the old valve.
4 With the engine off, shake the valve and make sure you hear a rattling noise. If not, the valve is defective and should be replaced.
5 To replace the PCV valve, simply pull it from the hose or vent module and install a new one.
6 Inspect the hose prior to installation to ensure that it isn't plugged or damaged. Compare the new valve with the old one to make sure they are the same. Chapter 1 contains additional PCV valve information.

7 Heated inlet air (HIA) system

Refer to illustrations 7.1, 7.6, 7.9, 7.11 and 7.14

General description

1 The heated inlet air system is used to keep the air entering the carburetor at a constant temperature. The carburetor can then be calibrated much leaner for emissions reductions, faster warm up and improved driveability **(see illustration)**.

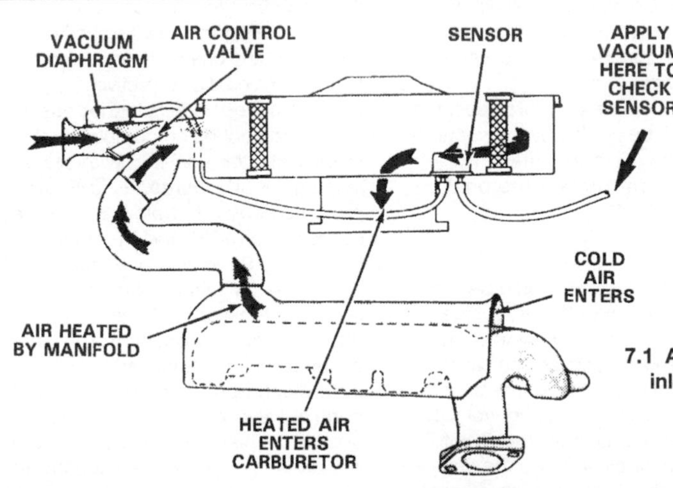

7.1 A typical heated inlet air system

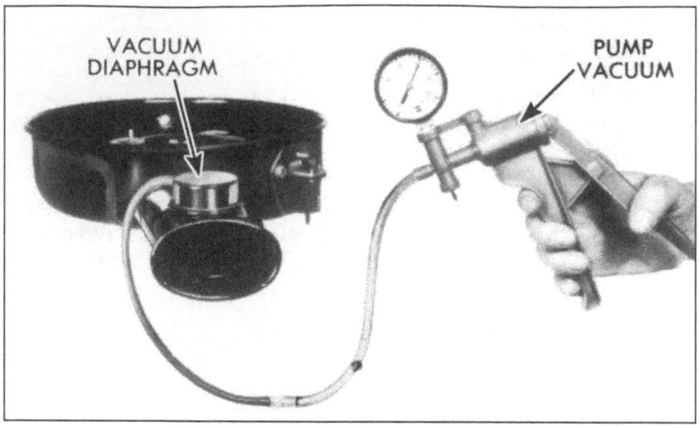

7.6 If the air control valve operates when vacuum is applied directly to the diaphragm, but not when vacuum is applied to the sensor, the sensor may be faulty - check the hoses and connections carefully for leaks before replacing it

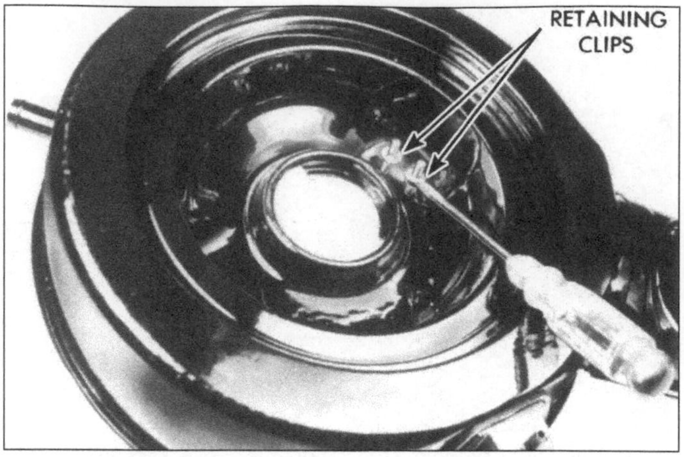

7.9 The sensor is held in place with two retaining clips - discard them and use new ones when the new sensor is installed

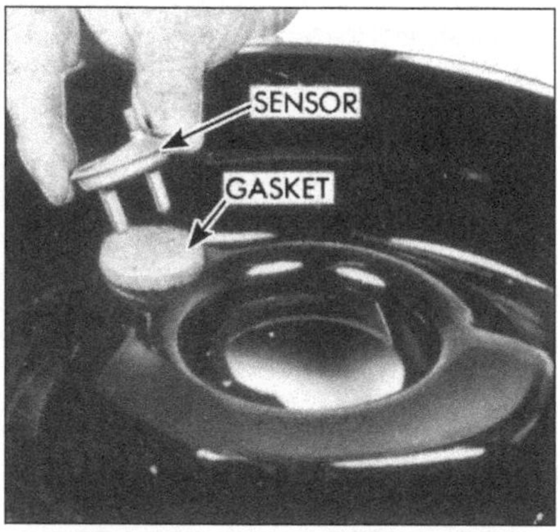

7.11 Make sure a new gasket is used and that it is compressed slightly when the sensor is fastened in place

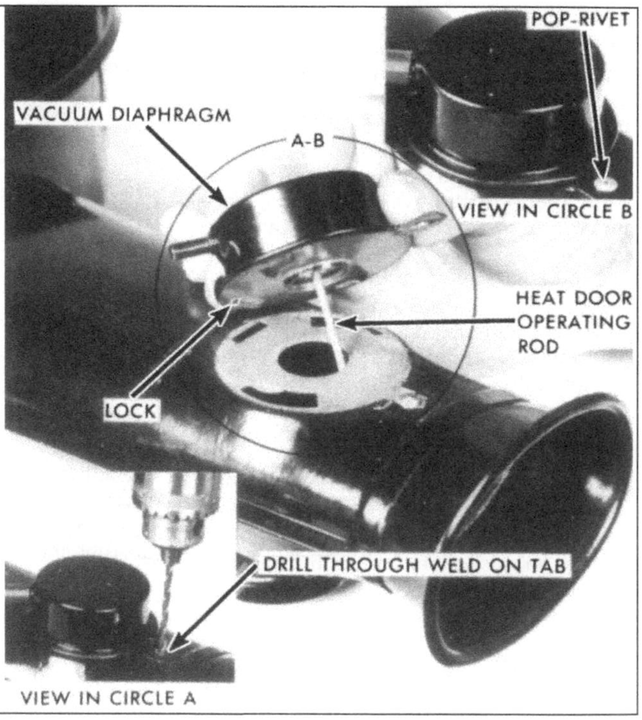

7.14 Removing the air control valve vacuum diaphragm (later model diaphragms are attached with a rivet)

2 Two air flow circuits are used. They are controlled by various carburetor vacuum and coolant temperature sensing valves. A vacuum diaphragm operates an air control valve in the air cleaner snorkel. When the underhood temperature is cold, air is drawn through the shroud which fits over the exhaust manifold, up through the air pipe and into the air cleaner. On dual-snorkel air cleaners, both air control valves open during hard acceleration (when intake manifold vacuum drops below 5-inches).

3 As the underhood temperature rises, the air control valve will be closed by the vacuum diaphragm and the air that enters the air cleaner will be drawn through the cold air snorkel or duct.

Checking

4 Checking of the system should be done while the engine is cold. If the engine is equipped with a cold air inlet snorkel duct, remove the clamp and the duct so you can view the air control valve through the air inlet snorkel. Start the engine and see if the air control valve moves up to allow hot air from the manifold shroud to enter the air cleaner. As the engine warms up, the vacuum diaphragm should allow the air control valve to close, eventually allowing cool air to enter through the snorkel. If this does not occur, any one of the many components in the system could be defective.

5 Remove the air cleaner assembly and allow it to cool down completely. **Note:** *Early models can be checked with the air cleaner just under 90-degrees F, but later models must be cooled to below 60-degrees F (50-degrees F on 1982 and later models).*

6 Apply 20-inches of vacuum to the sensor in the air cleaner housing with a hand pump. The air control valve should move to the up or closed position. If it doesn't, the vacuum should be applied directly to the vacuum diaphragm **(see illustration)**. If the air control valve now moves up, to the closed position, replace the sensor with a new one. If it doesn't, replace the vacuum diaphragm and repeat the check.

7 The diaphragm should not bleed down more than 10-inches of vacuum in five minutes. The air control valve should begin to move at 5 to 5-1/2-inches of vacuum minimum and be fully open at no more than 8-1/2 to 9-inches of vacuum.

8 On dual-snorkel air cleaners, check the second air control valve as described in Step 7.

Component replacement

Sensor

9 To replace the sensor, remove the air cleaner, detach the vacuum hoses from the sensor ports and pry up the retaining clips with a screwdriver **(see illustration)**.

10 Discard the old sensor and the gasket.

11 Position the new gasket and sensor on the housing, support the sensor at the outer

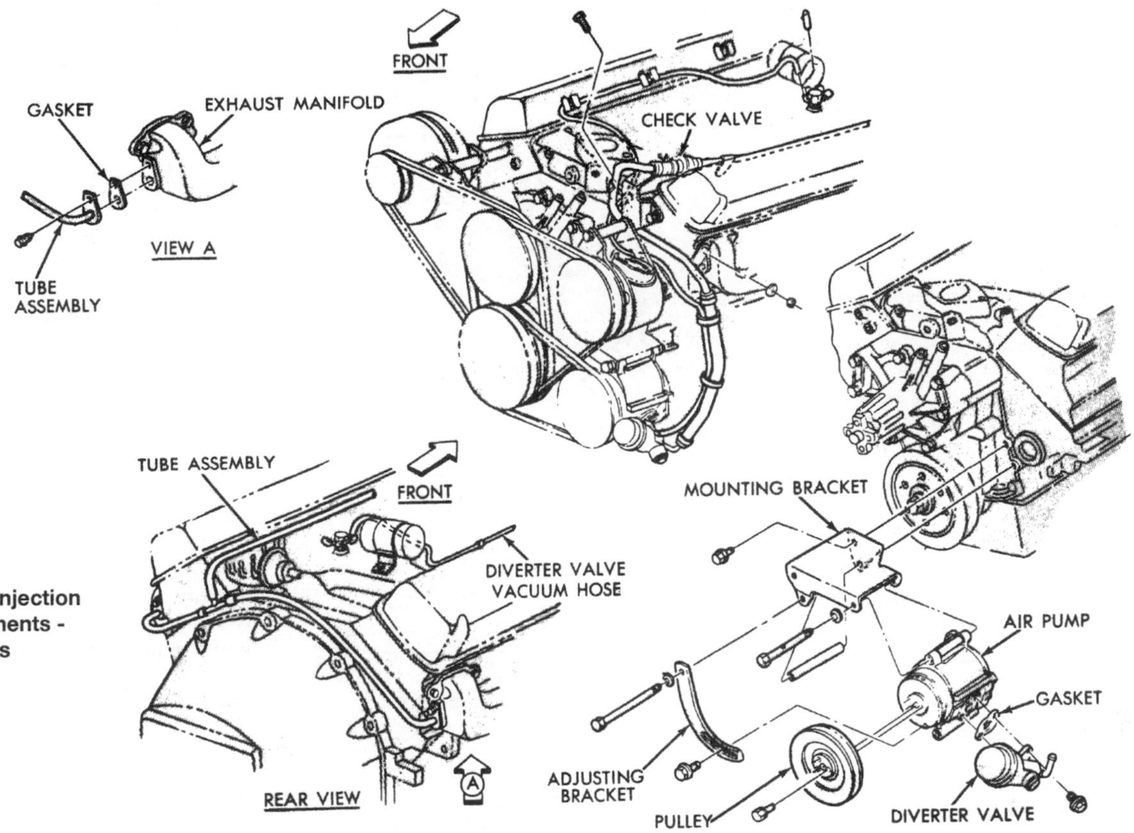

8.1a Typical air injection system components - V8 engines

edge and push the new retaining clips onto the hose ports **(see illustration)**. **Note:** *Make sure the gasket is compressed slightly to form an air-tight seal. Do not attempt to adjust the sensor.*

12 Hook up the vacuum hoses and install the air cleaner.

Vacuum diaphragm

13 Remove the air cleaner housing and disconnect the vacuum hose from the diaphragm.

14 On later models, drill through the welded tab on the air cleaner snorkel **(see illustration)**.

15 Tip the diaphragm slightly forward and rotate it counterclockwise.

16 Slide the diaphragm assembly to one side and disengage the operating rod from the heat control valve.

17 Check the heat control valve for free movement.

18 Installation is the reverse of removal. On later models, use a pop rivet tool to install a new rivet in the vacuum diaphragm tab. Do not manually operate the heat control valve - use vacuum only.

8 Air injection (AIR) system

Refer to illustrations 8.1a and 8.1b

General description

1 The air injection system is employed to reduce carbon monoxide and hydrocarbon emissions, The system consists of an air supply pipe, an air control valve, a check valve, injection tubes and connecting hoses **(see illustrations)**.

2 The air injection system continues combustion of unburned gases after they leave the combustion chamber by injecting fresh air into the hot exhaust stream leaving the exhaust ports. At this point, the fresh air mixes with hot exhaust gases to promote further oxidation of both hydrocarbons and carbon monoxide, thereby reducing their concentration and converting some of them into harmless carbon dioxide and water. During some modes of operation, such as higher engine speeds, the air is dumped into the atmosphere by the air control valve to prevent overheating of the exhaust system.

3 There are three types of air control valves used on the different air injection systems:

a) *The relief valve diverts excessive pump output at higher engine speeds to the atmosphere.*
b) *The diverter valve incorporates a similar relief valve, but also uses a vacuum sensitive valve, which dumps air into the*

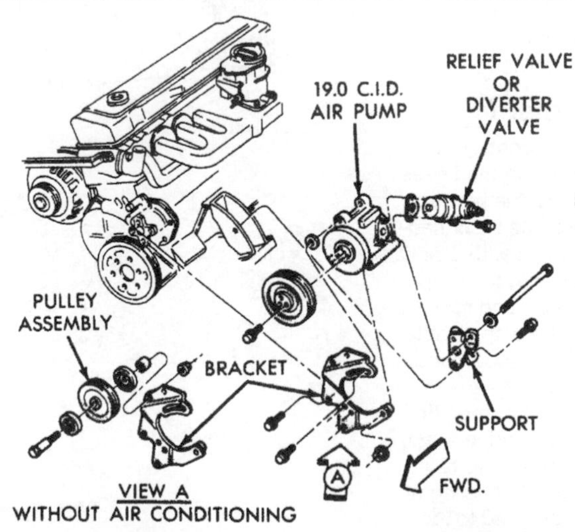

8.1b Typical air injection system components - six-cylinder engines

atmosphere upon sudden deceleration to prevent backfiring.

c) *The switch/relief valve, besides releasing excessive pressure, directs air injection flow to either the exhaust port location or to the downstream injection at the catalytic converter, depending on engine temperature.*

Checking

Air supply pump

4 Check and adjust the drivebelt tension (refer to Chapter 1).

5 Disconnect the air supply hose at the air bypass valve inlet.

6 The pump is operating satisfactorily if air flow is felt at the pump outlet with the engine running at idle, increasing as the engine speed is increased.

7 If the air pump does not successfully pass the above tests, replace it with a new or rebuilt unit.

Relief valve and diverter valve

8 Failure of the relief valve or diverter valve will cause excessive noise and air pump output at the valve.

9 There should be little or no air escaping from the silencer of the valve body at engine speed.

10 If these tests indicate a faulty air control valve, the unit should be replaced.

Switch/relief valve

11 If air injection flow is being directed to both the upstream and downstream inlets simultaneously, the switch/relief valve is malfunctioning.

12 There should be no air escaping from the relief valve silencer at idle speed.

13 If these tests indicate a faulty switch/relief valve the unit should be replaced.

Check valve

14 The check valve, located on the injection tube assembly, can be tested by removing the air hose from the valve inlet tube.

15 Start the engine and check for exhaust gas escaping from the valve inlet tube.

16 If there is an exhaust leak, the check valve is faulty and should be replaced.

Component replacement

17 The check valve and air control valve may be replaced by disconnecting the hoses leading to them (be sure to label the hoses as they are disconnected to facilitate reconnection), replacing the faulty element with a new one and reconnecting the hoses to the proper ports. Be sure to use new gaskets and remove all traces of the old gasket, where applicable. Make sure that the hoses are in good condition. If not, replace them with new ones.

18 To replace the air supply pump, first loosen the appropriate engine drivebelt(s) (refer to Chapter 1), then remove the faulty pump from the mounting bracket, labeling all wires and hoses as they are removed to facilitate installation of the new unit.

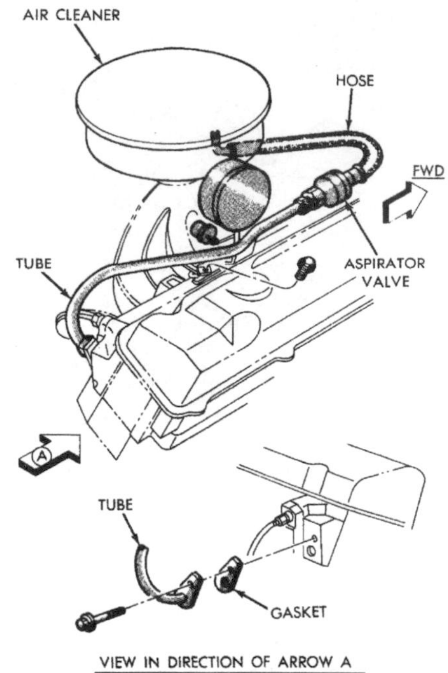

9.1 Typical air aspirator system components

19 After the new pump is installed, adjust the drivebelt(s) to the specified tension (refer to Chapter 1).

9 Air aspirator system

Refer to illustration 9.1

General description

1 Later model Canadian vehicles with a catalytic converter are equipped with an air aspirator system to reduce carbon monoxide and hydrocarbon emissions. The system draws fresh air from the air cleaner, past a one-way diaphragm, allowing fresh air to mix with the exhaust gases during negative pressure pulses in the exhaust system **(see illustration)**.

Checking

2 Failure of the aspirator valve results in excessive exhaust noise in the engine compartment at idle and hardening of the rubber hose from the valve to the air cleaner.

3 To check the aspirator valve, located midway in the hose between the air cleaner and the cylinder head, first disconnect the hose from the aspirator inlet.

4 With the engine at idle and the transmission in Neutral, the negative (vacuum) exhaust pulses should be felt at the aspirator inlet. If hot exhaust gases escape from the inlet, the valve is faulty and should be replaced.

5 If the aspirator tube is leaking at the exhaust manifold joint, the aspirator tube-to-cylinder head gasket should be replaced.

6 Inspect the rubber hose and replace it if it has cracked or hardened.

Component replacement

7 The aspirator valve can be replaced by disconnecting the air hose and unscrewing the valve from the tube assembly.

8 The aspirator tube-to-cylinder head gasket can be replaced by disconnecting the aspirator valve from the air hose and removing the aspirator tube mounting bolts. Be sure to scrape off all traces of the old gasket.

10 Electric assist choke system

Refer to illustration 10.1

General description

1 Most vehicles covered in this manual

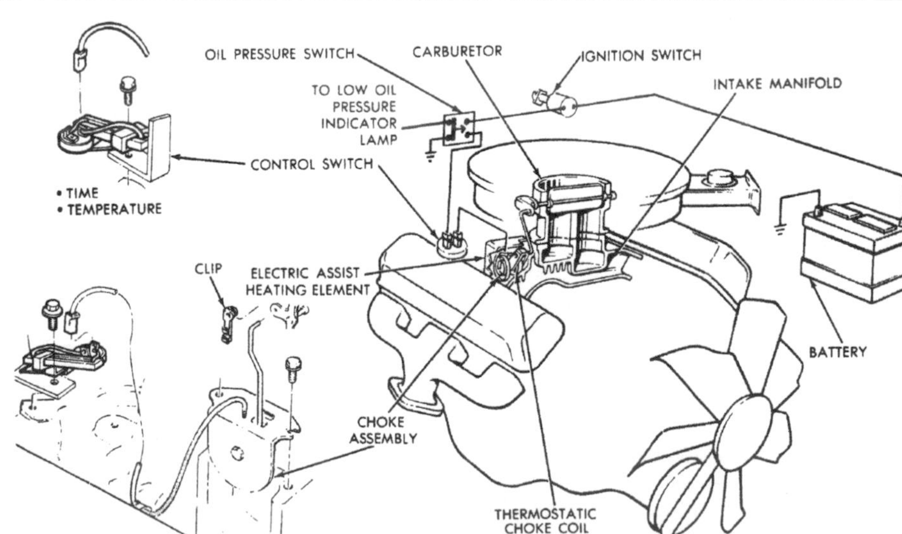

10.1 Typical electric assist choke system

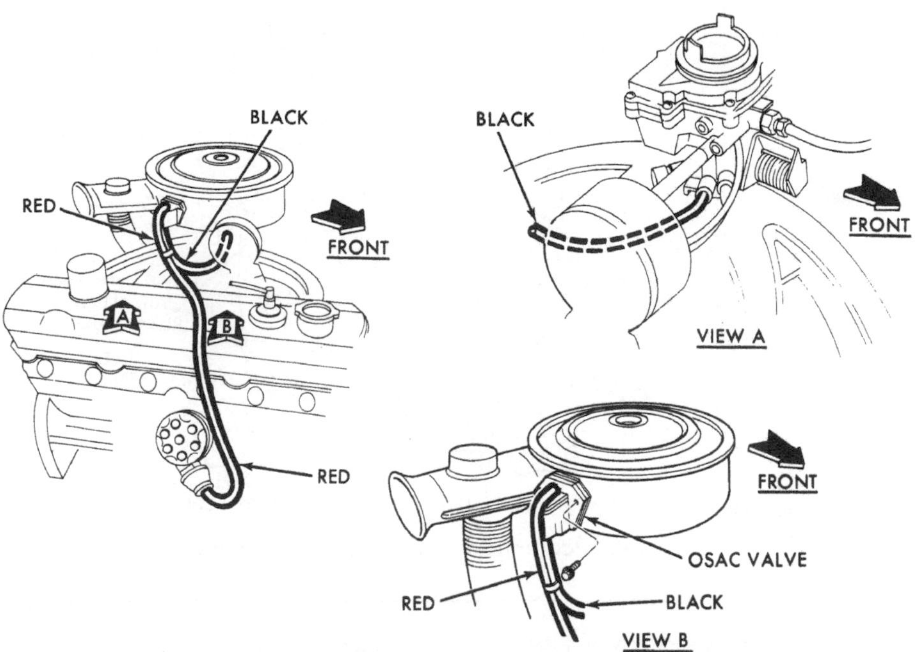

11.1a Typical OSAC system - six-cylinder engines

are equipped with a choke assist device to help reduce hydrocarbon and carbon monoxide emissions during engine warm up. An electric heat element, located next to the thermostatic coil spring in the choke well, assists engine heat to shorten choke duration **(see illustration)**.

2 Some models may employ a dual stage choke assist, which provides the choke heater with partial power below 58-degrees F and full power above 68 degrees F. The single stage system energizes the control switch above 68-degrees F. On both systems, the heater shuts off after the control switch reaches a temperature of 110 degrees F.

Checking

Single stage choke control switch

3 A test light will be needed to check the control switch.

4 With the engine off, remove the battery lead wire connector from the control switch.

5 Connect a test light to the load (small) terminal of the control switch.

6 Start the engine and allow it to warm up to operating temperature.

7 Apply 12-volts to the battery lead terminal of the control switch.

8 The test light should glow for a few seconds to several minutes. However, if the test light stays on for more than five minutes, the control switch is defective and must be replaced.

Dual stage choke control switch

9 The same procedure as above is used to check the dual stage control switch, except that two levels of bulb intensity should be observed.

10 Connect the test light directly to the battery terminals. Observe the intensity of the bulb.

11 Repeat Steps 4 through 8. When applying voltage to the circuit as described in Step 8, the test light bulb should initially show the same intensity as when connected directly to the battery.

Choke heating element

12 To test the choke heating element, an ohmmeter will be needed. Remove the battery lead connector from the control switch terminal (B+ terminal).

13 Connect one ohmmeter lead to the choke housing or choke retainer screw.

14 Touch the other ohmmeter lead to the metal portion of the choke wire connector at the control switch (not the B+ terminal).

15 A resistance of 4 to 12 ohms should be indicated. Meter readings indicating an open or short circuit are cause for installation of a new heating element.

11.1b Typical OSAC systems - V8 engines

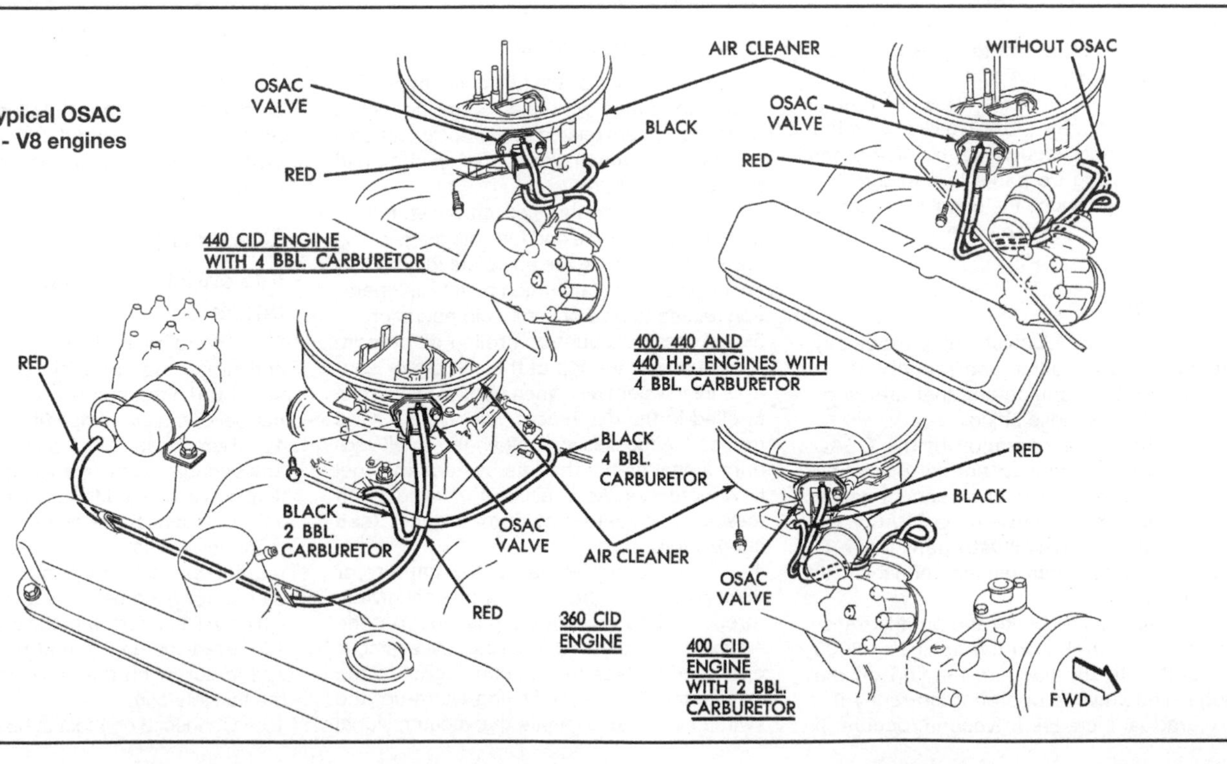

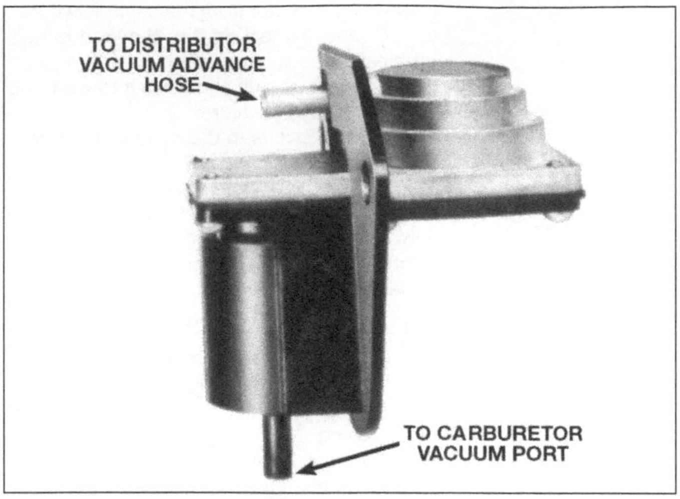

11.3 When checking OSAC valve operation, connect the vacuum gauge T-fitting to the hose between the upper OSAC valve port and the distributor vacuum advance unit

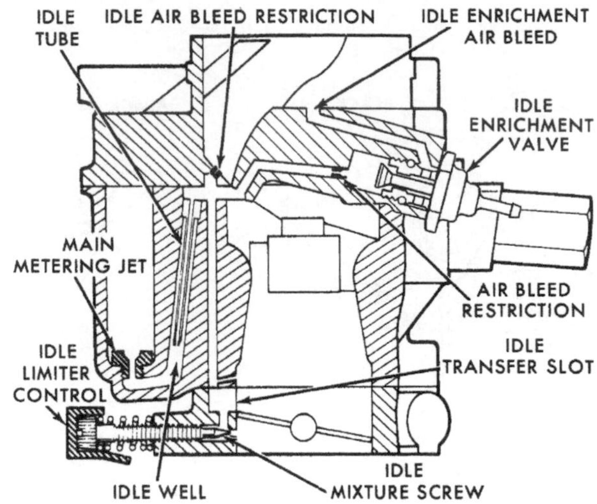

12.3 Idle enrichment control components (not all models)

Component replacement

16 The electric assist choke switch is mounted on the right side of the engine block, adjacent to the carburetor.

17 Disconnect the electrical lead and remove the two mounting bolts to remove the choke switch. Installation is the reverse of removal.

11 Orifice Spark Advance Control (OSAC)

Refer to illustrations 11.1a, 11.1b and 11.3

General description

1 The OSAC system was installed on some engines to aid in the control of NOx (oxides of nitrogen). This system controls the vacuum to the vacuum advance diaphragm of the distributor. The OSAC valve delays the change in ported vacuum to the distributor when transferring from idle to part throttle, and so helps to reduce the quantity of oxides of nitrogen produced by the engine **(see illustrations).**

Checking

2 Maintenance consists of periodically inspecting the condition and security of the vacuum hoses. Replace any that are hardened, cracked or loose fitting.

3 To check the operation of the OSAC valve, run the engine to normal operating temperature, then connect a vacuum gauge to the vacuum hose between the OSAC valve and the distributor **(see illustration)**. Use a T-fitting so the hose can remain connected at both components.

4 Set the parking brake and put the transmission in Neutral. Run the engine at 2000 rpm and note the vacuum reading on the gauge. The valve is operating correctly if a very gradual increase in vacuum occurs. If the vacuum increases suddenly or if there is no increase at all, replace the valve with a new one.

Component replacement

5 The vacuum hoses are simple to replace - just pull off the old one and install the new one. Replace one hose at a time to avoid mix-ups.

6 The OSAC valve is attached to the air cleaner housing with two screws.

12 Idle enrichment system

Refer to illustration 12.3

General description

1 The purpose of the idle enrichment system is to reduce cold engine stalling with a metering system related to the basic carburetor instead of the choke. It was used on mid to late 70's vehicles.

2 The system enriches carburetor mixtures in the curb idle and fast idle ranges. The carburetor will have the complete idle system enriched during that portion of vehicle operation related to cold or semi-cold operation.

3 A small vacuum-controlled diaphragm mounted near the top of the carburetor controls idle system air. When control vacuum is applied to the diaphragm, idle system air is reduced. Air losses strengthen the small vacuum signal within the idle system and fuel flow increases. As a result of more fuel and less air, the fuel/air mixture is richer **(see illustration)**.

4 The vacuum signal to the carburetor diaphragm is controlled by a thermal switch threaded into a coolant passage on the engine. When the engine is cold, the switch is open and passes the vacuum signal to the carburetor diaphragm. During warm-up, the switch closes to eliminate the vacuum signal and return carburetor metering to normal, lean levels. In one type system, the coolant vacuum switch also controls idle system enrichment duration.

5 In a second system, a similar coolant vacuum switch is used, but it receives its vacuum signal from a solenoid valve operated by an electric timer. Enrichment duration is approximately 35 seconds after the engine starts. The switch will prevent additional cycles of idle system enrichment after the engine reaches normal operating temperature, but while the engine is cold, each restart results in another 35 seconds of enrichment.

6 When the engine cools below the switch opening temperature, enriched idle will occur after restart. Duration of the enrichment will be controlled as described above for each type system.

7 The solenoid controlled system switch closes at approximately 150-degrees F, while others close at about 98-degrees F. The switches open approximately 12-degrees F below closing temperatures.

Checking

Idle enrichment system functional check

8 The following test is a check of the idle enrichment system within the carburetor.

9 Start the engine and allow it to reach normal operating temperature.

10 Remove the air cleaner but DO NOT cap any vacuum ports opened by hose removal (leakage is needed for this check).

11 Disconnect the hose to the idle enrichment valve diaphragm at the plastic connector (the connector must be removed from the carburetor hose because it is a filtered bleed which will interfere with the check).

12 Start the engine and place the fast idle speed screw on the slowest speed step of the fast idle cam.

13 Connect 3 or 4 feet of hose to the enrich-

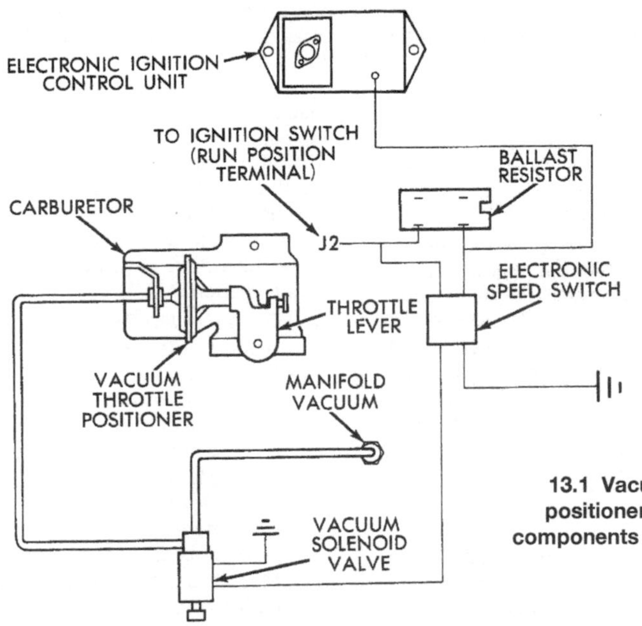

13.1 Vacuum throttle positioner and related components (not all models)

ment valve hose with a splice connector.

14 Apply vacuum to the other end of the hose with a hand vacuum pump and listen for an engine speed change. If speed can be controlled by vacuum, the diaphragm and idle enrichment valve are operating properly. If engine speed cannot be controlled, proceed to the next step (Carter carburetors) or replace the valve (Holley carburetors).

15 Place your finger over the idle air bleed passage and listen for an engine speed change. If engine speed can be controlled, the diaphragm is leaking or the valve is stuck open. If engine speed cannot be controlled, the valve is stuck closed. In either case, clean the valve and repeat Step 14. If engine speed still cannot be controlled, replace the idle enrichment valve.

Idle enrichment time delay check

16 In some applications, the timing module and solenoid valve serve the dual function of controlling both EGR delay and idle enrichment duration. Prior to testing the timer, the wires should be checked for proper connections and repairs made if necessary.

17 With the ignition switch off, detach the wiring connector from the time delay solenoid valve.

18 Connect a self-powered test light across the connector terminals. **Note:** *In order to avoid timer overload, test light current draw should not exceed 0.5 amps. A typical automotive 6-12 volt test light is satisfactory. If the light is shop built, instrument panel size bulbs or smaller should be used.*

19 Start the engine. The light should come on and remain on for 60 seconds after the engine starts (30 to 40 seconds on 1976 models).

20 If the light does not come on, remains on indefinitely or does not conform to delay times described in Step 19, the timer should be replaced.

21 If the timer is replaced, repeat the check on the new one.

Coolant Control Idle Enrichment (CCIE) valve check

22 Check all vacuum hoses for proper routing and valve installation. Check engine coolant level (Chapter 1).

23 With the valve installed properly, disconnect the molded connector from the valve. Attach a 1/8-inch ID hose to the bottom port of the valve.

24 With the radiator top tank warm to the touch (no warmer than 75-degrees F), blow through the hose. If it is not possible to blow through the hose/valve, it is defective and must be replaced.

25 Bring the engine to normal operating temperature. Attach a vacuum pump and gauge to the bottom port of the valve. Apply a 10-inch vacuum signal. If the vacuum level drops more than one inch in 15 seconds, replace the valve.

13 Vacuum throttle positioner

Refer to illustrations 13.1 and 13.11

General description

1 Some carburetors are equipped with a throttle positioner system which consists of an electronic speed switch, an electronically controlled vacuum solenoid valve and a vacuum-actuated throttle positioner **(see illustration)**.

2 The system's function is to prevent unburned hydrocarbons from entering the atmosphere through the vehicle's exhaust system when the throttle is closed after running at high RPM. The electronic speed switch receives ignition pulses from the five ohm ballast resistor terminal connected to the electronic ignition control unit. It then senses when the engine speed exceeds 2000 RPM and allows vacuum to energize the throttle positioner. When the positioner is energized, it prevents the throttle from returning to the idle position. When the throttle is released, it will return to the new stop position (positioner energized) calibrated for a steady-state engine speed of 1750 RPM. As the engine is decelerating, the electronic speed switch senses when the engine speed drops below 2000 RPM and de-energizes the throttle positioner. This allows the throttle to return to the normal idle stop position and the engine will continue to decelerate to idle speed. This operation positions the throttle partially open (1750 RPM) whenever the engine decelerates from a speed above 2000 RPM to a speed just below 2000 RPM. This provides sufficient air flow through the engine to adequately dilute the fuel/air mixture.

Checking

3 Check all wiring harness and hose connections to make sure they are secure.

4 Apply vacuum from an external source to the positioner port. If the plunger does not extend, replace the positioner. If it does extend, pinch off the vacuum hose and watch the plunger. If it remains extended for one minute or more, the positioner is in good condition.

5 Apply vacuum from an external source to the manifold vacuum hose connection on the solenoid valve. Disconnect the wiring harness from the solenoid and ground one solenoid terminal with a jumper wire. Apply battery voltage to the other terminal and watch the positioner plunger. If the plunger does not extend, replace the solenoid. If operation is normal, replace the speed switch.

Adjustment

6 It is essential that the curb idle speed adjustment be made using only the curb idle speed screw on the carburetor. Curb idle speed adjustment must not be made using the adjustment for the vacuum throttle positioner. Improper adjustment of the throttle positioner may result in excessive curb idle speed.

7 When finished with idle speed or throttle positioner adjustments, the engine speed must be increased to above 2500 RPM to check for proper idle return.

8 Start the engine and allow it to idle in Neutral, then increase engine speed to above 2000 RPM and verify that the vacuum positioner unit plunger extends. Try to retract it by hand - resistance should be encountered. If the positioner plunger does not extend, refer to the checking procedure above.

9 If the plunger operates normally, adjust the positioner as follows:

10 Increase engine speed to approximately 2500 RPM.

11 Loosen the positioner adjustment lock-

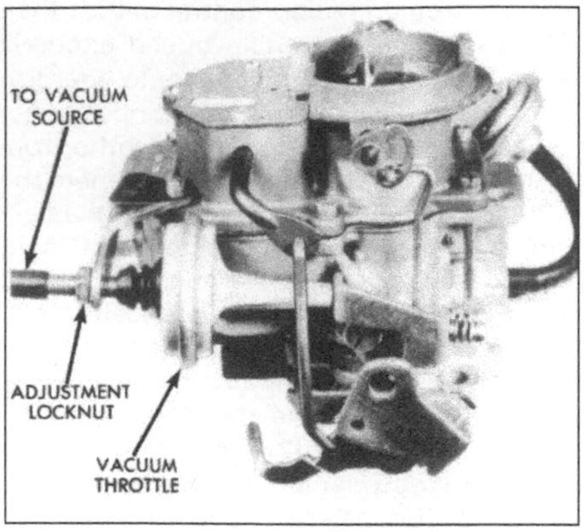

13.11 The vacuum throttle positioner can be turned after loosening the adjustment locknut

nut and rotate the throttle positioner assembly until the plunger just contacts the throttle lever **(see illustration)**.

12 Release the throttle, then slowly turn the positioner in the opposite direction (away from the throttle lever) until a sudden drop in speed occurs (over 1000 RPM). At this point, turn the positioner an additional 1/4-turn and tighten the locknut. Increase the engine speed to approximately 2500 RPM and release the throttle; the engine should return to normal idle.

14 Catalytic converter

General description

1 The catalytic converter is an emission control device added to the exhaust system to reduce pollutants from the exhaust gas stream. There are two types of converters. The conventional oxidation catalyst reduces the levels of hydrocarbon (HC) and carbon monoxide (CO). The three-way catalyst lowers the levels of nitrogen oxides (NOx) as well as hydrocarbons (HC) and carbon monoxide (CO).

Checking

2 The test equipment for a catalytic converter is expensive and highly sophisticated. If you suspect that the converter on your vehicle is malfunctioning, take it to a dealer or authorized emissions inspection facility for diagnosis and repair.

3 Whenever the vehicle is raised for servicing of underbody components, check the converter for leaks, corrosion, dents and other damage. Check the welds/flange bolts that attach the front and rear ends of the converter to the exhaust system. If damage is discovered, the converter should be replaced.

4 Although catalytic converters don't break too often, they do become plugged. The easiest way to check for a restricted converter is to use a vacuum gauge to diagnose the effect of a blocked exhaust on intake vacuum.

- a) *Open the throttle until the engine speed is about 2000 RPM.*
- b) *Release the throttle quickly.*
- c) *If there is no restriction, the gauge will quickly drop to not more than 2 in Hg or more above its normal reading.*
- d) *If the gauge does not show 5 in Hg or more above its normal reading, or seems to momentarily hover around its highest reading for a moment before it returns, the exhaust system, or the converter, is plugged (or an exhaust pipe is bent or dented, or the core inside the muffler has shifted).*

Component replacement

5 Refer to Chapter 4.

Chapter 7 Part A
Manual transmission

Contents

Specifications

General

Transmission lubricant type	See Chapter 1

Torque specifications

	Ft-lbs
Shift linkage (column shift models)	
Swivel bolts	
1971 through 1974	100 in-lbs
1975 and later	125 in-lbs
Transmission-to-clutch housing bolts	50

1 General information

Vehicles covered by this manual are equipped with a three or four-speed manual or a three-speed automatic transmission. Information on the manual transmission is included in this Part of Chapter 7. Information on the automatic transmission can be found in Part B of this Chapter.

The models covered by this Chapter use either a three or four-speed manual transmission. Several different three-speed units have been used: The A-230 three-speed is used on various 1971 through 1980 vehicles; the A-250 is used on some 1973 and 1974 vehicles; the A-903 is used on some 1971 and 1972 vehicles; the A-390 is used on some 1975 and 1976 vehicles. The A-250 and the A-903 are identical in design. There are also a couple of four-speeds: The A-833 is used on some 1971 through 1975 vehicles; the "Overdrive-4" is an option for some 1976 through 1980 vehicles. These two four-speed transmissions are virtually identical in design, except that the Overdrive-4 unit uses an overdrive fourth gear.

Depending on the expense involved in having a transmission overhauled, it may be a better idea to consider replacing it with either a used or rebuilt one. Your local dealer or transmission shop should be able to supply information concerning cost, availability and exchange policy. Regardless of how you decide to remedy a transmission problem, you can still save a lot of money by removing and installing the unit yourself.

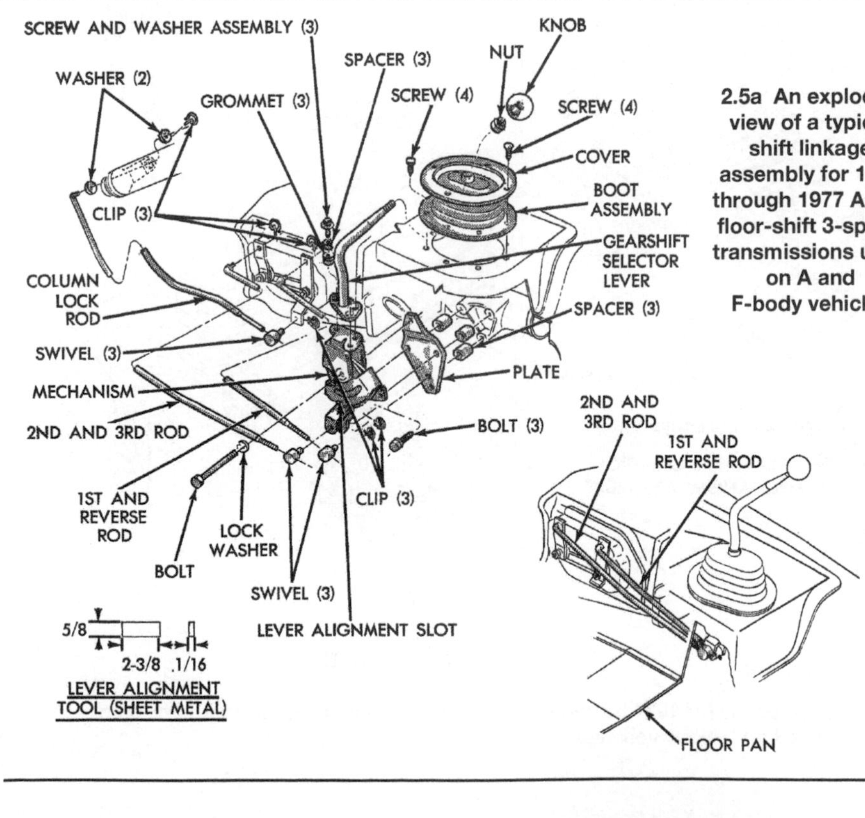

2.5a An exploded view of a typical shift linkage assembly for 1971 through 1977 A-230 floor-shift 3-speed transmissions used on A and F-body vehicles

2 Shift linkage - removal, installation and adjustment

Removal

Refer to illustrations 2.5a, 2.5b, 2.5c, 2.5d, 2.5e and 2.5f

Floor-shift models

1 Disconnect the negative cable from the battery.

2 On three-speed models, loosen the locknut and unscrew the shift knob.

3 Remove the console (see Chapter 11).

4 Remove the retaining screws from the floor pan boot and remove the boot.

5 Remove the shift lever mounting bolts and remove the shift lever **(see illustrations)**.

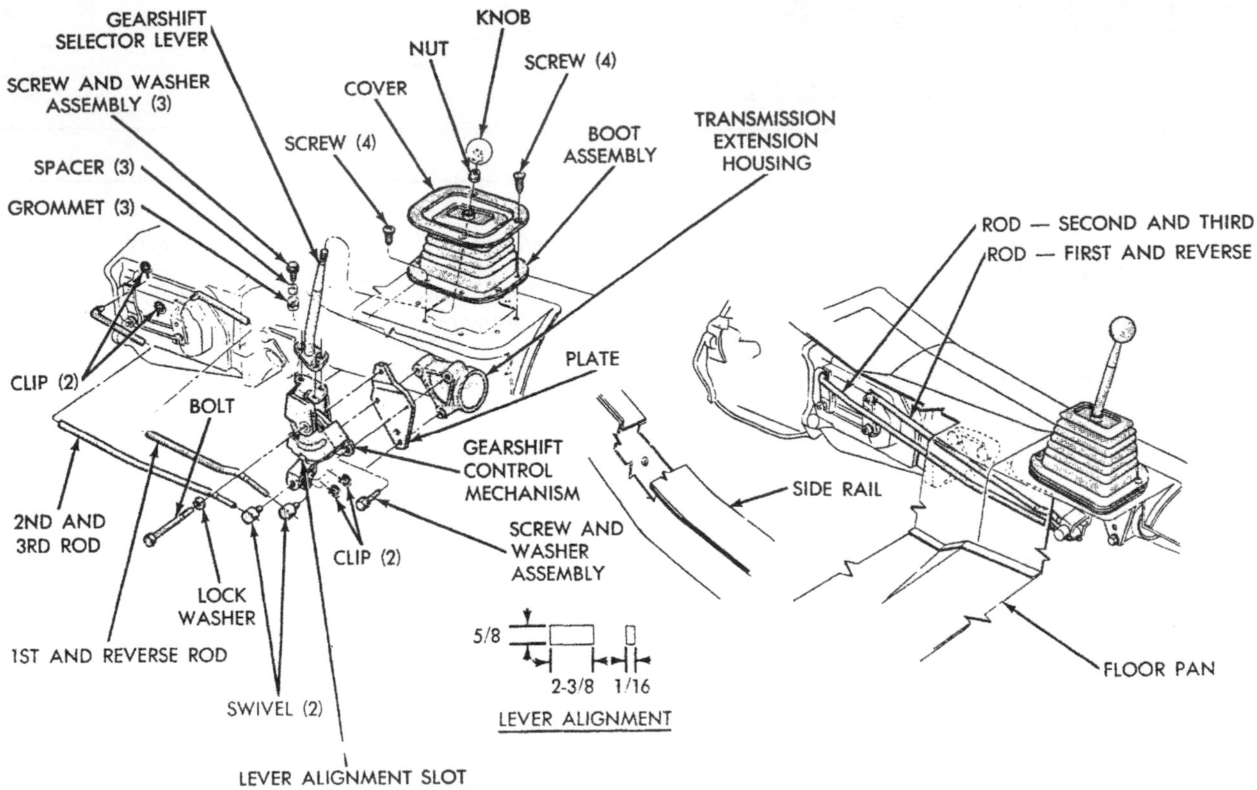

2.5b An exploded view of a typical shift linkage assembly for 1971 through 1974 A-230 floor-shift 3-speed transmission used on E and R-body vehicles

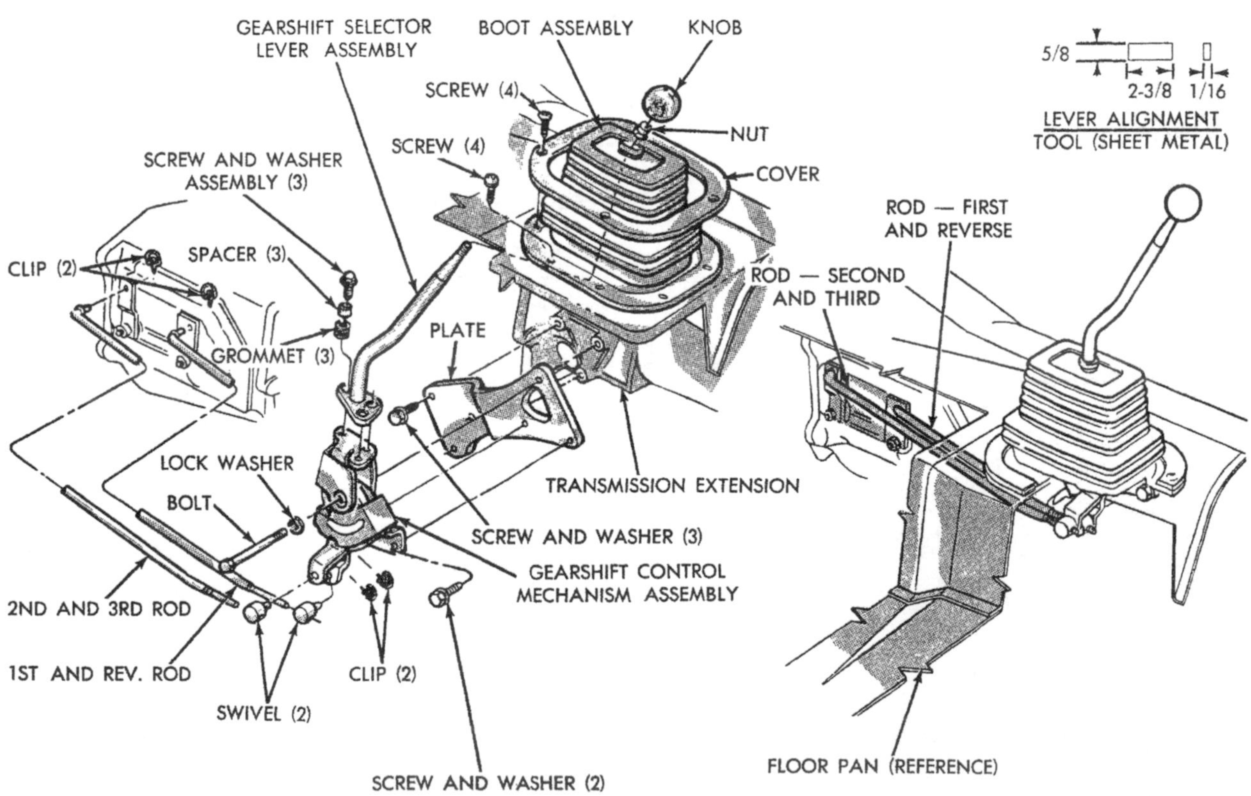

2.5c An exploded view of a typical shift linkage assembly for 1971 through 1974 A-230 floor-shift 3-speed transmission used on B-body vehicles

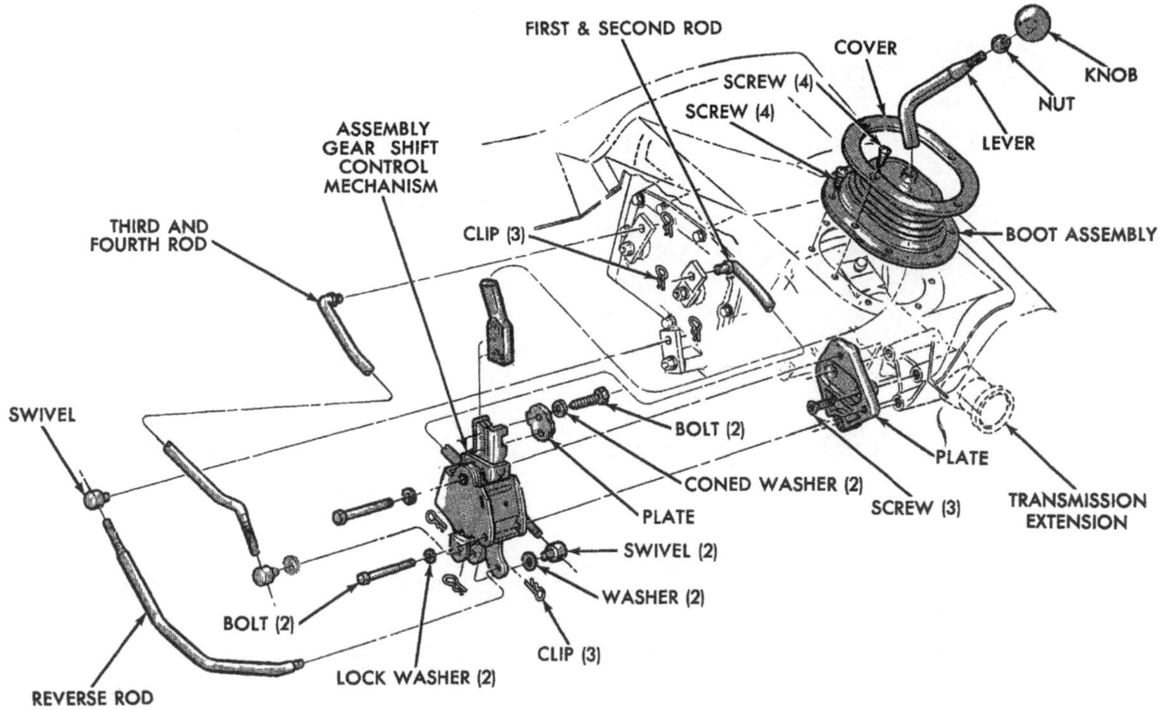

2.5d An exploded view of typical shift linkage assembly for 1971 through 1975 A-833 floor-shift 4-speed transmission used on A-body vehicles; also the linkage for Overdrive-4floor-shift 4-speed transmission, which was available as special equipment on some 1976 through 1980 vehicles

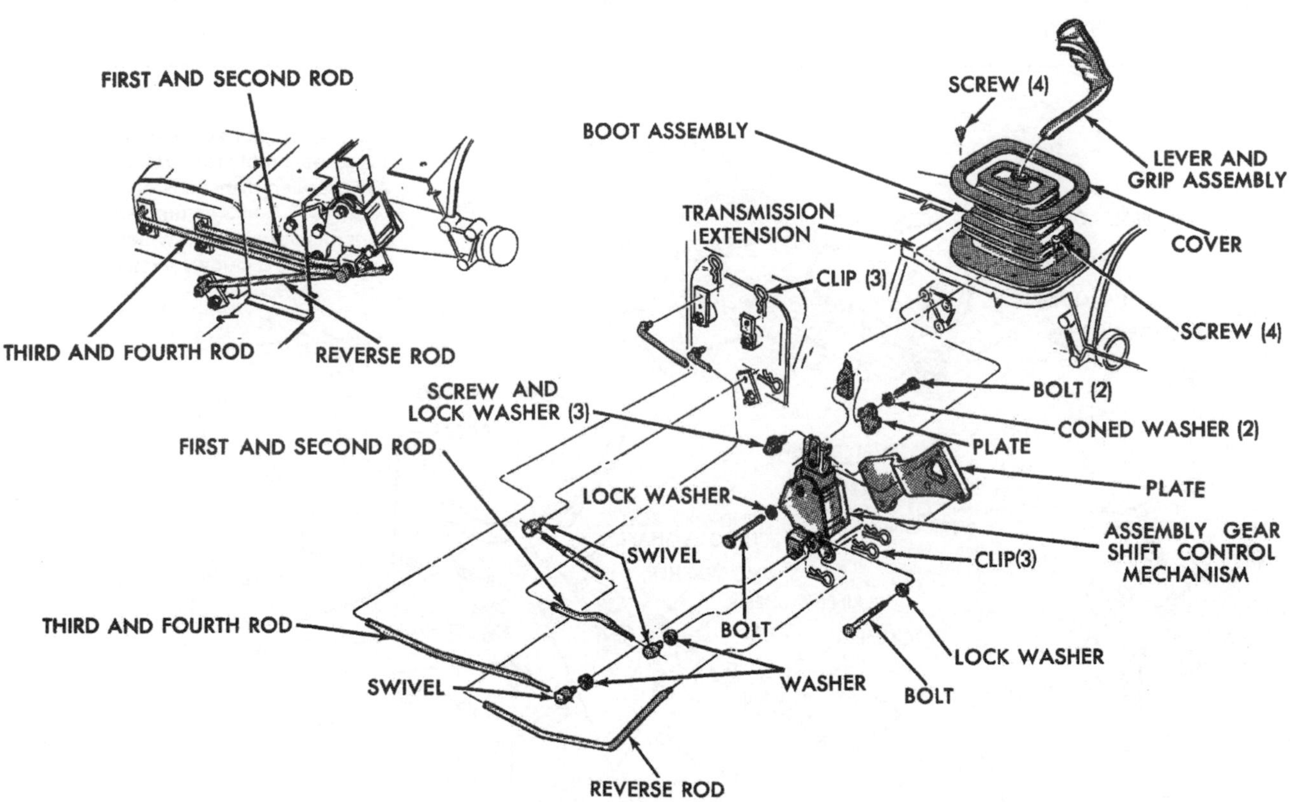

2.5e An exploded view of typical shift linkage assembly for 1971 through 1975 A-833 floor-shift 4-speed transmission used on B-body vehicles

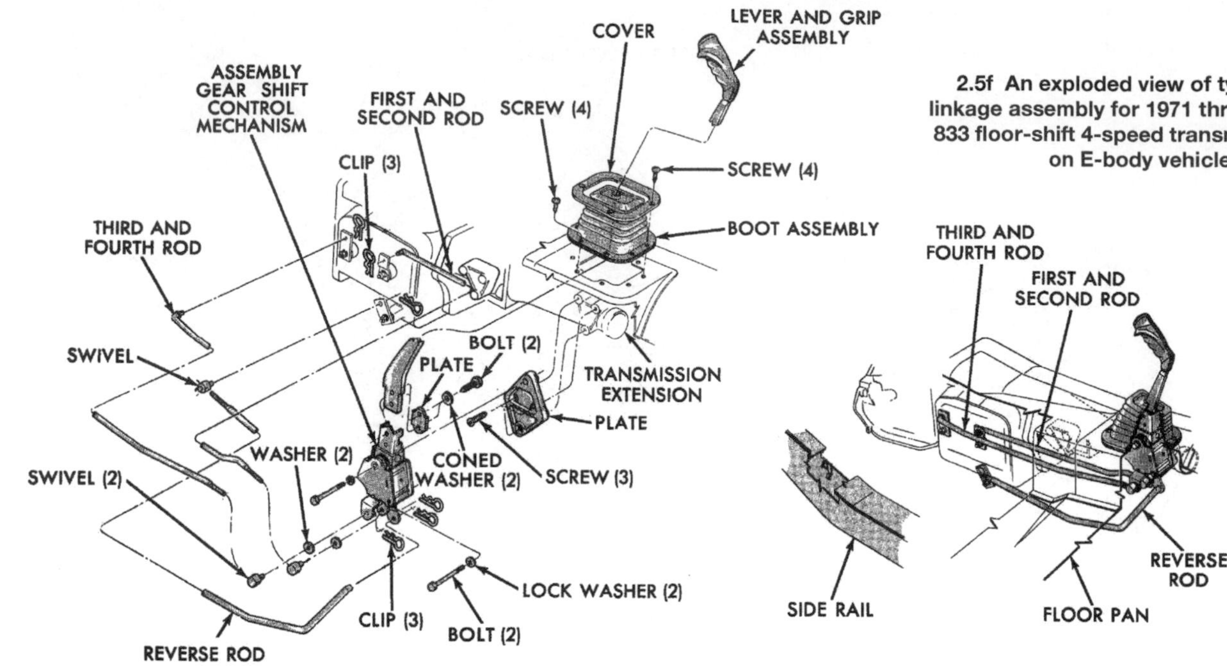

2.5f An exploded view of typical shift linkage assembly for 1971 through 1975 A-833 floor-shift 4-speed transmission used on E-body vehicles

6 Remove the retaining clips, washers and shift rods (three-speeds) or control rods (four-speeds) from the shift mechanism levers under the floor pan. Remove the bolts and washers which secure the shift mechanism to the transmission extension housing mounting plate and remove the unit.

Column-shift models

Refer to illustrations 2.7a, 2.7b, 2.7c, 2.7d, 2.7e and 2.7f

7 On column-shift models, the shifter assembly is integral with the steering column. The mechanism in the column should not normally require service; however, the bushings at the ends of the shift rods frequently become worn, causing difficult shifting. To replace the bushings, remove the clips (if equipped) and pull or pry the rods out of the levers at the steering column or transmission **(see illustrations)**. When removing the transmission, disconnect the shift rods at the transmission shift levers.

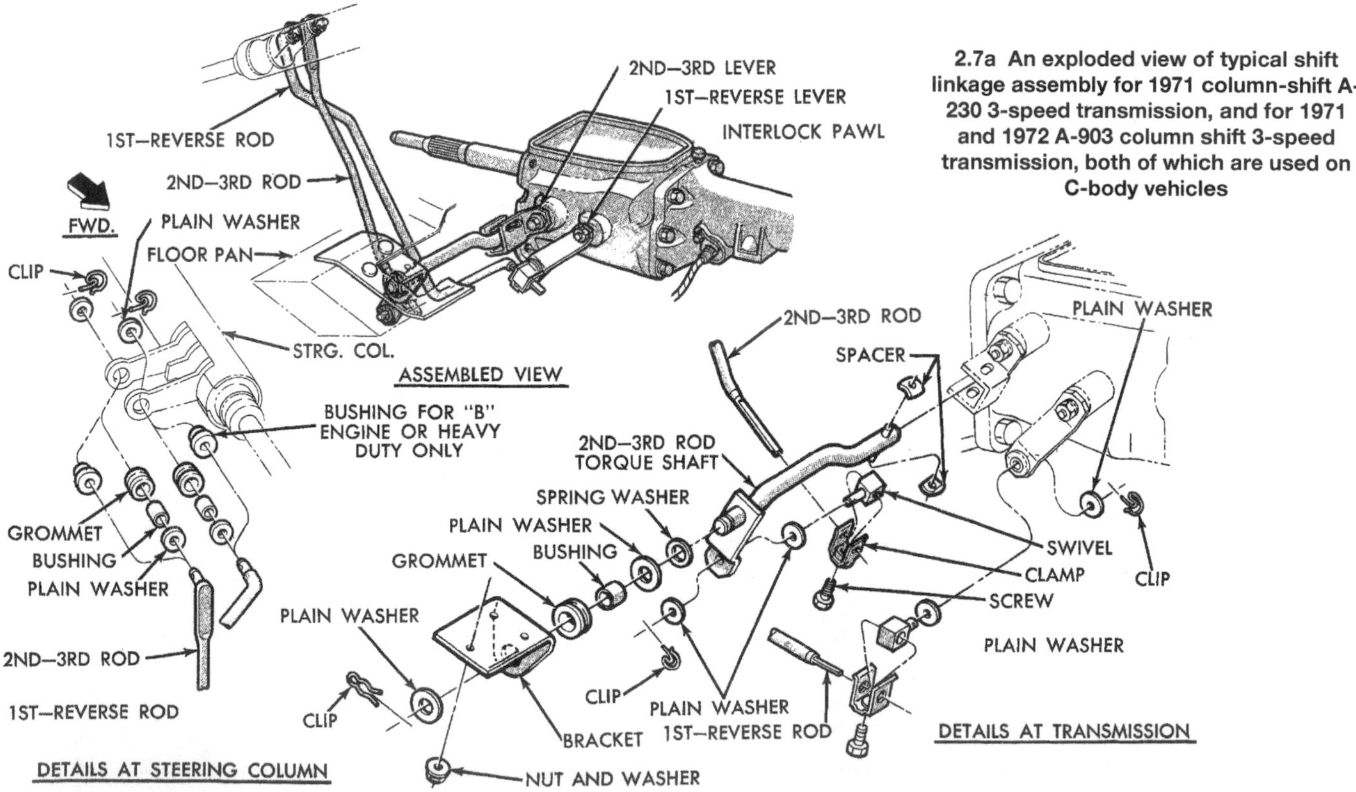

2.7a An exploded view of typical shift linkage assembly for 1971 column-shift A-230 3-speed transmission, and for 1971 and 1972 A-903 column shift 3-speed transmission, both of which are used on C-body vehicles

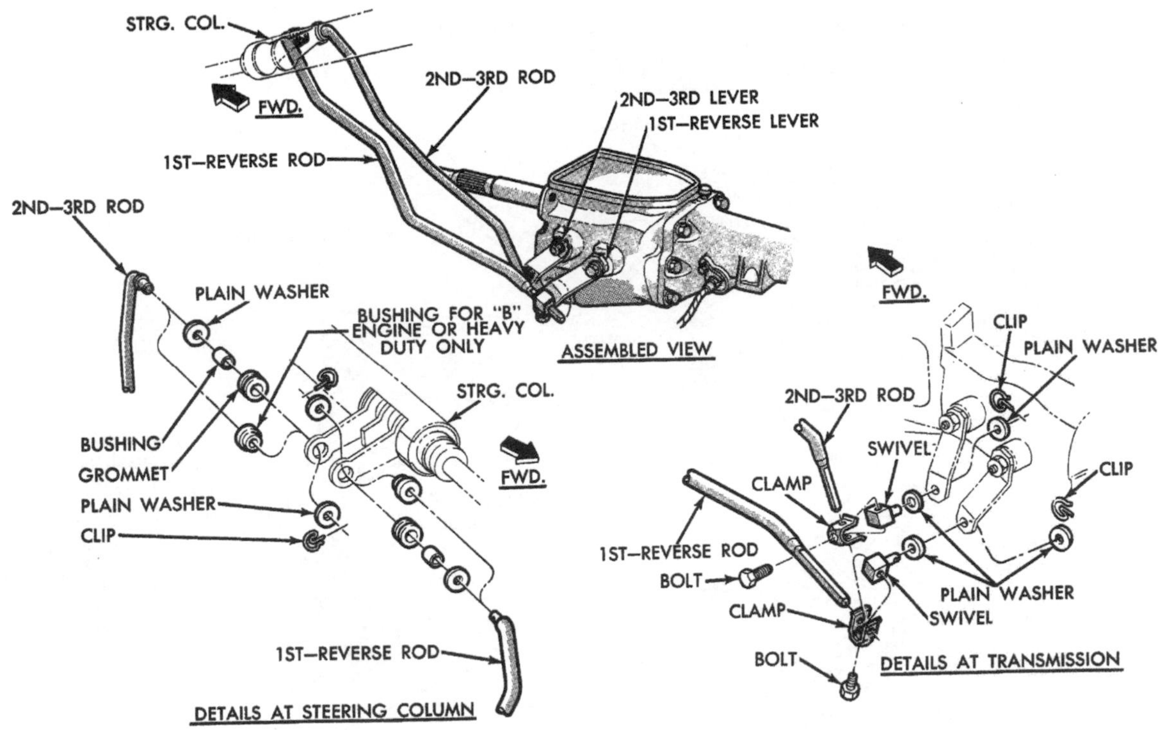

2.7b An exploded view of typical shift linkage assembly for 1971 column-shift A-230 3-speed transmission used on A-body vehicles

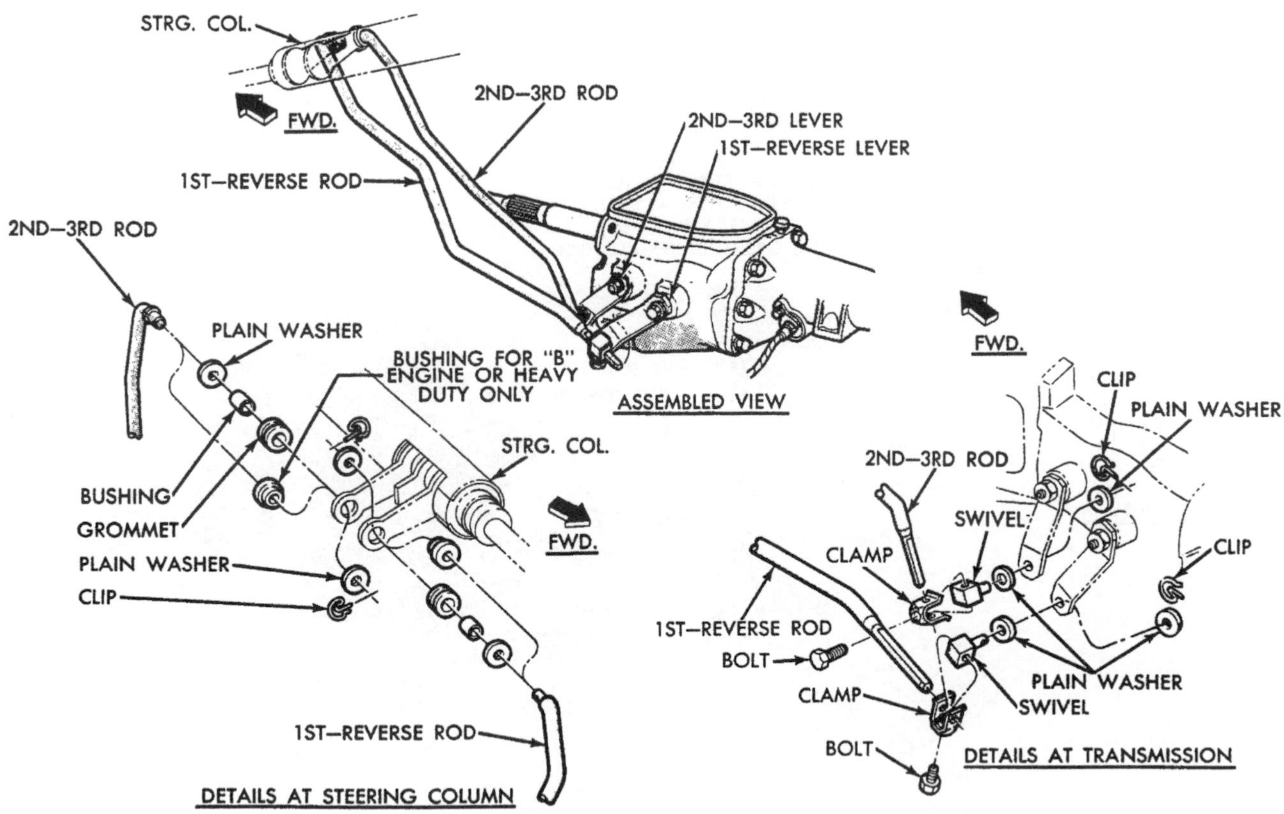

2.7c An exploded view of a typical shift linkage assembly for 1971 and 1972 A-230 column-shift 3-speed transmission, and for 1971 and 1972 A-903 column-shift 3-speed transmission, both of which are used on B-body vehicles

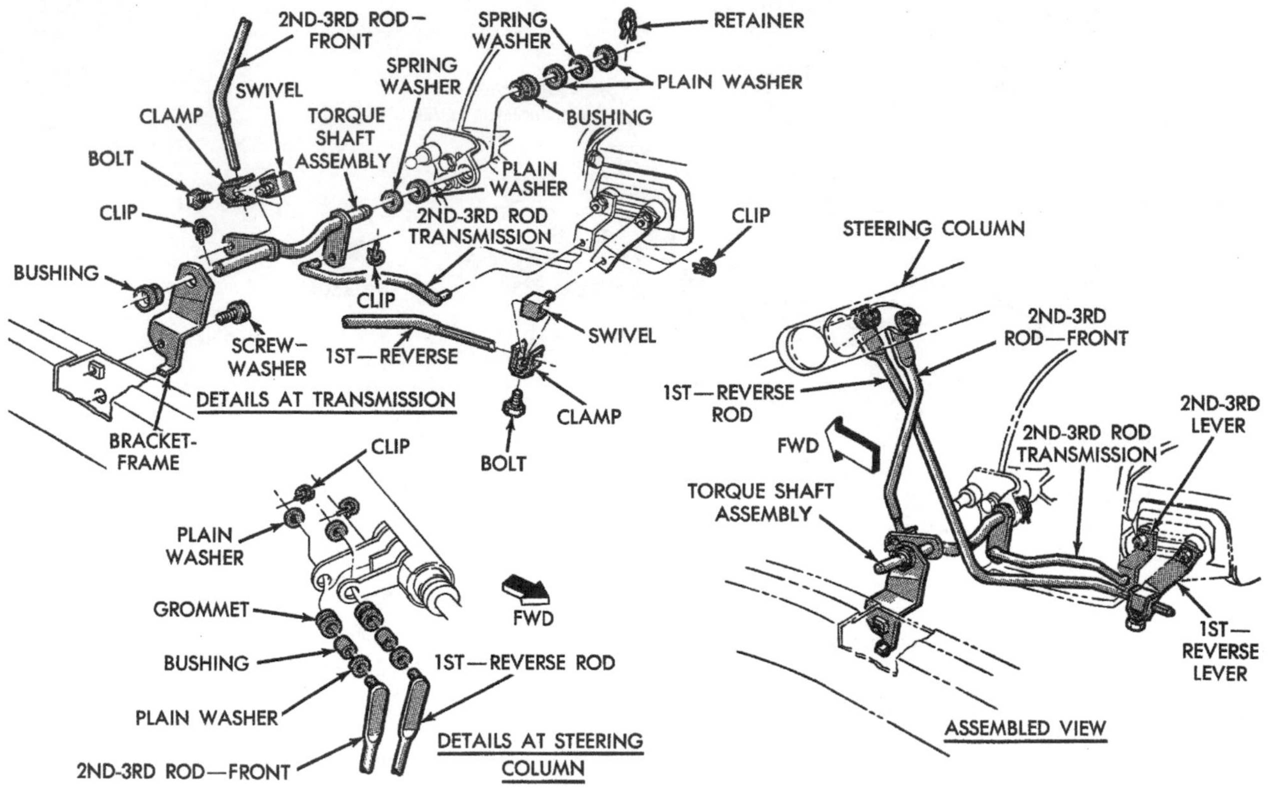

2.7d An exploded view of typical shift linkage assembly for 1972 through 1975 A-230 column-shift 3-speed transmission, and for 1973 and 1974 A-250 column-shift 3-speed transmission, both of which are used on A-body vehicles

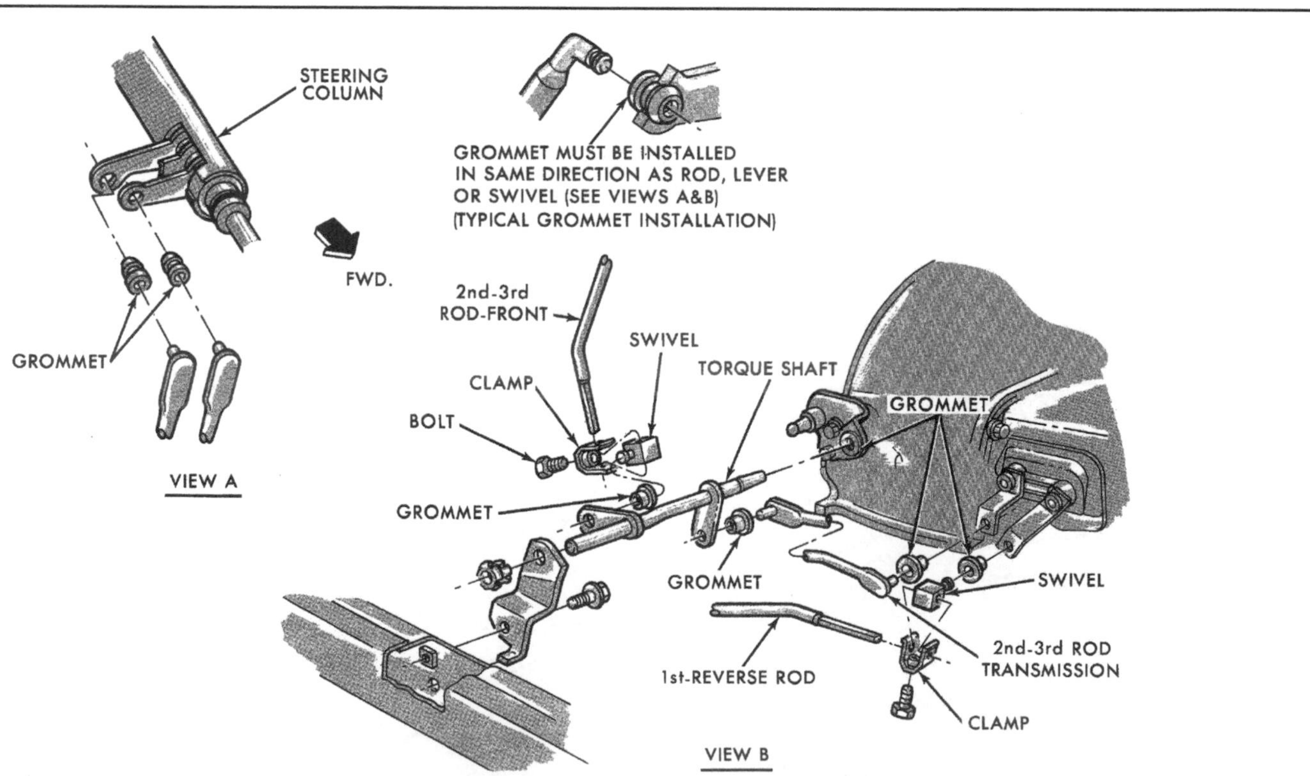

2.7e An exploded view of typical shift linkage assembly for 1976 through 1980 A-230 column-shift 3-speed transmission, and for 1975 and1976 A-390 column-shift 3-speed transmission,both of which are used on A and F-body vehicles

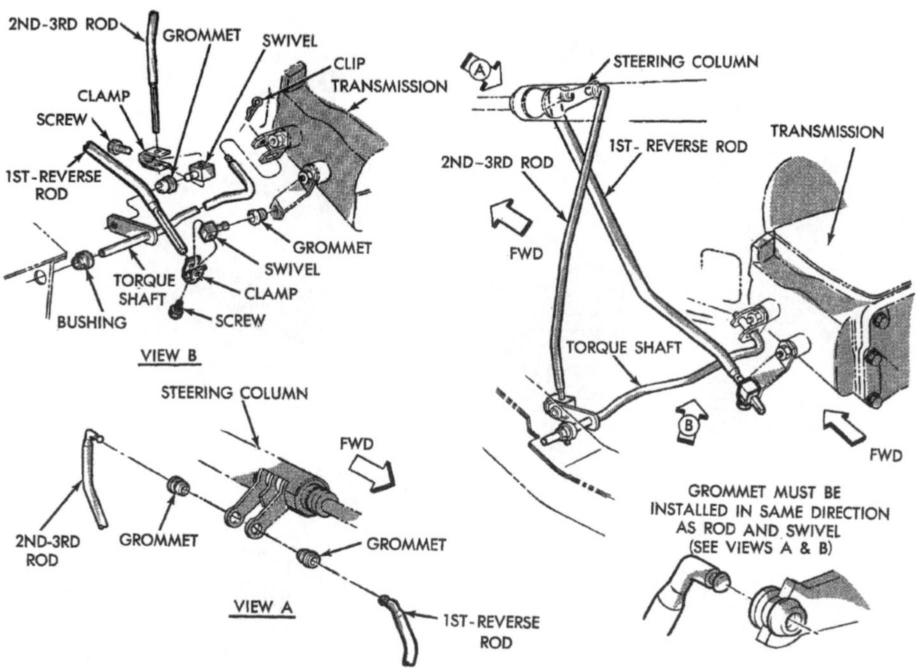

2.7f An exploded view of typical shift linkage assembly for 1973 and later A-230 column-shift 3-speed transmission used on B-body vehicles

Installation (column-shift models)

8 Position the shift mechanism on the transmission extension housing mounting plate and secure it with the three mounting bolts (three-speeds) or two bolts and lock washers (four-speeds). On three-speed models, if there's a lower boot, place it on the mechanism first, then insert the long mounting bolt through the unit and boot before attaching it to the mounting plate.

9 On four-speed models, install the shift rods, washers and clips.

10 Install the shift lever.

11 Adjust the linkage (see below).

Adjustment

Column-shift models

Refer to illustration 2.15

12 On 1971 through 1974 models, remove both shift rod swivels from the transmission shift levers; on 1975 and later models, loosen - but don't remove - the shift rod swivels.

13 Put the transmission shift levers in the Neutral (the middle detent) position.

14 Move the shift lever to align the locating slots in the bottom of the steering column shift housing and the bearing housing. Install a suitable tool in the slot and lock the ignition switch.

15 Place a screwdriver between the crossover blade and the 2nd-3rd lever at the steering column so that both lever pins are engaged by the crossover blade **(see illustration)**.

1971 through 1974 models

16 Set the 1st-Reverse lever on the transmission to the Reverse position by rotating it clockwise.

17 Adjust the 1st-Reverse rod swivel by loosening the clamp bolt and sliding swivel along the rod so it will enter the 1st-Reverse lever at the transmission. Install the washers and clip. Tighten the swivel bolt to the torque listed in this Chapter's Specifications.

18 Remove the gearshift housing locating tool and unlock the ignition switch and shift column lever to the Neutral position.

19 Adjust the 2nd-3rd rod swivel by loosening the clamp bolt and sliding swivel along the rod so it will enter the 2nd-3rd lever at the transmission. Install the washers and the clip. Tighten the swivel bolt to the torque listed in this Chapter's Specifications.

2.15 Place a screwdriver between the crossover blade and the 2nd-3rd lever at the steering column so that both lever pins are engaged by the crossover blade

1975 and later models

20 Tighten both shift rod swivel bolts to the torque listed in this Chapter's Specifications.

21 Remove the gearshift housing locating tool.

All models

22 Remove the tool from the crossover blade at the steering column and shift through all the gears to check the adjustment and crossover smoothness.

23 Check for proper operation of the steering column lock in Reverse and in the other gear positions. With proper linkage adjustment, the column should lock only in the Reverse position.

Floor-shift models

Three-speed models

24 Fabricate a lever alignment tool from 1/16-inch thick sheet metal to the dimensions shown in the accompanying illustrations.

25 Insert the lever alignment tool into the slots in the levers and the mechanism frame to hold the levers in the Neutral crossover position.

26 Place the transmission levers in the Neutral (the middle detent) position and adjust the shift rod swivels so the rod will install freely to the levers.

27 Secure the rods with the proper washers and clips and remove the alignment tool. If there's a shift boot, work the flange portion of the lower boot up above the floor pan opening.

Four-speed models

Refer to illustration 2.28

28 Install the floor shift lever aligning tool to hold the levers in the Neutral-crossover position. For A-833 units, you'll need a special tool, available at most auto parts stores, or its equivalent to do the job. For Overdrive-4 units, you can make your own aligning tool out of 1/4-inch diameter rod **(see illustration)**.

29 With all rods removed from the transmission shift levers, place the levers in the Neutral detent position.

30 Rotate the threaded shift rods until their length is correct for entering the transmission levers. Start with the 1-2 shift rod (you may

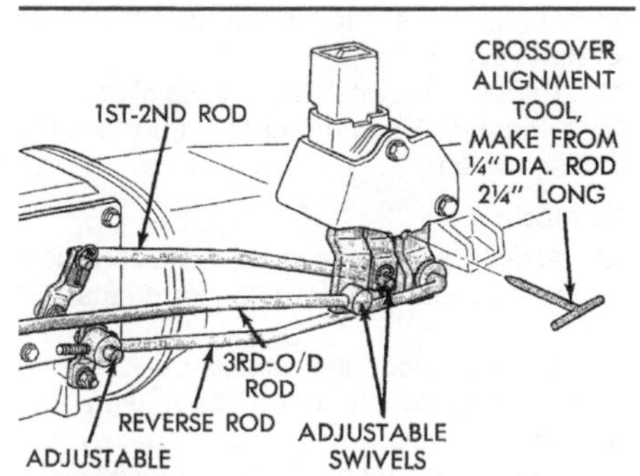

2.28 For Overdrive-4 units, you can make your own 2-1/4 inch long tool out of 1/4-inch diameter rod

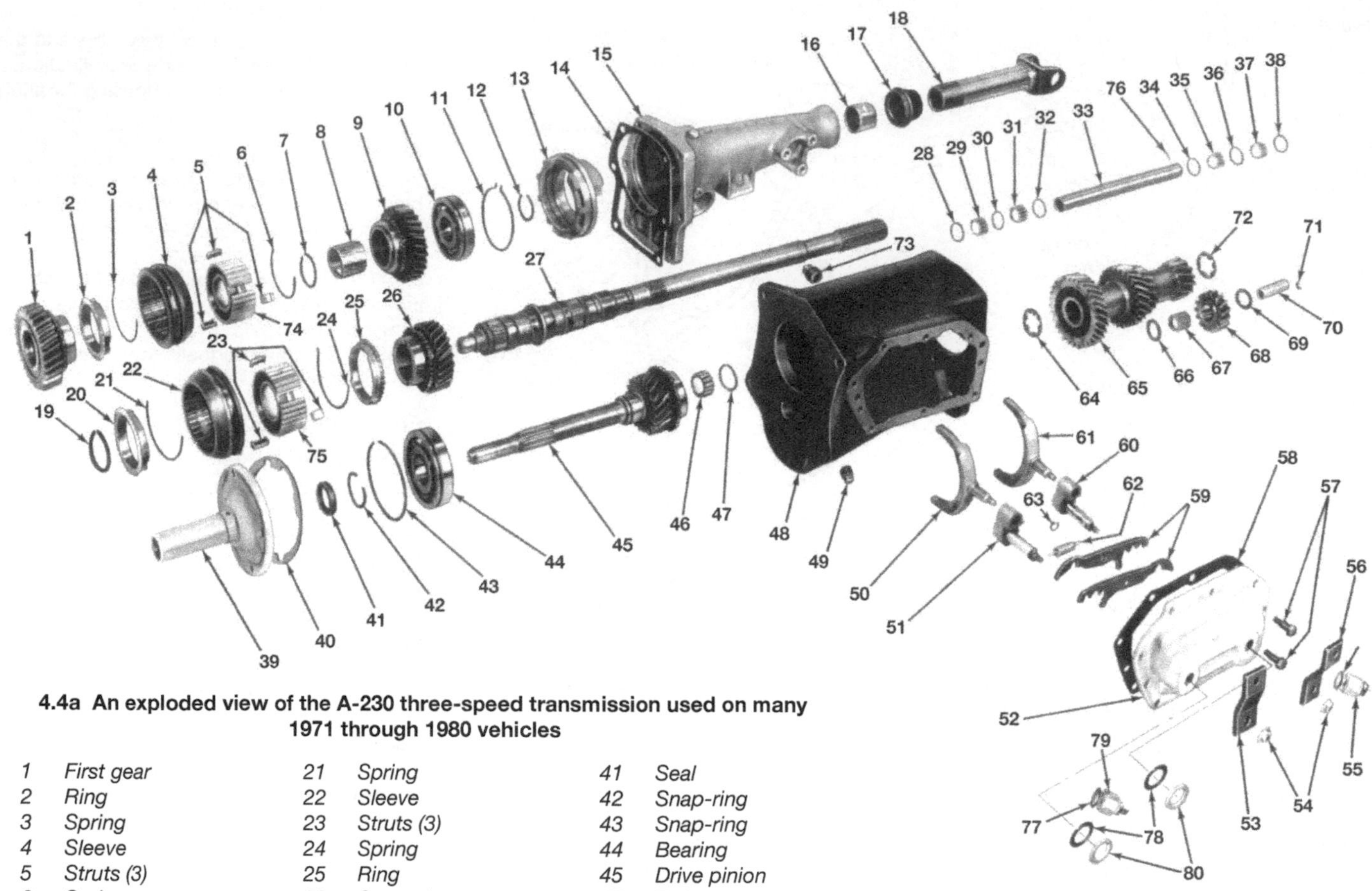

4.4a An exploded view of the A-230 three-speed transmission used on many 1971 through 1980 vehicles

1	*First gear*	21	*Spring*	41	*Seal*	61	*Fork*
2	*Ring*	22	*Sleeve*	42	*Snap-ring*	62	*Spring*
3	*Spring*	23	*Struts (3)*	43	*Snap-ring*	63	*Snap-ring*
4	*Sleeve*	24	*Spring*	44	*Bearing*	64	*Washer*
5	*Struts (3)*	25	*Ring*	45	*Drive pinion*	65	*Countershaft gear*
6	*Spring*	26	*Second gear*	46	*Roller*	66	*Washer*
7	*Snap-ring*	27	*Output shaft*	47	*Snap-ring*	67	*Roller*
8	*Bushing*	28	*Washer*	48	*Case*	68	*Idler gear*
9	*Reverse gear*	29	*Roller*	49	*Drain plug*	69	*Washer*
10	*Bearing*	30	*Washer*	50	*Fork*	70	*Shaft*
11	*Snap-ring*	31	*Roller*	51	*Lever*	71	*Key*
12	*Snap-ring*	32	*Washer*	52	*Housing*	72	*Washer*
13	*Retainer*	33	*Countershaft*	53	*Lever*	73	*Filler plug*
14	*Gasket*	34	*Washer*	54	*Locking nut*	74	*Clutch gear*
15	*Extension housing*	35	*Roller*	55	*Switch*	75	*Clutch gear*
16	*Bushing*	36	*Washer*	56	*Lever*	76	*Key*
17	*Seal*	37	*Roller*	57	*Bolt*	77	*Exhaust emission switch gasket (1971 and 1972 vehicles only)*
18	*Yoke*	38	*Washer*	58	*Gasket*	78	*O-rings (2)*
19	*Snap-ring*	39	*Retainer*	59	*Interlock lever*	79	*Exhaust emission switch (1971 and 1972 vehicles only)*
20	*Ring*	40	*Gasket*	60	*Lever*	80	*O-ring retainers (2)*

have to pull the clip at the shifter end to rotate this rod).

31 Replace the washers and the clips.

32 Remove the aligning tool and test the shifting action.

All models

33 Attach the shift lever to the mechanism with three mounting screws, grommets and spacers.

34 Slide the boot and retainer over the shift lever and fasten it to the floor with four screws.

35 Install the console (see Chapter 11).

36 On three-speed models, install the shift knob and lock nut on the shift lever.

37 Connect the negative battery cable.

38 Test the shifting action for smoothness.

3 Transmission - removal and installation

Removal

1 Disconnect the negative cable from the battery,

2 Raise the front end of the vehicle on a jack. Raise the vehicle sufficiently to provide clearance to easily remove the transmission.

3 Support the vehicle securely on jackstands.

4 Disconnect the shift levers or rods from the transmission levers (see Section 2).

5 Disconnect the speedometer cable (see Chapter 7B), then unplug the back-up light switch connector.

6 Drain the lubricant from the transmission (see Chapter 1).

7 Remove the driveshaft (see Chapter 8). Use a plastic bag to cover the end of the transmission to prevent fluid loss and contamination.

8 Remove the exhaust system components as necessary for clearance (see Chapter 4).

9 Support the engine. This can be done from above with an engine hoist, or by placing a jack (with a block of wood as an insulator)

under the engine oil pan. The engine should remain supported at all times while the transmission is out of the vehicle.

10 Support the transmission with a jack - preferably a special jack made for this pur-

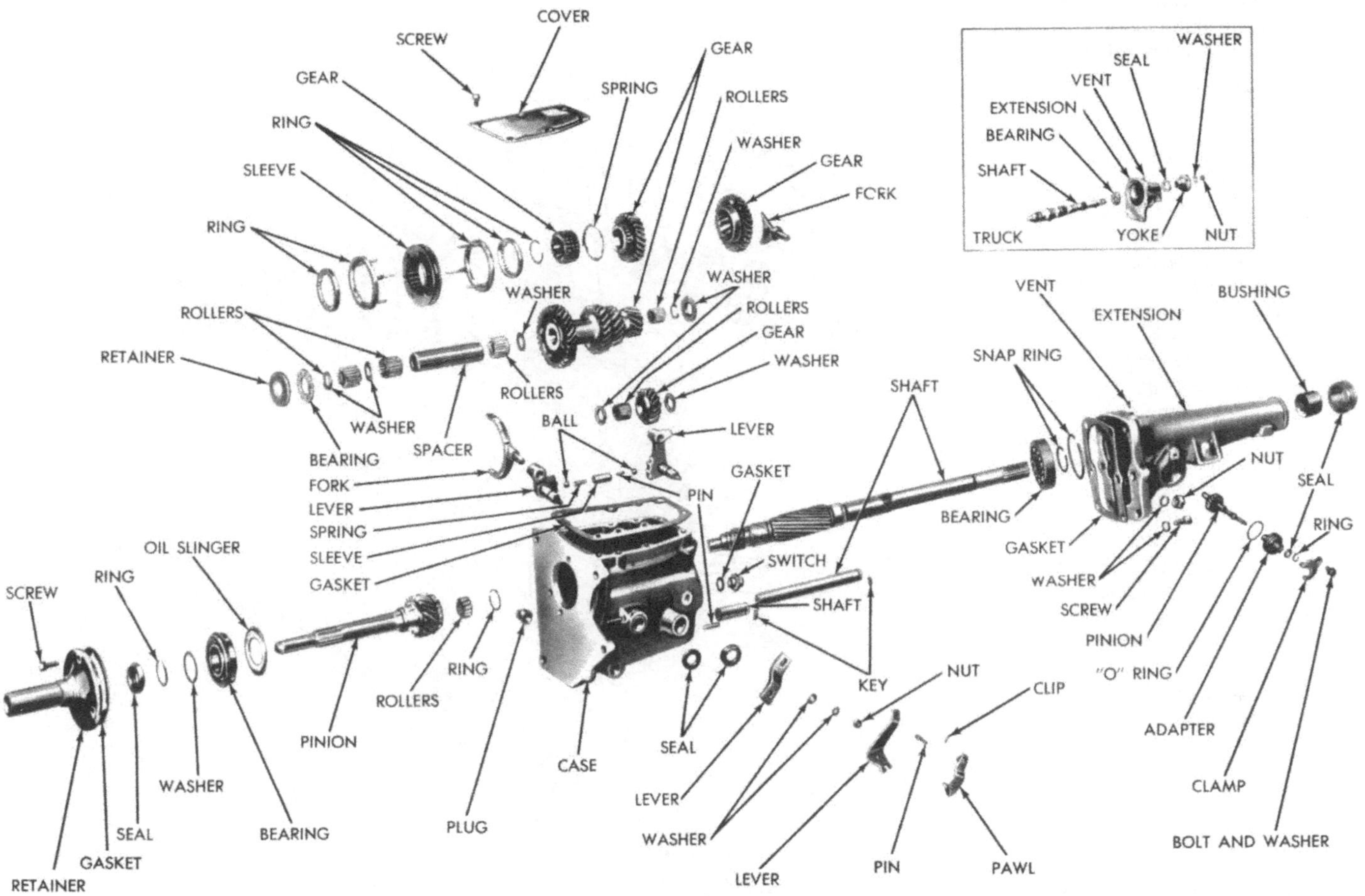

4.4b An exploded view of the A-250 three-speed transmission used on some 1973 and 1974 vehicles (also applies to A-903 3-speed unit used on some 1971 and 1972 vehicles)

pose. Safety chains will help steady the transmission on the jack.

11 Raise the engine slightly and disconnect the extension housing from the center cross-member.

12 Raise the transmission slightly and remove the center cross-member.

13 Remove the bolts securing the transmission to the clutch housing.

14 Make a final check that all wires and hoses have been disconnected from the transmission, then move the transmission and jack toward the rear of the vehicle until the transmission drive pinion shaft (input shaft) clears the splined hub in the clutch disc. Keep the transmission level as this is done.

15 Once the input shaft is clear, lower the transmission and remove it from under the vehicle.

16 The clutch components can be inspected by removing the clutch housing from the engine (see Chapter 8). In most cases, new clutch components should be routinely installed if the transmission is removed.

Installation

17 Insert a small amount of multi-purpose grease into the pilot bushing in the crankshaft and lubricate the inner surface of the bushing. Make sure no grease gets on the input shaft, clutch disc splines or the release lever.

18 If removed, install the clutch components (see Chapter 8).

19 If removed, attach the clutch housing to the engine and tighten the bolts to the torque listed in the Chapter 8 Specifications.

20 With the transmission secured to the jack as on removal, raise the transmission into position behind the clutch housing and then carefully slide it forward, engaging the input shaft with the clutch plate hub. Do not use excessive force to install the transmission - if the input shaft does not slide into place, readjust the angle of the transmission so it is level and/or turn the input shaft so the splines engage properly with the clutch.

21 Install the transmission-to-clutch housing bolts. Tighten the bolts to the torque listed in this Chapter's Specifications.

22 Install the crossmember and attach it to the extension housing. Tighten all nuts and bolts securely.

23 Remove the jacks supporting the transmission and the engine.

24 Install the various items removed previously, referring to Chapter 8 for the installation of the driveshaft and Chapter 4 for information regarding the exhaust system components.

25 Make a final check that all wires, hoses and the speedometer cable have been connected and that the transmission has been filled with lubricant to the proper level (see Chapter 1). Lower the vehicle.

26 Install the console, if equipped (see Chapter 11) and the shift lever boot and knob (see Section 2).

27 Connect the negative battery cable. Road test the vehicle for proper operation and check for leakage.

4 Transmission overhaul - general information

Refer to illustrations 4.4a, 4.4b, 4.4c and 4.4d

Overhauling a manual transmission is a difficult job for the do-it-yourselfer. It involves the disassembly and reassembly of many small parts. Numerous clearances must be precisely measured and, if necessary, changed with select fit spacers and snap-rings. As a result, if transmission problems arise, it can be removed and installed by a

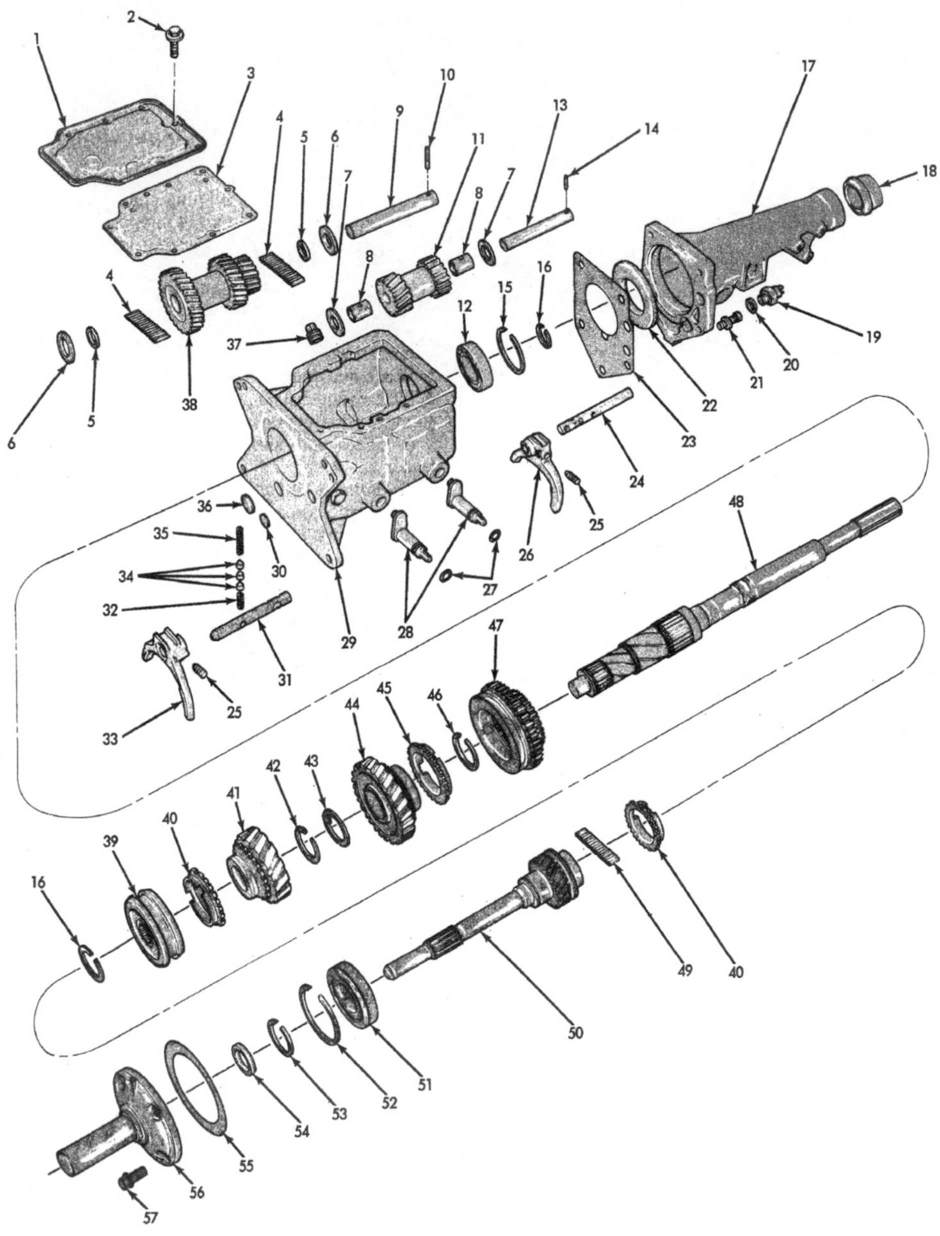

4.4c An exploded view of the A-390 three-speed transmission (some 1975 and 1976 A and F-body vehicles)

1 *Case cover*
2 *Case cover screw*
3 *Case cover gasket*
4 *Countershaft bearing roller*
5 *Countershaft bearing washer*
6 *Countershaft thrust washer*
7 *Reverse idler thrust washer*
8 *Reverse idler bushing*
9 *Countershaft*
10 *Countershaft roll pin*
11 *Reverse idler gear*
12 *Output shaft bearing*
13 *Reverse idler shaft*
14 *Reverse idler stop pin*
15 *Output shaft bearing outer snap-ring*
16 *Output shaft bearing inner snap-ring*
17 *Extension housing*
18 *Extension housing seal*
19 *Back-up light switch*
20 *Back-up light switch gasket*
21 *Extension housing screw and lock-washer*
22 *Output shaft bearing retainer*
23 *Extension gasket*
24 *Gearshift 1st and Reverse rail*
25 *Fork set screw*
26 *Gearshift 1st and Reverse fork*
27 *Gearshift lever shaft oil seal*
28 *Gearshift lever*
29 *Transmission case*
30 *Plug*
31 *Gearshift 2nd and 3rd rail*
32 *Gearshift detent-pin spring*
33 *Gearshift 2nd and 3rd fork*
34 *Gearshift detent pin*
35 *Gearshift detent-pin spring*
36 *Plug*
37 *Case filler plug*
38 *Countershaft gear*
39 *2nd and 3rd synchronizer ring*
40 *Synchronizer 2nd and 3rd stop ring*
41 *2nd-speed gear*
42 *Low-speed gear thrust washer snap-ring*
43 *Low-speed gear thrust washer*
44 *Low-speed gear*
45 *Synchronizer low stop ring*
46 *Synchronizer low and Reverse clutch-gear snap-ring*
47 *Low and Reverse synchronizer assembly*
48 *Output shaft*
49 *Output shaft pilot roller*
50 *Input shaft*
51 *Input shaft bearing*
52 *Bearing outer snap-ring*
53 *Bearing inner snap-ring*
54 *Bearing retainer oil seal*
55 *Bearing retainer gasket*
56 *Bearing retainer*
57 *Bearing retainer screw*

competent do-it-yourselfer, but overhaul should be left to a transmission repair shop. Rebuilt transmissions may be available - check with your dealer parts department and auto parts stores. At any rate, the time and money involved in an overhaul is almost sure to exceed the cost of a rebuilt unit.

Nevertheless, it's not impossible for an inexperienced mechanic to rebuild a transmission if the special tools are available and the job is done in a deliberate step-by-step manner so nothing is overlooked.

The tools necessary for an overhaul include internal and external snap-ring pliers, a bearing puller, a slide hammer, a set of pin punches, a dial indicator and possibly a hydraulic press. In addition, a large, sturdy workbench and a vise or transmission stand will be required.

During disassembly of the transmission, make careful notes of how each piece comes off, where it fits in relation to other pieces and what holds it in place. Exploded views are included **(see illustrations)** to show where the parts go - but actually noting how they are installed when remove the parts will make it much easier to get the transmission back together.

Before taking the transmission apart for repair, it will help if you have some idea what area of the transmission is malfunctioning. Certain problems can be closely tied to specific areas in the transmission, which can make component examination and replacement easier. Refer to the Troubleshooting section at the front of this manual for information regarding possible sources of trouble.

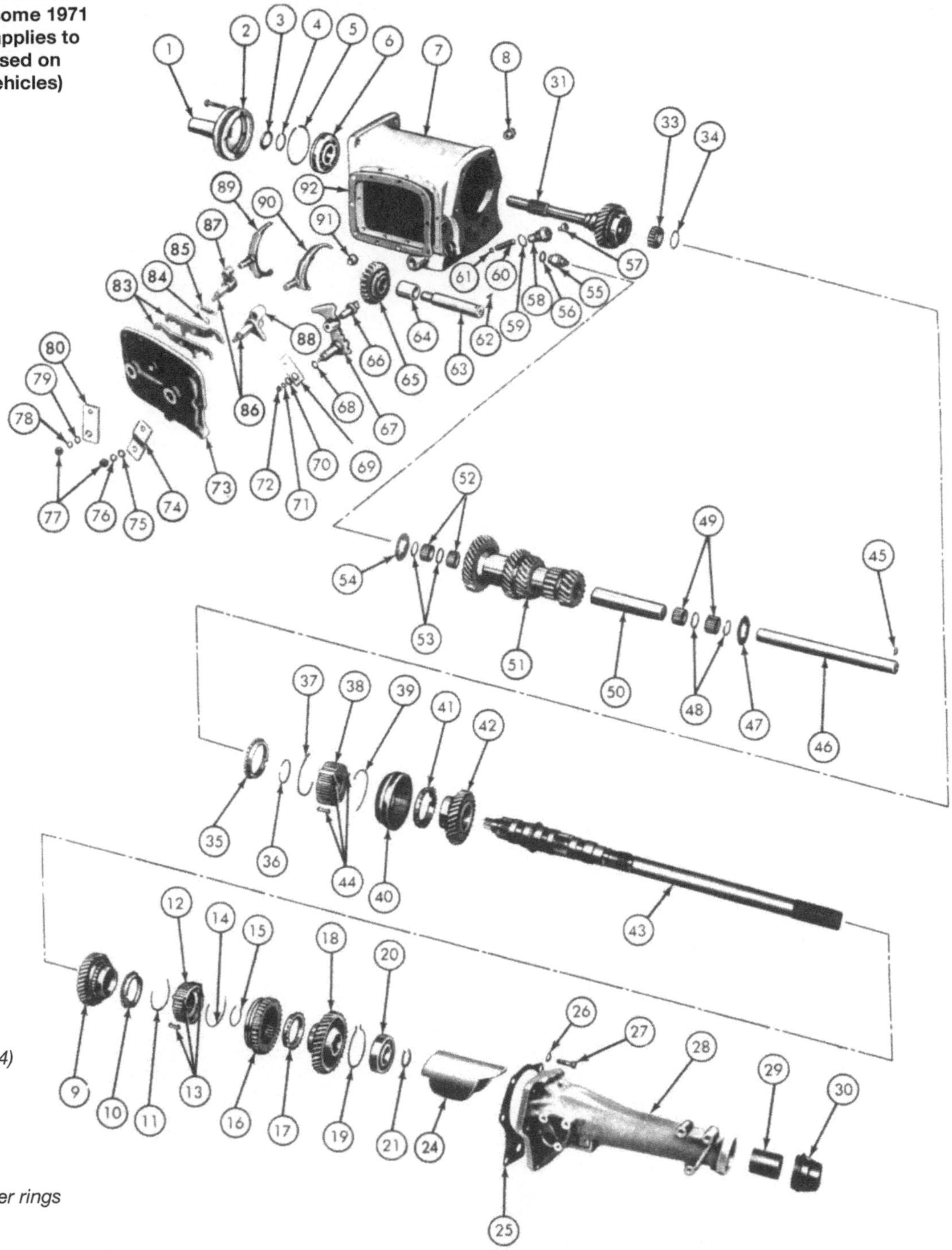

4.4d An exploded view of the A-833 four-speed transmission used on some 1971 through 1975 vehicles (also applies to optional Overdrive-4 unit used on some 1976 through 1980 vehicles)

1 *Bearing retainer*
2 *Bearing retainer gasket*
3 *Bearing retainer oil seal*
4 *Bearing inner snap-ring*
5 *Bearing outer snap-ring*
6 *Pinion bearing*
7 *Transmission case*
8 *Filler plug*
9 *2nd-speed gear*
10 *Snap-ring*
11 *Shift strut springs*
12 *Clutch gear*
13 *Shift struts (3)*
14 *Shift strut spring*
15 *Snap-ring*
16 *1st and 2nd clutch sleeve gear*
17 *Synchronizer ring*
18 *1st-speed gear*
19 *Bearing retainer ring*
20 *Rear bearing*
21 *Snap-ring*
24 *Baffle*
25 *Case-to-extension housing gasket*
26 *Lock-washer*
27 *Bolt*
28 *Extension housing*
29 *Mainshaft yoke bushing*
30 *Oil seal*
31 *Main drive pinion*
33 *Needle bearing rollers*
34 *Snap-ring*
35 *Synchronizer ring*
36 *Snap-ring*
37 *Shift-strut spring*
38 *Clutch gear*
39 *Shift-strut spring*
40 *Clutch sleeve*
41 *Synchronizer ring*
42 *3rd speed gear (A-833), Overdrive gear (Overdrive-4)*
43 *Mainshaft (output shaft)*
44 *Shift struts (3)*
45 *Woodruff key*
46 *Countershaft*
47 *Gear thrust washer*
48 *Needle roller bearing spacer rings*
49 *Needle roller bearing*
50 *Bearing spacer*
51 *Countershaft gear cluster*
52 *Needle bearing rollers*
53 *Needle roller bearing spacer rings*
54 *Gear thrust washer*
55 *Back-up light switch*
56 *Back-up light switch gasket*
57 *Plug*
58 *Reverse detent ball spring retainer*
59 *Gasket*
60 *Reverse detent ball spring*
61 *Reverse detent ball*
62 *Woodruff key*
63 *Reverse idler gear shaft*
64 *Reverse idler gear bushing*
65 *Reverse idler gear*
66 *Reverse shifter fork*
67 *Reverse lever*
68 *Reverse lever shaft oil seal*
69 *Reverse operating lever*
70 *Flat washer*
71 *Lock-washer*
72 *Nut*
73 *Gearshift control housing*
74 *1st and 2nd operating lever*
75 *Flat washer*
76 *Lock-washer lever*
77 *Lever nut*
78 *Lock-washer lever*
79 *Flat-washer lever*
80 *3rd and Overdrive operating lever*
83 *Interlock lever (2)*
84 *E-ring*
85 *Spring*
86 *Oil seal (2)*
87 *3rd and Overdrive lever*
88 *1st and 2nd lever*
89 *3rd and Overdrive-speed fork*
90 *1st and 2nd-speed fork*
91 *Drain plug*
92 *Shift control housing gasket*

Notes

Chapter 7 Part B
Automatic transmission

Contents

Specifications

Torque specifications

	Ft-lbs
Neutral start/backup light switch	24
Transmission fluid pan bolts	See Chapter 1
Transmission-to-engine bolts	30
Torque converter-to-driveplate bolts	23

1 General information

All vehicles covered in this manual come equipped with either a three or four-speed manual transmission or an automatic transmission. All information on the automatic transmission is included in this Part of Chapter 7. Information on the manual transmission can be found in Part A of this Chapter. You'll also find certain procedures common to both automatic and manual transmissions - such as oil seal replacement - here in Part B.

All automatic transmissions are three-speeds based on the basic Torqueflite design. The A-904 is the most common type; a variant is known as the A-904LA. The A-727 is a heavy-duty unit normally used behind big-block engines and early high-performance small-block engines (also, a few Police and Taxi six-cylinders). Many 1978 and later transmissions are equipped with a lock-up torque converter that engages in high gear. The lock-up torque converter provides a direct connection between the engine and the drive wheels for improved efficiency and economy. The lock-up converter consists of a solenoid-controlled clutch on the torque converter that engages to lock up the converter in high gear.

Due to the complexity of the automatic transmissions covered in this manual and the need for specialized equipment to perform most service operations, this Chapter contains only general diagnosis, routine maintenance, adjustment and removal and installation procedures.

If the transmission requires major repair work, it should be left to a dealer service department or an automotive or transmission repair shop. You can, however, remove and install the transmission yourself and save the expense, even if the repair work is done by a transmission shop.

2 Diagnosis - general

Note: *Automatic transmission malfunctions may be caused by five general conditions: poor engine performance, improper adjustments, hydraulic malfunctions, mechanical malfunctions or malfunctions in the computer or its signal network (later models). Diagnosis of these problems should always begin with a check of the easily repaired items: fluid level and condition (see Chapter 1), shift linkage adjustment and throttle rod linkage adjustment. Next, perform a road test to determine if the problem has been corrected or if more diagnosis is necessary. If the problem persists after the preliminary tests and corrections are completed, additional diagnosis should be done by a dealer service department or transmission repair shop. Refer to the* Troubleshooting *Section at the front of this manual for information on symptoms of transmission problems.*

Preliminary checks

1 Drive the vehicle to warm the transmission to normal operating temperature.

2 Check the fluid level as described in Chapter 1:

a) If the fluid level is unusually low, add enough fluid to bring the level within the designated area of the dipstick, then check for external leaks (see below).

b) If the fluid level is abnormally high, drain off the excess, then check the drained fluid for contamination by coolant. The presence of engine coolant in the automatic transmission fluid indicates that a failure has occurred in the internal radiator walls that separate the coolant from the transmission fluid (see Chapter 3).

c) If the fluid is foaming, drain it and refill the transmission, then check for coolant in the fluid or a high fluid level.

3 Check the engine idle speed. **Note:** *If the engine is malfunctioning, do not proceed with the preliminary checks until it has been repaired and runs normally.*

4 Check the throttle rod for freedom of movement. Adjust it if necessary (see Section 5). **Note:** *The throttle rod may function properly when the engine is shut off and cold, but it may malfunction once the engine is hot. Check it cold and at normal engine operating temperature.*

5 Inspect the shift linkage (see Section 4). Make sure it's properly adjusted and that the linkage operates smoothly.

Fluid leak diagnosis

6 Most fluid leaks are easy to locate visually. Repair usually consists of replacing a seal or gasket. If a leak is difficult to find, the following procedure may help.

7 Identify the fluid. Make sure it's transmission fluid and not engine oil or brake fluid

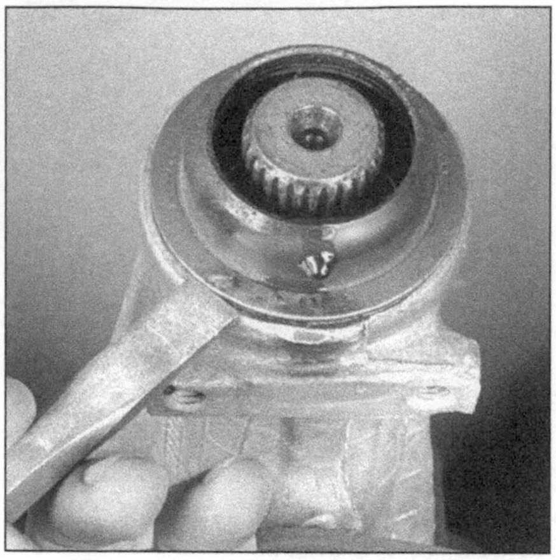

3.4 Use a hammer and chisel to dislodge the rear seal

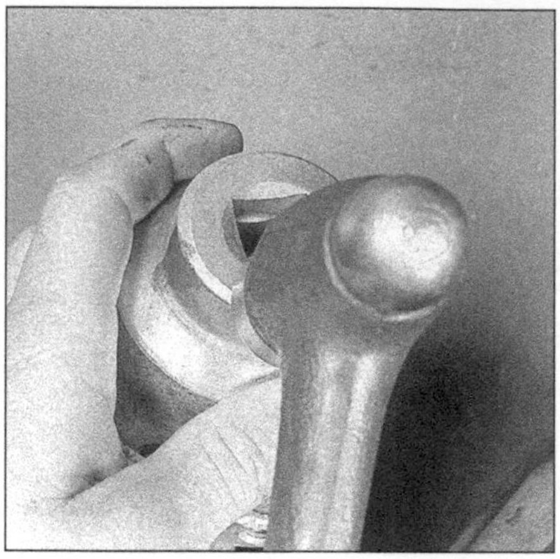

3.6 A large socket can be used to drive the new seal evenly into the bore

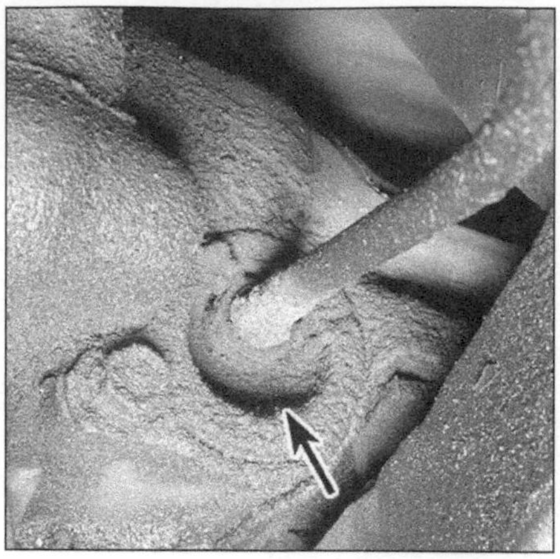

3.8 The adapter (arrow), or driven gear housing, as it's sometimes called, is located on the left side of the extension housing, right in front of the crossmember; it can be hard to find if it's coated with oil and dirt like this one

(automatic transmission fluid is a deep red color).

8 Try to pinpoint the source of the leak. Drive the vehicle several miles, then park it over a large sheet of cardboard. After a minute or two, you should be able to locate the leak by determining the source of the fluid dripping onto the cardboard.

9 Make a careful visual inspection of the suspected component and the area immediately around it. Pay particular attention to gasket mating surfaces. A mirror is often helpful for finding leaks in areas that are hard to see.

10 If the leak still cannot be found, clean the suspected area thoroughly with a degreaser or solvent, then dry it.

11 Drive the vehicle for several miles at normal operating temperature and varying speeds. After driving the vehicle, visually inspect the suspected component again.

12 Once the leak has been located, the cause must be determined before it can be properly repaired. If a gasket is replaced but the sealing flange is bent, the new gasket will not stop the leak. The bent flange must be straightened.

13 Before attempting to repair a leak, check to make sure that the following conditions are corrected or they may cause another leak. **Note:** *Some of the following conditions cannot be fixed without highly specialized tools and expertise. Such problems must be referred to a transmission repair shop or a dealer service department.*

Gasket leaks

14 Check the pan periodically. Make sure the bolts are tight, no bolts are missing, the gasket is in good condition and the pan is flat (dents in the pan may indicate damage to the valve body inside).

15 If the pan gasket is leaking, the fluid level or the fluid pressure may be too high, the vent may be plugged, the pan bolts may be too tight, the pan sealing flange may be warped, the sealing surface of the transmission housing may be damaged, the gasket may be damaged or the transmission casting may be cracked or porous. If sealant instead of gasket material has been used to form a seal between the pan and the transmission housing, it may be the wrong type sealant.

Seal leaks

16 If a transmission seal is leaking, the fluid level or pressure may be too high, the vent may be plugged, the seal bore may be damaged, the seal itself may be damaged or improperly installed, the surface of the shaft protruding through the seal may be damaged or a loose bearing may be causing excessive shaft movement.

17 Make sure the dipstick tube seal is in good condition and the tube is properly seated. Periodically check the area around the speedometer gear or sensor for leakage. If transmission fluid is evident, check the O-ring for damage.

Case leaks

18 If the case itself appears to be leaking, the casting is porous and will have to be repaired or replaced.

19 Make sure the oil cooler hose fittings are tight and in good condition.

Fluid comes out vent pipe or fill tube

20 If this condition occurs, the transmission is overfilled, there is coolant in the fluid, the case is porous, the dipstick is incorrect, the vent is plugged or the drain-back holes are plugged.

3 Oil seal replacement

1 Oil leaks frequently occur due to wear of the extension housing oil seal, and/or the speedometer drive gear oil seal and O-ring. Replacement of these seals is relatively easy, since the repairs can usually be performed without removing the transmission from the vehicle.

Extension housing oil seal

Refer to illustrations 3.4 and 3.6

2 The extension housing oil seal is located at the extreme rear of the transmission, where the driveshaft is attached. If leakage at the seal is suspected, raise the vehicle and support it securely on jackstands. If the seal is leaking, transmission fluid will be built up on the front of the driveshaft and may be dripping from the rear of the transmission.

3 Refer to Chapter 8 and remove the driveshaft.

4 Using a chisel and hammer, carefully pry the oil seal out of the rear of the transmission **(see illustration)**. Do not damage the splines on the transmission output shaft.

5 If the oil seal and bushing cannot be removed with a chisel, a special oil seal removal tool (available at auto parts stores) will be required.

6 Using a large section of pipe or a very large deep socket as a drift, install the new oil seal **(see illustration)**. Drive it into the bore squarely and make sure it's completely seated.

7 Lubricate the splines of the transmission output shaft and the outside of the driveshaft sleeve yoke with lightweight grease, then install the driveshaft. Be careful not to damage the lip of the new seal.

Speedometer driven gear seals

Refer to illustrations 3.8, 3.9, 3.10a, 3.10b, 3.11, 3.12, 3.13 and 3.15

Note: *Any time the speedometer pinion adapter is removed, you MUST install a new O-ring.*

8 The speedometer cable and driven gear housing **(see illustration)** is located on the left (driver's) side of the extension housing. Look for transmission fluid around the cable and housing to determine if the seal and O-ring are leaking.

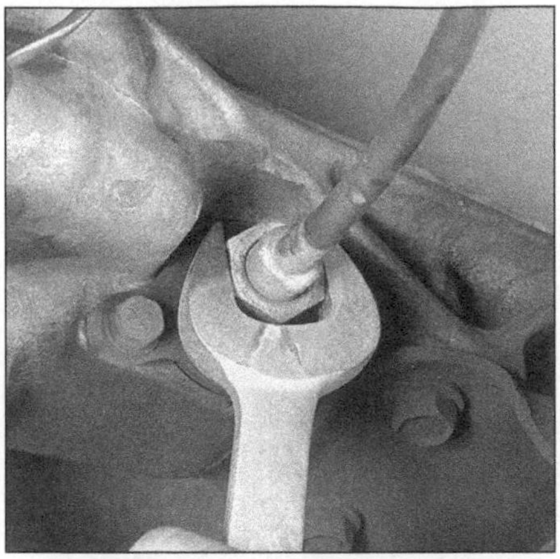

3.9 **Unscrew the speedometer cable collar with a wrench (1971 through 1977 models)**

3.10a To remove the driven gear housing, remove the retainer bolt . . .

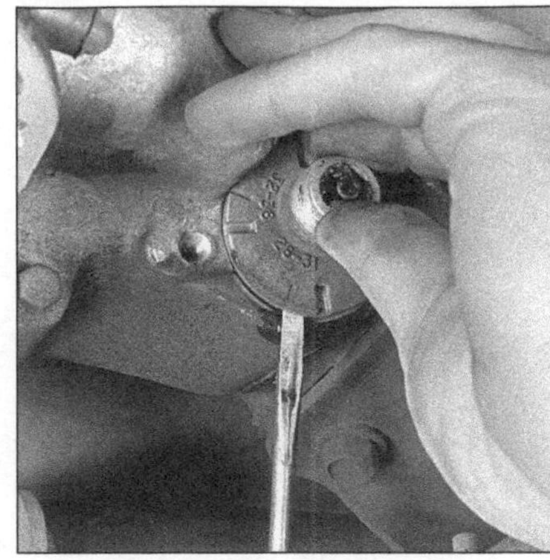

3.10b . . . and pry the housing out with a screwdriver

3.11 Use a hooked tool to remove the O-ring

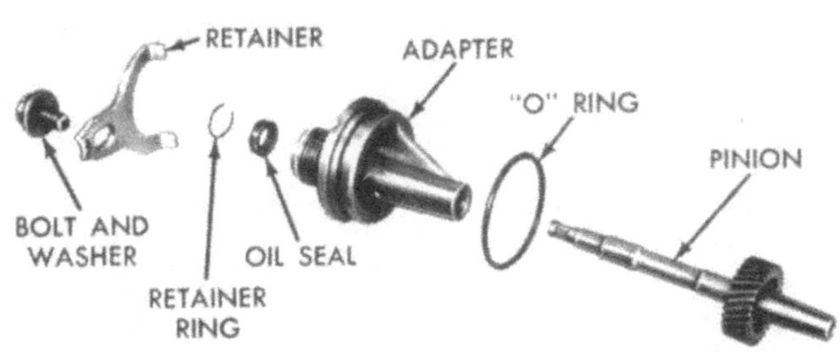

3.12 An exploded view of a typical speedometer pinion and adapter assembly (1971 through 1977 models)

9 On 1977 and earlier models, disconnect the speedometer cable **(see illustration)**.

10 Mark the relationship of the adapter to the transmission, then remove the bolt and retainer and pull out the adapter **(see illustrations)**. **Caution:** *Every combination of rear axle gear ratio and tire size requires a different pinion (driven gear), and each pinion is a different diameter, so the position of the (eccentric) adapter is critical. If you try to install the adapter in some position other than the one it was in before you removed it, the pinion will be stripped or will seize as soon as the vehicle is driven. That's why it's important to carefully mark the position of the adapter in relation to the transmission housing so that the adapter is installed in exactly the same orientation it was in prior to disassembly.*

11 Remove the old O-ring **(see illustration)** and install a new O-ring in the adapter.

12 On 1977 and earlier models, remove the retainer ring with a small screwdriver, then, using a hook, remove the seal **(see illustration)**.

13 On 1978 and later models, remove the adapter clip and pinion clip, then remove the pinion gear and speedometer cable from the adapter **(see illustration)**.

14 Using a small socket as a drift, install the new seal.

15 Installation is the reverse of removal. Make SURE you align the marks you made so the adapter will be in the same position. If you can't find the mark, there should be a number molded into the end of the plastic pinion gear. This number will fall into one of four ranges cast into the outside flange of the adapter. When reinstalling the adapter, rotate it until the correct range is in the six O'clock position **(see illustration)**. For example, if the

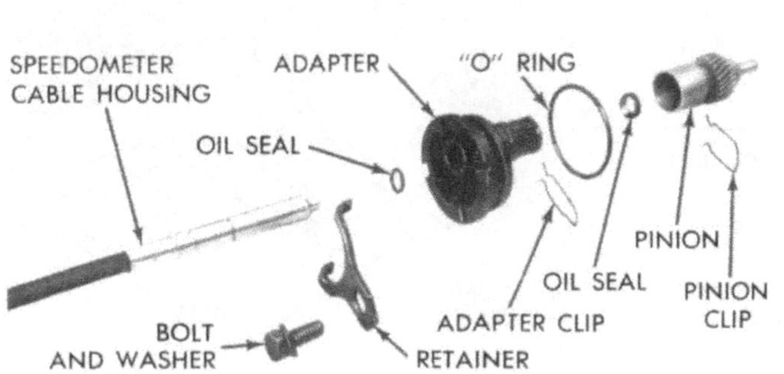

3.13 An exploded view of a typical speedometer pinion and adapter assembly (1978 and later models)

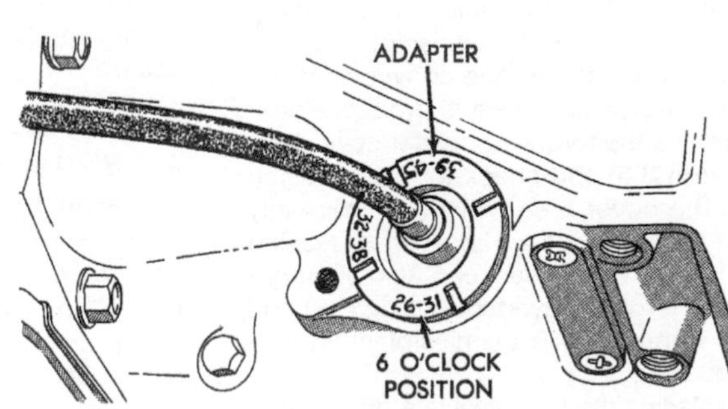

3.15 If you forgot to mark the adapter, look at the number molded into the end of the pinion gear, then find the number range it falls into on the adapter - position the correct range in the six o'clock position

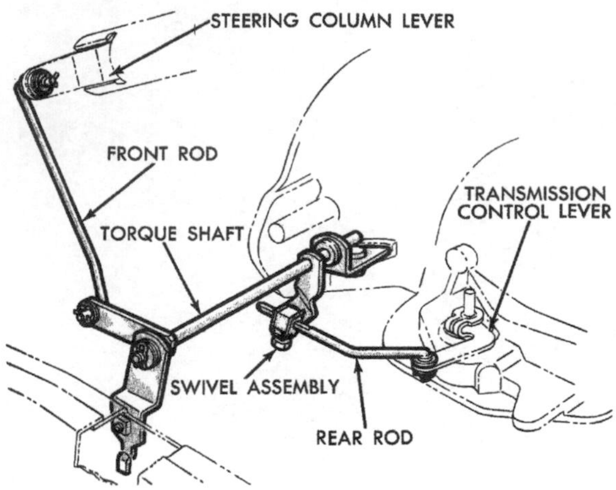

4.4a Shift linkage assembly for column-shift models (1971 through 1974 A-body)

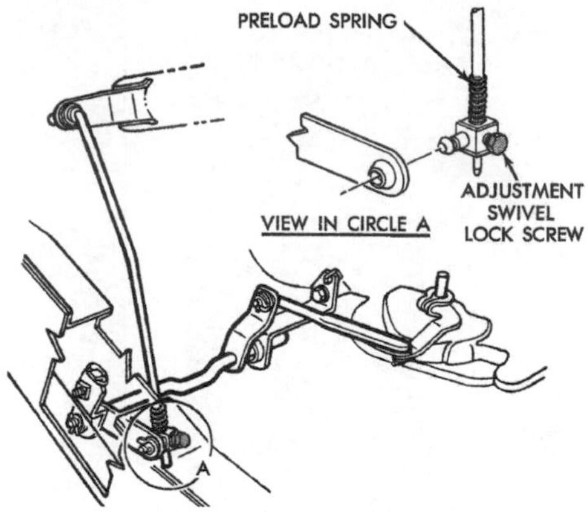

4.4b Shift linkage assembly for column-shift models (1971 and 1972 E and B-body; 1973 E, B and C-body; 1974 and 1975, all models except A-body; 1976, all models except A and F-body; 1977, all models except F-body; 1978 and 1979, all models except F and M-body)

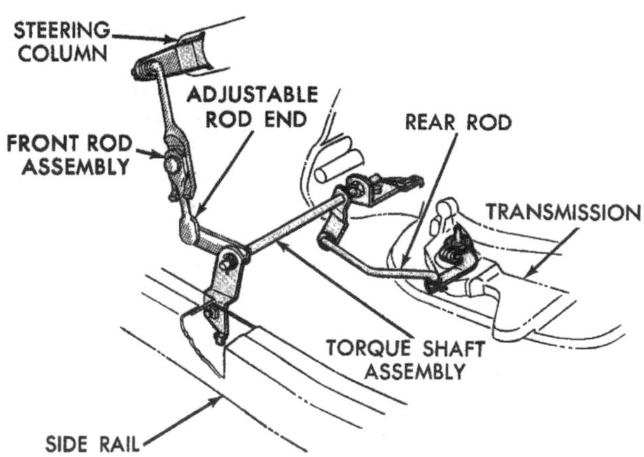

4.4c Shift linkage assembly for column-shift models (1971 and 1972 C-body)

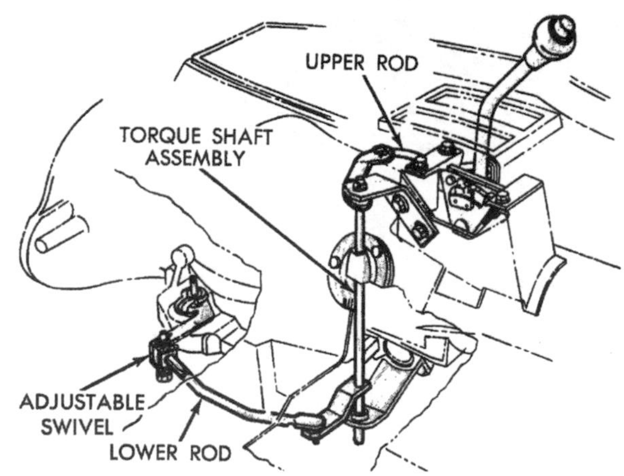

4.4d Shift linkage assembly for floor-shift models (1971 through 1976 A-body; 1973 C-body)

number on the end of the pinion gear is 28, position the 26-31 range down.

16 Reinstall the retainer and bolt, tightening the bolt securely.

Front seal

17 If you find transmission fluid leaking from the front of the transmission, coming out from around the torque converter cover, the transmission front seal is probably leaking. Remove the torque converter cover **(see illustration 8.6)** and check for transmission fluid in the cover area, which will normally confirm your diagnosis. However, oil leaking from the engine rear seal can also accumulate in the torque converter housing area and lead to confusion. Transmission fluid is a deep red color.

18 Replacing the front seal is a difficult job, since the transmission and torque converter must be removed first (see Section 8). Also, the torque converter bushing in the transmission front pump should also be replaced at this time, since front seal leaks are frequently caused by a worn bushing which causes increased movement of the torque converter hub and subsequent seal failure. We recommend taking the removed transmission to an automotive machine shop for replacement of the front seal and bushing - this way you'll make sure the seal will not fail on you again soon.

4 Shift linkage - check and adjustment

Refer to illustrations 4.4a through 4.4j

Note: *Should it be necessary to disassemble any linkage rods from levers which use plastic grommets as retainers, always replace the grommets.*

1 Check the operation of the shift linkage, making sure the lever is within the stops and the engine starts only in the Park or Neutral position.

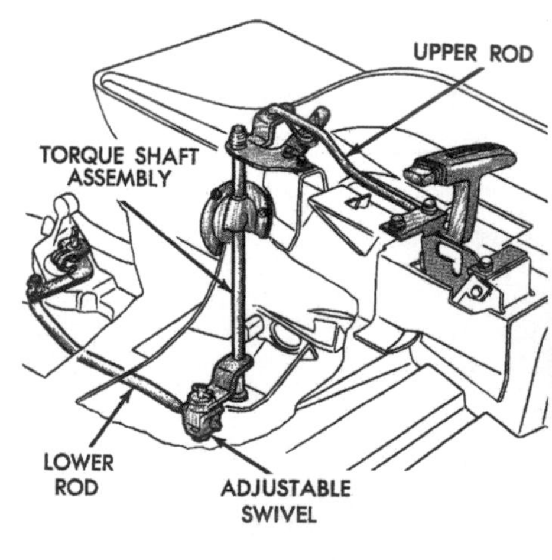

4.4e Shift linkage assembly for floor-shift models (1971 through 1974 B and E-body; 1975 B-body; 1976 and 1977 F and B-body; 1978 F, B and M-body)

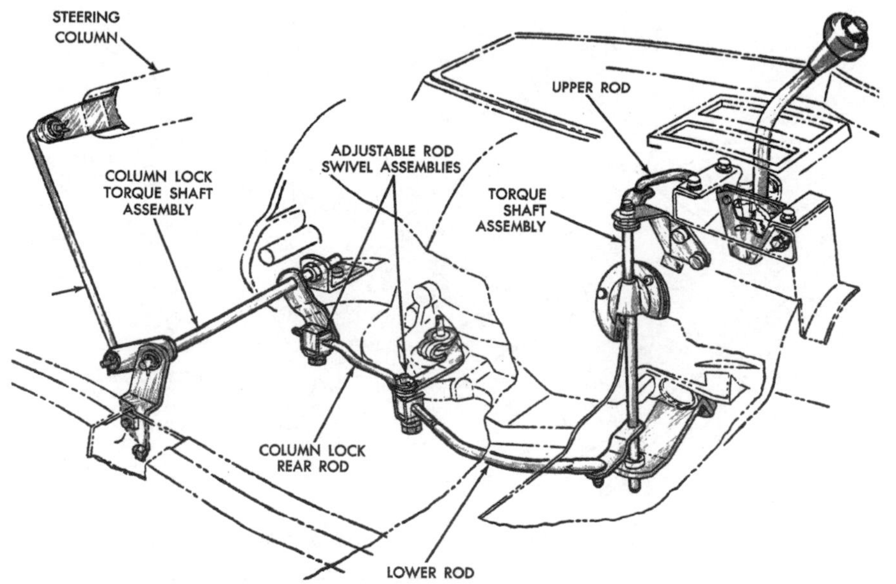

4.4f Shift linkage assembly for floor-shift models (1971 and 1972 C-body)

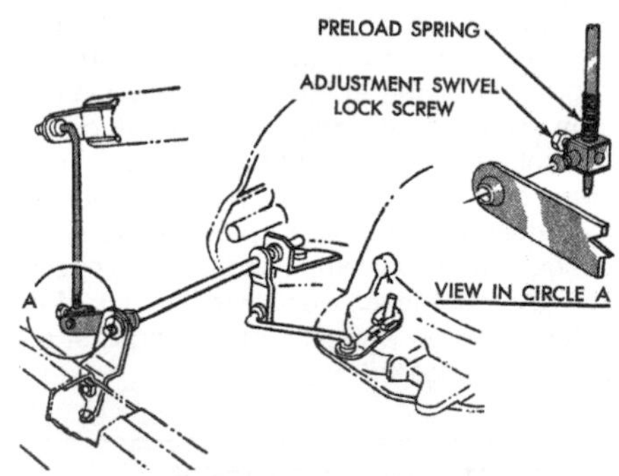

4.4g Shift linkage assembly for column-shift models (1975 and 1976 A-body; 1977 F-body; 1978 and 1979 F and M-body)

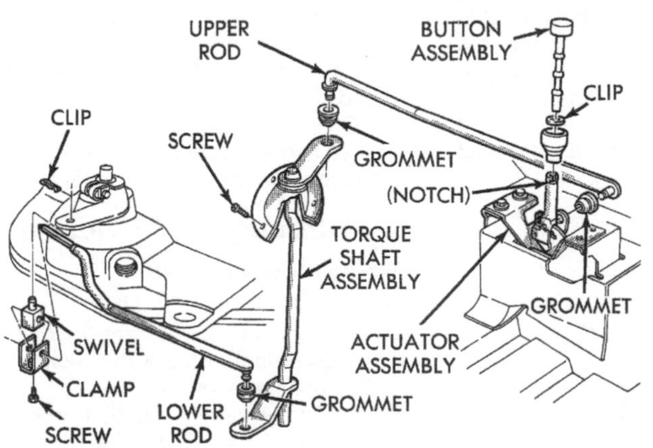

4.4h Typical console shift linkage assembly for floor-shift models (1979 through 1984)

2 Disconnect the negative battery cable.

3 Raise the front of the vehicle and support it securely on jackstands. In the passenger compartment, place the shift lever in Park.

4 Before proceeding with the adjustment, loosen the adjustment swivel lockscrew **(see illustrations)** and make sure the adjustable swivel block is free to turn on the shift rod. If the swivel block binds at all on the shift rod, clean off any corrosion, dirt or grease with solvent and a wire brush before proceeding.

5 Place the shift lever (the one inside the vehicle) in the Park position. On column-shift models, lock the steering column with the ignition key.

6 With all linkage assembled and the adjustment swivel lockscrew still loose, move the shift lever on the transmission all the way to its rear detent (the Park position).

7 With the shift lever inside the vehicle and the shift lever on the transmission in their Park positions, tighten the adjustment swivel lockscrew securely.

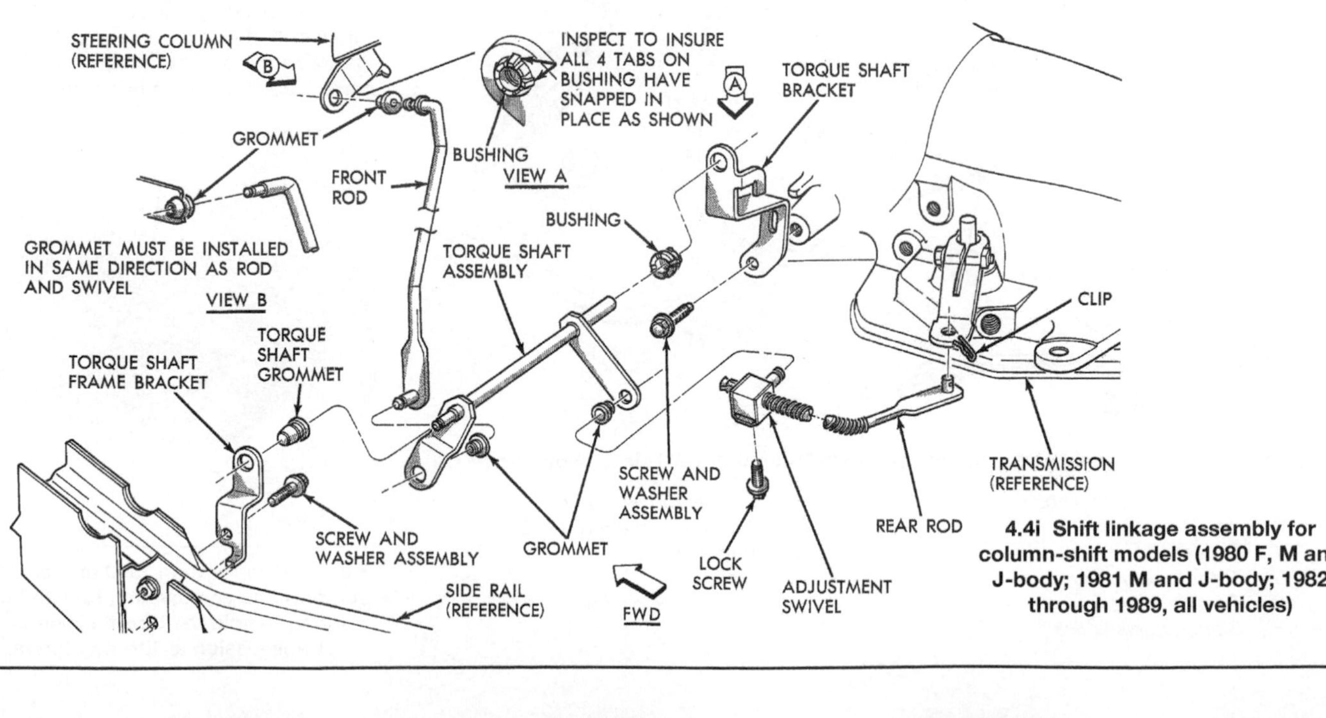

4.4i Shift linkage assembly for column-shift models (1980 F, M and J-body; 1981 M and J-body; 1982 through 1989, all vehicles)

8 Lower the vehicle and check the shift linkage operation, again making sure the engine starts only in Park or Neutral. On column-shift models, the detent position for Neutral and Drive should be within the limits of the shift lever stops. On floor-shift models, the detent position must be close enough to the gate stops in Neutral and Drive to ensure that the shift lever won't remain out of its detent position when it's moved against the gate, then released.

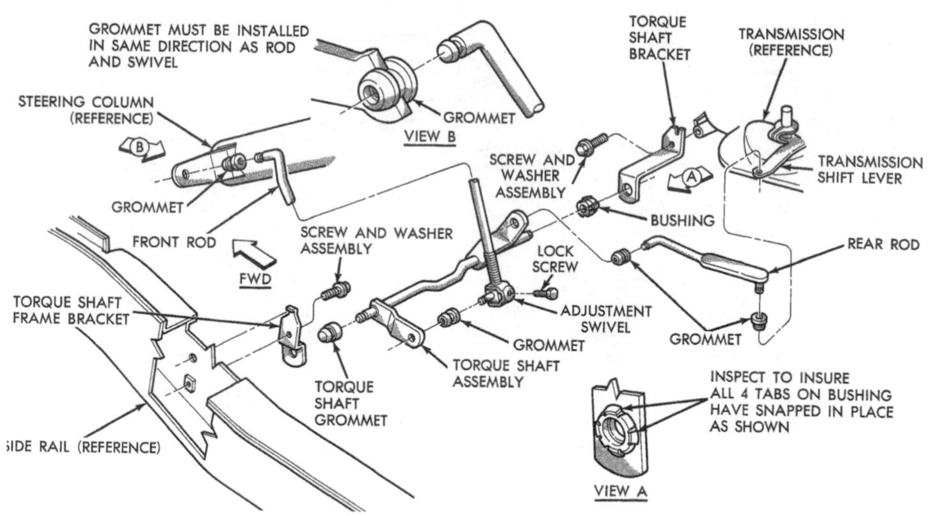

4.4j Shift linkage assembly for column-shift models (1980 and 1981 R-body)

5 Throttle rod adjustment

Caution: *This adjustment is critical and must be correct for the transmission to function correctly. Also, incorrect adjustment can lead to early transmission failure. When the adjustment is complete, the transmission should shift into third (high) gear at 20 to 30 mph under light throttle pressure. If it does not, take the vehicle to a dealer service department or qualified transmission shop to have the transmission checked.*

All models

1 The throttle rod (which is the only transmission shift control, since the Torqueflite has no vacuum modulator) controls a valve in the transmission which governs shift quality and speed. Under light throttle pressure, the transmission should shift into third gear at 20 to 30 mph. If shifting is harsh, erratic or not at the right speed, the throttle rod should be adjusted. You can quickly check the throttle rod's operation by moving the rod (near the carburetor) rearward, then slowly releasing it. If properly adjusted and in good condition, the throttle rod will return to its full forward position when released. If it doesn't, it should be adjusted and/or lubricated so that it does.

2 Before performing the following adjustment, start the engine and warm it up to its normal operating temperature. Remove the air cleaner assembly and make sure the carburetor is **off** the fast-idle cam (see Chapter 4) and the idle speed is correctly adjusted (see Chapter 1). Also, make sure all linkage is cleaned and lubricated, especially bellcranks and carburetor and throttle rod levers.

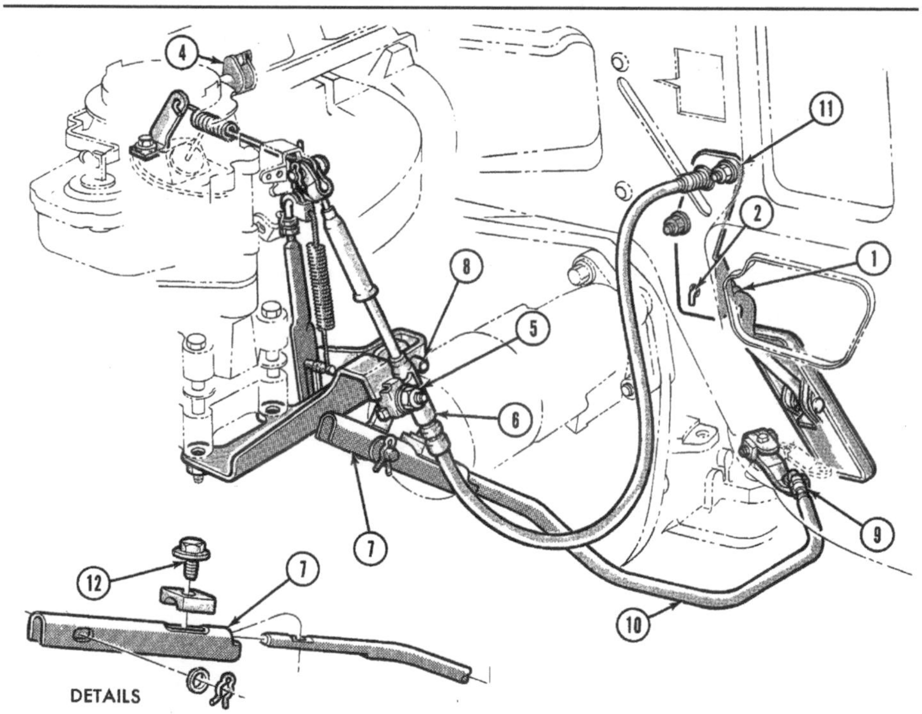

5.3 Throttle rod assembly adjustment details (1971 through 1973 six-cylinder models)

1 Accelerator pedal shaft
2 Anti-rattle spring
4 Choke
5 Throttle cable clamp nut
6 Throttle cable
7 Slotted adjuster link
8 Throttle rod lever pin
9 Transmission throttle lever
10 Throttle rod
11 Bracket
12 Slotted link lockbolt

1971 through 1973 models

Six-cylinder engines

Refer to illustration 5.3 and 5.5

3 Disconnect the choke at the carburetor **(see illustration)** or block the choke valve in its fully-open position.

4 Open the throttle slightly to release the fast-idle cam, then return the carburetor to curb idle.

5 The following adjustment must be made with the throttle lever on the transmission all the way forward against its stop. To insure correct adjustment, it is essential that the

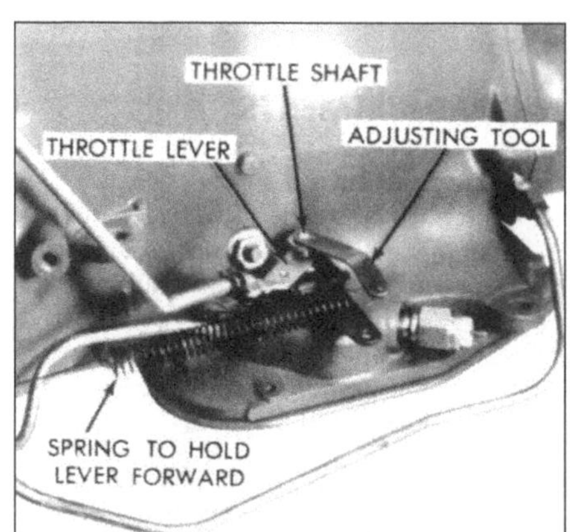

5.5 On 1971 through 1973 models, obtain or fabricate an adjusting tool and use a spring to hold the throttle lever on the transmission all the way forward

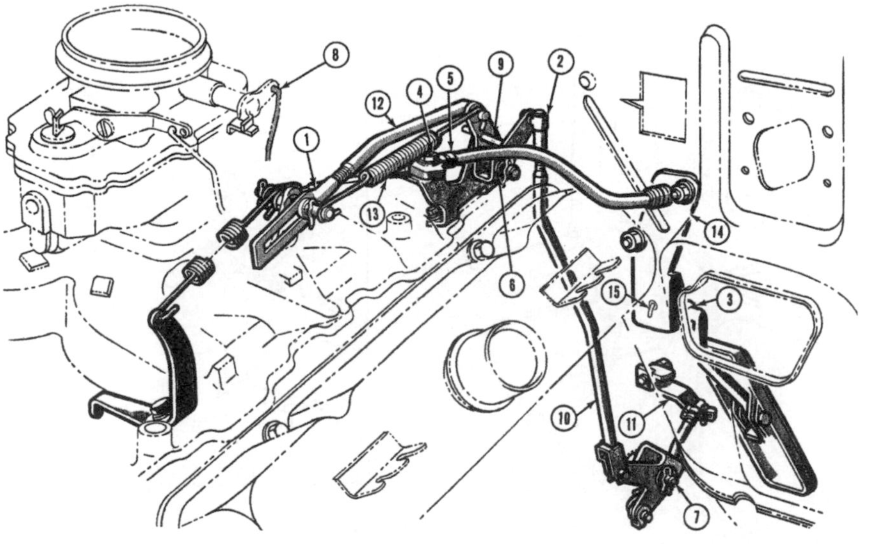

5.9 Throttle rod assembly adjustment details (1971 through 1973 V8 models with three-piece throttle rod)

1 *Slotted and threaded adjuster*
2 *Ball socket*
3 *Accelerator pedal shaft*
4 *Throttle cable clamp nut*
5 *Throttle cable*
6 *Upper bellcrank*
7 *Lower bellcrank*
8 *Choke rod*
9 *3/16-inch diameter adjustment rod*
10 *Intermediate transmission rod*
11 *Transmission throttle lever*
12 *Carburetor rod*
13 *Return spring*
14 *Bracket*
15 *Anti-rattle spring*

transmission lever remain firmly against the stop, so the factory recommends using a special adjusting tool and spring **(see illustration)**; however, if the tool is not available or cannot be fabricated, have an assistant hold the throttle lever all the way forward during the adjustment. While adjusting the linkage, make sure that no other force is exerted against the lower end of the rod.

6 Loosen the slotted link lock bolt and pull forward on the slotted adjuster link so that it contacts the carburetor lever pin and all backlash is removed. Tighten the lock bolt securely. Release the lever.

7 To check the linkage for freedom of operation, move the slotted adjuster link to its full rearward position, then allow it to return slowly, making sure it returns to the full forward position. Connect the choke rod or remove the blocking fixture. Reinstall the air cleaner

V8 engines

Three-section throttle rod

Refer to illustrations 5.9 and 5.11

8 First, clean and lubricate the linkage, then follow Steps 3, 4 and 5, making sure the throttle lever on the transmission is held all the way forward. On engines with solenoid idle stops, make sure the solenoid plunger is in its fully extended position.

9 Using a screwdriver, pry the intermediate transmission rod off the upper bellcrank **(see illustration)**. With a 3/16-inch diameter rod inserted into the holes provided in the upper bellcrank and lever, adjust the rod by means of the threaded adjustment at the upper end of the intermediate rod. The ball socket must line up with the ball end with a slight downward effort on the rod (enough to remove the play in the linkage, but not enough to move the throttle lever on the transmission).

10 Assemble the ball socket to the ball end and remove the 3/16-inch rod from the upper bellcrank and lever.

11 At the carburetor, disconnect the return

5.11 Adjusting the throttle rod at the carburetor (1971 through 1973 V8 models with a three-piece throttle rod)

spring, clip and washer, then adjust the length of the carburetor rod by pushing rearward on the rod with a slight effort (to remove all backlash) and turning the threaded adjustment **(see illustration)**. The rear end of the slot should contact the carburetor lever pin (i.e. there should be no backlash) without exerting any forward force on the pin when the slotted adjuster link is in its normal operating position.

12 Install the slotted adjuster onto the carburetor lever pin and install the washer and retainer clip. Install the return spring.

13 Release the transmission throttle lever. To verify that the throttle rod operates without binding, move the slotted link at the carburetor to its full rear position, then allow it to return slowly, making sure it returns to the full-forward position.

14 Connect the choke rod or remove the blocking fixture. Reinstall the air cleaner.

One-piece throttle rod

Refer to illustration 5.15

15 First, clean and lubricate the linkage,

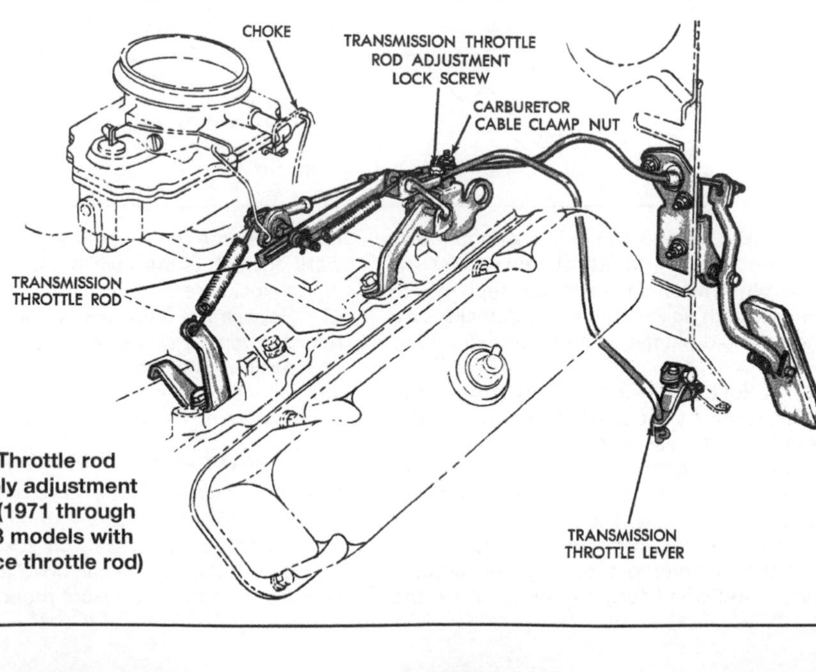

5.15 Throttle rod assembly adjustment details (1971 through 1973 V8 models with one-piece throttle rod)

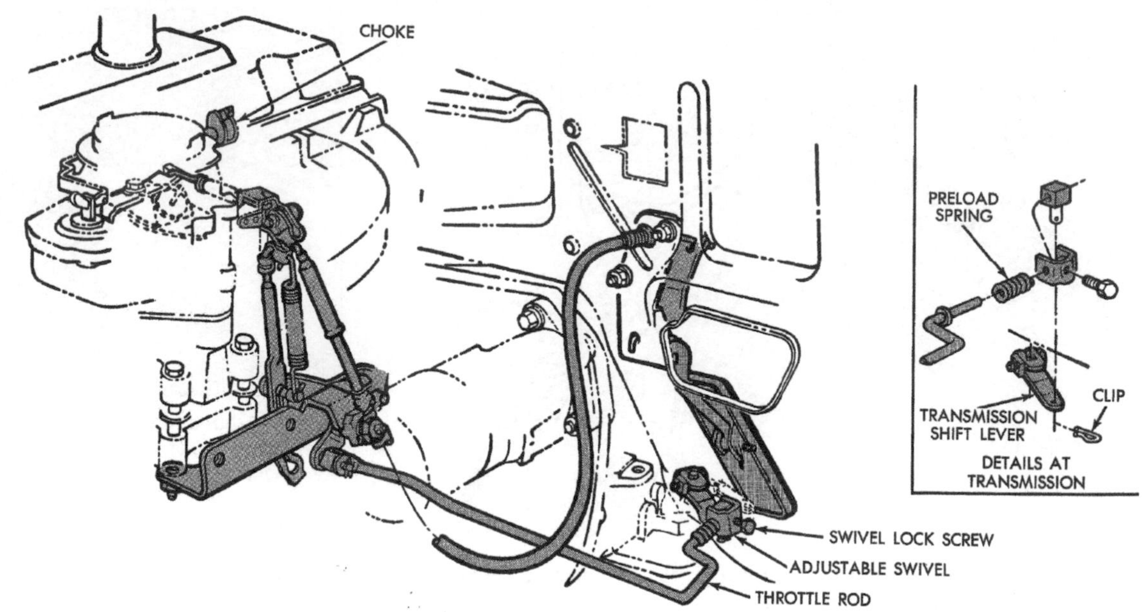

5.20a Throttle rod assembly adjustment details (1974 through 1983 six-cylinder models)

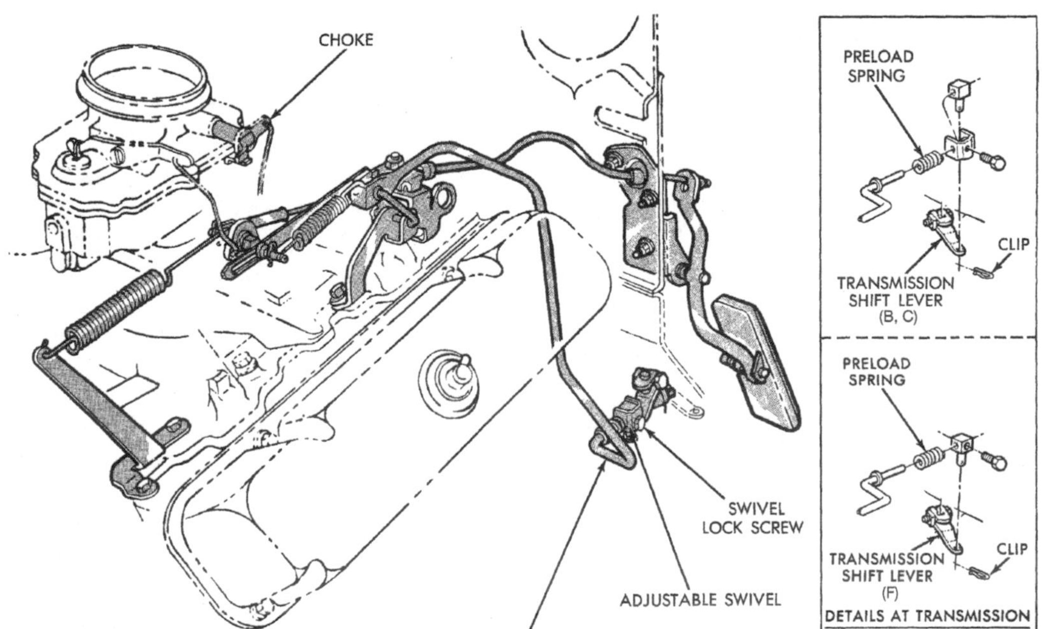

5.20b Throttle rod assembly adjustment details (1974 through 1976, all except A-body; 1977, all except F-body; 1978 and 1979, all except F and M-body; 1980 and 1981 R-body vehicles)

then follow Steps 3 and 4. Loosen the transmission throttle rod adjustment lockscrew **(see illustration)**, then follow Step 5. On engines with solenoid idle stops, make sure the solenoid plunger is in its fully extended position.

16 Adjust the transmission rod at the carburetor by pushing forward on the retainer and rearward on the rod (i.e. in opposite directions) with a slight force to remove all backlash. Tighten the transmission rod adjustment lockscrew. The rear edge of the link slot must be against the carburetor lever pin during this adjustment. And don't forget - the lever on the transmission must be held all the way forward against its stop while you're tightening the lockscrew.

17 Release the transmission throttle lever, refer to Step 13 and verify that the linkage operates freely as described.

18 Connect the choke rod or remove the blocking fixture. Reinstall the air cleaner

1974 and later (all models)

Refer to illustrations 5.20a, 5.20b and 5.20c

19 Refer to Steps 1 and 2, then raise the vehicle and place it securely on jackstands.

20 Loosen the adjustment swivel lockscrew **(see illustrations)**. To insure proper adjustment, the swivel must be free to slide along the flat end of the throttle rod so that the preload spring action is unrestricted. If necessary, disassemble, clean and lubricate all parts before proceeding.

21 Push the transmission lever forward as far as it will go and hold it there while tightening the swivel lockscrew securely. That's all there is to it! Linkage backlash is automatically removed by the preload spring.

22 Lower the vehicle, reattach the choke and verify that the linkage operates freely - move the throttle rod to the rear, slowly release it and verify that it returns fully to its forward position.

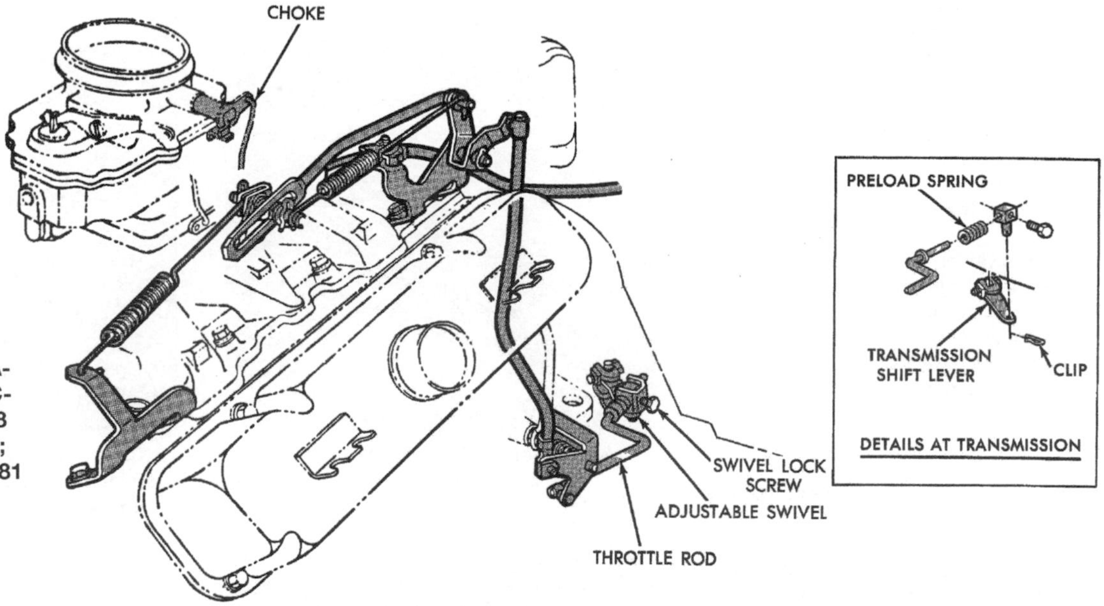

5.20c Throttle rod assembly adjustment details (1974 and 1975 A-body; 1976 A, F, B and C-body; 1977 F-body; 1978 and 1979 F and M-body; 1980 F, M and J-body; 1981 M and J-body)

6 Neutral start/backup light switch - check and replacement

1 The Neutral start/backup switch is threaded into the lower left front edge of the transmission case. The Neutral start and backup light switch functions are combined into one unit, with the center terminal of the switch grounding the starter solenoid circuit when the transmission is in Park or Neutral, allowing the engine to start. The outer terminals make up the backup light switch circuit.

Check

Refer to illustration 6.3

2 Prior to checking the switch, make sure the shift linkage is properly adjusted (see Section 4). Raise the vehicle and support it securely on jackstands.

3 Unplug the connector and use an ohmmeter or self-powered test light to check for continuity between the center terminal and the transmission case **(see illustration)**. Continuity should exist only when the transmission is in Park or Neutral.

4 Check for continuity between the two outer terminals. There should be continuity only when the transmission is in Reverse. There should be no continuity between either of the outer terminals and the transmission case.

Replacement

5 Place a container under the transmission to catch the fluid which will be released, then use a six-point socket to remove the switch.

6 Move the shift lever from Park to Neutral while checking to see that the switch operating fingers are centered in the opening. If they aren't, the shift linkage is incorrectly adjusted or there is an internal problem with transmission.

6.3 Unplug the electrical connector from the switch for access to the terminals (arrow)

7 Install the new switch and O-ring (be sure to use a new O-ring, even if for some reason you end up putting the same switch back in), tighten it to the torque listed in this Chapter's Specifications, then repeat the checks before plugging in the connector.

7 Transmission mount - check and replacement

Check

Refer to illustration 7.1

1 Raise the vehicle and support it securely on jackstands. Insert a large screwdriver or prybar into the space between the transmission extension housing and the crossmember and try to pry the transmission up slightly **(see illustration)**.

2 The transmission should not move much at all and the rubber in the center of the mount should fully insulate the center of the mount from the mount bracket around it.

7.1 To check a transmission mount, insert a prybar or large screwdriver between the mount and its bracket, then try to lever the mount sideways; it should move very little - excessive movement indicates that the mount is worn out and needs to be replaced

Replacement

3 To replace the mount, remove the through-bolt attaching the mount to the crossmember and the bolts attaching the mount to the transmission.

4 Raise the transmission slightly with a jack and remove the mount.

5 Installation is the reverse of the removal procedure. Be sure to tighten the bolts/nut securely.

8 Automatic transmission - removal and installation

Removal

Refer to illustrations 8.6 and 8.8

1 The transmission and converter must be

removed as a single assembly. If you try to leave the torque converter attached to the driveplate, the converter driveplate, pump bushing and oil seal will be damaged. The driveplate is not designed to support the load, so none of the weight of the transmission should be allowed to rest on the plate during removal.

2 Disconnect the negative cable from the battery.

3 Raise the vehicle and support it securely on jackstands, then remove the starter motor (see Chapter 5). **Note:** *On six-cylinder models, it will probably be easier to remove the starter before you raise the vehicle.*

4 Drain the transmission fluid (see Chapter 1), then reinstall the pan.

5 Unbolt and remove the engine-to-transmission struts, if so equipped.

6 Remove the torque converter cover **(see illustration)**.

7 Mark the relationship of the torque converter and driveplate with white paint so they can be installed in the same position.

8 Remove the torque converter-to-driveplate bolts **(see illustration)**. Turn the crankshaft for access to each bolt. Turn the crankshaft in a clockwise direction only (as viewed from the front).

9 Remove the driveshaft (see Chapter 8).

10 Disconnect the speedometer cable (see Section 3). On 1978 and later models, you can remove the adapter assembly from the transmission and use wire to tie it out of the way, with the speedometer cable still attached. Just make sure you tie a plastic bag around the adapter to prevent it from getting dirty.

11 Detach all wire harness connectors from the transmission.

12 Remove any exhaust components which will interfere with transmission removal (see Chapter 4).

13 Disconnect the throttle rod at the transmission (see Section 5).

14 Disconnect the shift linkage (see Section 4).

15 Support the engine with a jack. Use a block of wood under the oil pan to spread the load.

16 Support the transmission with a jack - preferably a jack made for this purpose (often available from rental yards). Safety chains will help steady the transmission on the jack.

17 Remove the transmission mount through-bolt, then remove the bolts holding the crossmember to the chassis.

18 Remove the bolts securing the transmission to the engine and also remove the transmission dipstick tube. **Note:** *The upper bolts are often easier to remove from above, in the engine compartment. Also, on many V8 models, you'll have to remove the oil filter (see Chapter 1) before you can remove the lower right (passenger's side) bolt.*

19 Lower the transmission slightly and disconnect the transmission cooler lines at the transmission, using a flare-nut wrench. **Note 1:** *You'll often need to use a back-up wrench on the transmission cooler line fitting so it doesn't unscrew and cause the line to twist.* **Note 2:** *Fluid will leak out of the lines after you disconnect them. To prevent this, connect the ends of the disconnected lines together with a piece of 3/8-inch diameter fuel hose.*

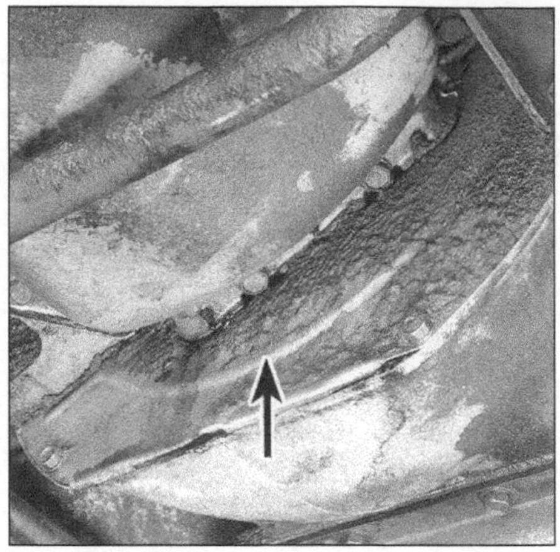

8.6 Unbolt and remove the torque converter cover . . .

8.8 . . . then remove the four torque converter-to-driveplate bolts - be sure to mark the relationship of the torque converter to the driveplate for reassembly reference

20 Remove the transmission dipstick tube.

21 Clamp a pair of Vise-Grip locking pliers onto the lower portion of the transmission case, just in front of the torque converter (but make sure they're not clamped in front of one of the torque converter-to-driveplate bolt locations, or the driveplate will hang up on the pliers). The pliers will prevent the torque converter from falling out while you're removing the transmission. Move the transmission to the rear to disengage it from the engine block dowel pins and make sure the torque converter is detached from the driveplate. Lower the transmission with the jack.

Installation

22 Prior to installation, make sure the torque converter hub is securely engaged in the pump. If you've removed the converter, spread transmission fluid on the torque converter rear hub, where the transmission front seal rides. With the front of the transmission facing up, rotate the converter back and forth. It should drop down into the transmission front pump in stages. To make sure the converter is fully engaged, lay a straightedge across the transmission-to-engine mating surface and make sure the converter hub is at least 1/2-inch below the straightedge. Reinstall the Vise-Grips to hold the converter in this position.

23 With the transmission secured to the jack, raise it into position. Connect the transmission fluid cooler lines.

24 Turn the torque converter to line up the holes with the holes in the driveplate. The white paint mark on the torque converter and driveplate made in Step 7 must line up.

25 Move the transmission forward carefully until the dowel pins and the torque converter are engaged. Make sure the transmission mates with the engine with no gap. If there's a gap, make sure there are no wires or other objects pinched between the engine and transmission and also make sure the torque converter is completely engaged in the transmission front pump. Try to rotate the converter - if it doesn't rotate easily, it's probably not fully engaged in the pump. If necessary, lower the transmission and install the converter fully.

26 Install the transmission-to-engine bolts and tighten them to the torque listed in this Chapter's Specifications. As you're tightening the bolts, check again that the engine and transmission mate completely at all points. If not, find out why. Never try to force the engine and transmission together with the bolts or you'll break the transmission case!

27 Install the torque converter-to-driveplate bolts. Tighten them to the torque listed in this Chapter's Specifications.

28 Install the transmission mount and crossmember.

29 Remove the jacks supporting the transmission and the engine.

30 Install the dipstick tube and engine-to-transmission bolts.

31 Install the starter motor (see Chapter 5).

32 Connect the shift and throttle rod linkage (see Sections 4 and 5, respectively).

33 Plug in the transmission wire harness connectors.

34 Install the torque converter cover and tighten the bolts securely.

35 Install the driveshaft (see Chapter 8).

36 Connect the speedometer cable (see Section 3).

37 Adjust the shift linkage.

38 Install any exhaust system components that were removed or disconnected (see Chapter 4).

39 Remove the jackstands and lower the vehicle.

40 Fill the transmission with the specified fluid (see Chapter 1), run the engine and check for fluid leaks.

Chapter 8
Clutch and driveline

Contents

Specifications

General

Clutch disc lining thickness	1/16 inch (above rivet)

Torque specifications

	Ft-lbs (unless otherwise indicated)
Clutch	
Pressure plate-to-flywheel bolts	
5/16-inch bolts	200 in-lbs
3/8-inch bolts	30
Release fork pivot bolts	200 in-lbs
Clutch housing-to-engine bolts	
3/8-inch bolts	30
7/16-inch bolts	50
Driveshaft	
Roller bushing-to-hub yoke clamp bolts	170 in-lbs
Flange-to-center socket yoke nuts (1972 and 1973 LeBaron)	25
Sliding yoke-to-centering yoke bolts (1972 and 1973 LeBaron)	25
Rear axle	
Pinion shaft lock bolt	100 in-lbs
Differential cover bolts	See Chapter 1

1 General information

The information in this Chapter deals with the components from the rear of the engine to the rear wheels, except for the transmission, which is dealt with in the previous Chapter. For the purposes of this Chapter, these components are grouped into three categories: clutch, driveshaft and axles. Separate Sections within this Chapter offer general descriptions and checking procedures for components in each of the three groups.

Since nearly all the procedures covered in this Chapter involve working under the vehicle, make sure it's securely supported on sturdy jackstands or on a hoist where the vehicle can be easily raised and lowered.

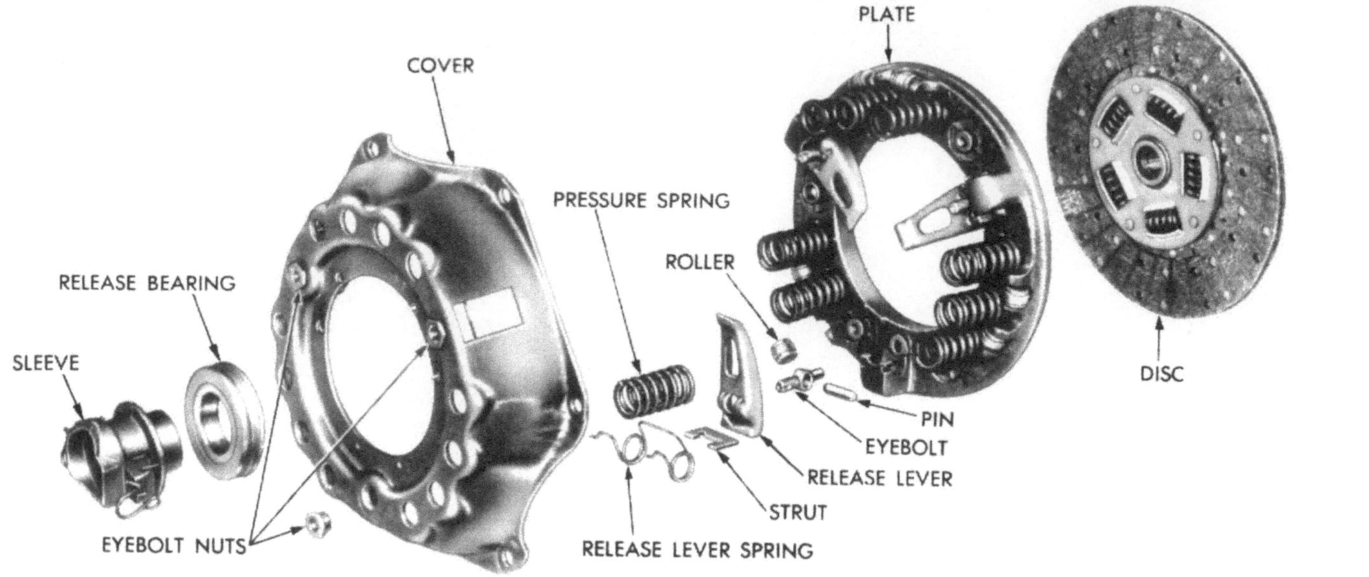

2.1 An exploded view of a typical clutch assembly

2 Clutch - description and check

Refer to illustration 2.1

1 All vehicles with a manual transmission use a single dry-plate type clutch **(see illustration)**. The clutch disc has a splined hub which allows it to slide along the splines of the transmission input shaft. The clutch and pressure plate are held in contact by spring pressure exerted by nine coil springs installed between the cover and the pressure plate.

2 The clutch release system consists of the clutch pedal, a rod connecting the pedal to the torque shaft, the torque shaft itself, another rod connecting the torque shaft to the clutch release lever, and the release, or throwout, bearing.

3 When the clutch pedal is depressed, the pedal rod pushes against the upper lever on the torque shaft, rotating the torque shaft the same direction as forward wheel rotation. This motion moves another lever on the bottom of the torque shaft to the rear, pushing against the rod connected to the outer end of the release lever. When the outer end of the release lever is pushed to the rear, its inner end slides the release bearing forward on the input shaft. The bearing pushes against the fingers of the release levers mounted on the pressure plate, which in turn relieves coil spring pressure against the clutch disc.

4 Terminology can be a problem when discussing the clutch components because common names are in some cases different from those used by the manufacturer. For example, the driven plate is also called the clutch plate or disc, the clutch release bearing is sometimes called a throwout bearing, the release fork is also called a release lever, and so on.

5 Before replacing any components with obvious damage, some preliminary checks should be performed to diagnose clutch problems.

a) The first check should be "clutch spin down time." Run the engine at normal idle speed with the transmission in Neutral (clutch pedal up - engaged). Disengage the clutch (pedal down), wait several seconds and shift the transmission into Reverse. No grinding noise should be heard. A grinding noise would most likely indicate a problem in the pressure plate or the clutch disc.

b) To check for complete clutch release, run the engine (with the parking brake applied to prevent movement) and hold the clutch pedal approximately 1/2-inch from the floor. Shift the transmission between First gear and Reverse several times. If the shift is hard or the transmission grinds, component failure is indicated. Make sure clutch pedal freeplay is adjusted correctly (see Chapter 1).

c) Visually inspect the pivot bushing at the top of the clutch pedal to make sure there is no binding or excessive play.

d) Crawl under the vehicle and make sure the clutch release lever is solidly mounted on the pivot.

3 Clutch components - removal, inspection and installation

Warning: *Dust produced by clutch wear and deposited on clutch components may contain asbestos, which is hazardous to your health. DO NOT blow it out with compressed air and DO NOT inhale it. DO NOT use gasoline or petroleum-based solvents to remove the dust. Brake system cleaner should be used to flush the dust into a drain pan. After the clutch components are wiped clean with a rag, dispose of the contaminated rags and cleaner in a covered, marked container.*

Removal

Refer to illustrations 3.4a, 3.4b and 3.10

1 Access to the clutch components is normally accomplished by removing the transmission and bellhousing, leaving the engine in the vehicle. If, of course, the engine is being removed for major overhaul, then check the clutch for wear and replace worn components as necessary. However, the relatively low cost of the clutch components compared to the time and trouble spent gaining access to them warrants their replacement anytime the engine or transmission is removed, unless they are new or in near-perfect condition. The following procedures are based on the assumption the engine will stay in place.

2 Referring to Chapter 7 Part A, remove the transmission from the vehicle. Support the engine while the transmission is out. Preferably, an engine hoist should be used to support it from above. However, if a jack is used underneath the engine, make sure a piece of wood is positioned between the jack and oil pan to spread the load. **Caution:** *The pickup for the oil pump is very close to the bottom of the oil pan. If the pan is bent or distorted in any way, engine oil starvation could occur.*

3 Remove the bellhousing access cover.

4 Remove the return spring from the clutch release fork and the clutch housing, or from the torque shaft lever **(see illustrations)**.

5 Remove the spring washer securing the fork rod to the torque shaft lever and remove the rod from the torque shaft and release fork **(see illustration 15.2 in Chapter 1)**.

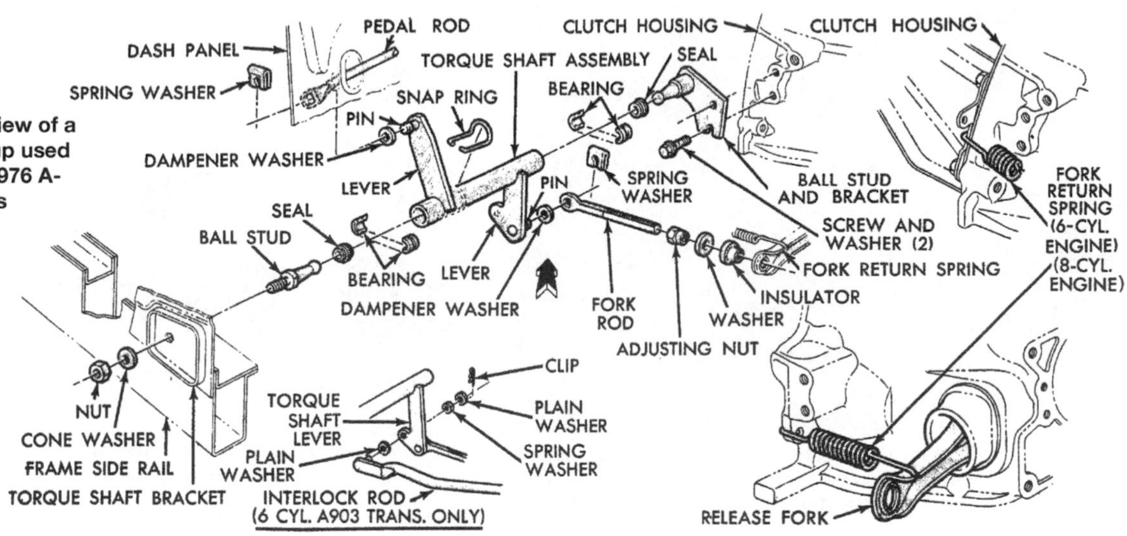

3.4a An exploded view of a typical linkage setup used on 1971 through 1976 A-body vehicles

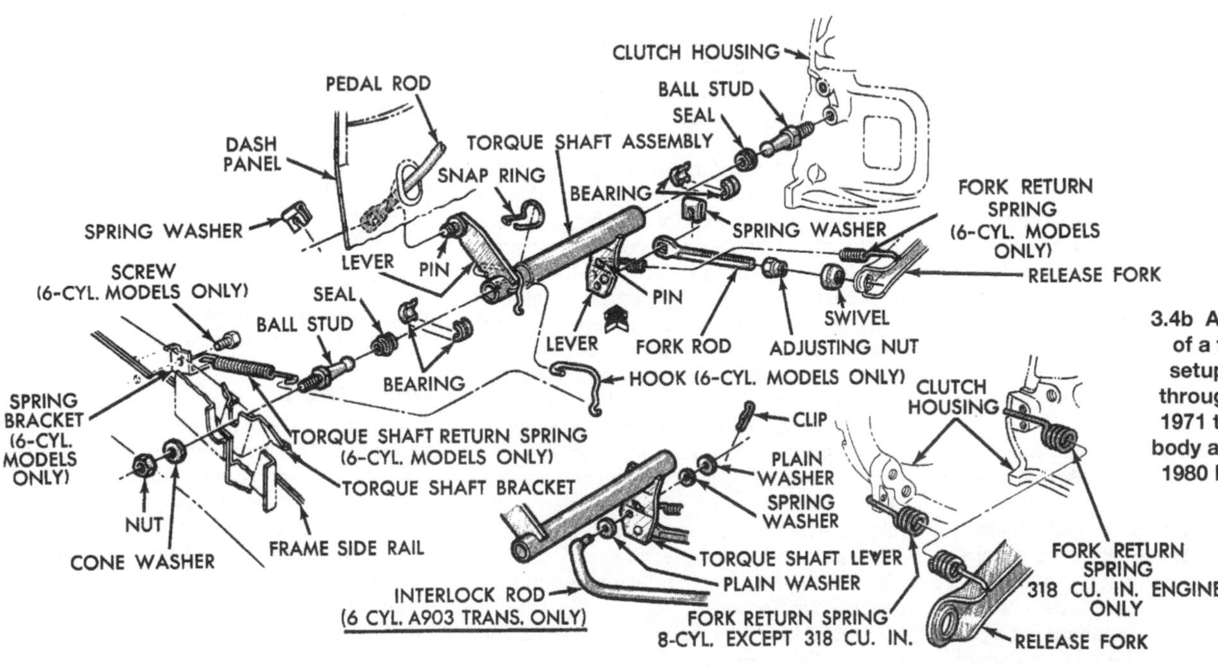

3.4b An exploded view of a typical linkage setup used on 1971 through 1974 E-body, 1971 through 1978 B-body and 1976 through 1980 F-body vehicles

6 On six-cylinder models with an A-903 or A-250 transmission, remove the clip and plain washer securing the interlock rod to the torque shaft lever and remove the spring washer, the plain washer and the rod from the torque shaft **(see illustration 3.4a)**.

7 Remove the bellhousing-to-engine bolts and then detach the housing. It may have to be gently pried off the alignment dowels with a screwdriver or prybar.

8 The clutch fork and release bearing can remain attached to the bellhousing for the time being.

9 To support the clutch disc during removal, install a clutch alignment tool through the clutch disc hub.

10 Carefully inspect the flywheel and pressure plate for indexing marks. The marks are usually an X, an O or a white letter. If they cannot be found, scribe marks yourself so the pressure plate and the flywheel will be in the same alignment during installation **(see illustration)**.

11 Turning each bolt only 1/4-turn at a time, loosen the pressure plate-to-flywheel bolts. Work in a criss-cross pattern until all spring pressure is relieved. Then hold the pressure plate securely and completely remove the bolts, followed by the pressure plate and clutch disc.

Inspection

Refer to illustration 3.13

12 Ordinarily, when a problem occurs in the clutch, it can be attributed to wear of the clutch driven plate assembly (clutch disc). However, all components should be inspected at this time.

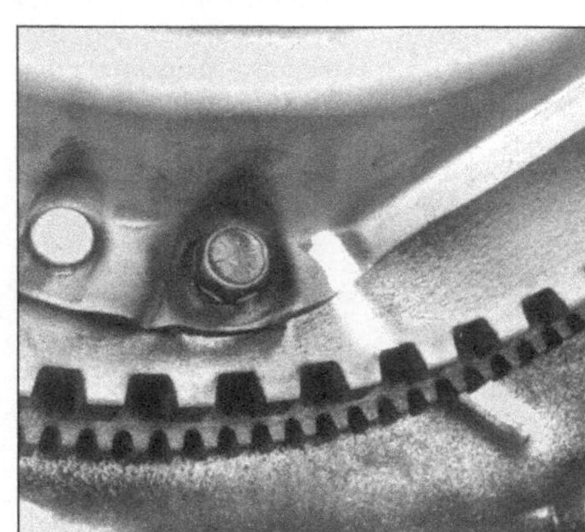
3.10 Be sure to mark the pressure plate and flywheel in order to insure proper alignment during installation (this won't be necessary if a new pressure plate is to be installed)

3.13 Check the flywheel for cracks, hot spots and other obvious defects (slight imperfections can be removed by resurfacing - see an automotive machine shop). This flywheel should be resurfaced

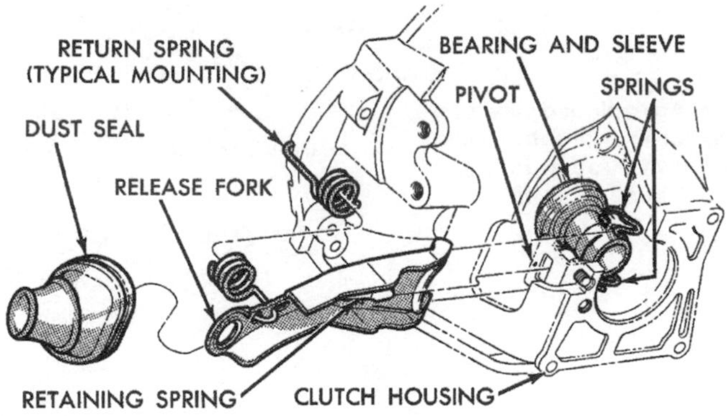

4.4 An exploded view of the release fork and release bearing assemblies

13 Inspect the flywheel for cracks, heat checking, grooves and other obvious defects **(see illustration)**. If the imperfections are slight, a machine shop can machine the surface flat and smooth, which is highly recommended regardless of the surface appearance. See Chapter 2 for the flywheel removal and installation procedure.
14 Inspect the pilot bushing (see Section 5).
15 Inspect the lining on the clutch disc. There should be at least 1/16-inch of lining above the rivet heads. Check for loose rivets, distortion, cracks, broken springs and other obvious damage. As mentioned above, ordinarily the clutch disc is routinely replaced, so if in doubt about the condition, replace it with a new one.
16 The release bearing should also be replaced along with the clutch disc (see Section 4).
17 Check the coil springs for bluing (which could indicate overheating, and loss of temper) and look for damage such as cracks. Check the ends of the release levers (where they contact the throwout bearing) for signs of excessive wear. Finally, be sure to inspect the pressure plate friction surface itself for obvious damage, distortion, cracking, etc. The friction surface of the pressure plate must be clean, dry, lightly sanded and free of scoring, warpage and heavy checking. Light glazing can be removed with sandpaper or emery cloth. If a new pressure plate is required, new and factory-rebuilt units are available.

Installation

18 Before installation, clean the flywheel and pressure plate machined surfaces with lacquer thinner or acetone. It's important that no oil or grease is on these surfaces or the lining of the clutch disc. Handle the parts only with clean hands.
19 Position the clutch disc and pressure plate against the flywheel with the clutch held in place with an alignment tool (available from auto parts stores). Make sure it's installed properly (most replacement clutch plates will be marked "flywheel side" or something similar - if not marked, install the clutch disc with the damper springs toward the transmission).
20 Tighten the pressure plate-to-flywheel bolts only finger tight, working around the pressure plate.
21 Center the clutch disc by ensuring the alignment tool extends through the splined hub and into the pilot bushing in the crankshaft. Wiggle the tool up, down or side-to-side as needed to bottom the tool in the pilot bushing. Tighten the pressure plate-to-flywheel bolts a little at a time, working in a criss-cross pattern to prevent distorting the cover. After all of the bolts are snug, tighten them to the torque listed in this Chapter's Specifications. Remove the alignment tool.
22 Using high-temperature grease, lubricate the inner groove of the release bearing (see Section 4). Also place grease on the release lever contact areas and the transmission input shaft bearing retainer.
23 Install the clutch release bearing as described in Section 4.
24 Install the clutch housing and tighten the bolts to the torque listed in this Chapter's Specifications.
25 Install the transmission, linkage, and any other components previously removed. Tighten all fasteners to the proper torque specifications.

4 Clutch release bearing - removal, inspection and installation

Warning: *Dust produced by clutch wear and deposited on clutch components may contain asbestos, which is hazardous to your health. DO NOT blow it out with compressed air and DO NOT inhale it. DO NOT use gasoline or petroleum-based solvents to remove the dust. Brake system cleaner should be used to flush the dust into a drain pan. After the clutch components are wiped clean with a rag, dispose of the contaminated rags and cleaner in a covered, marked container.*

Removal

Refer to illustration 4.4

1 Disconnect the negative cable from the battery.
2 Remove the transmission (see Chapter 7 Part A).
3 Remove the access cover from the bellhousing, disconnect the linkage and remove the clutch housing from the engine (see Section 3).
4 Remove the clutch release fork from the pivot, then remove the bearing and sleeve from the fork **(see illustration)**.

Inspection

5 Hold the center of the bearing and rotate the outer portion while applying pressure. If the bearing doesn't turn smoothly or if it's noisy, remove it from the sleeve and replace it with a new one. Wipe the bearing with a clean rag and inspect it for damage, wear and cracks. Don't immerse the bearing in solvent - it's sealed for life and to do so would ruin it. Also check the release fork for cracks and other damage. **Note:** *Because of the relative low expense of the release bearing compared with the amount of work it takes to remove it, we recommend replacing the bearing routinely whenever it is removed.*

Installation

6 Lightly lubricate the clutch fork crown and spring retention crown where they contact the bearing sleeve with high-temperature grease. Fill the inner groove of the bearing with the same grease.
7 Attach the release bearing sleeve to the clutch release fork.
8 Lubricate the fork pivot socket with high-temperature grease and push the fork onto the pivot until it's firmly seated. **Note:** *Be sure the retaining spring is positioned behind the pivot.*
9 Apply a light coat of high-temperature grease to the face of the release bearing, where it contacts the pressure plate diaphragm fingers.

10 Install the clutch housing and tighten the bolts to the torque listed in this Chapter's Specifications. Reattach the clutch release linkage (see Section 3).
11 Prior to installing the transmission, apply a light coat of grease to the transmission front bearing retainer.
12 The remainder of installation is the reverse of the removal procedure.
13 Be sure to adjust clutch pedal freeplay when you're done (see Chapter 1).

5 Pilot bushing - inspection and replacement

1 The clutch pilot bushing is an oil-impregnated bronze-type bushing which is pressed into the rear of the crankshaft. It is greased at the factory and does not require additional lubrication. Its primary purpose is to support the front of the transmission input shaft. The pilot bushing should be inspected whenever the clutch components are removed from the engine. Due to its inaccessibility, if you are in doubt as to its condition, replace it with a new one. **Note:** *If the engine has been removed from the vehicle, disregard the following steps which do not apply.*
2 Remove the transmission (see Chapter 7 Part A).
3 Remove the clutch components (see Section 3).
4 Inspect for any excessive wear, scoring, lack of grease, dryness or obvious damage. If any of these conditions are noted, the bushing should be replaced. A flashlight will be helpful to direct light into the recess.
5 The bushing must be pulled from its hole in the crankshaft, gripping the bushing at the rear. Special tools are available, but the do-it-yourselfer may be able to get by with a an alternative tool or fabricated tool. One method that works well is to use a slide-hammer with a tip that has two adjustable hooks, 180-degrees apart. Such tips are commonly available for slide-hammers, and are often included with better-quality slide-hammer kits. A slide-hammer is a tool with many uses, such as pulling dents from body parts and removing seals; you'll use it later for more things than just removing bushings. If a slide-hammer with the correct hooked tip is not available, try to find a hooked tool that will fit into the bushing hole and hook behind the bushing, then clamp a large pair of Vise-Grip locking pliers to the tool and strike the pliers, near the jaws, to pull the bushing out.
6 To install the new bushing, lightly lubricate the outside surface with multi-purpose grease, then drive it into the recess with a soft-face hammer. Most new bushings come already lubricated, but if it is dry, apply a thin coat of high-temperature grease to it.
7 Install the clutch components, transmission and all other components removed previously, tightening all fasteners properly.

6 Driveshaft and universal joints - general information

1 A driveshaft is a tube that transmits power between the transmission and the differential. Universal joints are located at either end of the driveshaft.
2 The driveshaft employs a splined yoke at the front, which slips into the extension housing of the transmission. This arrangement allows the driveshaft to alter its length during vehicle operation. An oil seal prevents leakage of fluid at this point and keeps dirt from entering the transmission. If leakage is evident at the front of the driveshaft, replace the oil seal (see Chapter 7 Part B).
3 The driveshaft assembly requires very little service. The universal joints are lubricated for life and must be replaced if problems develop. The driveshaft must be removed from the vehicle for this procedure.
4 Since the driveshaft is a balanced unit, it's important that no undercoating, mud, etc. be allowed to stay on it. When the vehicle is raised for service it's a good idea to clean the driveshaft and inspect it for any obvious damage. Also, make sure the small weights used to originally balance the driveshaft are in place and securely attached. Whenever the driveshaft is removed it must be reinstalled in the same relative position to preserve the balance.
5 Problems with the driveshaft are usually indicated by a noise or vibration while driving the vehicle. A road test should verify if the problem is the driveshaft or another vehicle component. Refer to the Troubleshooting section at the front of this manual. If you suspect trouble, inspect the driveline (see the next Section).

7 Driveline inspection

1 Raise the rear of the vehicle and support it securely on jackstands. Block the front wheels to keep the vehicle from rolling off the stands.
2 Crawl under the vehicle and visually inspect the driveshaft. Look for any dents or cracks in the tubing. If any are found, the driveshaft must be replaced.
3 Check for oil leakage at the front and rear of the driveshaft. Leakage where the driveshaft enters the transmission extension housing indicates a defective extension housing seal (see Chapter 7 Part B). Leakage where the driveshaft enters the differential indicates a defective pinion seal (see Section 14).
4 While under the vehicle, have an assistant rotate a rear wheel so the driveshaft will rotate. As it does, make sure the universal joints are operating properly without binding, noise or looseness.
5 The universal joints can also be checked with the driveshaft motionless, by gripping your hands on either side of a joint and attempting to twist the joint. Any movement at all in the joint is a sign of considerable wear. Lifting up on the shaft will also indicate movement in the universal joints.
6 Finally, check the driveshaft mounting bolts at the ends to make sure they're tight.

8 Driveshaft - removal and installation

Removal

Refer to illustrations 8.3 and 8.4

1 Disconnect the negative cable from the battery.
2 Raise the vehicle and support it securely on jackstands. Place the transmission in Neutral with the parking brake off.
3 On conventional driveshafts, make reference marks on the driveshaft and the pinion flange in line with each other **(see illustration)**. This is to make sure the driveshaft is reinstalled in the same position to preserve the balance. On 1972 and 1973 LeBaron models, make reference marks on the driveshaft, the rear constant velocity U-joint assembly and the pinion flange.
4 On conventional driveshafts, remove the rear universal joint bolts and clamps **(see**

8.3 Always make reference marks on the driveshaft and pinion flange to insure that the driveshaft's balance is maintained

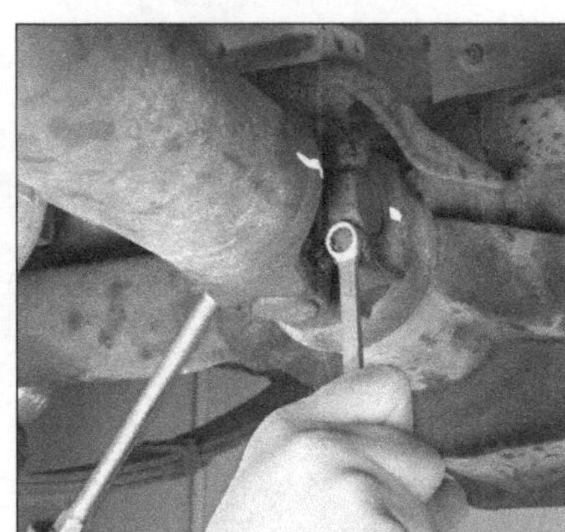

8.4 Use a large screwdriver to hold the flange while breaking the U-joint bolts loose

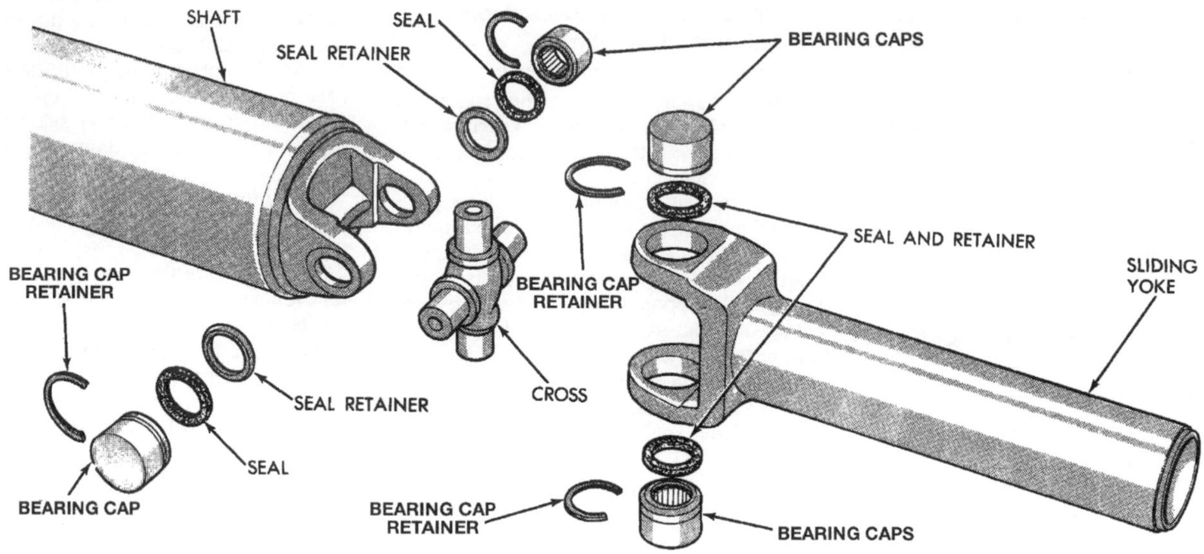

9.3a An exploded view of the front U-joint assembly

illustration). Turn the driveshaft (or wheels) as necessary to bring the bolts into the most accessible position. On 1972 and 1973 LeBaron models, remove the nuts and lock-washers attaching the rear constant velocity U-joint to the rear axle pinion flange **(see illustration 9.13)**.

5 On conventional driveshafts, tape the bearing caps to the cross to prevent the caps from coming off during removal.

6 Lower the rear of the driveshaft. Slide the front of the driveshaft out of the transmission extension housing.

7 Wrap a plastic bag over the transmission extension housing and hold it in place with a rubber band. This will prevent loss of fluid and protect against contamination while the driveshaft is out.

Installation

8 Remove the plastic bag from the transmission extension housing and wipe the area clean. Inspect the oil seal carefully. Procedures for replacement of this seal can be found in Chapter 7.

9 Slide the front of the driveshaft into the transmission.

10 Raise the rear of the driveshaft into position, checking to be sure the marks are in alignment. If not, turn the rear wheels to match the pinion flange and the driveshaft.

11 Remove the tape securing the bearing caps and install the clamps and bolts. Tighten all clamp bolts to the torque listed in this Chapter's Specifications. Lower the vehicle and connect the negative battery cable.

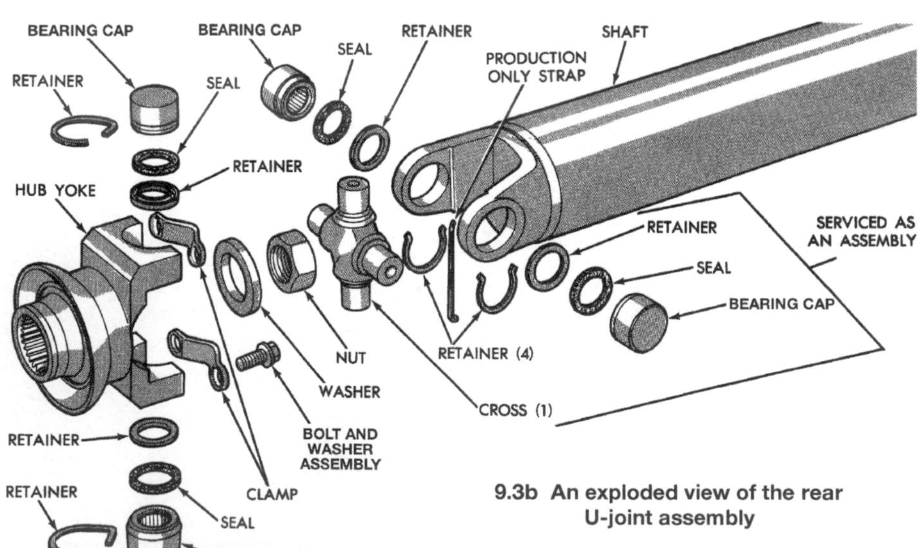

9.3b An exploded view of the rear U-joint assembly

9 Universal joints - replacement

Conventional U-joints

Refer to illustrations 9.3a, 9.3b, 9.3c, 9.5a, 9.5b, 9.9a and 9.9b

Note: *A press or large vise will be required for this procedure. It may be a good idea to take the driveshaft to a repair or machine shop where the universal joints can be replaced for you, normally at a reasonable charge.*

1 Remove the driveshaft (see Section 8).

2 Apply penetrating oil to the bearing caps.

3 Use a screwdriver and hammer to tap off the bearing cap retainers **(see illustrations)**.

4 Supporting the driveshaft, place it in position on either an arbor press or on a workbench equipped with a vise.

5 Place a piece of pipe or a large socket, having an inside diameter slightly larger than

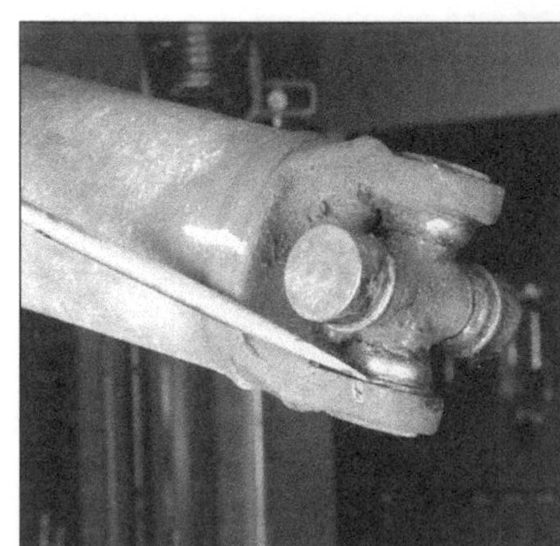

9.3c Remove the inner retainers from the U-joint by tapping them off with a screwdriver and hammer

9.5a Press the bearing caps from the yoke with sockets and a large vise

9.5b Pliers can be used to grip the bearing cap to detach it from the yoke after it has been pushed out

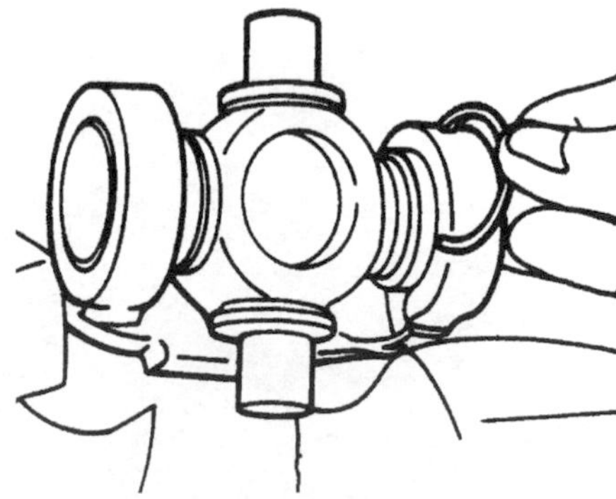

9.9a Press the bearing cap into place and install the retainer in its groove in the bearing cap

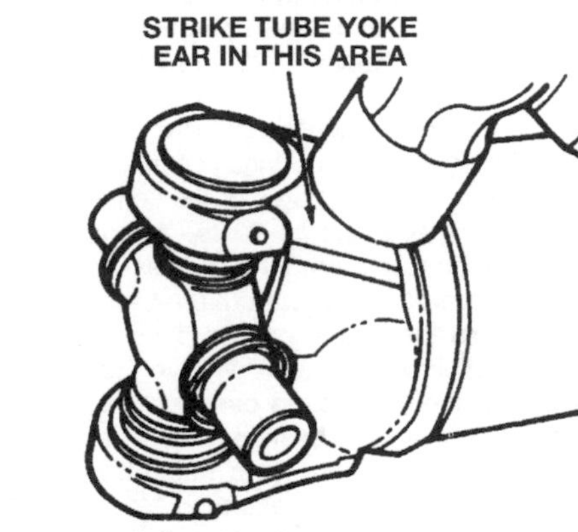

9.9b Strike the yoke with a hammer to reposition the retainer groove (this will also relieve stress in the joint caused by pressing the bearing caps in)

the outside diameter of the bearing caps, over one of the bearing caps. Position a socket with an outside diameter slightly smaller than that of the opposite bearing cap against the cap **(see illustration)** and use the vise or press to force the bearing cap out (inside the pipe or large socket), stopping just before it comes completely out of the yoke. Use the vise or large pliers to work the bearing cap the rest of the way out **(see illustration)**.

6 Transfer the sockets to the other side and press the opposite bearing cap out in the same manner.

7 Pack the new universal joint bearings with chassis grease. Position the spider in the yoke and partially install one bearing cap in the yoke. If the replacement spider is equipped with a grease fitting, be sure it's offset in the proper direction (toward the driveshaft).

8 Start the spider into the bearing cap and then partially install the other bearing cap. Align the spider and press the bearing caps into position, being careful not to damage the dust seals.

9 Install the bearing cap retainers. If difficulty is encountered in seating the retainers, strike the driveshaft yoke sharply with a hammer. This will spring the yoke ears slightly and allow the retainers to seat in their grooves **(see illustrations)**.

10 Install the grease fitting (if equipped) and fill the joint with grease. Be careful not to overfill the joint, as this could blow out the grease seals.

11 Install the driveshaft (see Section 8).

Constant-velocity universal joint (1972 and 1973 LeBaron models only)

Disassembly

Refer to illustrations 9.13, 9.15 and 9.16

12 Remove the driveshaft (see Section 8).

13 Remove the four screws and lockwashers which attach the spline yoke to the constant velocity joint and remove the spline yoke **(see illustration)**. Slide the two loose

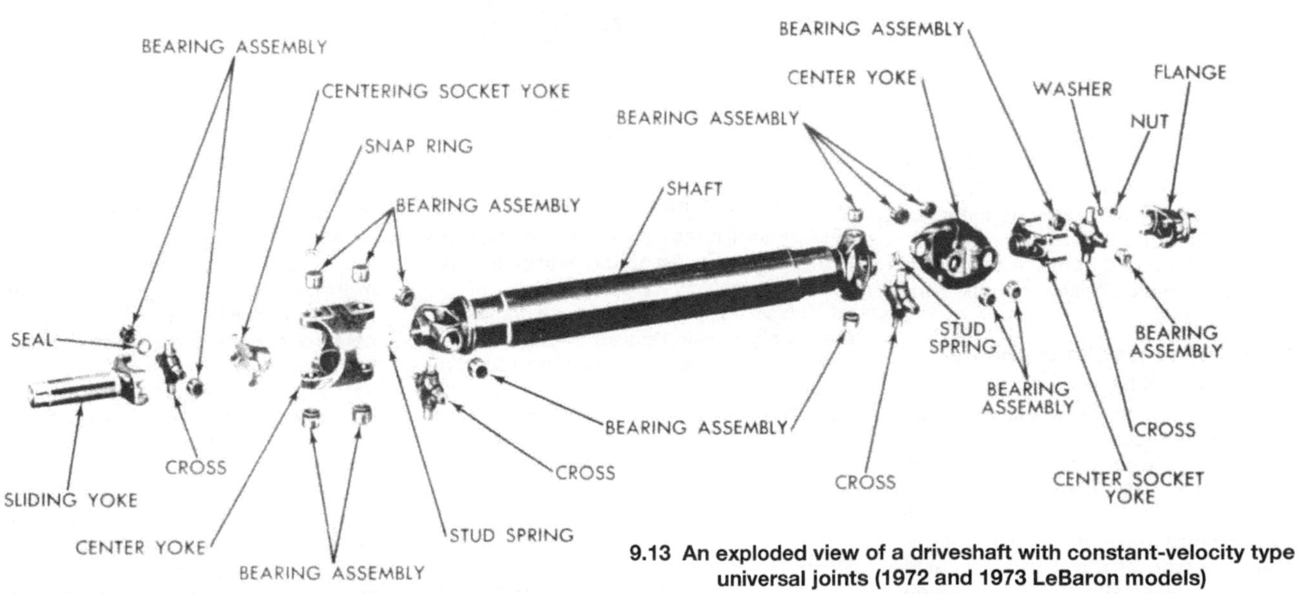

9.13 An exploded view of a driveshaft with constant-velocity type universal joints (1972 and 1973 LeBaron models)

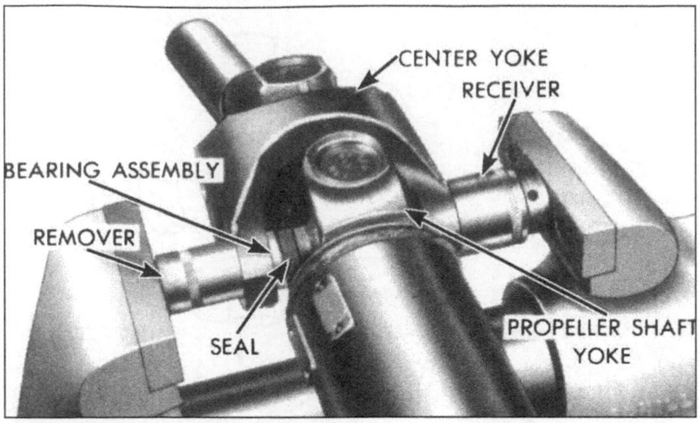

9.15 Press the bearing assemblies about 3/8-inch out of the the yokes with a "remover" (3/4-inch diameter piece of round bar stock or 3/4-inch socket) and a "receiver" (short length of pipe or socket with an inside diameter of not less than 1-1/16 inch) as shown (1972 and 1973 LeBaron models)

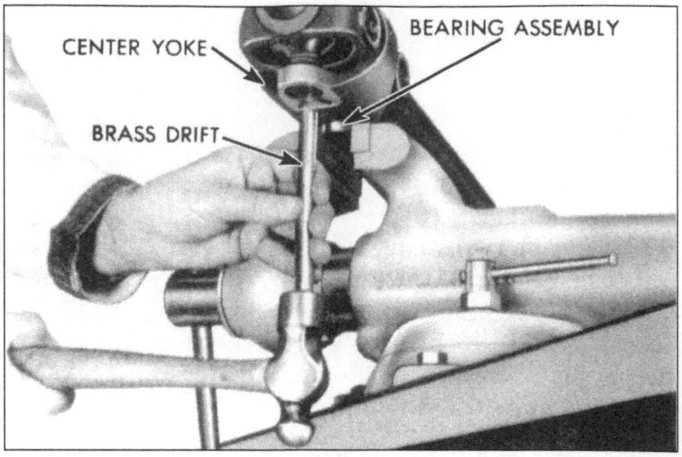

9.16 Clamp the exposed part of the bearing in the vise and drive the yoke from the bearing with a brass drift; use light blows - don't hit the drift too hard (1972 and 1973 LeBaron models)

bearings from the centering socket yoke.

14 Remove the snap-rings securing the two bearings in the front bores of the center yoke. **Note:** *If the bearings are rusted or corroded, apply penetrating lubricant in the bearing bores before attempting to press out the bearings.*

15 Press the bearing assemblies from the yokes as follows: Using a "remover" (3/4-inch diameter piece of round bar stock or a 3/4-inch socket) on one bearing, and a "receiver" (short length of pipe or a socket with an inside diameter of not less than 1-1/16 inch) on the opposite bearing, clamp up the joint, remover and receiver in a bench vise **(see illustration)** and press one of the rear yoke bearings about 3/8-inch out of the yoke.

16 Clamp the exposed part of the bearing with the vise and drive the yoke from the bearing with a brass drift **(see illustration)**. Use light blows - don't hit the drift too hard.

17 Using the same procedure, press the exposed end of the cross to force the bearing on the opposite end about 3/8-inch out of the yoke. Remove the bearing from the yoke with a brass drift as described in Step 15.

18 With the driveshaft clamped firmly in the vise, press in on the cross and the centering socket yoke and remove the cross and socket yoke assembly from the center yoke.

19 Remove the remaining four bearings from the rear bores of the center yoke and driveshaft yoke in the same manner as described in Steps 15, 16 and 17.

20 Remove the cross from the driveshaft yoke. Remove the stud spring from the centering stud.

Centering socket yoke assembly

Refer to illustration 9.24

21 Carefully pry the centering ball seal assembly from the socket yoke.

22 Remove the seal and bearing rollers from the centering ball assembly.

23 Fill the cavity behind the centering ball and inside the ball with multi-purpose grease.

24 Obtain a pipe or rod with an outside diameter slightly smaller than the inside diameter of the centering ball. Insert the rod into the ball and strike it sharply with a hammer **(see illustration)**. The force applied by this hammer blow should force the ball and retainer assembly from the yoke.

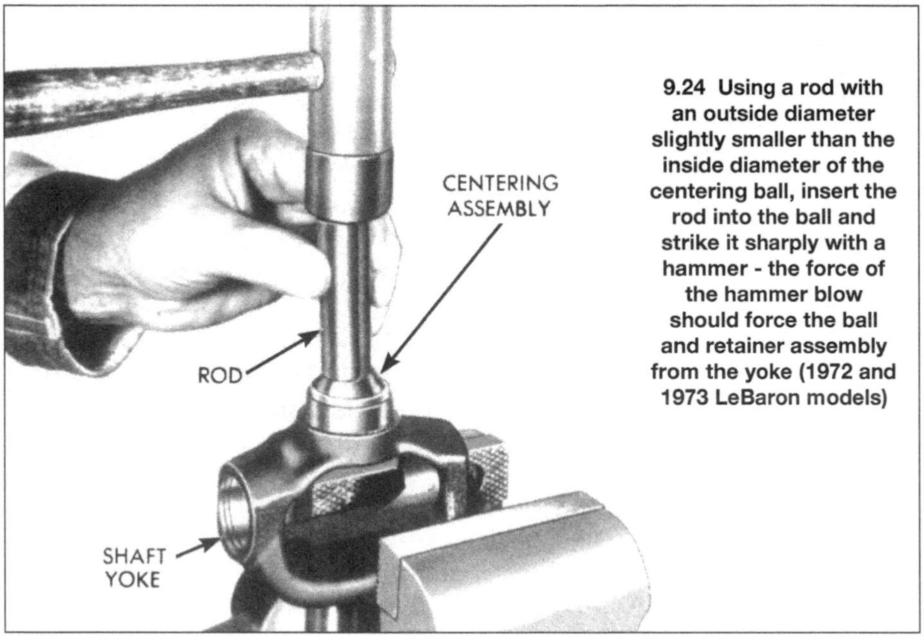

9.24 Using a rod with an outside diameter slightly smaller than the inside diameter of the centering ball, insert the rod into the ball and strike it sharply with a hammer - the force of the hammer blow should force the ball and retainer assembly from the yoke (1972 and 1973 LeBaron models)

Cleaning and inspection

25 Clean all parts with solvent and blow dry them with compressed air. **Warning:** *Wear eye protection!*

26 Inspect all parts for damage and excessive damage wear. Discard all damaged and worn parts. Inspect the bearing races for grooves and ridges. Don't reuse rollers that have been operating in damaged races. Inspect the seals on the cross assemblies for damage. These seals cannot be serviced separately. They're part of the cross assembly replacement kit.

Reassembly

27 Position the centering assembly in the yoke with the large-diameter hole facing up and press it firmly into its seat.

28 Apply a film of multi-purpose grease on the inside surface of the centering ball. Install the rollers (there should be 34 of them). Install the centering stud seal in the ball.

29 Install the centering ball seal assembly onto the yoke and press it firmly into place.

30 Coat the inside surfaces of the bearing races with multi-purpose grease and install the rollers (32 of these). Pack the reservoirs in the ends of the cross with the same lubricant.

31 Place the cross in the shaft yoke. Insert one bearing assembly in the bearing bore of the shaft yoke. Using the bar stock or socket you used as a remover when disassembling the joint, press the bearing into place and simultaneously guide the cross into the bearing. Press the bearing into the yoke far enough (about 3/16-inch) to install the snap-ring. Install the snap-ring, then reverse the position of the yoke

and install the bearing and snap-ring in the opposite bore in the same manner.
32 Install the two bearings in the rear bores of the center yoke and on the cross as previously described. Install the two snap-rings.
33 Install the centering stud spring on the centering stud (large end first). Apply a film of multi-purpose grease on the stud.
34 Install the two slip spline yoke bearing assemblies on the cross and assemble them in the bearing bores of the yoke.
35 Install the centering yoke and cross as an assembly in the center yoke, guiding the centering ball on the stud.
36 Apply slight pressure on the cross to align the cross in the front bores of the center yoke. Insert one bearing in the yoke and guide the end of the cross into the bearing.
37 Press the bearing into the bore and install the snap-ring.
38 Install the remaining bearing in the center yoke. Install the snap-ring.
39 Install the sliding yoke onto the center yoke, install the lock-washers and bolts and tighten the bolts to the torque listed in this Chapter's Specifications.
40 Install the driveshaft (see Section 8).

10 Rear axle - description and check

Description

1 Axles are referred to by the diameter (in inches) of the differential ring gear - "7-1/4," "8-1/4," "8-3/4" or "9-1/4." Two basic types of axles are used: On some units, the bearing is pressed onto the axleshaft; on others, the bearing is pressed into the axle tube. Although you can't actually see how the axle bearings are installed in either type, you can quickly identify which type you're servicing by looking for a retainer bolted to the inside of the brake backing plate. 1971 through 1982 7-1/4 inch and all 8-3/4 inch axles use flange-type retainers at the brake backing plate (so you know the bearing is pressed onto the axleshaft). 8-1/4 inch, 9-1/4 inch and later 7-1/4 inch axles use "C-washer" locks to secure the axles to the differential (there are no retainers, so you know the bearing is pressed into the axle tubes on these units). All rear axle assemblies are a *semi-floating* design, i.e. the axle supports the weight of the vehicle on the axleshaft in addition to transmitting driving forces to the rear wheels. All axles also use a *hypoid*-type pinion gear, i.e. the pinion gear teeth are cut in a spiral configuration to allow the centerline of the pinion to be lower than the centerline of the ring gear, so the vehicle floor can be lower. All differential housings are cast iron and the axle tubes are made of steel, pressed and welded into the differential housing. Except for the 8-3/4 inch axle, all other differentials can be inspected and serviced by removing a pressed-steel access cover from the back side of the differential housing. The 8-3/4 inch axle uses a "drop-out" style differential: The carrier, which is bolted to the front side of the axle housing, is also the differential housing; the entire differential assembly can be removed as a unit by removing the axleshafts, then unbolting and "dropping out" the differential from the front of the axle housing.
2 Most axles use a conventional differential. But a "Sure-Grip" model is an available option. A conventional differential allows the driving wheels to rotate at different speeds during cornering (the outside wheel must rotate faster than the inside wheel), while equally dividing the driving torque between the two wheels. In most normal driving situations, this function of the differential is appropriate. However, the total driving torque can't be more than double the torque at the lower-traction wheel. When traction conditions aren't the same for both driving wheels, some of the available traction is lost. An optional "Sure-Grip" differential is available for poor traction conditions - i.e. mud, ice or snow. During normal cornering and turning conditions, its controlled internal friction is easily overcome, allowing the driving wheels to turn at different speeds, just like a conventional differential. However, during extremely slippery conditions, the Sure-Grip differential allows the driving wheel with the better traction condition to develop more driving torque than the other wheel. Thus, the total driving torque can be significantly greater than with a conventional differential.
3 Obviously, with all these variations in axle size and type, proper identification is critical. If you're not sure what axle the vehicle has, make the following inspection:

a) *Does the axle have a bolt-on pressed-steel cover on the back side of the differential housing? If not, it must be an 8-3/4 inch axle.*
b) *Does it use flange-type axle retainers? If not, it could only be an 8-1/4 inch, 9-1/4 inch or 1983 or newer 7-1/4 inch axle.*
c) *Does it have a hooded breather on the left axle tube? If it does, It's a 7-1/4 inch axle.*
d) *Ring gear diameter can be determined by counting the number of cover screws and measuring the diameter of the axle tubes: 7-1/4 inch axles have a 10-screw cover and 1-1/2 inch diameter axle tubes; 8-1/4 inch axles have a 10-screw cover and 3-inch axle tubes; 9-1/4 inch axles have 12-screw covers.*
e) *Does it have a Sure-Grip differential? One easy way to tell is to raise the rear of the vehicle, place it securely on jackstands and rotate the rear wheels. If they both turn in the same direction, the axle is equipped with a Sure-Grip; if they turn in opposite directions, it's a conventional differential. Another way to tell - without raising the vehicle - is to shine a flashlight into the filler plug hole and look at the differential case. When you look into the "windows" of a Sure-Grip differential, you'll see coil springs.*
f) *Need to know the axle ratio? It's usually stamped onto a small metal tag attached to one of the bolts that secures the differential access cover.*

Check

4 Many times, a problem is suspected in an axle area when, in fact, it lies elsewhere. For this reason, a thorough check should be performed before assuming an axle problem.
5 The following noises are those commonly associated with axle diagnosis procedures:

a) *Road noise is often mistaken for mechanical faults. Driving the vehicle on different surfaces will show whether the road surface is the cause of the noise. Road noise will remain the same if the vehicle is under power or coasting.*
b) *Tire noise is sometimes mistaken for mechanical problems. Tires which are worn or low on pressure are particularly susceptible to emitting vibrations and noises. Tire noise will remain about the same during varying driving situations, where axle noise will change during coasting, acceleration, etc.*
c) *Engine and transmission noise can be deceiving because it will travel along the driveline. To isolate engine and transmission noises, make a note of the engine speed at which the noise is most pronounced. Stop the vehicle and place the transmission in Neutral and run the engine to the same speed. If the noise is the same, the axle is not at fault.*

6 Overhaul and general repair of the front or rear axle differential are beyond the scope of the home mechanic due to the many special tools and critical measurements required. Thus, the procedures listed here will involve axleshaft removal and installation, axleshaft oil seal replacement, axleshaft bearing replacement and removal of the entire unit for repair or replacement.

11 Rear axleshaft - removal and installation

8-1/4 inch, 9-1/4 inch and 1983 and later 7-1/4 inch axles

Refer to illustrations 11.2, 11.3a, 11.3b and 11.4

1 The usual reason for axleshaft removal is that the bearing is worn or the seal is leaking. To check the bearing, raise the vehicle, support it securely on jackstands and remove the wheel and brake drum (see Chapter 9). Move the axle flange up and down. If the axleshaft moves up and down, bearing wear is excessive. Also check for differential lubricant leaking out from below the axleshaft - this indicates the seal is leaking.
2 Remove the cover **(see illustration)** from the differential carrier and allow the oil to drain into a suitable container.

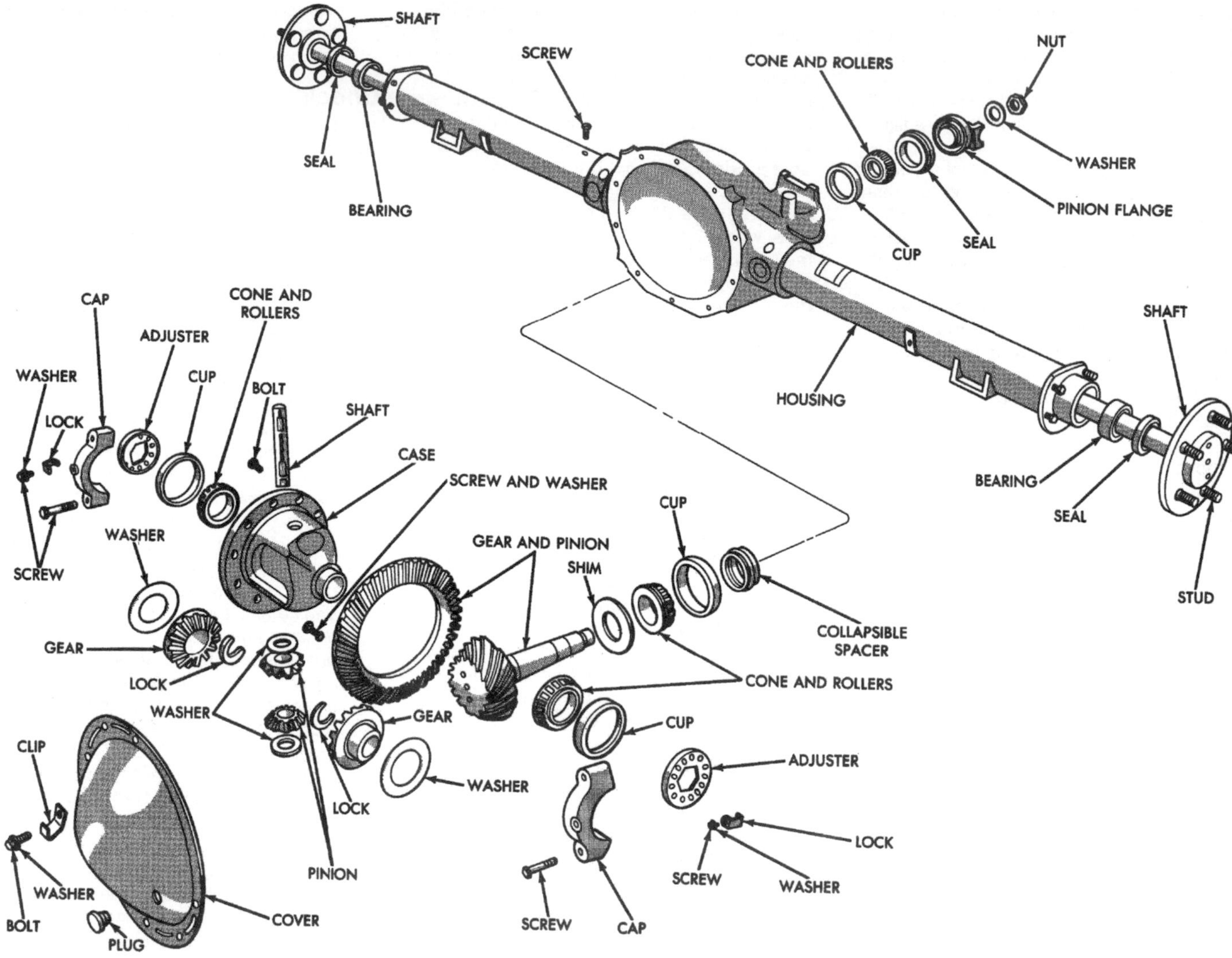

11.2 An exploded view of a typical rear axle assembly using C-clip type locks inside the differential (8-1/4 inch axle shown, 9-1/4 and 1983 and later 7-1/4 axles similar)

11.3a Remove the pinion shaft lock bolt . . .

11.3b . . .then carefully remove the pinion shaft from the differential carrier (don't turn the wheels or the carrier after the shaft has been removed, or the pinion gears may fall out)

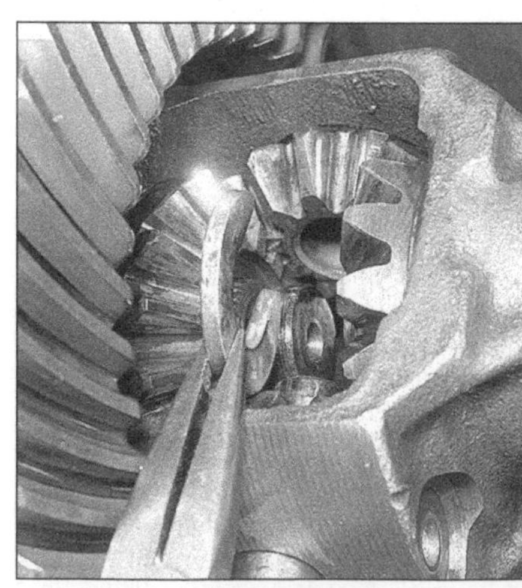

11.4 Push the axle flange in, then remove the C-lock from the inner end of the axleshaft

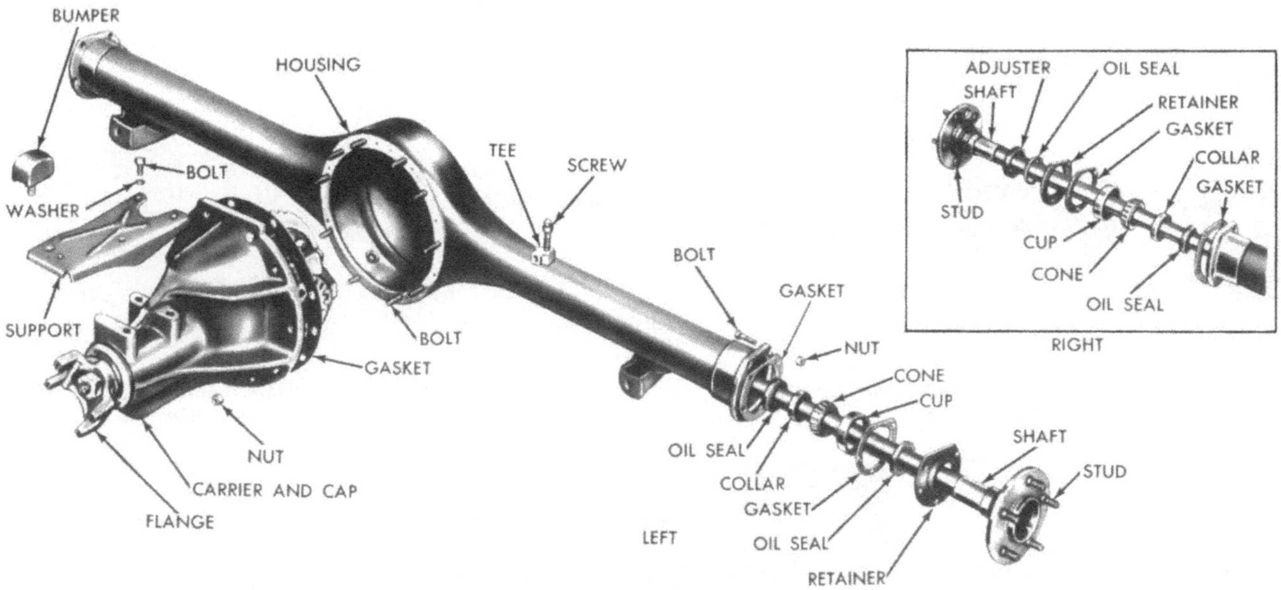

11.10a An exploded view of a typical 8-3/4 inch rear axle assembly (1971 through 1974 models)

3 Remove the lock bolt from the differential pinion shaft **(see illustration)**. Remove the pinion shaft **(see illustration)**.

4 Push the outer (flanged) end of the axleshaft in and remove the C-lock from the inner end of the shaft **(see illustration)**.

5 Withdraw the axleshaft, taking care not to damage the oil seal in the end of the axle housing as the splined end of the axleshaft passes through it.

6 Installation is the reverse of removal. Tighten the pinion shaft lock bolt to the torque listed in this Chapter's Specifications.

7 Always use a new cover gasket and tighten the cover bolts to the torque listed in the Chapter 1 Specifications.

8 Refill the axle with the correct quantity and grade of lubricant (see Chapter 1).

1971 through 1982 7-1/4 inch and all 8-3/4 inch axles

Refer to illustrations 11.10a, 11.10b, 11.11a and 11.11b

9 The usual reason for axleshaft removal is that the bearing is worn or the seal is leaking. To check the bearing, raise the vehicle, support it securely on jackstands and remove the wheel and brake drum. Move the axle flange up and down. If the axleshaft moves up and down, bearing wear is excessive. Also check for differential lubricant leaking out from below the axleshaft - this indicates the seal is leaking.

10 Unscrew and remove the nuts which attach the retainer to the brake backing plate **(see illustrations)**. A large hole is provided in the axle flange (where the brake drum attaches) for access to the nuts with a socket,

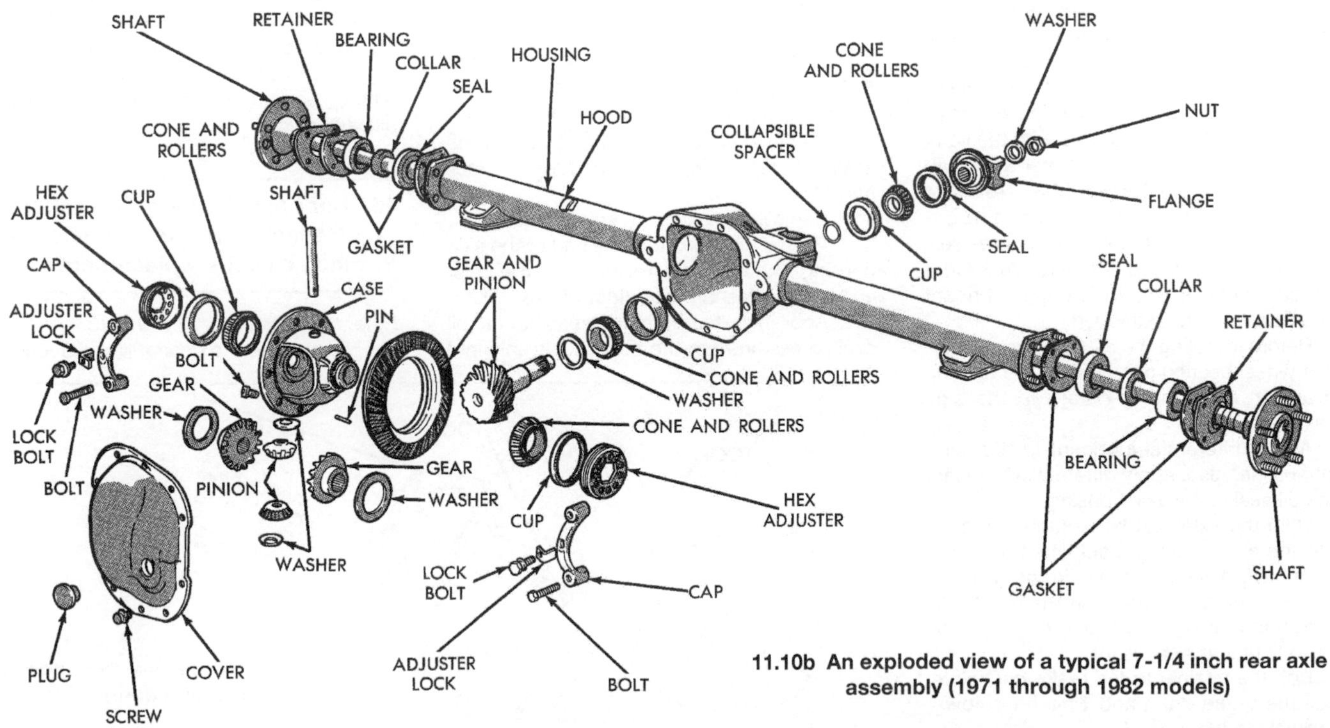

11.10b An exploded view of a typical 7-1/4 inch rear axle assembly (1971 through 1982 models)

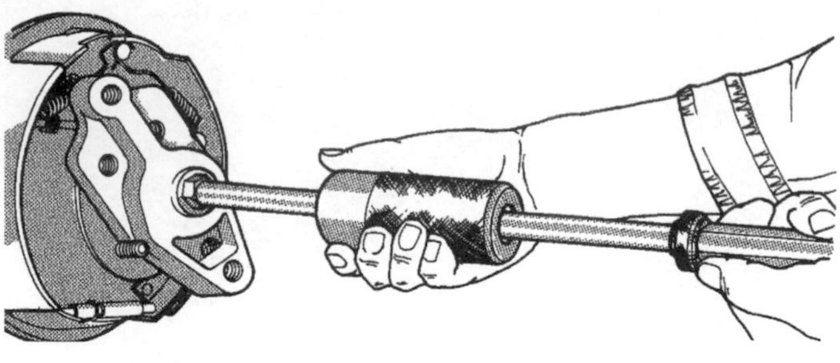

11.11a Attach a slide hammer to the wheel mounting studs and withdraw the axleshaft

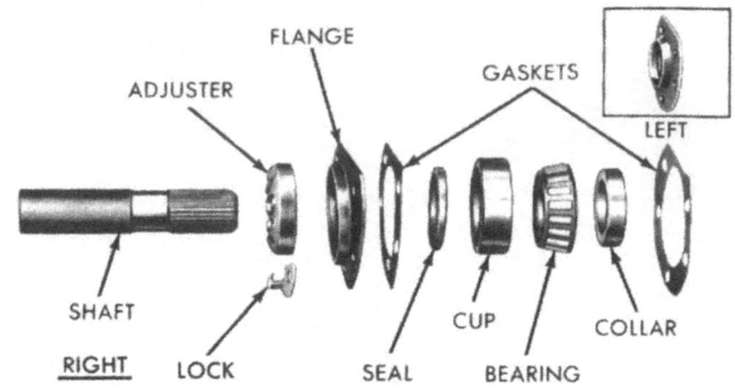

11.11b An exploded view of the axleshaft components (8-3/4 inch axle shown, 7-1/4 axle similar)

ratchet and extension. Rotate the axle to align this hole with each nut.

11 Attach a slide hammer to the wheel mounting studs and withdraw the axleshaft **(see illustrations)**. Sometimes, if the axleshaft is not too tight, you can pull it out by hand, but be very careful not to pull too hard or you'll pull the vehicle off the jackstands.

12 As the axleshaft is removed, it is possible that the bearing will come apart. This indicates the bearing has failed and will need to be replaced; be sure to remove all bearing parts from the axle housing. Take the axleshaft to an automotive machine shop to have the bearing replaced, since replacement requires pressing a new bearing and collar onto the shaft with a hydraulic press.

13 Considering the relative low cost of axle seals compared with the amount of work involved in removing the axle, we recommend replacing the axle seal whenever the axleshaft is removed. To replace the seal, use a slide-hammer-type tool or seal removal tool to pull the oil seal out of the axle housing. Clean the seal bore so there's no oil where the new seal will seat. Coat the outside diameter of the new seal with RTV sealant, then drive the seal into the seal bore with the lip of the seal facing in, using a hammer and seal-installer tool **(see illustration 12.3). Note:** *If a seal removal tool is not available, a large-diameter socket (slightly smaller than the seal bore) and an extension will also work.* Coat the rubber seal lip with differential lubricant so the seal won't be damaged.

14 Before installing the axleshaft assembly, smear wheel bearing grease onto the bearing end and in the bearing recess in the axle housing tube.

15 Apply differential lubricant to the axleshaft splines. Also, apply differential lubricant to the oil seal in the axle housing.

16 Hold the axleshaft horizontal and insert it into the axle housing. Feel when the shaft splines have picked up those in the differential side gears and then push the shaft fully into position, using a soft-faced hammer on the end flange as necessary.

17 Bolt the retainer to the brake backplate, install the brake drum and wheel and lower the vehicle to the ground.

12.2 The end of the axleshaft can be used to pry the old seal out of the axle housing

12 Rear axleshaft oil seal (8-1/4 inch, 9-1/4 inch and 1983 and later 7-1/4 inch axles) - replacement

Refer to illustrations 12.2 and 12.3

Note: *Oil seal replacement on 8-3/4 and earlier 7-1/4 inch axles is covered as part of the axleshaft replacement procedure in Section 11.*

1 Remove the axleshaft (see Section 11).

2 Pry the oil seal out of the end of the axle housing with a large screwdriver or the inner end of the axleshaft **(see illustration)**.

3 Apply high-temperature grease to the oil seal recess and tap the new seal evenly into place with a hammer and seal installation tool **(see illustration)**, large socket or piece of pipe so the lips are facing in and the metal face is visible from the end of the axle housing. When correctly installed, the face of the oil seal should be flush with the end of the axle housing.

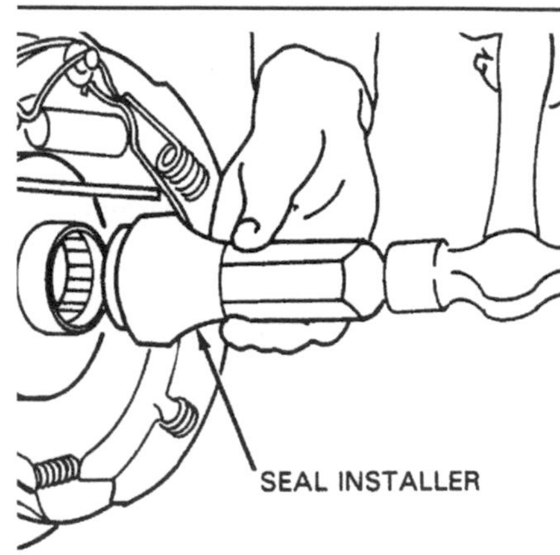

12.3 Use a seal installer tool, large socket or piece of pipe to tap the seal evenly into place

13 Rear axle bearing (8-1/4 inch, 9-1/4 inch and 1983 and later 7-1/4 inch axles) - replacement

Refer to illustrations 13.3 and 13.4

1 Remove the axleshaft (see Section 11)

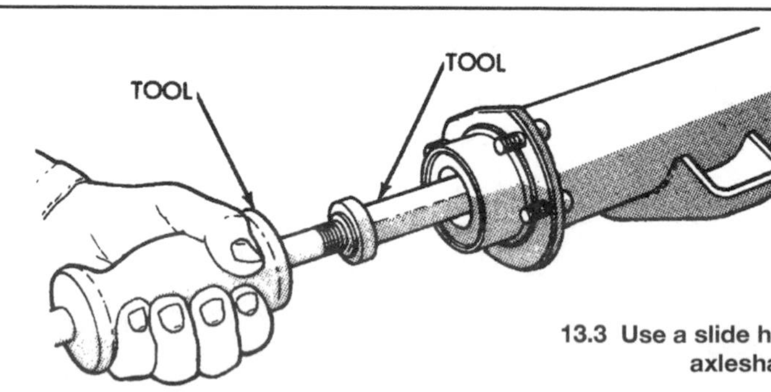

13.3 Use a slide hammer to remove the axleshaft bearing

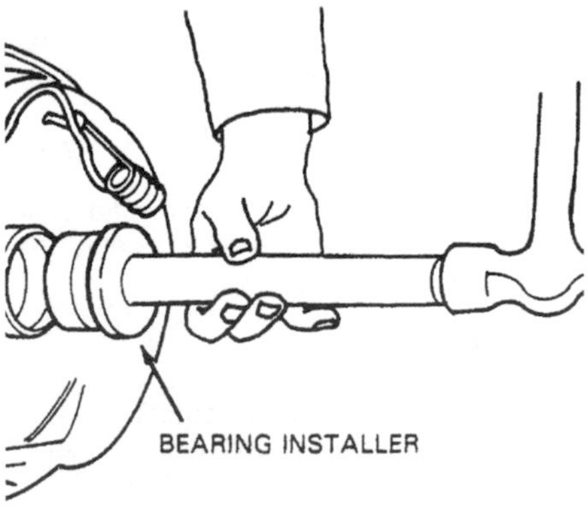

13.4 Use a special bearing installer tool, large socket or piece of pipe to tap the bearing evenly into the axle housing

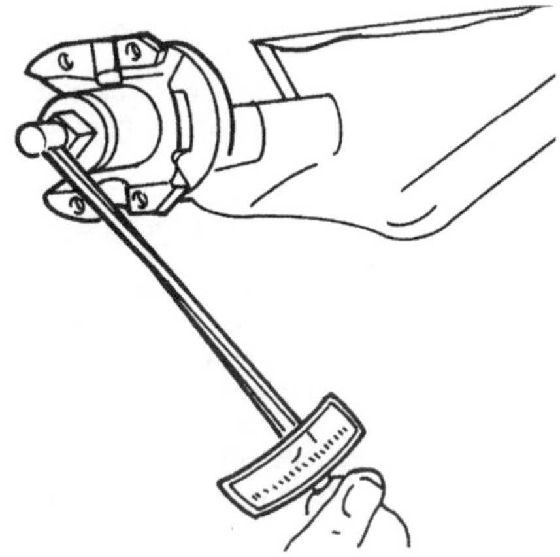

14.4 Use an in-lbs torque wrench to check the torque necessary to move the pinion shaft

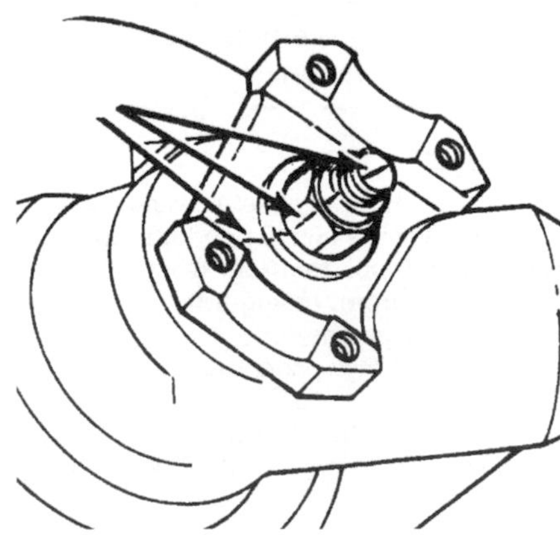

14.5 Mark the relative positions of the pinion, nut and flange (arrows) before removing the nut

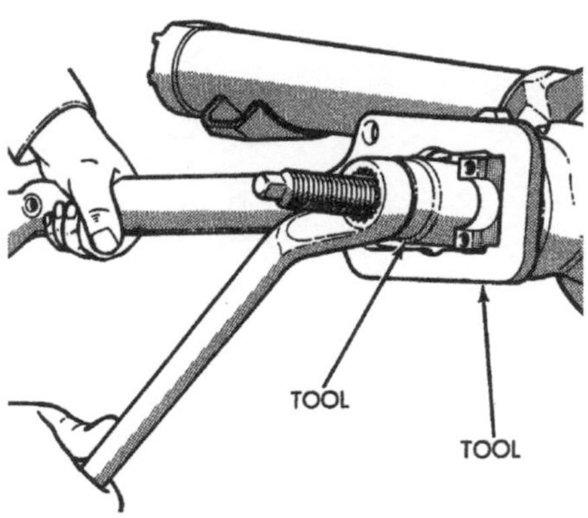

14.7a If you've got the factory tool or something similar, it's easy to hold the pinion flange while breaking loose the lock nut

14.7b If you don't have the factory tool or one like it, try holding the flange with a prybar or large screwdriver while you loosen the locknut

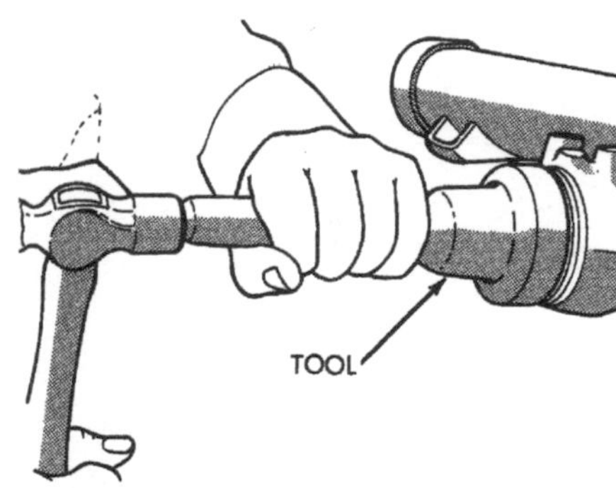

14.11 Use the proper-size driver to install the pinion seal

and the oil seal (see Section 12).

2 A slide-hammer-type puller which grips the bearing from behind will be required for this job.

3 Attach the puller and extract the bearing from the axle housing **(see illustration)**.

4 Clean out the bearing recess and drive in the new bearing with a bearing installer or a piece of pipe positioned against the outer bearing race **(see illustration)**. Make sure the bearing is tapped in to the full depth of the recess and the numbers on the bearing are visible from the outer end of the axle housing.

5 Install a new oil seal (see Section 12), then install the axleshaft (see Section 11).

14 Pinion oil seal - replacement

Refer to illustrations 14.4, 14.5, 14.7a, 14.7b and 14.11

1 Disconnect the cable from the negative battery terminal.

2 Raise the rear of the vehicle and support it securely on jackstands. Block the front wheels to keep the vehicle from rolling off the stands.

3 Disconnect the driveshaft and fasten it out of the way.

4 Use an inch-pound torque wrench to check the torque required to rotate the pinion. Record it for use later **(see illustration)**.

5 Scribe or punch alignment marks on the pinion shaft, nut and flange **(see illustration)**.

6 Count the number of threads visible between the end of the nut and the end of the pinion shaft and record it for use later.

7 A special flange holding tool can be used to keep the companion flange from moving while the self-locking pinion nut is loosened **(see illustration)**. If the special tool isn't available, try locking the flange with a prybar or a large screwdriver **(see illustration)**.

8 Remove the pinion nut.

9 Withdraw the companion flange. It may be necessary to use a two or three-jaw puller engaged behind the flange to draw it out. Do Not attempt to pry behind the flange or hammer on the end of the pinion shaft.

10 Pry out the old seal and discard it.

11 Lubricate the lips of the new seal with high-temperature grease and tap it evenly into position with a seal installation tool or a large socket. Make sure it enters the housing squarely and is tapped in to its full depth **(see illustration)**.

12 Align the mating marks made before disassembly and install the companion flange. If necessary, tighten the pinion nut to draw the flange into place. Do not try to hammer the flange into position.

13 Apply non-hardening sealant to the ends of the splines visible in the center of the flange so oil will be sealed in.

14 Install the washer (if equipped) and pinion nut. Tighten the nut carefully until the original number of threads are exposed and the marks are aligned.

15 Measure the torque required to rotate the pinion and tighten the nut in small increments until it matches the figure recorded in Step 4. In order to compensate for the drag of the new oil seal, the nut should be tightened more until the rotational torque of the pinion exceeds earlier recording by 10 in-lbs. with a torque reading of at least 210 ft-lbs. on the nut.

16 Connect the driveshaft and lower the vehicle.

15 Rear axle assembly - removal and installation

Removal

1 Loosen the rear wheel lug nuts, raise the rear of the vehicle and support it securely on jackstands. Block the front wheels to keep the vehicle from rolling off the stands. Remove the rear wheels.
2 Position a jack under the rear axle differential case.
3 Disconnect the driveshaft from the rear axle companion flange. Fasten the driveshaft out of the way with a piece of wire to the underbody.
4 Disconnect the shock absorbers at the lower mounts, then compress them to get them out of the way.
5 If equipped, disconnect the vent hose from the fitting on the axle housing and fasten it out of the way.
6 Disconnect the brake hose from the junction block on the axle housing, then plug the hose to prevent fluid leakage.
7 Remove the brake drums (see Chapter 9).
8 Disconnect the parking brake cables from the actuating levers and the backing plate (see Chapter 9).
9 Disconnect the spring U-bolts (see Chapter 10). Remove the spacers and clamp plates.
10 Disconnect any wiring interfering with removal.
11 Lower the jack under the differential, then remove the rear axle assembly from under the vehicle.

Installation

12 Installation is the reverse of removal. Lower the vehicle weight onto the wheels before tightening the U-bolt nuts completely.
13 Bleed the brakes (see Chapter 9).

Chapter 9 Brakes

Contents

Specifications

Disc brakes

Brake pad minimum thickness	See Chapter 1
Disc lateral runout limit	
1971 and 1972	0.0025 inch
1973 on	0.004 inch
Disc minimum thickness	Cast into disc
Thickness variation	0.0005 inch

Drum brakes

Minimum brake lining thickness	See Chapter 1
Maximum drum diameter	Cast into drum

Torque specifications

	Ft-lbs (unless otherwise indicated)
Brake booster mounting nuts	
1971	200 in-lbs
1972 through 1979	150 to 220 in-lbs
1980 on	200 to 250 in-lbs
Brake hoses and lines	
Brake hose banjo bolt-to-caliper	
1971 through 1975	25
1976 through 1987	19 to 29
1988 and 1989	30 to 40
Fixed caliper	
Caliper assembly bridge bolts	70 to 80
Caliper mounting bolts	50 to 80
Floating caliper	
Caliper guide pins (mounting bolts)	
1971	30 to 35
1972 through 1975	25 to 35
1976 and 1977	25 to 40
Caliper adapter-to-steering knuckle bolts	95 to 125
Master cylinder-to-brake booster mounting nuts	
1971 through 1977	80 to 200 in-lbs
1978 on	200 + 30 in-lbs
Sliding caliper	
Caliper retaining clip bolts	
1973 through 1975	220 in-lbs
1976 on	170 to 260 in-lbs
Caliper adapter-to-steering knuckle bolts	
1973 through 1975	180
1976 on	95 to 125
Wheel cylinder-to-brake backing plate mounting nuts	
1971 through 1975	
9-inch brakes	110 in-lbs
10-inch and 11-inch brakes	220 in-lbs
1976 on (all)	75 in-lbs (minimum)

1 General information

General

All models covered by this manual are equipped with hydraulically-operated, four wheel brake systems. Most, but not all, front brakes are discs (some early models use drum front brakes); most rear brakes are drum brakes, although some 1974 and 1975 Imperial models were equipped with optional rear disc brakes.

All brakes are self-adjusting. Disc brakes automatically compensate for pad wear, while drum brakes incorporate an adjustment mechanism which is activated as the brakes are applied.

The hydraulic system has separate circuits for the front and rear brakes. If one circuit fails, the other circuit will remain functional and a warning indicator will light up on the dashboard when a substantial amount of brake fluid is lost, showing that a failure has occurred.

Calipers

The vehicles covered by this manual have been equipped with either a *fixed*, *floating* or *sliding* caliper. The fixed caliper, which was used on some 1971 and 1972 A-body vehicles, uses two pairs of opposed pistons. Unlike most conventional automotive calipers, the fixed caliper assembly doesn't slide sideways when hydraulic pressure is applied.

The floating caliper, which was used on various vehicles from 1971 through 1978, is a single-piston design that "floats" on four rubber bushings (two on the outer side of the caliper housing, two on the inner side) and two steel guide pins which radially locate the brake pads and prevent them from falling out. When the brake pedal is depressed, hydraulic pressure pushing on the piston is transmitted to the inner brake pad and against the inner surface of the brake disc. As the force against the disc from the inner pad is increased, the caliper assembly moves in, sliding on the guide pins, providing a pinching force on the disc.

The sliding caliper, which is used on many vehicles from 1973 through 1989, is also a single-piston design. Unlike the floating caliper design, the sliding caliper uses no guide pins; instead, a pair of machined abutments on the caliper adapter position and align the caliper. Retaining clips keep the caliper on machined guides or "ways" on the adapter, allowing lateral movement of the caliper. When hydraulic pressure is applied, the force applied to the piston is transmitted to the inner brake pad and against the inner surface of the brake disc. As the force against the inner surface of the disc increases, the caliper assembly moves in, sliding on the machined ways on the adapter, providing a clamping force on the disc.

Master cylinder

The master cylinder is located under the hood on the driver's side, and can be identified by the large fluid reservoir on top. The master cylinder has separate primary and secondary piston assemblies for the front and rear circuits.

Power brake booster

The power brake booster uses engine manifold vacuum to provide assistance to the brakes. It is mounted on the firewall in the engine compartment, directly behind the master cylinder.

Sure Brake System

The Sure Brake System, which was used on some 1972 and 1973 Imperial models, is designed to prevent a wheel from locking up during braking above 5 mph. The system reduces the likelihood of a skid caused by a locked wheel while maintaining sufficient brake pressure for maximum stopping power. This improves the directional control and steerability of the vehicle, and in many cases reduces the distance required to bring the vehicle to a stop. For more general information regarding the Sure Brake System, refer to Section 16.

Parking brake

The parking brake mechanically operates the rear brakes only. The parking brake cables pull on a lever attached to the brake shoe assembly, causing the shoes to expand against the drum (or, on rear disc brake models, a pair of small brake shoes inside the disc/hub assembly).

Precautions

There are some general precautions and warnings related to the brake system:

a) *Use only brake fluid conforming to DOT 3 specifications.*
b) *The brake pads and linings may contain asbestos fibers which are hazardous to your health if inhaled. Whenever you work on brake system components, clean all parts with brake system cleaner or denatured alcohol. Do not allow the fine dust to become airborne.*
c) *Safety should be paramount whenever any servicing of the brake components is performed. Do not use parts or fasteners which are not in perfect condition, and be sure all clearances and torque specifications are adhered to. If you are at all unsure about a certain procedure, seek professional advice. Upon completion of any brake system work, test the brakes carefully in a controlled area before driving the vehicle in traffic.*

If a problem is suspected in the brake system, don't drive the vehicle until it's fixed.

2 Disc brake pads - replacement

Refer to illustration 2.3

Warning: *Disc brake pads must be replaced on both front wheels at the same time - never replace the pads on only one wheel. Also, the dust created by the brake system may contain asbestos, which is harmful to your health. Never blow it out with compressed air and don't inhale any of it. An approved filtering mask should be worn when working on the brakes. Do not, under any circumstances, use petroleum-based solvents to clean brake parts. Use brake system cleaner only!*

Note: *When servicing the disc brakes, use only high-quality, nationally-recognized, name-brand parts.*

2.3 On most calipers, a C-clamp can be used to depress the piston into its bore; this aids removal of the caliper and installation of the new pads (on fixed calipers, a C-clamp won't work - you'll have to lever the four pistons into their bores with a pair of screwdrivers)

1 Loosen the front wheel lug nuts, raise the front of the vehicle and support it securely on jackstands. Apply the parking brake. Remove the front wheels (or the rear wheels, if you're working on a vehicle with rear discs).

2 Remove about two-thirds of the fluid from the master cylinder reservoir and discard it. Position a drain pan under the brake assembly and clean the caliper and surrounding area with brake system cleaner.

3 Push the piston back into its bore to provide room for the new brake pads. A C-clamp can be used (on floating and sliding type calipers) to accomplish this **(see illustration)**. As the piston is depressed to the bottom of the caliper bore, the fluid in the master cylinder will rise. Make sure it doesn't overflow. If necessary, siphon off some of the fluid. If you're servicing a fixed type caliper, a C-clamp won't work - instead, depress the pistons by squeezing the pads to the caliper frame (away from the disc) with a pair of water pump pliers (otherwise, when you press down on one piston, the other pistons will rise!).

Fixed caliper (1971 and 1972 A-body models with disc brakes)

Refer to illustrations 2.4a, 2.4b and 2.4c

4 To change the pads on a fixed caliper, remove the brake pad retainer hold-down

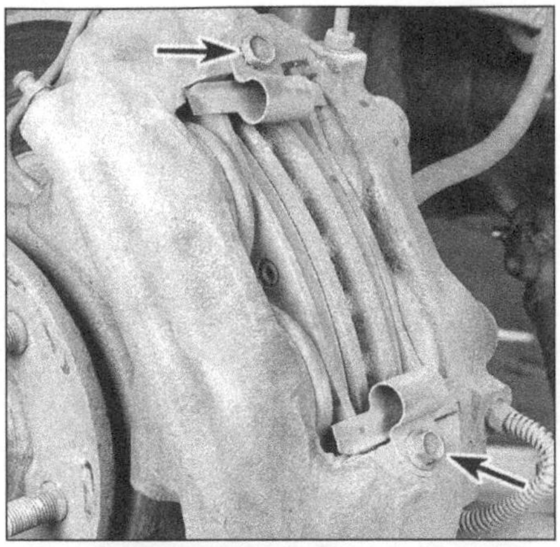

2.4a To remove the brake pad retainer hold-down clips, remove these bolts (arrows) (fixed caliper)

2.4b Pull out the old inner pad then install the new one . . . (fixed caliper)

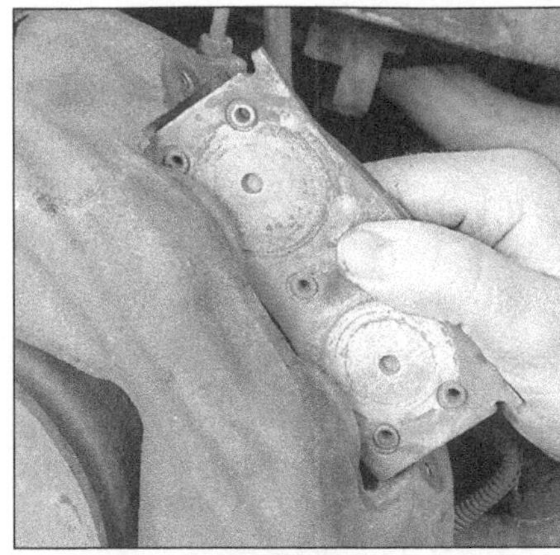

2.4c . . . then replace the outer pad (fixed caliper)

2.6a Remove the hold-down spring by depressing the ends and pulling the spring out (floating caliper)

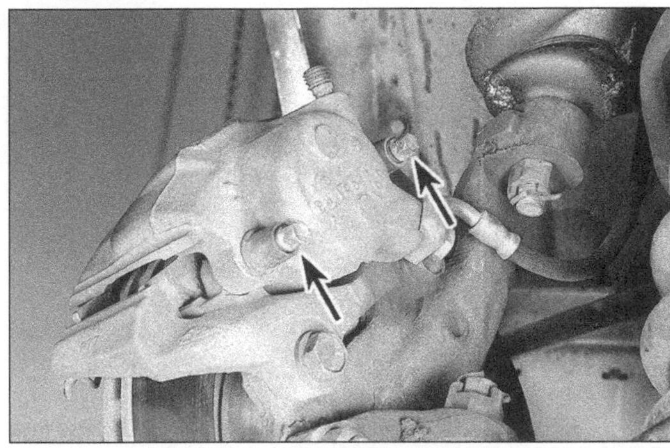

2.6b Unscrew the two caliper pins (arrows) far enough to remove the caliper - don't remove the pins completely unless the sleeves or bushings are to be replaced (floating caliper)

clips, depress one pad with two pairs of water pump pliers and remove the pad **(see illustrations)**. Install the new pad, then repeat the procedure to replace the remaining pad.

5 While the pads are removed, inspect the brake disc carefully as outlined in Section 4. If machining is necessary, follow the information in that Section to remove the disc. Inspect the caliper for seal leaks (evidenced by fluid moisture around the cavity) and for any damage to the piston dust boots. If excessive moisture is evident, install a new piston dust boot and piston seal (see Section 3).

Floating caliper

Refer to illustrations 2.6a, 2.6b, 2.6c, 2.6d and 2.6e

6 Follow the accompanying photos, beginning with **illustration 2.6a**, for the pad removal procedure. Be sure to stay in order

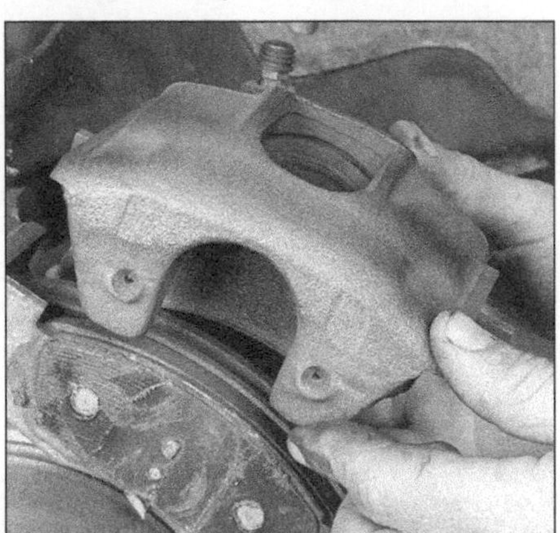

2.6c Lift the caliper straight up off the disc (floating caliper)

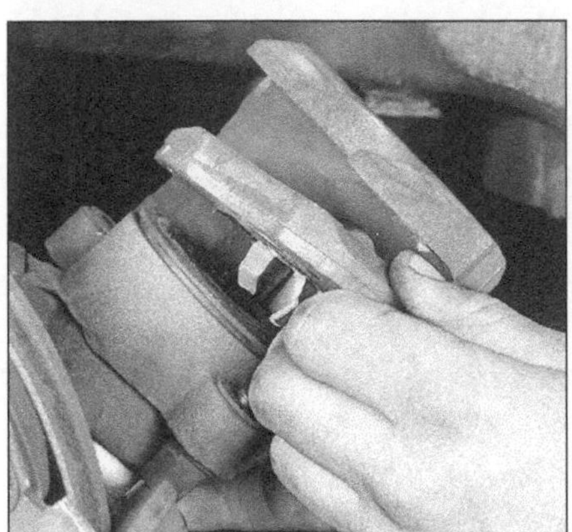

2.6d Pull the inner pad out of the caliper, then hang the caliper out of the way with a piece of wire to prevent damage to the brake hose . . .

2.6e . . . and remove the outer pad from the caliper adapter (floating caliper)

2.13a Remove the lower retainer clip and anti-rattle spring (shown), then remove the upper clip and spring (not shown) (sliding caliper)

2.13b Remove the caliper housing, detach the outer brake pad from the caliper and hang the caliper out of the way to prevent damage to the brake hose

2.13c Remove the inner brake pad from the caliper adapter

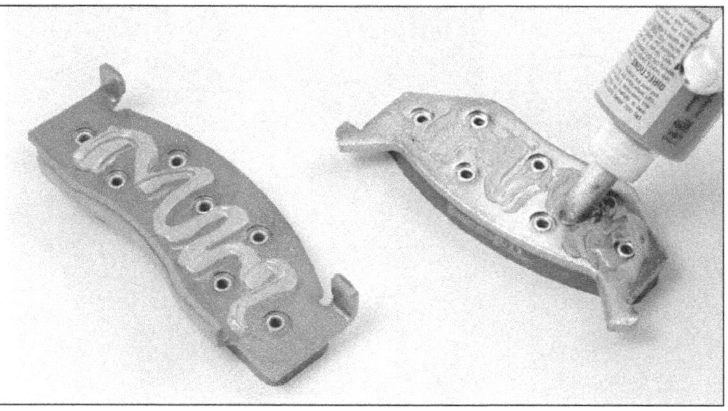

2.13d Apply anti-squeal compound to the backing plates of both the new pads

and read the caption under each illustration.

7 While the pads are removed, inspect the caliper for brake fluid leaks and ruptures of the piston boot. Overhaul or replace the caliper as necessary (see Section 3).

8 Clean the caliper and mounting bolts and check them for corrosion and damage. Replace the bolts with new ones if they're significantly corroded or damaged. Inspect the brake disc carefully as outlined in Section 4. If machining is necessary, follow the information in that Section to remove the disc.

9 If the caliper pin bushings appear worn, replace them as described in Section 3.

10 Remove the protective paper from the noise suppression gasket on the new brake pads (if equipped). Position the inner pad with the clip in the caliper.

11 Position the outer pad in the caliper adapter so the ears on the pad engage the cutouts in the caliper.

12 Hold the caliper in position over the disc, push the pins through the bushings and thread them into the bracket. Tighten them to the torque listed in this Chapter's Specifications. Install the hold-down spring.

Sliding caliper

Refer to illustrations 2.13a, 2.13b, 2.13c, 2.13d, 2.13e, 2.13f and 2.13g

Note: *The accompanying photo sequence depicts pad replacement on a* front *sliding caliper; the rear sliding caliper, used on some 1974 and 1975 Imperial models, is virtually identical to a front caliper. Both the caliper and the brake pads are removed and installed the same way as a front caliper.*

13 Follow the accompanying photos, beginning with **illustration 2.13a**, for the pad replacement procedure. Be sure to stay in order and read the caption under each illustration.

2.13e When you install the new inner brake pad, make sure the "ears" (arrows) at either end of the backing plate are properly seated onto the ways of the machined surfaces on the adapter

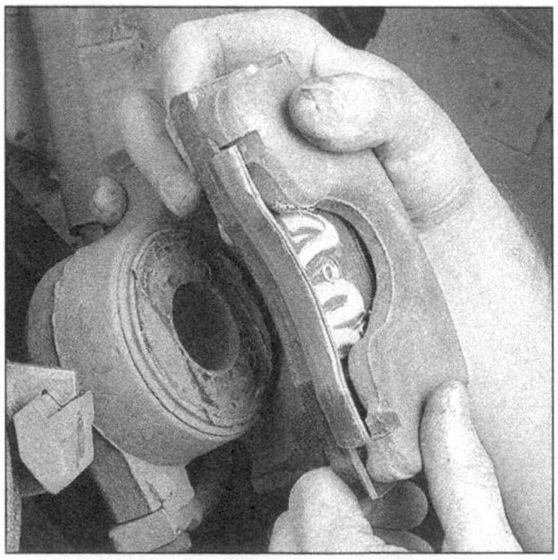

2.13f Pop the new outer brake pad into place in the caliper like this - make sure the new outer brake pad backing plate is properly seated in the caliper as shown, then install the caliper

2.13g Install the lower retainer clip and anti-rattle spring as shown, with the anti-rattle spring on top of the retainer clip, and tighten the bolt securely; install the upper clip, spring and bolt the same way

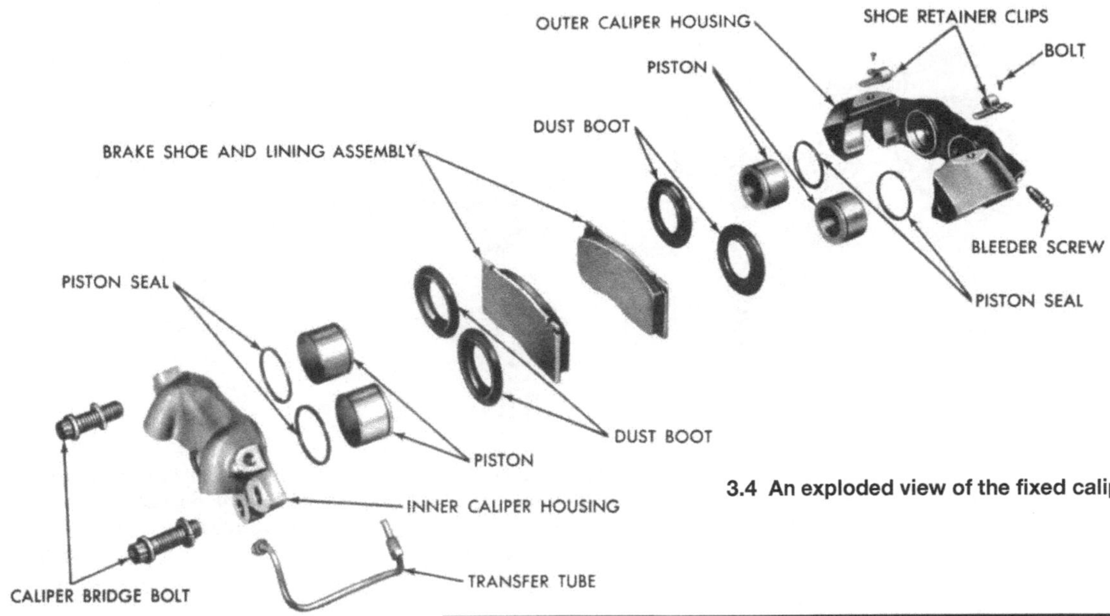

3.4 An exploded view of the fixed caliper assembly

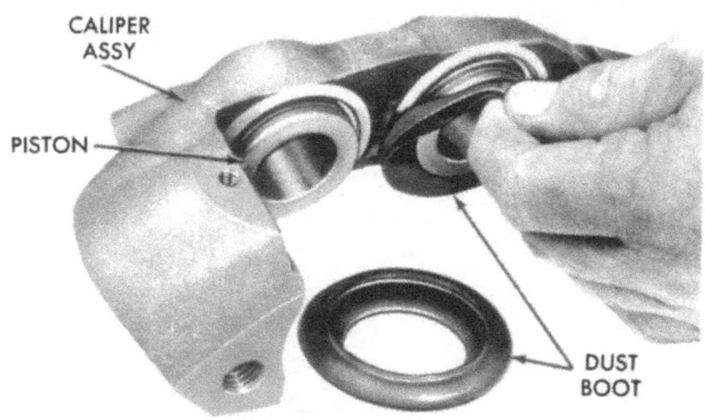

3.9 As you peel away the old dust boots, note how they're installed over both the caliper housing retainer and into the piston groove - that's exactly how the new boots must be installed during reassembly (fixed caliper)

14 While the pads are removed, inspect the caliper for seal leaks (evidenced by fluid moisture around the cavity) and for any damage to the piston dust boots. If excessive moisture is evident, install a new piston dust boot and piston seal (see Section 3). When you're done, make sure you tighten the bolts that secure the anti-rattle springs and retainers to the torque listed in this Chapter's Specifications.

All models

15 Install the brake pads on the opposite wheel, then install the wheels and lower the vehicle. Tighten the lug nuts to the torque listed in the Chapter 1 Specifications. Add brake fluid to the reservoir until it's full (see Chapter 1).

16 Pump the brakes several times to seat the pads against the disc, then check the fluid level again.

17 Check the operation of the brakes before driving the vehicle in traffic. Try to avoid heavy brake applications until the brakes have been applied lightly several times to seat the pads.

3 Disc brake caliper - removal, overhaul and installation

Warning: *Dust created by the brake system may contain asbestos, which is harmful to your health. Never blow it out with compressed air and don't inhale any of it. An approved filtering mask should be worn when working on the brakes. Do not, under any circumstances, use petroleum-based solvents to clean brake parts. Use brake system cleaner only.*

Note: *If an overhaul is indicated (usually because of fluid leaks, a stuck piston or broken bleeder screw) explore all options before beginning this procedure. New and factory rebuilt calipers are available on an exchange basis, which makes this job quite easy. If you decide to rebuild the calipers, make sure rebuild kits are available before proceeding. Always rebuild or replace the calipers in pairs - never rebuild just one of them.*

Removal

1 Loosen the front wheel lug nuts, raise the vehicle and support it securely on jackstands. Remove the front wheels.

2 On models where the brake hose threads directly into the caliper, disconnect the brake line fitting from the brake hose (see Section 13), then unscrew the hose from the caliper. (If the caliper is only being removed for access to other components, don't disconnect the hose.) On brake hoses with banjo fittings, unscrew the fitting bolt and detach the hose. Discard the two copper sealing washers on each side of the fitting and use new ones during installation. Wrap a plastic bag around the end of the hose to prevent fluid loss and contamination.

3 To remove a fixed caliper, remove the bolts that attach the caliper assembly to the steering knuckle. To remove a floating or sliding caliper, refer to the appropriate photo sequence in Section 2 (caliper removal is the first part of the brake pad replacement procedure for those calipers). Clean the caliper assembly with brake system cleaner. DO NOT use kerosene, gasoline or petroleum-based solvents. And while you've got the caliper off, be sure to check the pads as well and replace them if necessary (see Section 2).

Overhaul

Fixed caliper

Refer to illustrations 3.4, 3.9, 3.11 and 3.15

4 Remove the pad retainer spring clips **(see illustration)**.

5 Remove the transfer tube. Use a flare-nut wrench, if available.

6 Mount the caliper assembly in a bench vise equipped with brass protector jaws. **Caution:** *Clamp on the caliper mounting lugs, NOT the caliper itself!*

7 Remove the brake pads, if you haven't already done so.

8 Remove the bridge bolts that hold the two halves of the caliper together. Separate the two caliper halves.

9 Peel the dust boot out and away from the caliper housing retainer and out of the piston groove **(see illustration)**.

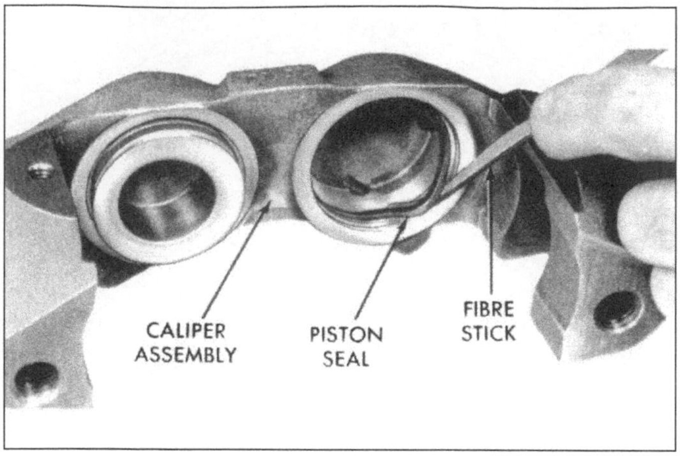

3.11 Use a small wooden or plastic tool to peel the old piston seals out of the piston bores - don't use a metal tool or you might damage the bore walls (fixed caliper)

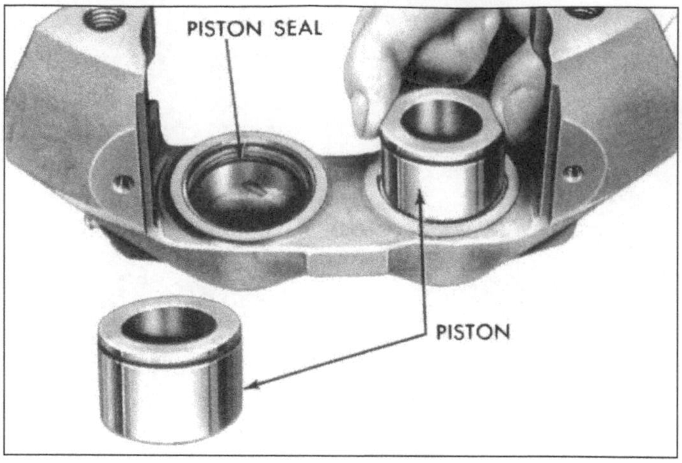

3.15 When you install the pistons into the caliper halves, coat them with fresh brake fluid, double-check the new seals to make sure they're properly seated into their grooves in the piston bore walls and make SURE the pistons are square to the bores (fixed caliper)

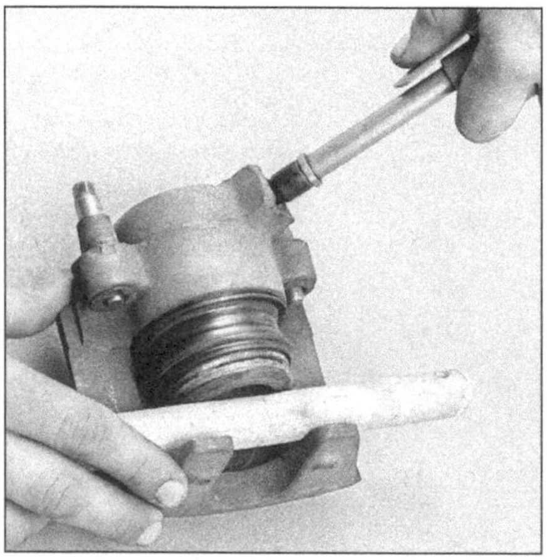

3.21 With a block of wood placed between the piston and the caliper frame, use compressed air to ease the piston out of the bore (fixed caliper)

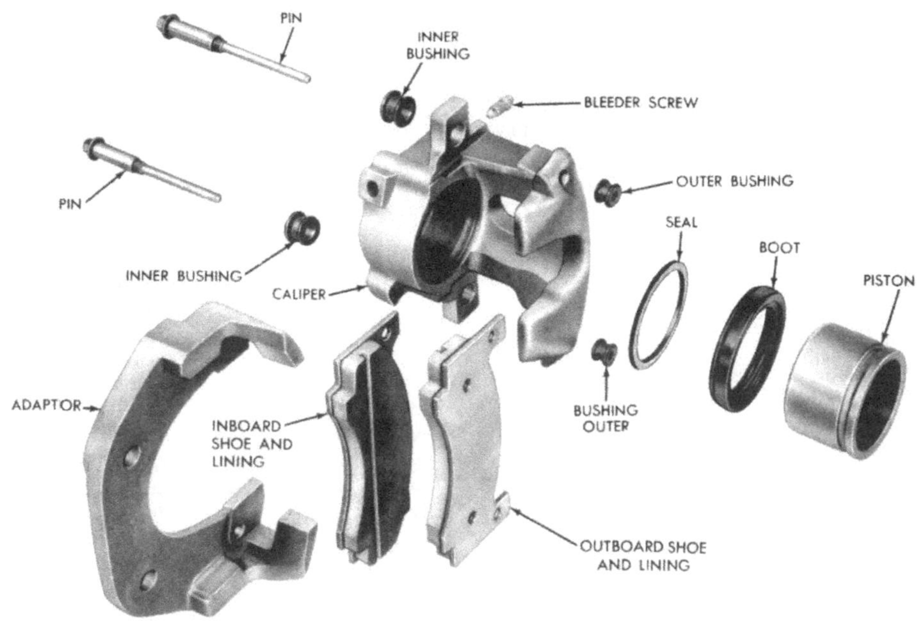

3.22a An exploded view of a typical floating caliper (sliding calipers similar internally)

10 Place each caliper-half face down (pistons facing down) on top of a block of wood and, while holding the caliper-half down against the wood with one hand, plug the hole for the brake hose, apply a brief burst of compressed air through the hole for the transfer tube and pop out the pistons. The outer caliper-half also has another hole, for the bleeder screw, which must be installed during this procedure, or the air will go in one hole and out the other, instead of pushing out the pistons. If a piston becomes cocked in its bore during removal, don't try to jerk it out with a pair of pliers, or you will damage the piston and the bore. Instead, carefully tap the piston back into its bore until it's square and try again.

11 Using a small pointed wooden or plastic tool, remove the piston seals from the groove in the cylinder bore **(see illustration)**. Discard the old seals. And be careful! Don't scratch or gouge the piston bore or the seal groove.

12 Clean everything in brake fluid and wipe it dry with a lint-free shop towel. Blow out the drilled passages in both caliper-halves with filtered, unlubricated compressed air. Inspect the dust boots for cracks, tears and deterioration. Replace as necessary. Inspect the piston bores in both housings for scratches, scoring and pitting. Black stains in the bore walls are caused by the piston seals and are harmless. Light scratches, scoring and pitting can be cleaned up with a crocus cloth. If the damage is deeper, the caliper should be replaced (rebuilts are available). The pistons should be similarly inspected and cleaned up as necessary. If a piston is severely damaged - pitted, scored or the chrome plating worn off - replace it.

13 Clamp the mounting lugs of the inner half of the caliper in a bench vise with protector jaws.

14 Dip the new piston seals in clean brake fluid and install them in the caliper grooves. Position each seal into its groove at one spot and gently work it around the piston bore with your finger until it's properly seated. **Warning:** *Do NOT use old seals!*

15 Coat the outside diameter of the pistons with clean brake fluid and install them in the cylinder bores with the open end of each piston facing away from the caliper **(see illustration)**. Position the piston squarely in the bore and apply a slow, steady pressure until the piston is fully seated. If you encounter resistance, remove the piston and make sure the seal is properly installed.

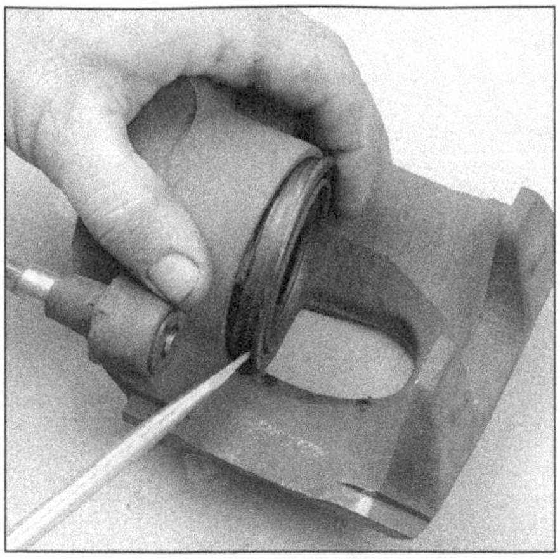

3.22b Carefully pry the dust boot out of the caliper (floating caliper shown, sliding caliper similar)

3.23 The piston seal should be removed with a plastic or wooden tool to avoid damage to the bore and the seal groove (a pencil will do the job) (floating caliper shown, sliding caliper similar)

3.27 Position the new seal in the cylinder groove - make sure it isn't twisted (floating caliper shown, sliding caliper similar)

3.28a Slip the boot over the piston (floating caliper shown, sliding caliper similar)

3.28b Push the piston straight into the cylinder - make sure it doesn't become cocked in the bore (floating caliper shown, sliding caliper similar)

3.29 If you don't have a boot installation tool, gently seat the boot with a drift punch (floating caliper shown, sliding caliper similar)

16 Install the new dust boots into their grooves in the caliper and the pistons. Make sure the dust boots are properly seated.

17 Remove the inner caliper-half from the bench vise and install the outer caliper-half the same way. Install the seals, pistons and dust boots as described in Steps 14, 15 and 16.

18 Install the inner caliper-half onto the outer caliper-half that's still clamped into the vise, install the bridge bolts and tighten them to the torque listed in this Chapter's Specifications.

19 Install the transfer tube and tighten the fittings securely.

20 Install the bleeder screw, if removed.

Floating and sliding calipers

Refer to illustrations 3.21, 3.22a, 3.22b, 3.23, 3.27, 3.28a, 3.28b and 3.29

Note: *The accompanying photos depict overhaul of a front sliding caliper; the rear sliding caliper, used on some 1974 and 1975 Imperial models, is virtually identical to a front caliper.*

21 Place several shop towels or a block of wood in the center of the caliper to act as a cushion, then use compressed air, directed into the fluid inlet, to remove the piston **(see illustration)**. Use only enough air pressure to ease the piston out of the bore. If the piston is blown out, even with the cushion in place, it may be damaged. **Warning:** *Never place your fingers in front of the piston in an attempt to catch or protect it when applying compressed air, as serious injury could occur.*

22 Pry the dust boot from the caliper bore **(see illustrations)**.

23 Using a wood or plastic tool, remove the piston seal from the groove in the caliper bore **(see illustration)**. Metal tools may cause bore damage.

24 Remove the bleeder screw, then remove and discard the caliper pin bushings and sleeves.

25 Clean the remaining parts with brake system cleaner or clean brake fluid, then blow them dry with filtered, unlubricated compressed air.

26 Inspect the surfaces of the piston for nicks and burrs and loss of plating. If surface defects are present, the caliper must be replaced. Check the caliper bore in a similar way. Light polishing with crocus cloth is permissible to remove slight corrosion and stains. Discard the caliper pins if they're severely corroded or damaged.

27 Lubricate the new piston seal with clean brake fluid and position the seal in the cylinder groove using your fingers only **(see illustration)**.

28 Install the new dust boot in the groove in the end of the piston **(see illustration)**. Dip the piston in clean brake fluid and insert it squarely into the cylinder. Depress the piston to the bottom of the cylinder bore **(see illustration)**.

29 Seat the boot in the caliper counterbore using a boot installation tool or a blunt punch **(see illustration)**.

30 Install the new mounting pin bushings and sleeves.

31 Install the bleeder screw.

32 Clean the sliding surfaces of the caliper and the caliper adapter.

4.3 Use a dial indicator to check disc runout - if the reading exceeds the specified runout limit, the disc will have to be machined or replaced

4.4a On some models, the minimum thickness is cast into the inside of the disc - on others, it's located on the outside of the disc

4.4b Use a micrometer to measure the thickness of the disc at several points

Installation

33 If you're servicing a fixed caliper, install the caliper assembly and the caliper mounting bolts, then tighten the bolts to the torque listed in this Chapter's Specifications and install the brake pads (see Section 2). If you're servicing a floating or sliding caliper, caliper installation is a part of brake pad installation in Section 2. Make sure you tighten the caliper pins/bolts to the torque listed in this Chapter's Specifications.

34 On models where the brake hose threads directly into the caliper, connect the brake hose to the caliper and tighten it securely. Position the other end of the hose in its bracket and connect the brake line fitting. Tighten the fitting securely. On models that use banjo fittings, connect the hose to the caliper using new sealing washers. Tighten the banjo bolt to the torque listed in this Chapter's Specifications.

35 Bleed the brakes as outlined in Section 11.

36 Install the wheels and lug nuts. Lower the vehicle and tighten the lug nuts to the torque listed in the Chapter 1 Specifications.

37 After the job has been completed, firmly depress the brake pedal a few times to bring the pads into contact with the disc.

38 Check the operation of the brakes before driving the vehicle in traffic.

4 Brake disc - inspection, removal and installation

Inspection

Refer to illustrations 4.3, 4.4a and 4.4b

1 Loosen the wheel lug nuts, raise the front of the vehicle and support it securely on jackstands. Apply the parking brake. Remove the front wheels.

2 Remove the brake caliper as described in Section 3. Visually inspect the disc surface for score marks and other damage. Light

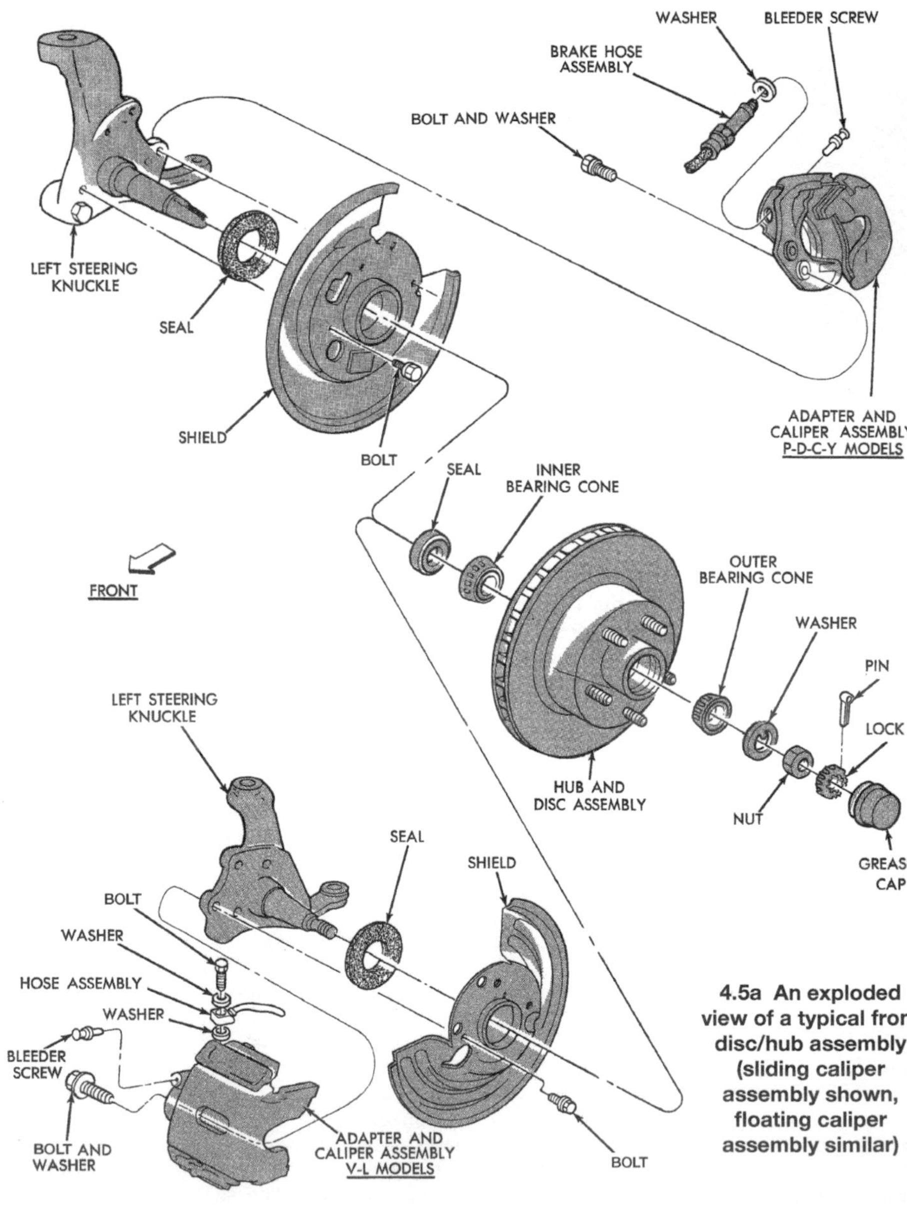

4.5a An exploded view of a typical front disc/hub assembly (sliding caliper assembly shown, floating caliper assembly similar)

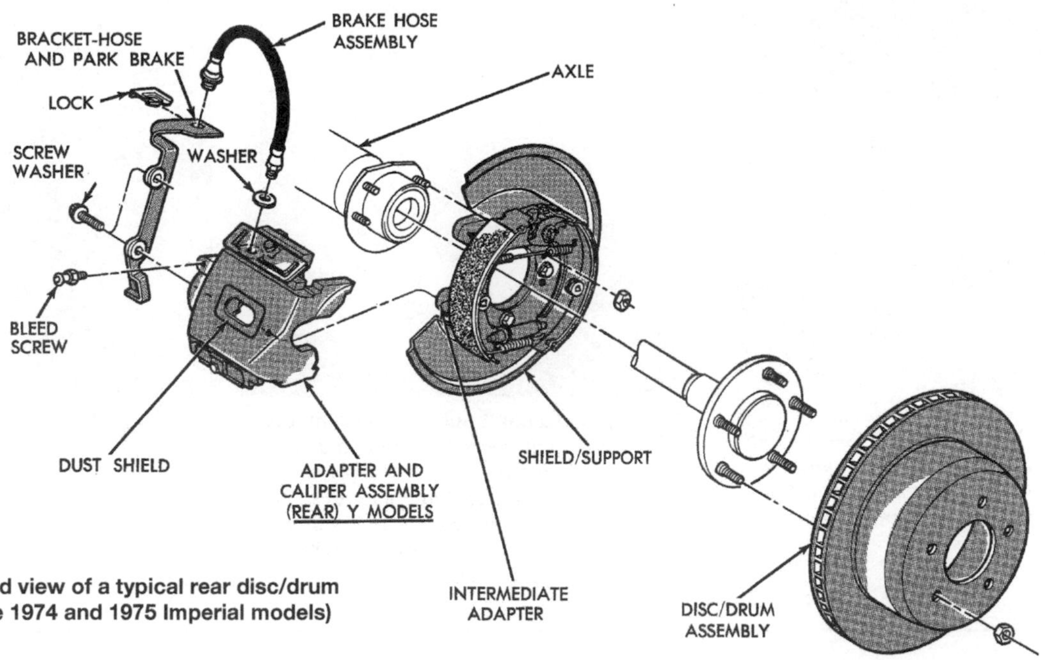

4.5b An exploded view of a typical rear disc/drum assembly (some 1974 and 1975 Imperial models)

scratches and shallow grooves are normal after use and won't affect brake operation. Deep grooves - over 0.015-inch (0.38 mm) deep - require disc removal and refinishing by an automotive machine shop. Be sure to check both sides of the disc.

3 To check disc runout, place a dial indicator at a point about 1/2-inch from the outer edge of the disc **(see illustration)**. Set the indicator to zero and turn the disc. The indicator reading should not exceed the runout limit listed in this Chapter's Specifications. If it does, the disc should be refinished by an automotive machine shop. **Note:** *Professionals recommend resurfacing the brake discs regardless of the dial indicator reading (to produce a smooth, flat surface that will eliminate brake pedal pulsations and other undesirable symptoms related to questionable discs). At the very least, if you elect not to have the discs resurfaced, deglaze them with sandpaper or emery cloth.*

4 The disc must not be machined to a thickness less than the specified minimum thickness. The minimum (or discard) thickness is cast into the disc **(see illustration)**. The disc thickness can be checked with a micrometer **(see illustration)**.

Removal and installation

Refer to illustrations 4.5a and 4.5b

5 The disc is an integral part of the hub **(see illustrations)**. Front hub removal and installation is done as part of the front wheel bearing maintenance procedure (see Chapter 1). The following procedure applies only to the rear brake disc used on some 1974 and 1975 Imperial models.

6 Loosen the rear wheel lug nuts, raise the rear of the vehicle and place it securely on jackstands. Remove the rear wheel.

7 Remove the rear caliper (see Section 3).

8 Remove the plug from the parking brake adjusting access hole. Engage the notches in the adjusting star wheel with a suitable tool (such as a long screwdriver) and retract the parking brake shoes (see Section 10 for an exploded view of the parking brake shoe components).

9 Remove the retaining clips, if present, from the wheel studs.

10 Remove the disc/drum assembly.

11 Installation is the reverse of removal. Make sure you tighten the bolts that secure the anti-rattle springs and retainer clips to the torque listed in this Chapter's Specifications.

12 Be sure to adjust the parking brake shoes when you're done (see Section 10).

5 Drum brake shoes - replacement

Refer to illustrations 5.4a through 5.4r and 5.5

Warning: *Drum brake shoes must be replaced on both wheels at the same time - never replace the shoes on only one wheel. Also, the dust created by the brake system may contain asbestos, which is harmful to your health. Never blow it out with compressed air and don't inhale any of it. An approved filtering mask should be worn when working on the brakes. Do not, under any circumstances, use petroleum-based solvents to clean brake parts. Use brake system cleaner only!*

Caution: *Whenever the brake shoes are replaced, the retractor and hold-down springs should also be replaced. Due to the continuous heating/cooling cycle that the springs are subjected to, they lose their tension over a period of time and may allow the shoes to drag on the drum and wear at a much faster rate than normal. When replacing the rear brake shoes, use only high quality, nationally recognized brand-name parts.*

Note: *The photo sequence accompanying the following procedure depicts a typical* rear *brake assembly. Early vehicles with* front *drum brakes are virtually identical, except that they don't have a parking brake lever.*

1 Loosen the wheel lug nuts, raise the rear of the vehicle and support it securely on jackstands. Block the front wheels to keep the vehicle from rolling.

2 If you're servicing the rear brakes, release the parking brake.

3 Remove the wheel. **Note:** *All four front or rear brake shoes must be replaced at the same time, but to avoid mixing up parts, work on only one brake assembly at a time.*

4 Follow **illustrations 5.4a through 5.4r**

5.4a If you can't pull off the brake drum fairly easily, apply some penetrating oil at the hub-to-drum joint and allow it to soak in, lightly tap the drum to break it loose . . .

5.4b . . . then carefully tap around the outer edge of the drum to drive it off the studs - don't use excessive force!

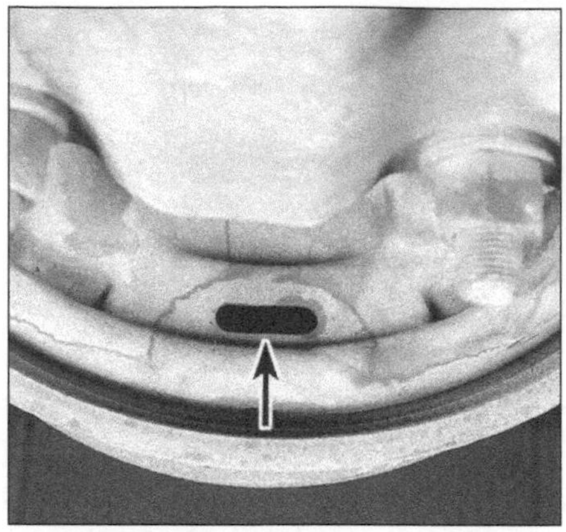

5.4c If you still can't pull off the drum, the shoes have worn into the drum and will have to be retracted - insert a screwdriver into the slot in the backing plate to hold the adjusting lever away from the star wheel, then turn the star wheel with another screwdriver

5.4d Before beginning work, wash away all traces of dust with brake system cleaner - DO NOT use compressed air

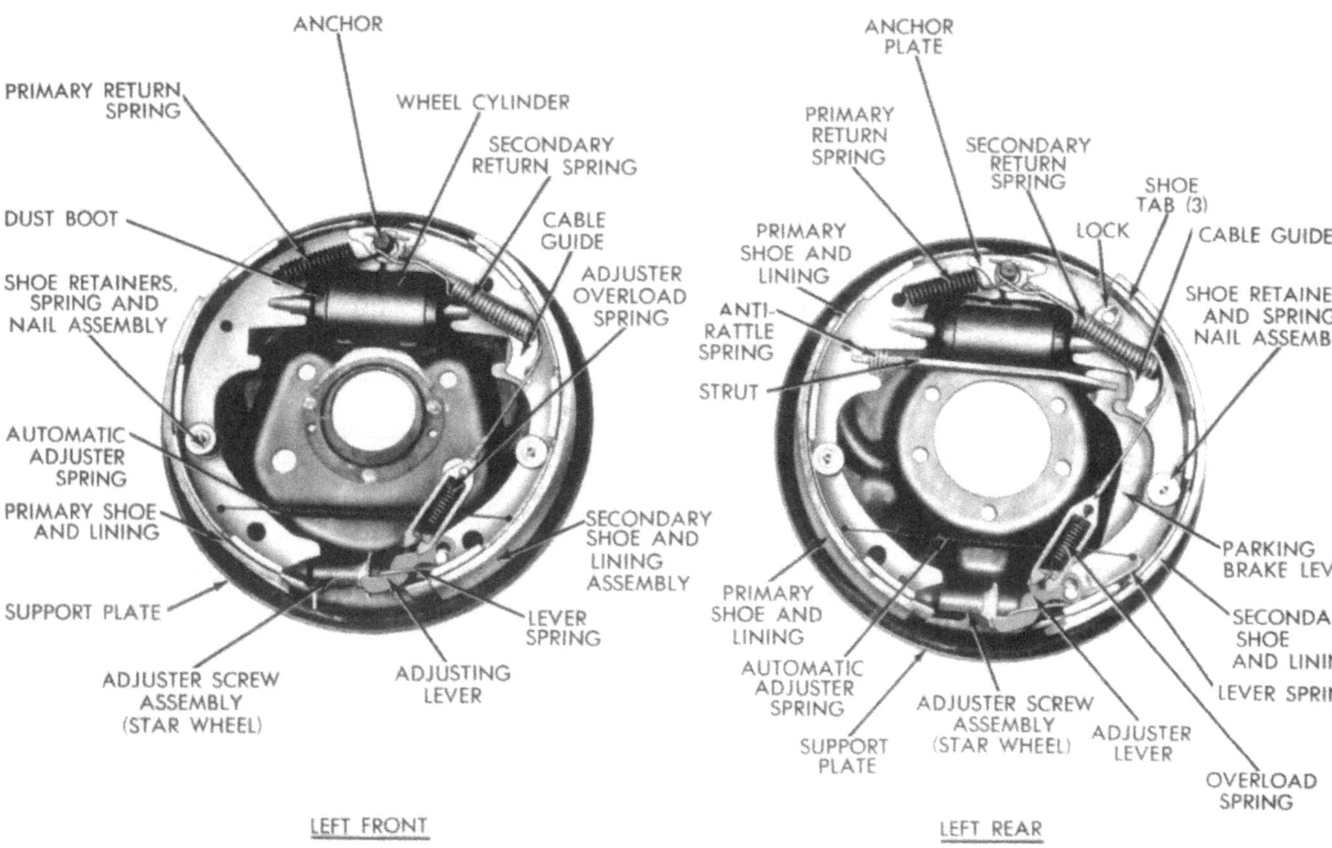

5.4e Details of typical front and rear drum brake assemblies (10-inch models shown, 9 and 11-inch units similar)

for the inspection and replacement of the brake shoes. Be sure to stay in order and read the caption under each illustration.

5 Before reinstalling the drum it should be checked for cracks, score marks, deep scratches and hard spots, which will appear as small discolored areas. If the hard spots cannot be removed with fine emery cloth or if any of the other conditions listed above exist, the drum must be taken to an automotive machine shop to have it turned. **Note:** *Professionals recommend resurfacing the drums whenever a brake job is done. Resurfacing will eliminate the possibility of out-of-round drums. If the drums are worn so much that they can't*

5.4f Remove the shoe return springs - the spring removal tool shown here can be purchased at most auto parts stores and greatly simplifies this step

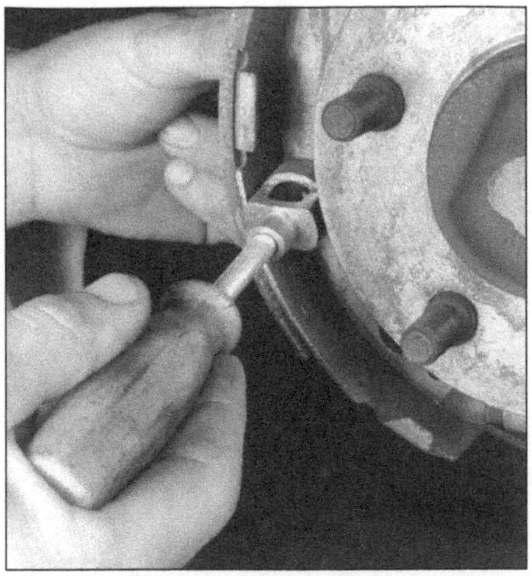

5.4g Remove the hold-down springs

5.4h Remove the adjuster cable

5.4i Remove the cable guide and spring

5.4j Separate the shoes at the top, lift them away from the backing plate, then detach the parking brake lever (rear brakes only) from the secondary shoe - the shoes, adjuster and spring can now be separated

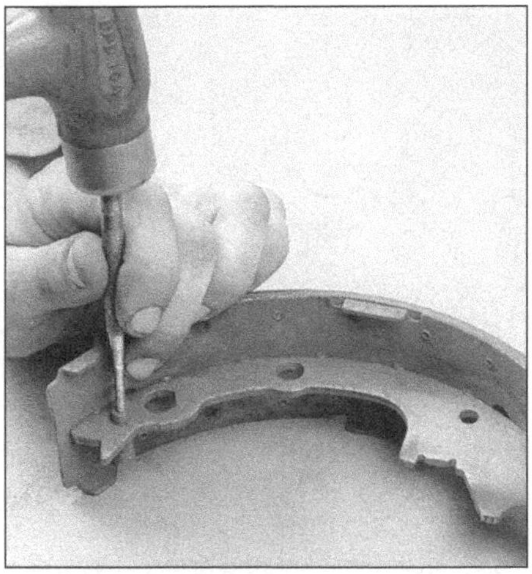

5.4k Install a new adjuster lever pin to the new shoe

5.4l Apply high temperature grease to the areas on the backing plate that support the shoes (don't use too much grease or get it on the shoes or the drum)

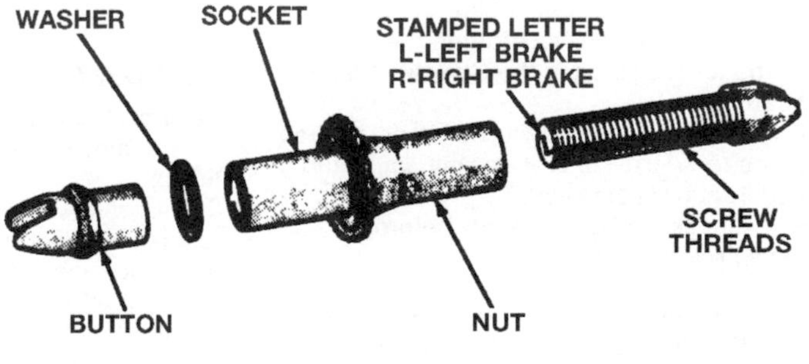

5.4m Clean the adjuster screw, then lubricate the threads and the sliding surface of the button with high temperature grease

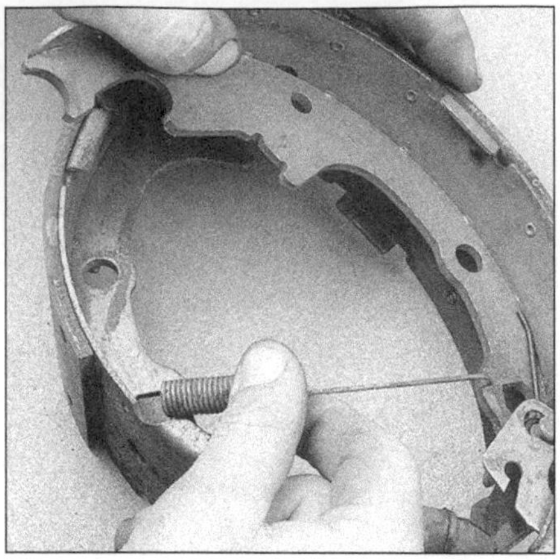

5.4n Install the shoe-to-shoe spring and adjuster assembly

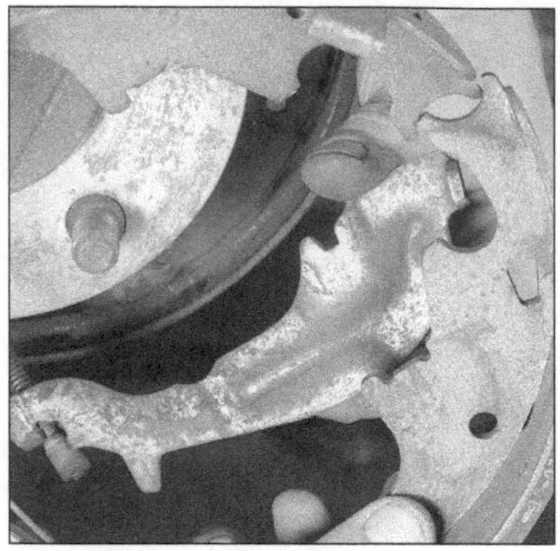

5.4o Raise the shoe assembly up to the backing plate and connect the tang on the parking brake lever with its slot in the secondary shoe

5.4p Fit the shoes over the axle flange, then install the hold-down pins, springs and retainers - make sure the shoes engage with the wheel cylinder pistons (or, on some models, the wheel cylinder pushrods)

5.4q Install the parking brake strut (rear brakes only) and anti-rattle spring

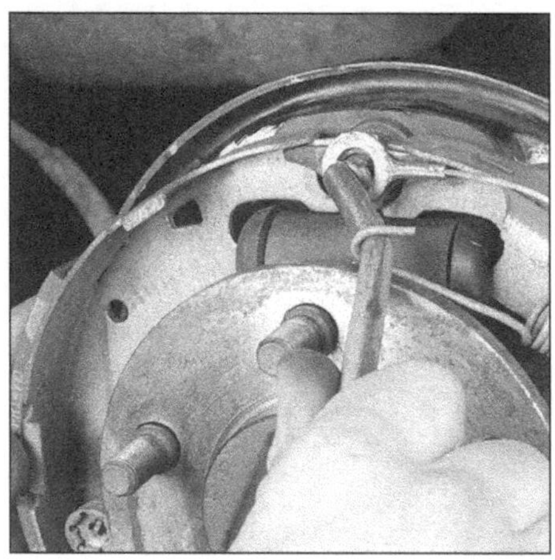

5.4r Install the return springs, route the adjuster cable around the cable guide and hook it to the adjuster lever - the brake drum can now be installed

5.5 The maximum allowable inside diameter is cast into the outer edge of the drum

be resurfaced without exceeding the maximum allowable diameter (stamped into the drum) **(see illustration)**, *then new ones will be required. At the very least, if you elect not to have the drums resurfaced, remove the glazing from the surface with emery cloth or sandpaper using a swirling motion.*

6 Install the brake drum on the axle flange.

7 Mount the wheel, install the lug nuts, then lower the vehicle. Tighten the lug nuts to the torque listed in the Chapter 1 Specifications.

8 Make a number of forward and reverse stops to adjust the brakes until satisfactory pedal action is obtained.

9 Check brake operation before driving the vehicle in traffic.

6 Wheel cylinder - removal, overhaul and installation

Note: *If an overhaul is indicated (usually because of fluid leakage or sticky operation) explore all options before beginning the job. New wheel cylinders are available, which makes this job quite easy. If you decide to rebuild the wheel cylinder, make sure a rebuild kit is available before proceeding. Never overhaul only one wheel cylinder. Always rebuild both of them at the same time.*

Removal

Refer to illustration 6.2

1 Remove the brake shoes (see Section 5).

2 Unscrew the brake line fitting from the rear of the wheel cylinder **(see illustration)**. If available, use a flare-nut wrench to avoid rounding off the corners on the fitting. Don't pull the metal line out of the wheel cylinder - it could bend, making installation difficult.

3 Remove the two bolts securing the wheel cylinder to the brake backing plate.

4 Remove the wheel cylinder.

5 Plug the end of the brake line to prevent the loss of brake fluid and the entry of dirt.

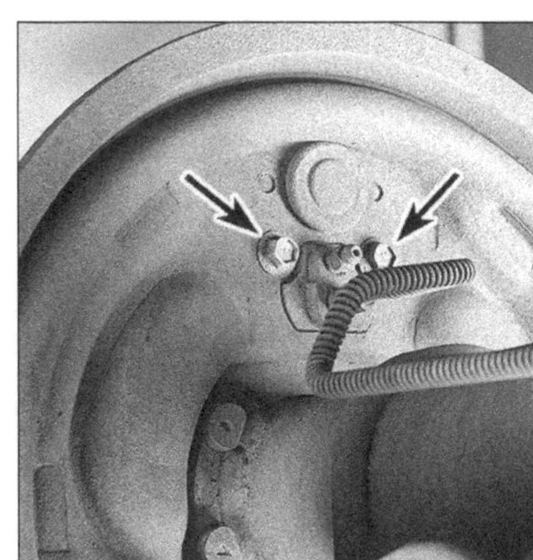

6.2 Unscrew the brake line fitting from the rear of the wheel cylinder with a flare-nut wrench, then remove the two bolts (arrows) securing the wheel cylinder to the brake backing plate

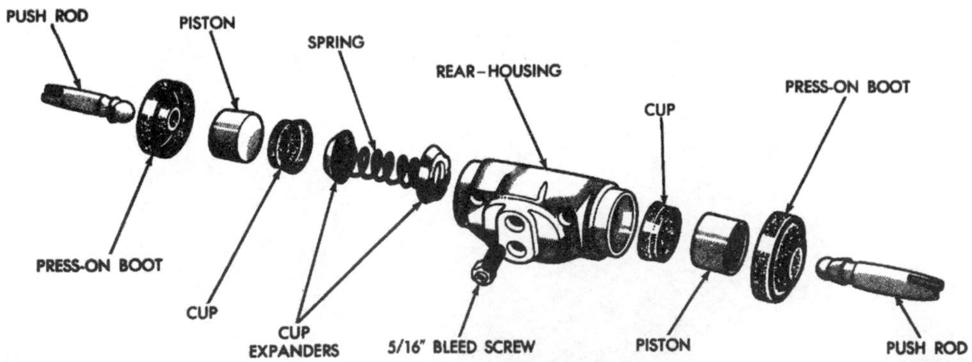

6.6a An exploded view of a typical wheel cylinder assembly (not all units are equipped with pushrods)

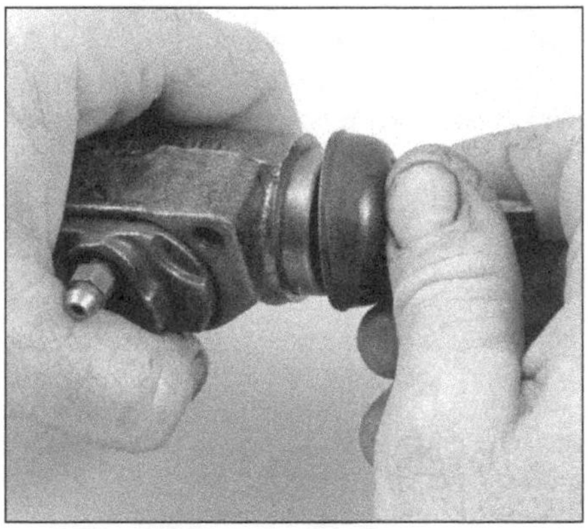

6.6b Remove the wheel cylinder dust seals

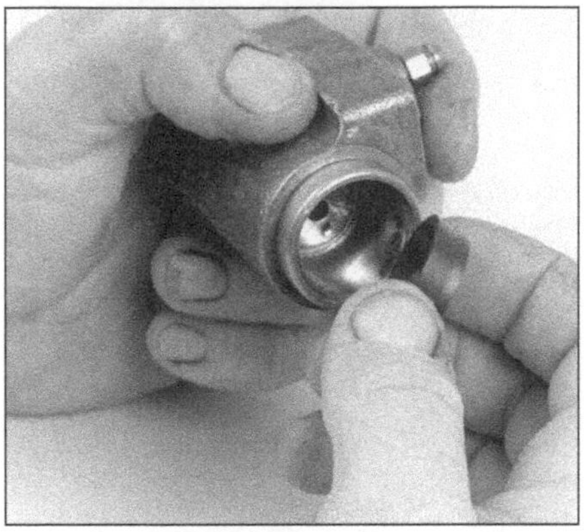

6.11 Install the piston cups with their open ends facing in, toward the cup expanders

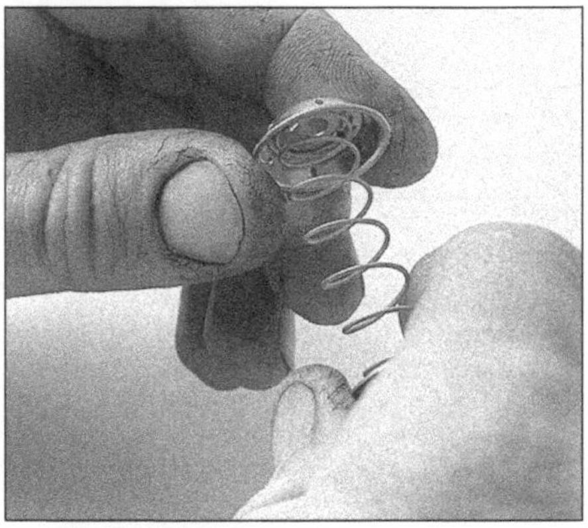

6.12a Attach the cup expanders to the spring . . .

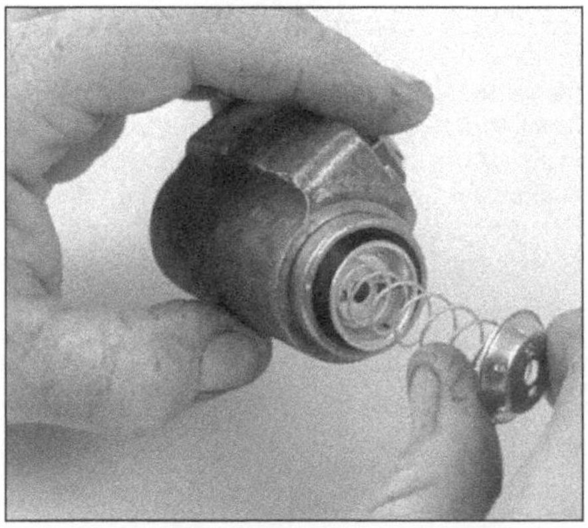

6.12b . . . then insert the expander/spring assembly into one end of the wheel cylinder housing (expanders are used only on 1974 and later vehicles; on earlier vehicles, simply install the spring)

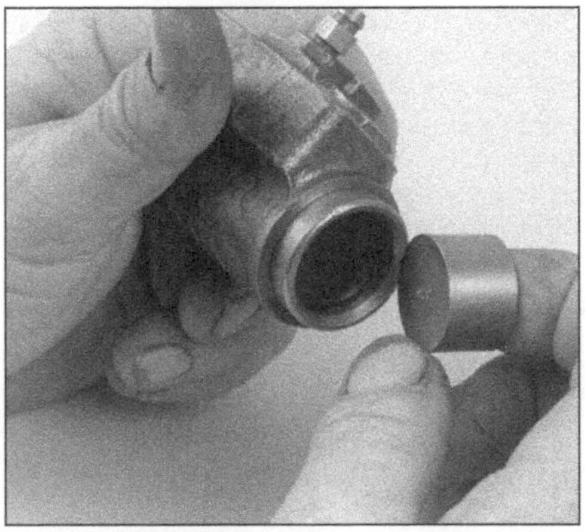

6.14 Install a piston into each end end of the wheel cylinder housing

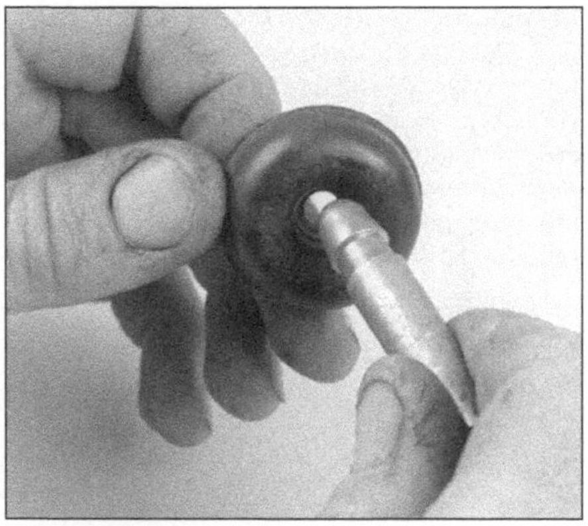

6.15 Insert the wheel cylinder pushrods into the new dust boots, then install the dust boots and pushrods onto both ends of the wheel cylinder

Overhaul

Refer to illustration 6.6a, 6.6b, 6.11, 6.12a, 6.12b, 6.14 and 6.15

6 To disassemble the wheel cylinder, remove the rubber dust boot from each end of the cylinder, then push out the two pistons, the cups (seals) and the expander/spring assembly **(see illustrations)**. Discard the rubber parts and use new ones from the rebuild kit when reassembling the wheel cylinder.

7 Inspect the pistons for scoring and scuff marks. If present, the pistons should be replaced with new ones.

8 Examine the inside of the cylinder bore for score marks and corrosion. If these conditions exist, the cylinder can be honed slightly to restore it, but replacement is recommended.

9 If the cylinder is in good condition, clean it with brake system cleaner or denatured alcohol. **Warning:** *DO NOT, under any circumstances, use gasoline or petroleum-based solvents to clean brake parts!*

10 Remove the bleeder screw and make sure the hole is clean.

11 Lubricate the cylinder bore with clean brake fluid, then insert one of the new rubber cups into the bore **(see illustration)**. Make sure the lip on the rubber cup faces in.

12 Install the cup expanders, if equipped, onto the ends of the spring, insert the expander/spring assembly into the opposite end of the bore and push it in until it contacts the rear of the rubber cup **(see illustrations)**. **Note:** *Cup expanders were introduced in 1974 - you won't find them on pre-1974 vehicles.*

13 Install the remaining cup in the cylinder bore.

14 Install the pistons **(see illustration)**.

15 Insert the pushrods (if equipped) into the boots **(see illustration)**, then install the boots and pushrods.

Installation

16 Installation is the reverse of removal. Attach the brake line to the wheel cylinder before installing the mounting bolts and tighten the line fitting after the wheel cylinder mountings bolts have been tightened. If available, use a flare-nut wrench to tighten the line fitting. Make sure you tighten the line fitting securely and the wheel cylinder mounting bolts to the torque listed in this Chapter's Specifications.
17 Install the brake shoes and brake drum, (see Section 5).
18 Bleed the brakes (see Section 11). Don't drive the vehicle in traffic until the operation of the brakes has been thoroughly tested.

7 Master cylinder - removal, overhaul and installation

Removal

Refer to illustrations 7.2 and 7.4

1 Place rags under the brake line fittings and prepare caps or plastic bags to cover the ends of the lines once they're disconnected. **Caution:** *Brake fluid will damage paint. Cover all painted surfaces and avoid spilling fluid during this procedure.*
2 Loosen the tube nuts at the ends of the brake lines where they enter the master cylinder. To prevent rounding off the flats on these nuts, a flare-nut wrench, which wraps around the nut, should be used **(see illustration)**.
3 Pull the brake lines away from the master cylinder slightly and plug the ends to prevent contamination. Also plug the openings in the master cylinder to prevent fluid spillage.
4 Remove the two master cylinder mounting nuts **(see illustration)** and unbolt the bracket. Move the bracket aside slightly, tak-

7.2 Disconnect the brake line fittings with a flare-nut wrench

ing care not to kink the hydraulic lines. Remove the master cylinder from the vehicle.
5 Remove the reservoir caps, or cover and gasket, then discard any fluid remaining in the reservoir.

Overhaul

Refer to illustrations 7.6, 7.8, 7.9, 7.10, 7.11, 7.13, 7.14a and 7.14b

Warning: *A number of different master cylinders have been used on the vehicles covered by this manual. NOT ALL OF THEM ARE REBUILDABLE! On 1971 through 1975 vehicles, the units on models with front drum brakes are different from those used on models with front disc brakes. Up through the mid-Seventies, there are also some minor differences between master cylinders used with different body lines. And, finally, two different materials - cast iron and aluminum - have been used for master cylinder castings: The master cylinders used on 1971 through 1977 vehicles*

7.4 To detach the master cylinder from the power brake booster, remove these nuts (arrows), pull the support bracket off the right (passenger-side) stud and pull the master cylinder assembly straight off the mounting studs

are all cast iron; master cylinders used on 1979 and later vehicles are all aluminum. 1978 vehicles use both - some are cast-iron units, some are aluminum. With all these different master cylinder units, it's essential that you obtain the correct rebuild kit for the master cylinder you're servicing. Be sure to state the year, make and model of the vehicle when buying a kit. If you're still unsure, take the master cylinder with you when buying parts. DO NOT try to use the rebuild kit for a cast iron master cylinder in an aluminum unit, or vise versa. And don't open a new parts kit until you've compared the contents to the old parts you're removing from the master cylinder you intend to rebuild. Regarding the availability of parts for various master cylinders: According to the manufacturer, the aluminum master cylinders on

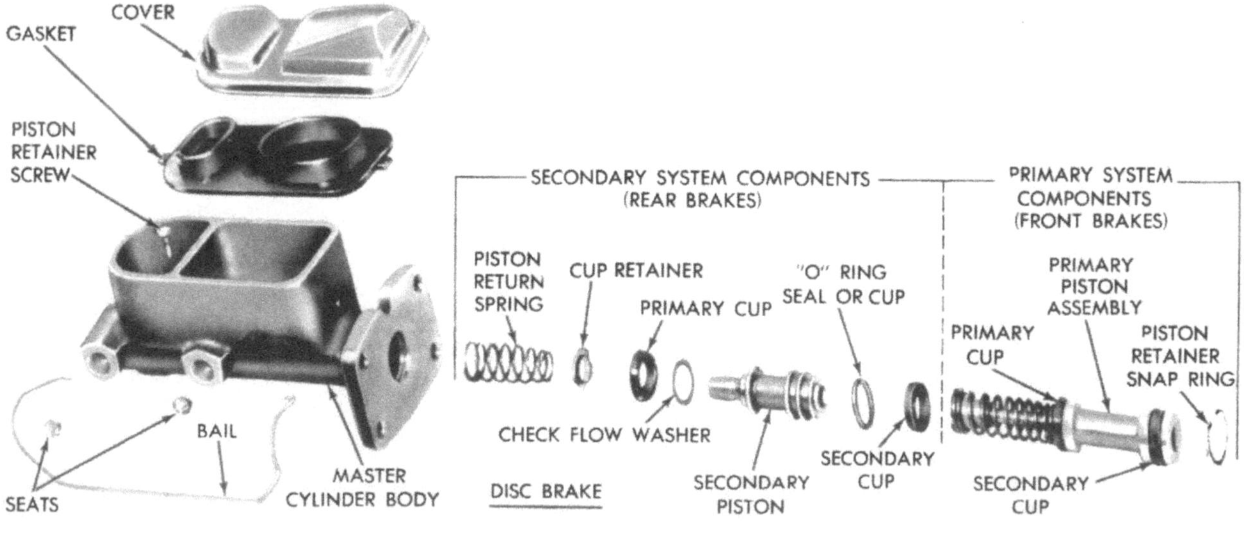

7.6 An exploded view of a typical cast iron master cylinder assembly

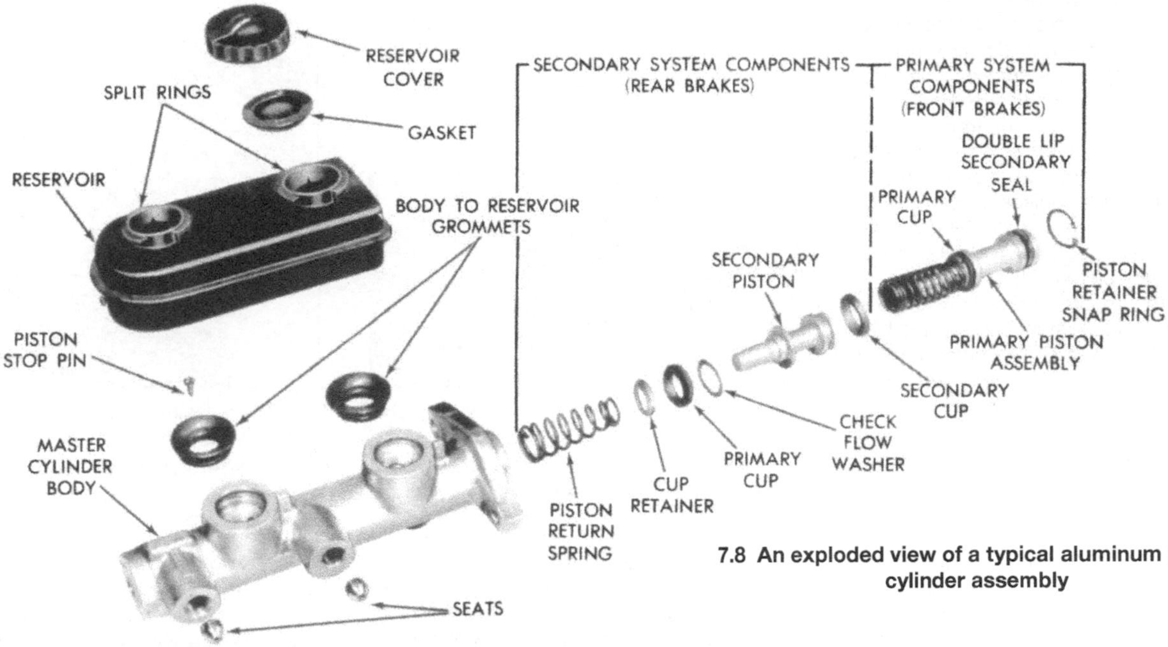

7.8 An exploded view of a typical aluminum master cylinder assembly

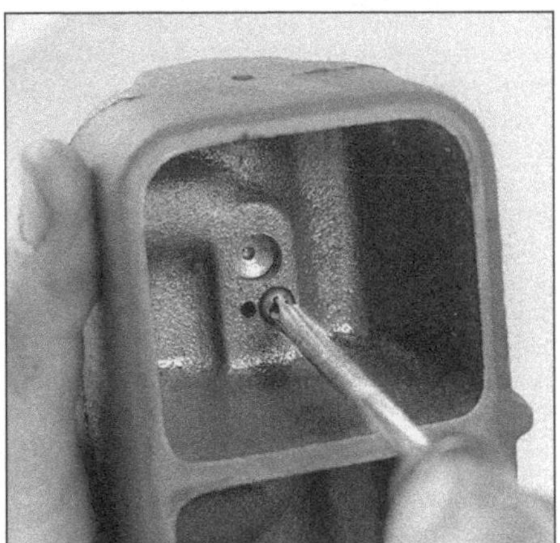

7.9 Remove the piston retainer screw (also referred to as a piston retainer pin or piston stop pin on aluminum master cylinders)

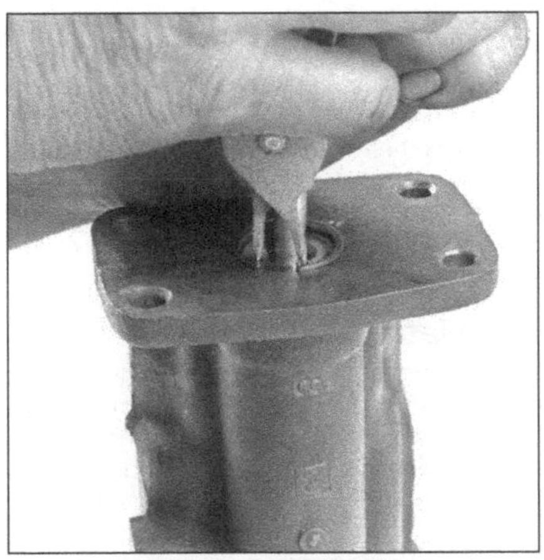

7.10 Depress the primary piston assembly with a Phillips screwdriver and remove the snap-ring with a pair of snap-ring pliers

7.11 As you remove the primary piston assembly, note the relationship of the piston, spring and cups to one another - the new parts must be assembled EXACTLY like this during installation

1983 and later vehicles are NOT rebuildable. At dealerships, there may be no parts available for ANY aluminum master cylinders (i.e. even 1978 through 1982 units). If one of these aluminum units is leaking or malfunctioning, you must replace it with a new or rebuilt unit (unless, of course, you can obtain the correct kit from an aftermarket source). Finally, it should be noted that rebuild kits are not always available for all older cast-iron master cylinders either, particularly those from the early Seventies. Always verify parts availability before beginning a master cylinder rebuild.

6 Remove the caps or the cover and gasket **(see illustration)** and drain the brake fluid into a container, if you haven't already done so. Remove the plugs from the master cylinder outlets and drain the remaining brake fluid.

8 On aluminum-body master cylinders, remove the plastic reservoir **(see illustration)** by rocking it back-and-forth and pulling up on it. Remove the reservoir-to-housing grommets.

9 Remove the piston retainer screw (cast iron models) **(see illustration)** or piston retainer pin/piston stop pin (aluminum models).

10 Push in on the primary piston and remove the piston snap-ring **(see illustration)**.

11 Remove the primary piston assembly from the master cylinder bore **(see illustration)**.

12 Tap the open end of the master cylinder on the workbench to dislodge the secondary piston assembly.

13 Remove the secondary piston and spring from the master cylinder bore **(see illustration)**.

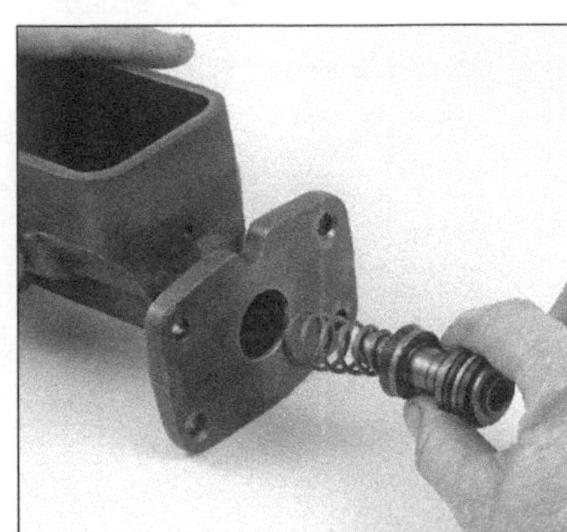

7.13 Remove the secondary piston assembly; again, the same comments apply

14 On cast iron units, install an EZ out or a metal screw in the tube seats as shown **(see illustration)**. Remove and discard the tube seats **(see illustration)**. Remove the residual pressure valves and springs, if equipped.

15 Remove the rubber cups from the ends of the pistons. However, don't remove the primary cup of the primary piston. The primary piston must be replaced as an assembly. On cast iron master cylinders, remove the O-ring or seal cup from the second land of the secondary piston.

16 Clean the master cylinder with fresh brake fluid or brake system cleaner and blow dry it with compressed air. **Warning:** *DO NOT, under any circumstances, use petroleum-based solvents to clean brake parts.*

17 Clean the cylinder bore with fresh brake fluid and inspect the bore for scoring and pitting. Light scratches and corrosion on the cylinder bore walls can be usually be removed with crocus cloth or with a hone. **Warning:** *Never attempt to hone an aluminum master cylinder.* However, deep scratches or score marks mean the cylinder must be replaced with a new unit. If the pistons or bore are severely corroded, replace them. Always use new piston cups and seals (or piston assemblies) when overhauling a master cylinder.

18 All components should be assembled wet after dipping them in clean brake fluid.

19 Carefully insert the complete secondary piston and return spring assembly into the master cylinder bore, easing the seals into the bore. Push the assembly all the way in.

20 Insert the primary piston assembly into the master cylinder bore.

21 Depress the primary piston assembly and install and tighten the secondary piston retainer screw or retainer pin.

22 Push in on the piston again and install the snap-ring.

7.14a To remove the master cylinder tube seats, tap a metal screw into them; this will give you something to pry on during removal

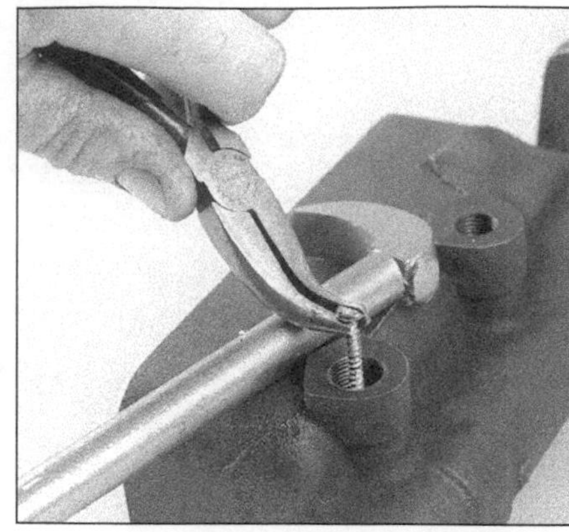

7.14b Using a prybar as a fulcrum, remove the tube seats by prying them out with a pair of long or curved needle-nose pliers

23 Install the residual pressure valves and springs, if equipped, in the outlet ports.

24 The best way to drive the seats into place is with a spare section of brake line (a flared end) with a tube fitting. Place the flared portion of the line over the seat, slide the tube nut into place and thread it into the outlet. Tighten the fitting to push the seat into place.

25 On aluminum master cylinders, lubricate a pair of new grommets with clean brake fluid and install them into their bores in the master cylinder. Install the reservoir into the grommets - make sure the reservoir is completely seated and the rounded portion of the reservoir is facing forward.

Installation

26 Whenever the master cylinder is removed, the entire hydraulic system must be bled. The time required to bleed the system can be reduced is the master cylinder is filled with fluid and bench bled before it's installed on the vehicle. Since you'll have to apply pressure to the master cylinder piston and, at the same time, control flow from the brake line outlets, the master cylinder should be mounted in a vise, with the jaws of the vise clamping on the mounting flange.

27 Insert threaded plugs into the brake line outlet holes and snug them down so that air won't leak past them - but not so tight that they can't be easily loosened.

28 Fill the reservoir with brake fluid of the recommended type (see Chapter 1).

29 Remove one plug and push the piston assembly into the bore to expel the air from the master cylinder. A large Phillips screwdriver can be used to push on the piston assembly.

30 To prevent air from being drawn back into the master cylinder, the plug must be replaced and snugged down before releasing the pressure on the piston.

31 Repeat the procedure until only brake

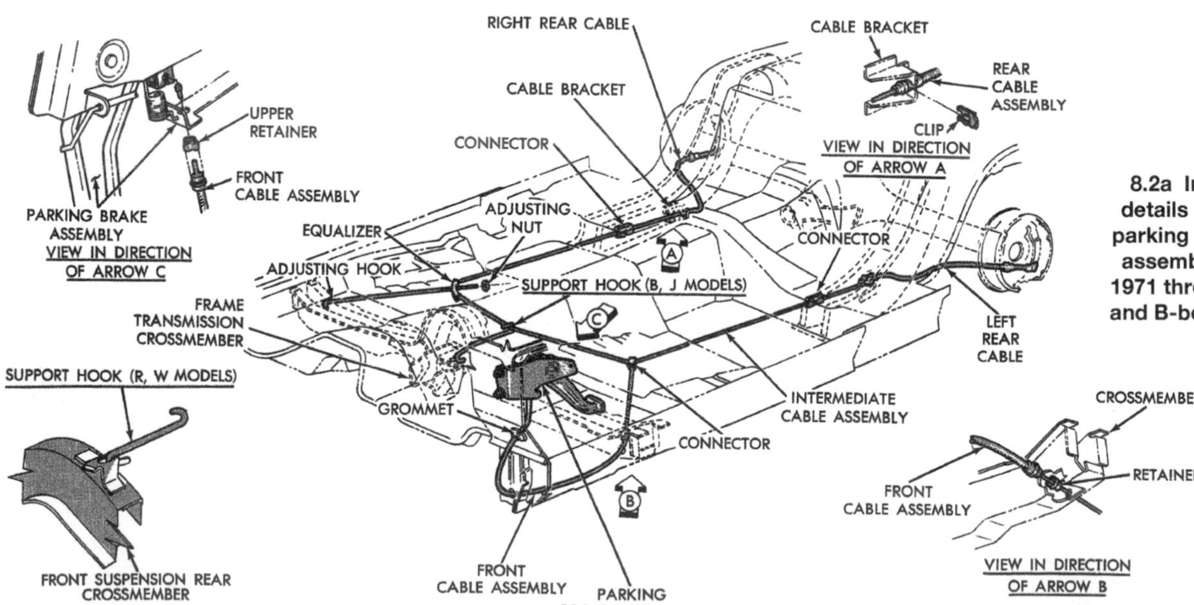

8.2a Installation details of a typical parking brake cable assembly used on 1971 through 1974 E and B-body vehicles

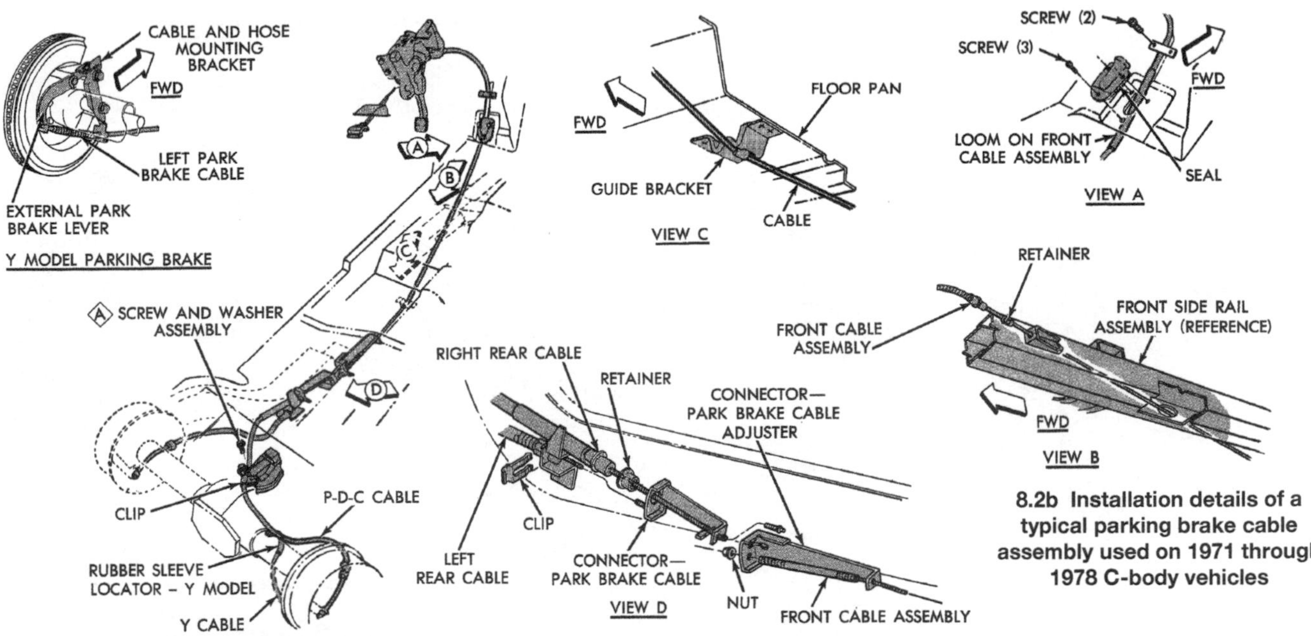

8.2b Installation details of a typical parking brake cable assembly used on 1971 through 1978 C-body vehicles

fluid is expelled from the brake line outlet hole. When only brake fluid is expelled, repeat the procedure at the other outlet hole and plug. Be sure to keep the master cylinder reservoir filled with brake fluid to prevent the introduction of air into the system.

32 Since high pressure isn't involved in the bench bleeding procedure, an alternative to the removal and replacement of the plugs with each stroke of the piston assembly is available. Before pushing in on the piston assembly, remove the plug as described in Step 29. Before releasing the piston, however, instead of replacing the plug, simply put your finger tightly over the hole to keep air from being drawn back into the master cylinder. Wait several seconds for brake fluid to be drawn from the reservoir into the bore, then depress the piston again, removing your finger as brake fluid is expelled. Be sure to put your finger back over the hole each time before releasing the piston, and when the bleeding procedure is complete for that outlet, replace the plug and tighten it before going on to the other port.

33 Carefully install the master cylinder by reversing the removal steps. Make sure you tighten the master cylinder mounting nuts to the torque listed in this Chapter's Specifications.

34 Bleed the brake system as described in Section 11.

8 Parking brake cables - replacement

Refer to illustrations 8.2a, 8.2b, 8.2c, 8.2d, 8.2e, 8.4, 8.5, 8.6a and 8.6b

1 Release the parking brake.

2 Loosen the adjuster nut on the adjuster hook and disconnect the cable from the equalizer **(see illustrations)**.

3 To remove the front cable, slip the ball at the end of the cable out of the clevis on the

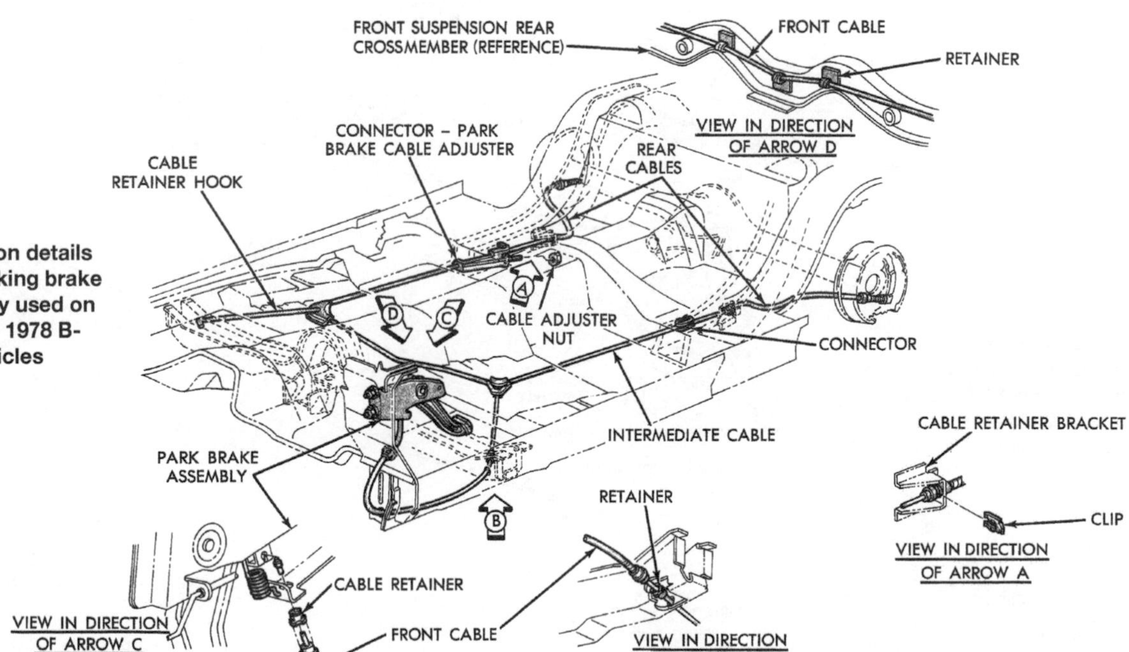

8.2c Installation details of a typical parking brake cable assembly used on 1975 through 1978 B-body vehicles

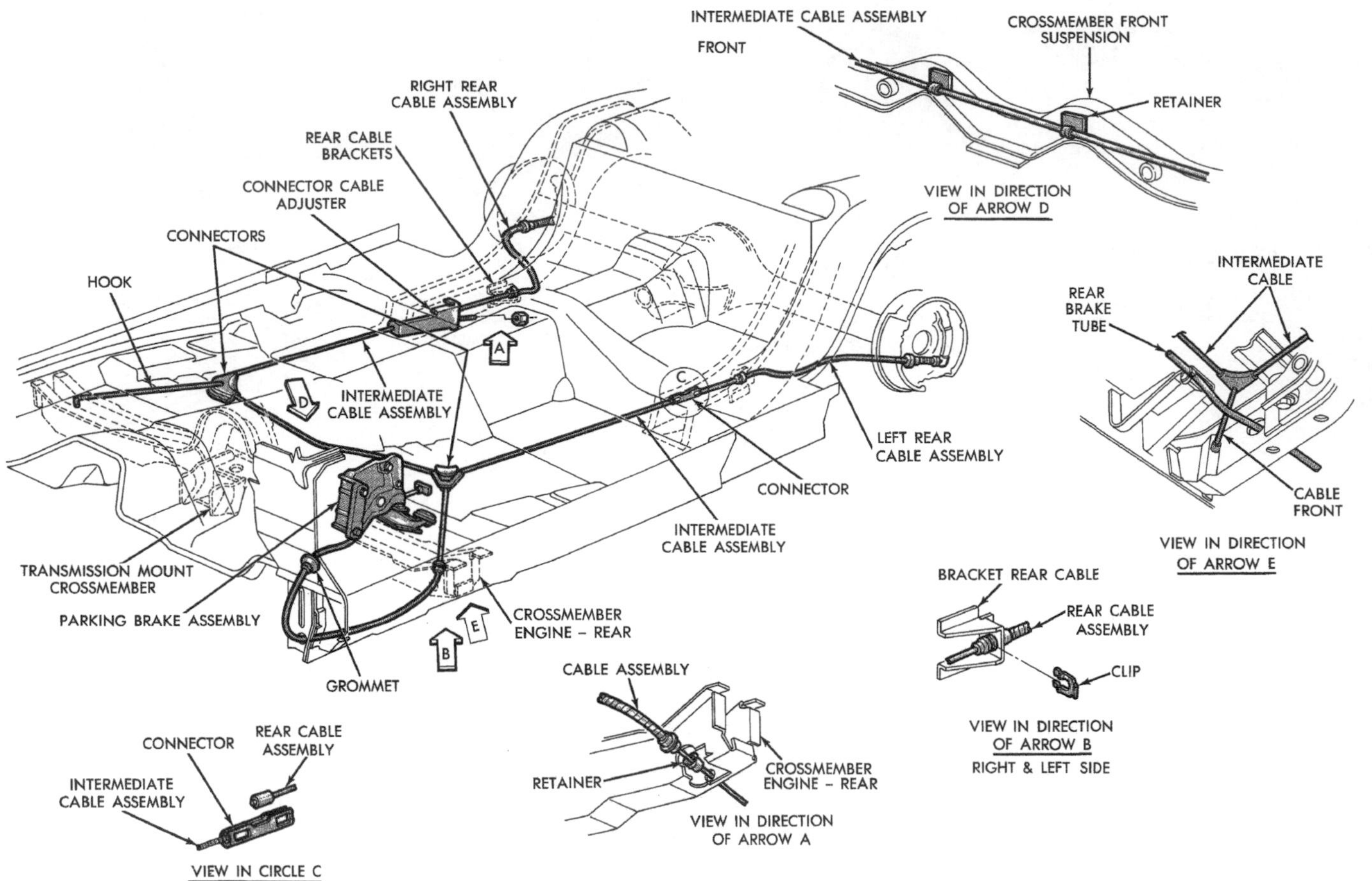

8.2d Installation details of a typical parking brake cable assembly used on 1979 through 1981 R-body vehicles

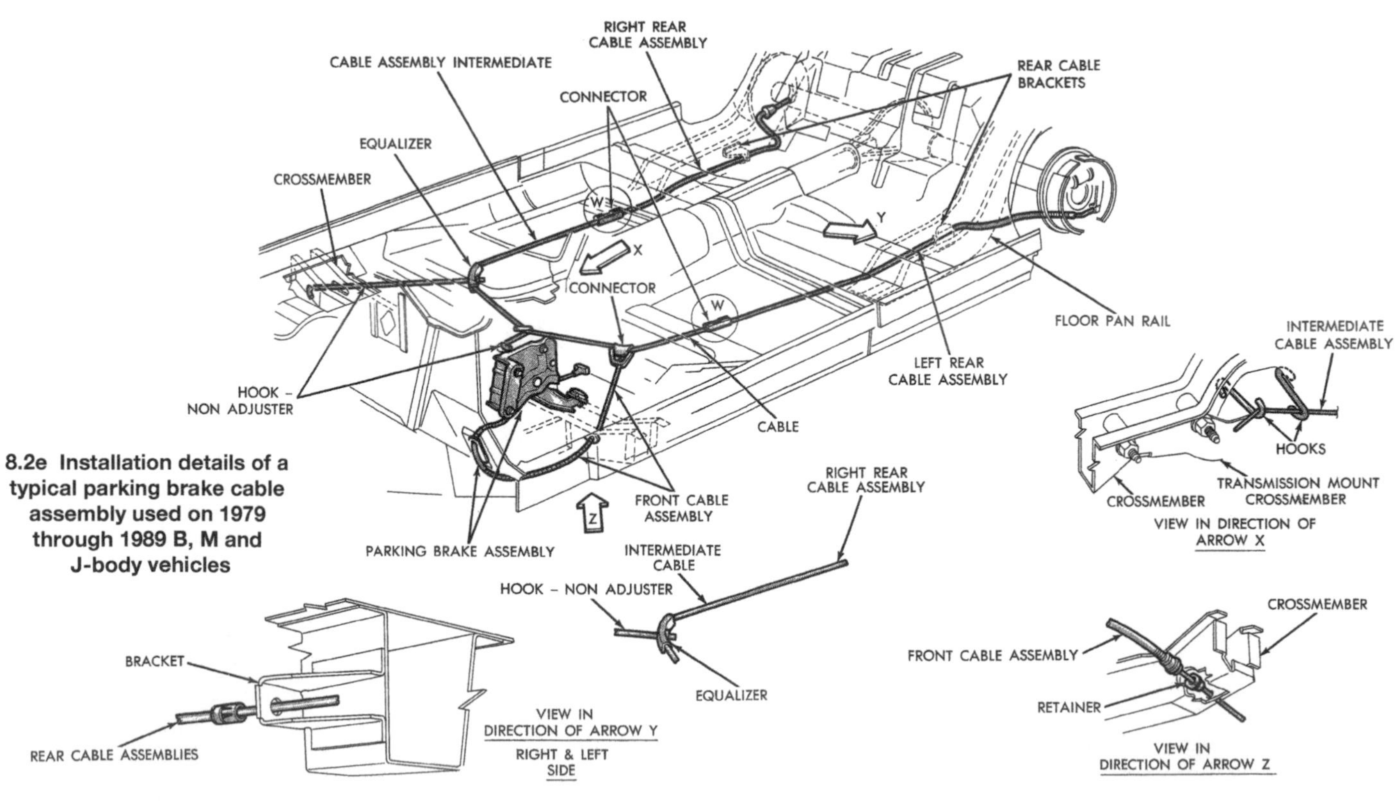

8.2e Installation details of a typical parking brake cable assembly used on 1979 through 1989 B, M and J-body vehicles

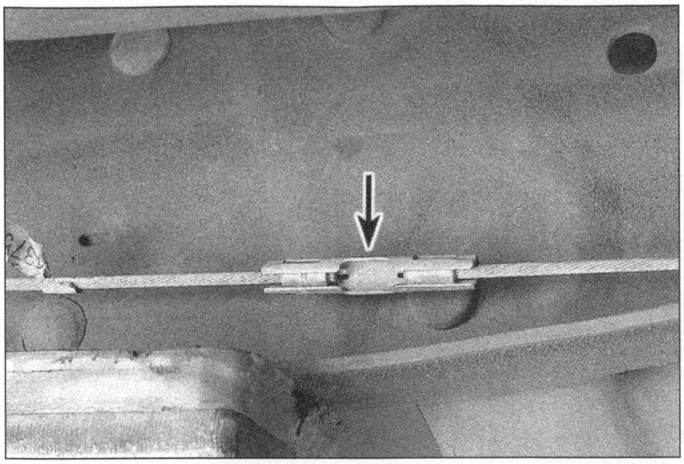

8.4 To detach a cable from a connector (arrow), push the cable toward the center opening and pull it out

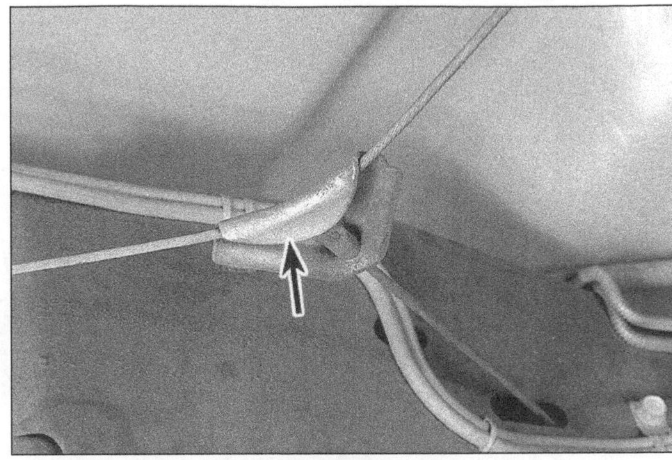

8.5 To remove the intermediate cable, loosen the adjuster nut and pull the cable out of the connector (arrow)

8.6a Pinch the retainer tangs together with needle-nose pliers, then push the cable through the backing plate

8.6b Remove the rear cable retaining clip (arrow)

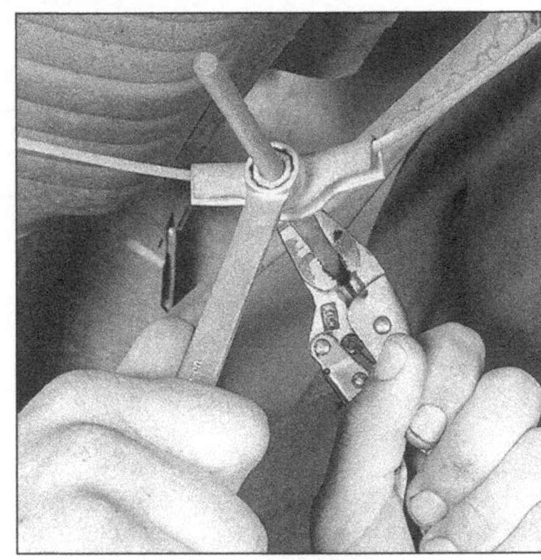

9.3 Hold the cable with locking pliers and turn the adjuster nut with a wrench

parking brake pedal.

4 Slip the front cable out of the connector located below the driver's side of the cab **(see illustration)** and remove the cable from the vehicle.

5 To remove the intermediate cables, loosen the adjuster nut and detach the cable at the connectors **(see illustration)**.

6 To remove the rear cable, remove the rear brake shoes as described in Section 5. Disengage the rear cable from the parking brake lever. Compress the tangs of the cable retainer and withdraw the cable assembly from the brake backing plate **(see illustration)**. Remove the retaining clip **(see illustration)** and lift the cable from the vehicle.

7 Installation is the reverse of removal. Adjust the parking brake as described in Section 9.

9 Parking brake - adjustment

Refer to illustration 9.3

1 The parking brake is pedal operated and is normally self-adjusting through the automatic adjusters in the rear brake drums. However, supplementary adjustment may be needed in the event of cable stretch, wear in the linkage or after installation of new components.

2 Raise the rear of the vehicle until the wheels are clear of the ground, support it securely on jackstands and block the front wheels. Release the parking brake pedal by pulling on the release lever.

3 The adjuster is located on the inside of the frame under catalytic converter. Hold the cable from turning with locking pliers, then loosen the locknut and tighten the adjuster nut **(see illustration)** until a slight drag is felt when the rear wheels are turned. **Note:** *If the threads appear rusty, apply penetrating oil before attempting adjustment.*

4 Loosen the parking brake adjustment until there's no longer any drag when the rear wheels are turned, then loosen the cable adjusting nut two turns.

5 Tighten the locknut, then apply and release the parking brake several times. Confirm that the brake does not drag.

6 Lower the vehicle to the ground.

10 Parking brake shoes (1974 models with rear disc brakes) - removal, inspection and installation

Warning: *Dust created by the brake system may contain asbestos, which is harmful to your health. Never blow it out with compressed air and don't inhale any of it. An approved filtering mask should be worn when working on the brakes. Do not, under any circumstances, use petroleum-based solvents to clean brake parts. Use brake system cleaner only.*

Note: *Parking brake shoes should be replaced on both wheels at the same time - never replace the shoes on only one wheel.*

Removal

Refer to illustration 10.5

1 The parking brake system should be checked as a normal part of driving. With the vehicle parked on a hill, apply the brake, place the transmission in Neutral and verify that the parking brake alone will hold the

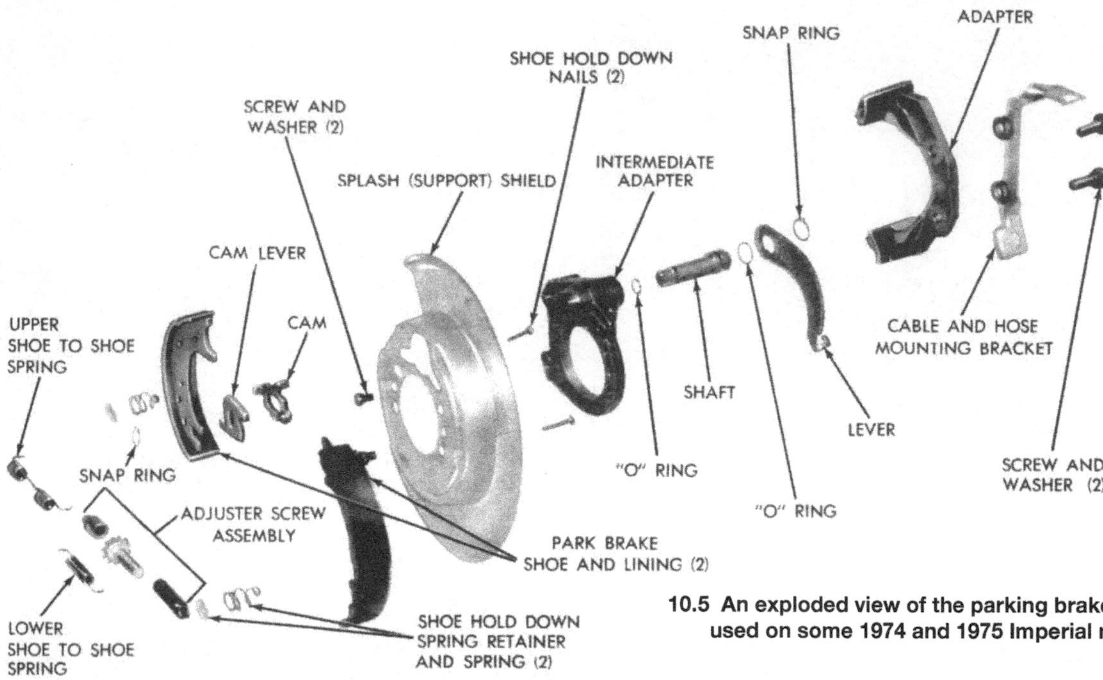

10.5 An exploded view of the parking brake assembly used on some 1974 and 1975 Imperial models

vehicle (be sure to stay in the vehicle during this check). Additionally, every 24 months - and any time a fault is suspected - the assembly itself should be visually inspected.

2 With the vehicle raised and supported on jackstands, remove the rear wheels.

3 Remove the rear calipers (see Section 3). Support the caliper assemblies with a coat hanger or heavy wire and don't disconnect the brake line from the caliper.

4 Remove the rear discs (see Section 4). Clean the parking brake assembly with brake system cleaner.

5 Remove the lower shoe-to-shoe return spring **(see illustration)**.

6 Spread the brake shoes slightly and remove the adjusting star wheel.

7 Remove the upper shoe-to-shoe return spring.

8 Move the brake shoes off of the support platform and remove the brake shoe retainers, springs and nails.

9 Disconnect the parking brake cable from the inner operating lever, remove the snap-ring retainer from the shaft and remove the operating lever.

10 Remove the shaft.

Inspection

11 Inspect the lining contact pattern to determine whether the shoes are bent or have been improperly adjusted. The lining should show contact across the entire width, extending from head to toe. Shoes showing contact only on one side should be replaced.

12 Clean the splash (support) shield with a suitable solvent, inspect it for burrs and scrub them off if necessary.

13 Clean and inspect the adjusting screws for pulled or stripped threads, then apply a thin film of lubricant to the threads, socket and washer. Replace the adjuster screw if corrosion of any part inhibits its free operation.

14 Clean the cam lever, cam, shaft and operating lever, inspect them for damage and replace as necessary. Inspect all O-rings for distortion and replace as necessary.

15 Inspect the drum (see Section 5).

Installation

16 Lubricate the shaft with a thin film of multi-purpose grease and install the shaft in the intermediate adapter.

17 Install the outer cam and cam lever on the shaft and install the snap-ring.

18 Install the inner operating lever on the shaft and install the snap-ring.

19 Install the inner operating lever on the shaft and install the snap-ring. Connect the parking brake cable to the operating lever.

20 Lubricate the shoe contact area on the splash (support) shield with a thin film of multi-purpose grease.

21 Install the upper shoe-to-shoe return spring.

22 Install the adjuster star wheel. The star adjuster is forward on the left (driver's) side and toward the rear of the vehicle on the right (passenger's) side.

23 Install the lower shoe-to-shoe return spring.

24 Install the disc/drum (see Section 4 or 5).

25 To adjust the parking brake shoes, insert a long thin screwdriver through the hole in the splash (support) shield, rotate the star wheel until the brake shoes contact the surface of the disc/drum, then back the shoes off 12 clicks of the star wheel.

26 Install the caliper (see Section 3).

11 Brake system bleeding

Refer to illustration 11.8

Warning: *Wear eye protection when bleeding the brake system. If the fluid comes in contact with your eyes, immediately rinse them with water and seek medical attention.*

Note: *Bleeding the brake system is necessary to remove any air that's trapped in the system when it's opened during removal and installation of a hose, line, caliper, wheel cylinder or master cylinder.*

1 It will probably be necessary to bleed the system at all four brakes if air has entered the system due to low fluid level, or if the brake lines have been disconnected at the master cylinder.

2 If a brake line was disconnected only at a wheel, then only that caliper or wheel cylinder must be bled.

3 If a brake line is disconnected at a fitting located between the master cylinder and any of the brakes, that part of the system served by the disconnected line must be bled.

4 Remove any residual vacuum from the brake power booster (if equipped) by applying the brake several times with the engine off.

5 Remove the master cylinder reservoir cover and fill the reservoir with brake fluid. Reinstall the cover. **Note:** *Check the fluid level often during the bleeding operation and add fluid as necessary to prevent the fluid level from falling low enough to allow air bubbles into the master cylinder.*

6 Have an assistant on hand, as well as a supply of new brake fluid, an empty clear plastic container, a length of 3/16-inch plastic, rubber or vinyl tubing to fit over the bleeder valve and a wrench to open and close the bleeder valve.

11.8 **To bleed the brakes, attach a clear piece of tubing to the bleeder screw fitting and submerge the other end of the tubing in brake fluid - you can easily see any air bubbles in the tube and container (when no more bubbles appear, the air has been purged from the caliper or wheel cylinder)**

7 Beginning at the right rear wheel, loosen the bleeder screw slightly, then tighten it to a point where it's snug but can still be loosened quickly and easily.

8 Place one end of the tubing over the bleeder screw fitting and submerge the other end in brake fluid in the container **(see illustration)**.

9 Have the assistant pump the brakes a few times to get pressure in the system, then hold the pedal firmly depressed.

10 While the pedal is held depressed, open the bleeder screw just enough to allow a flow of fluid to leave the valve. Watch for air bubbles to exit the submerged end of the tube. When the fluid flow slows after a couple of seconds, tighten the screw and have your assistant release the pedal.

11 Repeat Steps 9 and 10 until no more air is seen leaving the tube, then tighten the bleeder screw and proceed to the left rear wheel, the right front wheel and the left front wheel, in that order, and perform the same procedure. Be sure to check the fluid in the master cylinder reservoir frequently.

12 Never use old brake fluid. It contains moisture which will deteriorate the brake system components.

13 Refill the master cylinder with fluid at the end of the operation.

14 Check the operation of the brakes. The pedal should feel solid when depressed, with no sponginess. If necessary, repeat the entire process. Bleed the height sensing proportioning valve or anti-lock hydraulic valve, if equipped. **Warning:** *Do not operate the vehicle if you are in doubt about the effectiveness of the brake system.*

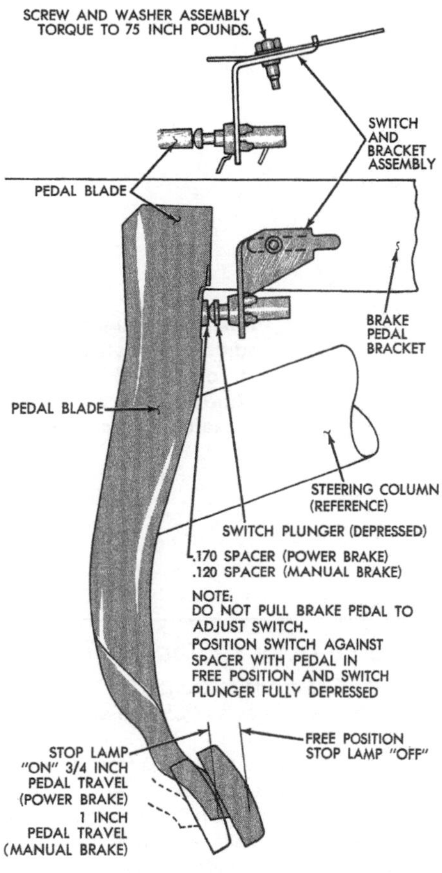

12.2a Brake light switch adjustment details (1971 through 1976 A-body vehicles)

12 Brake light switch - replacement and adjustment

Replacement

Refer to illustrations 12.2a, 12.2b, 12.2c and 12.2d

1 Disconnect the negative cable from the battery.

2 The brake light switch is located at the top of the brake pedal **(see illustrations)**.

3 To remove the switch, unplug the electrical connector and unbolt the switch and

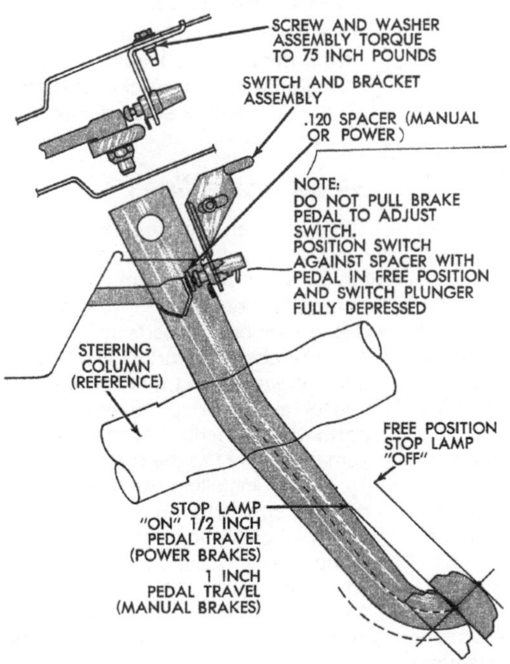

12.2b Brake light switch adjustment details (1971 through 1973 C-body vehicles)

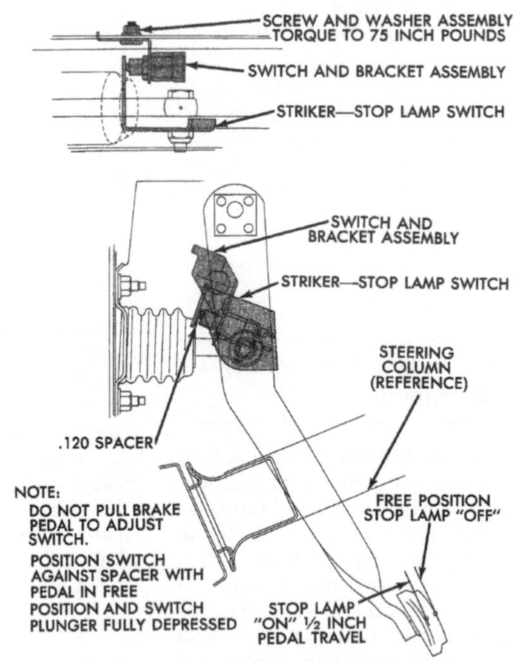

12.2c Brake light switch adjustment details (1974 through 1978 C-body vehicles)

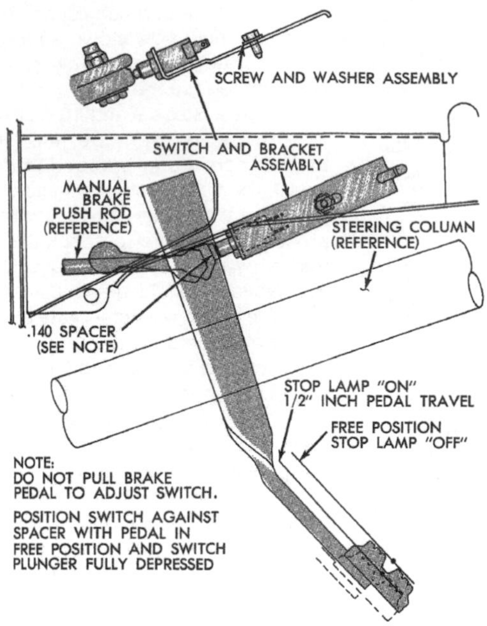

12.2d Brake light switch adjustment details (1971 through 1976 E, B and F-body vehicles; 1977 and later, all models)

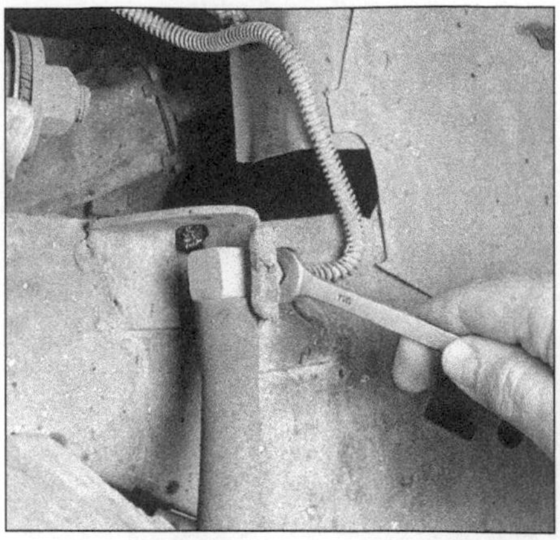

13.3 Disconnect the brake line from the hose fitting; although it's not necessary on this particular fitting, on some fittings you will need a back-up wrench; if necessary, soak the connection with penetrating oil; be careful not to bend the frame bracket or line

bracket assembly.

4 Installation is the reverse of removal. Reconnect the electrical connector and have an assistant check the brake lights.

Adjustment

5 Loosen the switch assembly-to-pedal bracket mounting screw and slide the assembly away from the pedal blade or brake pushrod.

6 Push the brake pedal down and allow it to return to its free position. Do NOT pull the brake pedal back at any time!

7 Place the proper spacer gauge on the pedal blade or pushrod **(see illustration 12.2a, 12.2b, 12.2c or 12.2d)**.

8 Slide the switch assembly toward the pedal blade or pushrod until the switch plunger is fully depressed against the spacer gauge (on heavy duty or stop light/speed control switches, depress the plunger until the switch body contacts the spacer gauge).

9 Retighten the switch bracket screw securely.

10 Remove the spacer. Make sure the stop light switch doesn't hang up on the pedal blade after the adjustment.

13 Brake hoses and lines - check and replacement

1 About every six months, with the vehicle raised and placed securely on jackstands, the flexible hoses which connect the steel brake lines with the front and rear brake assemblies should be inspected for cracks, chafing of the outer cover, leaks, blisters and other damage. These are important and vulnerable parts of the brake system and inspection should be complete. A light and mirror will be needed for a thorough check. If a hose exhibits any of the above defects, replace it with a new one.

Flexible hose replacement

Refer to illustrations 13.3 and 13.4

2 Clean all dirt away from the ends of the hose.

3 Disconnect the brake line from the hose fitting **(see illustration)**. you'll need a back-up wrench on some fittings. Be careful not to bend the frame bracket or line. If necessary, soak the connections with penetrating oil.

4 Remove the U-clip (lock) from the female fitting at the bracket **(see illustration)** and remove the hose from the bracket.

5 On models that use banjo fittings, unscrew the banjo bold and disconnect the hose from the caliper, discarding the copper washers on either side of the fitting. On models where the brake hose threads directly into the caliper, unscrew the hose from the caliper.

6 Using new copper washers, attach the new brake hose to the caliper (models that use banjo fittings). Tighten the banjo bolt to the torque listed this Chapter's Specifications. On models where the brake hose threads directly into the caliper, tighten the hose fitting securely.

7 Pass the female fitting through the frame or frame bracket. With the least amount of twist in the hose, install the fitting in this position. **Note:** *The weight of the vehicle must be on the suspension, so the vehicle should not be raised while positioning the hose.*

8 Install the U-clip (lock) in the female fitting at the frame bracket.

9 Attach the brake line to the hose fitting using a back-up wrench on the fitting. Tighten the fitting securely.

10 Carefully check to make sure the suspension or steering components don't make contact with the hose. Have an assistant push on the vehicle and also turn the steering wheel lock-to-lock during inspection.

11 Bleed the brake system as described in Section 11.

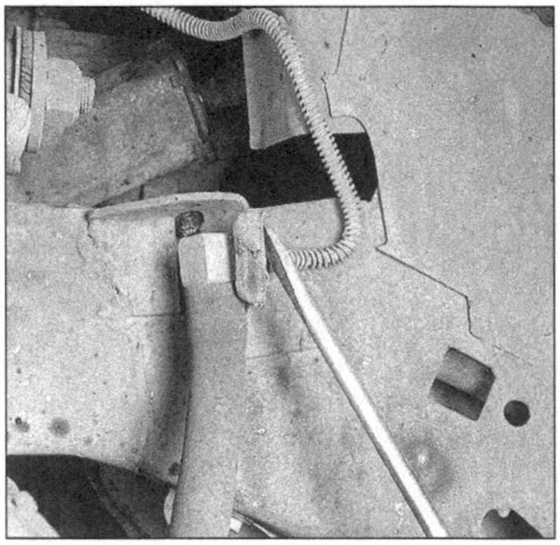

13.4 Before you can disconnect the hose from the line, you'll have to pry a U-clip (lock) off the bracket and fitting with a screwdriver

Metal brake lines

12 When replacing brake lines, be sure to use the correct parts. Don't use copper tubing for any brake system components. Purchase steel brake lines from a dealer parts department or auto parts store.

13 Prefabricated brake line, with the tube ends already flared and fittings installed, is available at auto parts stores and dealer parts departments. These lines are also sometimes bent to the proper shapes.

14 When installing the new line make sure it's well supported in the brackets and has plenty of clearance between moving or hot components. Make sure you tighten the fittings securely.

15 After installation, check the master cylinder fluid level and add fluid as necessary. Bleed the brake system as outlined in Section 11 and test the brakes carefully before placing the vehicle into normal operation.

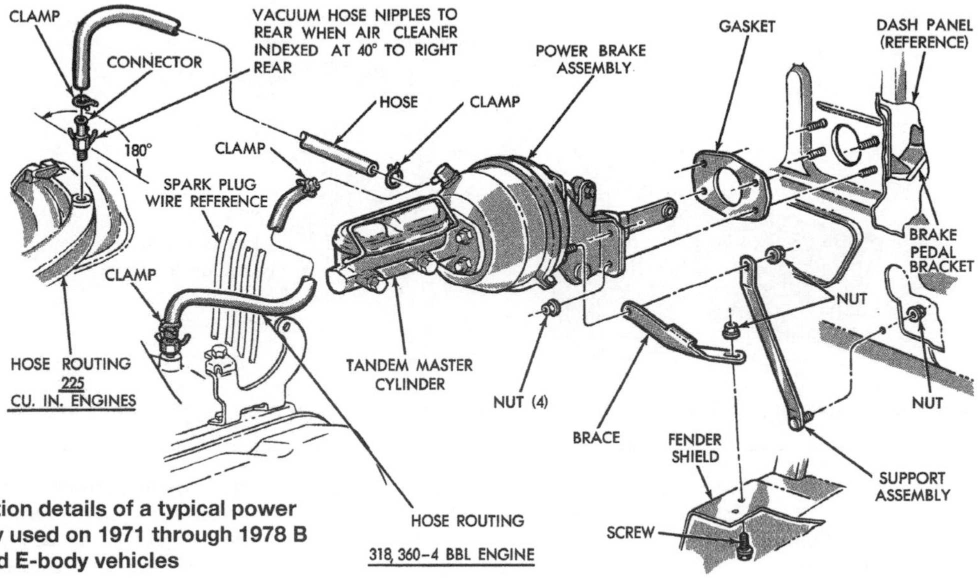

14.8a Installation details of a typical power brake assembly used on 1971 through 1978 B and E-body vehicles

14 Power brake booster - check, removal and installation

Operating check

1 Depress the brake pedal several times with the engine off and make sure that there is no change in the pedal reserve distance.

2 Depress the pedal and start the engine. If the pedal goes down slightly, operation is normal.

Airtightness check

3 Start the engine and turn it off after one or two minutes. Depress the brake pedal several times slowly. If the pedal goes down farther the first time but gradually rises after the second or third depression, the booster is airtight.

4 Depress the brake pedal while the engine is running, then stop the engine with the pedal depressed. If there is no change in the pedal reserve travel after holding the pedal for 30 seconds, the booster is airtight.

Removal

Refer to illustrations 14.8a, 14.8b, 14.8c and 14.8d

5 The power brake booster unit requires no special maintenance apart from periodic inspection of the vacuum hose and the case.

6 Disassembly of the power unit requires special tools and is not ordinarily performed by the home mechanic. If a problem develops, it's recommended that a new or factory rebuilt unit be installed.

7 Remove the master cylinder (see Section 7).

8 Disconnect the vacuum hose where it attaches to the power brake booster **(see illustrations)**.

9 Working under the dash, disconnect the power brake pushrod from the top of the brake pedal by prying the clip off.

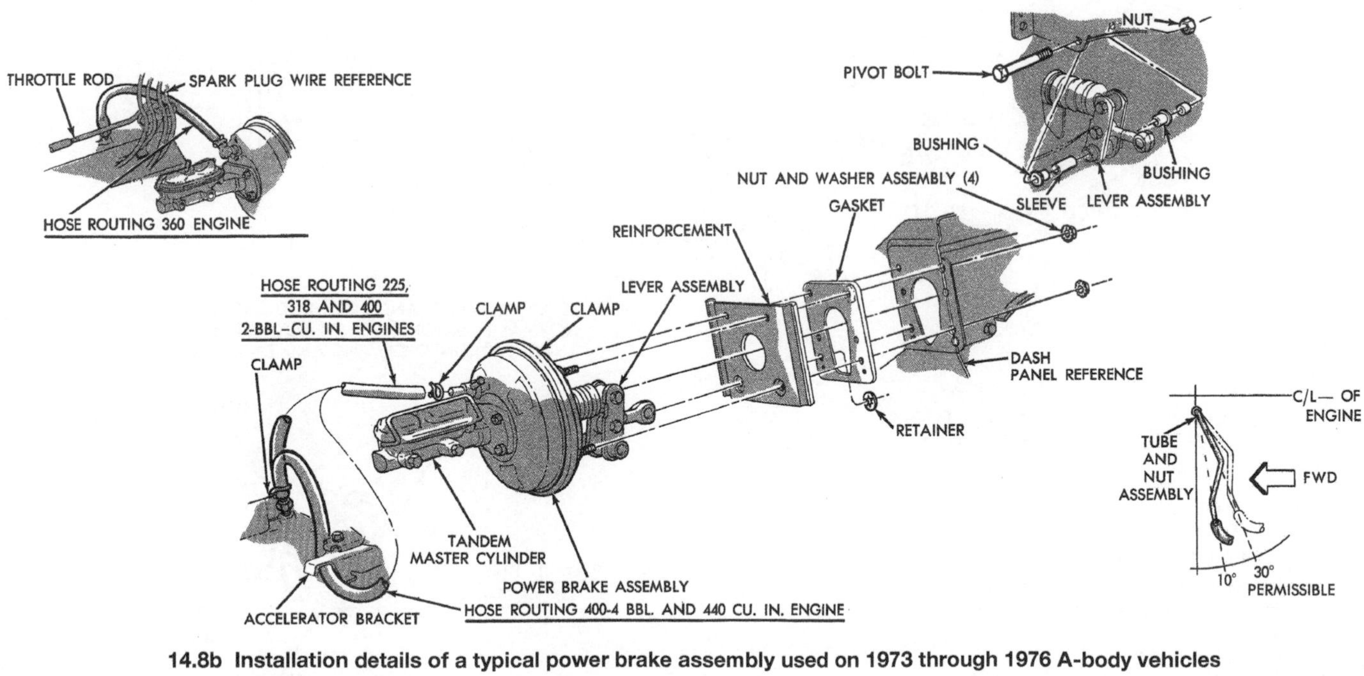

14.8b Installation details of a typical power brake assembly used on 1973 through 1976 A-body vehicles

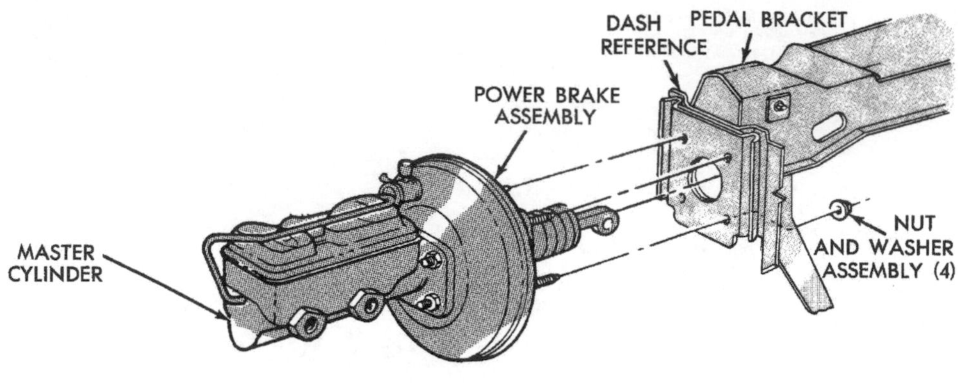

14.8c Installation details of a typical power brake assembly used on 1974 through 1978 C-body vehicles

10 Remove the nuts attaching the booster to the firewall.

11 Carefully lift the booster unit away from the firewall and out of the engine compartment.

Installation

12 To install the booster, place it into position and tighten the retaining nuts to the torque listed in this Chapter's Specifications. Connect the brake pedal. **Warning:** *Use a new clip.*

13 Install the master cylinder and vacuum hose.

14 Carefully test the operation of the brakes before placing the vehicle in normal operation.

15 Hydraulic system control valve - check, resetting and replacement

Refer to illustration 15.1

1 All models have a hydraulic system control valve **(see illustration)** that incorporates a brake warning switch to warn if one half of the hydraulic system fails and a proportioning valve which restricts - at a given ratio - hydraulic pressure to the rear wheels when system hydraulic pressure reaches a certain point. Models with disc brakes in the front and drum brakes in the rear also have a metering valve (or hold-off valve) which limits the front brake pressure until the rear brake shoes have overcome the return springs and contacted the drums.

Check

2 The brake warning light normally comes on during the engine start sequence, then goes off again when the engine starts. It can

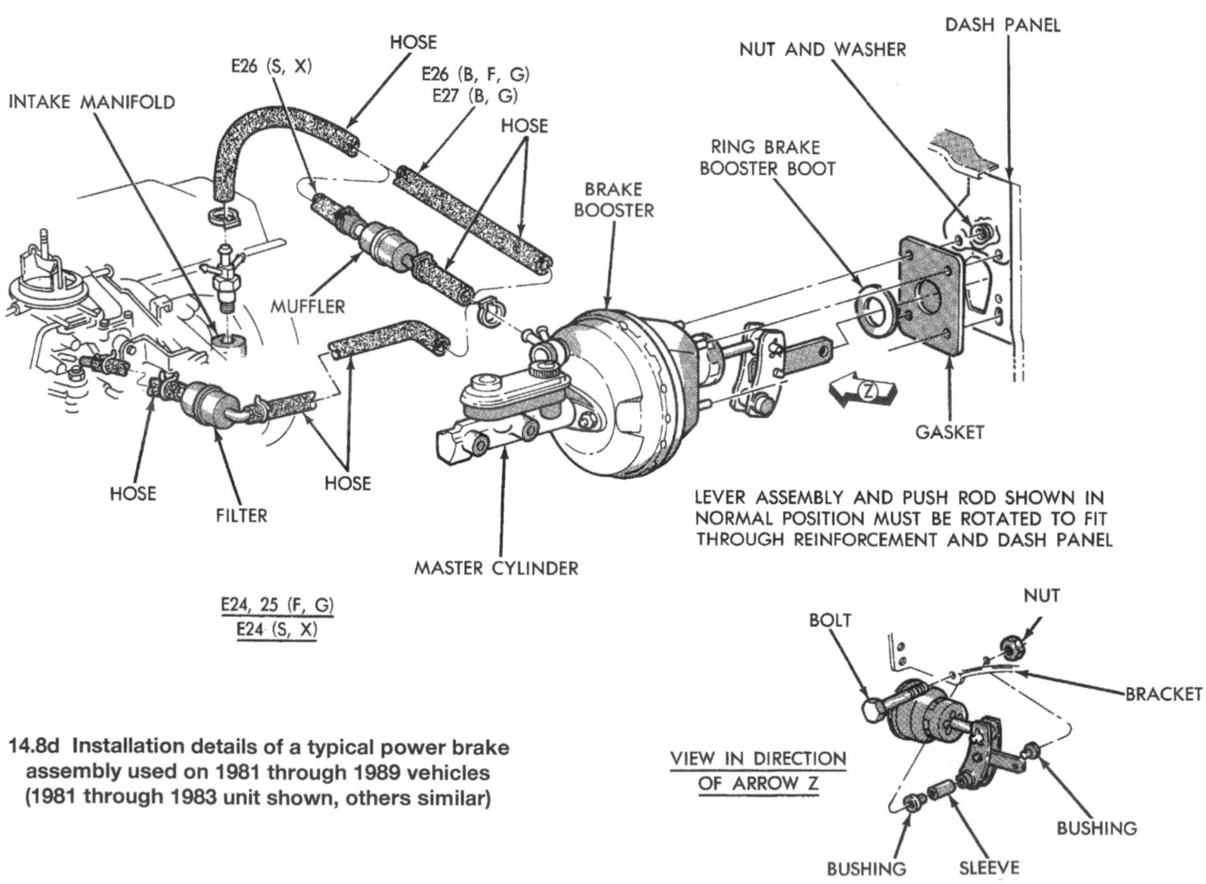

14.8d Installation details of a typical power brake assembly used on 1981 through 1989 vehicles (1981 through 1983 unit shown, others similar)

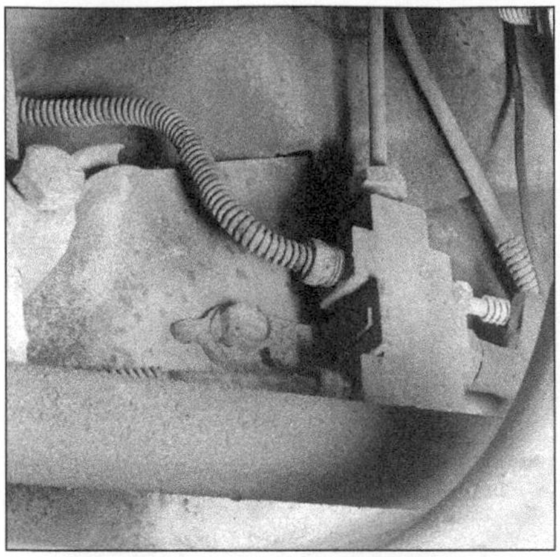

15.1 The hydraulic control valve is probably located on the left side of the engine compartment, somewhere below the master cylinder - to find the valve, simply follow the lines from the master cylinder

be checked functionally by opening one bleeder screw slightly while an assistant presses down on the brake pedal. The light should come on with the ignition switch On.

Resetting

3 If the light for the brake warning system comes on, a leak or problem has occurred in the system and must be corrected. Once the leak or problem has been repaired or if the hydraulic system has been opened up for brake cylinder overhaul or a similar repair, the pressure differential warning switch must be centered. Once the valve is centered, the brake light on the dash will go out.

4 To center the valve, first fill the master cylinder reservoir and make sure the hydraulic system has been bled.

5 Turn the ignition switch to On or Accessory. Slowly press the brake pedal down and the piston will center itself, causing the light to go out.

6 Check the brake pedal for firmness and proper operation.

Replacement

Caution: *Brake fluid will damage paint. Cover all painted surfaces and avoid spilling fluid during this procedure.*

7 If a hydraulic system control valve is defective or if it is leaking, it must be replaced. It is a non-serviceable unit and no repair operations are possible by the home mechanic.

8 Disconnect the brake warning light connector from the warning light switch.

9 Disconnect the brake line fittings from the valve assembly. Plug the ends of the lines to prevent loss of brake fluid and the entry of dirt.

10 Remove the bolts and nuts securing the valve assembly to the chassis.

11 Remove the valve assembly and bracket.

12 Installation is the reverse of removal.

13 Bleed the system after the replacement valve has been installed.

14 Center the valve as described in Steps 3 through 6 above.

16 Sure Brake System - general information

1 The Sure Brake System was an available option on some 1972 and 1973 Imperial models. This system prevents any wheel from locking up during brake applications above a speed of about 5 mph. The Sure Brake System reduces the likelihood of a skid caused by a locked wheel, yet still maintains enough brake pressure for maximum stopping power. This improves the directional control and steerability of the vehicle, and often even reduces the distance required to bring the vehicle to a stop.

2 There are three major components in the Sure Brake System. They include a mechanically driven speed sensor at each wheel; a logic controller (LC) located inside right rear quarter-panel and three pressure modulators (one under each front fender ahead of the wheels and one next to the radiator on the right side of the engine compartment).

3 Because of the complexity of this system and the special tools needed to diagnose and service it, we don't recommend tampering with this system. If it malfunctions, take the vehicle to an authorized dealer service department or other qualified repair shop to have it repaired.

Notes

Chapter 10
Suspension and steering systems

Contents

Specifications

Ride height

1971 and 1972*	
A-body	
2-door	1-5/8 inches
4-door	2-1/8 inches
B-body	1-5/8 inches
C-body	
Chrysler	1-1/8 inches
Imperial	1-3/4 inches
All others	1-3/8 inches
E-body	1 inch
1973*	
A-body	
2-door	1-7/8 inches
4-door	2-1/8 inches
B-body	1-7/8 inches
C-body	
Chrysler	1-1/4 inches
Plymouth Fury and Dodge Monaco	1-1/2 inches
Imperial	1-3/4 inches
E-body	1-1/8 inches
1974*	
A-body	1-7/8 inches
B-body	1-7/8 inches
C-body	1 inch
E-body	1-1/8 inches
1975 through 1979	
A-body (1975 and 1976)	10-15/16 inches
B-body (1975 through 1977)	
Sedan	10-3/4 inches
Station wagon	11-1/4 inches
C-body (1975 through 1978)	10-1/8 inches
F-body(1976 through 1979)	
M-body (1978 through 1980)	
J-body (1980)	10-1/4 inches
R-body	10-3/4 inches

Ride height (continued)

1981	
M-body and J-body	
Acceptable range	12 to 13 inches
Service reset	12-1/2 inches + 1/4-inch
R-body	
Acceptable range	16-1/4 to 17-1/4 inches
Service reset	16-3/4 inches + 1/4-inch
1982 on (all models)	
Acceptable range	12 to 13 inches
Service reset	12-1/2 inches + 1/4-inch

** The dimensions listed here for 1971 through 1974 vehicles are for the* difference *in height between points A and B (see illustrations in Sections 4 and 5)*

Torque specifications

	Ft-lbs (unless otherwise indicated)

Steering

Idler arm nut	
1971 through 1975	65
1976 on	70
Pitman arm nut	175
Steering wheel nut	
1971 and 1972	28
1973 through 1982	60
1983 on	45
Tie-rods	
Tie-rod end	40
Tie-rod sleeve clamps	
1971 through 1973	
All except C-body	115 in-lbs
C-body	150 in-lbs
1974	
All except C-body	132 in-lbs
C-body	150 in-lbs
1975	150 in-lbs
1976 through 1978	200 in-lbs
1979 on	150 in-lbs

Front suspension

Balljoints

Balljoint-to-upper control arm	
1971 through 1974	
All except Imperial	125
Imperial	150
1975 through 1978	
All except C-body	125
C-body	150
1979 on	125
Lower balljoint stud nut	
1971 through 1973	
A, B and 1971 and 1972 B-body	85
1973 B-body	150
All C-bodies except Imperial	115
Imperial	135
1974 through 1978	
All except C-body	100
C-body	135
1979 on	100
Upper balljoint stud nut	
1971 and 1972	
A-body	55
B and E body	100
C-body	125
1973	
All except Imperial	100
Imperial	125
1974 through 1978	
All except C-body	100
C-body	135
1979 on	100

Isolator	
1972 (Imperial)	
Isolator-to-crossmember	150
Rear isolator-to-frame	75
Isolator bracket and sleeve assembly-to-frame	75
Front isolator assembly nut	30
Lower control arm pivot shaft	
1971 through 1974	
All except C-body	145
C-body	190
1975	
A-body	145
All except A-body	190
1976 through 1981	
A-body (1976 only)	145
F, M and J-body (pivot bolt nut)	75
B and R-body	145
C-body	190
1982 on	75
Rear anchor crossmember insulator bolt (1974 through 1979 B-body; 1979 through 1981 R-body)	75
Steering knuckle	
Lower steering knuckle bolts	
1971 and 1972	
A-body	100
All except A-body	120
1973	
A, B and E-body and Imperial	160
C-body except Imperial	120
1974 on	160
Upper support plate steering knuckle bolts (A-body)	55
Strut	
Strut-to-frame	
1971 and 1972	
A-body	45
All except A-body	52
1973 through 1975	52
1976 through 1981	70
Strut-to-lower control arm bolt/nut	
1971 and 1972	
All except Imperial	105
Imperial	115
1973	
A, C (except Imperial) and E-body	105
B-body	95
Imperial	115
1974 and 1975	
A and E-body	105
B and C-body	95
1976 through 1981	
A-body (1976 only)	105
B, C and R-body	100
Torsion bar	
Torsion bar bushing-to-lower control arm bolt/nut	
1976 F-body	50
1977 F-body	70
1978 and 1979 F and M-body	70
1980 F, M and J-body	70
1981 J and M-body	70
1982 on	70
Torsion bar pivot cushion retainer nut/bolt	
1976 and 1977 F-body	75
1978 F and M-body	75
1979 F and M-body	85
1980 F, M and J-body	85
1981 J and M-body	85
1982 on	85

Front suspension (continued)

Upper control arm cam bolt (adjusting bolt)	
1971 and 1972	
All except Imperial	65
Imperial	160
1973 and 1974	
All except B-body and Imperial	65
B-body (pivot shaft bolt)	160
Imperial	160
1975	
A-body	70
B and C-body (pivot shaft bolt)	175
1976	
A-body	70
F, B and C-body (pivot shaft bolt	150
1977	
B and F-body (pivot shaft bolt)	150
C-body	175
1978	
B, F and M-body	150
C-body	175
1979 through 1989 (pivot shaft bolt)	150
Upper control arm pivot bushing nut (1979 on)	110

Rear suspension

Rear leaf springs	
Center bolt/nut	
1971	10
1972 through 1976	
All except A-body	40
A-body	25
1977 on	40
Front hanger nut	
1971 through 1978	30
1979 on	35
Pivot bolt and nut	
1971 through 1975	
A-body	85
All except A-body	125
1976	
A-body	85
All except C and F-body	125
C and F-body	105
1977 and 1978	105
1979	
F and M-body	105
B and R-body	125
1980 on	105
Rear hanger bolt	
1971 through 1975	30
1976 through 1978	40
1979 on (rear hanger screw)	35
Shackle nut	
1971 through 1976	
All except A-body	40
A-body	30
1977 and 1978	40
1979 on	35
U-bolt nut	
1971 through 1975	
All except 7-1/4 inch axle	45
7-1/4 inch axle	40
1976 on	45

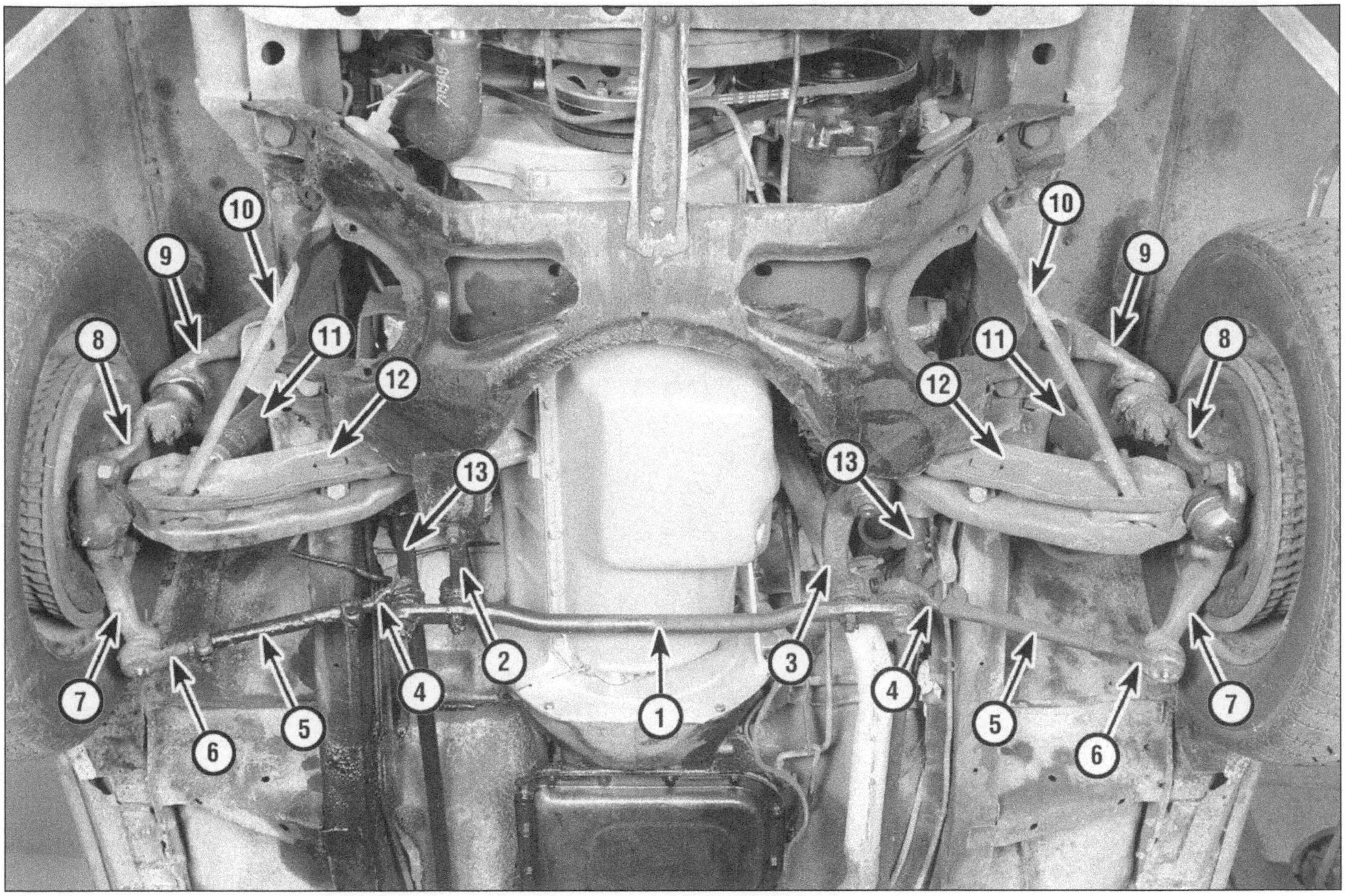

1.2a Steering and front suspension components (vehicles with longitudinal torsion bars)

1. *Center link*
2. *Idler arm*
3. *Pitman arm*
4. *Inner tie-rod (tie-rod end)*
5. *Tie-rod adjuster sleeve*
6. *Outer tie-rod (tie-rod end)*
7. *Steering arm*
8. *Steering knuckle*
9. *Upper control arm*
10. *Strut rod*
11. *Shock absorber*
12. *Lower control arm*
13. *Torsion bar*

1 General information

Refer to illustrations 1.2a, 1.2b and 1.3

The front suspension **(see illustrations)** is fully independent. Each wheel is connected to the frame by a steering knuckle, upper and lower balljoints and upper and lower control arms. All models use shock absorbers and torsion bars. The upper end of each front shock is attached to a bracket on the frame. The lower end of each shock is attached to the lower control arm. Longitudinal torsion bars connect the lower control arms to the frame on all 1971 through 1975 vehicles; transverse torsion bars connect the frame to the control arms on all 1982 and later vehicles. Both longitudinal and transverse types of torsion bars are used on 1976 through 1981 vehicles, depending on body type: F, M and J-body vehicles use transverse torsion bars; all other vehicles use longitudinal bars.

The rear suspension **(see illustration)** consists of a pair of multi-leaf springs and two shock absorbers. The rear axle assembly is attached to the leaf springs by U-bolts. The front ends of the springs are attached to the frame at the front hangers, through rubber bushings. The rear ends of the springs are attached to the frame by shackles which allow the springs to alter their length when the vehicle is in operation.

The steering linkage consists of a Pitman (steering gear) arm, an idler arm, a center link, two adjustable tie-rods and a steering damper (not on all models). When the steering wheel is turned, the gear rotates the Pitman arm which forces the center link to one side. The tie-rods, which are connected to the center link by ball studs, transfer steering force to the wheels. The tie-rods are adjustable and are used for toe-in adjustments. The center link is supported by the Pitman arm and idler arm. The idler arm pivots on a support attached to the frame rail. The steering damper is attached to the frame and center link.

Frequently, when working on the suspension or steering system components, you may come across fasteners which seem impossible to loosen. These fasteners on the underside of the vehicle are continually subjected to water, road grime, mud, etc., and can become rusted or "frozen," making them extremely difficult to remove. In order to unscrew these stubborn fasteners without damaging them (or other components), be sure to use lots of penetrating oil and allow it to soak in for a while. Using a wire brush to clean exposed threads will also ease removal of the nut or bolt and prevent damage to the threads. Sometimes a sharp blow with a hammer and punch is effective in breaking the bond between a nut and bolt threads, but care must be taken to prevent the punch from slipping off the fastener and ruining the

1.2b Steering and front suspension components (vehicles with transverse torsion bars)

1. *Right torsion bar*
2. *Right torsion bar anchor*
3. *Right torsion bar pivot cushion bushing*
4. *Left torsion bar*
5. *Left torsion bar anchor*
6. *Left torsion bar pivot cushion bushing*
7. *Stabilizer bar*
8. *Lower control arm*
9. *Pitman arm*
10. *Idler arm*
11. *Center link*
12. *Tie-rod*

threads. Heating the stuck fastener and surrounding area with a torch sometimes helps too, but isn't recommended because of the obvious dangers associated with fire. Long breaker bars and extension, or "cheater," pipes will increase leverage, but never use an extension pipe on a ratchet - the ratcheting mechanism could be damaged. Sometimes, turning the nut or bolt in the tightening (clockwise) direction first will help to break it loose. Fasteners that require drastic measures to unscrew should always be replaced with new ones.

Since most of the procedures that are dealt with in this Chapter involve jacking up the vehicle and working underneath it, a good pair of jackstands will be needed. A hydraulic floor jack is the preferred type of jack to lift the vehicle, and it can also be used to support certain components during various operations. **Warning:** *Never, under any circumstances, rely on a jack to support the vehicle while working on it. Also, whenever any of the suspension or steering fasteners are loosened or removed they must be inspected and, if necessary, replaced with new ones of the same part number or of original equipment quality and design. Torque specifications must be followed for proper reassembly and component retention. Never attempt to heat or straighten any suspension or steering component. Instead, replace bent or damaged parts with new ones.*

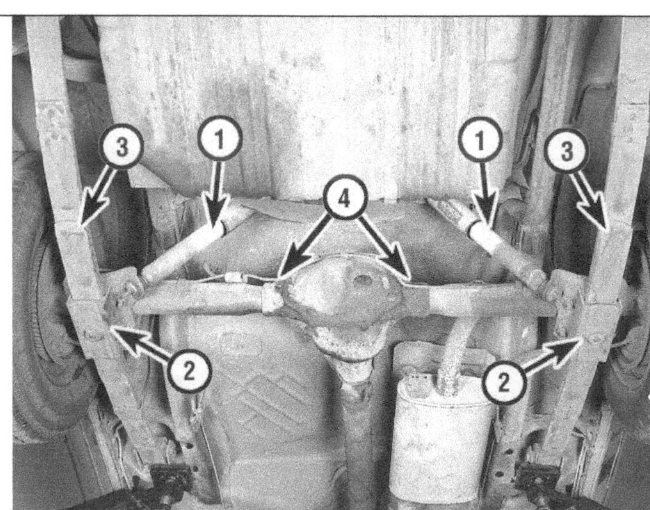

1.3 Rear suspension components (typical)

1. *Shock absorber*
2. *Rear spring isolator clamp*
3. *Leaf springs*
4. *Rear axle*

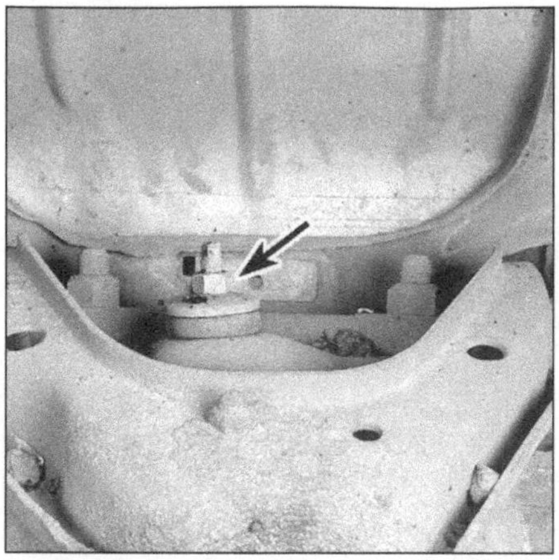

2.1 Remove this nut (arrow) from the upper end of the shock absorber - note the order in which the retainer washer and rubber bushing are installed

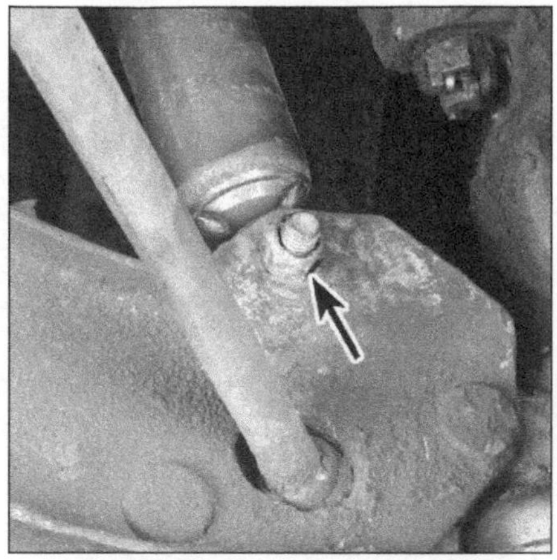

2.4a On most earlier models with longitudinal torsion bars, the lower end of the shock absorber is attached to the lower arm with a nut (arrow) and horizontal through-bolt - be sure to note the direction in which the bolt head and the nut are facing, so you can install them facing in the same direction during reassembly

2.4b On vehicles with transverse torsion bars, the lower end of the shock absorber is attached to the lower control arm with a nut - again, note the order in which the retainer washer and rubber bushing are installed, with the washer between the nut and the bushing

2 Shock absorber (front) - removal and installation

Refer to illustrations 2.1, 2.4a, 2.4b, 2.6a and 2.6b

1 Using a backup wrench on the stem, remove the upper mounting nut **(see illustration)**.

2 Remove the retainer and grommet.

3 Raise the vehicle and place it securely on jackstands.

4 Working underneath the vehicle, remove the nut and through bolt or the nut which attaches the lower end of the shock absorber to the lower control arm **(see illustrations)**. Pull the shock out from below.

5 Remove the lower grommet and retainer from the stem.

6 Fully extend the shock absorber prior to installation. The remaining installation steps are the reverse of removal **(see illustrations)**. Be sure to tighten the upper mounting nut and the lower mounting bolts securely.

3 Stabilizer bar (front) - removal and installation

Refer to illustrations 3.2a, 3.2b, 3.2c, 3.2d, 3.2e, 3.2f and 3.3

1 Raise the vehicle and place it securely

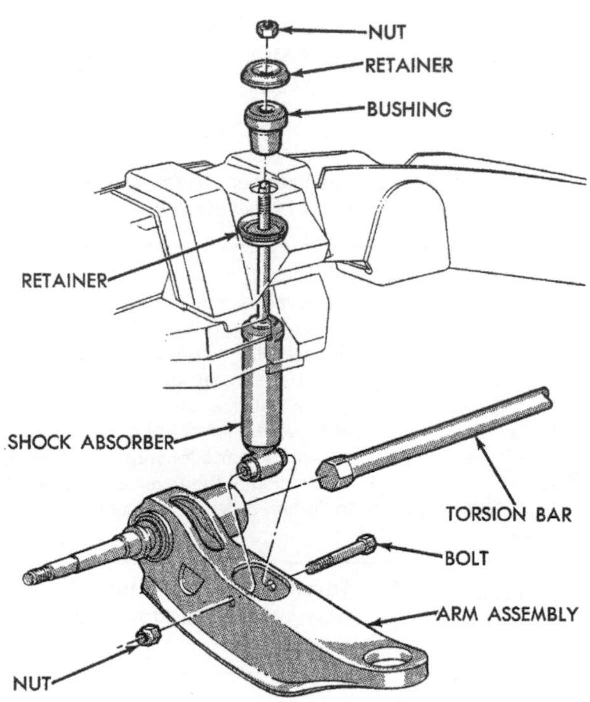

2.6a An exploded view of a typical front shock absorber installation (lower shock eye and through-bolt)

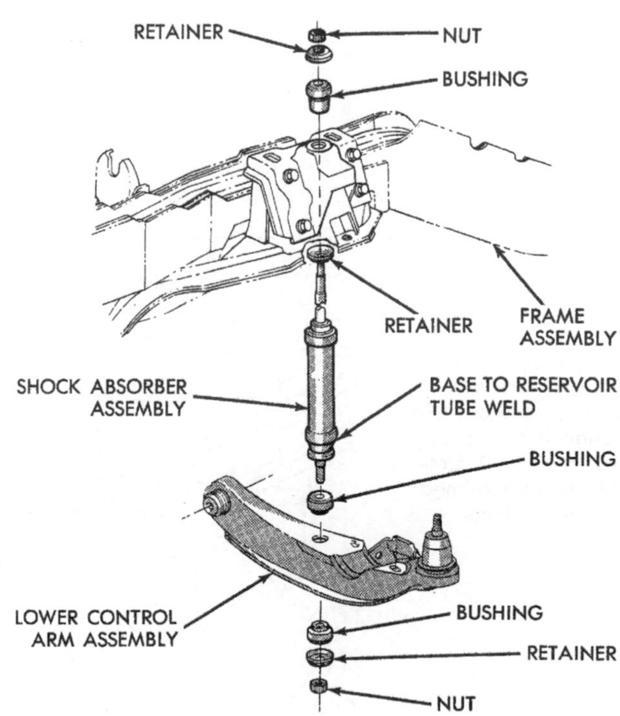

2.6b An exploded view of a typical front shock absorber installation (threaded lower extension with nut)

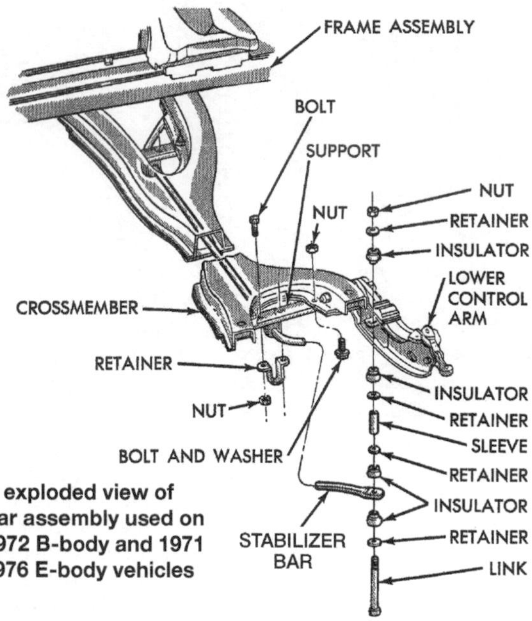

3.2a An exploded view of stabilizer bar assembly used on 1971 and 1972 B-body and 1971 through 1976 E-body vehicles

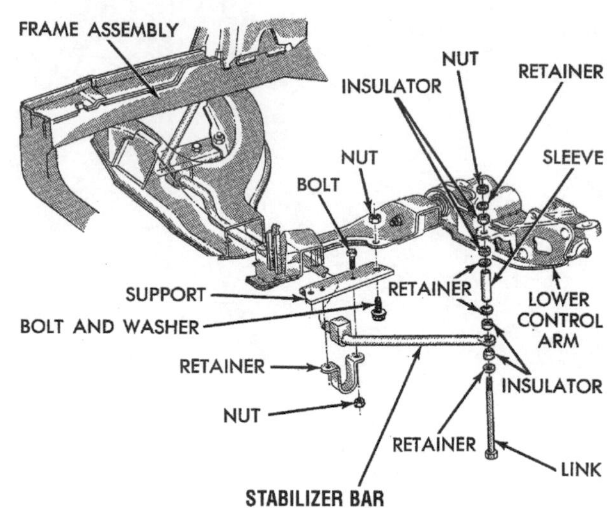

3.2b An exploded view of stabilizer bar assembly used on 1973 through 1976 A-body vehicles

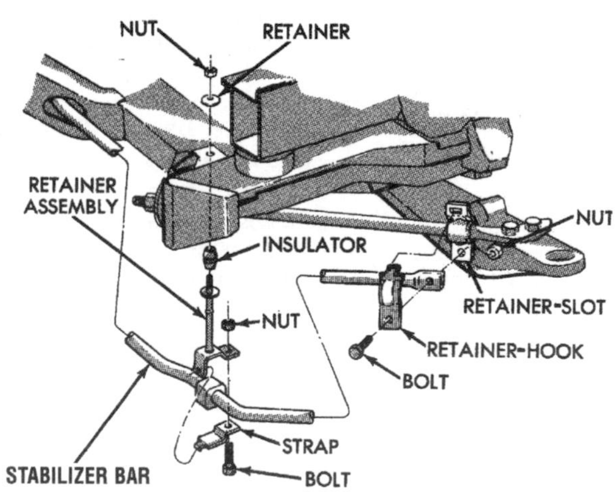

3.2c An exploded view of stabilizer bar assembly used on 1973 through 1979 B-body and 1979 through 1981 R-body vehicles

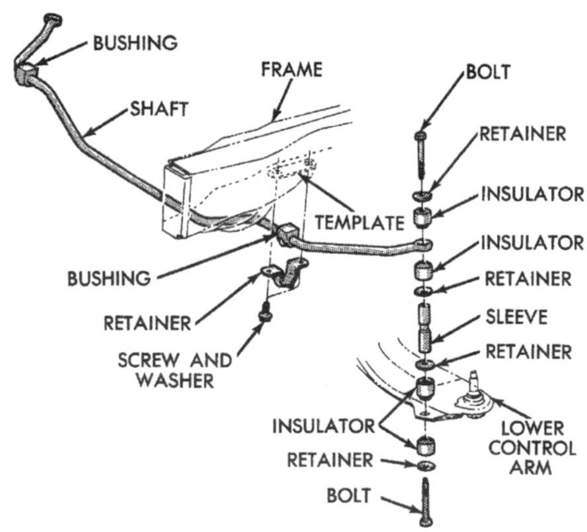

3.2d An exploded view of stabilizer bar assembly used on 1974 through 1978 C-body vehicles

3.2e An exploded view of stabilizer bar assembly used on 1976 through 1979 F-body, 1978 through 1981 M-body and all 1982 and later vehicles

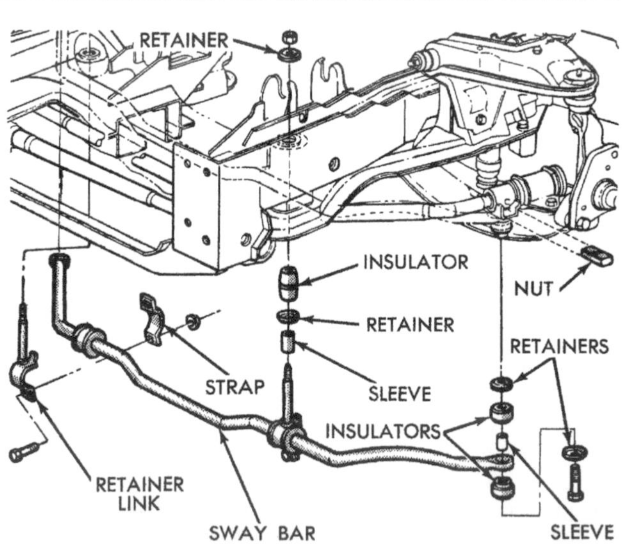

on jackstands.

2 Remove the nuts from the link bolts and remove the link bolts **(see illustrations)**.

3 Unscrew the bracket bolts **(see illustration)** from the frame or crossmember and remove the stabilizer bar.

4 Remove the rubber bushings.

5 Inspect all parts for wear and damage.

6 Installation is the reverse of removal. Be sure to tighten all fasteners securely.

4 Torsion bar (longitudinal) - removal, installation and adjustment

Removal

Refer to illustrations 4.3, 4.4a and 4.4b

Caution: *Don't interchange torsion bars between vehicles. They're identified for use*

3.2f To disconnect the ends of the stabilizer bar from the control arms, remove these link bolts (arrow) (left link bolt on 1985 Dodge Diplomat shown, link setup on 1976 through 1979 F-body, 1979 through 1981 M-body and all 1982 and later vehicles similar)

3.3 To detach the stabilizer bar assembly from the frame or crossmember, remove these bracket bolts (arrow) (left bracket on 1985 Dodge Diplomat shown, brackets on 1976 through 1979 F-body, 1979 through 1981 M-body and all 1982 and later vehicles similar)

4.3 To release tension on the torsion bar, turn this adjusting bolt (arrow) counterclockwise (1974 A-body shown, other models similar)

by length and thickness (depending on carline, body, engine, etc.) Don't even interchange torsion bars from one side to the other. They're marked either "L" or "R" on the ends to indicate the side they must be used on.

1 Remove the upper control arm rebound bumper on 1971 through 1976 vehicles.

2 Raise the front of the vehicle and support it securely on jackstands positioned under the frame. **Note:** *Don't place jackstands under the control arms - put them under the frame. The front suspension must be in full rebound (under no load) for this procedure.*

3 Release tension on the torsion bar by turning the adjusting bolt counterclockwise **(see illustration)**. **Warning:** *On 1971 through 1973 Imperials, the load on both torsion bars must be released before proceeding. Otherwise, the rubber isolator in the rear crossmember remains under a load - and could cause severe damage or personal injury.*

4 Remove the lock-ring from the torsion bar rear crossmember **(see illustrations)**.

5 On 1974 and 1975 A and B-body vehicles with an automatic transmission and floor-mounted shifter, remove the automatic transmission torque shaft.

6 Wrap the torsion bar with a rag to protect the bar from damage, then clamp up a pair of vise-grips to the protected part of the bar and tap on the vise-grips with a hammer to disengage the front end of the torsion bar from the control arm. Remove the vise-grips and rag from the torsion bar, slide the rear dust boot or seal off the front end of the torsion bar, then pull the torsion bar through its hole in the crossmember.

Installation

7 Clean and inspect the seal. If the seal is cracked or torn, replace it.

8 Put a little chassis grease in the hole in the crossmember and slide the torsion bar through the hole.

9 Coat the inside walls of the seal bore with a little grease, install the seal onto the front end of the torsion bar and slide it toward the rear end of the torsion bar (you can't place the seal in its final position until the torsion bar is installed).

10 Wrap the torsion bar with a rag again, clamp it up with a pair of vise-grips and drive the front end of the bar into place in the hex opening of the control arm by tapping against the vise-grips. Remove the vise-grips and shop rag from the torsion bar. Slide the new seal all the way back so that its lip is fully engaged with the groove in the crossmember hole.

11 On 1974 and 1975 A and B-body vehicles with an automatic transmission and floor-mounted shifter, install the automatic transmission torque shaft.

12 Install a new lock-ring on the rear of the torsion bar. Make sure it's fully seated.

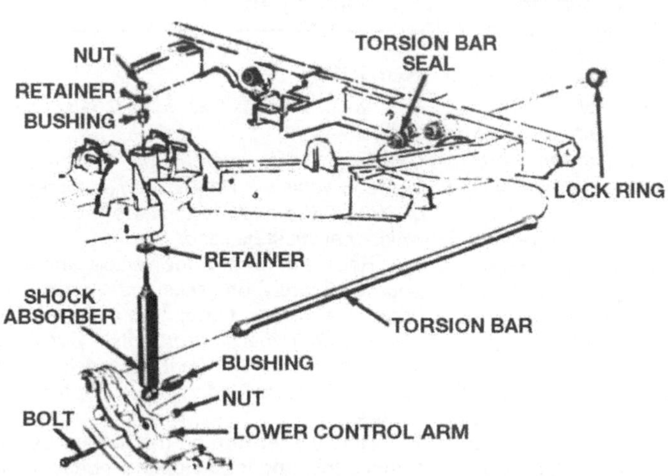

4.4a An exploded view of a typical torsion bar assembly used on 1971 through 1973 vehicles, except Imperials

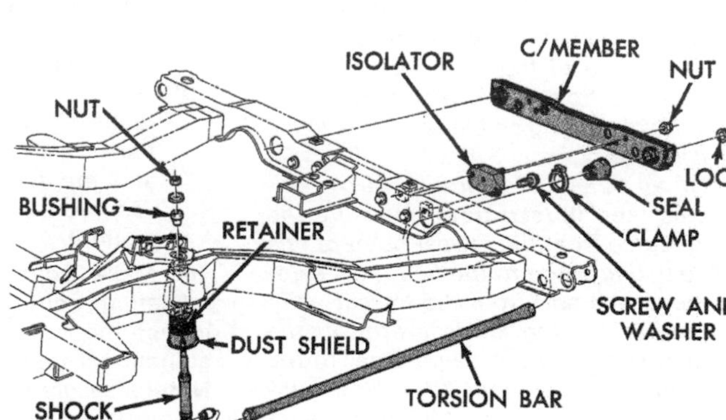

4.4b An exploded view of a typical torsion bar assembly used on early Imperials

4.17a On 1971 through 1973 Chryslers and on 1974 B and C-body vehicles, measure the distance from the torsion bar anchor to the floor (measurement "A") and from the balljoint housing to the floor (measurement "B") - the difference is the ride height; on 1975 and later vehicles, measure the distance from the torsion bar anchor to the floor (measurement "A")

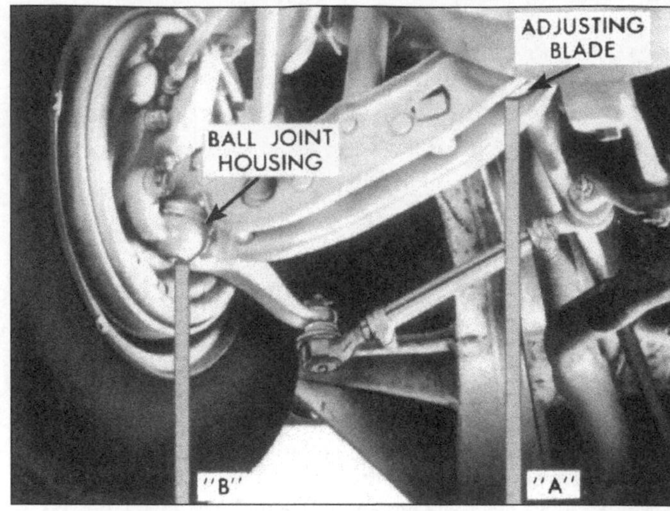

4.17b On all other 1971 through 1974 vehicles, measure the distance from the lowest point of one adjusting blade to the floor (measurement "A") and from the lowest point of the steering knuckle arm, at the centerline, on the same side, to the floor (measurement "B") - the difference is the ride height

13 After installation, the vehicle ride height must be set.

Adjustment

Refer to illustrations 4.17a, 4.17b and 4.17c

Note: *Ride height adjustment must always be performed on a level surface.*

14 Check and adjust the tire pressure (see Chapter 1).

15 Place the vehicle on level, solid pavement. It must be unloaded with a full tank of fuel.

16 Jounce the front bumper up and down several times at least four inches to settle the suspension.

17 The procedure for measuring the ride height - and where to measure it - differs slightly on various vehicles:

a) *On 1971 through 1973 Chryslers and on 1974 B and C-body vehicles* **(see illustration)**, *measure the distance from the lowest point of the front torsion bar anchor at the rear of the lower control arm flange to the floor (measurement "A") and from the lowest point of the balljoint housing on the same side to the floor (measurement "B").*

b) *On all other 1971 through 1974 vehicles* **(see illustration)**, *measure the distance from the lowest point of one adjusting blade to the floor (measurement "A") and from the lowest point of the steering knuckle arm, at the centerline, on the same side, to the floor (measurement "B"). Now subtract measurement "B" from measurement "A." In other words, the ride height dimension listed in this Chapter's Specifications for 1971 through 1974 vehicles is actually the difference between the higher point, (measurement "A") and the lower point (measurement "B").*

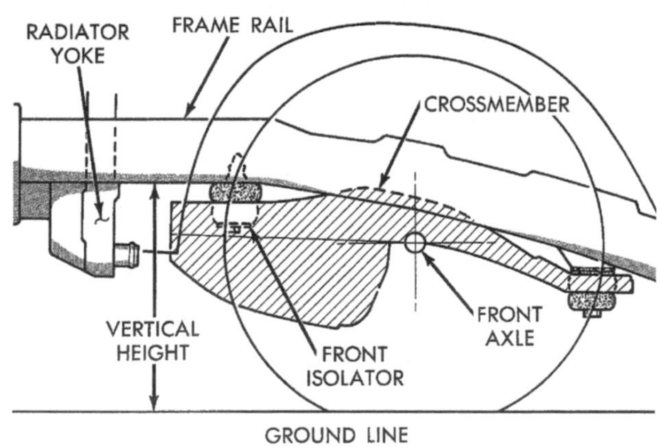

4.17c On 1981 vehicles, measure the distance from the bottom of the front frame rail, between the radiator yoke and the forward edge of the front suspension crossmember

c) *On 1975 through 1980 vehicles, measure the distance from the lowest point of the lower control arm torsion bar anchor at a point one inch forward of the rear face of the anchor, to the floor* **(see illustration 4.17a)**.

d) *On 1981 vehicles, measure the distance from the bottom of the front frame rail, between the radiator yoke and the forward edge of the front suspension crossmember* **(see illustration)**.

18 Make adjustments by turning the adjusting bolt. After each adjustment, jounce the front bumper again. The final adjustment must be in the up direction. **Note:** *Measure both sides even if only one side was replaced.*

19 Front vehicle height should not vary more than 1/4-inch from the dimension listed in this Chapter's Specifications, and should be within 1/8 to 1/4-inch side-to-side (i.e. there should be no more than 1/8-inch difference in vehicle height at the left and right sides).

5 Torsion bar (transverse) - removal, installation and adjustment

Removal

Refer to illustrations 5.2a, 5.2b, 5.3, 5.5, 5.6 and 5.7

Caution: *Don't interchange torsion bars between vehicles. They're marked either "L" or "R" on the ends to indicate the side on which they must be used.*

1 Raise the front of the vehicle and support it securely on jackstands positioned under the frame. **Note:** *Don't place jackstands under the control arms - put them under the frame. The front suspension must be in full rebound (under no load) for this procedure.*

2 Release tension on the torsion bar by turning the anchor adjusting bolts in the frame crossmember counterclockwise **(see illustrations)**. Remove the anchor adjusting bolt on the torsion bar to be removed.

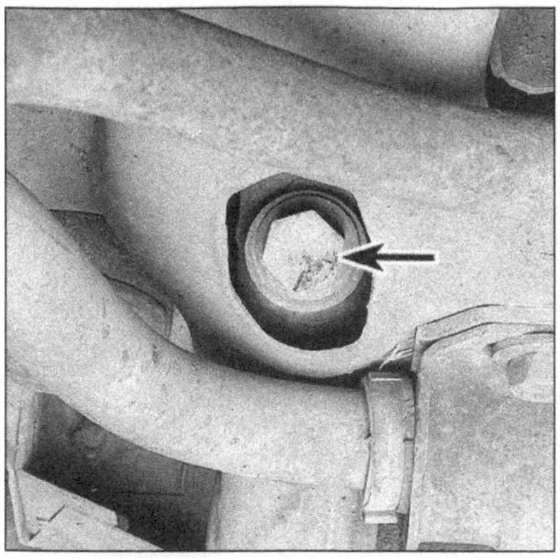

5.2a The left (driver's side) torsion bar anchor adjusting bolt . . .

5.2b . . . and the right (passenger's side) torsion bar anchor adjusting bolt; both adjusting bolts, which are hidden in the crossmember, face toward the rear of the vehicle

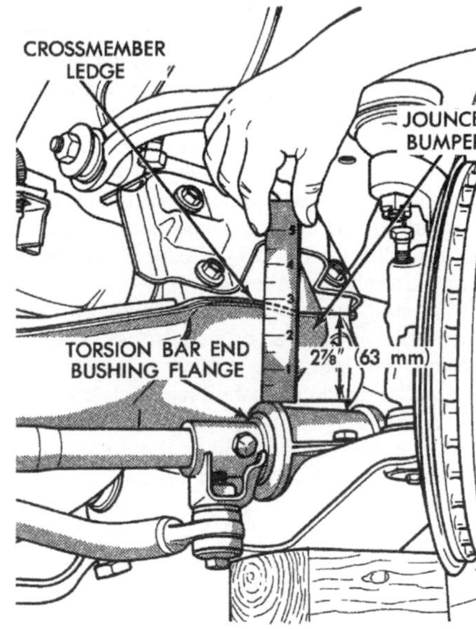

5.3 Before disconnecting a torsion bar, align the stabilizer bar and lower control arm attaching points by raising the lower control arms until the clearance between the crossmember ledge (at the jounce bumper) and the torsion bar end-bushing is 2-7/8 inches - support the control arms at this "design height" until the torsion bar has been reinstalled

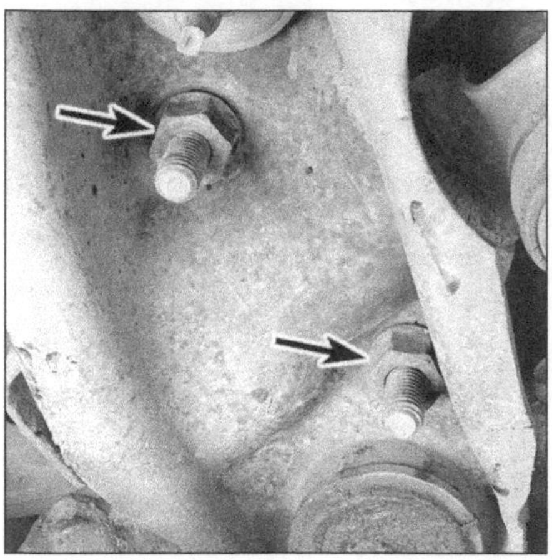

5.5 Remove these two nuts (arrows) and the two bolts that connect the torsion bar end-bushing to the lower control arm

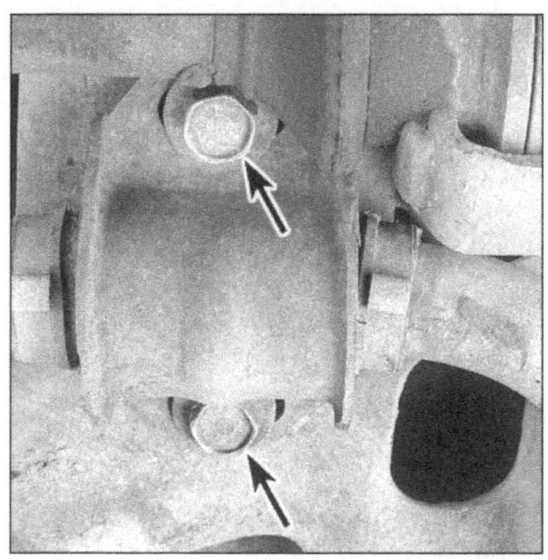

5.6 Remove the two bolts (arrows) that attach the torsion bar pivot cushion bushing to the crossmember and remove the torsion bar and anchor assembly from the crossmember

3 Before disconnecting a torsion bar, you must align the stabilizer bar and lower control arm attaching points to facilitate disassembly, component realignment and attachment during reassembly. To do so, raise the lower control arms until the clearance between the crossmember ledge (at the jounce bumper) and the torsion bar end-bushing is 2-7/8 inches **(see illustration)**. Support the lower control arms at this "design height".

4 Remove the stabilizer bar-to-control arm attaching bolt and retainers **(see illustration 3.2f)**.

5 Remove the two bolts that connect the torsion bar end-bushing to the lower control arm **(see illustration)**.

6 Remove the two bolts that attach the torsion bar pivot cushion bushing to the crossmember **(see illustration)** and remove the torsion bar and anchor assembly from the crossmember.

7 Separate the anchor from the torsion bar **(see illustration)**.

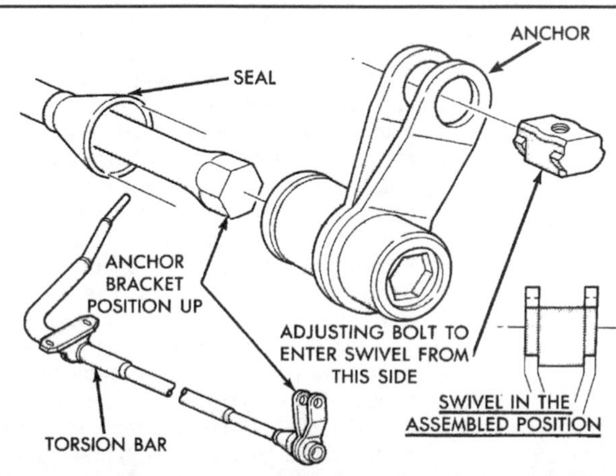

5.7 An exploded view of the hex end of the torsion bar and the anchor bracket and swivel assembly

Inspection

8 Inspect the boots for damage - cracks, tears and deterioration - and replace as necessary.

9 Inspect the lower control arm and pivot cushion bushings for cuts, tears, deterioration and other damage and replace as necessary.

10 Make sure that the seals at either end of the pivot cushion bushings are also in good shape; if they're damaged, moisture can get into the cushion. If the seals are damaged, inspect the torsion bar area under the bushing very carefully. If corrosion is evident in this area, replace the torsion bar.

11 Inspect the torsion bar paint for chips. If the paint is damaged, touch it up. Don't allow a torsion bar to remain uncoated or it will rust.

12 Inspect the torsion bar adjusting bolt and swivel for signs of corrosion or other damage. Replace as necessary.

Installation

Refer to illustration 5.16

13 Carefully slide the seal over the end of the torsion bar (the cupped end of the seal

should face toward the hex).

14 Make sure the hex end of the torsion bar and the hex-shaped receptacle in the anchor are clean, then coat the hex end of the bar with multi-purpose grease.

15 Install the hex end of the torsion bar into the anchor bracket. With the torsion bar in a horizontal position, the ears of the anchor bracket should be positioned nearly straight up **(see illustration 5.7)**. Position the swivel into the anchor bracket ears.

16 Install the torsion bar anchor bracket assembly into the crossmember anchor retainer and install the anchor adjusting bolt and thrust bearing (washer) **(see illustration)**. Install the two bolts and washers that secure the pivot cushion bushing to the crossmember and tighten the bolts finger-tight.

17 With the lower control arms at their "design height," install the two bolts and nuts that connect the torsion bar bushing to the lower control arm. Tighten them to the torque listed in this Chapter's Specifications.

18 Make sure the anchor bracket is fully seated in the crossmember, then tighten the pivot cushion retainer-to-crossmember bolts to the torque listed in this Chapter's Specifications.

19 Position the seal over the anchor bracket.

20 Install a new stabilizer bar link bolt (see Section 3) and tighten it securely.

21 Load the torsion bar by turning the adjusting bolt clockwise.

22 Lower the vehicle and adjust the vehicle ride height.

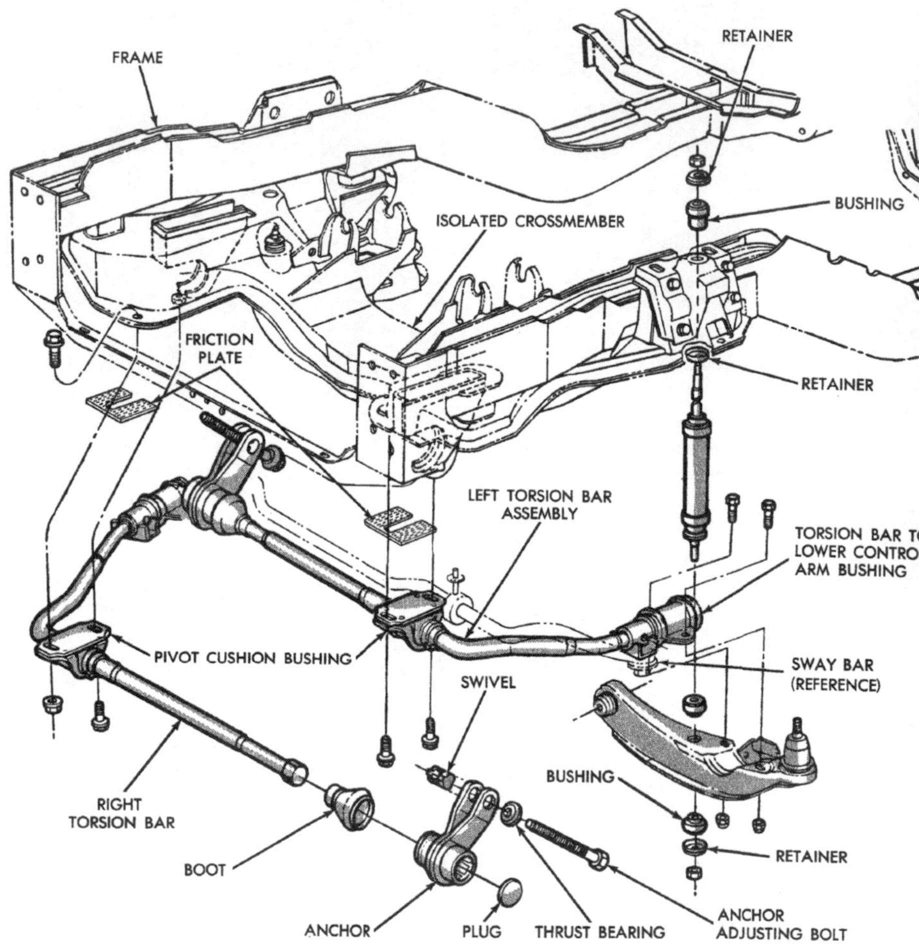

5.16 An exploded view of the transverse torsion bar assembly

Adjustment

Refer to illustration 5.26

23 Check and adjust tire pressure (see Chapter 1).

24 Place the vehicle on a level, solid surface. The vehicle must be unloaded, but with a full tank of fuel.

25 Jounce the front bumper up and down several times at least four inches to settle the suspension.

26 On 1976 through 1980 vehicles, measure the distance from the lowest point of the lower control arm inner pivot bushing to the floor **(see illustration 4.17a)**. On 1981 and later vehicles, measure the distance from the head of the front suspension front crossmember isolator bolt to the ground **(see illustration)**.

27 Compare your measurements with the dimensions listed in this Chapter's Specifications and adjust as necessary. To increase ride height, turn the adjusting bolts clockwise; to decrease height, turn the bolts counterclockwise. After each adjustment, jounce the vehicle as described in Step 25. Both sides must be measured even if only one side has been adjusted or is to be adjusted. Ride heights shouldn't vary more than 1/4-inch from the dimensions listed in this Chapter's Specifications, and they should be within 1/4-inch from side-to-side.

5.26 On 1981 and later vehicles, measure the distance from the head of the front suspension front crossmember isolator bolt to the ground

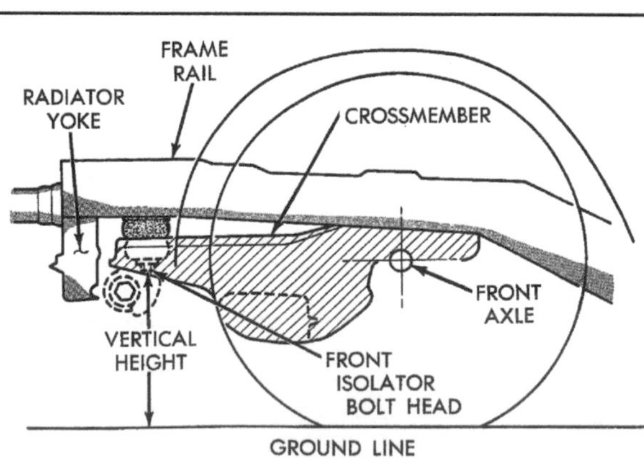

6 Steering knuckle - removal and installation

Refer to illustration 6.7

1 Loosen the wheel lug nuts, raise the front of the vehicle and support it securely on jackstands. Support the lower control arm with a jack so that the suspension is in full rebound (under no load). **Warning:** *The jack must remain in this position throughout the entire procedure*. Remove the wheel.

2 Remove all load from the torsion bar by turning the adjusting bolt counterclockwise (longitudinal torsion bars, see Section 4; transverse torsion bars, see Section 5).

3 On drum brake models, remove the brake drum, brake shoe assembly and brake support (backing plate) (see Chapter 9).

4 On disc brake models, remove the brake caliper (see Chapter 9) and hang it out

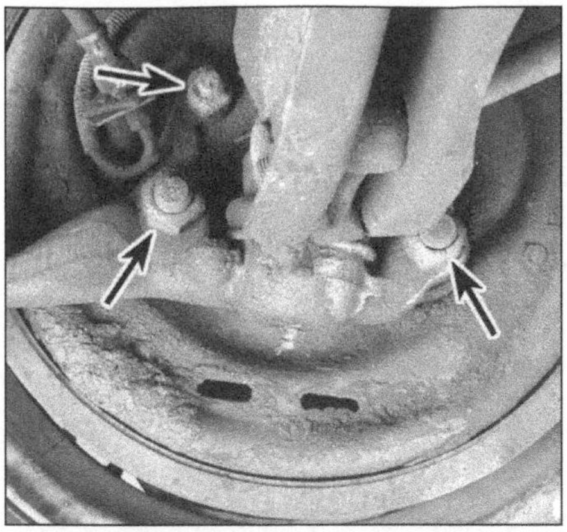

6.7 To disconnect the steering arm from the steering knuckle, remove these two bolts (arrows); to disconnect the brake support (backing plate) from the steering knuckle, remove the two upper bolts (left bolt arrowed, other bolt not visible in this photo)

of the way with a piece of wire. Remove the brake disc/hub assembly (see Chapter 1, *Front wheel bearing check, repack and adjustment*). Remove the disc splash shield, if equipped.

5 Disconnect the tie-rod end from the steering knuckle arm (see Section 15).

6 Disconnect the upper balljoint stud from the steering knuckle (see Section 10).

7 Remove the two upper bolts attaching the steering knuckle to the brake support and the two lower bolts attaching the steering arm to the steering knuckle **(see illustration)**.

8 Remove the steering knuckle.

9 Installation is the reverse of removal.

10 Adjust the front wheel bearings (see Chapter 1) and have the front wheel alignment checked by a dealer service department or alignment shop.

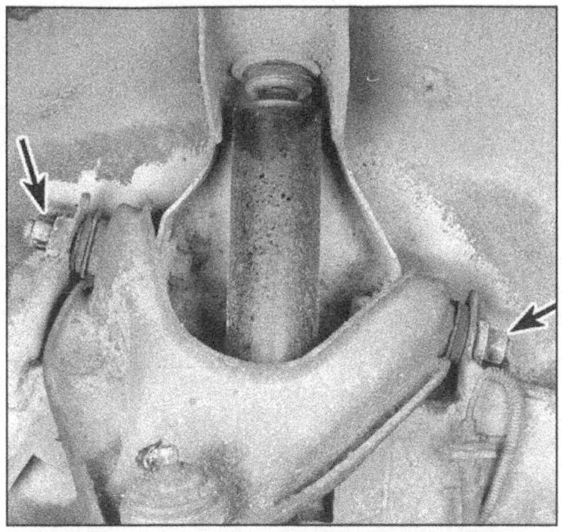

7.4a To disconnect the inner end of the upper control on an A-body vehicle, remove these two nuts (arrows) and pull out the pivot shaft (1974 A-body vehicle shown, 1971 through 1976 A-body vehicles similar)

7 Upper control arm - removal and installation

Removal

Refer to illustrations 7.4a, 7.4b, 7.4c, 7.4d and 7.4e

1 Loosen the wheel lug nuts, raise the front of the vehicle and support it securely on jackstands. Remove the wheel.

2 Position a floor jack, with a wood block on the jack head (to act as a cushion), under the lower control arm as close to the balljoint as

possible. Raise the jack slightly to take the spring pressure off the upper control arm. **Warning:** *The jack must remain in this position throughout the entire procedure.*

3 Disconnect the upper balljoint stud from the steering knuckle (see Section 6). Using a

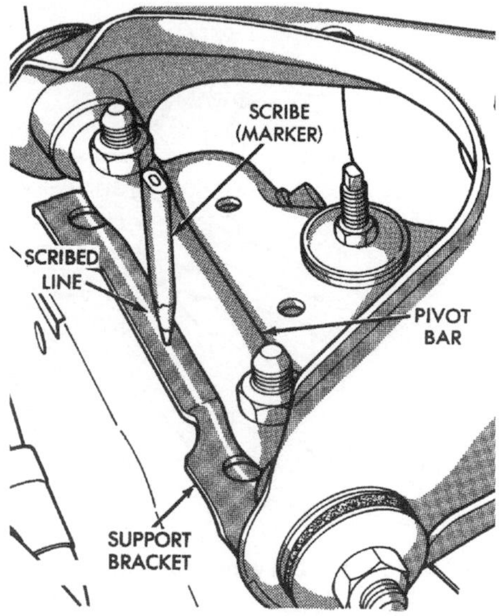

7.4b To insure basic suspension alignment after reassembly, scribe a line on the support bracket along the inner edge of the pivot bar

piece of wire or rope tied to the frame rail or some other component, support the top of the steering knuckle so it doesn't fall and stretch the brake hose.

4 If the upper control arm is equipped with an eccentric (cam-type) pivot bolt, make matchmarks on the cams and the frame to facilitate reassembly. Remove the upper control arm nut, washer, cams (if equipped), bushings and pivot bolt **(see illustration)**. On 1973 through 1978 B and F-body vehicles and on all 1979 and later vehicles, remove the splash shield, scribe a line on the support bracket along the inner edge of the pivot bar to insure that the suspension is still basically aligned after reassembly **(see illustration)**, remove the pivot bolts or nuts **(see illustra-**

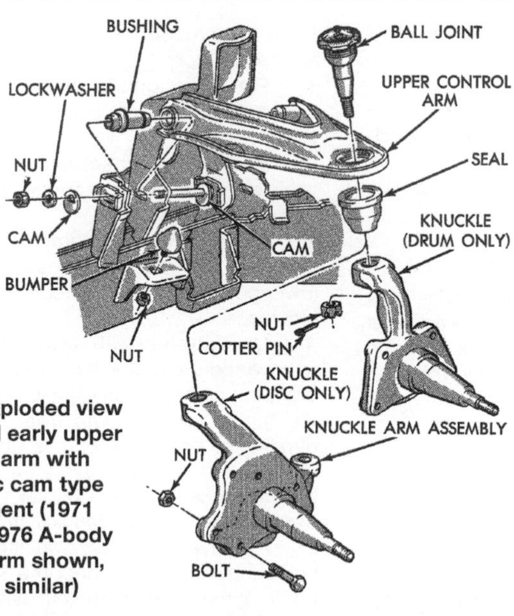

7.4c An exploded view of a typical early upper control arm with eccentric cam type adjustment (1971 through 1976 A-body control arm shown, others similar)

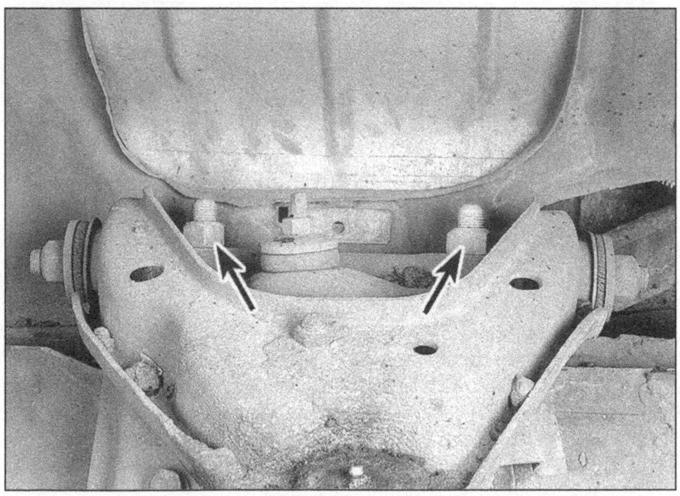

7.4d To disconnect the inner end of the upper control arm from a later vehicle, remove these two nuts (arrows) - or, on some models, bolts - from the pivot bar

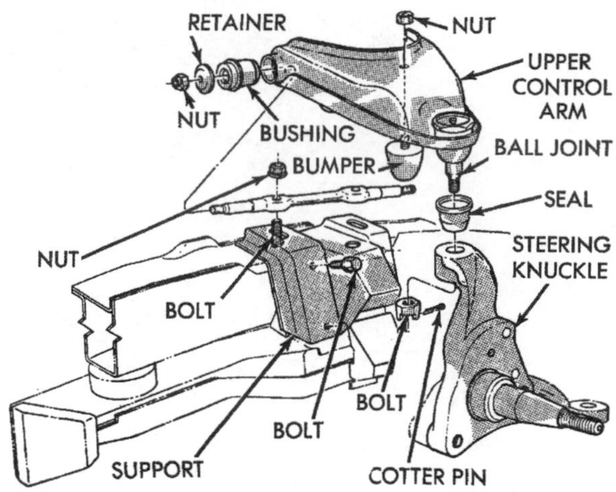

7.4e An exploded view of the upper control arm used on 1973 through 1978 B and F-body vehicles and on all 1979 and later vehicles

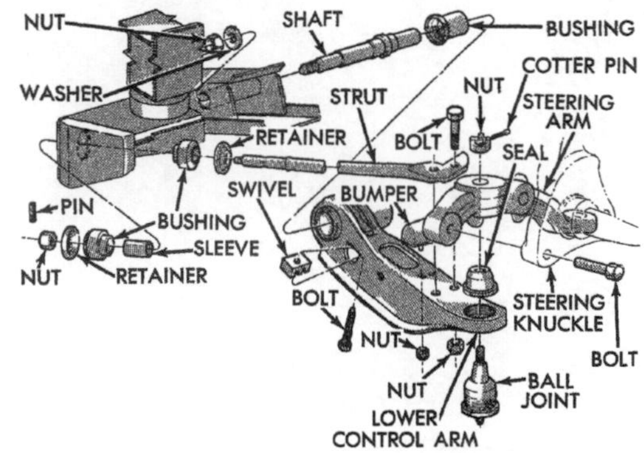

8.6 An exploded view of a typical lower control arm assembly with a strut that bolts to the top of the arm (mid-Seventies B-body shown, others similar)

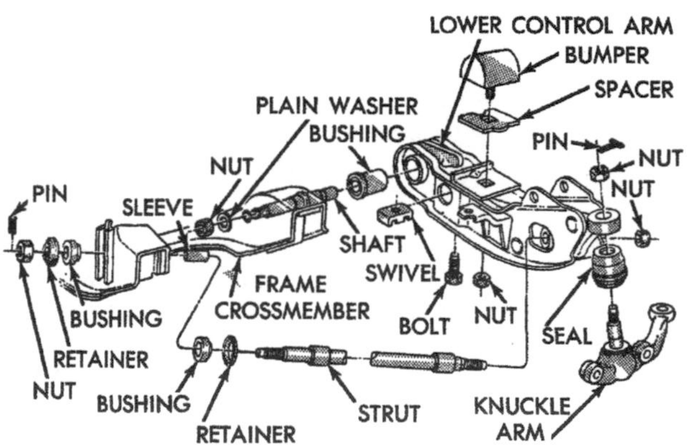

8.8 An exploded view of a typical lower control arm assembly with a strut that's inserted through the lower arm (mid-Seventies A-body shown, others similar)

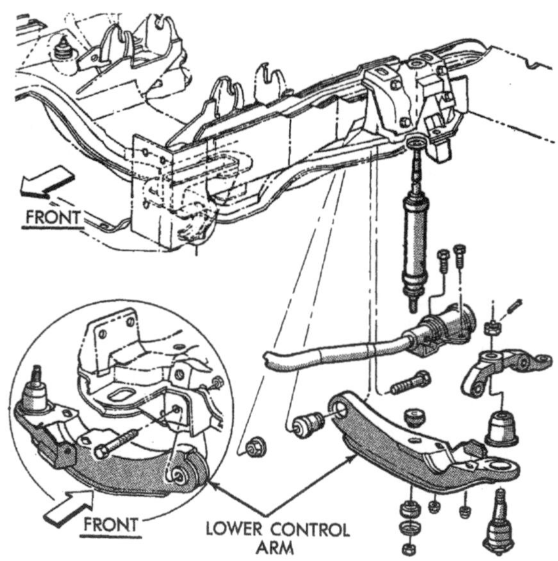

8.9 An exploded view of a typical lower control arm assembly without a strut bar (i.e. a transverse torsion bar design)

tions) and remove the upper control arm.

5 Inspect the control arm bushings for wear. Replace them if necessary. A hydraulic press may be required to remove and install the bushings, in which case you may have to take them to a dealer service department or other repair shop to have this done.

Installation

6 Position the arm in the frame brackets and install the bolt, washers, cams (if equipped) and nut.

7 Line up the previously applied match-marks (if applicable). Tighten the mounting bolts and nuts to the torque listed in this Chapter's Specifications.

8 Attach the balljoint to the steering knuckle (see Section 6). Tighten the balljoint nut to the torque listed in this Chapter's Specifications.

9 Install the wheel and lug nuts, then lower the vehicle. Tighten the lug nuts to the torque listed in the Chapter 1 Specifications.

10 Have the front end alignment checked and, if necessary, adjusted.

8 Lower control arm - removal and installation

Removal

Refer to illustrations 8.6, 8.8 and 8.9

Note: *Vehicles with longitudinal torsion bars use struts between the lower control arm and the front crossmember. You can remove the lower control arm without removing the strut on those vehicles which use struts that are bolted to the top of the lower control arm. However, on vehicles which use struts that are inserted through bushings in the lower control arms, you'll have to remove the arm and strut as a single assembly, then disassemble them off the vehicle. If you're removing the control arm on a vehicle with transverse torsion bars, disregard the Steps in the following procedure which refer to control arms with struts.*

1 Place the ignition switch in the Off or Unlocked position (so you can turn the steering knuckle).

2 Loosen the wheel lug nuts, raise the vehicle and support it securely on jackstands. Remove the wheel.

3 On vehicles so equipped, remove the rebound bumper. If necessary, raise the lower control arm with a floor jack to provide clearance.

4 On vehicles with front drum brakes, remove the drum, the shoe assembly and the brake support (brake backing plate) (see Chapter 9). On vehicles with front disc brakes, remove the brake caliper (see Chapter 9) and hang it out of the way, then remove the hub/disc assembly (see Chapter 1, *Front wheel bearing check, repack and adjustment*).

5 Remove the disc splash shield, if equipped, then disconnect the shock absorber from the lower control arm (see Section 2).

6 On vehicles with struts that are bolted to the top of the lower control arm, remove the two bolts **(see illustration)**.

7 Disconnect the stabilizer bar from the lower control arm (see Section 3).

8 On vehicles with struts that are inserted through the control arm **(see illustration)**, measure the torsion bar anchor bolt depth into the lower control arm before unwinding the torsion bar anchor bolt (see Section 4).

9 Disconnect the torsion bar from the lower control arm (longitudinal torsion bar, see Section 4; transverse torsion bar, see Section 5) **(see illustration)**.

10 Disconnect the tie-rod end from the steering arm (see Section 15). Remove the steering knuckle arm-to-brake support bolts and remove the steering knuckle arm **(see illustration 6.7)**.

11 On vehicles with integral balljoint/steering arm assemblies located *below* the control arm, disconnect the balljoint stud from the lower control arm (see Section 10). On vehicles with steering arms *above* the lower control arm, disconnect the balljoint stud from the steering arm.

12 Remove the spring pin, nut and bushing retainer from the forward end of the crossmember (see Section 9).

13 Remove the nut and washer from the lower control arm pivot shaft.

14 Tap the end of the lower control arm shaft with a soft-blow hammer and knock it out of the crossmember bracket. Remove the lower control arm, or lower control arm/strut assembly. Place the strut in a bench vise, remove the nut and detach the arm from the strut.

15 Inspect the control arm pivot shaft bushings. If they're worn or damaged, have them replaced at an automotive machine shop.

16 Inspect the front and rear strut bushings (if equipped) in the crossmember for cracks and tears. If they're damaged, replace them (see Section 9).

Installation

17 If you're installing the type of lower control arm that uses a strut bar inserted through the arm, install the strut now (see Section 9, if necessary), but don't tighten the nut yet.

18 If you removed the strut with the control arm, it's a good idea to install new strut bushings in the crossmember. Coat the bushings with water (not oil or grease) and use a twisting motion to work each bushing into its hole in the crossmember.

19 Coat the control arm pivot shaft with multi-purpose grease, install the lower control arm and install the pivot shaft washer and nut. Don't tighten the nut yet. If you're installing a control arm which must be installed with the strut already attached, place the strut bushing inner retainer on the strut and install the lower control arm, shaft and strut assembly. Install the strut bushing outer retainer and nut finger tight.

20 On drum-brake models, position the brake support (backing plate) on the steering knuckle and install the two upper bolts and nuts finger tight. Position the steering arm on the steering knuckle, install the two bolts and nuts and tighten them finger tight (the upper bolts go through the steering knuckle and through the brake support; the lower bolts go through the steering arm, the steering knuckle and the brake support). Tighten the upper, then the lower, bolt nuts to the torque listed in this Chapter's Specifications.

21 On vehicles with integral balljoint/steering arm assemblies, insert the balljoint stud into the lower control arm; on vehicles with the balljoint pressed or screwed into the lower arm, insert the balljoint stud into the steering arm (see Section 10). On all vehicles, tighten the balljoint stud nut to the torque listed in this Chapter's Specifications. Install a new cotter pin.

22 Inspect the tie-rod end seal and replace it if damaged. Connect the tie-rod end to the steering knuckle arm and tighten the nut to the torque listed in this Chapter's Specifications.

23 Connect the shock absorber to the control arm (see Section 2) and tighten it finger tight.

24 Install the torsion bar (longitudinal, see Section 4; transverse, see Section 5).

25 On drum-brake models, install the brake shoe assembly and the drum; on disc-brake models, install the brake disc (see Chapter 1, *Front wheel bearing check, repack and adjustment*), the caliper support and the caliper (see Chapter 9).

26 Raise the lower control arm with a floor jack to simulate normal ride height. Tighten the strut nut at the crossmember to the torque listed in this Chapter's Specifications, and install the strut pin. Tighten the lower control arm shaft nut bushings to the torque listed in this Chapter's Specifications.

27 Install the wheel and tighten the wheel lug nuts hand tight.

28 Lower the vehicle and tighten the wheel lug nuts to the torque listed in the Chapter 1 Specifications.

29 Adjust the vehicle ride height (see Section 4).

30 Drive the vehicle to an alignment shop and have the alignment checked, and if necessary, adjusted.

9 Strut bars - removal and installation

1 Loosen the wheel lug nuts, raise the front of the vehicle and support it securely on jackstands. Remove the wheel.

2 Remove the nut that secures the front end of the strut to the crossmember.

3 On vehicles with struts that are bolted to the top of the lower control arm, simply remove the two bolts, then remove the strut from the crossmember.

4 On some vehicles, the struts are inserted *through* the lower control arm, then attached with a nut. If you're removing the strut on one of these vehicles, the control arm and strut must be removed as a single assembly (see Section 8), then the strut can be detached from the arm.

5 Inspect the front and rear bushings for cracks and tears and other deterioration. If they're damaged or worn, replace them. In general, it's always a good idea to install new. strut bushings in the crossmember whether they look damaged or not. Coat the bushings with water (not oil or grease) and use a twisting motion to work each bushing into its hole in the crossmember.

6 Installation is the reverse of removal. Be sure to tighten all fasteners to the torque listed in this Chapter's Specifications.

10 Balljoints - check and replacement

Check

Upper balljoint

1 Loosen the wheel lug nuts, raise the front of the vehicle and support it securely on jackstands. Remove the wheel.

2 Place a floor jack under the lower control arm and raise it slightly. Using a large screwdriver or prybar, pry up on the upper control arm and watch for movement at the balljoint. Any movement indicates a worn balljoint. Now grasp the steering knuckle and attempt to move the top of it in-and-out - if any play is felt, the balljoint will have to be replaced.

Lower balljoint

3 Place a floor jack under the lower control arm.

4 Grasp the wheel and tire and attempt to move the bottom of it in-and-out - if any play is felt, the balljoint will have to be replaced.

Replacement

5 Loosen the wheel lug nuts, raise the front of the vehicle and support it securely on jackstands. Remove the wheel.

Upper balljoint

Refer to illustrations 10.9a, 10.9b, 10.9c and 10.9d

6 Remove the brake caliper and tie it out of the way (see Chapter 9).

7 Remove the cotter pin from the upper balljoint stud and back off the nut several turns.

8 Place a jack or jackstand under the outer end of the lower control arm. **Warning:** *The jack or jackstand must remain under the control arm during removal and installation of the balljoint to hold the spring and control arm in position.*

9 Separate the balljoint from the steering knuckle as follows:

a) *Remove the cotter pin and loosen the nut on the upper balljoint stud a few turns, but don't remove it.*

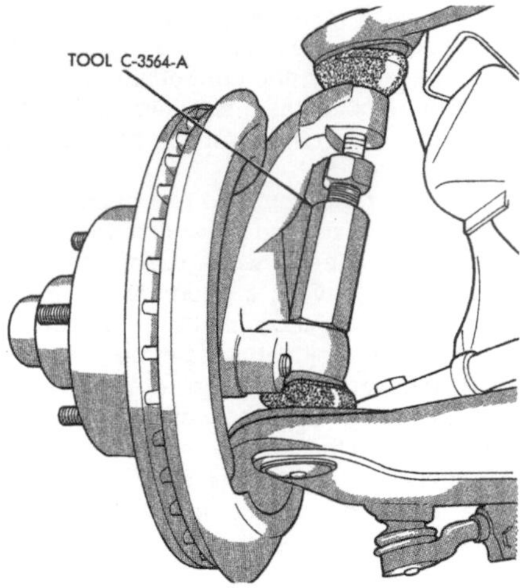

10.9a The best way to separate an upper balljoint stud from the steering knuckle (or to split a lower balljoint stud from the steering knuckle arm or the lower control arm, depending on the design) is with this special tool, available at most auto parts stores

b) *Install a special tool available at most auto parts stores, or its equivalent* **(see illustrations)** *shown. Position the balljoint stud remover over the lower balljoint stud, allowing the tool to rest on the knuckle arm. Set the tool nut securely against the upper stud. Tighten the tool to apply pressure to the upper stud and strike the knuckle sharply with a hammer to loosen the stud. Don't try to force out the stud with the tool alone.*

c) *If you you're not worried about damaging the balljoint boot, i.e. you're planning to replace the balljoint anyway, it's okay to use a picklefork type balljoint separator* **(see illustration)**.

d) *You can also loosen the balljoint stud nut and give the knuckle or steering arm a few sharp raps with a hammer* **(see illustration)**, *but we don't recommend this method unless you're working on a vehicle which makes the use of any other means difficult because of the shape of the knuckle, its close proximity to the ball stud nut, etc.*

10.9d If you don't have a picklefork tool, you can also rap the knuckle sharply with a hammer to loosen the upper balljoint stud

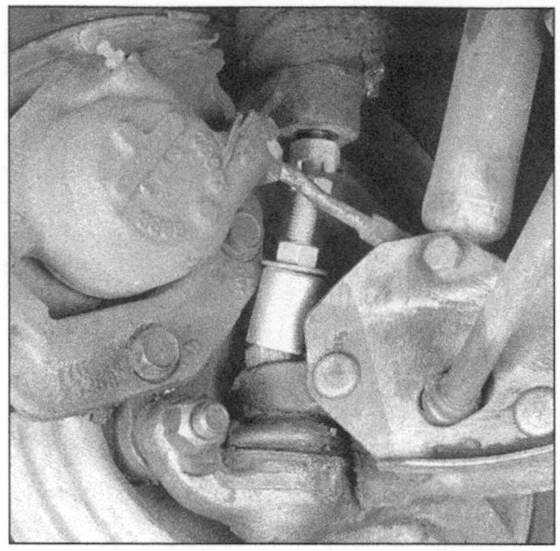

10.9b You can make your own version of the factory tool with a large bolt and nut (same diameter and thread pitch as bolt), a large washer and a big socket - this photo shows our homemade tool set up for removing an upper balljoint stud on an A-body

10 If the upper balljoint has a hex on top, unscrew it from the control arm using a special tool available at most auto parts stores, or equivalent (a large socket of the correct size will work). After you remove the balljoint, clean the balljoint bore in the control arm. If the balljoint is pressed in, i.e. there is no hex on top with which to unscrew it, remove the upper control arm (see Section 7) and have the old balljoint pressed out and a new unit pressed in at an automotive machine shop. **Note:** *If you can rent or borrow a balljoint removal/installation press, the control arm doesn't have to be removed from the vehicle. Follow the tool manufacturer's instructions.*

10.13a This is the same homemade setup we showed you in illustration 10.9b, except that the whole assembly has been flipped around so that it now presses on the lower balljoint stud (this particular vehicle has a balljoint in the lower control arm; the balljoint stud is attached to the steering knuckle arm)

10.9c On some models, like this old A-body, you won't be able to use the factory tool or the homemade setup because the balljoint studs don't line up - on these models, you'll have to split the upper balljoint stud from the knuckle with a "picklefork" tool (but remember, it will damage the balljoint boot)

Lower balljoint

Refer to illustrations 10.13a, 10.13b and 10.13c

Note: *Lower balljoints on the vehicles covered by this manual can be in the steering knuckle arm, with the* stud facing down *(lower control arm below); or in the steering knuckle arm, with the* stud facing up *(lower control arm above); or in the lower control arm, with the stud facing up (steering knuckle arm above).*

11 If the balljoint is an integral part of the steering knuckle arm, disconnect the tie-rod end from the steering knuckle arm (see Section 15).

10.13b The procedure shown in illustration 10.13a for splitting a lower balljoint just won't work on one of these old A-bodies, because the upper and lower balljoint studs don't line up; instead, separate the balljoint from the lower control arm with a picklefork (note the balljoint-in-the-steering arm, with the stud facing up, through the lower control arm)

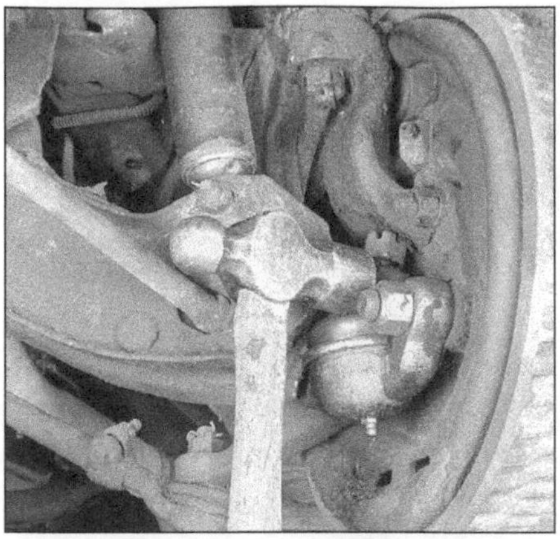

10.13c Or, if you don't have a picklefork, use a hammer - this is very tricky on some old A-bodies like this one because there's not a lot of knuckle surface to hit! Turn the brake and knuckle assembly all the way to the left to give yourself a clear target

12 Remove the cotter pin from the lower balljoint stud and back off the nut several turns.

13 Separate the lower balljoint stud from the steering knuckle arm **(see illustration)** or from the lower control arm **(see illustrations)**. Refer to Step 9 if necessary.

14 If the lower balljoint is an integral part of the steering knuckle arm, unbolt the steering arm from the steering knuckle **(see illustration 6.7)**, install a new steering arm and reattach it to the lower control arm.

15 If the balljoint is an integral part of the lower control arm, remove the lower control arm (see Section 8), then take the arm to a dealer device department or other repair shop to have the old balljoint pressed out and the new one installed. **Note:** *If you can rent or borrow a balljoint removal/installation press, the control arm doesn't have to be removed from the vehicle. Follow the tool manufacturer's instructions.*

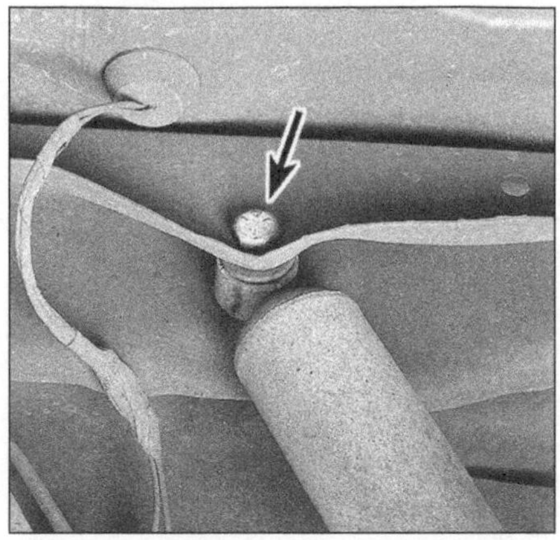

11.2 Remove the shock absorber upper mounting nut and bolt (arrow) from the frame crossmember

All balljoints

16 While the balljoint stud is disconnected from the lower control arm, steering knuckle and/or steering arm, inspect the tapered holes in the control arm, steering knuckle or steering arm. Remove any accumulated dirt. If out-of-roundness, deformation or other damage is noted, the lower control arm, steering knuckle and/or steering arm must be replaced with a new one.

17 Inspect the balljoint boot(s) for cracks and tears. If a boot is damaged, replace it. You can install a new balljoint boot and retainer using a large socket and a hammer.

18 Installation is the reverse of removal. After the suspension has been reassembled, raise the lower control arm with a floor jack to simulate normal ride height, then tighten all fasteners to the torque listed in this Chapter's Specifications.

19 If a cotter pin doesn't line up with an opening in the castle nut, tighten (never loosen) the nut just enough to allow installation of the cotter pin.

20 Install the grease fittings and lubricate the new balljoints (see Chapter 1).

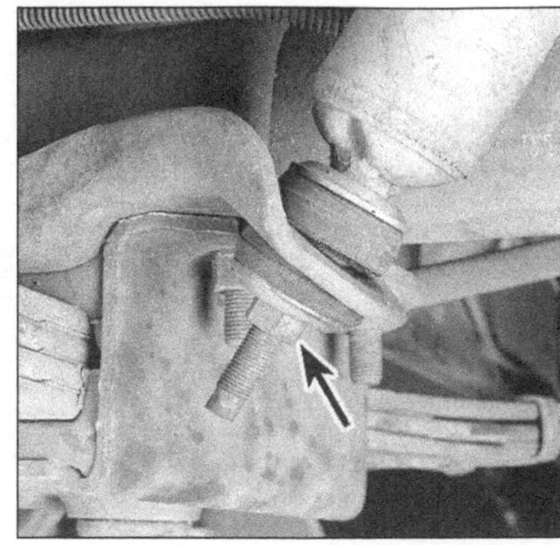

11.3 Remove the lower mounting nut (arrow), washer and bushing from the axle bracket

21 Install the wheels and lower the vehicle. Tighten the lug nuts to the torque listed in the Chapter 1 Specifications.

22 The front end alignment should be checked, and if necessary, adjusted, by a dealer service department or alignment shop.

11 Shock absorber (rear) - removal and installation

Refer to illustrations 11.2, 11.3, 11.5a and 11.5b

1 Raise the rear of the vehicle and place it securely on jackstands. Support the rear axle assembly with a floor jack.

2 Remove the shock absorber upper mounting nuts and bolts from the frame crossmember **(see illustration)**.

3 Remove the lower mounting nut, washer and bolt from the axle bracket **(see illustration)**.

4 Remove the shock absorber.

5 Installation is the reverse of removal **(see illustrations)**. Make sure you install the nuts and bolts facing in the proper direction and

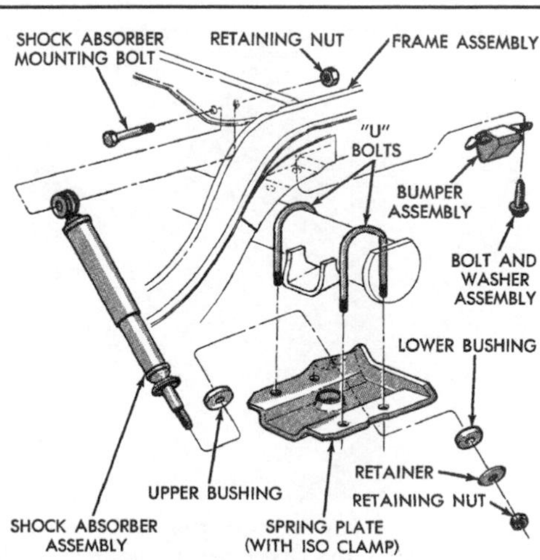

11.5a Shock absorber mounting details - threaded lower extension with nut

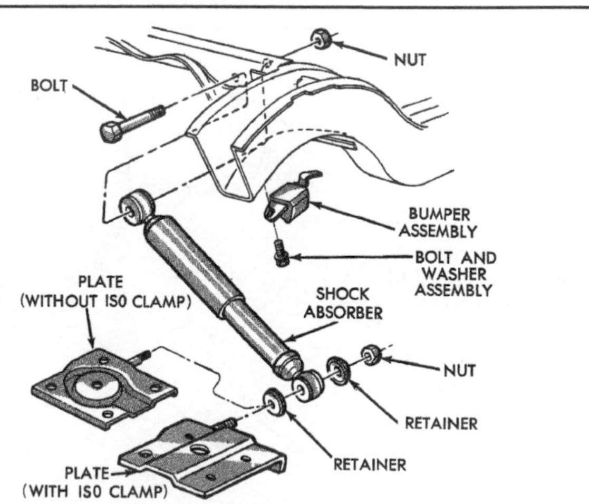

11.5b Shock absorber mounting details - lower mounting stud integral with isolator spring plate

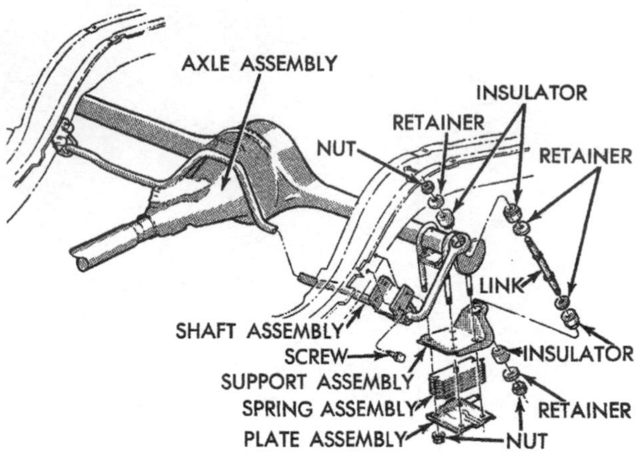

12.2 An exploded view of the rear stabilizer bar assembly used on E-body vehicles

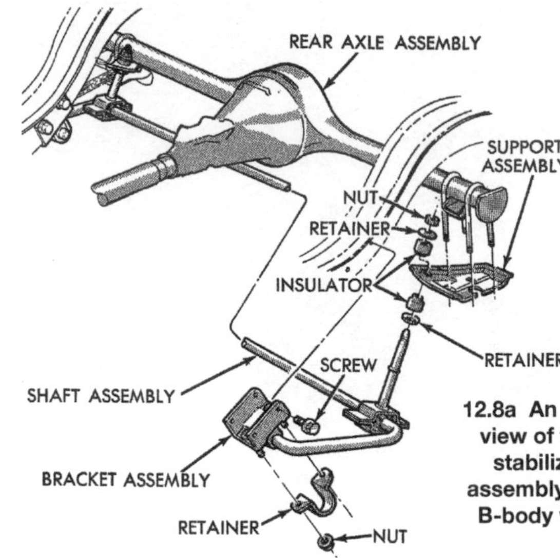

12.8a An exploded view of the rear stabilizer bar assembly used on B-body vehicles

tighten them securely. On shock absorbers with a threaded lower extension, don't tighten the nut so much that the rubber bushings are flattened.

12 Stabilizer bar (rear) - removal and installation

1 Loosen the rear wheel lug nuts, raise the rear of the vehicle and place it securely on jackstands. Remove the rear wheels.

E-body vehicles

Refer to illustration 12.2

2 Loosen and remove the upper link nut, retainers and rubber insulators on both sides **(see illustration)**.

3 Loosen and remove the screws attaching each bracket assembly to each rail.

4 Remove the stabilizer bar from the vehicle.

5 Loosen and remove the nuts, retainers and rubber insulators, then remove the links from the support assembly.

6 Inspect the rubber insulators for excessive wear and deterioration. If they're damaged or worn, replace them.

7 Installation is the reverse of removal. Be sure to tighten all fasteners securely.

All other vehicles

Refer to illustrations 12.8a and 12.8b

8 Loosen and remove the two stabilizer bar link retaining nuts, both retainers and both insulators **(see illustrations)**.

9 Loosen and remove the two nuts and the retainer off each bracket fastened to each rail.

10 Remove the stabilizer bar assembly from the vehicle.

11 Remove the insulator and retainer from each link.

12 Inspect the insulators and bracket bushings for cracks, tears and deterioration. If you find any, replace those parts.

13 Installation is the reverse of removal. Be sure to tighten all fasteners securely.

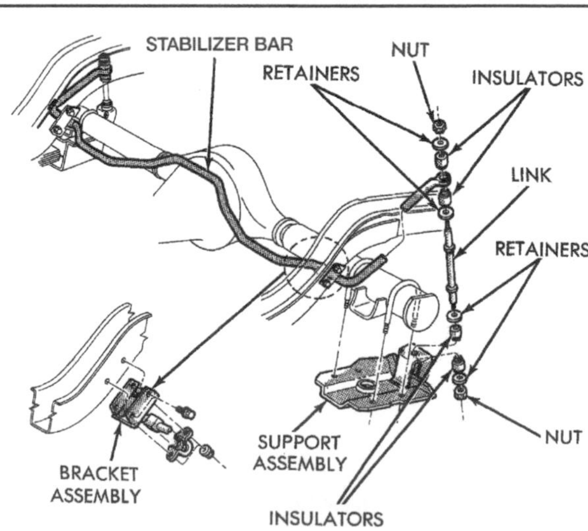

12.8b An exploded view of the rear stabilizer bar assembly used on all except E and B-body vehicles

13 Leaf spring/shackle - removal and installation

Removal

Refer to illustrations 13.3, 13.4, 13.5a, 13.5b and 13.7

Note: *Change one spring at a time, using the other spring to hold the rear axle in position.*

1 Raise the rear of the vehicle and place the frame securely on jackstands. Support the axle with a floor jack and raise it slightly to relieve the tension on the leaf springs.

2 Disconnect the lower end of the shock absorber from the spring plate (see Section 11). On vehicles so equipped, disconnect the stabilizer bar (see Section 12).

3 On all vehicles except 1975 through 1978 C-body vehicles, remove the nuts and washers from the U-bolts **(see illustration)**. Remove the U-bolts and the spring plate.

4 **Warning:** *1975 through 1978 C-body vehicles are equipped with preloaded "tension" type leaf springs - never attempt to*

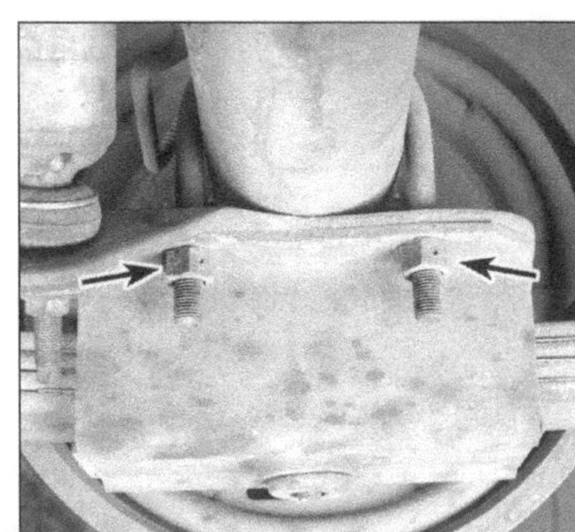

13.3 Remove the nuts and washers from the U-bolts, remove the U-bolts, then remove the spring plate (all except 1975 through 1978 C-body vehicles)

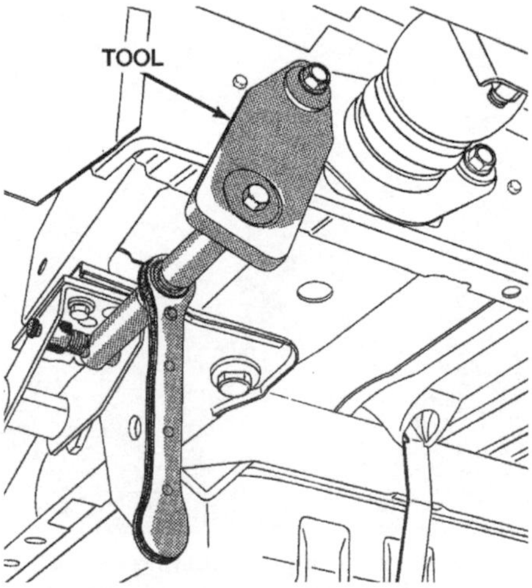

13.4 **Install a spring stretcher tool, as illustrated, available at auto parts stores that sell specialty tools, or similar**

remove springs without the use of a spring stretcher. The procedure for using the stretcher on these vehicles is as follows:

a) *Install the spring stretcher at the rear of the spring and apply tension* **(see illustration)**.
b) *Remove the rear spring hanger bolts and hanger* **(see illustration 13.7)**.
c) *Relieve spring tension and remove the spring stretcher; remove the front hanger as described in the next Step.*

5 On all other vehicles, loosen and remove the front pivot bolt nut and remove the bolt from the front spring hanger bracket **(see illustration)**. **Note:** *On some vehicles (early A-bodies, for example), it will be necessary to remove the nuts holding the front spring hanger bracket to the body because the front spring bolt can't be removed until the hanger bracket is lowered to provide enough clearance to remove the bolt* **(see illustration)**.

6 Loosen and remove the shackle plate nuts, remove the shackle plate, push the threaded shackle studs through their bushings and remove the spring.

13.5a To detach the front end of the spring from the front hanger bracket, loosen and remove the front pivot nut (arrow) and remove the bolt from the hanger

7 Loosen and remove the shackle nuts and remove the shackle from the rear spring **(see illustration)**.

8 Inspect all rubber bushings for cracks, tears and other deterioration. If any bushings are damaged or worn, have the old bushings removed and the new ones installed by an automotive machine shop.

Installation

9 Installation is basically the reverse of removal. Again, you'll need a spring stretcher tool to install the springs on 1975 through 1978 C-body vehicles. If you're working on one of these vehicles, perform the following Steps first. If you servicing anything else, proceed to Step 11.

a) *Assemble the front spring hanger to the front spring eye and install the pivot bolt (but don't tighten it yet).*
b) *Raise the spring and start the spring hanger bolts in the mounting bracket holes. Install the nuts and tighten them to the torque listed in this Chapter's Specifications.*
c) *Assemble the shackle and bushings to the spring rear eye and rear spring hanger. Don't lubricate the rubber bushings. Start the shackle bolt nuts but don't tighten them yet.*
d) *Raise the spring into position, install the spring stretcher tool and apply tension to facilitate installation of the hanger-to-frame bolts. Tighten all hanger bolts to the torque listed in this Chapter's Specifications. Remove the spring stretcher and proceed to Step 15.*

10 On all other vehicles, install the shackle and bushings in the rear of the spring and attach the spring and shackle to the rear spring hanger bracket. Don't lubricate the rubber bushings. Loosely tighten the shackle stud nuts, but don't tighten them yet.

11 On vehicles with insufficient clearance for the front spring bolt, position the spring in the spring hanger, install the spring bolt and nut (don't tighten the nut yet), reattach the spring hanger bracket and tighten the bracket mounting nuts securely.

12 On other vehicles, raise the front end of the spring, align the bushing eye with the holes in the front spring hanger bracket, install the pivot bolt, but don't tighten it completely yet.

13 Lower the axle assembly into the correct position with the axle centered over the spring center bolt.

14 Position the spring clamp and upper isolator over the center bolt followed by the bottom isolator and plate on vehicles so equipped. Install the U-bolts and nuts and tighten the nuts, in a diagonal sequence, to the torque listed in this Chapter's Specifications.

15 Reattach the shock absorber (see Section 11) and the stabilizer bar (see Section 12).

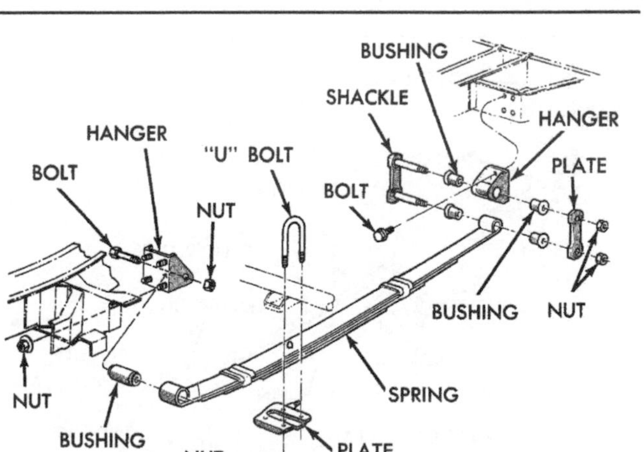

13.5b On some vehicles, you'll have to remove the nuts which attach the front spring hanger bracket to the body because the pivot bolt can't be removed until the hanger is lowered (A-body shown, some others similar)

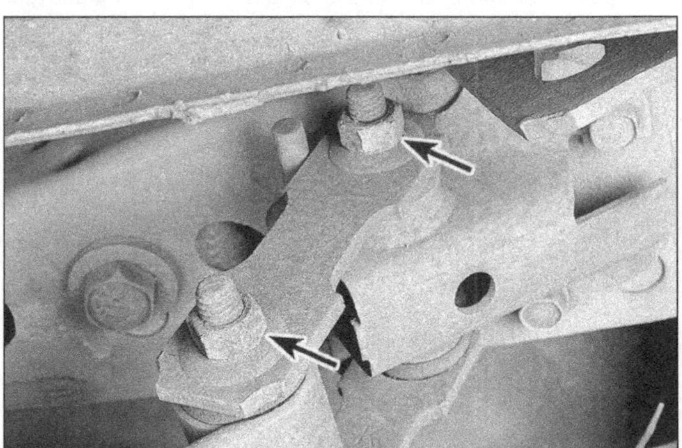

13.7 To detach the rear end of the spring from the rear hanger bracket, remove the shackle nuts (arrows), remove the shackle plate and push the two threaded shackle studs out of their bushings

14.2a To detach the horn pad, carefully pry it off like this with a small screwdriver (some pads may be secured to the steering wheel with screws installed through the side of the wheel facing toward the instruments, so always feel for screws before prying off a pad)

16 Install the wheels, if removed, remove the jackstands and lower the vehicle. Tighten the wheel lug nuts to the torque listed in the Chapter 1 Specifications.

17 With the full weight of the vehicle on the wheels, tighten all fasteners to the torque listed in this Chapter's Specifications.

18 After a rear spring has been replaced, it's a good idea to check the front suspension ride height and correct it as necessary.

14 Steering wheel - removal and installation

Refer to illustrations 14.2a, 14.2b, 14.3 and 14.4

1 Disconnect the cable from the negative terminal of the battery.

2 Remove the horn pad from the steering wheel **(see illustration)**. Lift the pad off and

14.2b Flip the horn pad over and disconnect the horn lead from its spade terminal

disconnect the horn switch wires **(see illustration)**.

3 Remove the steering wheel retaining nut and mark the position of the steering wheel to the shaft, if marks don't already exist or don't line up **(see illustration)**.

4 Use a puller to detach the steering wheel from the shaft **(see illustration)**. Don't hammer on the shaft to dislodge the wheel.

5 To install the wheel, align the mark on the steering wheel hub with the mark on the shaft and slide the wheel onto the shaft. Install the nut and tighten it to the torque listed in this Chapter's Specifications.

6 Connect the horn wire and install the horn pad.

7 Connect the cable to the negative terminal of the battery.

15 Steering linkage - inspection, removal and installation

Refer to illustration 15.1

1 The steering linkage **(see illustration)**

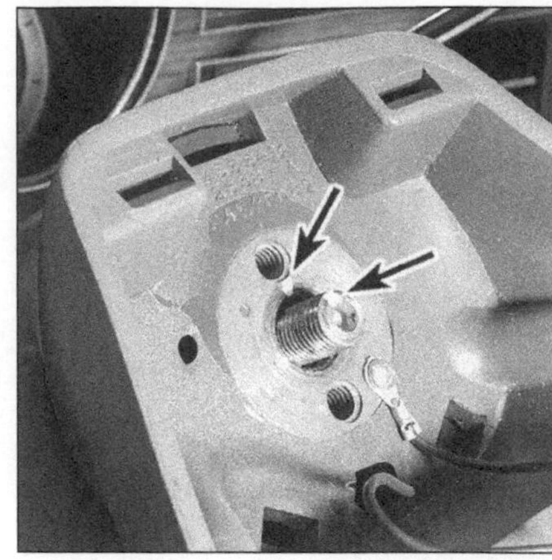

14.3 Before pulling the steering wheel off the steering shaft, mark the alignment of the wheel to the shaft to insure proper realignment during reassembly

connects the steering gear to the front wheels and keeps the wheels in proper relation to each other. The linkage consists of the Pitman (or steering gear) arm, the idler arm, the center link and two adjustable tie-rods. The Pitman arm, which is fastened to the steering gear shaft, moves the center link back-and-forth. The center link is supported on the other end by a frame-mounted idler arm. The back-and-forth motion of the center link is transmitted to the steering knuckles through a pair of tie-rod assemblies. Each tie-rod is made up of an inner and outer tie-rod end, a threaded adjuster tube and two clamps.

Inspection

2 Set the wheels in the straight ahead position and lock the steering wheel.

3 Raise one side of the vehicle until the tire is approximately 1-inch off the ground.

4 Grasp the front and rear of the tire and using light pressure, wiggle the wheel back-and-forth. There should be no noticeable

14.4 Sometimes you can pull the steering wheel off with your hands, but it's usually too tight a fit for that, so you'll have to use a steering wheel puller to get it off

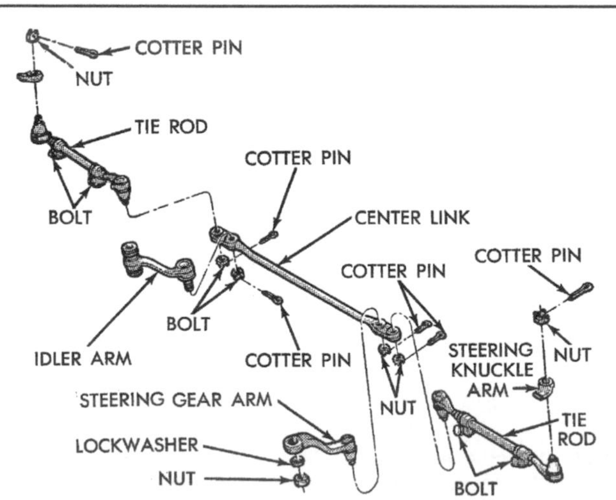

15.1 An exploded view of a typical steering linkage assembly

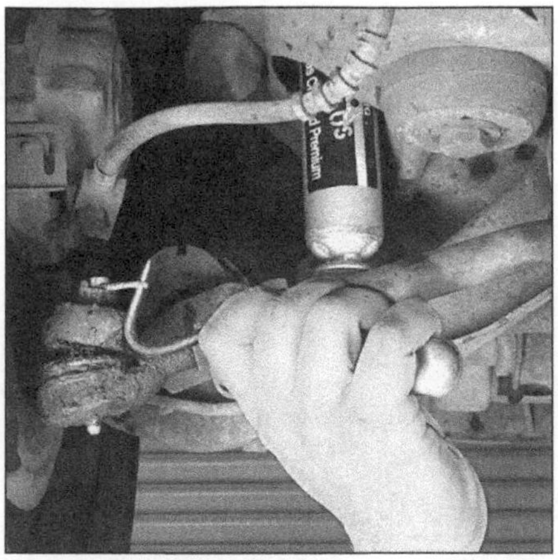

15.8 Remove the cotter pin and loosen, but do not remove, the castle nut from the ballstud

play. If the play in the steering system is noticeable, inspect each steering linkage pivot point and ballstud for looseness and replace parts if necessary.

5 Raise the front of the vehicle and support it on jackstands. Push up, then pull down on the center link end of the idler arm, exerting a force of approximately 25 pounds each way. If there is any looseness, replace the idler arm.

6 Check for frozen balljoints and bent or damaged linkage components.

Removal and installation

Tie-rod

Refer to illustrations 15.8, 15.9, 15.10, 15.12 and 15.13

7 Loosen the wheel lug nuts, raise the vehicle and support it securely on jackstands. Apply the parking brake. Remove the wheel.

8 Remove the cotter pin and loosen, but do not remove, the castle nut from the ballstud **(see illustration)**.

9 Using a small puller, separate the tie-rod end from the steering knuckle **(see illustration)**. Remove the castle nut and pull the tie-rod end from the knuckle.

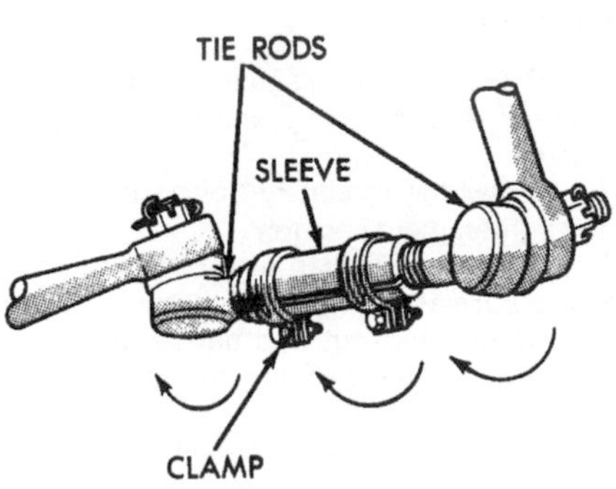

15.12 Make sure the tie-rod can pivot back and forth freely - otherwise, it may bind

15.9 Using a small puller, separate the tie-rod end from the steering arm

10 If a tie-rod end must be replaced, measure the distance from the end of the adjuster tube to the center of the ballstud and record it **(see illustration)**. Loosen the adjuster tube clamp and unscrew the tie-rod end.

11 Remove the nut securing the inner tie-rod end to the center link. Separate the inner tie-rod end from the center link in the same manner as in Step 9.

12 Lubricate the threaded portion of the tie-rod with chassis grease. Screw the new tie-rod end into the adjuster sleeve and adjust the distance from the tube to the ballstud to the previously measured dimension. The number of threads showing on the inner and outer tie-rod ends should be equal within three threads. Don't tighten the clamp nut yet **(see illustration)**.

13 To install the tie-rod, connect the outer tie-rod end to the steering knuckle and install the castle nut. Tighten the nut to the torque listed in this Chapter's Specifications and install a new cotter pin. If the ballstud spins when attempting to tighten the nut, force it into the tapered hole with a large pair of pliers **(see illustration)**. If necessary, tighten the nut slightly to align a slot in the nut with the cotter pin hole in the ballstud. Install the cotter pin.

15.13 If the balljoint stud spins when the nut is tightened, press the tie-rod end into the steering arm

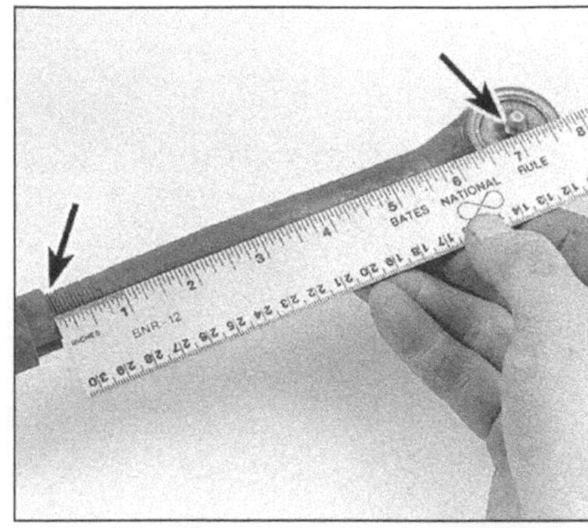

15.10 If a tie-rod end must be replaced, measure the distance from the end of the adjuster tube to the center of the ballstud and record the measurement before loosening the adjuster tube clamp and unscrewing the tie-rod end

14 Insert the inner tie-rod end ballstud into the center link until it's seated. Install the nut and tighten it to the torque listed in this Chapter's Specifications.

15 Tighten the clamp nuts. The center of the bolt should be nearly horizontal and the adjuster tube slot must not line up with the gap in the clamps.

16 Install the wheel and lug nuts, lower the vehicle and tighten the lug nuts to the torque listed in the Chapter 1 Specifications. Drive the vehicle to an alignment shop to have the front end alignment checked and, if necessary, adjusted.

Idler arm

Refer to illustration 15.18

17 Raise the vehicle and support it securely on jackstands. Apply the parking brake.

18 Loosen but do not remove the idler arm-to-center link nut **(see illustration)**.

15.18 To remove the idler arm, remove the cotter pin and loosen the nut on the balljoint stud that's attached to the center link (upper arrow), install a two-jaw puller, separate the idler arm ball stud from the center link, then remove the retaining nut (lower arrow) and pivot bolt

15.32 Use a puller to draw the Pitman arm off the shaft

19 Separate the idler arm from the center link with a two-jaw puller. Remove the nut.

20 Remove the idler arm-to-frame nut and bolt **(see illustration 15.18)**.

21 To install the idler arm, position it on the frame and install the bolts, tightening them to the torque listed in this Chapter's Specifications.

22 Insert the idler arm ballstud into the center link and install the nut. Tighten the nut to the torque listed in this Chapter's Specifications. If the ballstud spins when attempting to tighten the nut, force it into the tapered hole with a large pair of pliers **(see illustration 15.13)**.

Center link

23 Raise the vehicle and support it securely on jackstands. Apply the parking brake.

24 Separate the two inner tie-rod ends from the center link.

25 Separate the center link from the Pitman arm.

26 Separate the center link from the idler arm.

27 Installation is the reverse of the removal procedure. If the ballstuds spin when attempting to tighten the nuts, force them into the tapered holes with a large pair of pliers **(see illustration 15.13)**. Be sure to tighten all of the nuts to the torque listed in this Chapter's Specifications.

Pitman arm

Refer to illustration 15.32

28 Raise the vehicle and place it securely on jackstands.

29 Remove the center link nut from the Pitman arm ballstud. Discard the nut - don't re-use it.

30 Using a puller, separate the center link from the Pitman arm ballstud.

31 Remove the Pitman arm nut and washer.

32 Remove the Pitman arm with a Pitman arm puller or a two-jaw puller **(see illustration)**. Notice that the Pitman arm will only fit on one way - the unsplined area on the steering gear shaft must mesh with the unsplined area in the Pitman arm hole.

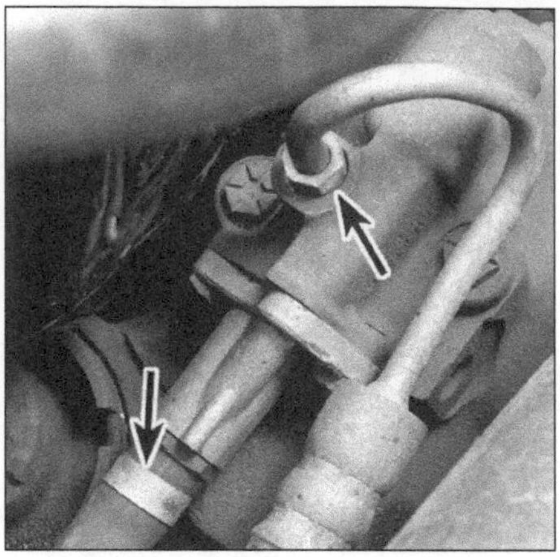

16.1 If the vehicle has power steering, place a drain pan under the steering gear, disconnect the hose fittings (arrows) , hang them up so the fluid doesn't run out and cap the ends to prevent excessive fluid loss and contamination

33 Inspect the ballstud threads for damage. Inspect the ballstud seals for excessive wear. Clean the threads of the ballstud.

34 Installation is the reverse of removal. Make sure the marks you made on the Pitman arm and steering gear shaft are aligned. **Note:** *If a clamp-type Pitman arm is used, spread the arm just enough, with a wedge, to slip the arm onto the shaft. Don't spread the arm more than necessary to slip it over the shaft with hand pressure. Do not hammer the arm onto the shaft or you may damage the steering gear.*

16 Steering gear - removal and installation

Removal

Refer to illustrations 16.1, 16.2 and 16.5

1 If the vehicle has power steering, place a drain pan under the steering gear. Disconnect the hose fittings **(see illustration)**, hang them up so the fluid doesn't run out and cap the ends to prevent excessive fluid loss and contamination. **Note:** *On some models with big-block engines it may be easier to detach the power steering lines from underneath, after the vehicle has been raised and supported.*

2 Remove the roll pin **(see illustration)** from the steering shaft lower coupling with a punch and hammer. **Note:** *On some models with big-block engines it may be easier to drive out the roll pin from underneath, after the vehicle has been raised and supported.*

3 Raise the front of the vehicle and support it securely on jackstands. Apply the parking brake.

4 Remove the Pitman arm (see Section 15).

5 Support the steering gear and remove

16.2 Remove the roll pin (arrow) from the steering shaft lower coupling with a punch and hammer

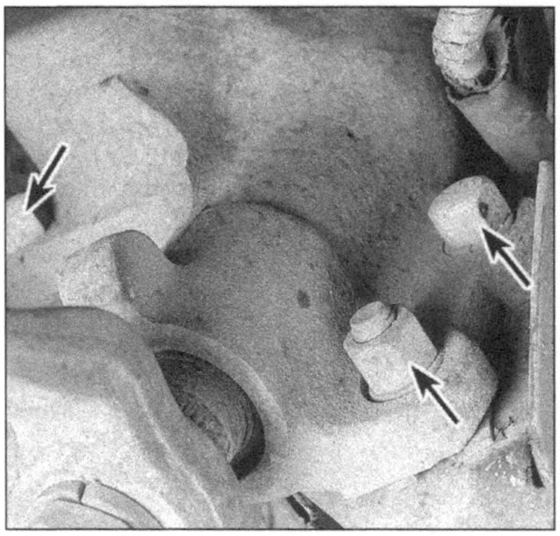

16.5 Support the steering gear and remove the mounting bolts (arrows), lower the unit, disconnect the steering shaft from the steering gear input shaft and remove the steering gear from the vehicle

the mounting bolts **(see illustration)**. Lower the unit, separate the steering shaft from the steering gear input shaft and remove the steering gear from the vehicle.

Installation

6 Raise the steering gear into position and connect the steering shaft, aligning the marks.

7 Install the mounting bolts and washers and tighten them securely.

8 Slide the Pitman arm onto the shaft. Make sure the marks are aligned. Install the washer and nut and tighten the nut to the torque listed in this Chapter's Specifications.

9 Install a *new* steering shaft roll pin.

10 Connect the power steering hose fittings to the steering gear and fill the power steering pump reservoir with the recommended fluid (see Chapter 1).

11 Lower the vehicle and bleed the steering system (see Section 18).

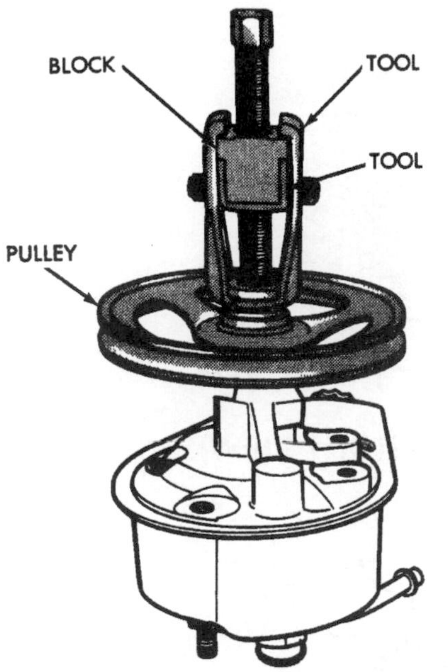

17.6 Remove the pulley with a special tool

17 Power steering pump - removal and installation

Removal

Refer to illustration 17.6

1 Disconnect the cable from the negative terminal of the battery.
2 Remove the power steering drivebelt (see Chapter 1).
3 Unscrew the cap on the pump and siphon off the fluid in the pump reservoir.
4 Position a drain pan under the power steering pump. Disconnect the pressure and return hoses from the backside of the pump. Plug the hoses to prevent contaminants from entering.
5 Remove the pump mounting fasteners and lift the pump from the vehicle, taking care not to spill fluid on the painted surfaces.
6 Using a special pulley removal tool, available at most auto parts stores, remove the pulley from the pump **(see illustration)**.

Installation

Refer to illustration 17.9

7 Position the pump in the mounting bracket and install the bolts/nuts. Tighten the fasteners securely.
8 Connect the hoses to the pump. Tighten the fittings securely and reinstall the cover.
9 Press the pulley onto the shaft using a special pulley installer tool, available at most auto parts stores, **(see illustration)**. An alternative tool can be fabricated from a long bolt, nut, washer and a socket of the same diameter as the pulley hub. Push the pulley onto the shaft until the front of the hub is flush with the shaft, but no further.
10 Install the drivebelt.
11 Fill the power steering reservoir with the recommended fluid (see Chapter 1) and bleed the system following the procedure described in the next Section.

18 Power steering system - bleeding

1 Following any operation in which the power steering fluid lines have been disconnected, the power steering system must be bled to remove all air and obtain proper steering performance.
2 With the front wheels in the straight ahead position, check the power steering fluid level and, if low, add fluid until it reaches the Cold mark on the dipstick.
3 Start the engine and allow it to run at fast idle. Recheck the fluid level and add more if necessary to reach the Cold mark on the dipstick.
4 Bleed the system by turning the wheels from side-to-side, without hitting the stops. This will work the air out of the system. Keep the reservoir full of fluid as this is done.
5 When the air is worked out of the system, return the wheels to the straight ahead position and leave the vehicle running for several more minutes before shutting it off.
6 Road test the vehicle to be sure the steering system is functioning normally and noise free.
7 Recheck the fluid level to be sure it is up to the Hot mark on the dipstick while the engine is at normal operating temperature. Add fluid if necessary (see Chapter 1).

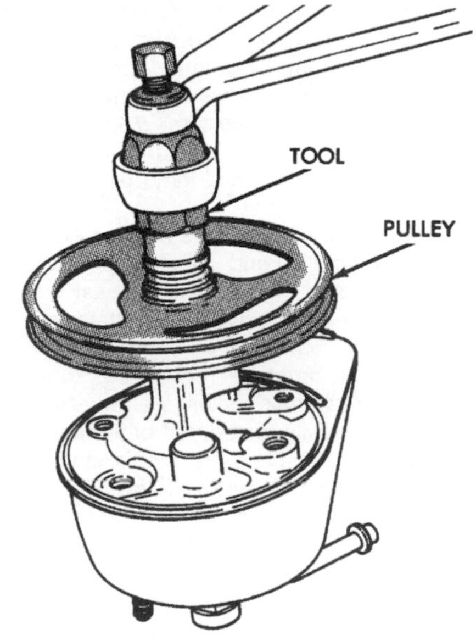

17.9 Use a pulley installation tool - do not hammer the pulley onto the shaft

19 Wheels and tires - general information

Refer to illustration 19.1

Most vehicles covered by this manual are equipped with metric-sized radial tires **(see illustration)**. Use of other size or type of tires may affect the ride and handling of the

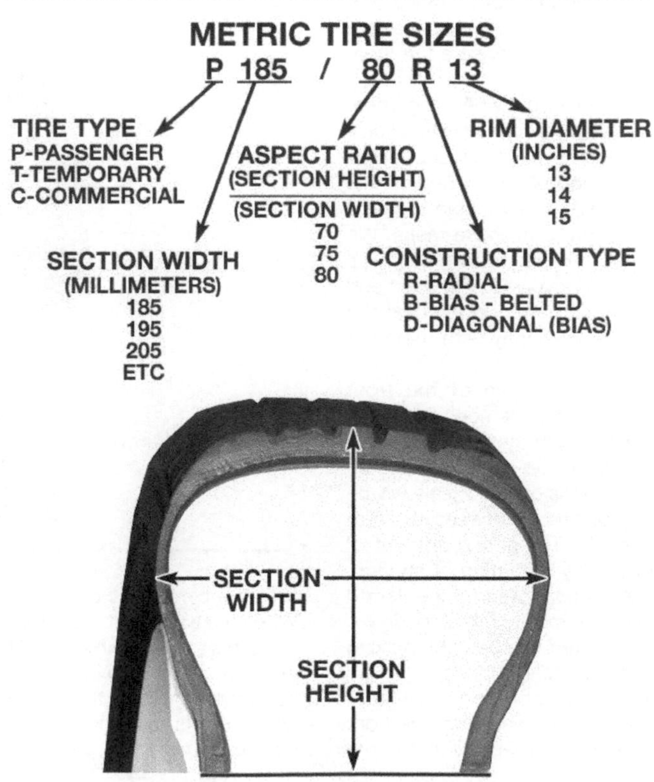

19.1 Metric tire size code

vehicle. Don't mix different types of tires, such as radials and bias belted, on the same vehicle as handling may be seriously affected. It's recommended that tires be replaced in pairs on the same axle, but if only one tire is being replaced, be sure it's the same size, structure and tread design as the other.

Because tire pressure has a substantial effect on handling and wear, the pressure on all tires should be checked at least once a month or before any extended trips (see Chapter 1).

Wheels must be replaced if they are bent, dented, leak air, have elongated bolt holes, are heavily rusted, out of vertical symmetry or if the lug nuts won't stay tight. Wheel repairs that use welding or peening are not recommended.

Tire and wheel balance is important to the overall handling, braking and performance of the vehicle. Unbalanced wheels can adversely affect handling and ride characteristics as well as tire life. Whenever a tire is installed on a wheel, the tire and wheel should be balanced by a shop with the proper equipment.

20 Front end alignment - general information

Refer to illustration 20.1

A front end alignment refers to the adjustments made to the front wheels so they are in proper angular relationship to the suspension and the ground. Front wheels that are out of proper alignment not only affect steering control, but also increase tire wear **(see illustration)**.

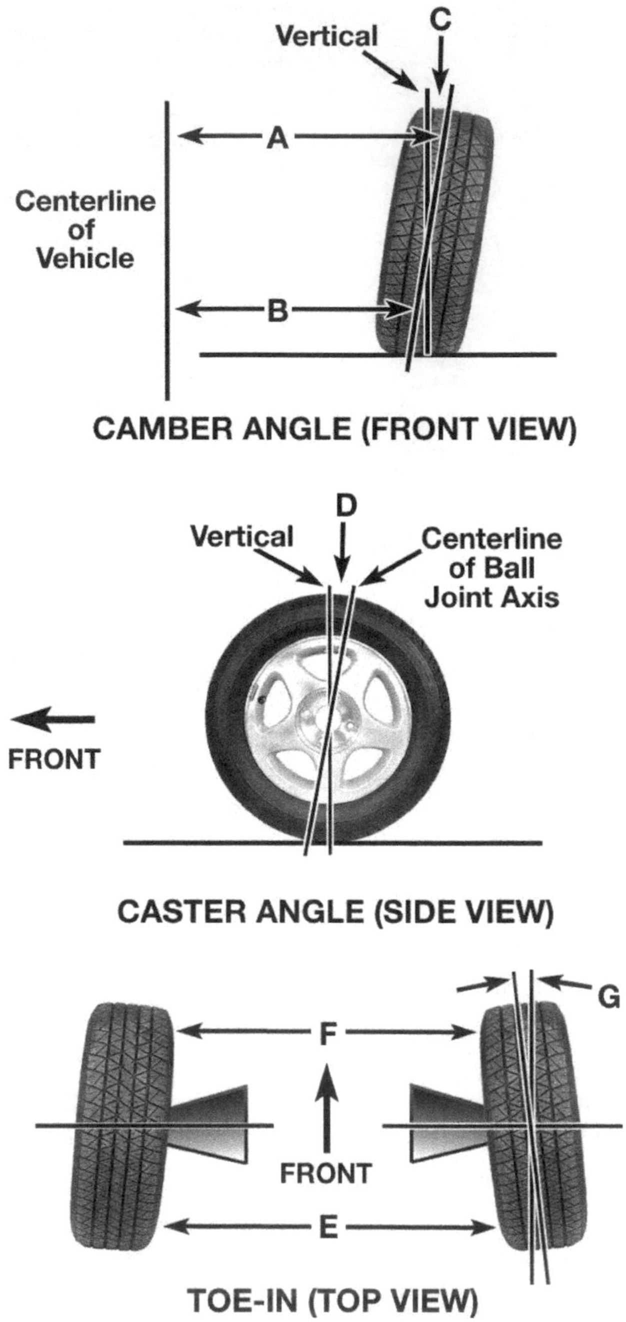

20.1 Front end alignment details

1 *A minus B = C (degrees camber)*
2 *E minus F = toe-in (measured in inches)*
3 *G - toe-in (expressed in degrees)*

Getting the proper front wheel alignment is a very exacting process, one in which complicated and expensive machines are necessary to perform the job properly. Because of this, you should have a technician with the proper equipment perform these tasks. We will, however, use this space to give you a basic idea of what is involved with front end alignment so you can better understand the process and deal intelligently with the shop that does the work.

Toe-in is the turning in of the front wheels. The purpose of a toe specification is to ensure parallel rolling of the front wheels. In a vehicle with zero toe-in, the distance between the front edges of the wheels will be the same as the distance between the rear edges of the wheels. The actual amount of toe-in is normally only a fraction of an inch. Toe-in adjustment is controlled by the tie-rod end position on the tie-rod. Incorrect toe-in will cause the tires to wear improperly by making them scrub against the road surface.

Camber is the tilting of the front wheels from the vertical when viewed from the front of the vehicle. When the wheels tilt out at the top, the camber is said to be positive (+). When the wheels tilt in at the top the camber is negative (-). The amount of tilt is measured in degrees from the vertical and this measurement is called the camber angle. This angle affects the amount of tire tread which contacts the road and compensates for changes in the suspension geometry when the vehicle is cornering or traveling over an undulating surface.

Caster is the tilting of the top of the front steering axis from the vertical. A tilt toward the rear is positive caster and a tilt toward the front is negative caster.

Chapter 11 Body

Contents

1 General information

The vehicles covered by this manual are of unitized construction. The body is designed to provide rigidity so that a separate frame is not necessary. Front and rear frame side rails integral with the body support the front end sheet metal, front and rear suspension systems and other mechanical components. Due to this type of construction, it is very important that, in the event of collision damage, the underbody be thoroughly checked by a facility with the proper equipment.

Certain components are particularly vulnerable to accident damage and can be unbolted and repaired or replaced. Among these parts are the body moldings, bumpers, the hood, doors and all glass.

Only general body maintenance practices and body panel repair procedures within the scope of the do-it-yourselfer are included in this Chapter.

2 Body - maintenance

1 The condition of your vehicle's body is very important, because the resale value depends a great deal on it. It's much more difficult to repair a neglected or damaged body than it is to repair mechanical components. The hidden areas of the body, such as the wheel wells, the frame and the engine compartment, are equally important, although they don't require as frequent attention as the rest of the body.

2 Once a year, or every 12,000 miles, it's a good idea to have the underside of the body steam cleaned. All traces of dirt and oil will be removed and the area can then be inspected carefully for rust, damaged brake lines, frayed electrical wires, damaged cables and other problems. The front suspension components should be greased after completion of this job.

3 At the same time, clean the engine and the engine compartment with a steam cleaner or water soluble degreaser.

4 The wheel wells should be given close attention, since undercoating can peel away and stones and dirt thrown up by the tires can cause the paint to chip and flake, allowing rust to set in. If rust is found, clean down to the bare metal and apply an anti-rust paint.

5 The body should be washed about once a week. Wet the vehicle thoroughly to soften the dirt, then wash it down with a soft sponge and plenty of clean soapy water. If the surplus dirt is not washed off very carefully, it can wear down the paint.

6 Spots of tar or asphalt thrown up from the road should be removed with a cloth soaked in solvent.

7 Once every six months, wax the body and chrome trim. If a chrome cleaner is used to remove rust from any of the vehicle's plated parts, remember that the cleaner also removes part of the chrome, so use it sparingly.

3 Vinyl trim - maintenance

Don't clean vinyl trim with detergents, caustic soap or petroleum-based cleaners. Plain soap and water works just fine, with a soft brush to clean dirt that may be ingrained. Wash the vinyl as frequently as the rest of the vehicle.

After cleaning, application of a high quality rubber and vinyl protectant will help prevent oxidation and cracks. The protectant can also be applied to weather-stripping, vacuum lines and rubber hoses, which often fail as a result of chemical degradation, and to the tires.

4 Upholstery and carpets - maintenance

1 Every three months remove the carpets or mats and clean the interior of the vehicle (more frequently if necessary). Vacuum the upholstery and carpets to remove loose dirt and dust.

2 Leather upholstery requires special care. Stains should be removed with warm water and a very mild soap solution. Use a clean, damp cloth to remove the soap, then wipe again with a dry cloth. Never use alcohol, gasoline, nail polish remover or thinner to clean leather upholstery.

3 After cleaning, regularly treat leather upholstery with a leather wax. Never use car wax on leather upholstery.

4 In areas where the interior of the vehicle is subject to bright sunlight, cover leather seats with a sheet if the vehicle is to be left out for any length of time.

5 Body repair - minor damage

See photo sequence

Repair of minor scratches

1 If the scratch is superficial and does not penetrate to the metal of the body, repair is very simple. Lightly rub the scratched area with a fine rubbing compound to remove loose paint and built up wax. Rinse the area with clean water.

2 Apply touch-up paint to the scratch, using a small brush. Continue to apply thin layers of paint until the surface of the paint in the scratch is level with the surrounding paint. Allow the new paint at least two weeks to harden, then blend it into the surrounding paint by rubbing with a very fine rubbing compound. Finally, apply a coat of wax to the scratch area.

These photos illustrate a method of repairing simple dents. They are intended to supplement *Body repair - minor damage* in this Chapter and should not be used as the sole instructions for body repair on these vehicles.

1 If you can't access the backside of the body panel to hammer out the dent, pull it out with a slide-hammer-type dent puller. In the deepest portion of the dent or along the crease line, drill or punch hole(s) at least one inch apart . . .

2 . . . then screw the slide-hammer into the hole and operate it. Tap with a hammer near the edge of the dent to help 'pop' the metal back to its original shape. When you're finished, the dent area should be close to its original contour and about 1/8-inch below the surface of the surrounding metal

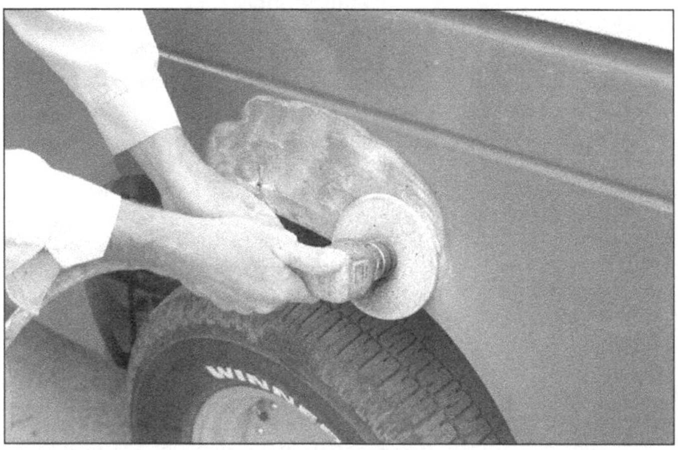

3 Using coarse-grit sandpaper, remove the paint down to the bare metal. Hand sanding works fine, but the disc sander shown here makes the job faster. Use finer (about 320-grit) sandpaper to feather-edge the paint at least one inch around the dent area

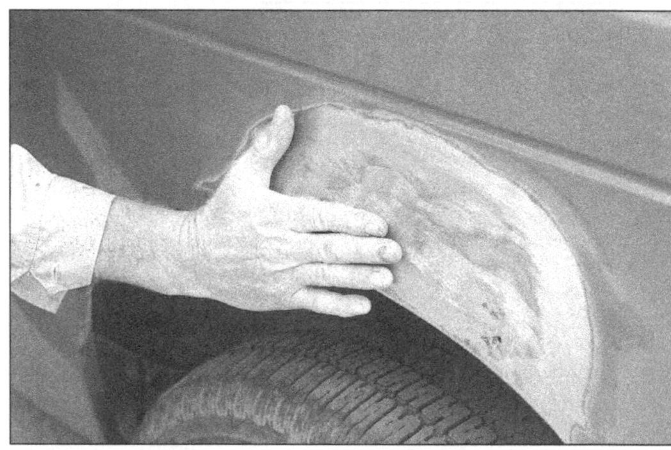

4 When the paint is removed, touch will probably be more helpful than sight for telling if the metal is straight. Hammer down the high spots or raise the low spots as necessary. Clean the repair area with wax/silicone remover

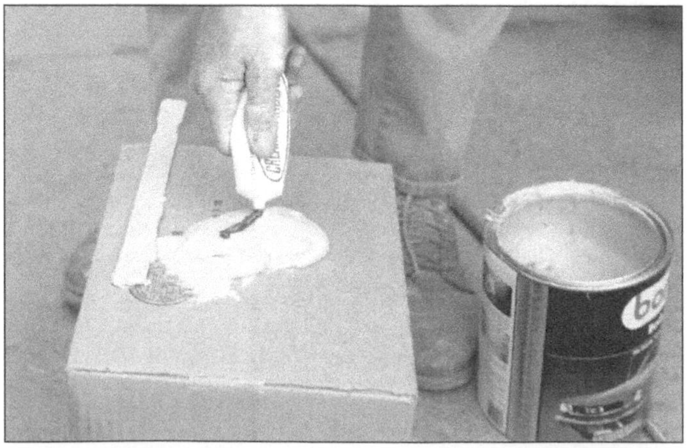

5 Following label instructions, mix up a batch of plastic filler and hardener. The ratio of filler to hardener is critical, and, if you mix it incorrectly, it will either not cure properly or cure too quickly (you won't have time to file and sand it into shape)

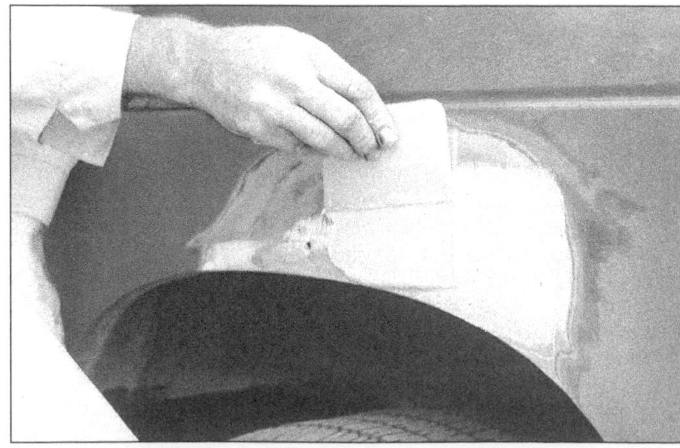

6 Working quickly so the filler doesn't harden, use a plastic applicator to press the body filler firmly into the metal, assuring it bonds completely. Work the filler until it matches the original contour and is slightly above the surrounding metal

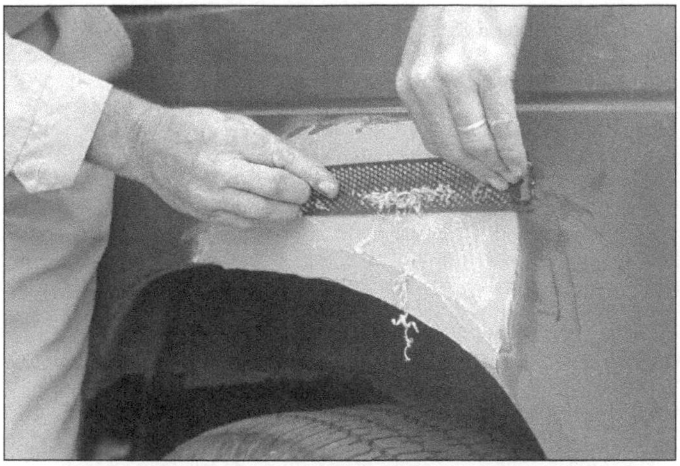

7 Let the filler harden until you can just dent it with your fingernail. Use a body file or Surform tool (shown here) to rough-shape the filler

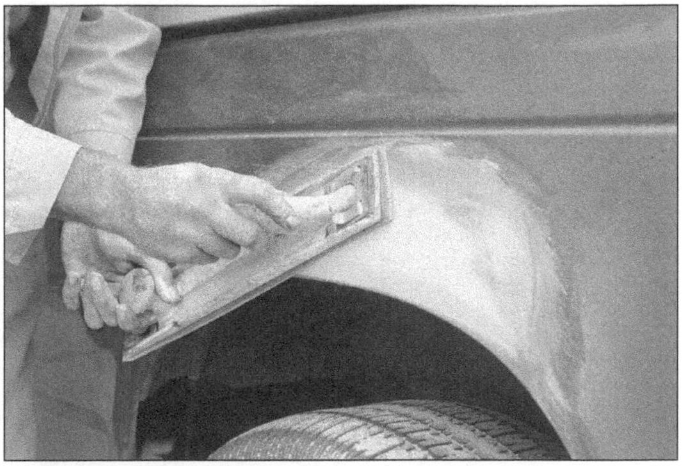

8 Use coarse-grit sandpaper and a sanding board or block to work the filler down until it's smooth and even. Work down to finer grits of sandpaper - always using a board or block - ending up with 360 or 400 grit

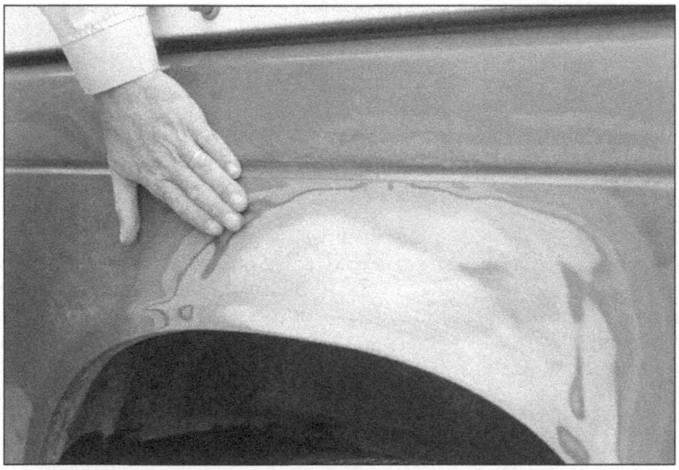

9 You shouldn't be able to feel any ridge at the transition from the filler to the bare metal or from the bare metal to the old paint. As soon as the repair is flat and uniform, remove the dust and mask off the adjacent panels or trim pieces

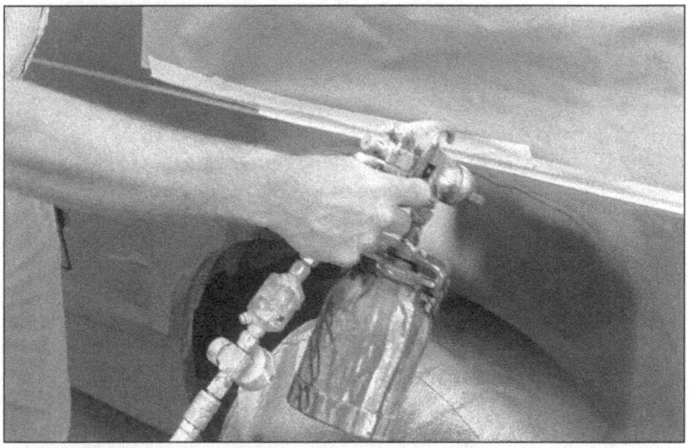

10 Apply several layers of primer to the area. Don't spray the primer on too heavy, so it sags or runs, and make sure each coat is dry before you spray on the next one. A professional-type spray gun is being used here, but aerosol spray primer is available inexpensively from auto parts stores

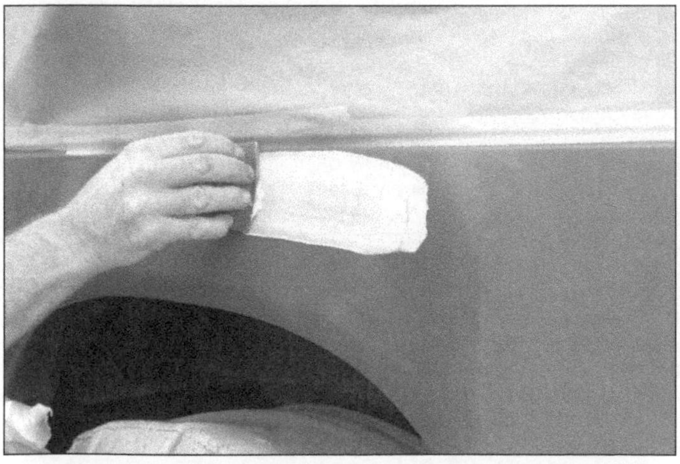

11 The primer will help reveal imperfections or scratches. Fill these with glazing compound. Follow the label instructions and sand it with 360 or 400-grit sandpaper until it's smooth. Repeat the glazing, sanding and respraying until the primer reveals a perfectly smooth surface

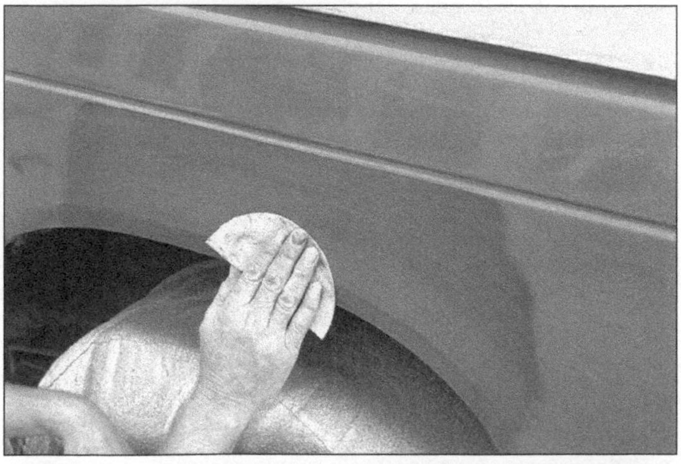

12 Finish sand the primer with very fine sandpaper (400 or 600-grit) to remove the primer overspray. Clean the area with water and allow it to dry. Use a tack rag to remove any dust, then apply the finish coat. Don't attempt to rub out or wax the repair area until the paint has dried completely (at least two weeks)

3 If the scratch has penetrated the paint and exposed the metal of the body, causing the metal to rust, a different repair technique is required. Remove all loose rust from the bottom of the scratch with a pocket knife, then apply rust inhibiting paint to prevent the formation of rust in the future. Using a rubber or nylon applicator, coat the scratched area with glaze-type filler. If required, the filler can be mixed with thinner to provide a very thin paste, which is ideal for filling narrow scratches. Before the glaze filler in the scratch hardens, wrap a piece of smooth cotton cloth around the tip of a finger. Dip the cloth in thinner and then quickly wipe it along the surface of the scratch. This will ensure that the surface of the filler is slightly hollow. The scratch can now be painted over as described earlier in this section.

Repair of dents

4 When repairing dents, the first job is to pull the dent out until the affected area is as close as possible to its original shape. There is no point in trying to restore the original shape completely as the metal in the damaged area will have stretched on impact and cannot be restored to its original contours. It is better to bring the level of the dent up to a point which is about 1/8-inch below the level of the surrounding metal. In cases where the dent is very shallow, it is not worth trying to pull it out at all.

5 If the back side of the dent is accessible, it can be hammered out gently from behind using a soft-face hammer. While doing this, hold a block of wood firmly against the opposite side of the metal to absorb the hammer blows and prevent the metal from being stretched.

6 If the dent is in a section of the body which has double layers, or some other factor makes it inaccessible from behind, a different technique is required. Drill several small holes through the metal inside the damaged area, particularly in the deeper sections. Screw long, self tapping screws into the holes just enough for them to get a good grip in the metal. Now the dent can be pulled out by pulling on the protruding heads of the screws with locking pliers.

7 The next stage of repair is the removal of paint from the damaged area and from an inch or so of the surrounding metal. This is easily done with a wire brush or sanding disk in a drill motor, although it can be done just as effectively by hand with sandpaper. To complete the preparation for filling, score the surface of the bare metal with a screwdriver or the tang of a file or drill small holes in the affected area. This will provide a good grip for the filler material. To complete the repair, see the Section on filling and painting.

Repair of rust holes or gashes

8 Remove all paint from the affected area and from an inch or so of the surrounding metal using a sanding disk or wire brush mounted in a drill motor. If these are not available, a few sheets of sandpaper will do the job just as effectively.

9 With the paint removed, you will be able to determine the severity of the corrosion and decide whether to replace the whole panel, if possible, or repair the affected area. New body panels are not as expensive as most people think and it is often quicker to install a new panel than to repair large areas of rust.

10 Remove all trim pieces from the affected area except those which will act as a guide to the original shape of the damaged body, such as headlight shells, etc. Using metal snips or a hacksaw blade, remove all loose metal and any other metal that is badly affected by rust. Hammer the edges of the hole in to create a slight depression for the filler material.

11 Wire brush the affected area to remove the powdery rust from the surface of the metal. If the back of the rusted area is accessible, treat it with rust inhibiting paint.

12 Before filling is done, block the hole in some way. This can be done with sheet metal riveted or screwed into place, or by stuffing the hole with wire mesh.

13 Once the hole is blocked off, the affected area can be filled and painted. See the following subsection on filling and painting.

Filling and painting

14 Many types of body fillers are available, but generally speaking, body repair kits which contain filler paste and a tube of resin hardener are best for this type of repair work. A wide, flexible plastic or nylon applicator will be necessary for imparting a smooth and contoured finish to the surface of the filler material. Mix up a small amount of filler on a clean piece of wood or cardboard (use the hardener sparingly). Follow the manufacturer's instructions on the package, otherwise the filler will set incorrectly.

15 Using the applicator, apply the filler paste to the prepared area. Draw the applicator across the surface of the filler to achieve the desired contour and to level the filler surface. As soon as a contour that approximates the original one is achieved, stop working the paste. If you continue, the paste will begin to stick to the applicator. Continue to add thin layers of paste at 20-minute intervals until the level of the filler is just above the surrounding metal.

16 Once the filler has hardened, the excess can be removed with a body file. From then on, progressively finer grades of sandpaper should be used, starting with a 180-grit paper and finishing with 600-grit wet-or-dry paper. Always wrap the sandpaper around a flat rubber or wooden block, otherwise the surface of the filler will not be completely flat. During the sanding of the filler surface, the wet-or-dry paper should be periodically rinsed in water. This will ensure that a very smooth finish is produced in the final stage.

17 At this point, the repair area should be surrounded by a ring of bare metal, which in turn should be encircled by the finely feathered edge of good paint. Rinse the repair area with clean water until all of the dust produced by the sanding operation is gone.

18 Spray the entire area with a light coat of primer. This will reveal any imperfections in the surface of the filler. Repair the imperfections with fresh filler paste or glaze filler and once more smooth the surface with sandpaper. Repeat this spray-and-repair procedure until you are satisfied that the surface of the filler and the feathered edge of the paint are perfect. Rinse the area with clean water and allow it to dry completely.

19 The repair area is now ready for painting. Spray painting must be carried out in a warm, dry, windless and dust free atmosphere. These conditions can be created if you have access to a large indoor work area, but if you are forced to work in the open, you will have to pick the day very carefully. If you are working indoors, dousing the floor in the work area with water will help settle the dust which would otherwise be in the air. If the repair area is confined to one body panel, mask off the surrounding panels. This will help minimize the effects of a slight mismatch in paint color. Trim pieces such as chrome strips, door handles, etc., will also need to be masked off or removed. Use masking tape and several thicknesses of newspaper for the masking operations.

20 Before spraying, shake the paint can thoroughly, then spray a test area until the spray painting technique is mastered. Cover the repair area with a thick coat of primer. The thickness should be built up using several thin layers of primer rather than one thick one. Using 600-grit wet-or-dry sandpaper, rub down the surface of the primer until it is very smooth. While doing this, the work area should be thoroughly rinsed with water and the wet-or-dry sandpaper periodically rinsed as well. Allow the primer to dry before spraying additional coats.

21 Spray on the top coat, again building up the thickness by using several thin layers of paint. Begin spraying in the center of the repair area and then, using a circular motion, work out until the whole repair area and about two inches of the surrounding original paint is covered. Remove all masking material 10 to 15 minutes after spraying on the final coat of paint. Allow the new paint at least two weeks to harden, then use a very fine rubbing compound to blend the edges of the new paint into the existing paint. Finally, apply a coat of wax.

6 Body repair - major damage

1 Major damage must be repaired by an auto body shop specifically equipped to perform body and frame repairs. These shops have the specialized equipment required to do the job properly.

2 If the damage is extensive, the body and frame must be checked for proper alignment or the vehicle's handling characteristics may be adversely affected and other components may wear at an accelerated rate.

3 Due to the fact that all of the major body components (hood, fenders, etc.) are separate and replaceable units, any seriously damaged components should be replaced rather than repaired. Sometimes the components can be found in a wrecking yard that specializes in used vehicle components, often at considerable savings over the cost of new parts.

7 Hinges and locks - maintenance

Once every 3000 miles, or every three months, the hinges and latch assemblies on the doors, hood and trunk should be given a few drops of light oil or lock lubricant. The door latch strikers should also be lubricated with a thin coat of grease to reduce wear and ensure free movement. Lubricate the door locks with spray-on graphite lubricant.

8 Windshield and fixed glass - replacement

Replacement of the windshield and fixed glass requires the use of special fast-setting adhesive/caulk materials and some specialized tools and techniques. These operations should be left to a dealer service department or a shop specializing in glass work.

9 Door trim panel - removal and installation

Refer to illustrations 9.3 and 9.4

Removal

1 Disconnect the negative cable from the battery.

9.3 Use an Allen wrench to remove the window crank

2 Remove all door trim panel retaining screws and door pull/armrest assemblies.
3 On models equipped with manual window regulators, remove the window crank **(see illustration)**. On power regulator models, pry out the control switch assembly and disconnect the electrical connector.
4 Insert a putty knife or flat screwdriver between the trim panel and the door and disengage the retainers **(see illustration)**. Work around the outer edge until the panel is free.
5 Once all of the clips are disengaged, detach the trim panel, unplug any other electrical connectors and remove the trim panel from the vehicle.
6 For access to the inner door, carefully peel back the plastic watershield.

Installation

7 Prior to installation of the door panel, be sure to reinstall any clips in the panel which may have come out during the removal procedure and remain in the door itself.
8 Plug in the electrical connectors and place the panel in position on the door. Press the door panel into place until the clips are

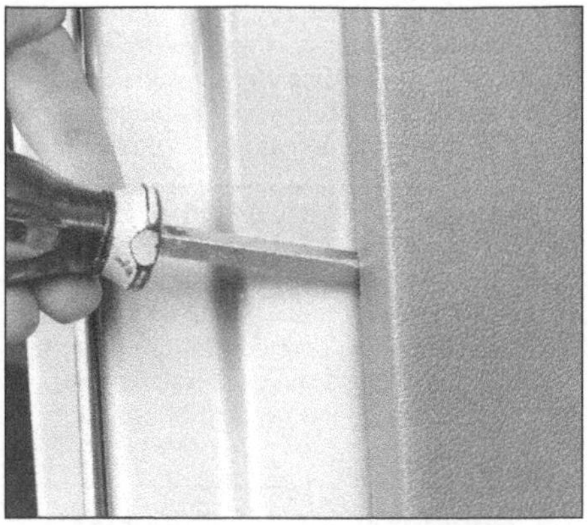

9.4 Pry carefully, working around the door trim panel until all of the clips are disengaged

seated and install the armrest/door pulls. Install the manual regulator window crank or power window switch assembly.

10 Door window glass - removal and installation

Refer to illustrations 10.3, 10.4, 10.7a and 10.7b

1 Remove the door trim panel and plastic watershield (see Section 9).
2 If necessary for clearance, remove the door belt moulding weather-strip.
3 On some models, the track is mounted to the glass and is removed along with it. On these, lower the window regulator to the full down position and remove the lift bracket screws **(see illustration)**.
4 Remove the front track upper and lower screws, then raise the glass and track up and out of the door as an assembly **(see illustration)**. After removal, separate the track from the glass.

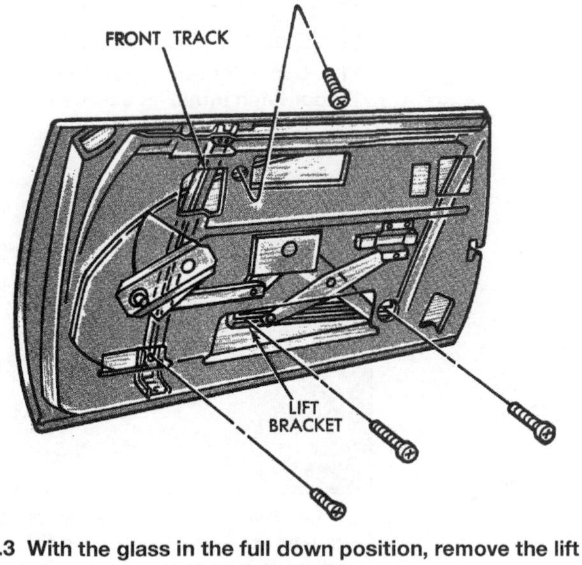

10.3 With the glass in the full down position, remove the lift bracket screws

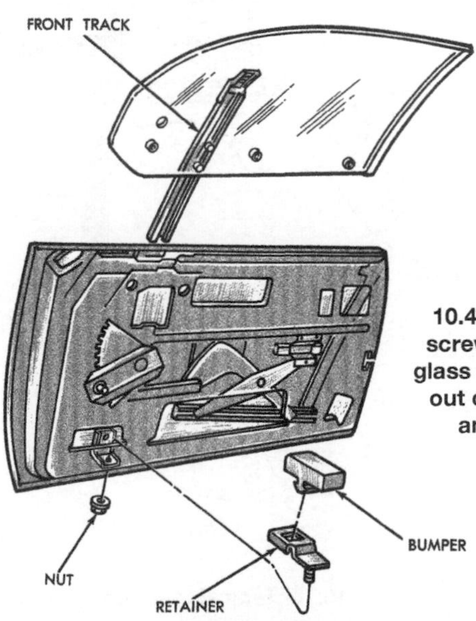

10.4 Remove the screws and lift the glass and front track out of the door as an assembly

5 Installation is the reverse of removal.
6 On other models, the glass is fastened directly to the lift channel by screws, or by glass stabilizers that clamp it in place.
7 On these other models, position the glass so you can access and remove the screws and nuts, or loosen the clamps. Detach the glass and lift it up and out of the door **(see illustration)**. On rear doors with glass division channels, it will be necessary first to remove the screws, detach the channel and lift the glass out **(see illustration)**.
8 Installation is the reverse of removal.

11 Door latch, lock cylinder and outside handle - removal and installation

Refer to illustrations 11.3, 11.4, 11.5a, 11.5b and 11.5c

1 Raise the glass fully and remove the door trim panel and plastic watershield.

Latch

2 Disconnect the operating links from the latch.
3 Remove the retaining screws **(see illustration)** and lift the latch out of the door through the inside panel opening.

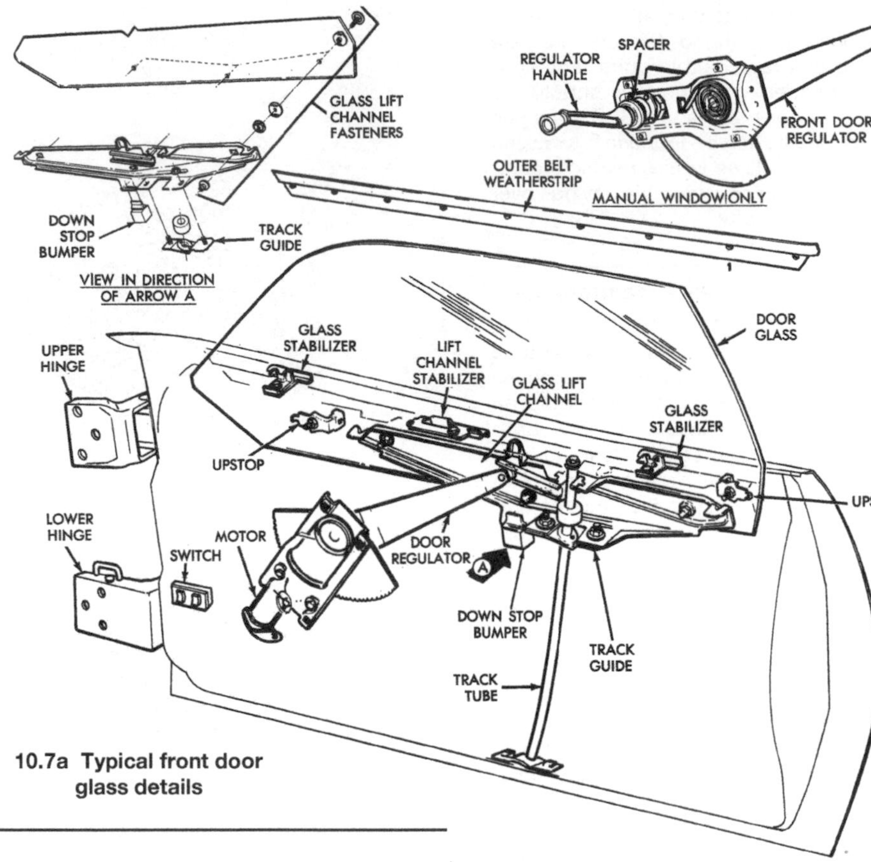

10.7a Typical front door glass details

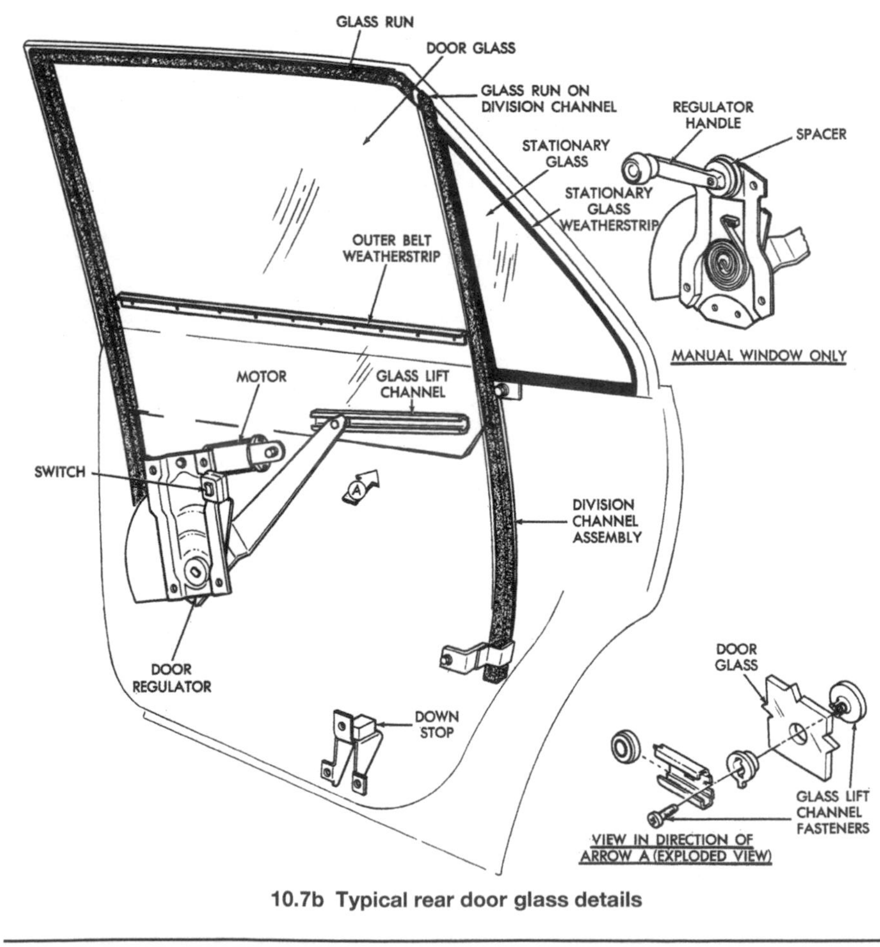

10.7b Typical rear door glass details

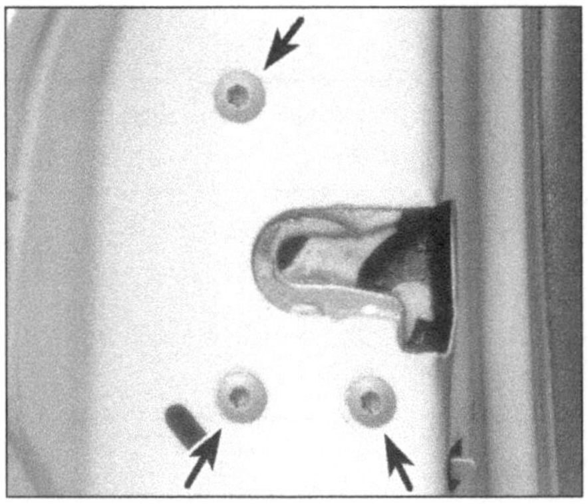

11.3 Use a Torx-head tool to remove the door latch retaining screws (arrows)

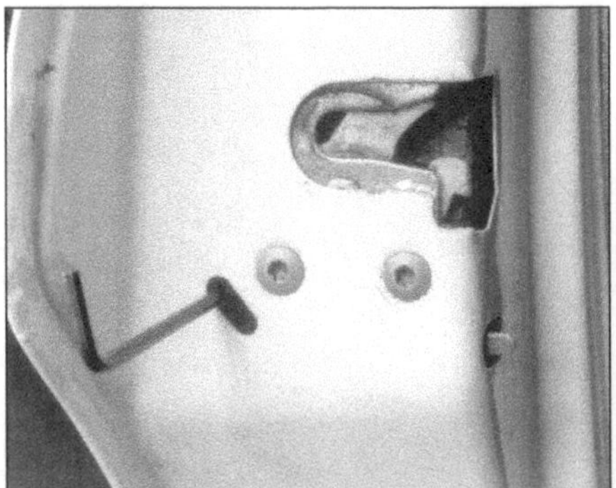

11.4 Insert an Allen wrench into the adjustment hole, then loosen the screw and move it up in the slot to remove any slack from the linkage

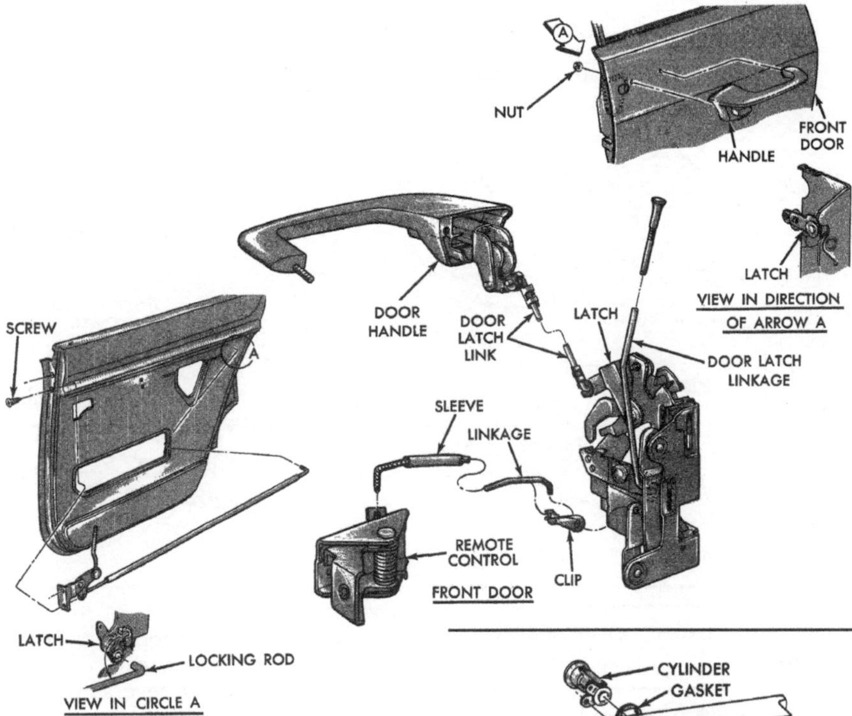

11.5a Early model push button-type outside handle details

4 Installation is the reverse of removal. After installation, loosen the linkage adjusting screw by inserting an Allen wrench into the access hole in the end of the door **(see illustration)**. Push up on the latch lever through the access hole to remove all slack from the linkage, then tighten the screw while holding the latch lever in position.

Lock cylinder

5 Disconnect the operating link at the lock cylinder. Use a screwdriver to push the retaining clip off, then withdraw the lock cylinder from the door **(see illustrations)**.

6 Installation is the reverse of removal.

Outside handle

7 Disconnect the operating links from the outside handle.

8 Remove the nuts and lift the handle from the door **(see illustration 11.5a and 11.5c).**

9 Installation is the reverse of removal.

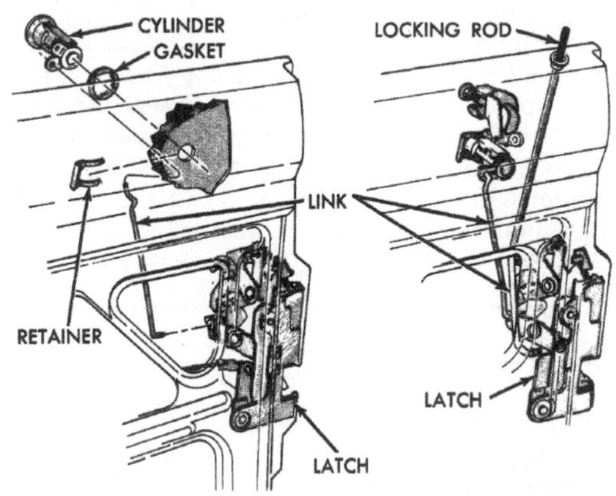

11.5b Early model lock cylinder and latch details

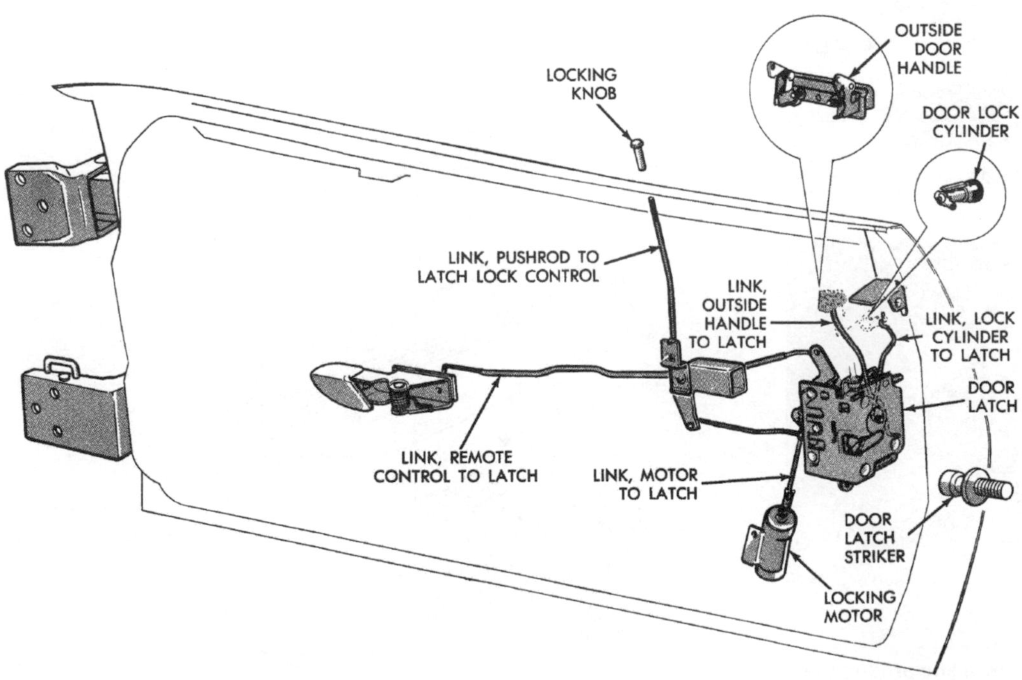

11.5c Typical later model latch, lock cylinder and outside handle details

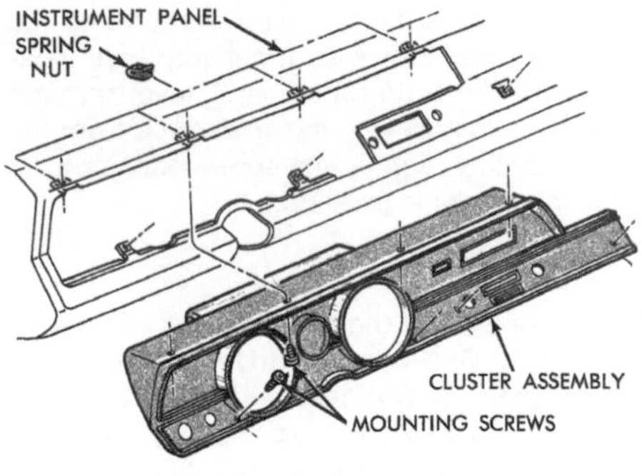

12.4a Typical earlier model instrument cluster bezel details

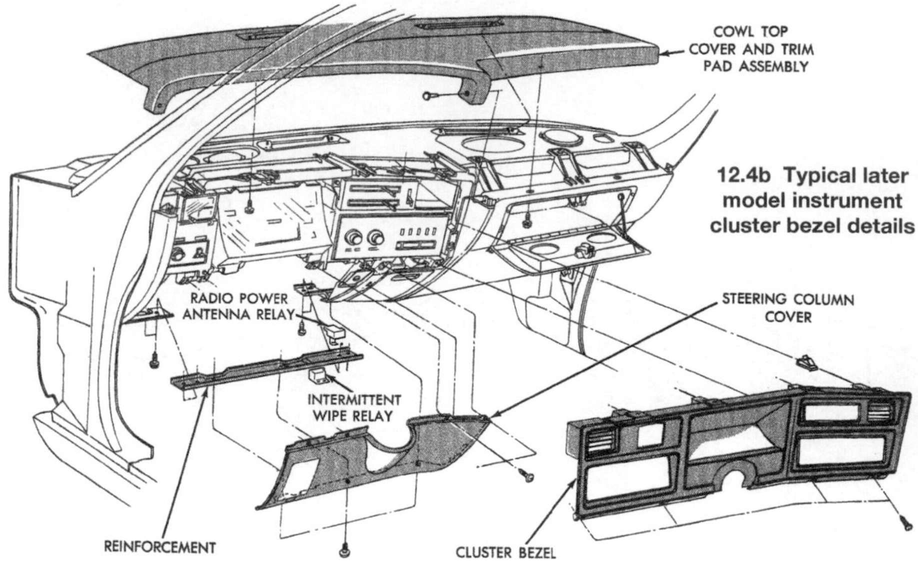

12.4b Typical later model instrument cluster bezel details

12 Instrument cluster bezel - removal and installation

Refer to illustrations 12.4a and 12.4b

1 Disconnect the negative battery cable.
2 Remove any knobs or other components which would interfere with removal.
3 On later automatic transmission models, shift the column shift lever into Low.
4 Remove the retaining screws, grasp the bezel securely, then pull it out and detach it from the instrument panel **(see illustrations)**.
5 Installation is the reverse of removal.

13 Center console - removal and installation

Refer to illustrations 13.1a and 13.1b

1 Remove the screws retaining the console assembly to the mounting brackets **(see illustrations)**. **Note:** *On some models it may be necessary to peel the edge of the carpet back to gain access to screws or bolts on the side. All other screws are accessible from inside the console.*
2 Remove the shift knob, if equipped, and remove the console assembly.
3 Installation is the reverse of removal. Using trim adhesive, re-glue the carpet to the console, if necessary.

14 Hood - removal, installation and adjustment

Refer to illustrations 14.2 and 14.11

Note: *The hood is heavy and somewhat awkward to remove and install - at least two people should perform this procedure.*

Removal and installation

1 Use blankets or pads to cover the cowl area of the body and the fenders. This will protect the body and paint as the hood is lifted off.

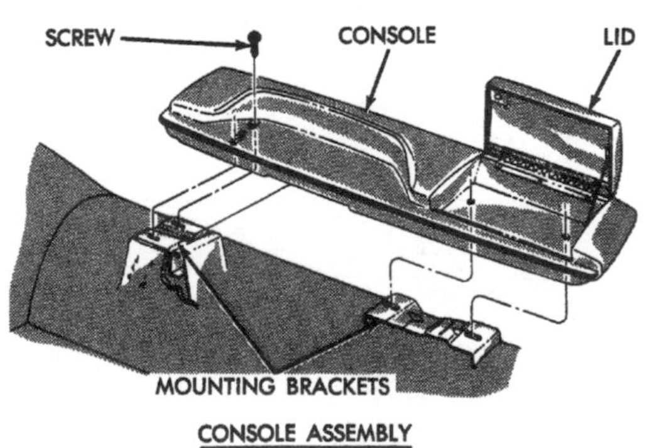

13.1a Typical early model console assembly

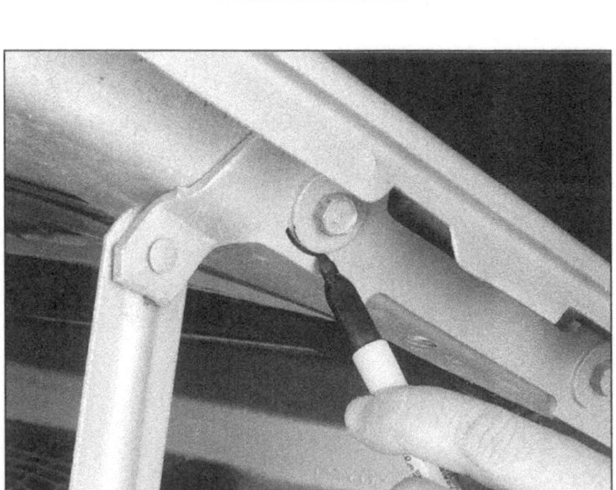

14.2 Use a marking pen to draw a line around the hood bolts

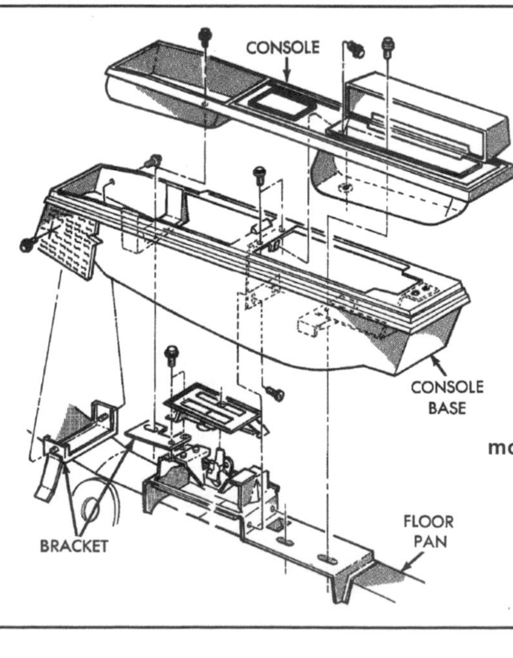

13.1b Typical later model console assembly

14.11 Screw the rubber bumper in-or-out to adjust the hood closed height

2 Scribe or draw alignment marks around the bolt heads to insure proper alignment during installation **(see illustration)**.
3 Disconnect any cables or electricals which will interfere with removal.
4 Have an assistant support the weight of the hood. Remove the hinge-to-hood bolts.
5 Lift off the hood.
6 Installation is the reverse of removal.

Adjustment

7 Fore-and-aft and side-to-side adjustment of the hood is done by moving the hood in relation to the hinge plate after loosening the bolts or nuts.
8 Scribe a line around the entire hinge plate so you can judge the amount of movement **(see illustration 14.2)**.
9 Loosen the bolts or nuts and move the hood into correct alignment. Move it only a little at a time. Tighten the hinge bolts or nuts and carefully lower the hood to check the alignment.
10 If necessary after installation, the entire hood latch assembly can be adjusted up-and-down as well as from side-to-side on the radiator support so the hood closes securely and is flush with the fenders. To do this, scribe a line around the hood latch and striker mounting bolts to provide a reference point. Then loosen the bolts and reposition the latch assembly and striker assemblies as necessary. Following adjustment, retighten the mounting bolts.
11 Finally, adjust the hood bumpers on the radiator support so the hood, when closed, is flush with the fenders **(see illustration)**.
12 The hood latch assembly, as well as the hinges, should be periodically lubricated with white lithium-base grease to prevent sticking and wear.

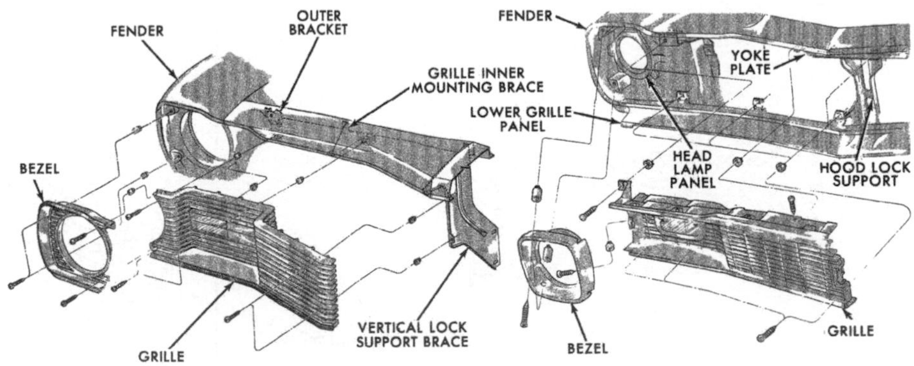

15.1a Typical early model grille assembly

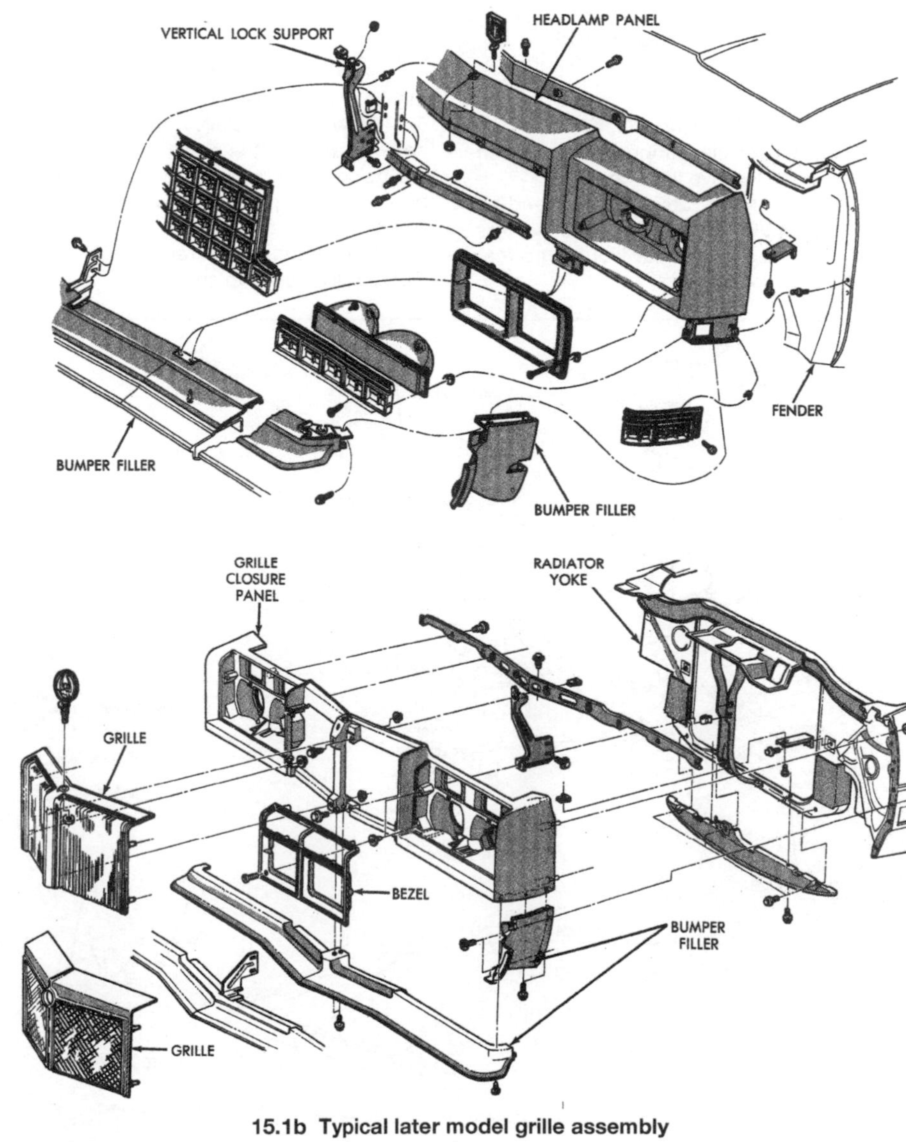

15.1b Typical later model grille assembly

15 Radiator grille - removal and installation

Refer to illustrations 15.1a and 15.1b

1 Remove the screws retaining the headlight bezels and grille **(see illustrations)** and remove the bezels.
2 Disconnect the electrical connectors to the lamp assemblies, if necessary.
3 Remove the grille assembly.
4 Installation is the reverse of removal.

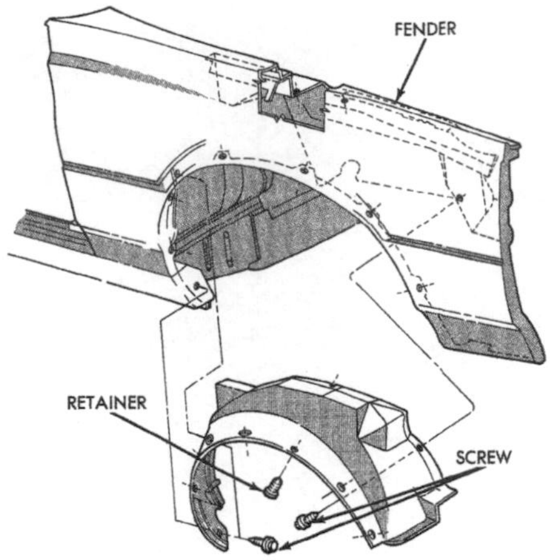

16.4a Typical front fender splash shield details

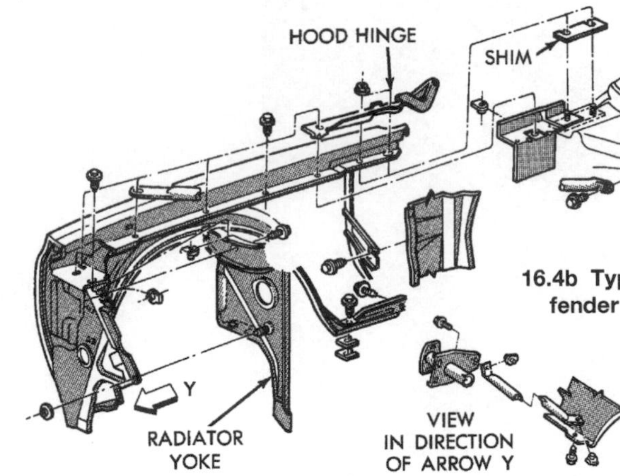

16.4b Typical front fender details

16 Front fender - removal and installation

Refer to illustrations 16.4a and 16.4b

1 Raise the vehicle, support it securely on jackstands and remove the front wheel.

2 Disconnect the radio antenna and all lighting assembly wiring harness connectors and other components that would interfere with fender removal.

3 Remove the front bumper (see Section 17).

4 Remove the splash shield and fender mounting bolts **(see illustrations)**.

5 Detach the fender. It is a good idea to have an assistant support the fender while it's being moved away from the vehicle to prevent damage to the surrounding body panels.

6 Installation is the reverse of removal. Tighten all nuts, bolts and screws securely.

17 Bumpers - removal and installation

Refer to illustrations 17.3a, 17.3b, 17.5a and 17.5b

1 Disconnect any wiring harnesses, bumper covers or fascia (if equipped), or any other components that would interfere with bumper removal.

2 Support the bumper with a jack or jack-

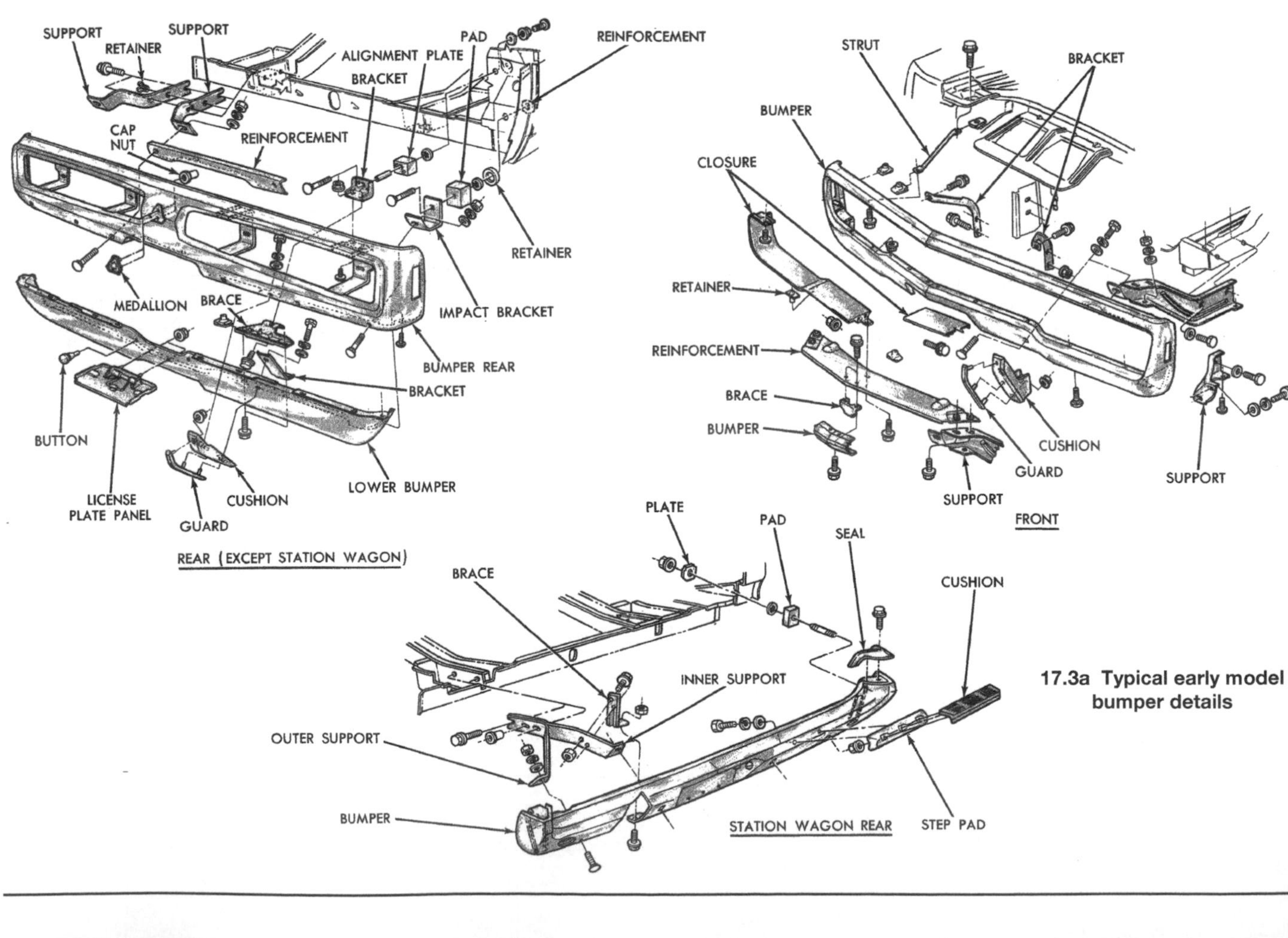

17.3a Typical early model bumper details

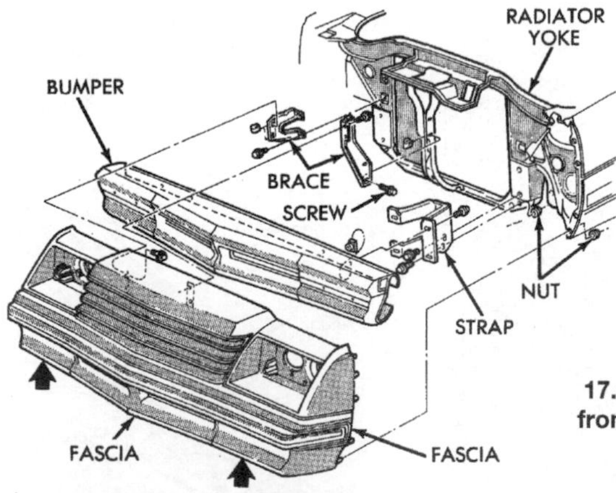

17.3b Typical early model front bumper fascia details

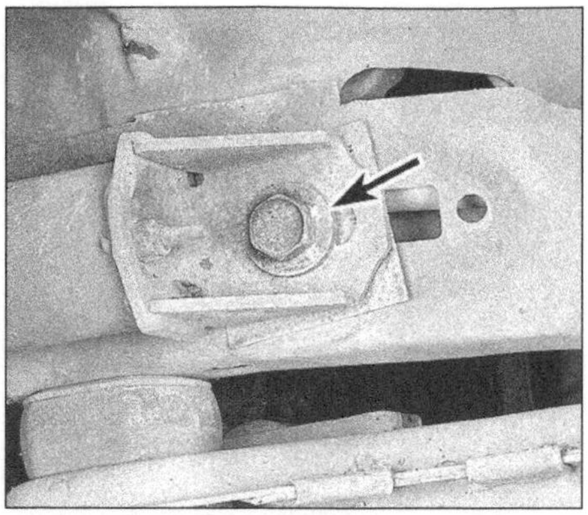

17.5a Energy absorber-to-frame bolt (arrow)

17.5b Energy absorber-to-chassis bolts (arrows)

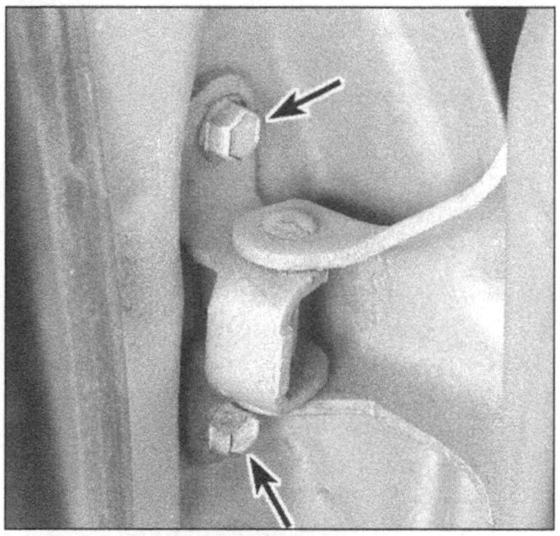

18.4 After marking their locations, remove the hinge bolts (arrows) and lift the door off

18.6 Use a Torx-head tool to loosen the door striker, then move it as necessary to adjust the door closed position

stand. Scribe or draw alignment marks around the bolt heads, then loosen the retaining bolts. Have an assistant support the bumper as the bolts are loosened.

3 Remove the retaining bolts and detach the bumper **(see illustrations)**.

4 Installation is the reverse of removal. Tighten the retaining bolts securely.

5 Later models are equipped with energy absorber units which bolt to the frame and chassis. On these models, the bumper and energy absorbers can be unbolted and removed as a unit **(see illustrations)**.

6 Installation is the reverse of removal. Tighten the retaining bolts securely.

18 Door - removal, installation and adjustment

Refer to illustrations 18.4 and 18.6

Removal and installation

1 Remove the door trim panel. Disconnect any electrical connectors and push them through the door opening so they won't interfere with door removal.

2 Place a jack or jackstand under the door and have an assistant on hand to support the door when the hinge bolts are removed. **Note:** *If a jack or jackstand is used, place a wood block and rag between it and the door to protect the door's painted surfaces.*

3 Scribe around the door hinges.

4 Remove the hinge-to-door bolts and carefully lift off the door **(see illustration)**.

5 Installation is the reverse of removal.

Adjustment

6 Following installation of the door, check the alignment and adjust, if necessary, as follows:

- *a) Up/down and forward/backward adjustments are made by loosening the hinge-to-body bolts and moving the door as necessary to gain proper alignment with the surrounding body panels (fender, roof, sill plate, etc.).*
- *b) The door lock striker can also be adjusted both up-and-down and sideways to provide positive engagement with the lock mechanism. This is done by loosening the striker, moving it as necessary, then retightening the striker* **(see illustration)**.

19 Trunk lid - removal, installation and adjustment

Refer to illustrations 19.3 and 19.7

1 Open the trunk lid and cover the edges of the trunk compartment and quarter panels with pads or cloths to protect the painted surfaces when the lid is removed.

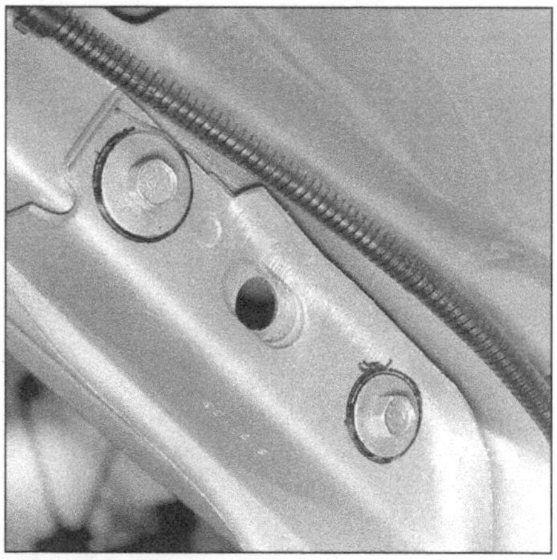

19.3 Draw alignment marks around the hinge bolts with a marking pen

2 Disconnect any cables or electrical connectors attached to the trunk lid that would interfere with removal.

3 Use a marking pen to make alignment marks around the hinge bolts **(see illustration)**.

4 With the help of an assistant, support

19.7 Loosen the bolts (arrows) and move the latch to adjust the trunk lid height

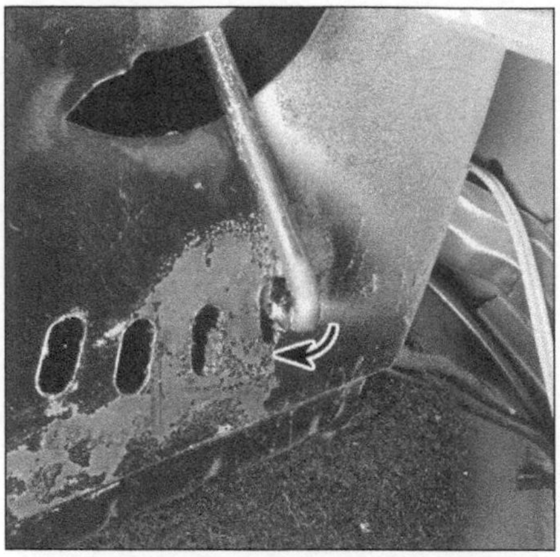

20.1 Pull the torsion bar out of the upper slot (it's under pressure, so be careful) and move it to a lower one to increase the support tension of the trunk lid

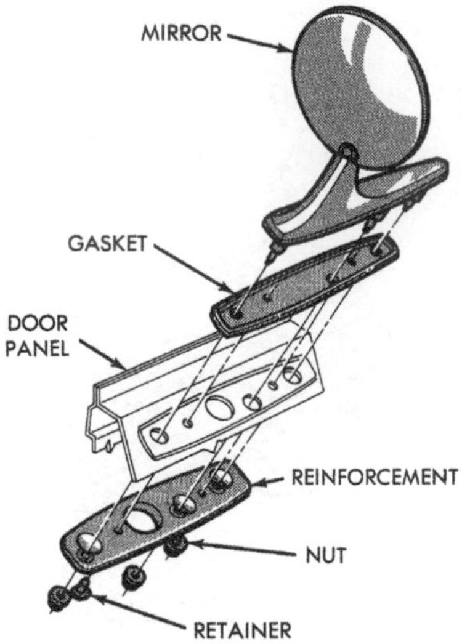

21.2a Typical outside mirror details

the trunk lid, remove the hinge bolts on both sides and lift the trunk lid off.

5 Installation is the reverse of the removal procedure. When reinstalling the trunk lid, align the heads of the bolts with the marks made during removal.

6 After installation, close the trunk lid and make sure the lid is aligned properly with the surrounding body panels. Fore/aft and up/down adjustments of the trunk lid are determined by the position of the bolts in the hinge arms. To adjust, loosen the hinge bolts, reposition the lid and retighten the bolts.

7 The height of the trunk lid in relation to the body can be adjusted by loosening the latch bolts, repositioning the latch and retightening the bolts **(see illustration)**.

20 Trunk lid torsion bar - adjustment

Refer to illustration 20.1

1 The support tension of the trunk lid can be adjusted by moving the end of the torsion bar to a different slot in the mount **(see illustration)**.

2 After adjustment, close the trunk lid to make sure the latch action is smooth and solid. Repeat the adjustment procedure until the trunk lid stays in the open position without falling, but closes easily without too much force.

21 Outside rear view mirror - removal and installation

Refer to illustration 21. 2a, 21.2b and 21.2c

1 Remove the door trim panel (see Section 9).

2 Remove the retaining nuts/bolts and detach the mirror **(see illustrations)**. If you're removing a right (passenger's) side remote-controlled mirror, remove any dash panels required to detach the control stalk end of the cables and route the new control cable into place.

3 Installation is the reverse of removal.

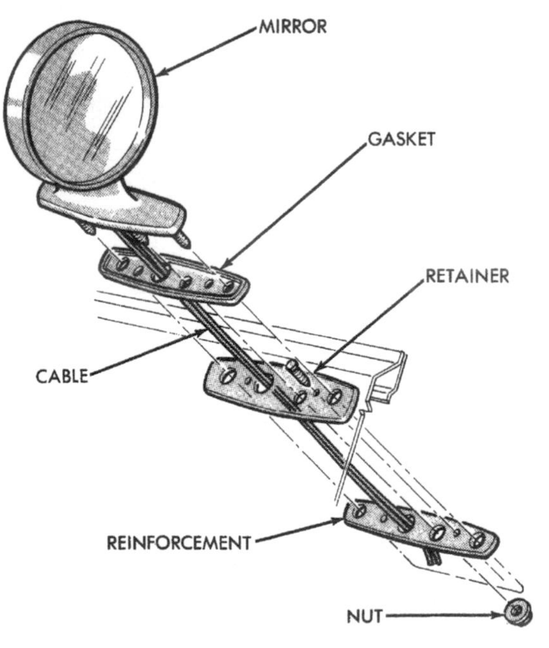

21.2b Remote-controlled mirror details

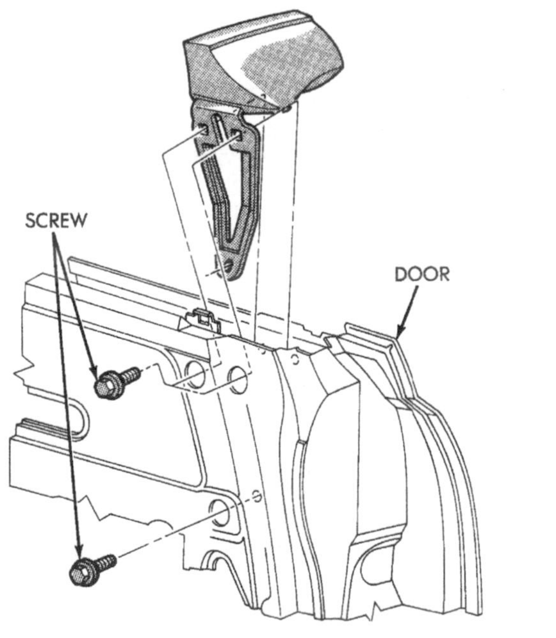

21.2c Sport mirror details

Chapter 12
Chassis electrical system

Contents

1 General information

The electrical system is a 12-volt, negative ground type. Power for the lights and all electrical accessories is supplied by a lead/acid-type battery which is charged by the alternator.

This Chapter covers repair and service procedures for the various electrical components not associated with the engine. Information on the battery, alternator, distributor and starter motor can be found in Chapter 5.

It should be noted that when portions of the electrical system are serviced, the negative cable should be disconnected from the battery to prevent electrical shorts and/or fires.

2 Electrical troubleshooting - general information

A typical electrical circuit consists of an electrical component, any switches, relays, motors, fuses, fusible links or circuit breakers related to that component and the wiring and connectors that link the component to both the battery and the chassis. To help you pinpoint an electrical circuit problem, wiring diagrams are included at the end of this book.

Before tackling any troublesome electrical circuit, first study the appropriate wiring diagrams to get a complete understanding of what makes up that individual circuit. Trouble spots, for instance, can often be narrowed down by noting if other components related to the circuit are operating properly. If several components or circuits fail at one time, chances are the problem is in a fuse or ground connection, because several circuits are often routed through the same fuse and ground connections.

Electrical problems usually stem from simple causes, such as loose or corroded connections, a blown fuse, a melted fusible link or a bad relay. Visually inspect the condition of all fuses, wires and connections in a problem circuit before troubleshooting it.

If testing instruments are going to be utilized, use the diagrams to plan ahead of time where you will make the necessary connections in order to accurately pinpoint the trouble spot.

The basic tools needed for electrical troubleshooting include a circuit tester or voltmeter (a 12-volt bulb with a set of test leads can also be used), a continuity tester, which includes a bulb, battery and set of test leads, and a jumper wire, preferably with a circuit breaker incorporated, which can be used to bypass electrical components. Before attempting to locate a problem with test instruments, use the wiring diagram(s) to decide where to make the connections.

Voltage checks

Voltage checks should be performed if a circuit is not functioning properly. Connect one lead of a circuit tester to either the negative battery terminal or a known good ground. Connect the other lead to a connector in the circuit being tested, preferably nearest to the battery or fuse. If the bulb of the tester lights, voltage is present, which means that the part of the circuit between the connector and the battery is problem free. Continue checking the rest of the circuit in the same fashion. When you reach a point at which no voltage is present, the problem lies between that point and the last test point with voltage. Most of the time the problem can be traced to a loose connection. **Note:** *Keep in mind that some circuits receive voltage only when the ignition key is in the Accessory or Run position.*

Finding a short

One method of finding shorts in a circuit is to remove the fuse and connect a test light or voltmeter in its place to the fuse terminals. There should be no voltage present in the circuit. Move the wiring harness from side-to-side while watching the test light. If the bulb goes on, there is a short to ground somewhere in that area, probably where the insulation has rubbed through. The same test can be performed on each component in the circuit, even a switch.

Ground check

Perform a ground test to check whether a component is properly grounded. Disconnect the battery and connect one lead of a self-powered test light, known as a continuity tester, to a known good ground. Connect the other lead to the wire or ground connection being tested. If the bulb goes on, the ground is good. If the bulb does not go on, the ground is not good.

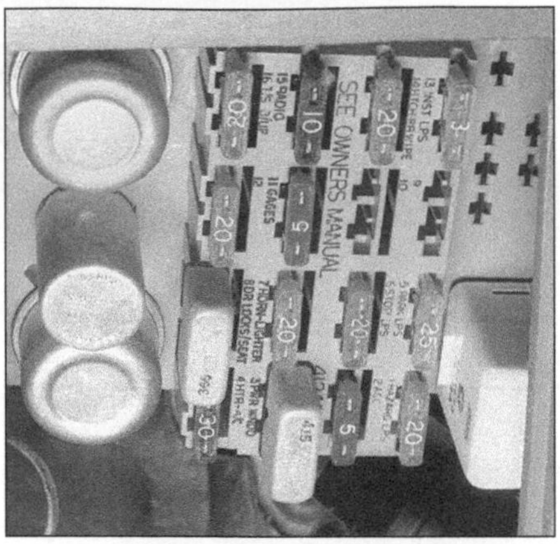

3.1 The fuse block is located on the driver's side of the instrument panel under a cover on this later model and under the dash on earlier models

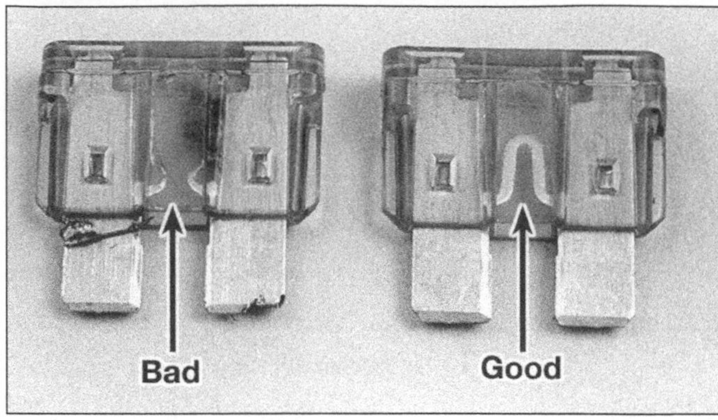

3.3 The fuses can be checked visually to determine if they are blown (later model fuse shown)

Continuity check

A continuity check is done to determine if there are any breaks in a circuit - if it is passing electricity properly. With the circuit off (no power in the circuit), a self-powered continuity tester can be used to check the circuit. Connect the test leads to both ends of the circuit (or to the "power" end and a good ground), and if the test light comes on the circuit is passing current properly. If the light doesn't come on, there is a break somewhere in the circuit. The same procedure can be used to test a switch, by connecting the continuity tester to the switch terminals. With the switch turned On, the test light should come on.

Finding an open circuit

When diagnosing for possible open circuits, it is often difficult to locate them by sight because oxidation or terminal misalignment are hidden by the connectors. Merely wiggling a connector on a sensor or in the wiring harness may correct the open circuit condition. Remember this when an open circuit is indicated when troubleshooting a circuit. Intermittent problems may also be caused by oxidized or loose connections.

Electrical troubleshooting is simple if you keep in mind that all electrical circuits are basically electricity running from the battery, through the wires, switches, relays, fuses and fusible links to each electrical component (light bulb, motor, etc.) and to ground, from which it is passed back to the battery. Any electrical problem is an interruption in the flow of electricity to and from the battery.

3 Fuses - general information

Refer to illustrations 3.1 and 3.3

The electrical circuits of the vehicle are protected by a combination of fuses, circuit breakers and fusible links. The fuse block is located under the left side of instrument panel **(see illustration)**.

Each of the fuses is designed to protect a specific circuit, and the various circuits are identified on the fuse panel itself.

On early models the fuses are tubular glass body types and on later models plastic miniaturized fuses are used. If an electrical component fails, always check the fuse first. A blown fuse is easily identified through the glass tube or clear plastic body. Visually inspect the element for evidence of damage **(see illustration)**. If the fuse looks good, but you still suspect it's faulty, perform a continuity check at the metal ends on glass tube fuses or the blade terminal tips exposed in the fuse body.

Be sure to replace blown fuses with the correct type. Fuses of different ratings are physically interchangeable, but only fuses of the proper rating should be used. Replacing a fuse with one of a higher or lower value than specified is not recommended. Each electrical circuit needs a specific amount of protection. The amperage value of each fuse is stamped on the metal end cap (glass-type) or molded into the fuse body (miniature-type).

If the replacement fuse immediately fails, don't replace it again until the cause of the problem is isolated and corrected. In most cases, the cause will be a short circuit in the wiring caused by a broken or deteriorated wire.

4 Fusible links - general information

Refer to illustrations 4.2 and 4.3

Some circuits are protected by fusible links. The links are used in circuits which are not ordinarily fused, such as the ignition circuit.

Although the fusible links appear to be a heavier gauge than the wire they are protecting, the appearance is due to the thick insulation. All fusible links are several wire gauges smaller than the wire they are designed to protect. The fusible links are color coded and a link should always be replaced with one of the same color obtained from a dealer parts department or auto parts store **(see illustration)**.

FUSIBLE LINK CHART

Wire Gauge	Color Code	Color
12 Ga.	BK	Black
14 Ga.	RD	Red
16 Ga.	DB	Dark Blue
18 Ga.	GY	Gray
20 Ga.	OR	Orange
30 Ga.	LG	Light Green

4.2 Always replace a burned out fusible link with one of the same color code

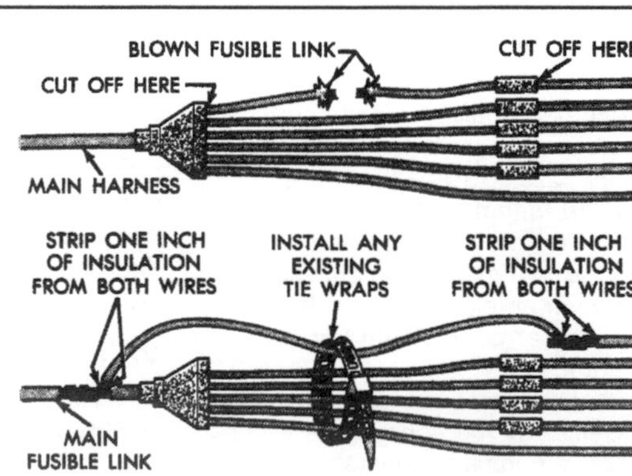

4.3 Fusible link repair details

8.3 Remove only the retainer screws (arrows) - do not disturb the headlight aim adjusting screws unless you're adjusting the headlights

Fusible links cannot be repaired, but a new link of the same size wire can be put in its place. The procedure is as follows:

a) Disconnect the negative cable from the battery.

b) Disconnect the fusible link from the wiring harness.

c) Cut the damaged fusible link out of the wiring just behind the connector.

d) Strip the insulation back approximately 1-inch **(see illustration)**.

e) Position the connector on the new fusible link and twist or crimp it into place.

f) Use rosin core solder at each end of the new link to obtain a good solder joint.

g) Use plenty of electrical tape around the soldered joint. No wires should be exposed.

h) Connect the battery ground cable. Test the circuit for proper operation.

5 Circuit breakers - general information

Circuit breakers protect components such as power windows, power door locks and headlights. Some circuit breakers are located in the fuse box.

On some models the circuit breaker resets itself automatically, so an electrical overload in a circuit breaker protected system will cause the circuit to fail momentarily, then come back on. If the circuit does not come back on, check it immediately. Once the condition is corrected, the circuit breaker will resume its normal function.

6 Relays - general information

Several electrical accessories in the vehicle use relays to transmit the electrical signal to the component. If the relay is defective, that component will not operate properly.

9.1a The vertical adjusting screw (arrow) is located at the top of the headlight

The various relays are grouped together in the fuse block on most models, as well as in other locations.

If a faulty relay is suspected, it can be removed and tested by a dealer service department or a repair shop. Defective relays must be replaced as a unit.

7 Turn signal and hazard flashers - check and replacement

Turn signal flasher

1 The turn signal flasher, a small canister-shaped unit located in the fuse block, flashes the turn signals.

2 When the flasher unit is functioning properly, an audible click can be heard during its operation. If the turn signals fail on one side or the other and the flasher unit does not make its characteristic clicking sound, a faulty turn signal bulb is indicated.

3 If both turn signals fail to blink, the problem may be due to a blown fuse, a faulty flasher unit, a broken switch or a loose or open connection. If a quick check of the fuse box indicates that the turn signal fuse has blown, check the wiring for a short before installing a new fuse.

4 To remove the flasher, simply pull it out of the fuse block or wiring harness.

5 Make sure that the replacement unit is identical to the original. Compare the old one to the new one before installing it.

6 Installation is the reverse of removal.

Hazard flasher

7 The hazard flasher, a small canister-shaped unit located in the fuse block, flashes all four turn signals simultaneously when activated.

8 The hazard flasher is checked in a fashion similar to the turn signal flasher (see Steps 2 and 3).

9 To replace the hazard flasher, pull it from the back of fuse block.

10 Make sure the replacement unit is identical to the one it replaces. Compare the old

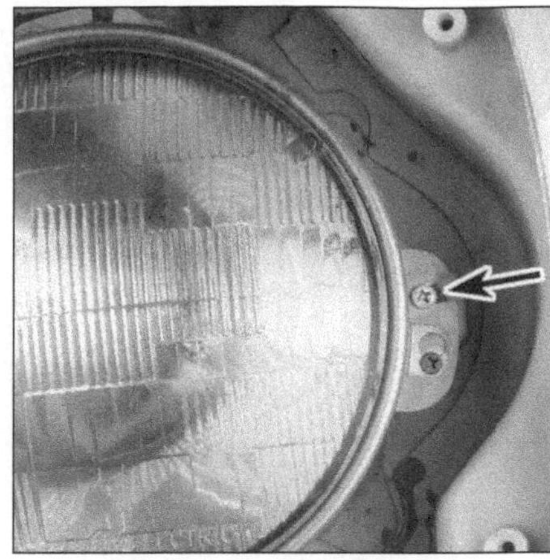

9.1b The horizontal adjusting screw (arrow) is located on the side of the headlight

one to the new one before installing it.

11 Installation is the reverse of removal.

8 Headlights - removal and installation

Refer to illustration 8.3

1 Disconnect the negative cable from the battery.

2 Remove the retaining screws and detach the headlight bezel.

3 Remove the retainer screws, taking care not to disturb the adjusting screws **(see illustration)**.

4 Remove the retainer and pull the headlight out enough to allow the connector to be unplugged.

5 Remove the headlight.

6 To install the headlight, plug the connector in, place the headlight in position and install the retainer and screws. Tighten the screws securely.

7 Place the bezel in position and install the retaining screws.

9 Headlights - adjustment

Refer to illustrations 9.1a and 9.1b

Note: *The headlights must be aimed correctly. If adjusted incorrectly they could blind the driver of an oncoming vehicle and cause a serious accident or seriously reduce your ability to see the road. The headlights should be checked for proper aim every 12 months and any time a new headlight is installed or front end body work is performed. It should be emphasized that the following procedure is only an interim step which will provide temporary adjustment until the headlights can be adjusted by a properly equipped shop.*

1 Headlights have two spring loaded adjusting screws, one on the top controlling up-and-down movement and one on the side controlling left-and-right movement **(see illustrations)**.

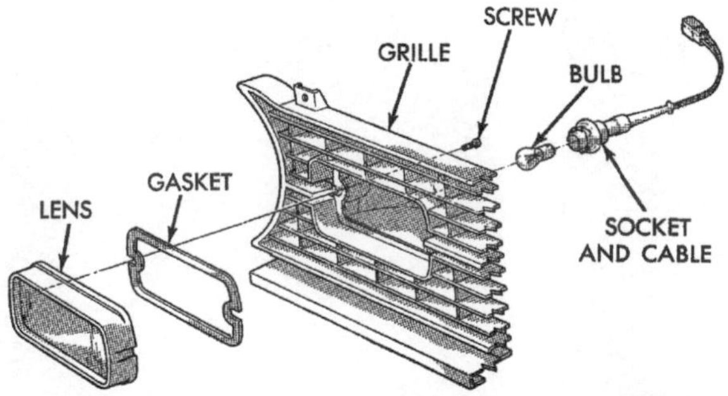

10.1a On early models, the turn signal bulb is accessible after removing the screws and detaching the lens

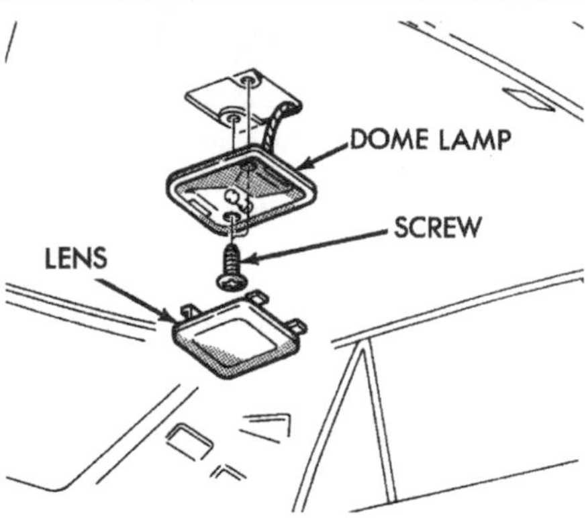

10.2 After prying out the lens, the dome lamp bulb can be pulled from the socket

2 There are several methods of adjusting the headlights. The simplest method requires a blank wall 25 feet in front of the vehicle and a level floor.

3 Position masking tape vertically on the wall in reference to the vehicle centerline and the centerlines of both headlights.

4 Position a horizontal tape line in reference to the centerline of all the headlights. **Note:** *It may be easier to position the tape on the wall with the vehicle parked only a few inches away.*

5 Adjustment should be made with the vehicle sitting level, the gas tank half-full and no unusually heavy load in the vehicle.

10.3b Push the bulb in and rotate it counterclockwise, then remove it from the holder

10.3a Rotate the parking light bulb holder counterclockwise and withdraw it from the housing

6 Starting with the low beam adjustment, position the high intensity zone so it is two inches below the horizontal line and two inches to the right of the headlight vertical line. Adjustment is made by turning the top adjusting screw clockwise to raise the beam and counterclockwise to lower the beam. The adjusting screw on the side should be used in the same manner to move the beam left or right.

7 With the high beams on, the high intensity zone should be vertically centered with the exact center just below the horizontal line. **Note:** *It may not be possible to position the headlight aim exactly for both high and low beams. If a compromise must be made, keep in mind that the low beams are the most used and have the greatest effect on driver safety.*

8 Have the headlights adjusted by a dealer service department or service station at the earliest opportunity.

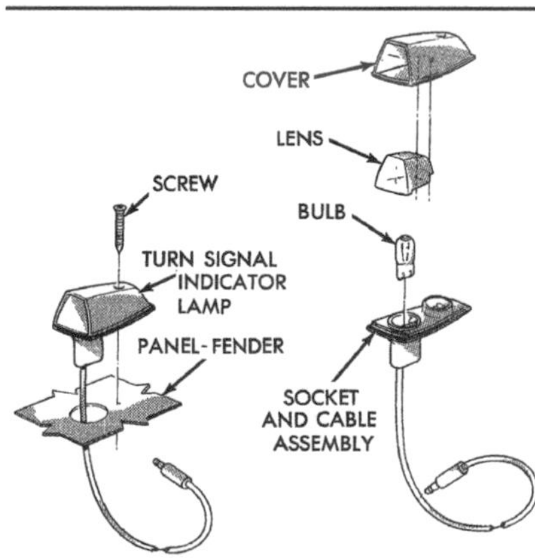

10.3c After removing the screw and detaching the turn signal lamp cover and lens, pull this type of bulb straight out of the socket

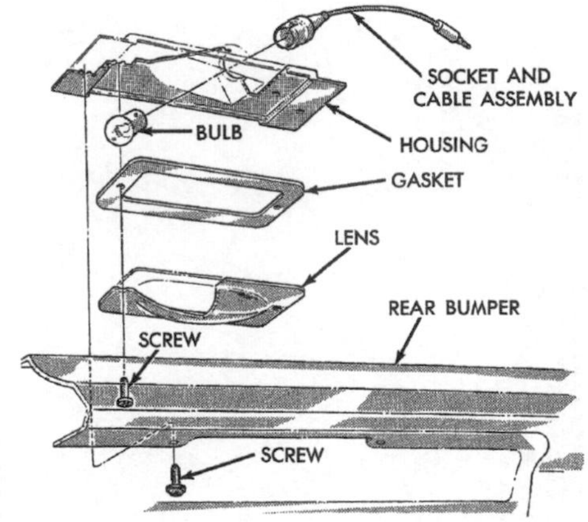

10.1b After removing the license plate light screws and lens assembly, the bulb can be removed

10 Bulb replacement

Refer to illustrations 10.1a, 10.1b, 10.2, 10.3a, 10.3b, 10.3c and 10.4

1 The lenses of many lights are held in place by screws, which makes it a simple procedure to gain access to the bulbs **(see illustrations).**

2 On some lights the lenses are held in place by clips. The lenses can be removed either by unsnapping them or by using a small screwdriver to pry them off **(see illustration).**

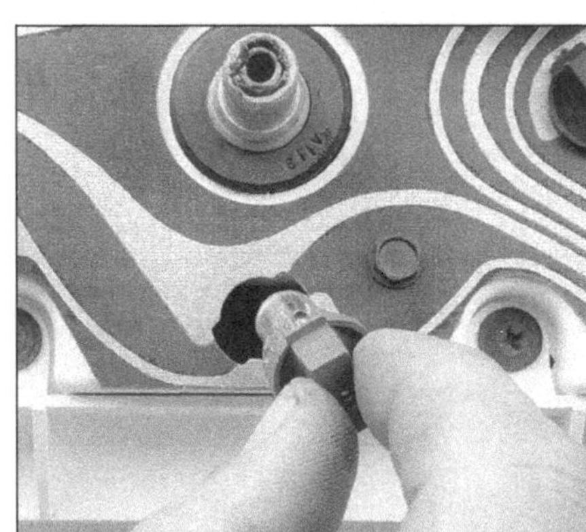

10.4 After removing the instrument cluster, turn the bulb holders counterclockwise, lift them out and pull out the bulbs

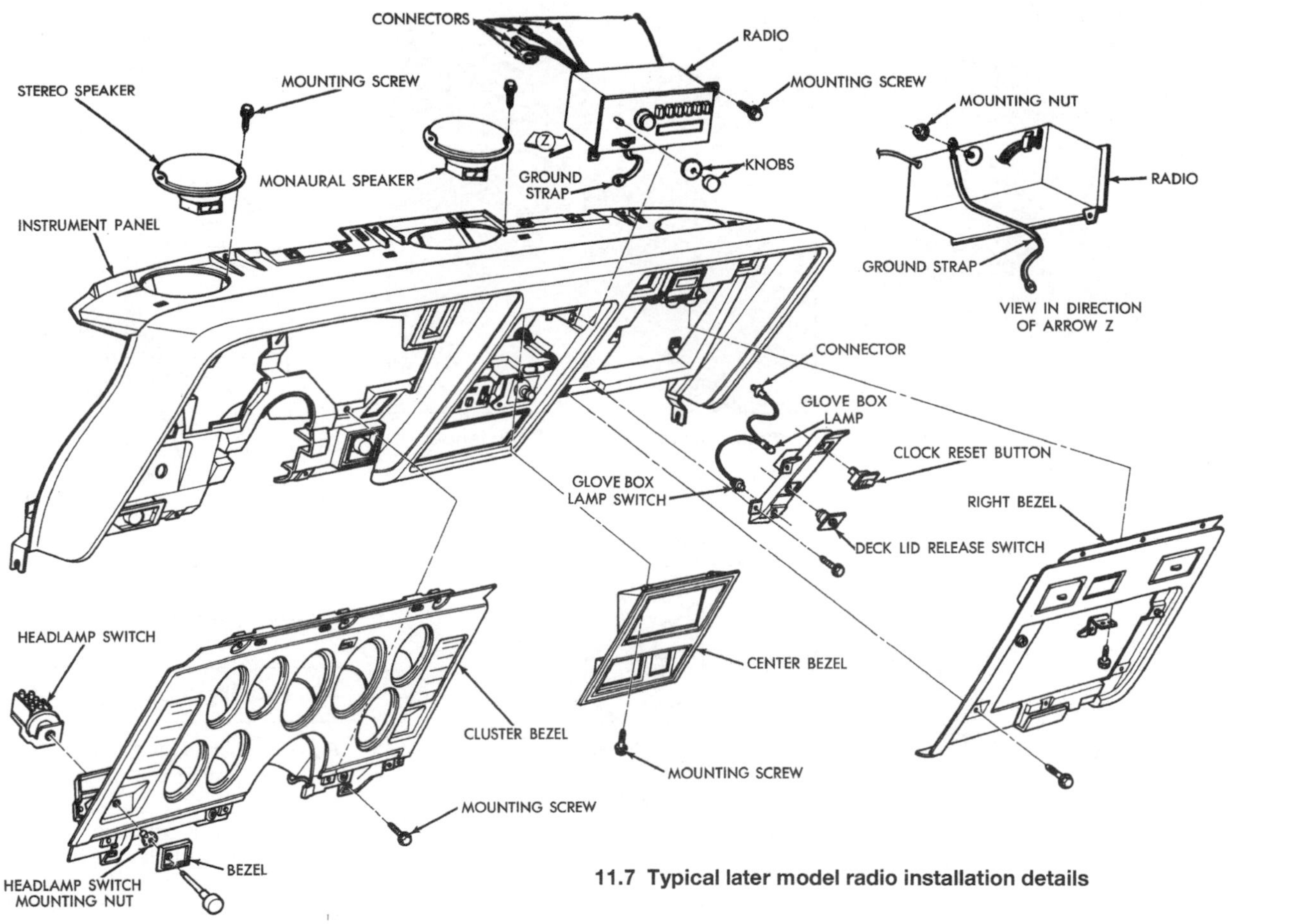

11.7 Typical later model radio installation details

3 Several types of bulbs are used. Some are removed by pushing in and turning them counterclockwise **(see illustration)**. Others can simply be unclipped from the terminals or pulled straight out of the socket **(see illustrations)**.

4 To gain access to the instrument panel lights **(see illustration)**, the instrument cluster will have to be removed first (see Section 18).

11 Radio - removal and installation

Refer to illustration 11.7

1 Disconnect the negative cable at the battery.

Early models

2 On some earlier models, the radio must be removed from below the instrument panel.

3 Pull off the control knobs and remove the retaining nuts retaining the radio to the instrument panel. On some models it may be necessary to pull out the ashtray for access and remove the bracket-to-radio screws.

4 Reach up under the instrument panel and unplug the antenna, power and speaker electrical connectors.

5 Push the radio back into the opening and lower it from under the dash.

6 Installation is the reverse of removal.

Later models

7 Remove the center bezel **(see illustration)**.

8 Remove the radio mounting screws.

9 Slide the radio out and support it, disconnect the ground cable, antenna and electrical connectors, then remove the assembly from the instrument panel.

10 Installation is the reverse of removal.

12 Antenna - removal and installation

Antenna mast

Refer to illustration 12.2

1 The antenna mast can be unscrewed and replaced with a new one in the event it is damaged, following the procedure below. Follow the procedure beginning with Step 4 to replace the antenna and cable as an assembly.

2 Use a small wrench to unscrew the mast **(see illustration)**.

3 Install the new antenna mast finger tight and tighten it securely with the wrench.

Antenna and cable

Refer to illustrations 12.8a and 12.8b

4 Disconnect the negative cable at the battery.

5 On earlier models reach around behind the radio and unplug the antenna lead. On later models it will probably be necessary to remove the radio and unplug the antenna.

6 Remove the fender splash shield (see Chapter 11).

7 Working through the fender opening, detach the antenna cable from the clips and push the grommet through the firewall. Connect a long piece of string or thin wire to the antenna cable.

12.2 Use a small open-end wrench to unscrew the antenna

12.8a Unscrew the cap nut with needle-nose pliers - use care so the pliers don't slip and scratch the fender

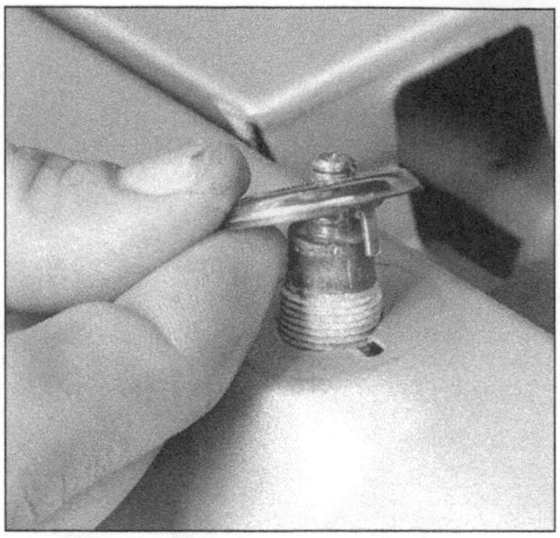

12.8b Lift the adapter off the fender

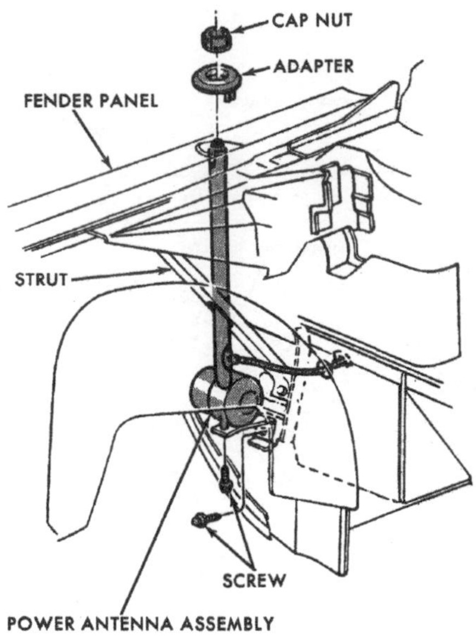

12.18 Power antenna mounting details

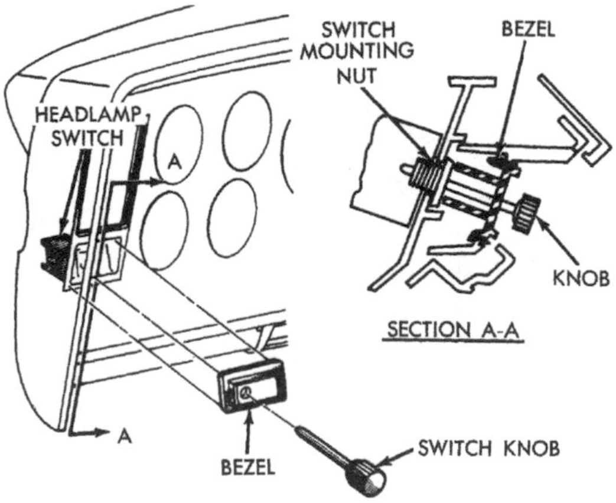

13.2 Early model headlight switch details

8 Remove the antenna cap nut, using needle-nose pliers, then lift off the adapter **(see illustrations)**. Be very careful - the pliers could slip and scratch the fender. It's a good idea to surround the base of the antenna with rags to prevent against scratching.

9 Install the antenna mast temporarily and push the antenna down far enough so that you can grasp the cable end through the fender opening.

10 Remove the antenna mast, then pull the cable through the firewall and into the fenderwell.

12 Push the new cable end up through the fender, then install the adapter, cap nut and antenna mast.

13 Connect the string or wire to the radio end of the new cable and pull it through into the passenger compartment.

14 Connect the cable to the radio, secure it with the clips, then push the grommet into place in the firewall.

15 Install the fender splash shield and connect the battery negative cable.

Power antenna motor

Refer to illustration 12.18

16 Remove the antenna cap nut as described in Step 8.

17 Turn the front wheels all the way to the right for access, then remove the front fender splash shield (see Chapter 11).

18 Disconnect the electrical and antenna lead connectors. Remove the screws/nuts and detach the antenna motor **(see illustration)**.

19 Push the new antenna motor up into place and install the screws/nuts and the cap nut.

20 Plug in the connectors and install the fender splash shield.

13 Headlight switch - replacement

Refer to illustrations 13.2 and 13.7

1 Disconnect the negative cable at the battery.

2 On early models, the switch is removed from under the instrument panel. On some models it may be necessary to remove the fuse block screw and move the block out of the way for access to the switch. Press the release button on the bottom and withdraw the switch knob and shaft. Remove the retaining nut and lower the switch from the back of the instrument panel **(see illustration)**.

3 On later models, if necessary, remove the instrument cluster bezel (see Chapter 11).

4 Remove the switch screws, pull the switch and bracket assembly out, then unplug the electrical connector.

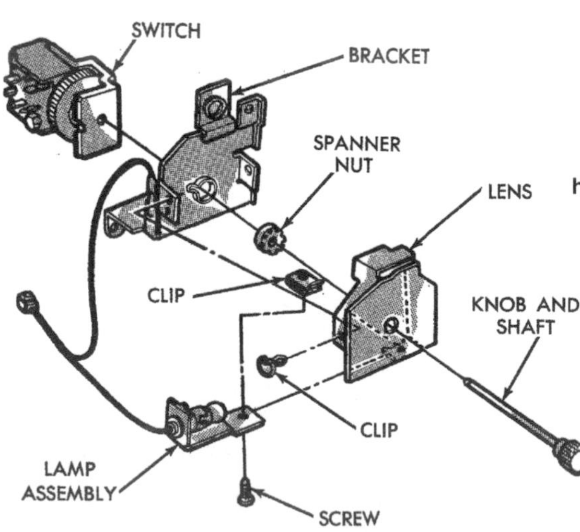

13.7 Later model headlight switch details

14.1 Remove the two bolts and detach the connector (arrows), then remove the headlight dimmer switch

5 Remove the nut retaining the bracket to the switch (some models).
6 Press the release button on the bottom of the switch and withdraw the switch knob and shaft.
7 Detach the switch lens or escutcheon, then remove the spanner nut and separate the switch from the bracket **(see illustration)**.
8 Place the new switch on the bracket and secure it with the spanner nut. Insert the switch knob shaft into the switch. The remainder of installation is the reverse of removal.

14 Headlight dimmer switch - check and replacement

Refer to illustrations 14.1, 14.2, 14.6 and 14.9

Floor-mounted switch

Check

1 Pull back the carpet, remove the bolts, then unplug the connector and detach the switch **(see illustration)**.
2 Use an ohmmeter to check the continuity between the switch terminals. Connect one lead to the center terminal and the other lead to the low beam terminal. There should be continuity with the switch in either position **(see illustration)**. Repeat the test with the high beam terminal - it should only have continuity in one position. Replace the switch if it fails the test.

Replacement

3 Plug the electrical connector into the new switch, place the switch in position and install the bolts. Tighten the bolts securely.

Steering column-mounted switch

Check

4 Remove the screws and detach the steering column cover.

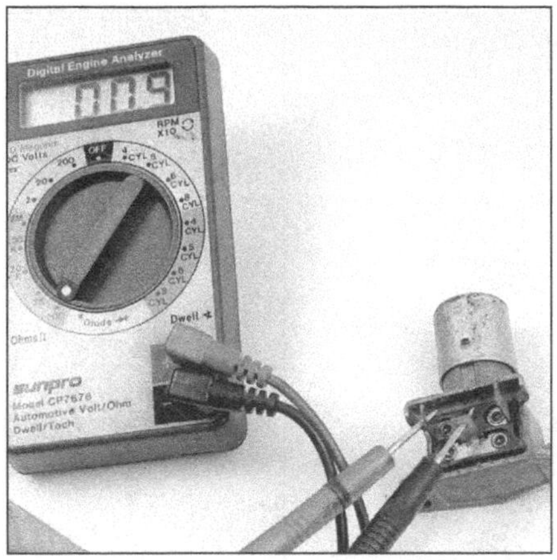

14.2 Checking a floor-mounted dimmer switch with an ohmmeter

5 Unplug the electrical connector from the switch.
6 Use an ohmmeter to check the continuity between the switch terminals. There should be continuity between the B and H terminals with the switch in the high beam position and the B and L terminals in the low beam position **(see illustration)**. If there isn't, replace the switch.

Replacement

7 Remove the two bolts and lower the switch from the steering column.
8 Place the new switch in position and install the mounting bolts finger tight.
9 Insert an adjustment pin fabricated from a piece of wire. Push the switch to the rear to take up the slack in the actuator control rod, then tighten the bolts securely **(see illustration)**.
10 Remove the pin, plug in the connector and install the steering column cover.

15 Turn signal/hazard switch - replacement

Refer to illustrations 15.9 and 15.11

1 The turn signal switch is located at the top end of the steering column and is operated by the multi-function control stalk. The hazard warning switch is mounted under the turn signal switch on most later models.
2 Disconnect the negative cable from the battery.
3 Remove the screws and detach steering column lower cover.
4 Unplug the turn signal switch electrical connector.
5 Remove the steering wheel (see Chapter 10).
6 Detach the wiring harness trough from the steering column (if equipped).
7 Unplug the electrical connector.

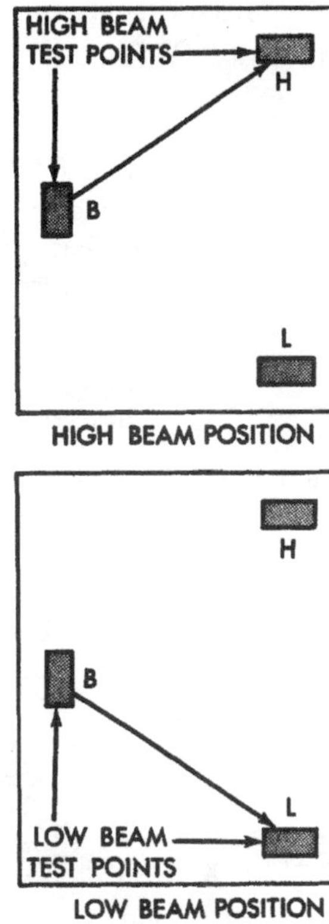

14.6 Dimmer switch continuity check points

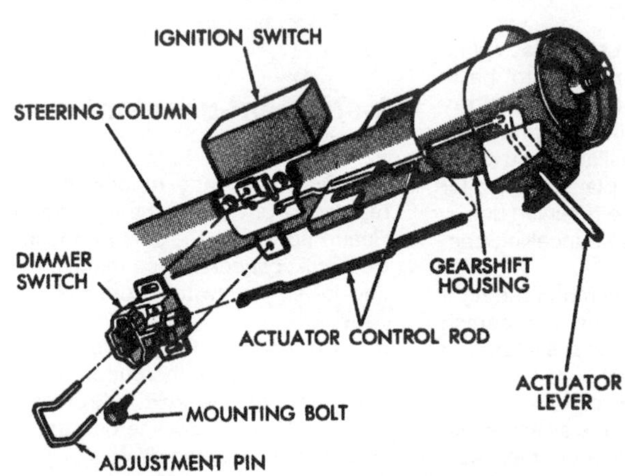

14.9 Dimmer switch adjustment details - you can fabricate your own adjustment pin with a paper clip

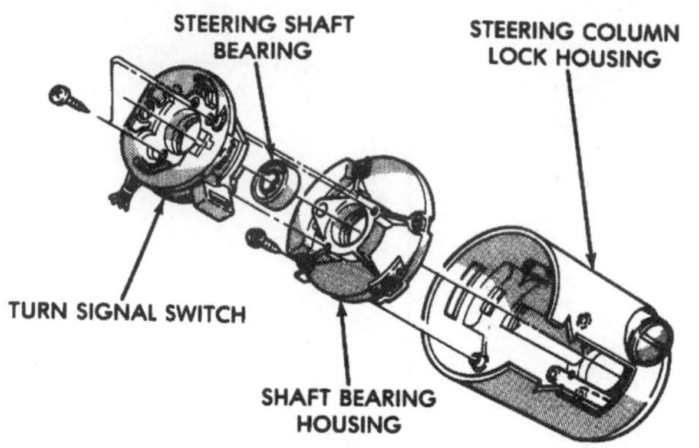

15.9 Turn signal switch mounting details (standard steering column)

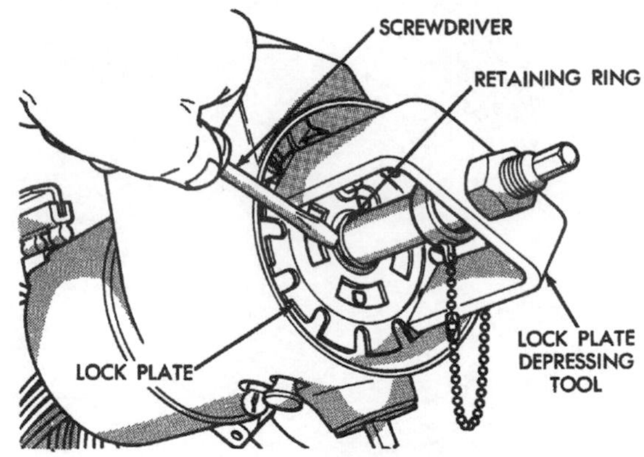

15.11 The retaining ring on tilt steering models can be removed with a screwdriver after depressing the lock plate

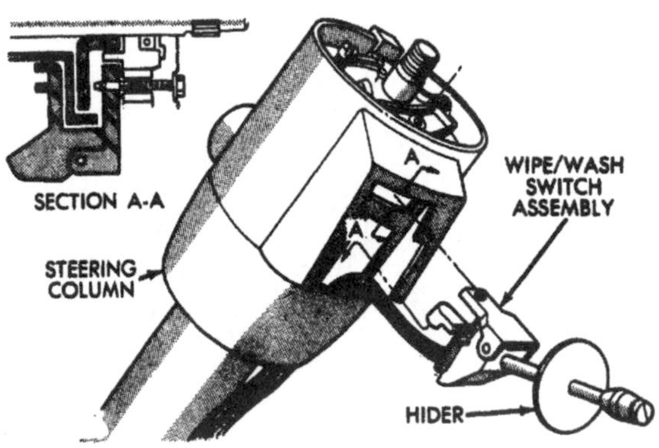

16.11 Typical wiper/washer switch removal details

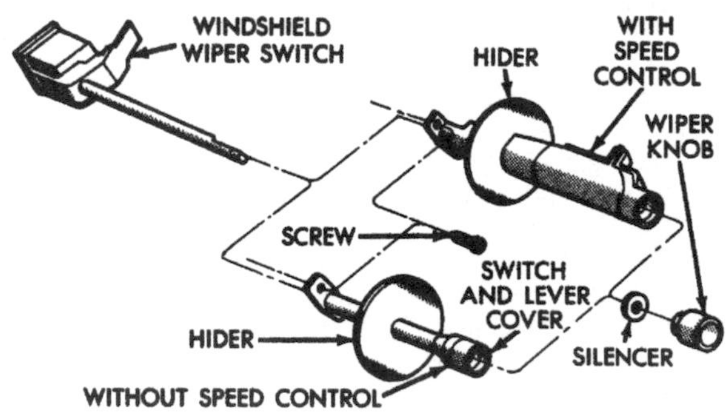

16.12 Wiper/washer stalk details

Standard steering column

8 Remove the wiper/washer switch pivot screw, leaving the control stalk in place.

9 Remove the three screws and detach the turn signal switch **(see illustration)**. Carefully pull the wiring harness out through the top of the steering column.

10 Installation is the reverse of removal, remembering to lubricate the steering shaft contact surfaces with multi-purpose grease and to make sure the dimmer switch rod is located securely in the pocket of the control stalk.

Tilt steering column

11 Depress the lock plate using a lock plate depressor (available at most auto parts stores), then use a small screwdriver to pry out the retaining ring **(see illustration)**. If the recommended tool is not available, you might be able to depress the lock plate by hand far enough so you can get at the retaining ring.

12 Remove the lock plate, cancelling cam and upper bearing spring.

13 Place the turn signal switch in the right turn position, remove the screw and detach the link between the turn signal switch and the wiper/washer switch pivot.

14 Unscrew the hazard switch knob.

15 Remove the screws, lift the switches out and guide the wiring harness out of the steering column opening.

16 Installation is the reverse of removal, noting the following: Place the turn signal switch in position in the right turn position, install the screws, then pull the link into position and install the wiper/washer pivot screw. Be sure to install a new lock plate retaining ring.

16 Wiper/washer switch - replacement

Refer to illustrations 16.11, 16.12, 16.13, 16.22 and 16.24

1 Disconnect the negative cable from the battery.

Instrument panel-mounted switch

2 On some early models the switch is removed from under the instrument panel. Reach up under the dash and unplug the electrical connector, then remove the screws and lower the switch from the instrument panel.

3 On other models, unplug the electrical connector, depress the button on the switch and withdraw the switch knob and shaft. Remove the nut retaining nut and lower the switch from under the dash.

4 Installation is the reverse of removal.

Column-mounted switch

5 These models are equipped with a multi-function lever located on the left side of the steering column which controls the wiper/washer, turn signal and dimmer switches.

6 Remove the steering wheel (see Chapter 10).

7 Remove screws and detach the steering column lower cover.

8 Detach the wiring harness trough from the steering column (some later models).

9 Unplug the electrical connector.

Standard steering column

10 Remove the lock housing cover screws and pull the cover off.

11 Pull the switch hider out for access **(see illustration)**.

12 Remove the two screws and detach the switch from the stalk, then remove the wiper knob from the end of the stalk **(see illustration)**.

13 Rotate the control stalk fully clockwise, align the slot and pin, then pull the stalk out **(see illustration)**.

14 Installation is the reverse of removal.

Tilt steering column

15 Compress the lock plate and remove the retaining ring (see Step 11 in Section 15).

16 Remove the lock plate, cancelling cam

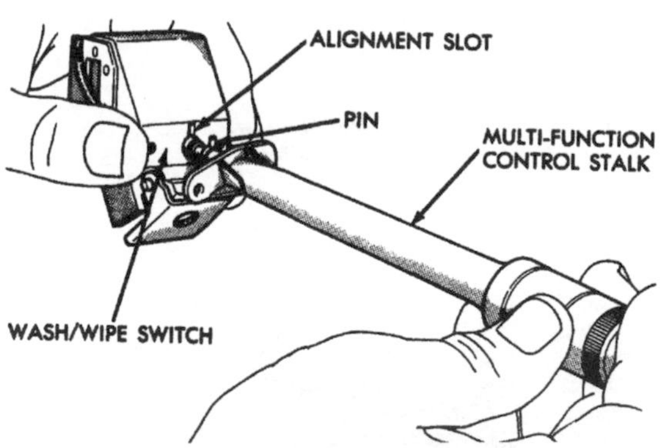

16.13 Line up the pin with the alignment slot and withdraw the multi-function stalk

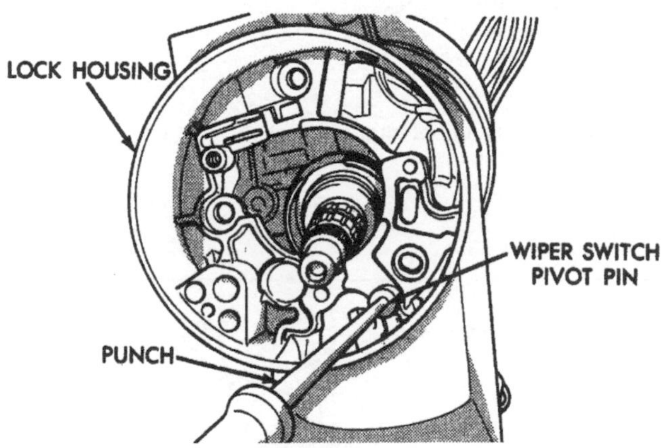

16.24 Use a thin punch or screwdriver to press the pivot pin out

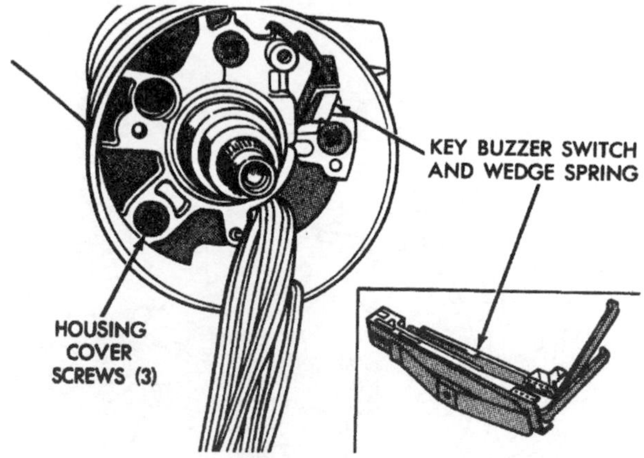

16.22 Hook a piece of wire (a paper clip works well) under the wedge spring and remove the key buzzer - the housing cover is retained by three screws

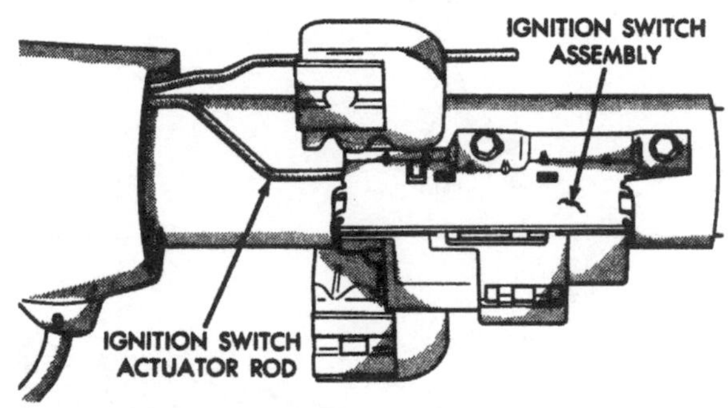

17.1 The ignition switch is located on the right side of the steering column

and upper bearing spring.

17 Remove the multi-function stalk retaining screw.

18 Remove the hazard flasher knob by pushing it in and unscrewing it.

19 Unplug the wiper/washer switch electrical connector and remove the turn signal switch (see Section 15).

20 Remove the ignition key lamp.

21 Remove the lock cylinder (see Section 21).

22 Bend the end of a straightened paper clip into a hook, insert it into the loop of the buzzer switch wedge spring, then lift the switch out **(see illustration)**.

23 Remove the three screws and lift off the column housing cover **(see illustration 16.22)**.

24 Press the wiper switch pivot pin out with a punch **(see illustration)**.

25 Remove the wiper/washer switch **(see illustration 16.11)**.

26 Use tape to secure the dimmer switch rod, pull the switch hider up the control stalk, then remove the two screws **(see illustration 16.12)**.

27 Pull the knob off the end of the multi-function stalk, turn the stalk fully clockwise, align the pin with the slot and pull the stalk straight out **(see illustration 16.13)**.

28 Installation is the reverse of removal.

17 Ignition switch - replacement

Refer to illustration 17.1

1 The ignition switch is located on the steering column and is actuated by a rod attached to the key lock cylinder **(see illustration)**.

2 Disconnect the negative cable from the battery.

3 Insert the key into the lock cylinder and turn it to the Acc position.

4 Remove the steering column cover.

5 Unplug the electrical connector, remove the bolts, then detach the switch from the actuator rod and lower it from the steering column.

6 Installation is the reverse of removal. As the switch is engaged to the actuator rod, push up on the switch to remove any slack from the rod before fully tightening the bolts.

18 Instrument cluster - removal and installation

Refer to illustrations 18.4, 18.6a and 18.6b

1 Disconnect the negative cable from the battery.

2 On some models it will be necessary to remove the bracket bolts and lower the steering column for access.

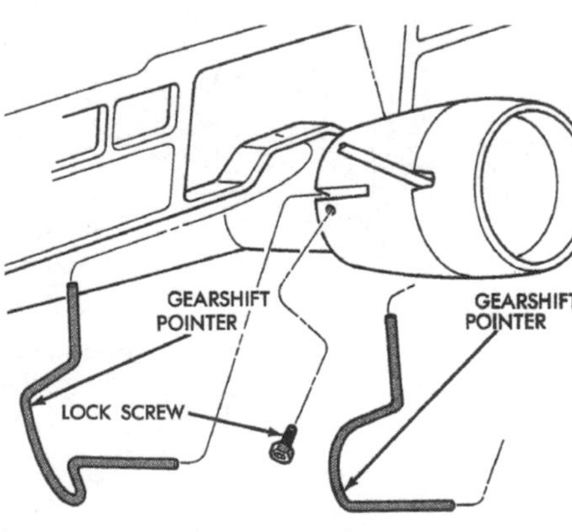

18.4 Typical later model gearshift pointer installation details

3 Remove the instrument cluster bezel (see Chapter 11).

4 On some later automatic transmission models, it will be necessary to remove the lock screw and detach the gearshift pointer **(see illustration)**.

5 Reach behind the cluster and disconnect the speedometer cable.

6 Remove the screws and pull the cluster out for access, then unplug the electrical

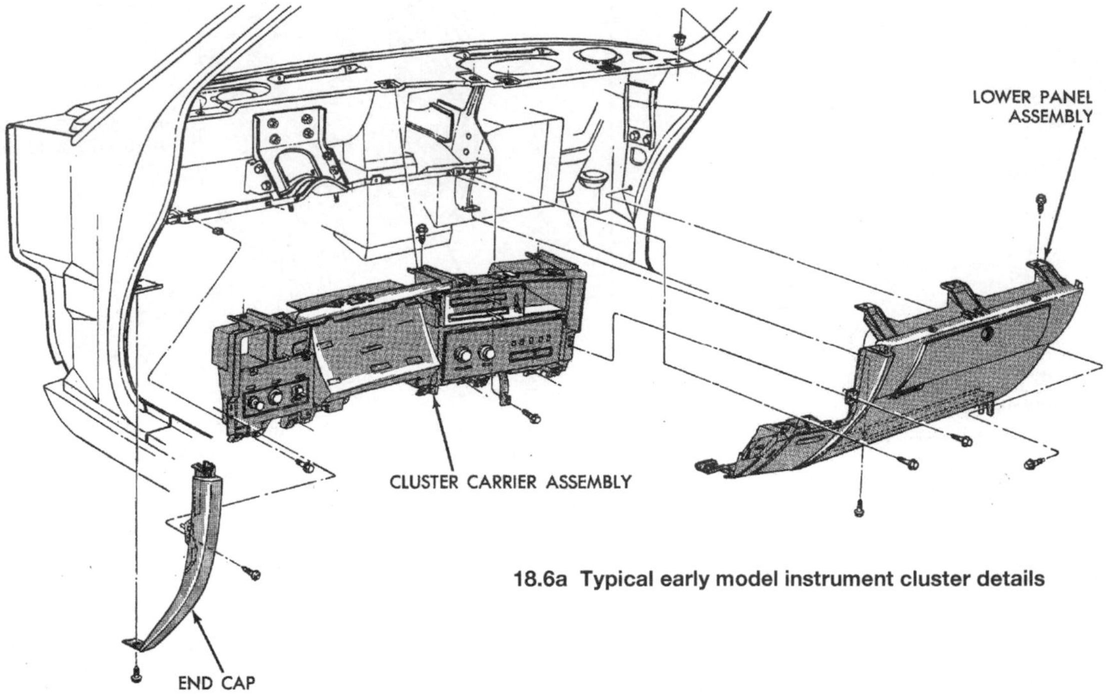

18.6a Typical early model instrument cluster details

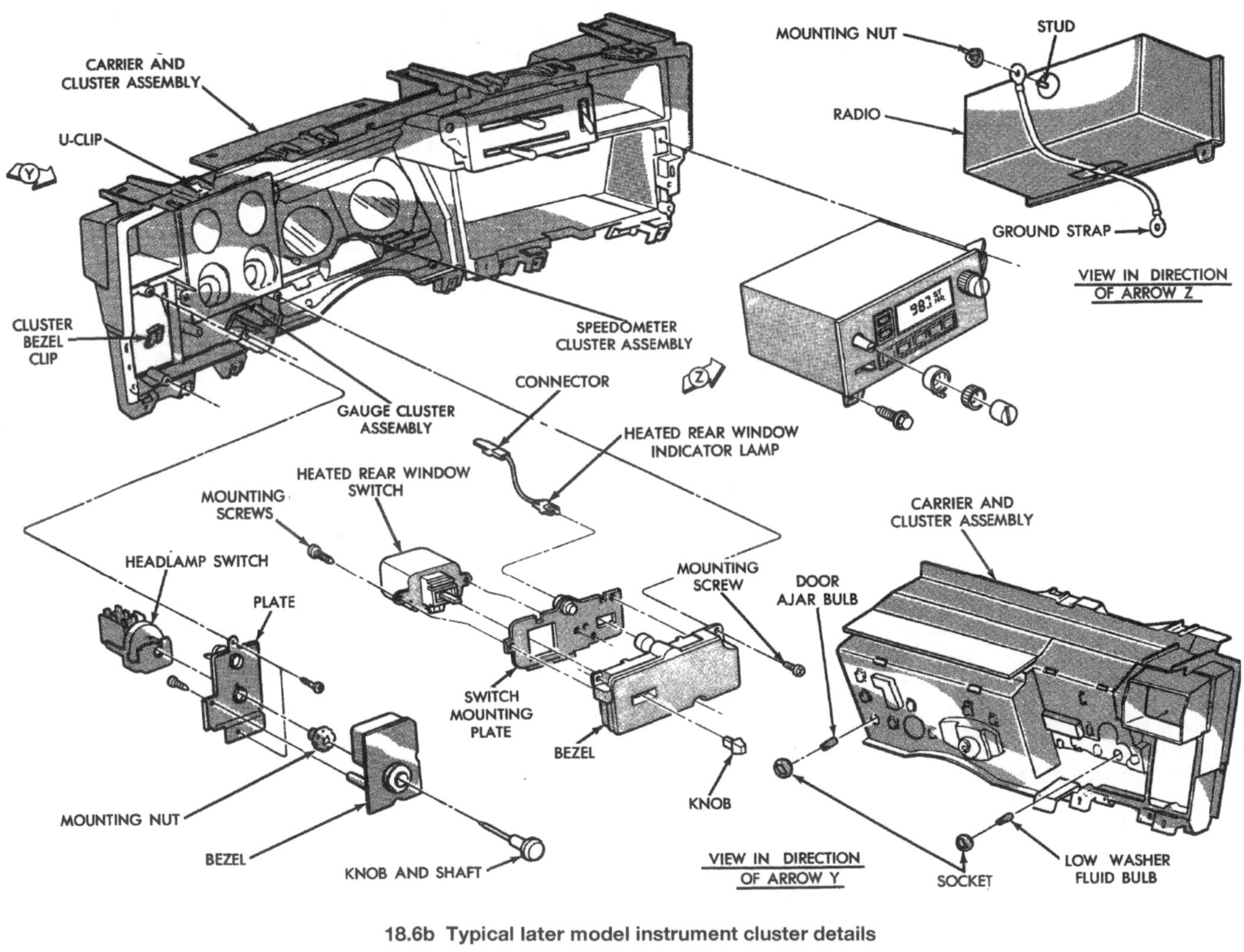

18.6b Typical later model instrument cluster details

19.2a Unscrew the collar and pull the speedometer cable out of the cruise control adapter

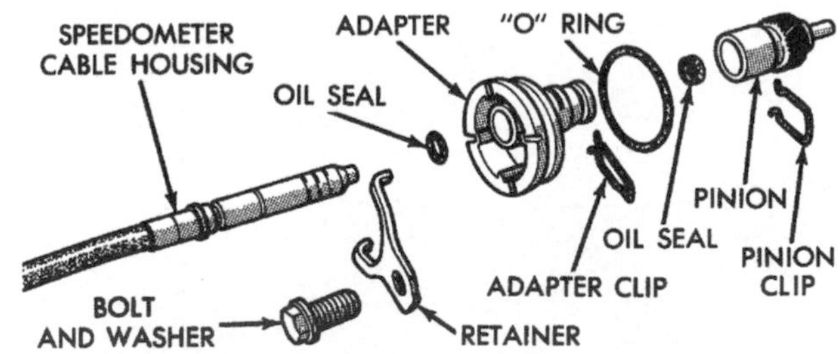

19.2b Remove the bolt and washer, pull the retainer and adapter out of the transmission, then detach the cable

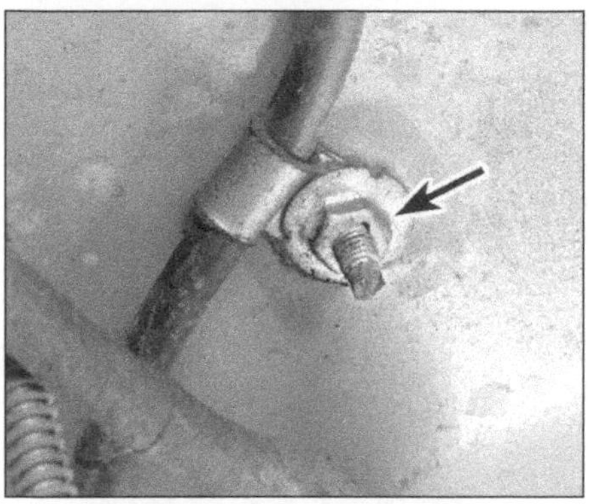

19.3 Remove the nut (arrow) to detach the speedometer cable from the clip

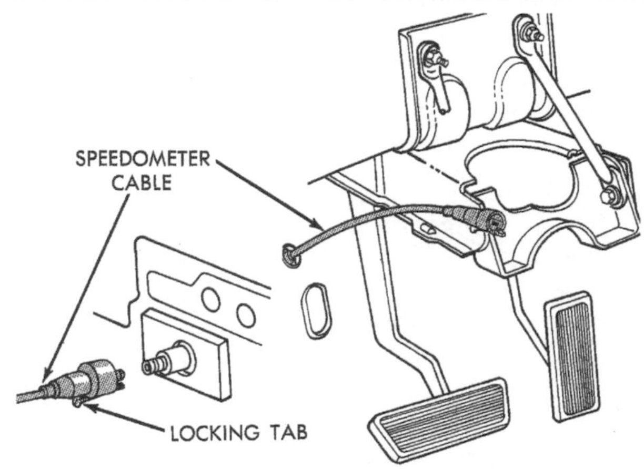

19.4a On some models the cable is secured to the speedometer with a locking tab

connectors and remove the cluster **(see illustrations)**.

7 Installation is the reverse of removal.

19 Speedometer cable - removal and installation

Refer to illustrations 19.2a, 19.2b, 19.3, 19.4a and 19.4b

1 Disconnect the negative cable from the battery.

2 Disconnect the speedometer cable from the transmission or cruise control adapter **(see illustrations)**.

3 Detach the cable from the routing clips in the engine compartment and pull the cable up to provide enough slack to allow disconnection from the speedometer **(see illustration)**.

4 Reach behind the instrument cluster screws and disconnect the speedometer cable from the back of the cluster **(see illustrations)**.

5 Remove the cable from the vehicle.

6 Prior to installation, lubricate the speedometer end of the cable with spray-on speedometer cable lubricant (available at auto parts stores).

7 Installation is the reverse of removal.

20 Windshield wiper motor - check and replacement

Refer to illustrations 20.2 and 20.3

1 Disconnect the negative cable from the battery.

2 Unplug the wiper motor electrical connector and remove the mounting nuts **(see illustration)**.

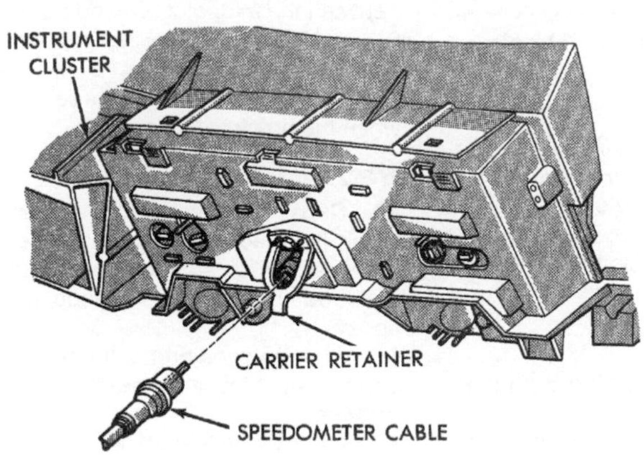

19.4b On other models, a retainer holds the speedometer cable in place

20.2 Unplug the connector (A) and remove the three nuts (B)

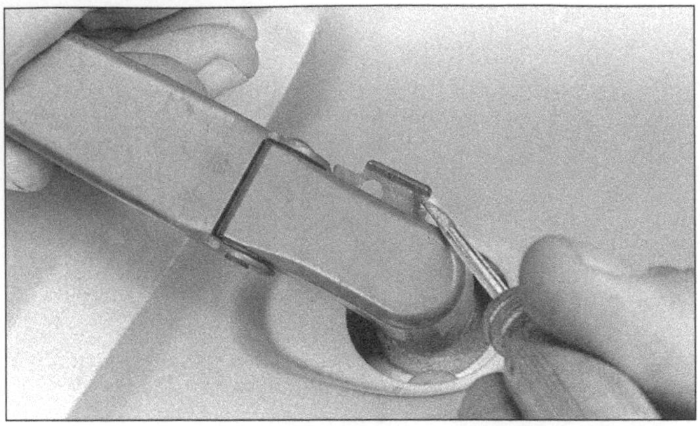

20.3 Pry the release lever out with a screwdriver and lift the wiper arm off

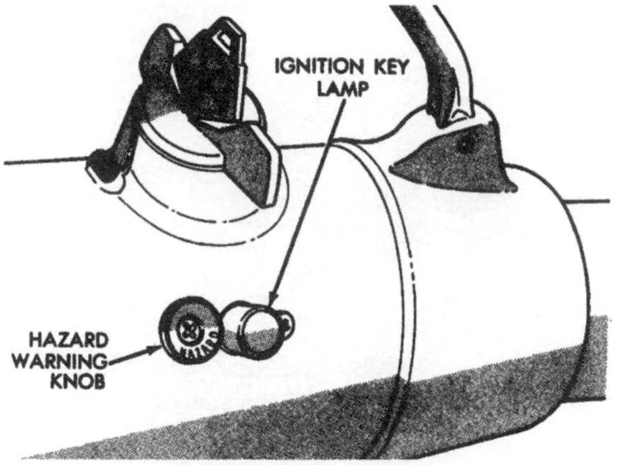

21.3 Typical ignition lamp details

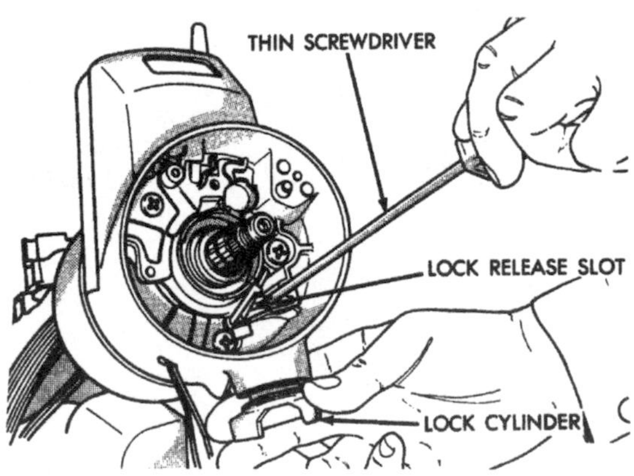

21.5 Insert a thin screwdriver into the lock release slot and press the spring latch while pulling the lock cylinder out

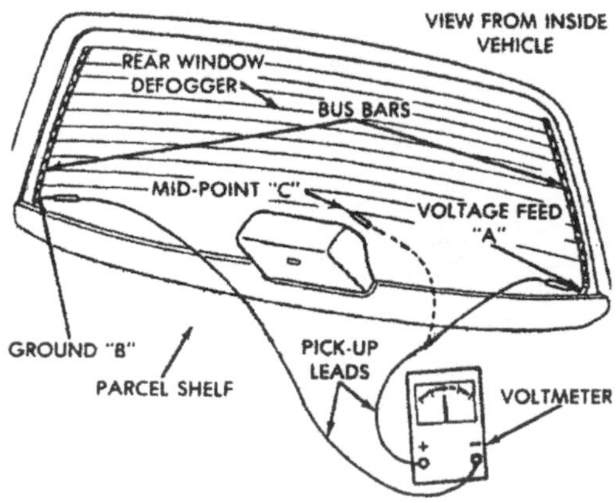

22.4 Rear window defogger test points

3 Remove the wiper arms if necessary and remove the cowl grille **(see illustration)**.

4 Hold the motor drive crank with a wrench, remove the crank nut and detach the crank, then lift the motor out. On some earlier models it may be easier to work from under the instrument panel to detach the crank.

5 Installation is the reverse of removal.

21 Ignition lock cylinder - removal and installation

Refer to illustrations 21.3 and 21.5

1 Disconnect the negative cable from the battery.

2 Remove the turn signal/hazard warning switches (later models) (see Section 15).

3 Remove the ignition key lamp assembly **(see illustration)**.

4 Place the lock cylinder in the Lock position.

5 Insert a thin screwdriver or punch into the lock release slot located next to the switch mounting screw boss, press down on the spring latch and withdraw the cylinder from the steering column **(see illustration)**.

6 Insert the new lock cylinder (without the key) into position until it contacts the ignition switch actuating rod. Move the rod up and down to remove any slack and align the components. Insert the key and push the lock cylinder in. The lock cylinder will snap into place automatically when the components align themselves, locking it in the housing.

7 The remainder of installation is the reverse of removal.

22 Rear window defogger - check and repair

Refer to illustrations 22.4 and 22.13

1 The rear window defogger consists of a number of horizontal elements baked onto the glass surface.

2 Small breaks in the element can be repaired without removing the rear window.

Check

3 Turn the ignition switch and defogger system switches On.

4 Ground the negative lead of the voltmeter to point "B" and the positive lead to point "A" **(see illustration)**.

5 The voltmeter should read between 10 and 15 volts. If the reading is lower, there is a poor ground connection.

6 Contact the negative lead to a good body ground. The reading should stay the same.

7 Connect the negative lead to point "B", then touch each grid line at the midpoint with the positive lead.

8 The reading should be approximately six volts. If the reading is 0, there is a break between the midpoint "C" and point "A".

9 A 10 to 14 volt reading is an indication of a break between midpoint "C". Move the lead toward the break: the voltage will change when the break is crossed.

Repair

10 Repair the break in the element using a repair kit specifically recommended for this purpose, available at most auto parts stores. Included in this kit is plastic conductive epoxy.

11 Prior to repairing a break, turn of the system and allow it to de-energize for a few minutes.

12 Lightly buff the element area with fine steel wool, then clean it thoroughly with rubbing alcohol.

13 Use masking tape to mask off the area

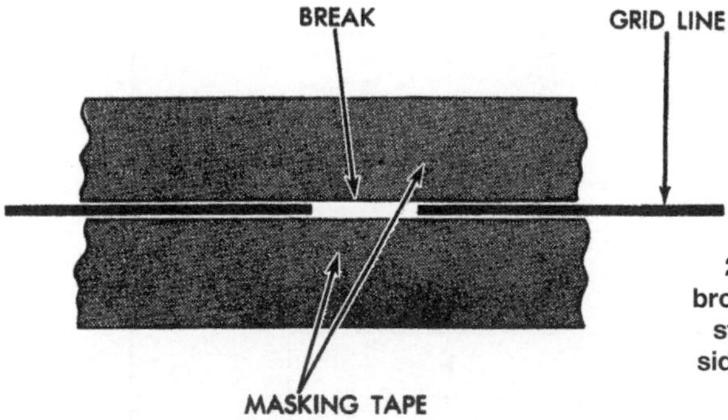

22.13 To repair the broken grid, first apply a strip of tape to either side of the grid to mask off the area

being repaired **(see illustration)**

14 Mix the epoxy thoroughly, following the instructions provided with the repair kit.

15 Apply the epoxy material to the slit in the masking tape, overlapping the undamaged area about 3/4-inch on either end.

16 Allow the repair to cure for 24 hours before removing the tape and using the system.

23 Power door lock system - description and check

The power door lock system operates the door lock actuators mounted in each door. The system consists of the switches, actuators and associated wiring. Diagnosis can usually be limited to simple checks of the wiring connections and actuators for minor faults which can be easily repaired. These include:

a) *Check the system fuse and/or circuit breaker.*
b) *Check the switch wires for damage and loose connections. Check the switches for continuity.*
c) *Remove the door panel(s) and check the actuator wiring connections to see if they're loose or damaged. Inspect the actuator rods (if equipped) to make sure they aren't bent or damaged. Inspect the actuator wiring for damaged or loose connections. The actuator can be checked by applying battery power momentarily. A discernible click indicates that the solenoid is operating properly.*

24 Power window system - description and check

The power window system operates the electric motors mounted in the doors which lower and raise the windows. The system consists of the control switches, the motors (regulators), glass mechanisms and associated wiring.

Diagnosis can usually be limited to simple checks of the wiring connections and motors for minor faults which can be easily repaired. These include:

a) *Inspect the power window actuating switches for broken wires and loose connections.*
b) *Check the power window fuse/and or circuit breaker.*
c) *Remove the door panel(s) and check the power window motor wires to see if they're loose or damaged. Inspect the glass mechanisms for damage which could cause binding.*

25 Cruise control system - description and check

The cruise control system maintains vehicle speed with a vacuum actuated servo motor located in the engine compartment, which is connected to the throttle linkage by a cable. The system consists of the servo motor, clutch switch, brake switch, control switches, a relay and associated vacuum hoses.

Because of the complexity of the cruise control system and the special tools and techniques required for diagnosis, repair should be left to a dealer service department or a repair shop. However, it is possible for the home mechanic to make simple checks of the wiring and vacuum connections for minor faults which can be easily repaired. These include:

a) *Inspect the cruise control actuating switches for broken wires and loose connections.*
b) *Check the cruise control fuse.*
c) *The cruise control system is operated by vacuum so it's critical that all vacuum switches, hoses and connections are secure. Check the hoses in the engine compartment for tight connections, cracks and obvious vacuum leaks.*

26 Wiring diagrams - general information

Refer to illustration 26.4

Since it isn't possible to include all wiring diagrams for every year and model covered by this manual, the following diagrams are those that are typical and most commonly needed.

Prior to troubleshooting any circuits, check the fuse and circuit breakers (if equipped) to make sure they're in good condition. Make sure the battery is properly charged and check the cable connections (see Chapter 1).

When checking a circuit, make sure that all connectors are clean, with no broken or loose terminals. When unplugging a connector, do not pull on the wires. Pull only on the connector housings themselves.

Refer to the accompanying chart **(see illustration)** for the wire color codes applica-

WIRE COLOR CODE CHART

COLOR CODE	COLOR	STANDARD TRACER COLOR	COLOR CODE	COLOR	STANDARD TRACER CODE
BK	BLACK	WT	PK	PINK	BK OR WH
BR	BROWN	WT	RD	RED	WT
DB	DARK BLUE	WT	TN	TAN	WT
DG	DARK GREEN	WT	VT	VIOLET	WT
GY	GRAY	BK	WT	WHITE	BK
LB	LIGHT BLUE	BK	YL	YELLOW	BK
LG	LIGHT GREEN	BK	*	WITH TRACER	
OR	ORANGE	BK			

26.4 Wiring diagram color codes

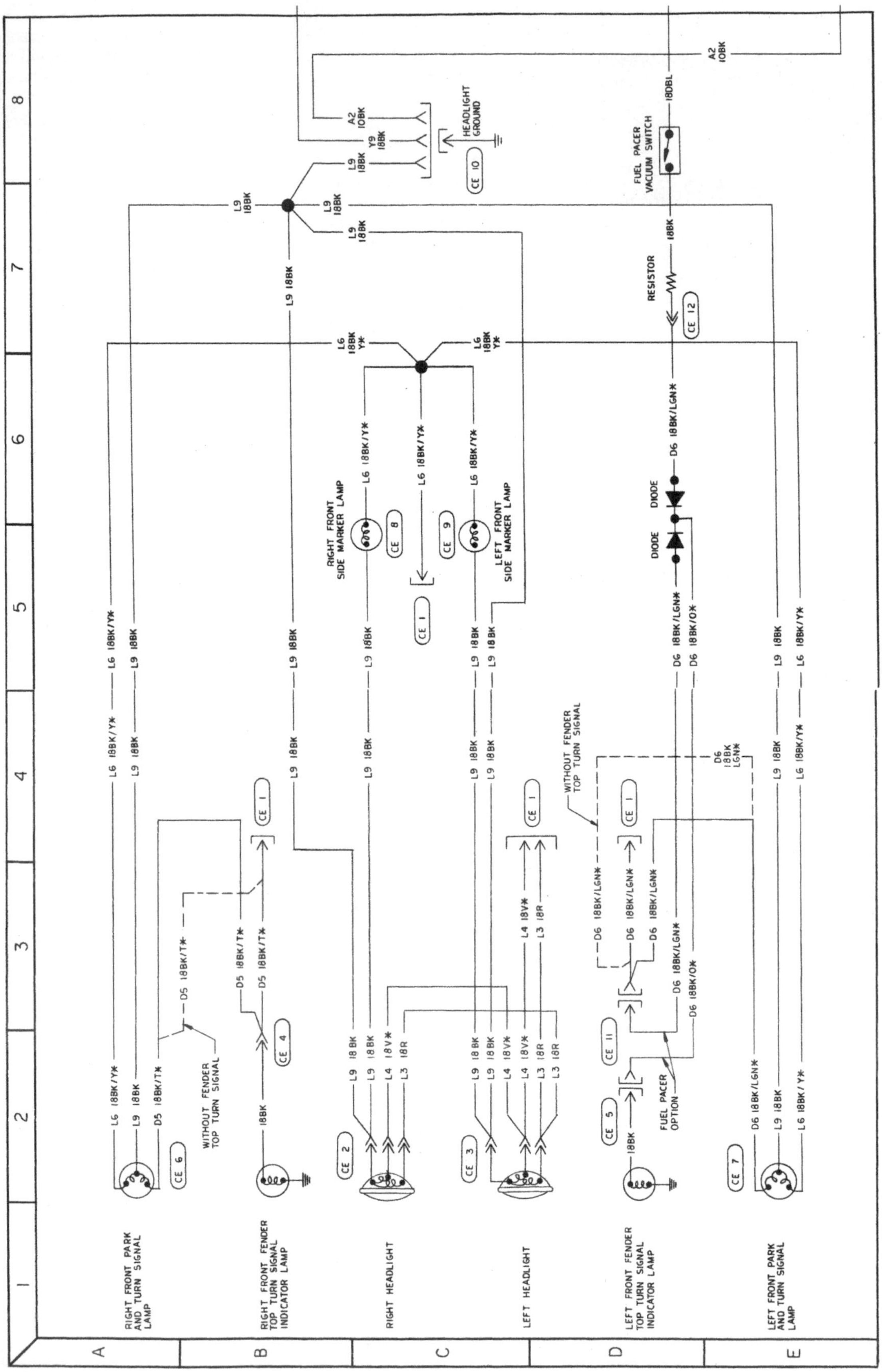

Typical A-body front lighting wiring diagram (E-body similar)

ble to your vehicle.

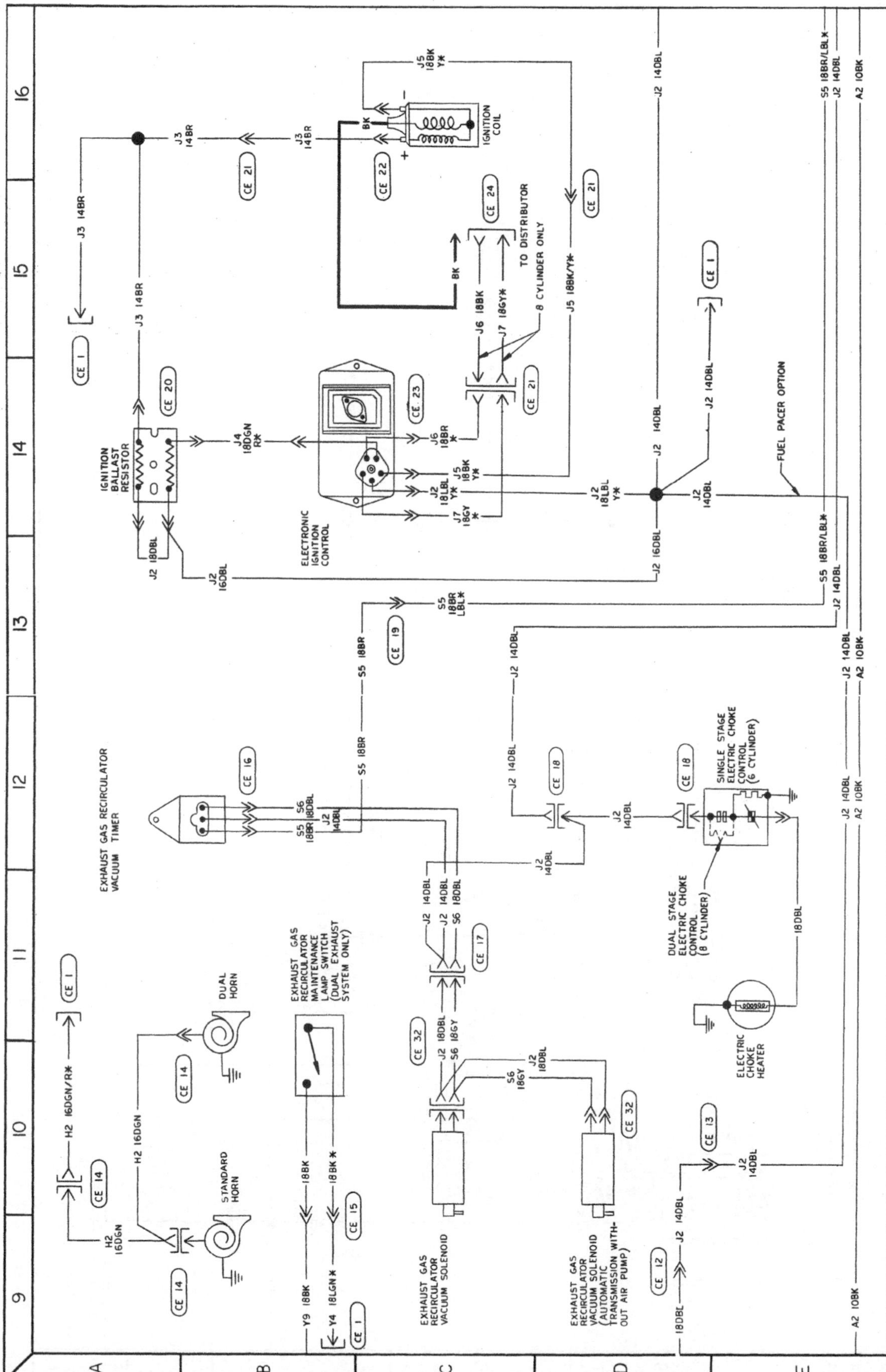

Typical A-body ignition system wiring diagram (E-body similar)

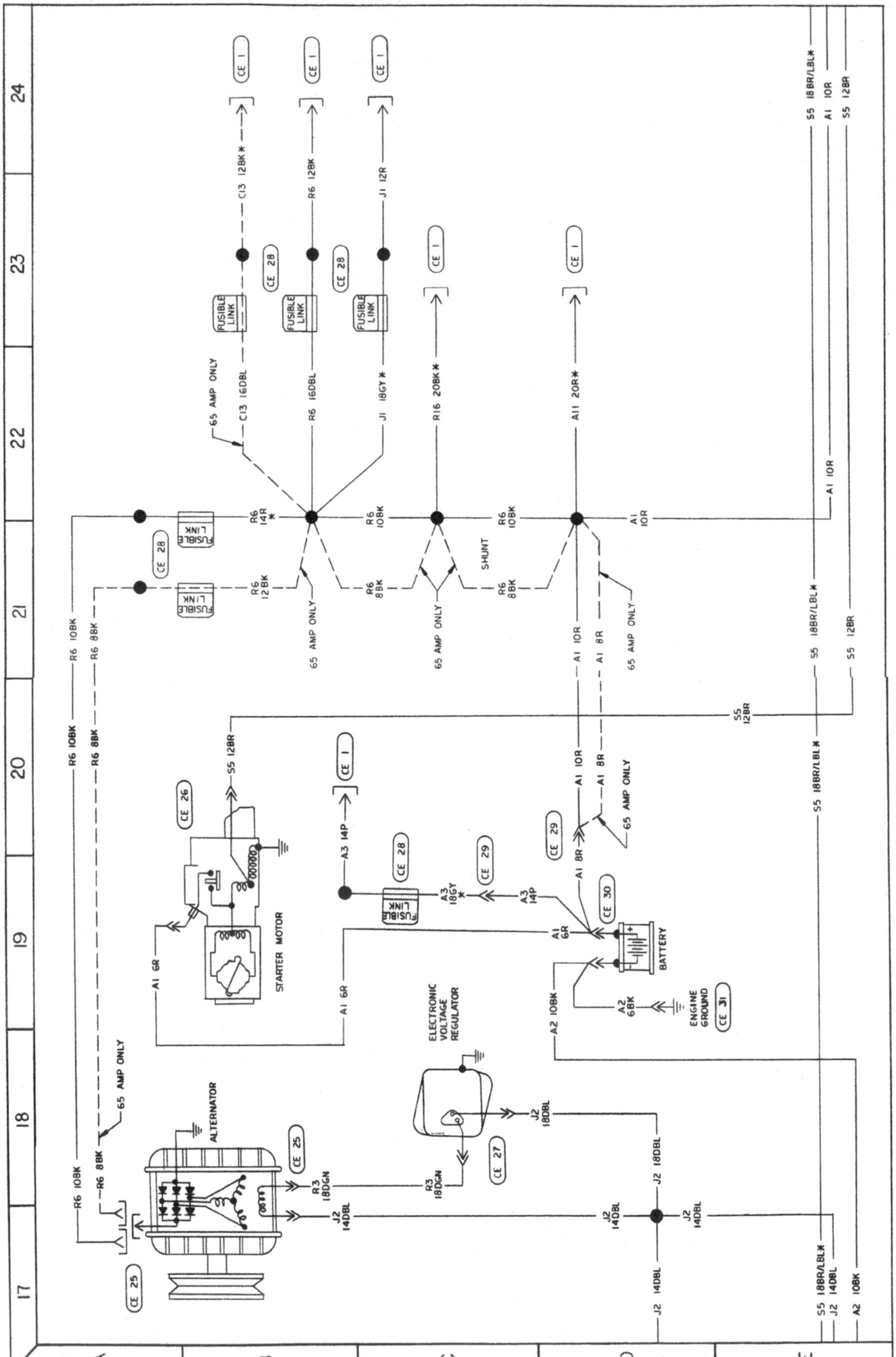

Typical A-body charging and starting system wiring diagram (E-body similar) (1 of 2)

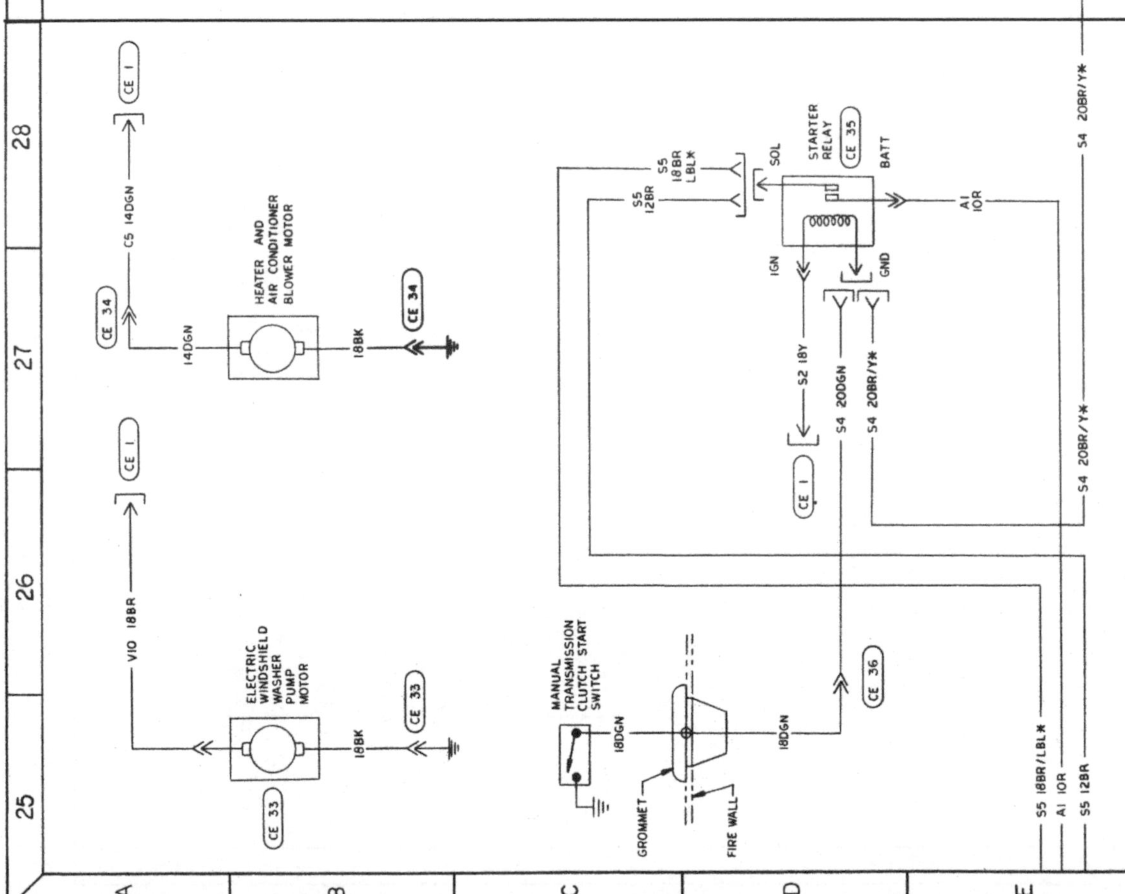

Typical A-body charging and starting system wiring diagram (E-body similar) (2 of 2)

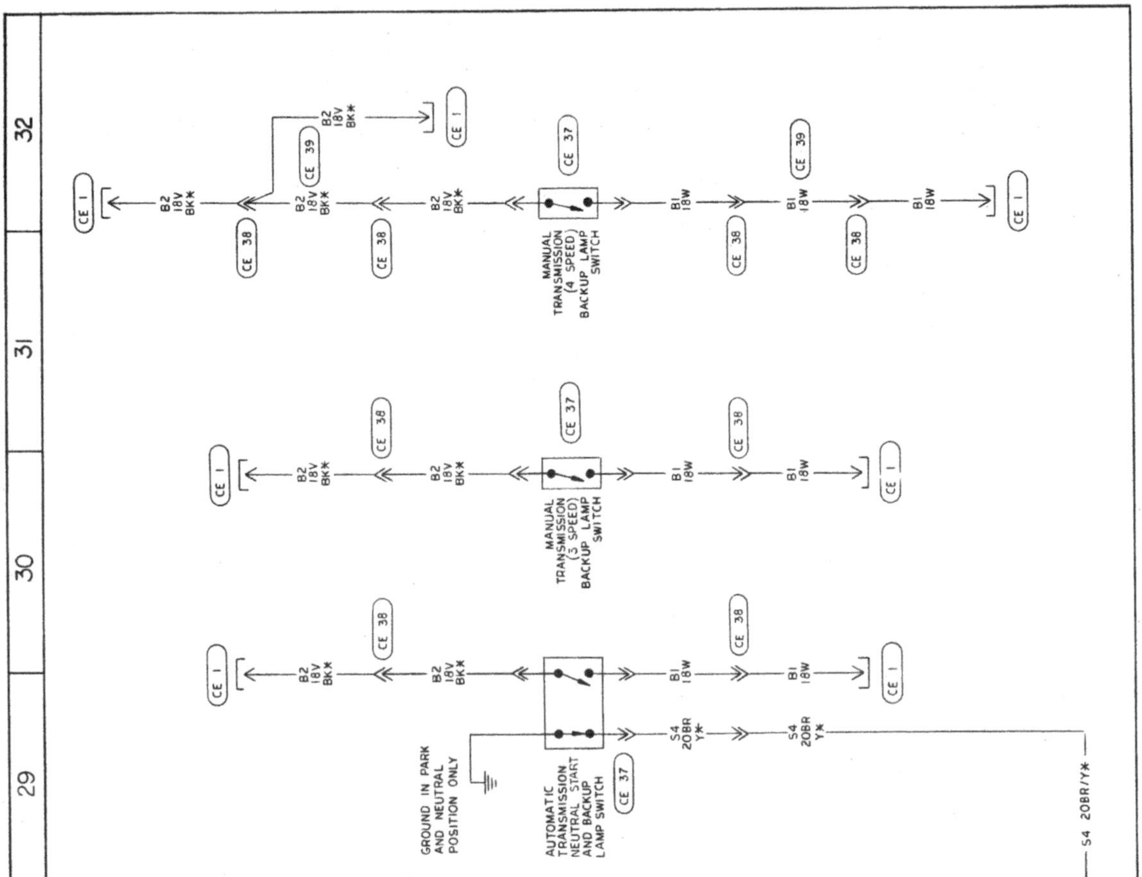

Typical A-body backup light wiring diagram (E-body similar)

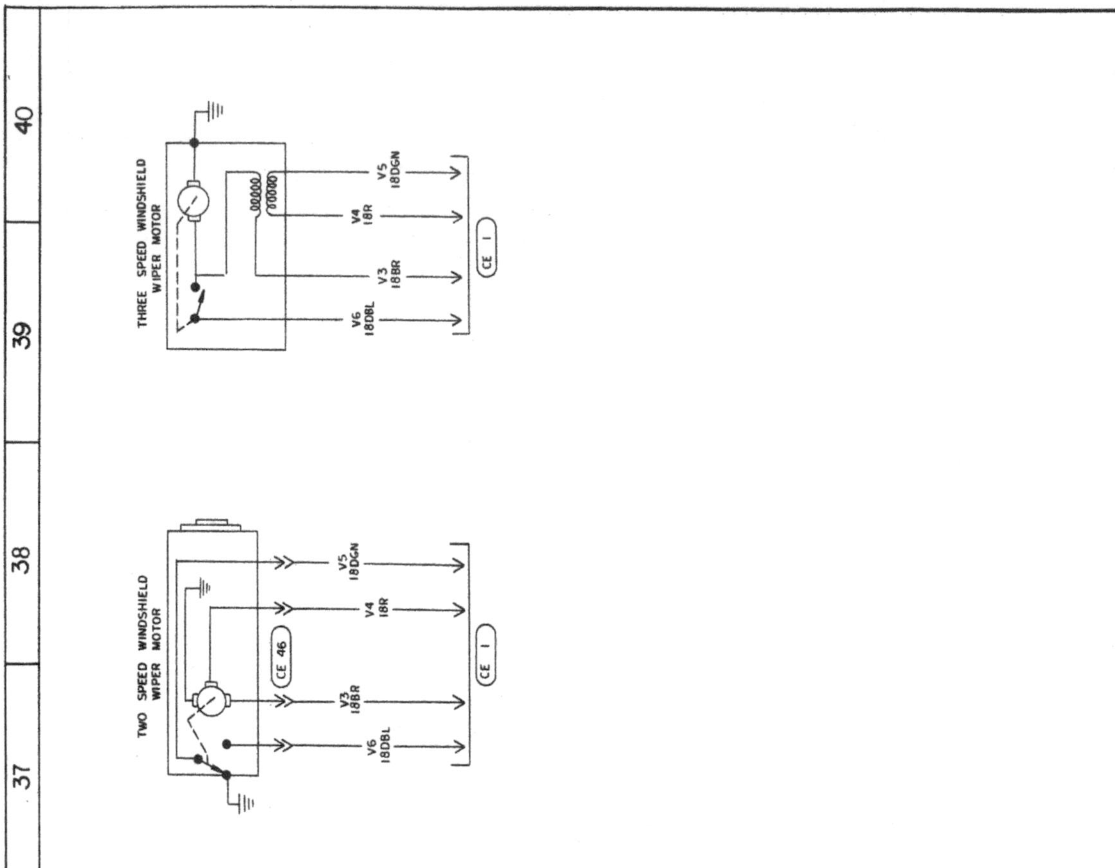

Typical A-body wiper switch wiring diagram (E-body similar)

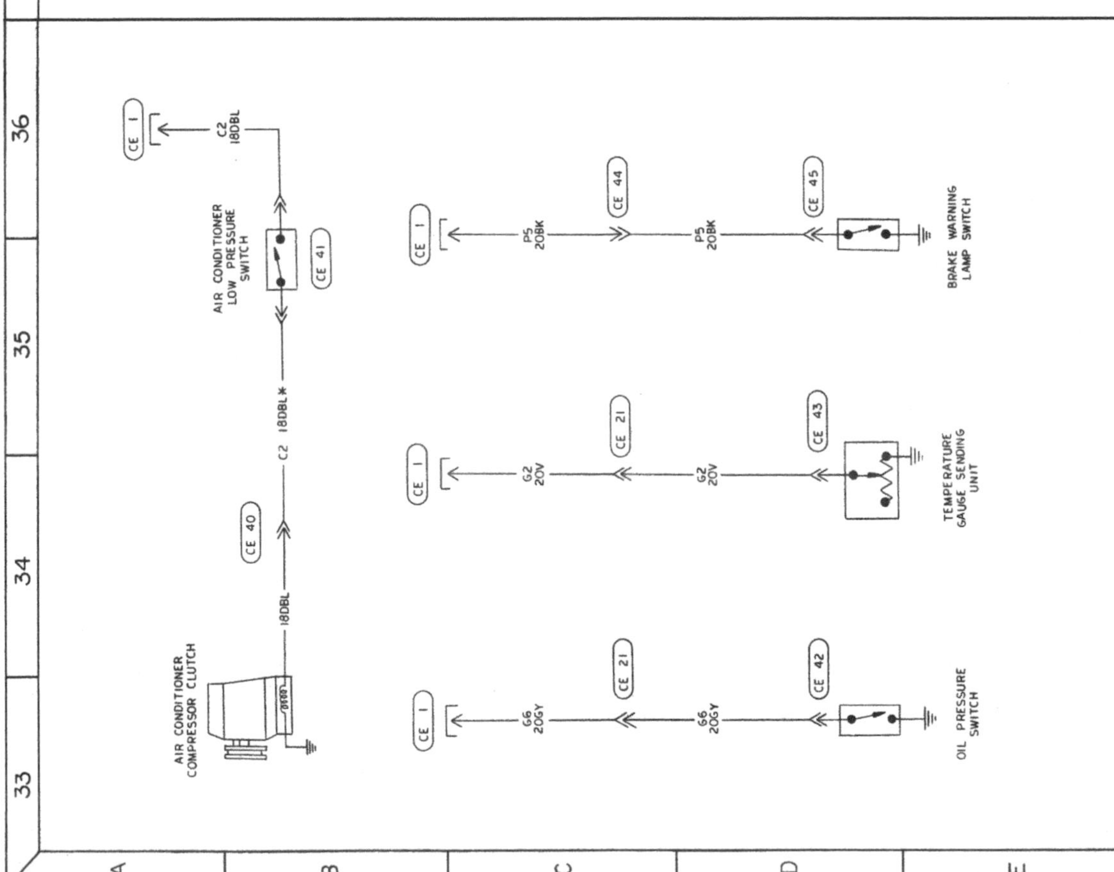

Typical A-body air conditioning compressor, oil pressure, temperature sending unit and brake switch wiring diagram (E-body similar)

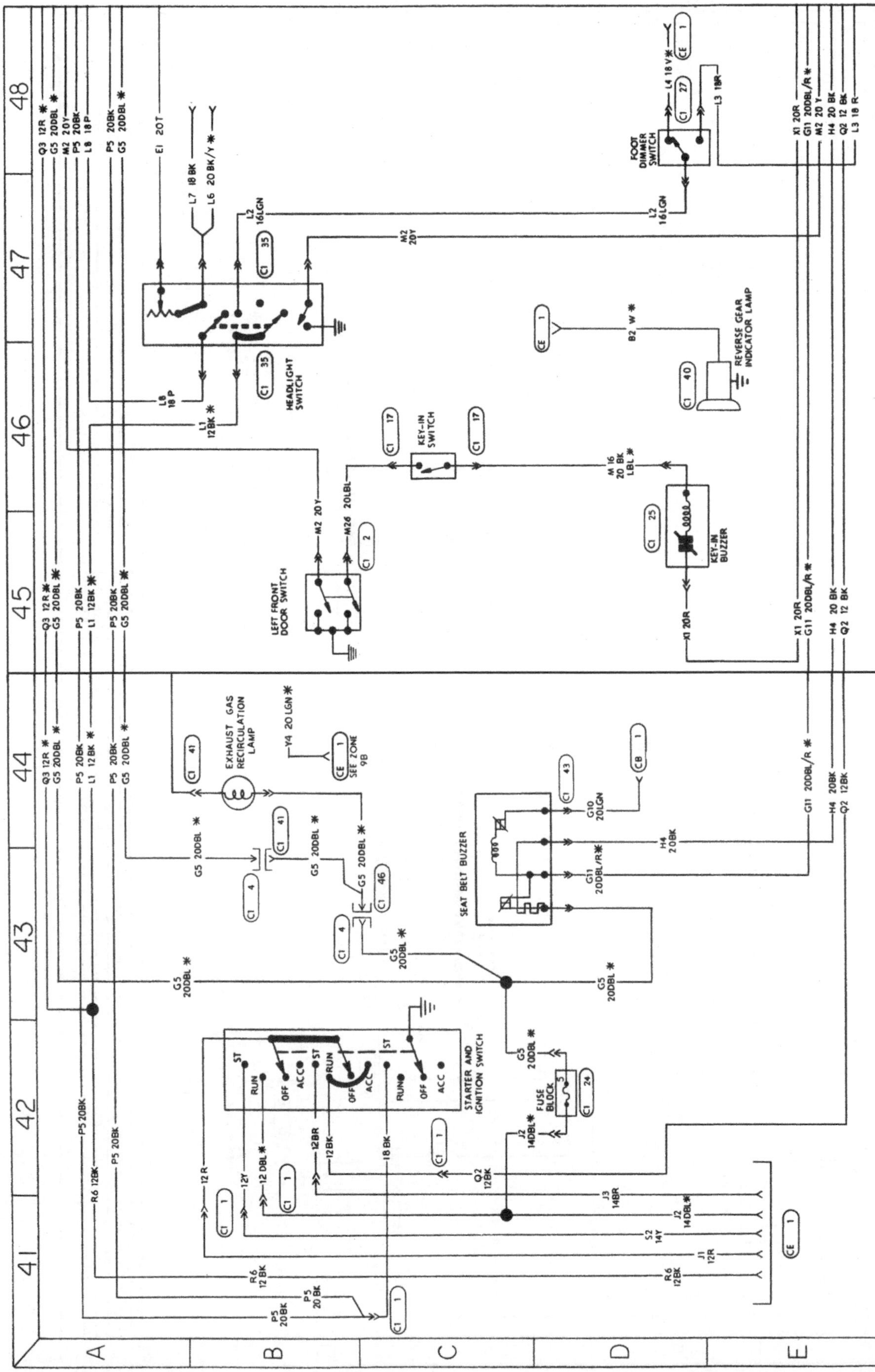

Typical A-body headlight switch wiring diagram (E-body similar)

Typical A-body ignition switch wiring diagram (E-body similar)

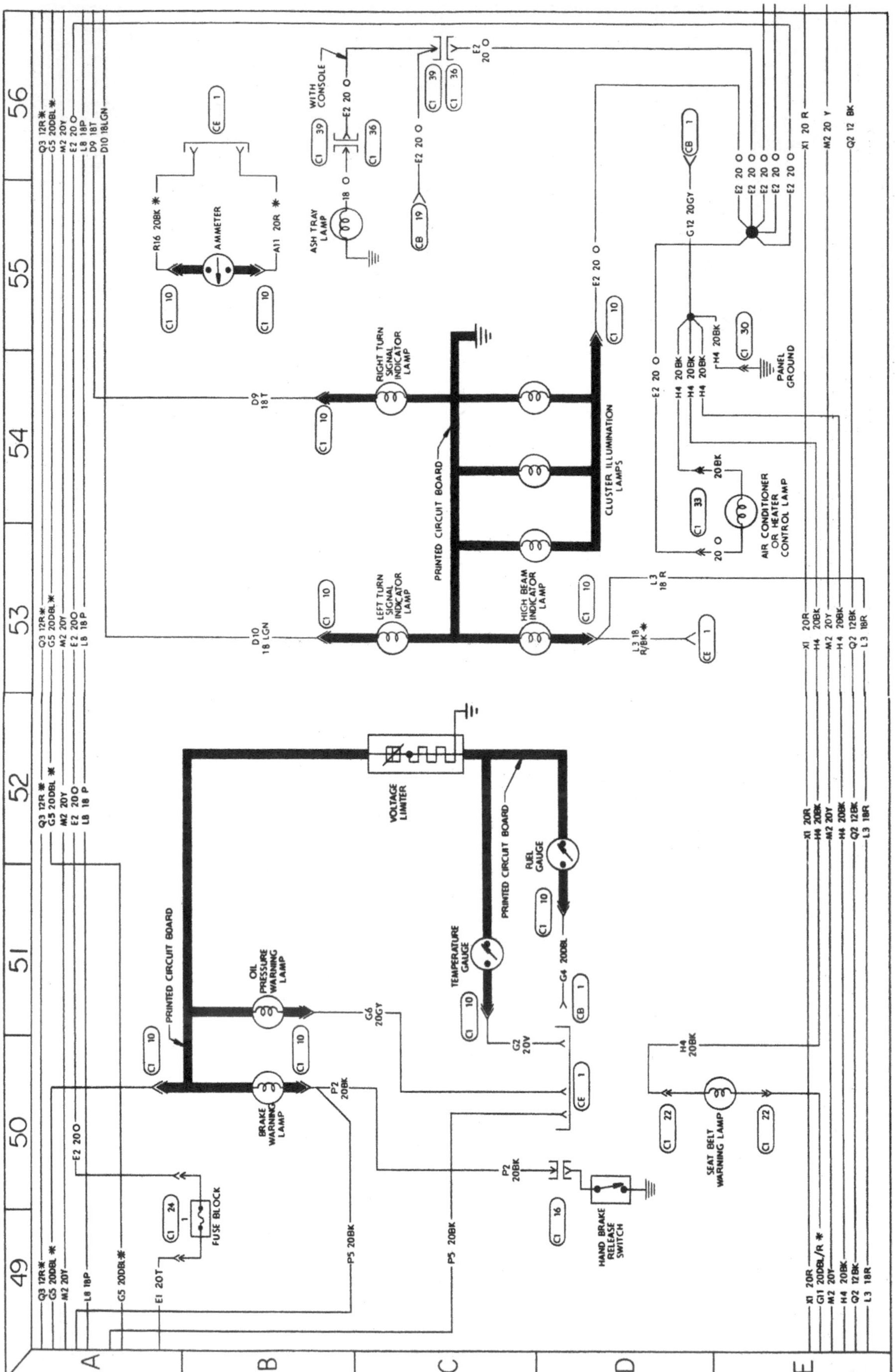

Typical A-body instrument panel wiring diagram (E-body similar)

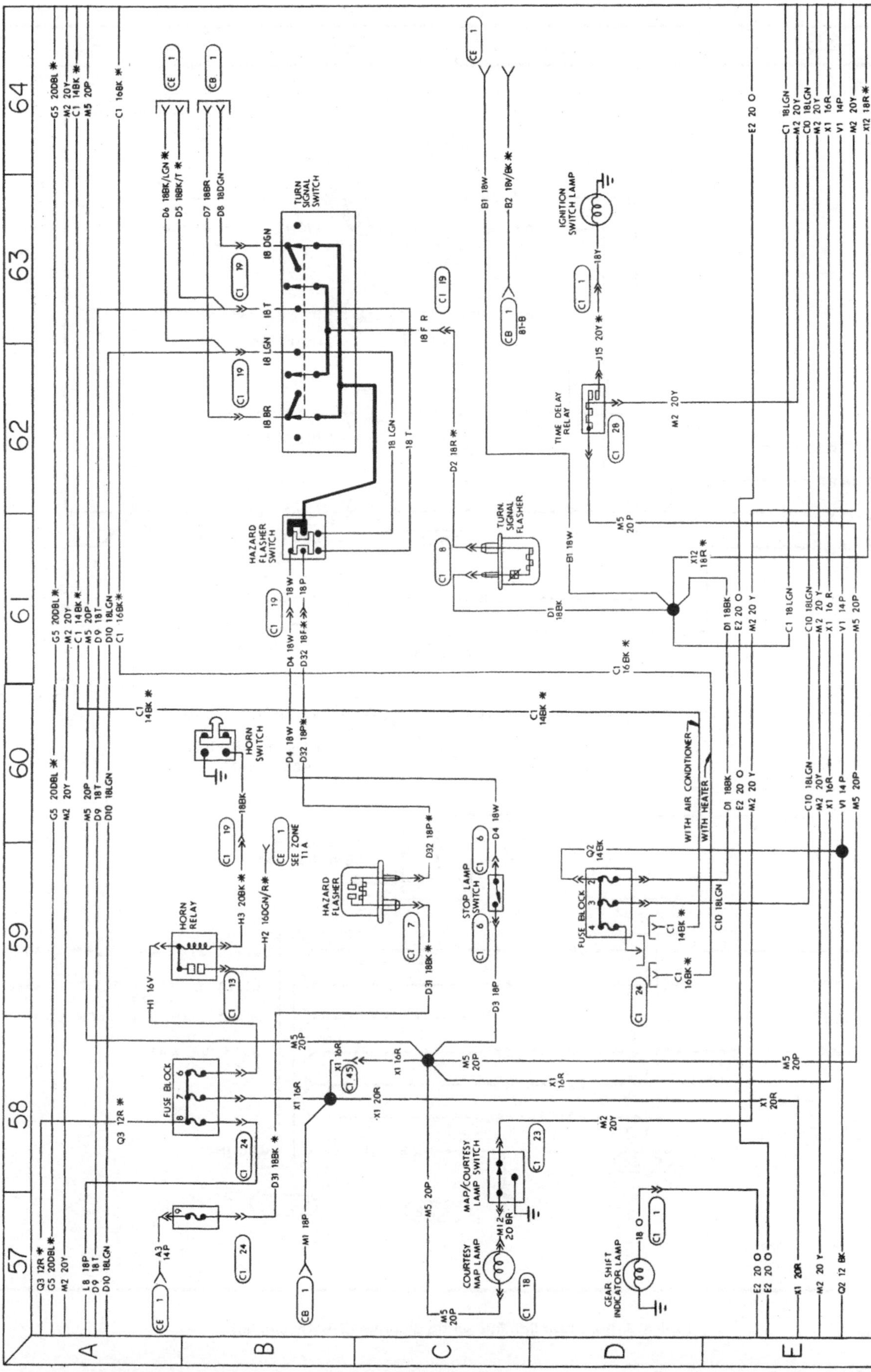

Typical A-body hazard and turn signal wiring diagram (E-body similar)

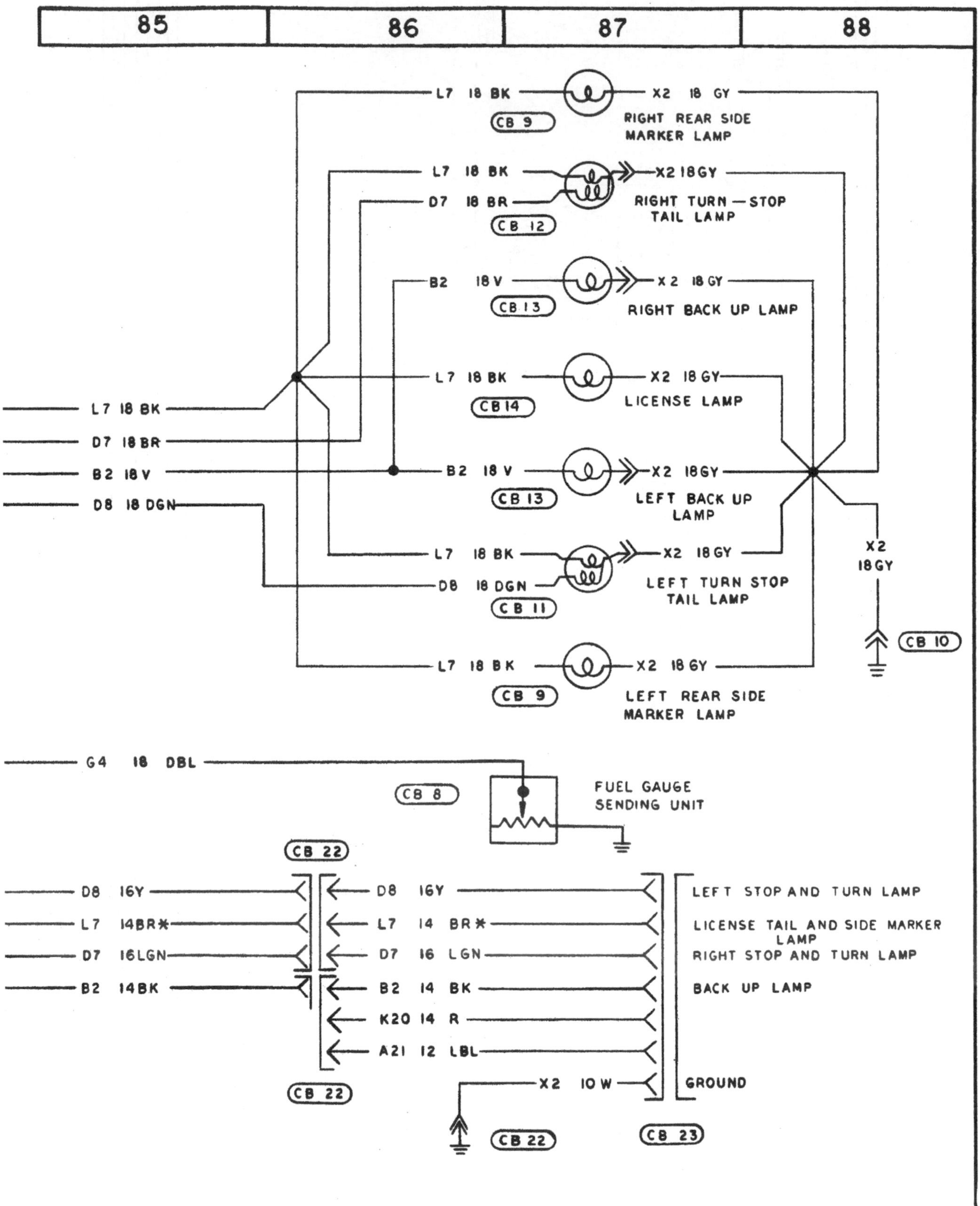

Typical A-body rear lighting wiring diagram (E-body similar)

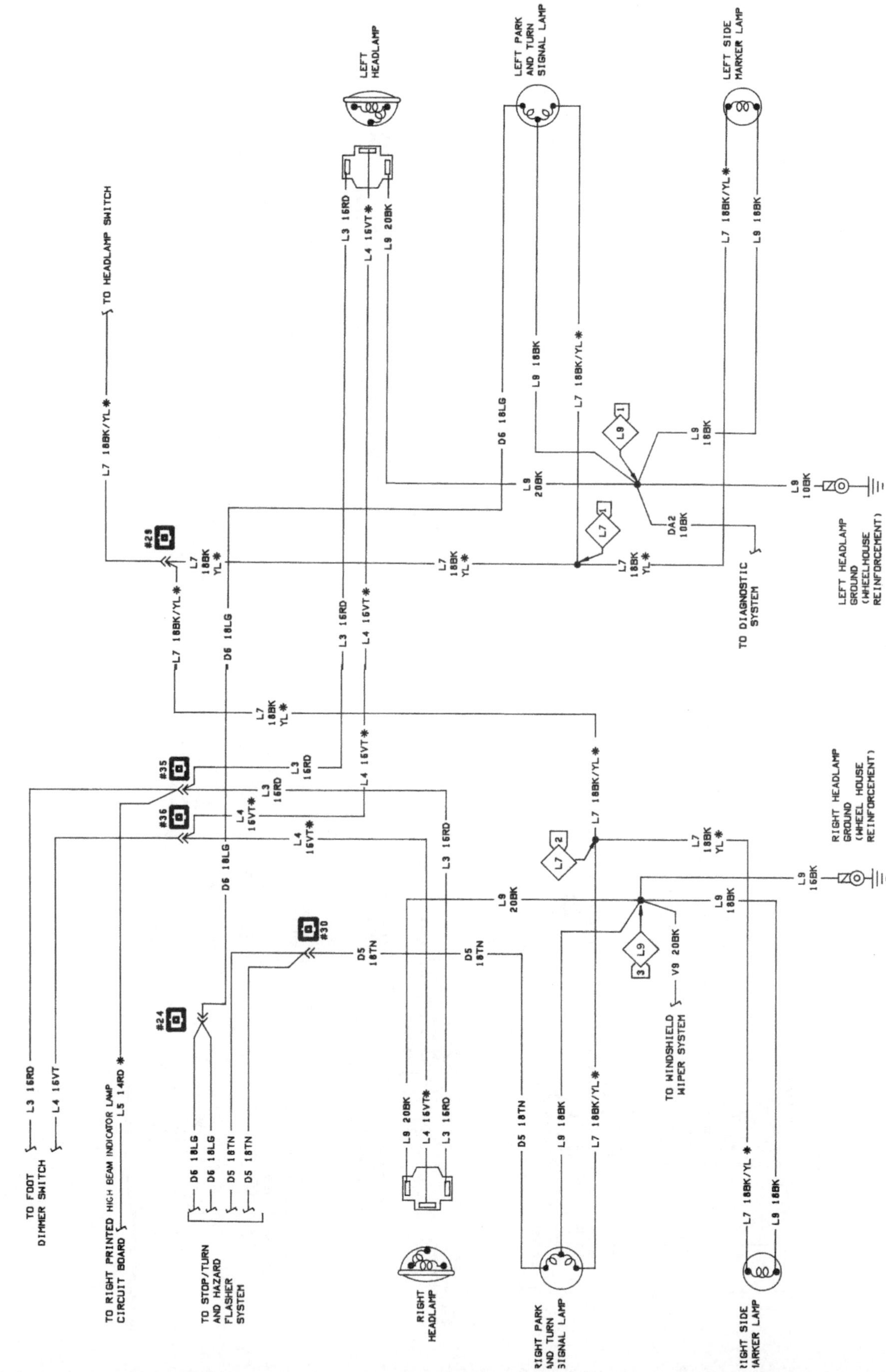

Typical F-body front lighting wiring diagram (early M-body similar)

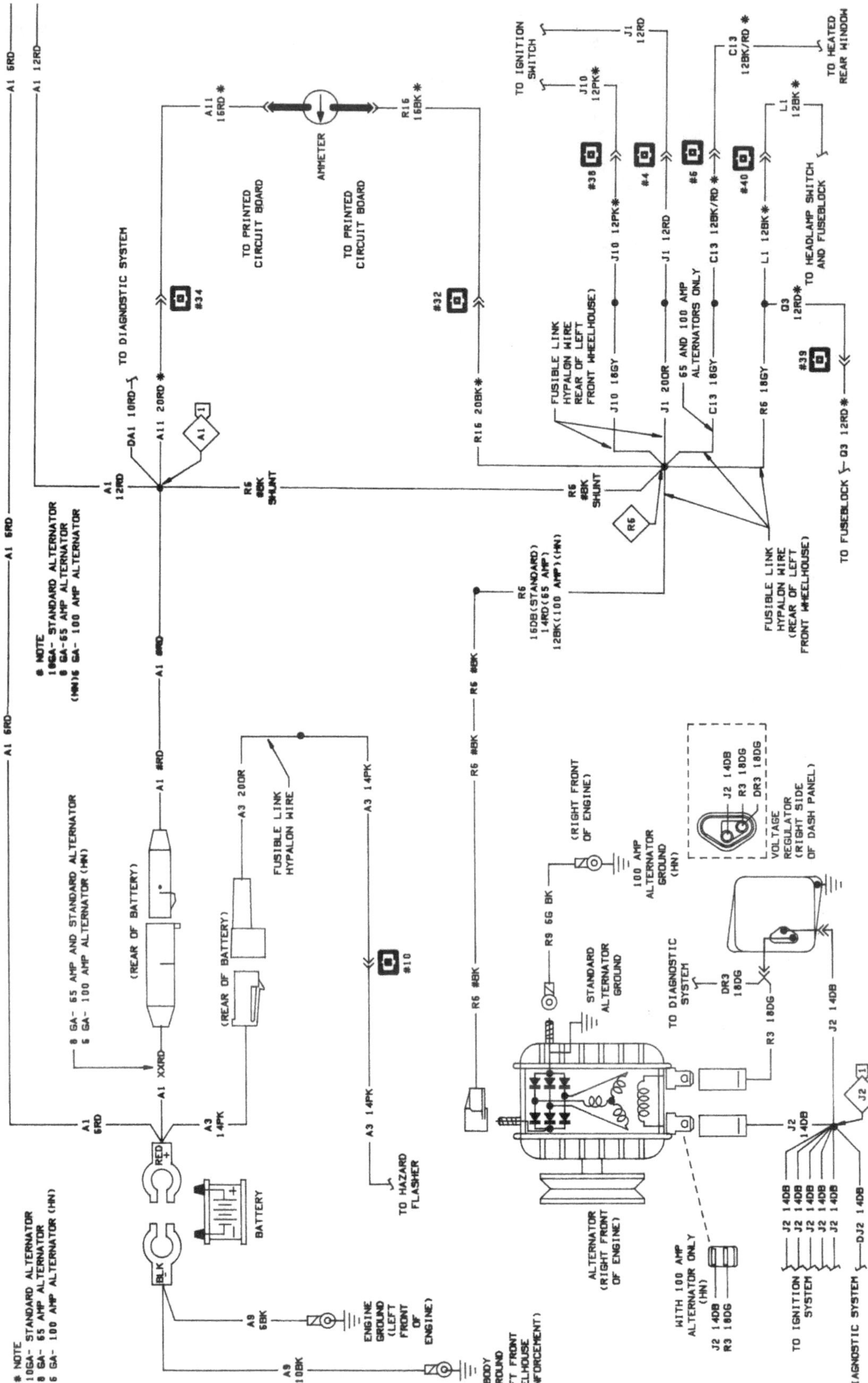

Typical F-body charging system wiring diagram (early M-body similar) (1 of 2)

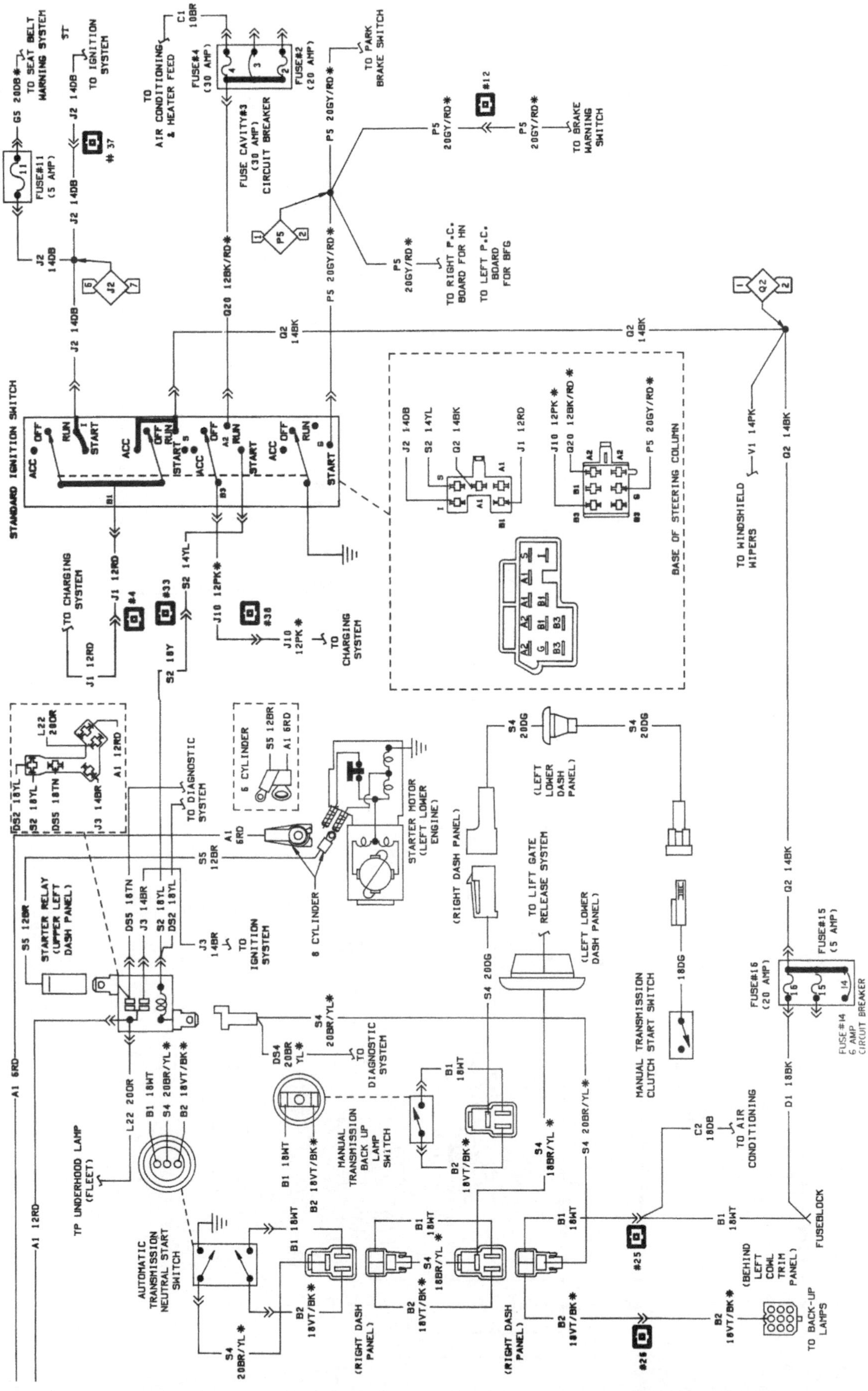

Typical F-body starting system and ignition switch wiring diagram (early M-body similar)

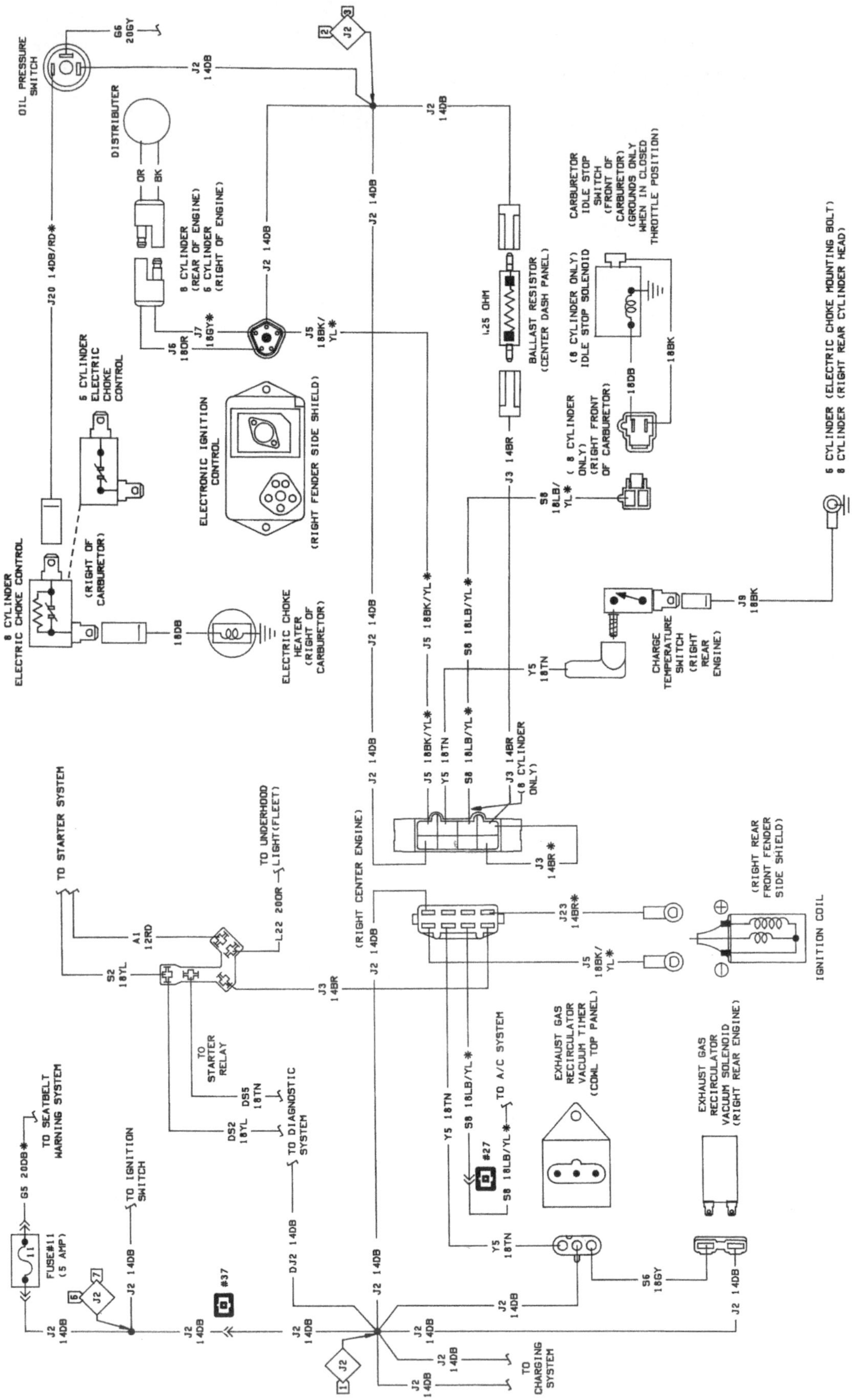

Typical F-body ignition system wiring diagram (early M-body similar)

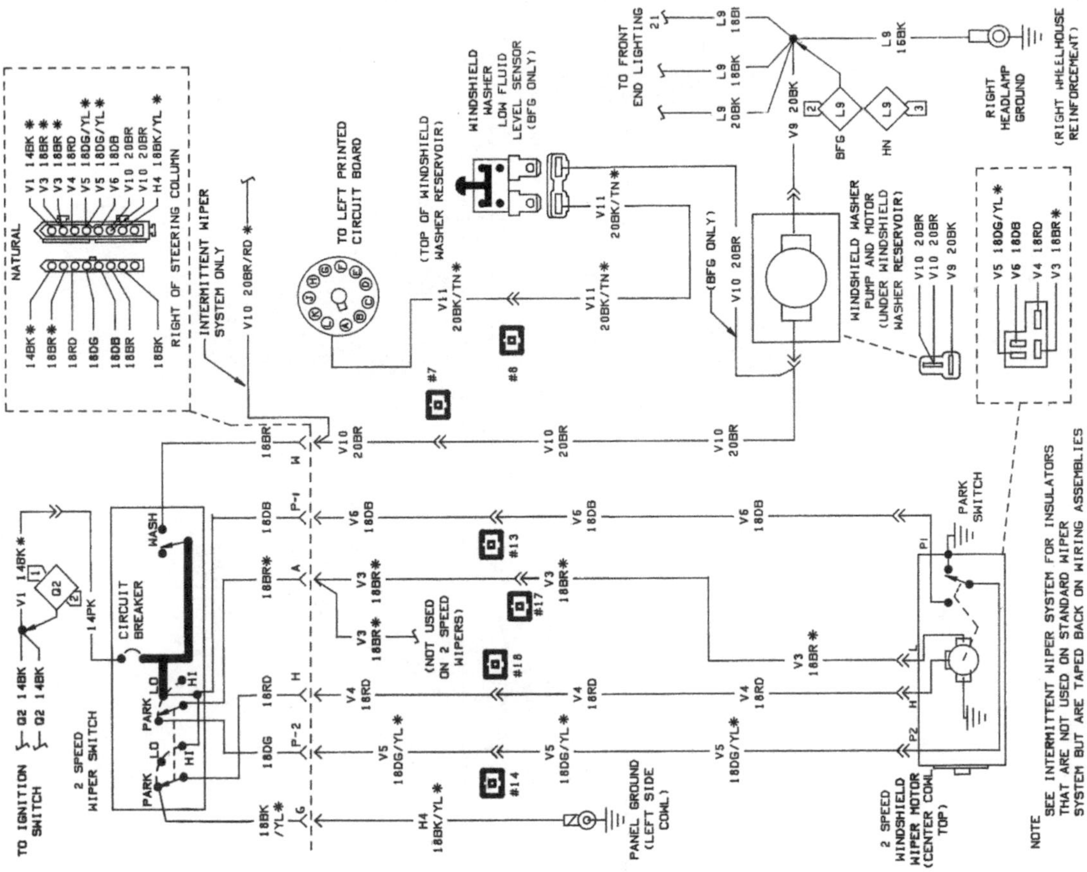

Typical F-body headlight switch wiring diagram (early M-body similar)

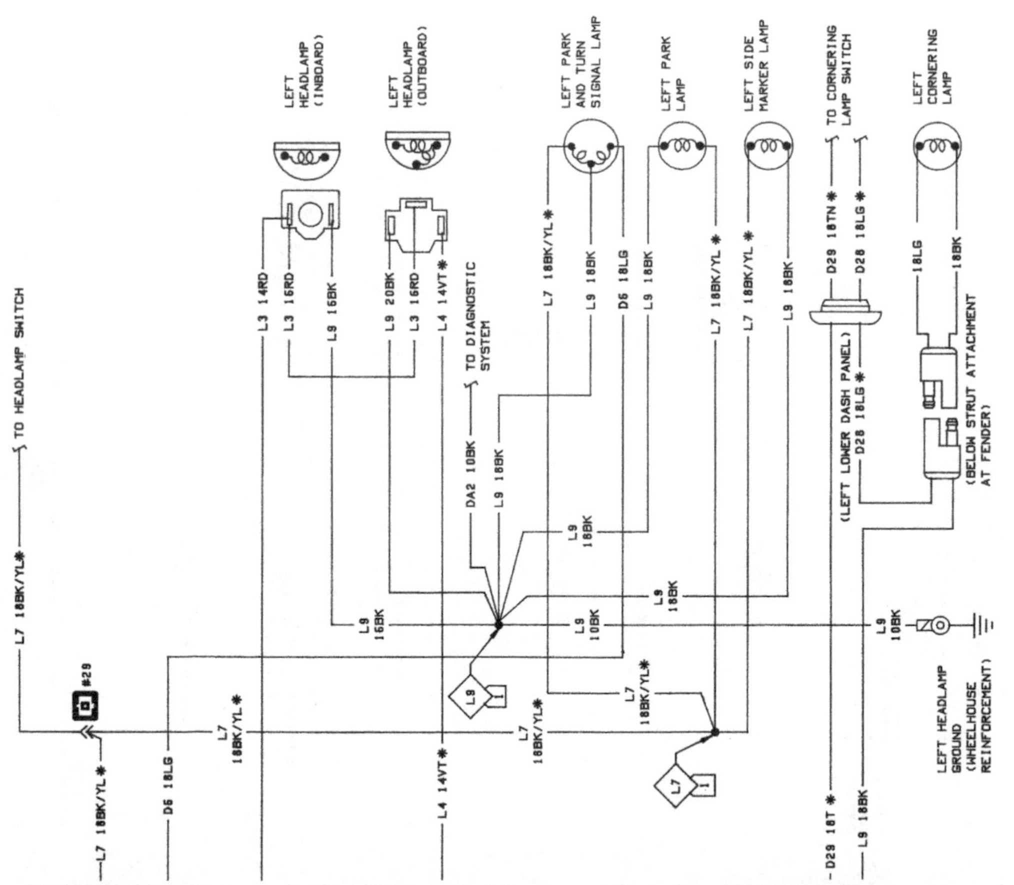

Typical F-body front lighting wiring diagram (early M-body similar)

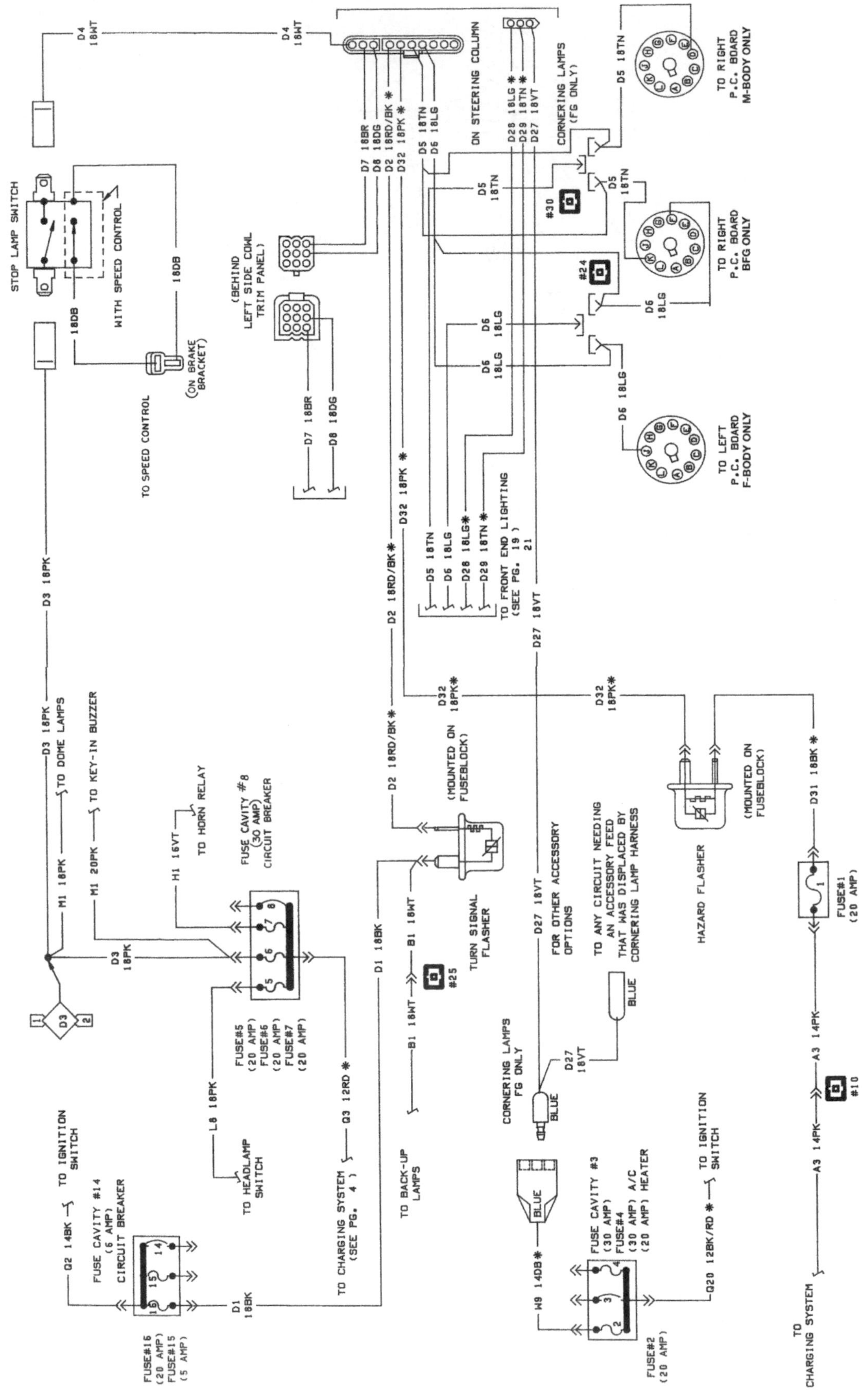

Typical F-body stop/turn/hazard flasher wiring diagram (early M-body similar)

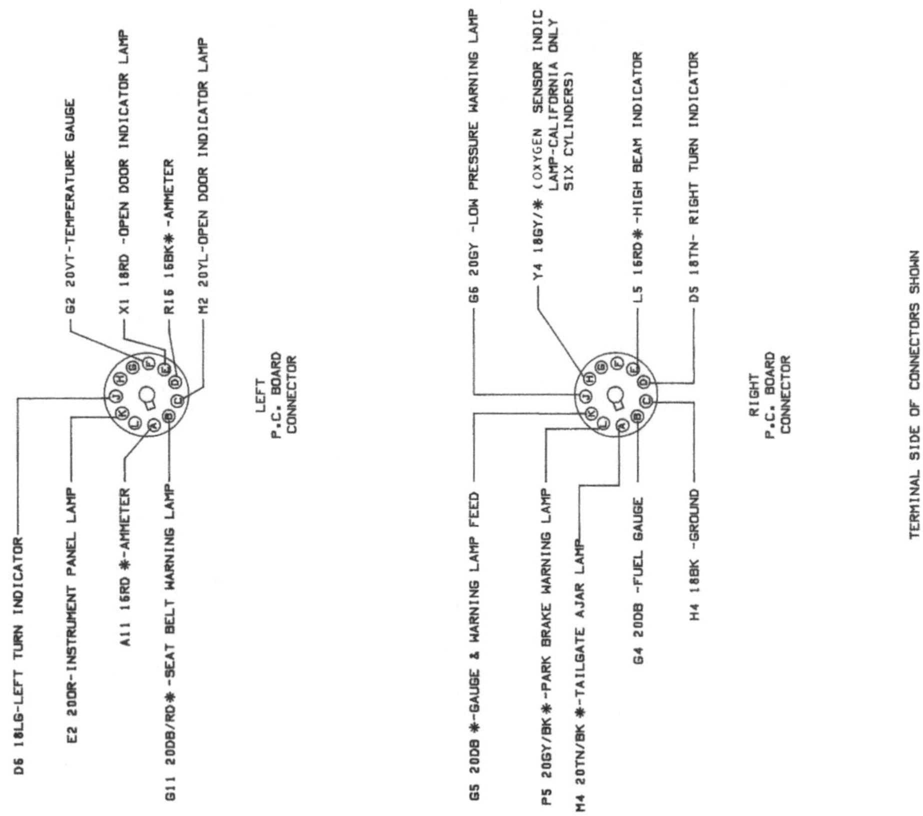

Typical F-body instrument panel circuit board connector diagram (early M-body similar)

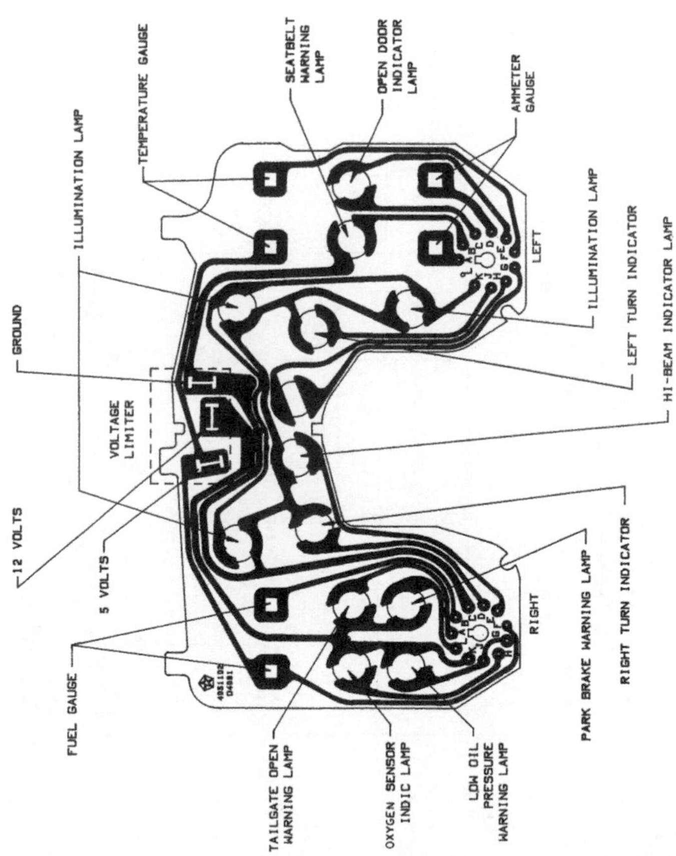

Typical F-body instrument panel circuit board diagram

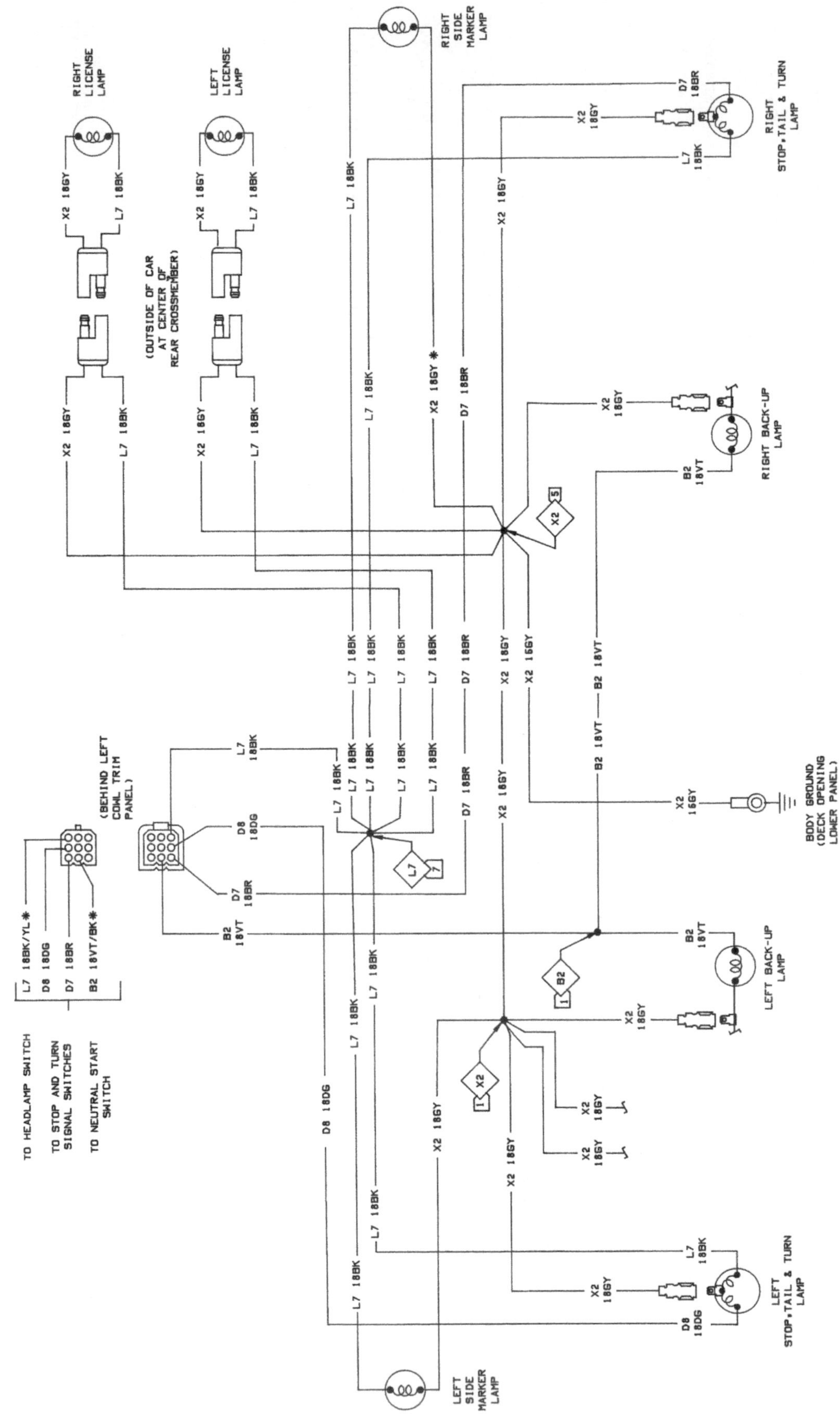

Typical F-body rear lighting wiring diagram (early M-body similar)

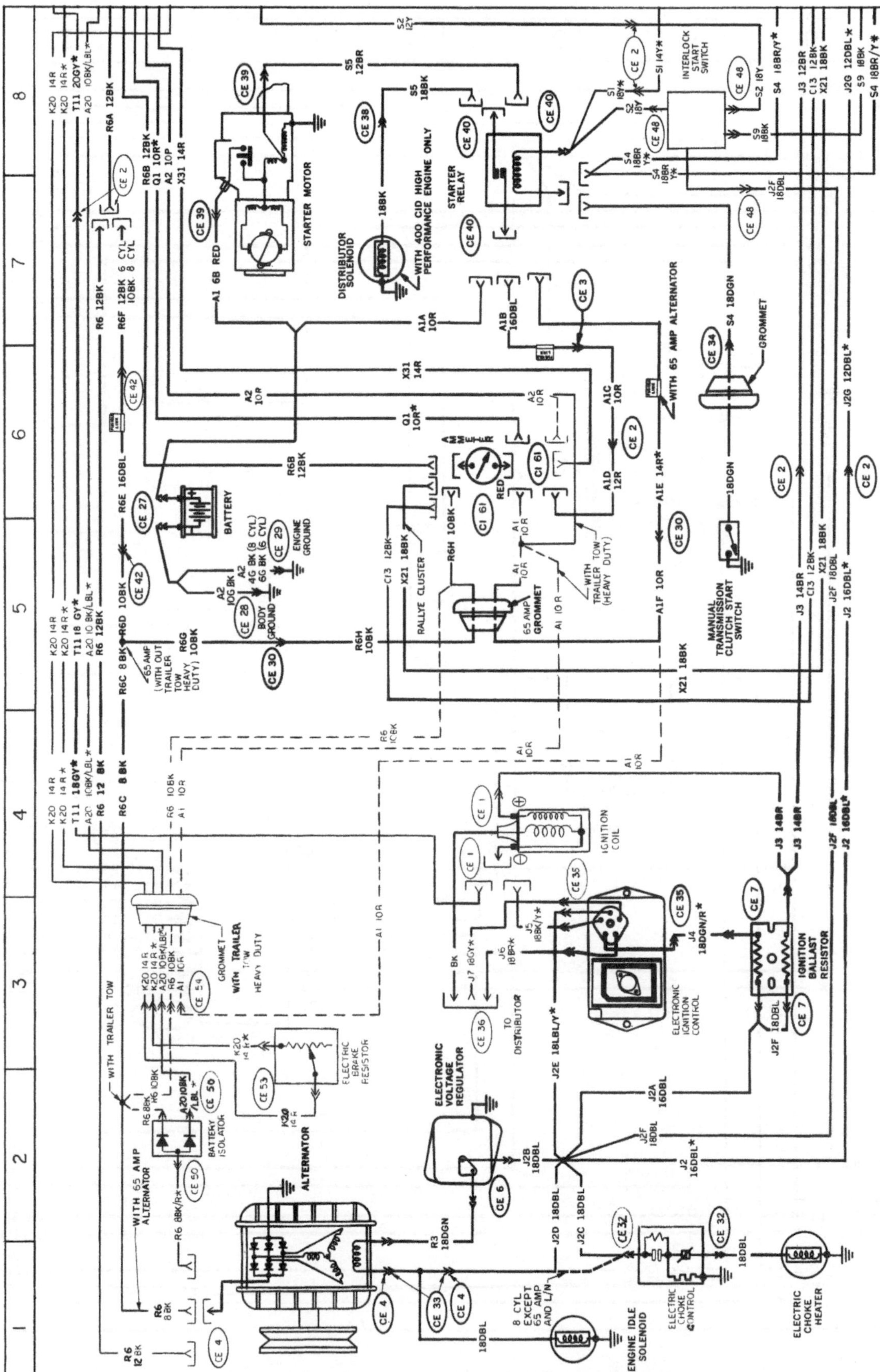

Typical B-body charging/starting system wiring diagram

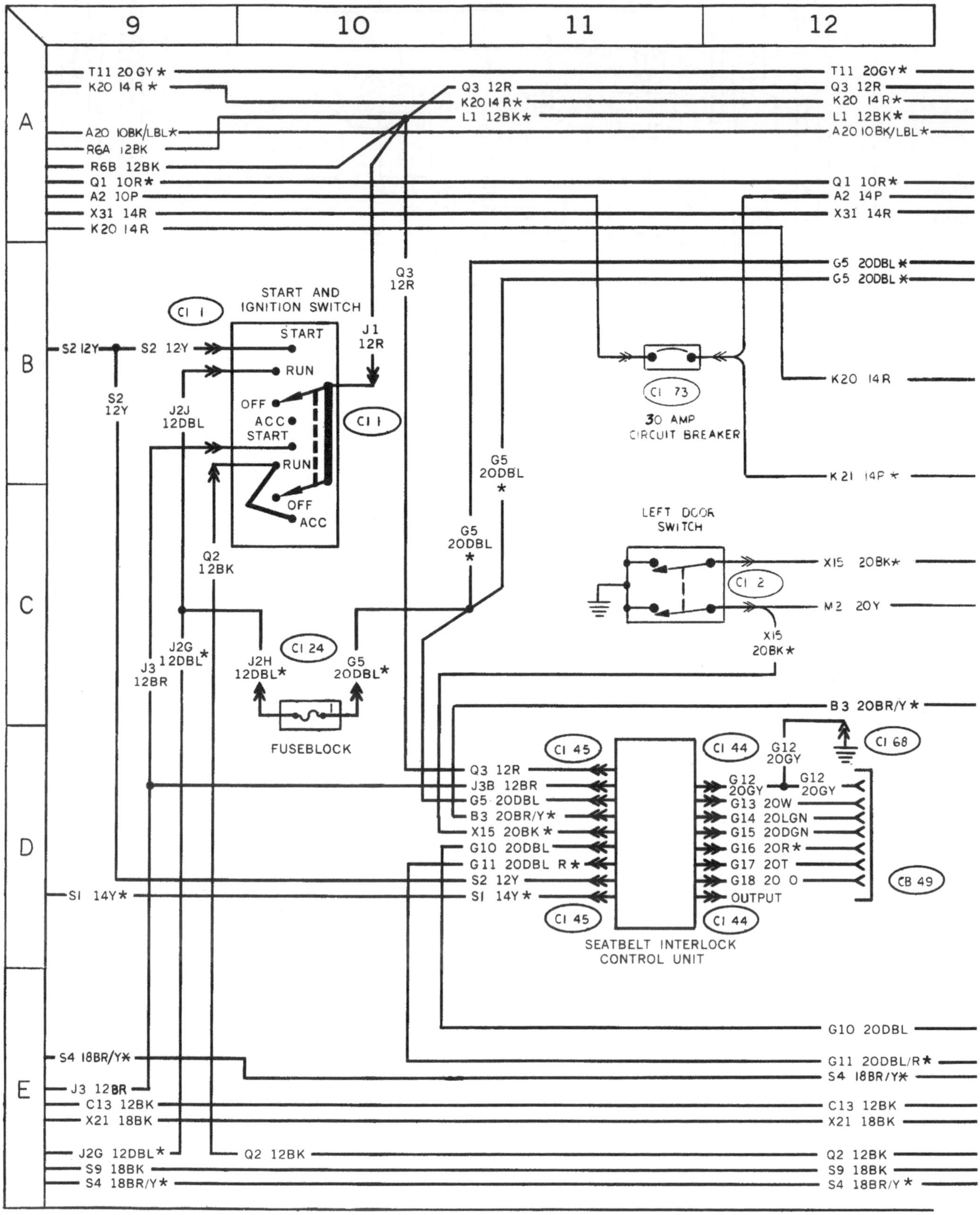

Typical B-body ignition switch wiring diagram

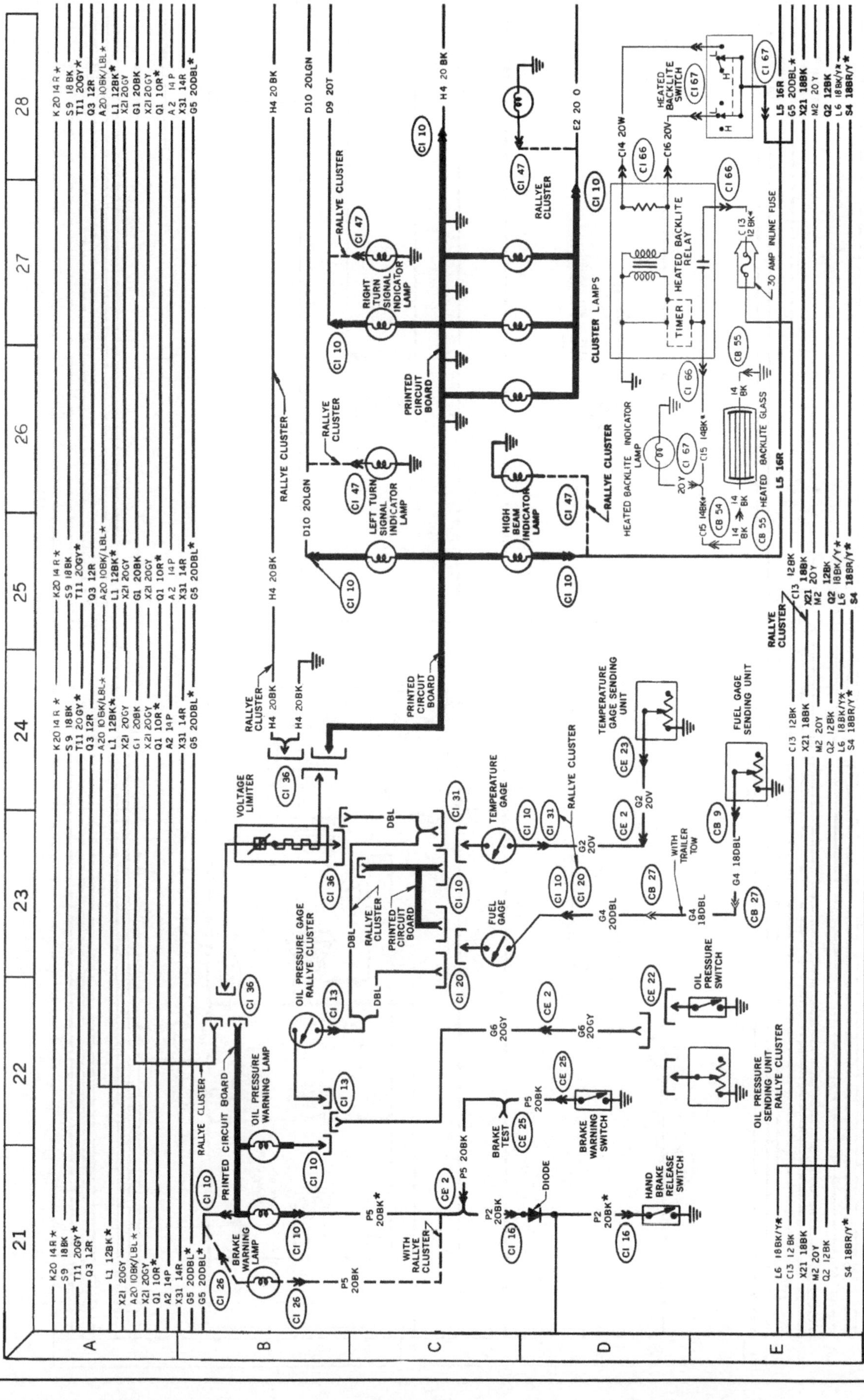

Typical B-body instrument panel wiring diagram

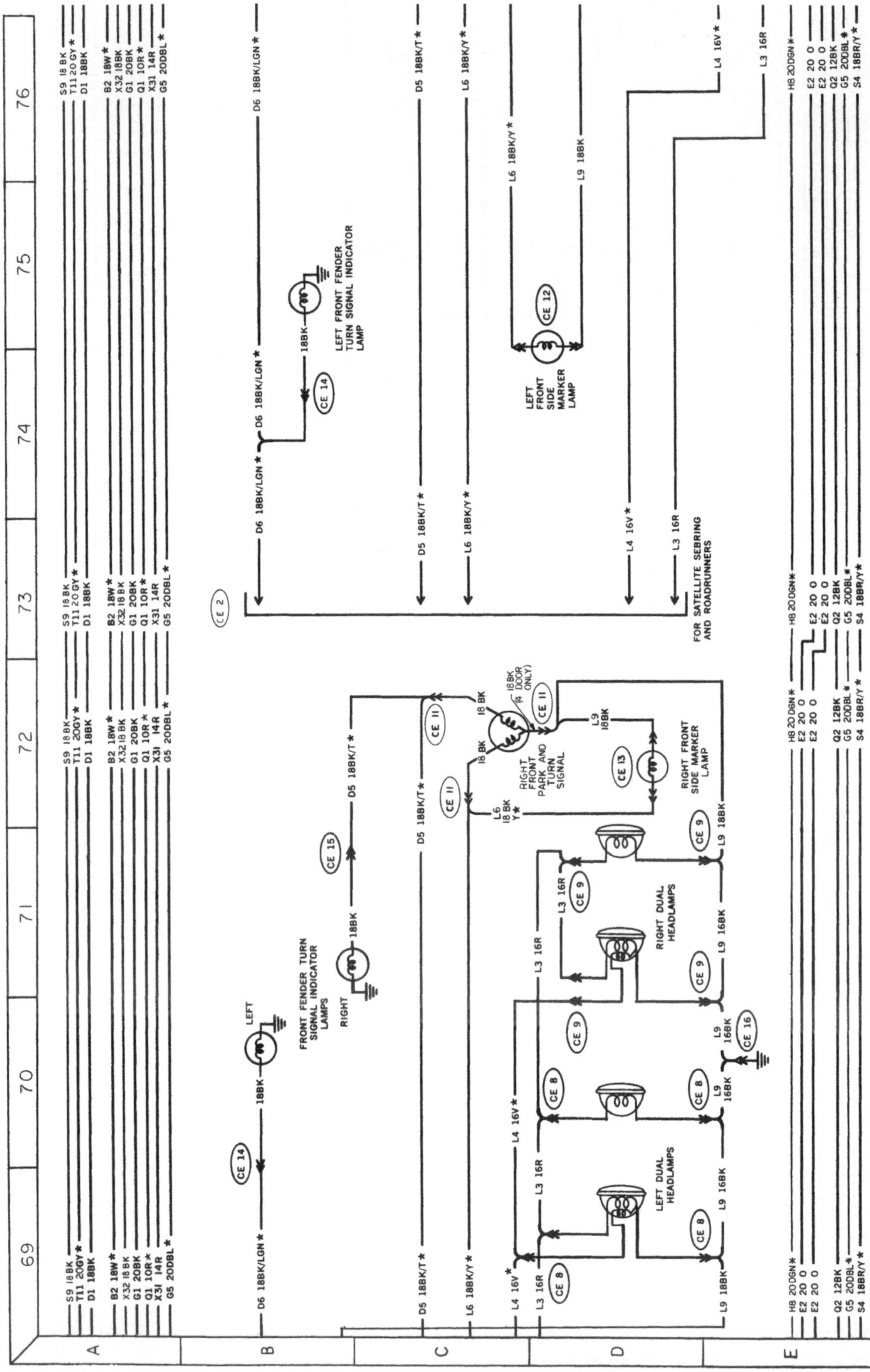

Typical B-body front lighting wiring diagram

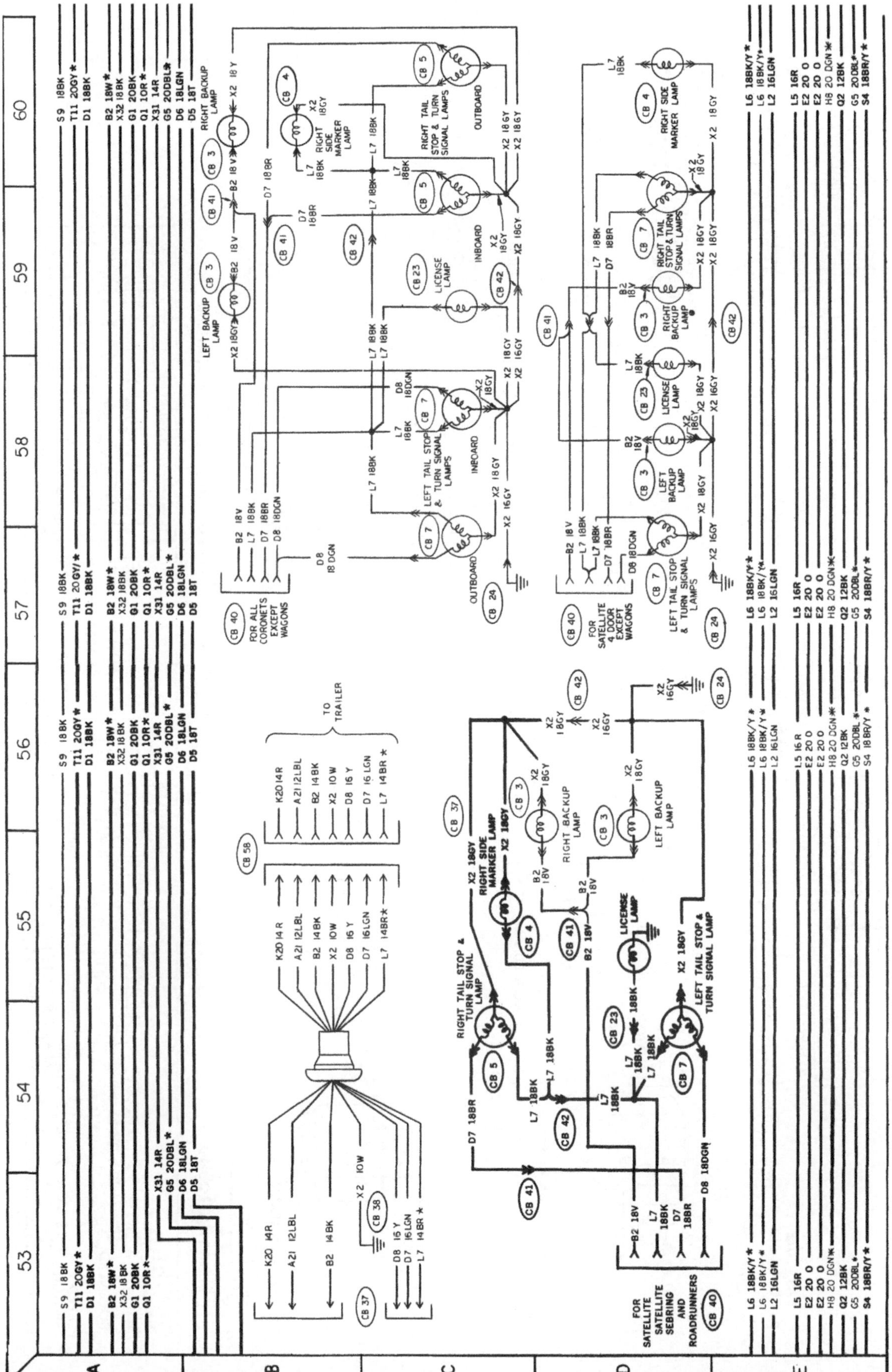

Typical B-body rear lighting wiring diagram

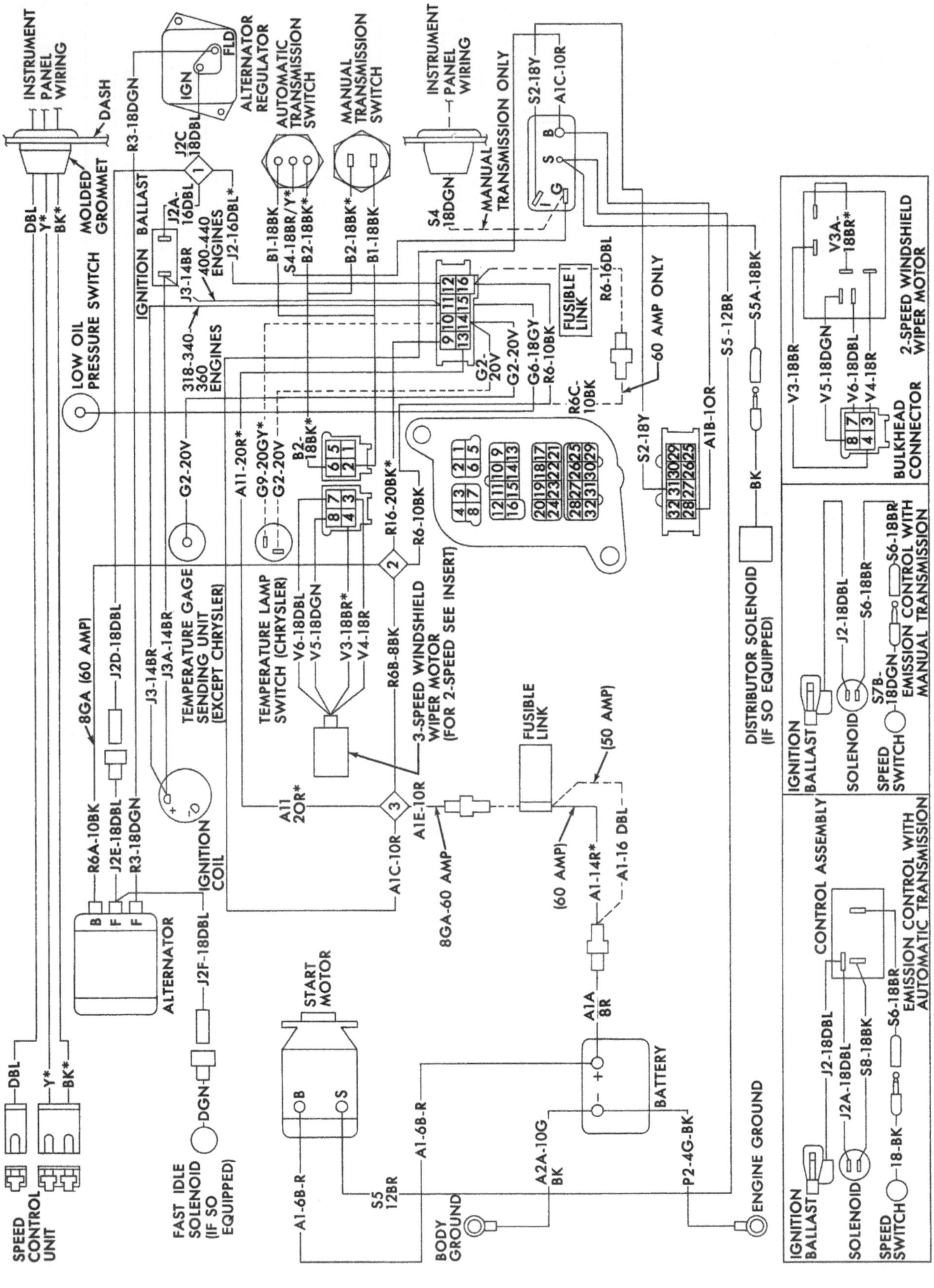

Typical C-body engine compartment wiring diagram

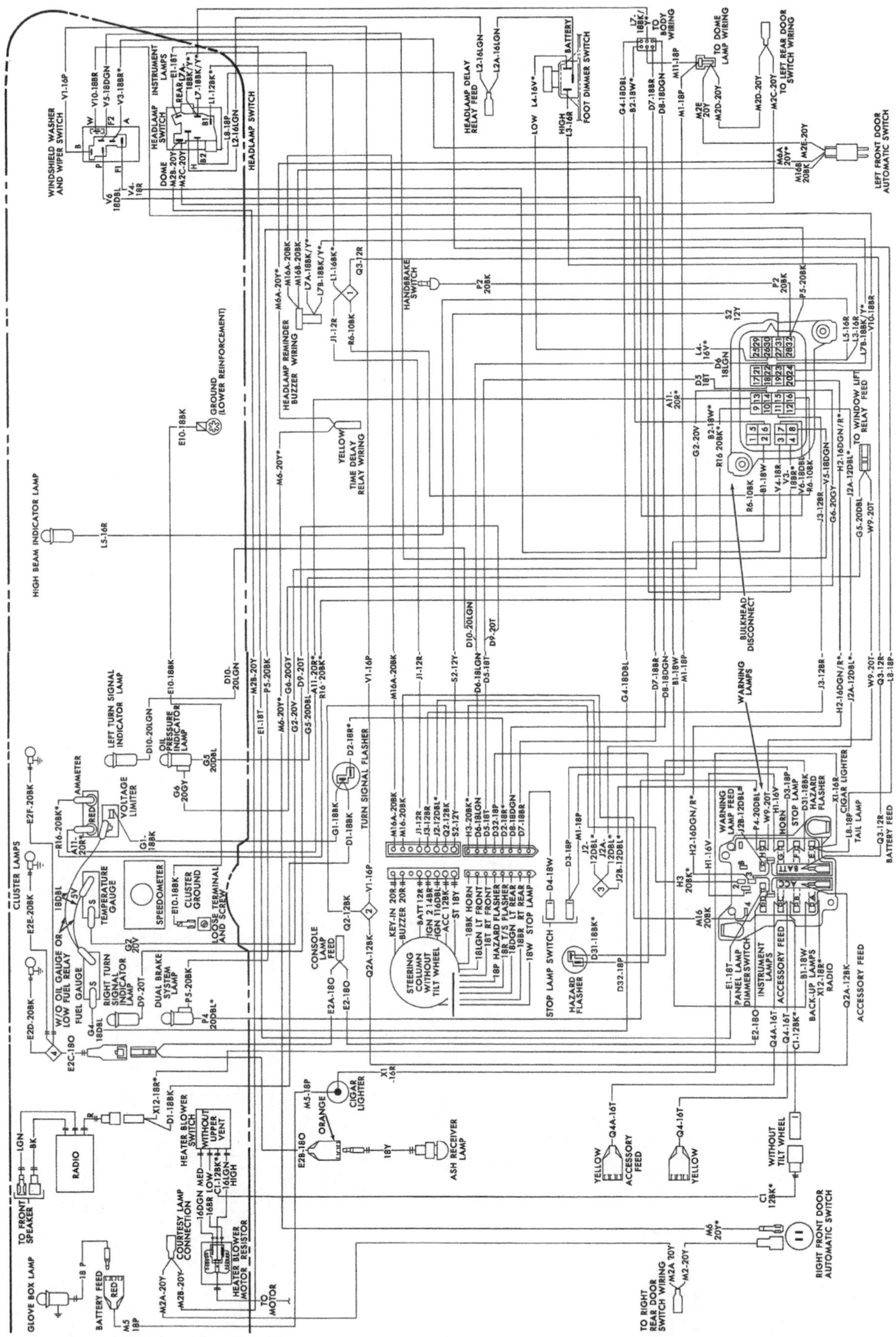

Typical C-body instrument panel wiring diagram

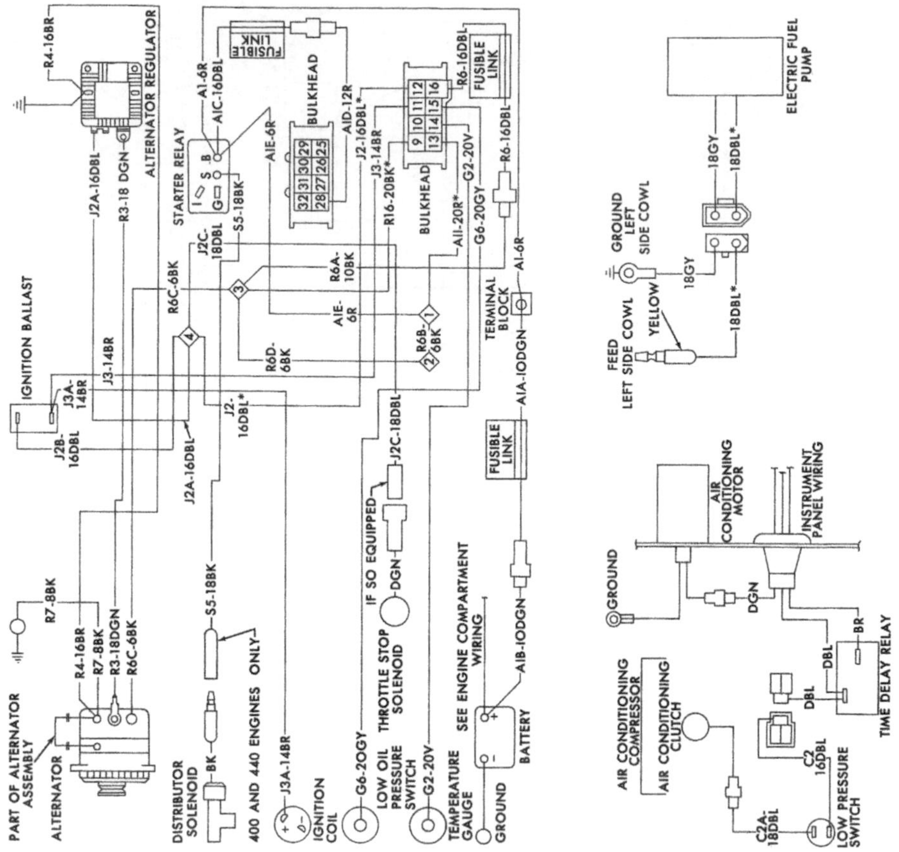

Typical C-body 65 amp alternator, air conditioning and electric fuel pump wiring diagram

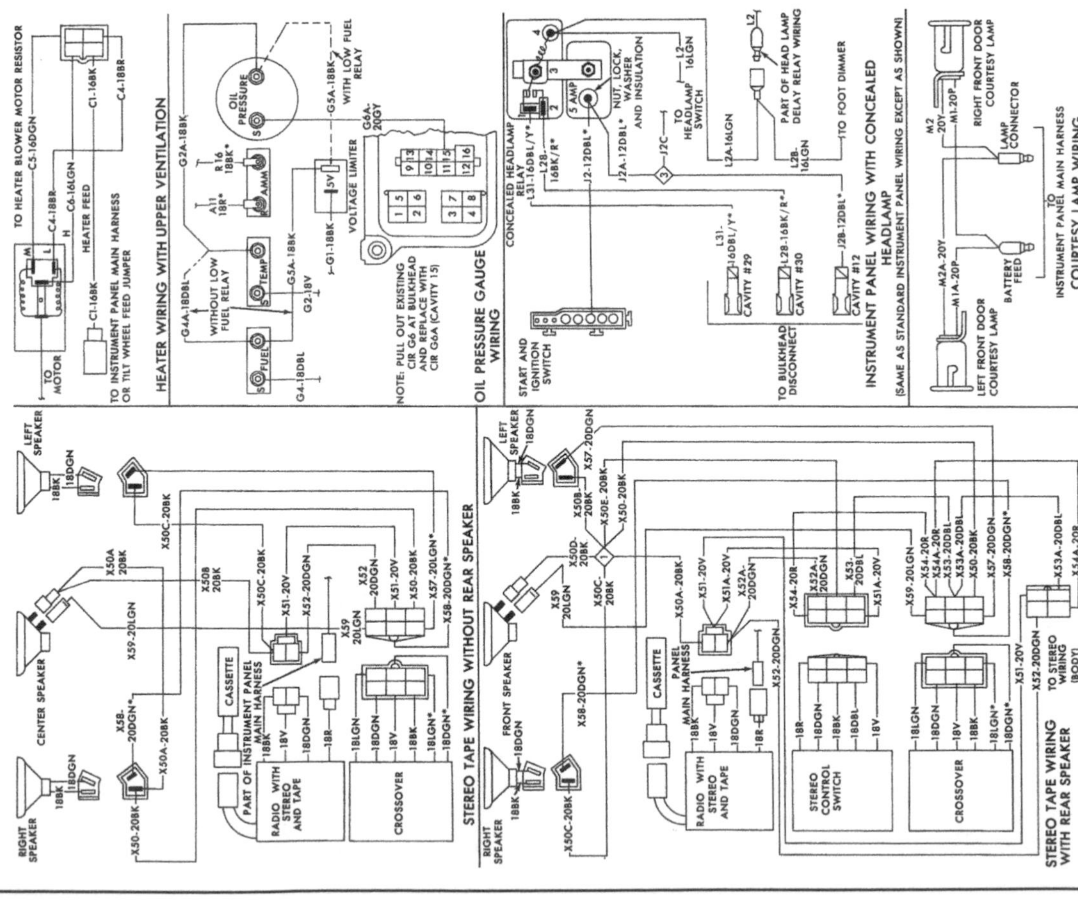

Typical C-body instrument panel accessory wiring diagram

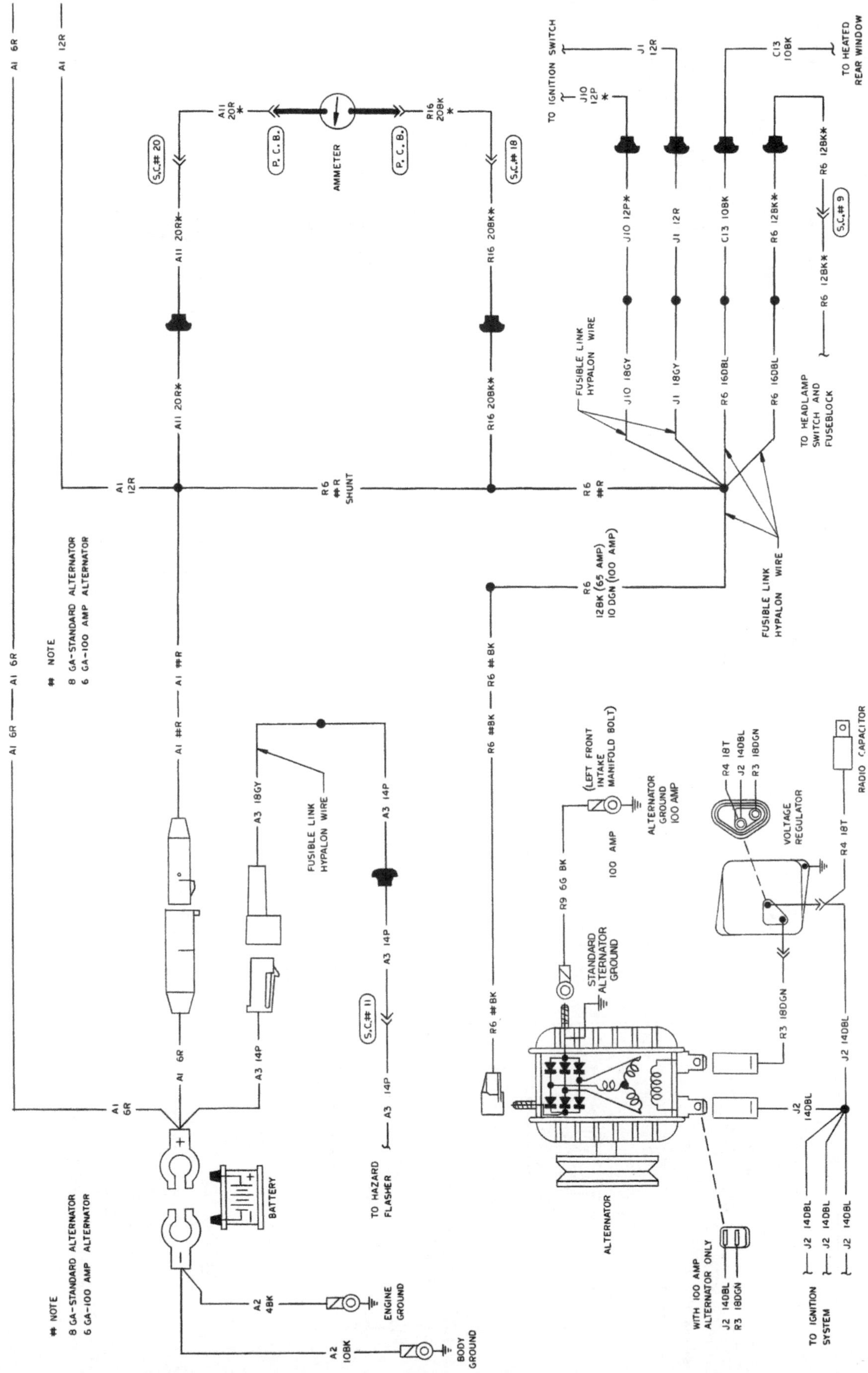

Typical C-body charging system wiring diagram

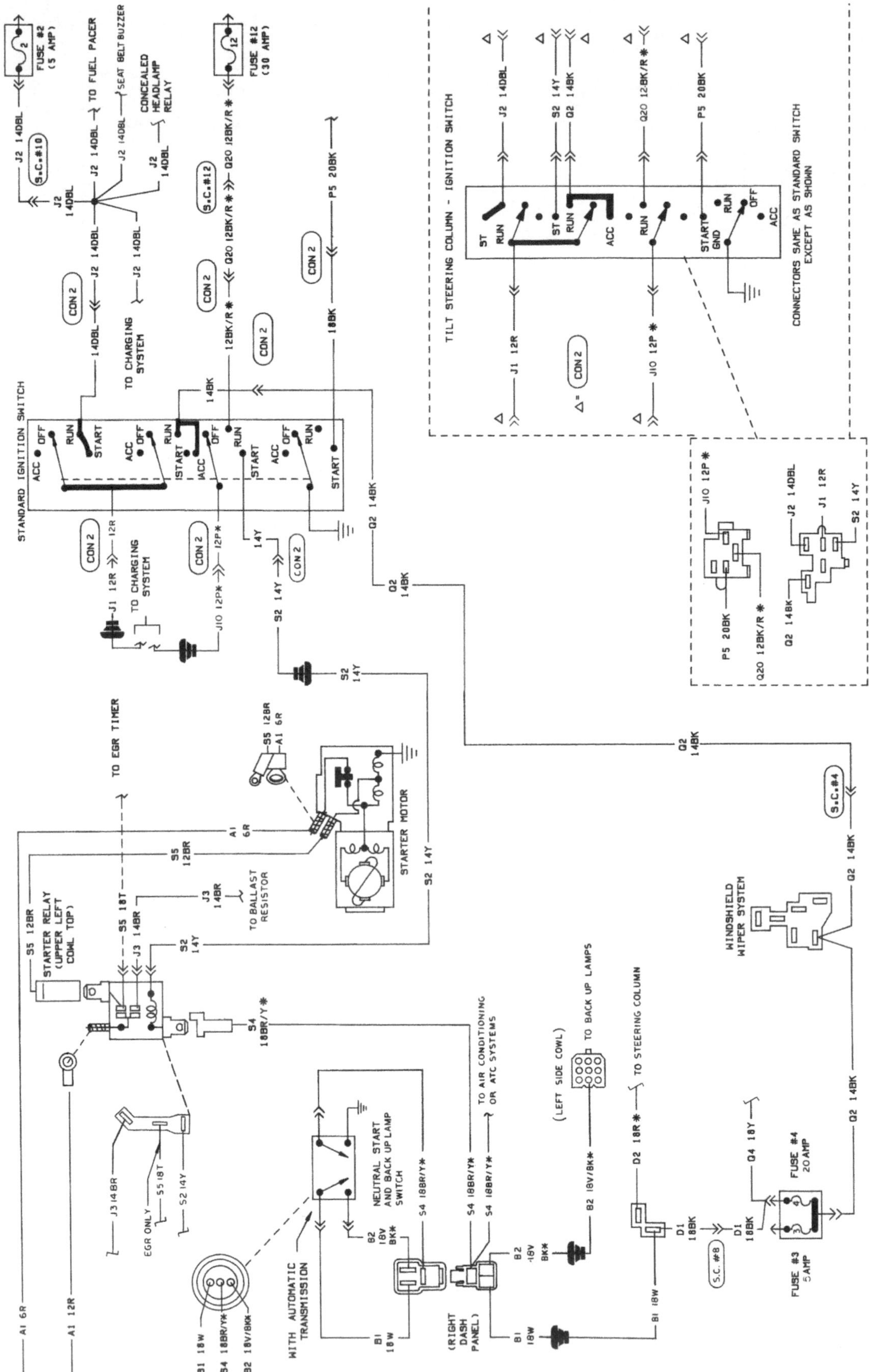

Typical C-body starting system and ignition switch wiring diagram

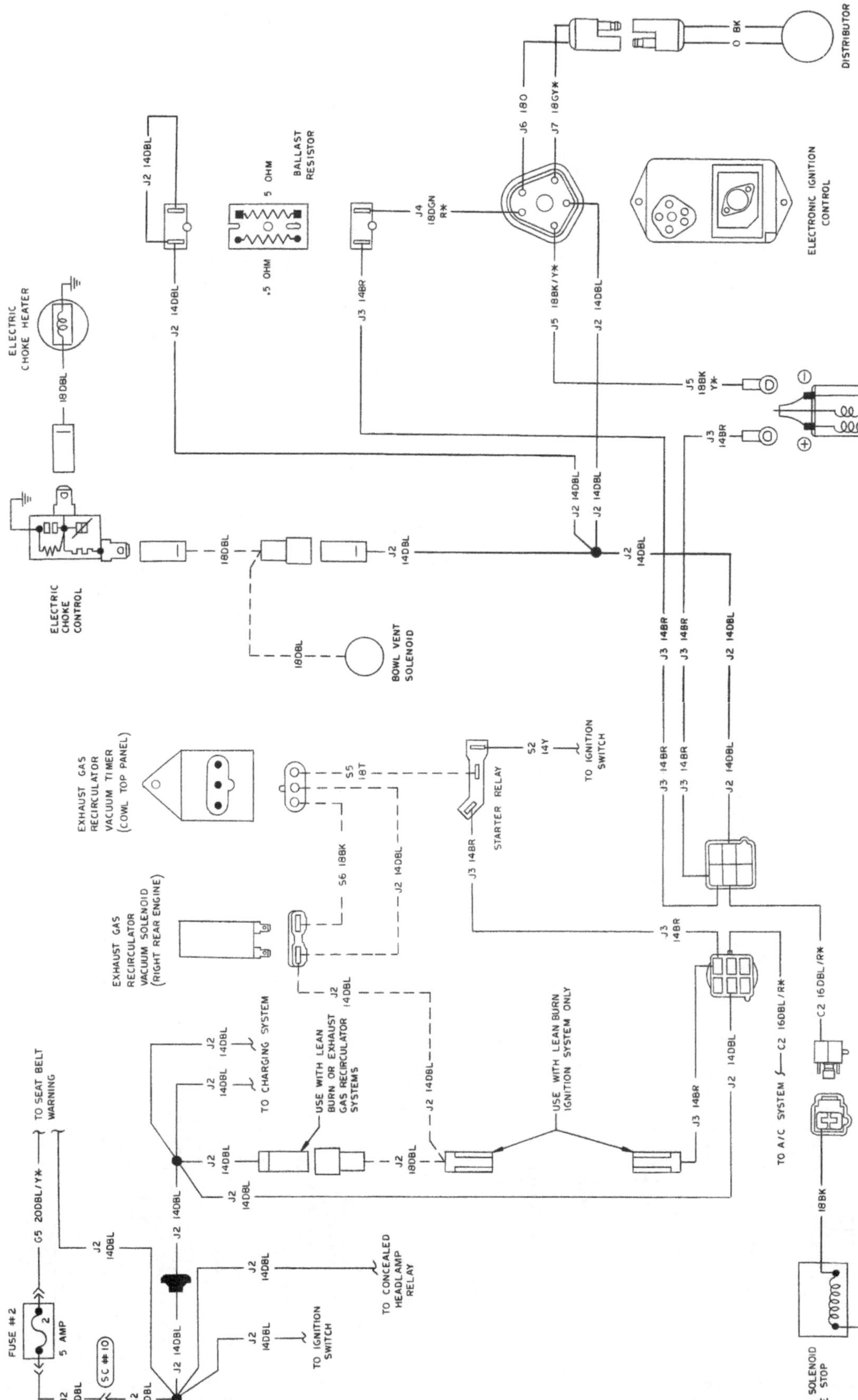

Typical C-body ignition system wiring diagram

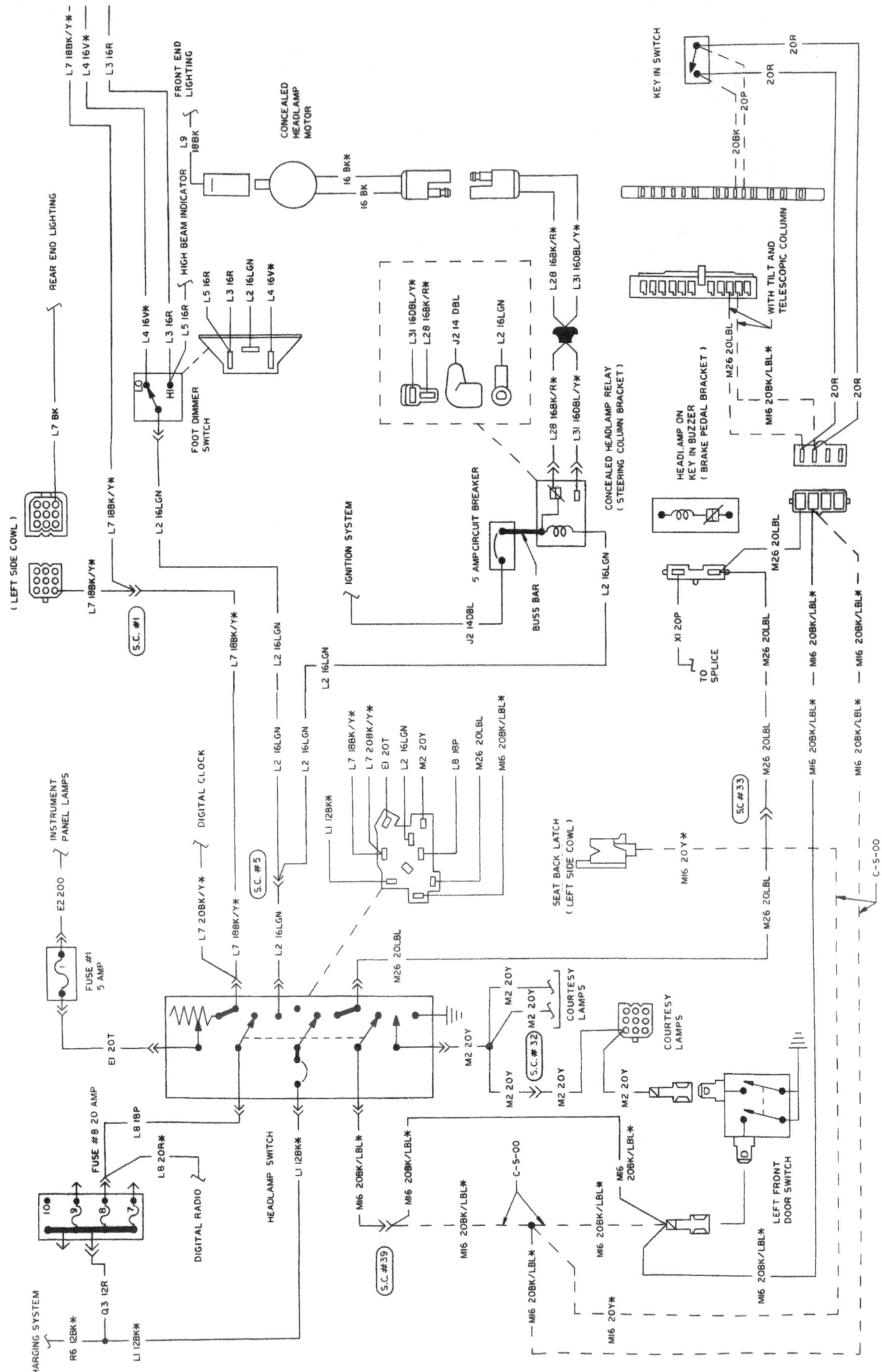

Typical C-body headlight switch wiring diagram

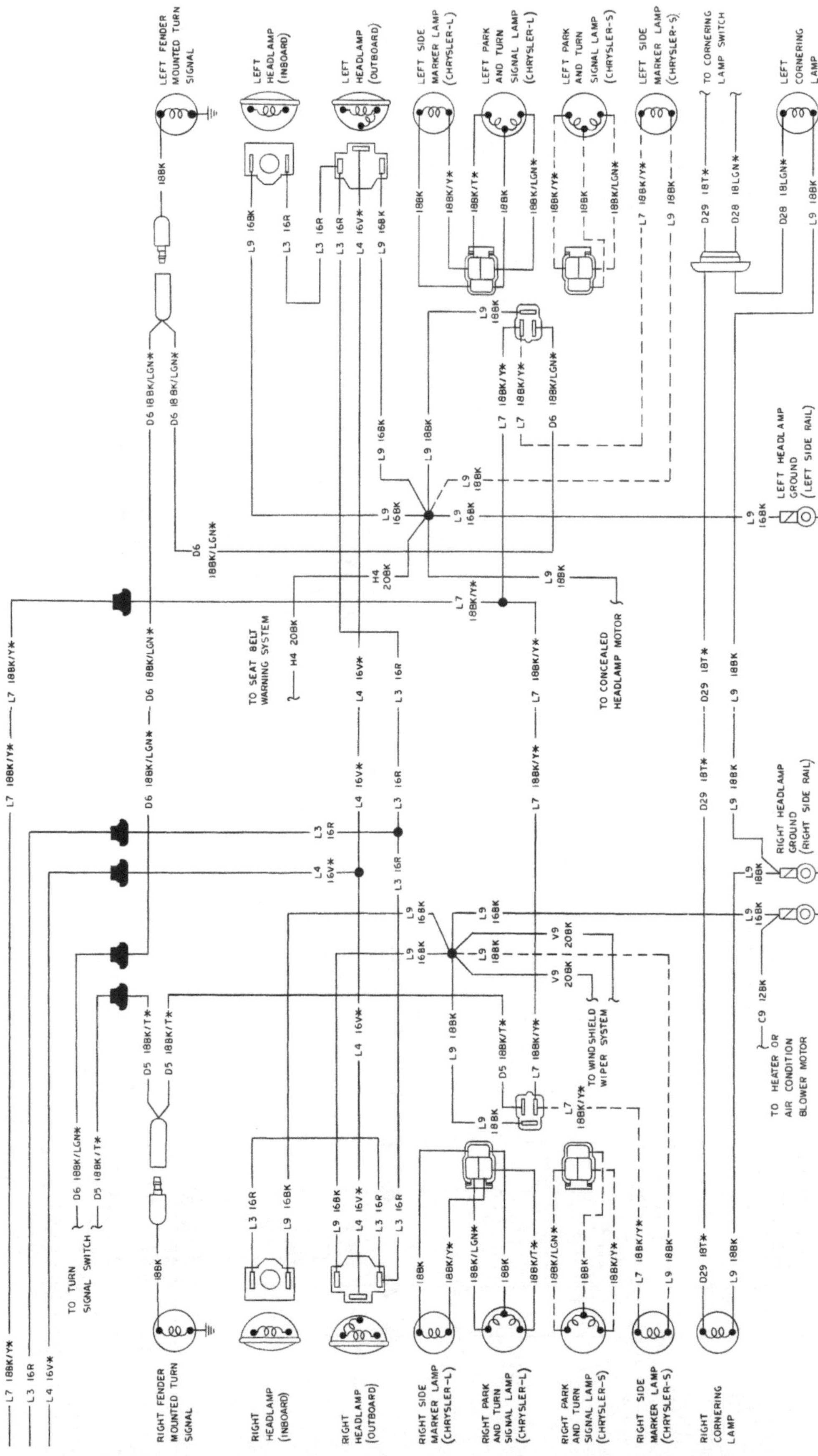

Typical C-body front end lighting wiring diagram

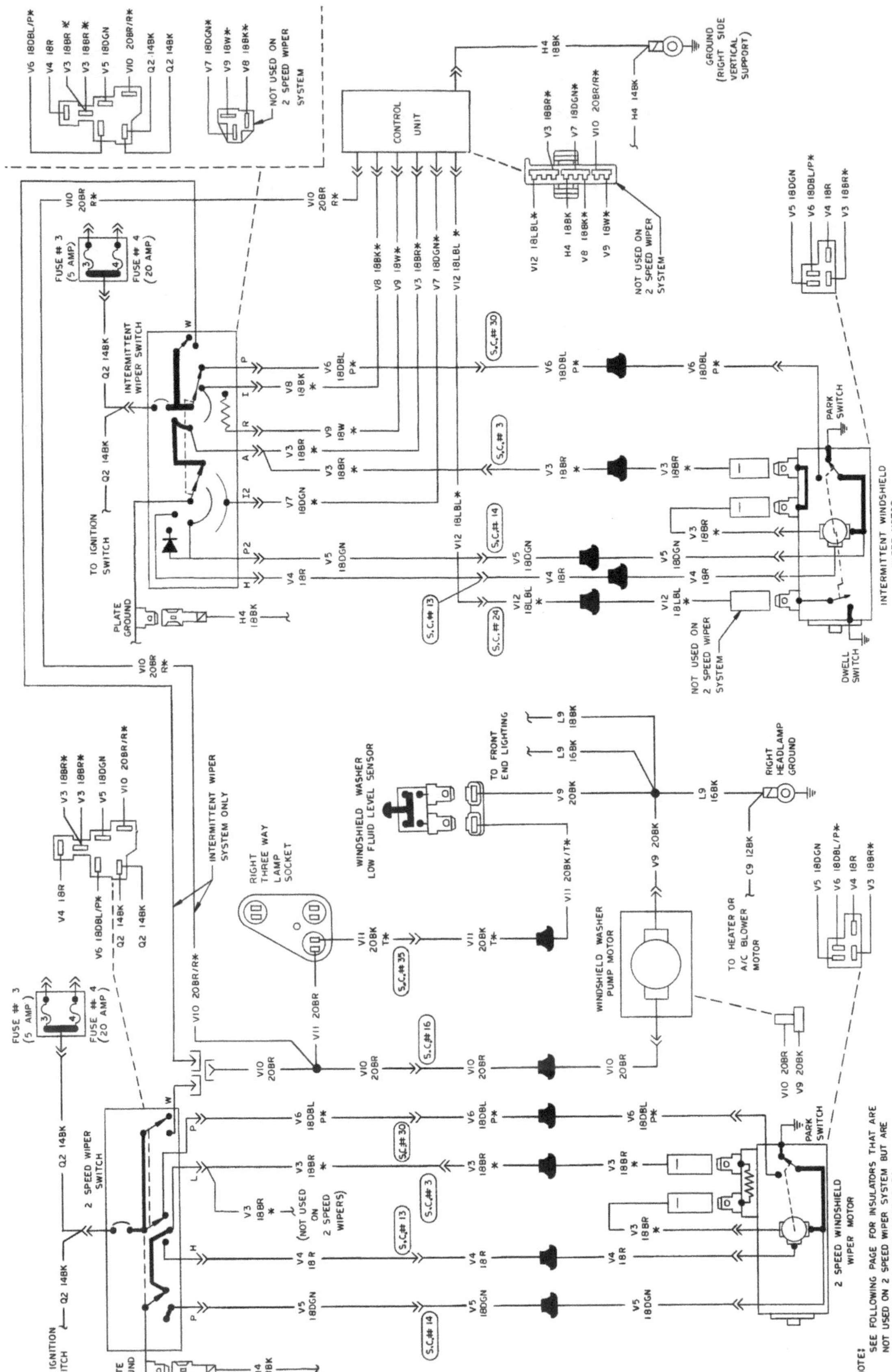

Typical C-body windshield wiper system wiring diagram

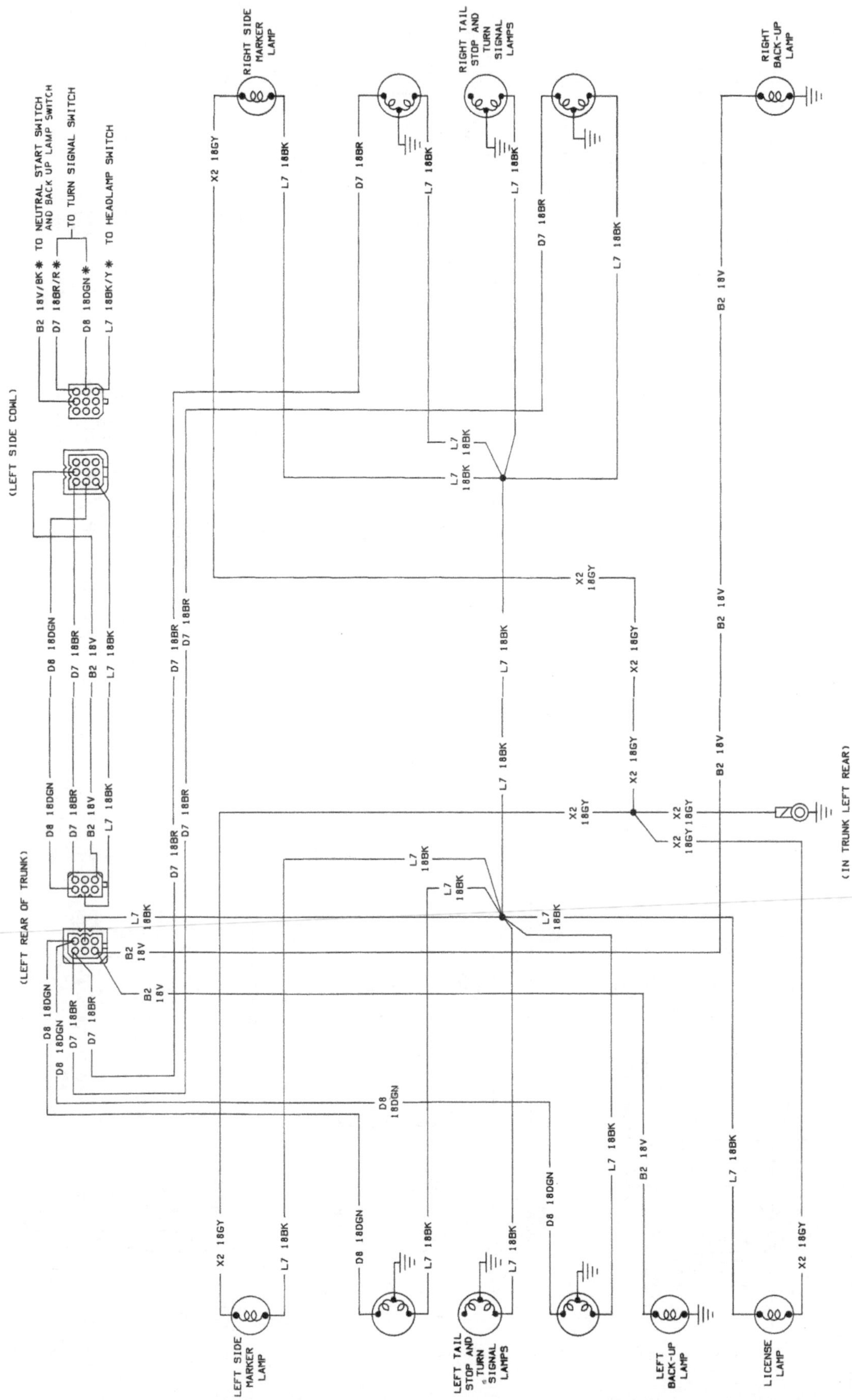

Typical C-body front end lighting wiring diagram

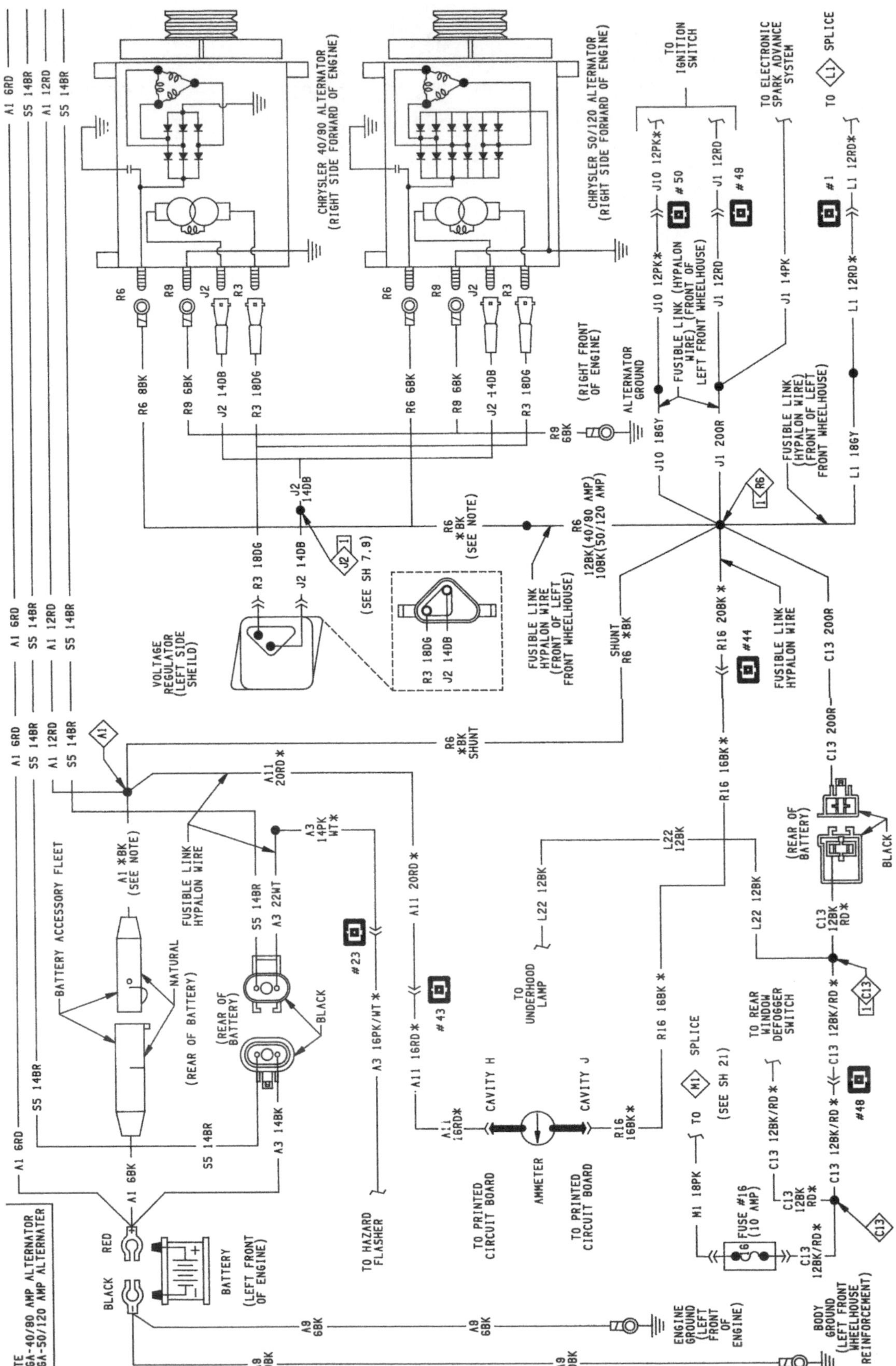

Typical later M-body charging system wiring diagram (J and R-body similar)

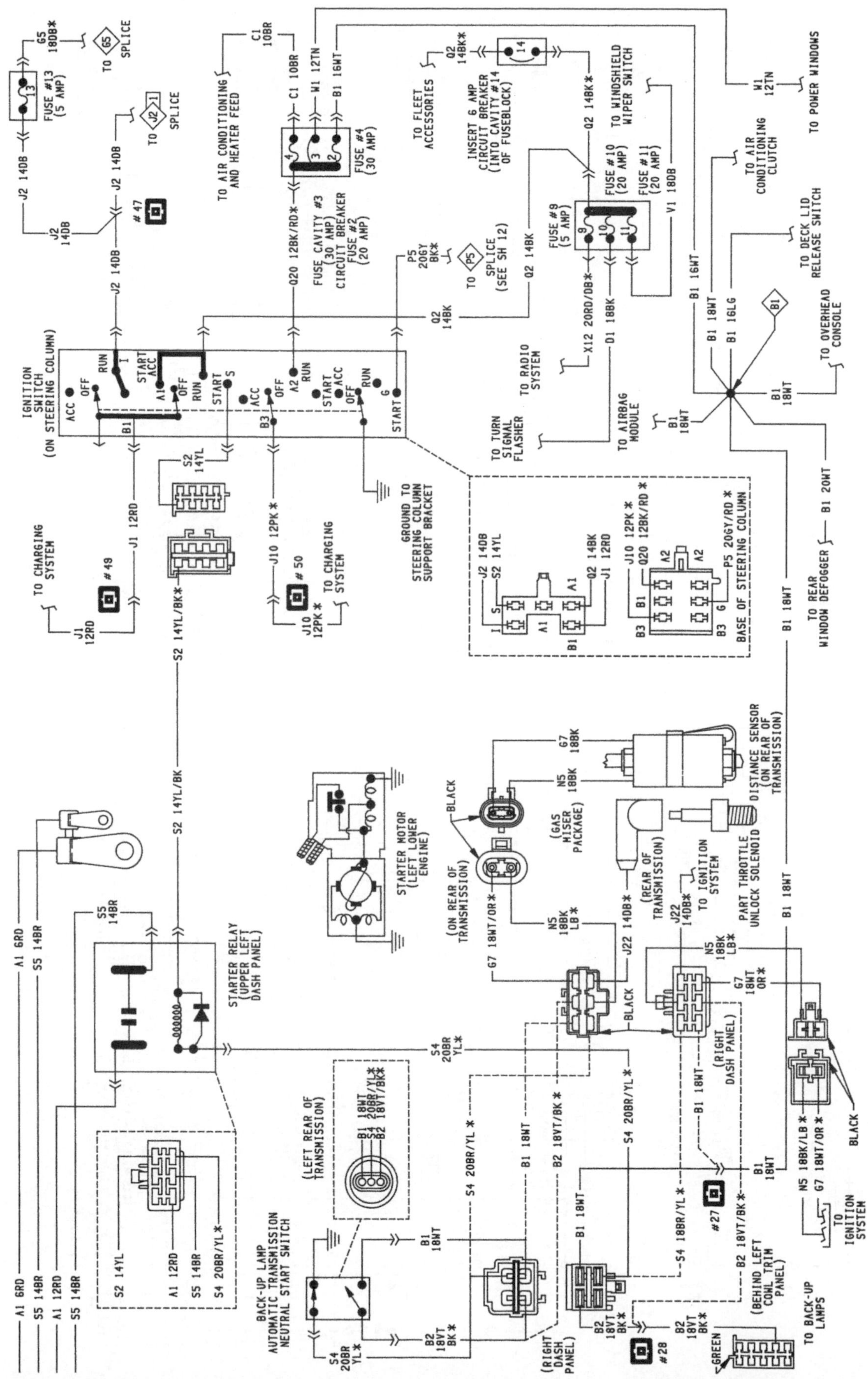

Typical later M-body charging system and ignition switch wiring diagram (J and R-body similar)

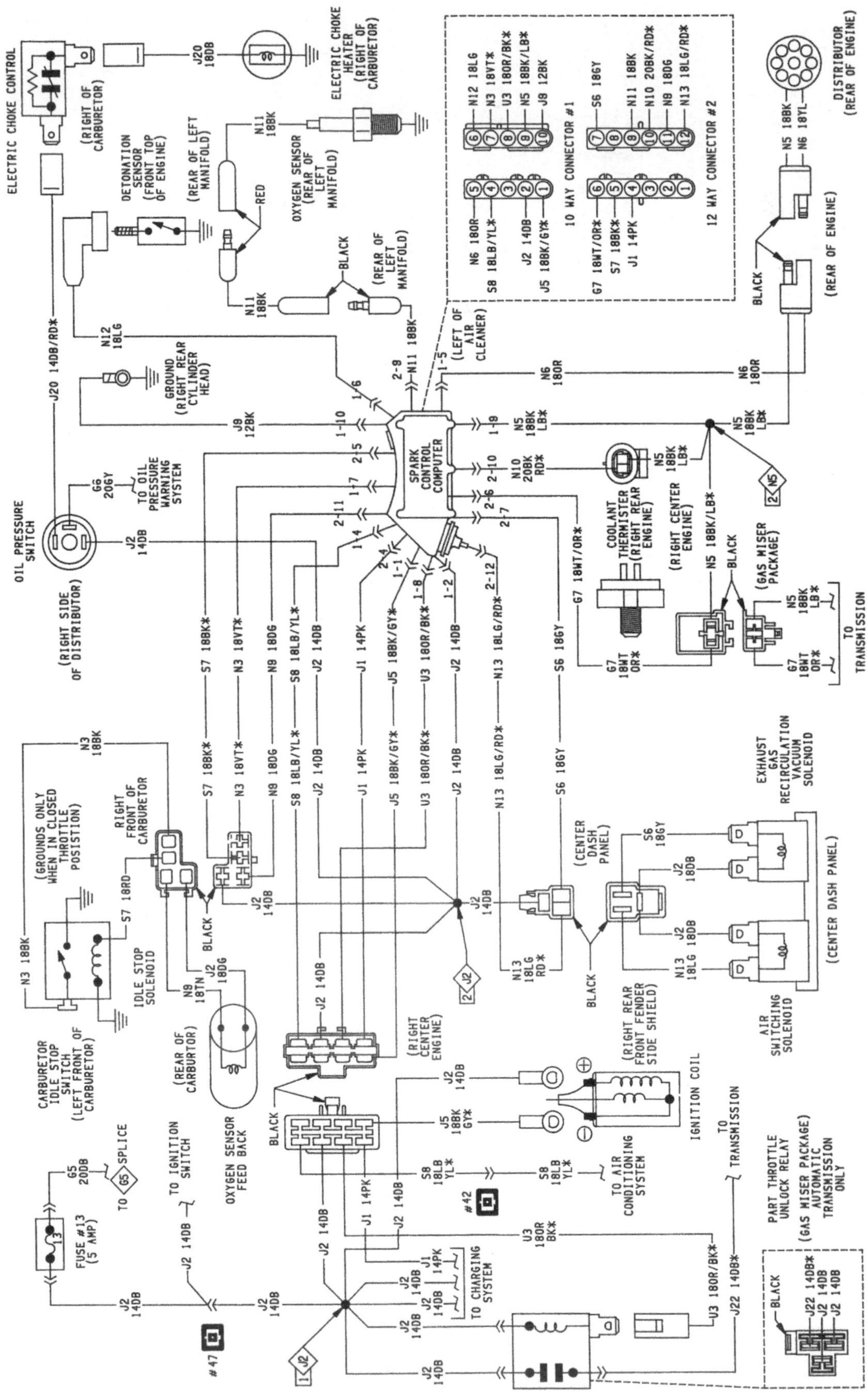

Typical later M-body electronic spark advance ignition (V8 with 2-barrel carburetor) system wiring diagram (J and R-body similar)

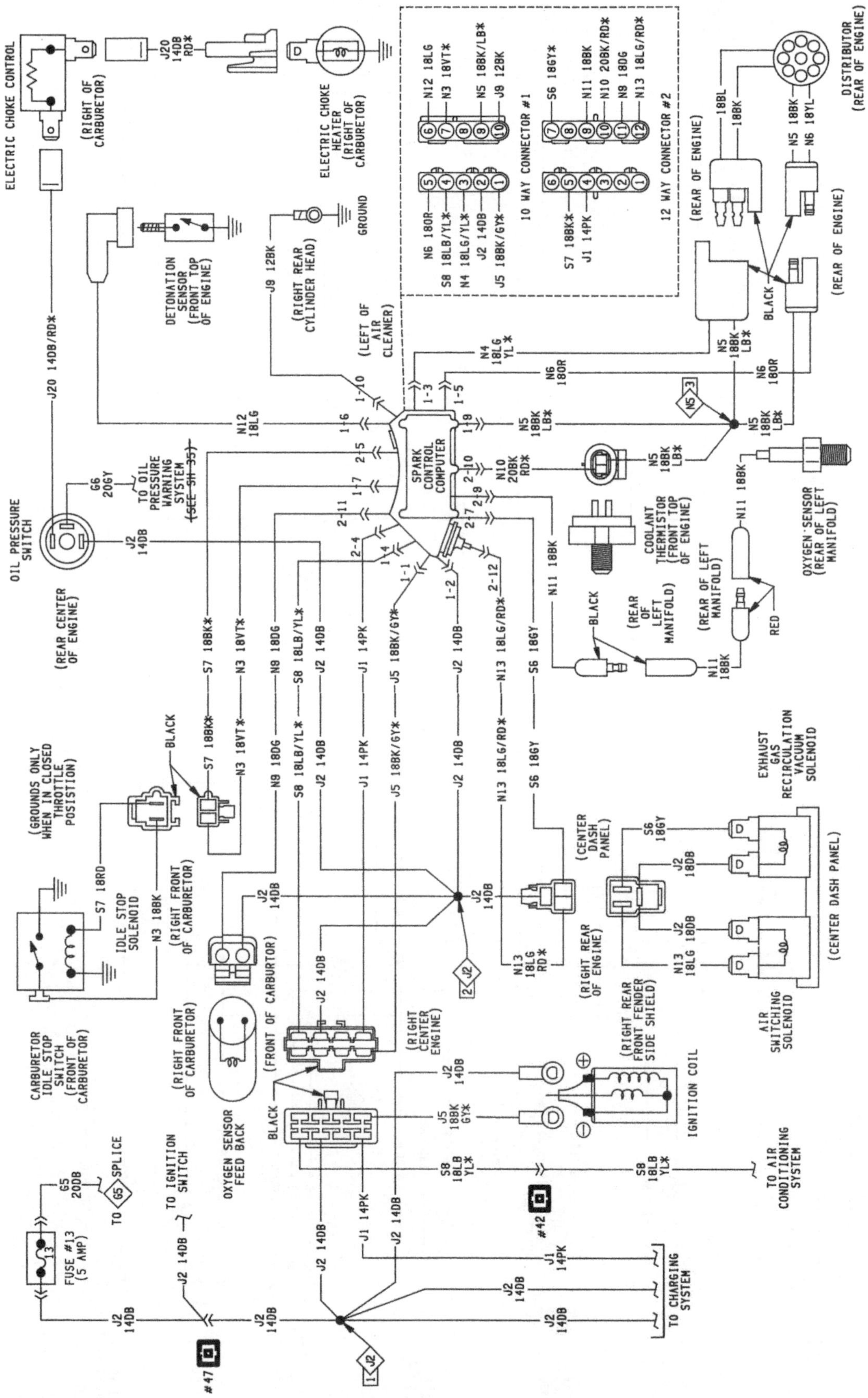

Typical later M-body electronic spark advance ignition (V8 with 4-barrel carburetor) system wiring diagram (J and R-body similar)

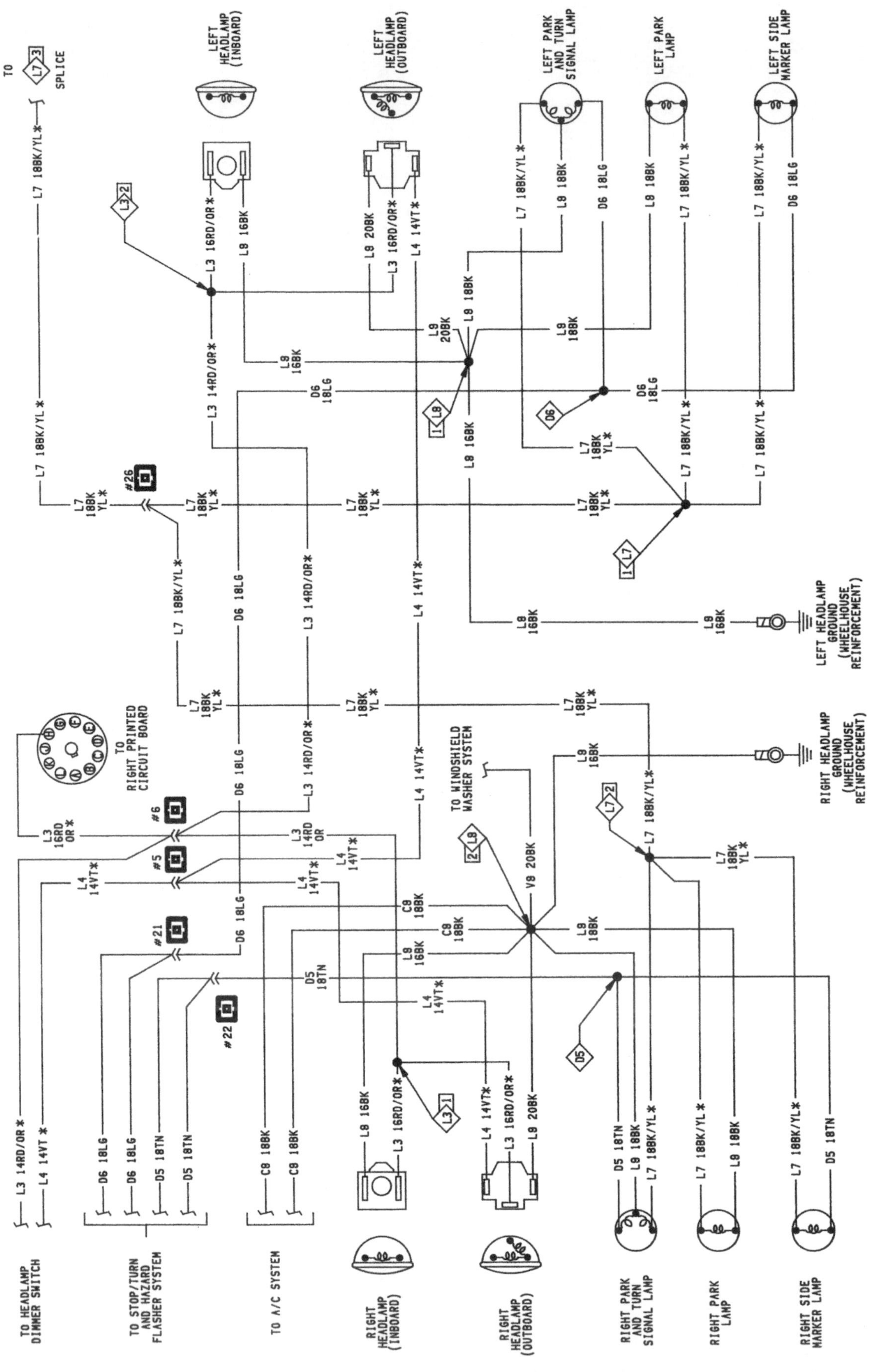

Typical later M-body front end lighting wiring diagram (J and R-body similar)

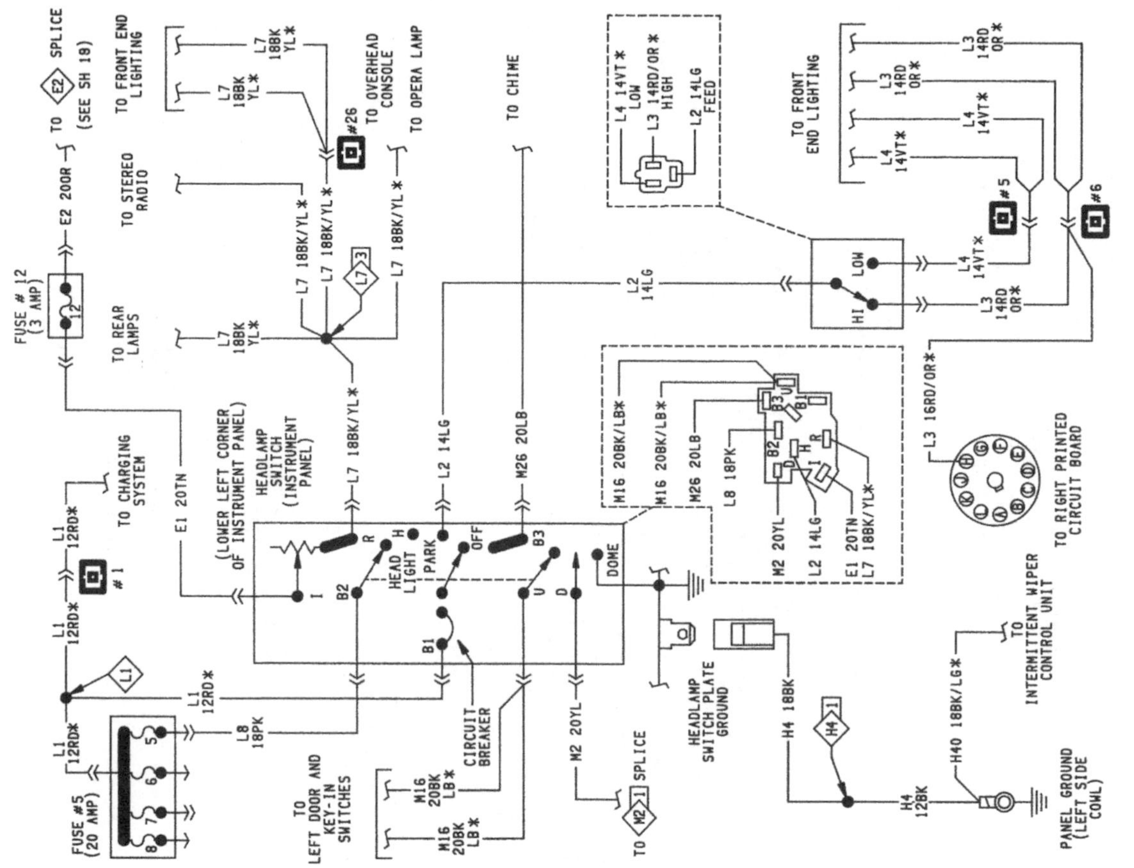

Typical later M-body headlight switch wiring diagram (J and R-body similar)

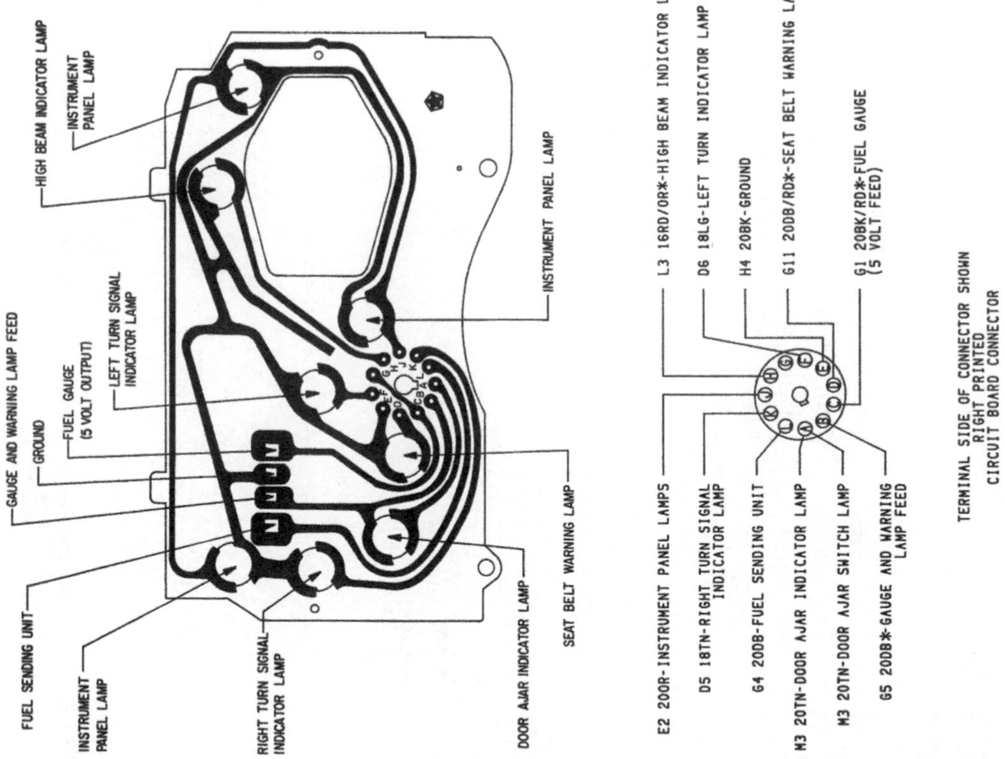

Typical later M-body printed circuit board connector diagram (J and R-body similar)

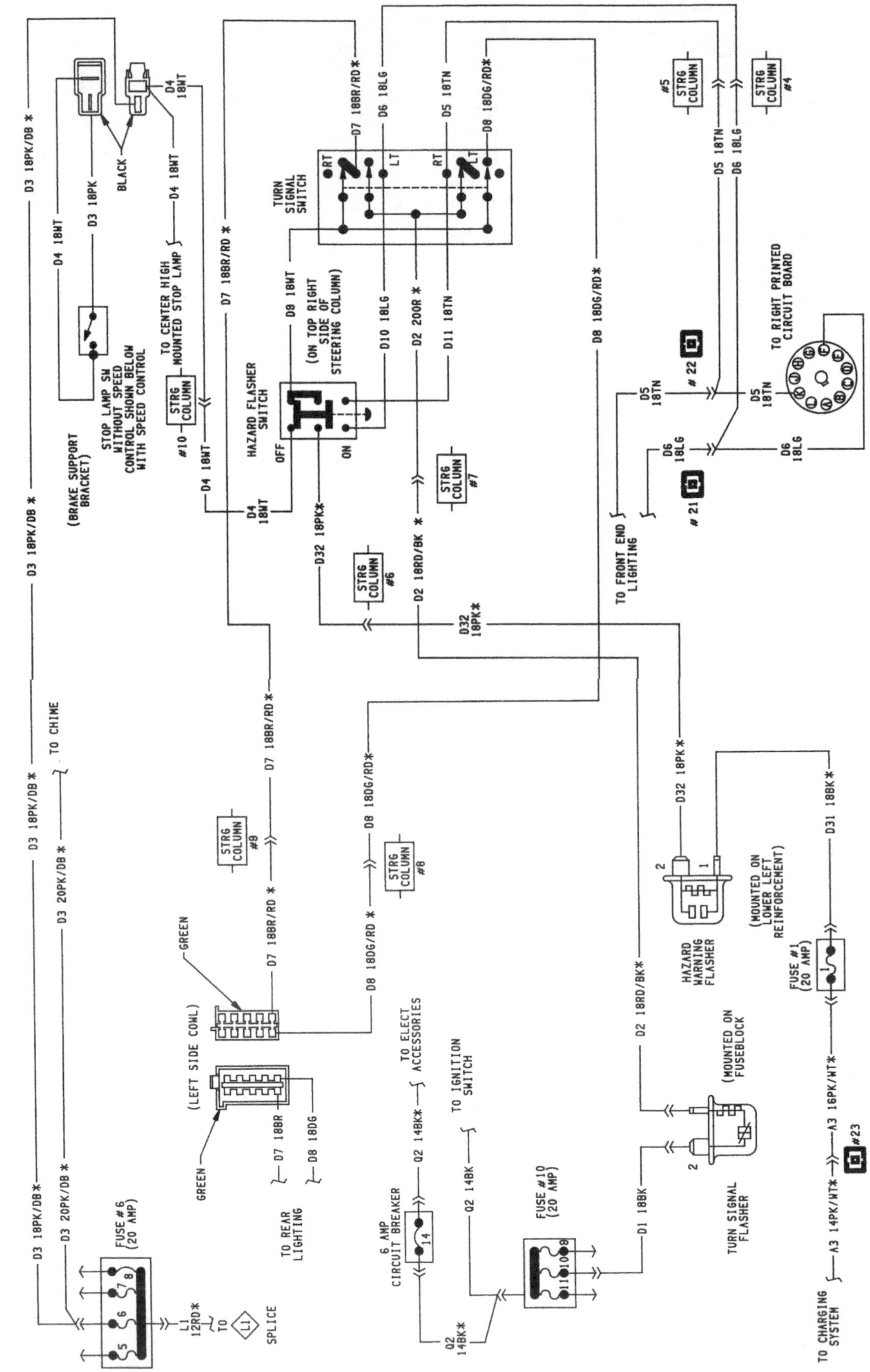

Typical later M-body turn signal/hazard flasher system wiring diagram (J and R-body similar)

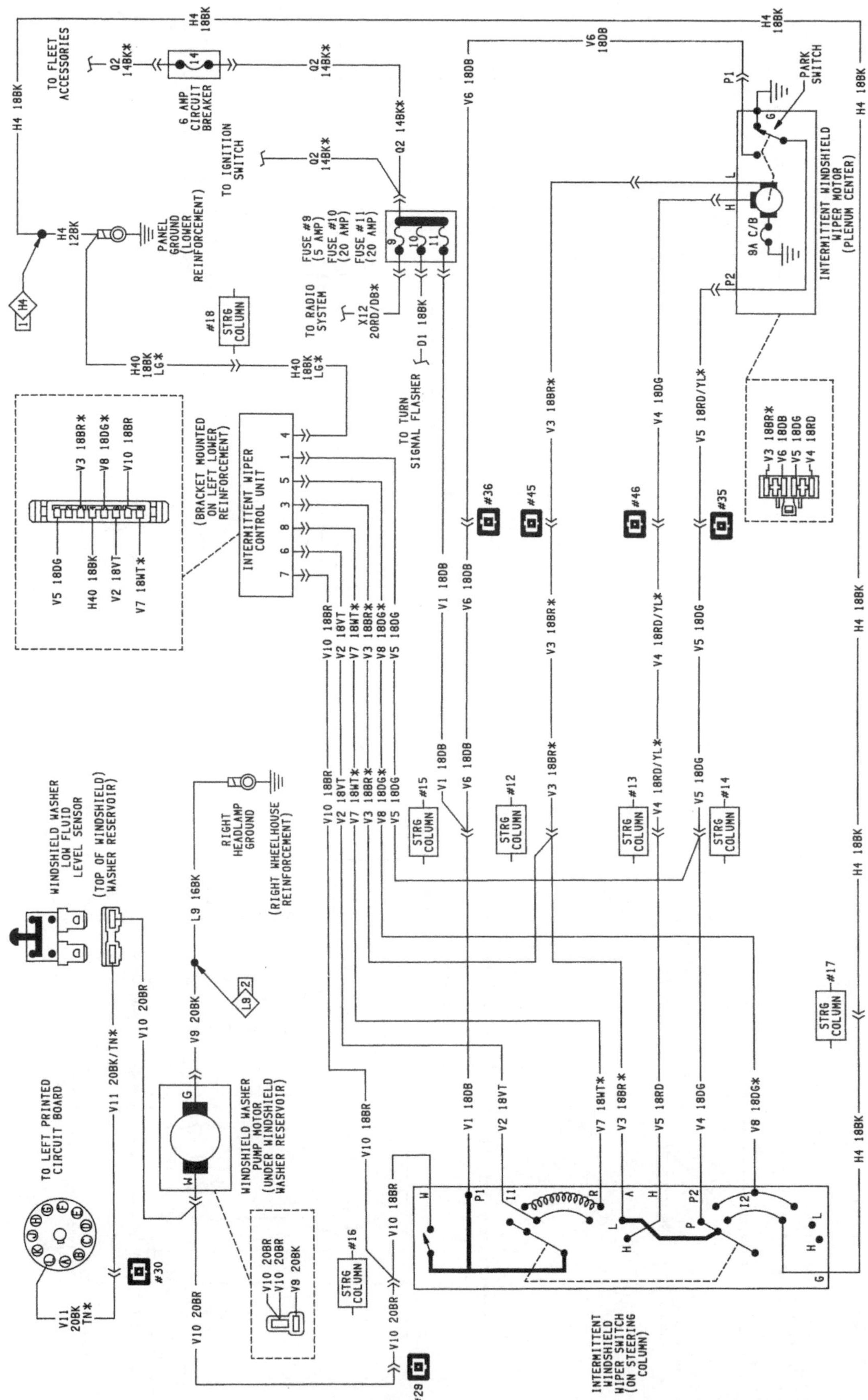

Typical later M-body windshield wiper system wiring diagram (J and R-body similar)

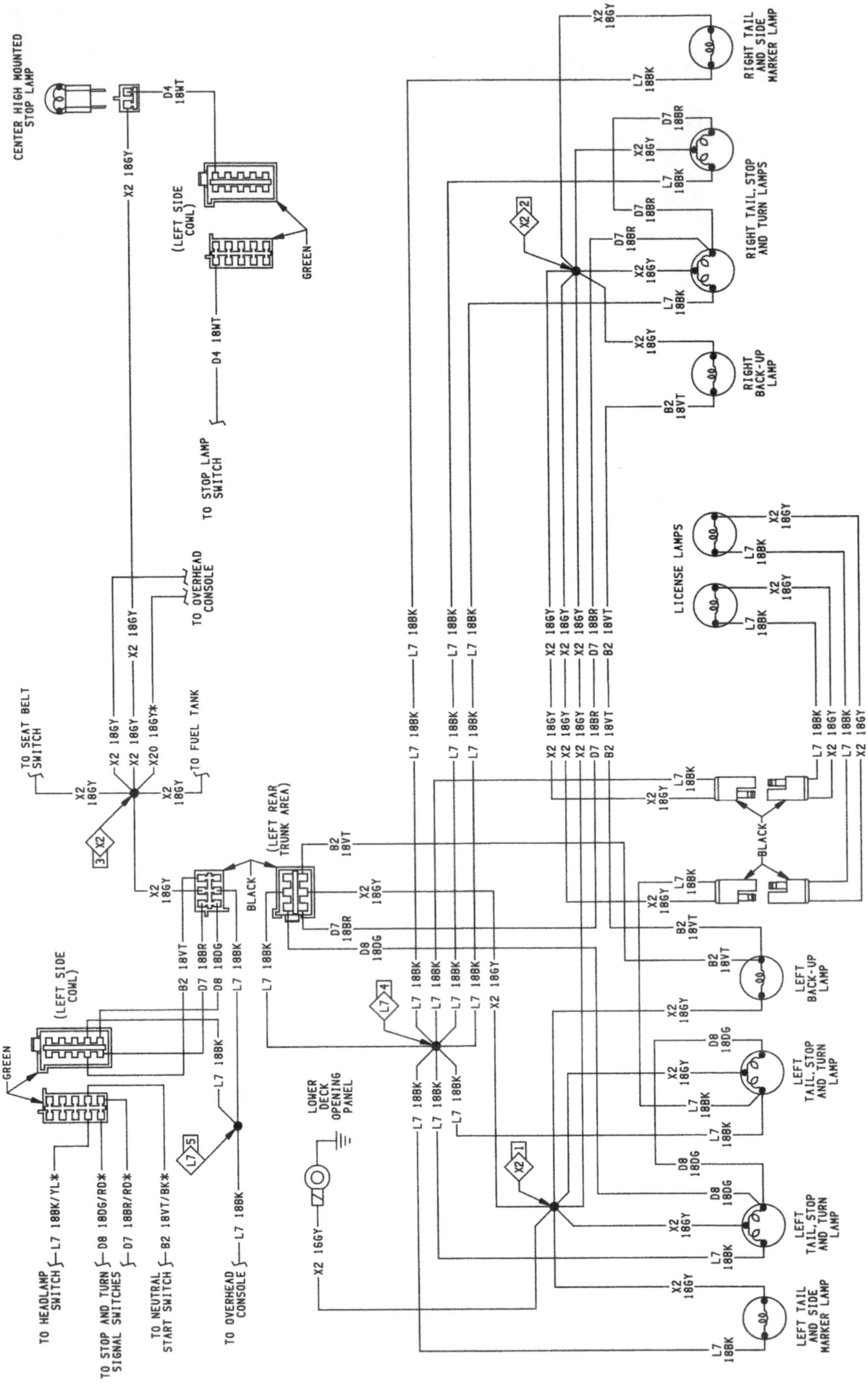

Typical later M-body rear lighting system wiring diagram (J and R-body similar)

Index

A

B

C

D

E

F

G

H

I

T

U

V

W

Notes

Haynes Automotive Manuals

NOTE: If you do not see a listing for your vehicle, please visit **haynes.com** *for the latest product information and check out our* ***Online Manuals!***

ACURA

12020 **Integra** '86 thru '89 **& Legend** '86 thru '90
12021 **Integra** '90 thru '93 **& Legend** '91 thru '95
Integra '94 thru '00 - *see HONDA Civic (42025)*
MDX '01 thru '07 - *see HONDA Pilot (42037)*
12050 **Acura TL** all models '99 thru '08

AMC

14020 **Mid-size models** '70 thru '83
14025 **(Renault) Alliance & Encore** '83 thru '87

AUDI

15020 **4000** all models '80 thru '87
15025 **5000** all models '77 thru '83
15026 **5000** all models '84 thru '88
Audi A4 '96 thru '01 - *see VW Passat (96023)*
15030 **Audi A4** '02 thru '08

AUSTIN-HEALEY

Sprite - *see MG Midget (66015)*

BMW

18020 **3/5 Series** '82 thru '92
18021 **3-Series** incl. Z3 models '92 thru '98
18022 **3-Series** incl. Z4 models '99 thru '05
18023 **3-Series** '06 thru '14
18025 **320i all 4-cylinder models** '75 thru '83
18050 **1500 thru 2002** except Turbo '59 thru '77

BUICK

19010 **Buick Century** '97 thru '05
Century (front-wheel drive) - *see GM (38005)*
19020 **Buick, Oldsmobile & Pontiac Full-size (Front-wheel drive)** '85 thru '05
Buick Electra, LeSabre and Park Avenue; **Oldsmobile** Delta 88 Royale, Ninety Eight and Regency; **Pontiac** Bonneville
19025 **Buick, Oldsmobile & Pontiac Full-size (Rear wheel drive)** '70 thru '90
Buick Estate, Electra, LeSabre, Limited, **Oldsmobile** Custom Cruiser, Delta 88, Ninety-eight, **Pontiac** Bonneville, Catalina, Grandville, Parisienne
19027 **Buick LaCrosse** '05 thru '13
Enclave - *see GENERAL MOTORS (38001)*
Rainier - *see CHEVROLET (24072)*
Regal - *see GENERAL MOTORS (38010)*
Riviera - *see GENERAL MOTORS (38030, 38031)*
Roadmaster - *see CHEVROLET (24046)*
Skyhawk - *see GENERAL MOTORS (38015)*
Skylark - *see GENERAL MOTORS (38020, 38025)*
Somerset - *see GENERAL MOTORS (38025)*

CADILLAC

21015 **CTS & CTS-V** '03 thru '14
21030 **Cadillac Rear Wheel Drive** '70 thru '93
Cimarron - *see GENERAL MOTORS (38015)*
DeVille - *see GENERAL MOTORS (38031 & 38032)*
Eldorado - *see GENERAL MOTORS (38030)*
Fleetwood - *see GENERAL MOTORS (38031)*
Seville - *see GM (38030, 38031 & 38032)*

CHEVROLET

10305 **Chevrolet Engine Overhaul Manual**
24010 **Astro & GMC Safari Mini-vans** '85 thru '05
24013 **Aveo** '04 thru '11
24015 **Camaro V8** all models '70 thru '81
24016 **Camaro** all models '82 thru '92
24017 **Camaro & Firebird** '93 thru '02
Cavalier - *see GENERAL MOTORS (38016)*
Celebrity - *see GENERAL MOTORS (38005)*
24018 **Camaro** '10 thru '15
24020 **Chevelle, Malibu & El Camino** '69 thru '87
Cobalt - *see GENERAL MOTORS (38017)*
24024 **Chevette & Pontiac T1000** '76 thru '87
Citation - *see GENERAL MOTORS (38020)*
24027 **Colorado & GMC Canyon** '04 thru '12
24032 **Corsica & Beretta** all models '87 thru '96
24040 **Corvette** all V8 models '68 thru '82
24041 **Corvette** all models '84 thru '96
24042 **Corvette** all models '97 thru '13
24044 **Cruze** '11 thru '19
24045 **Full-size Sedans** Caprice, Impala, Biscayne, Bel Air & Wagons '69 thru '90
24046 **Impala SS & Caprice and Buick Roadmaster** '91 thru '96
Impala '00 thru '05 - *see LUMINA (24048)*
24047 **Impala & Monte Carlo** all models '06 thru '11
Lumina '90 thru '94 - *see GM (38010)*
24048 **Lumina & Monte Carlo** '95 thru '05
Lumina APV - *see GM (38035)*
24050 **Luv Pick-up** all 2WD & 4WD '72 thru '82
24051 **Malibu** '13 thru '19
24055 **Monte Carlo** all models '70 thru '88
Monte Carlo '95 thru '01 - *see LUMINA (24048)*
24059 **Nova** all V8 models '69 thru '79
24060 **Nova and Geo Prizm** '85 thru '92
24064 **Pick-ups** '67 thru '87 - Chevrolet & GMC
24065 **Pick-ups** '88 thru '98 - Chevrolet & GMC
24066 **Pick-ups** '99 thru '06 - Chevrolet & GMC
24067 **Chevrolet Silverado & GMC Sierra** '07 thru '14
24068 **Chevrolet Silverado & GMC Sierra** '14 thru '19
24070 **S-10 & S-15 Pick-ups** '82 thru '93, **Blazer & Jimmy** '83 thru '94,
24071 **S-10 & Sonoma Pick-ups** '94 thru '04, including **Blazer, Jimmy & Hombre**
24072 **Chevrolet TrailBlazer, GMC Envoy & Oldsmobile Bravada** '02 thru '09
24075 **Sprint** '85 thru '88 **& Geo Metro** '89 thru '01
24080 **Vans - Chevrolet & GMC** '68 thru '96
24081 **Chevrolet Express & GMC Savana** Full-size Vans '96 thru '19

CHRYSLER

10310 **Chrysler Engine Overhaul Manual**
25015 **Chrysler Cirrus, Dodge Stratus, Plymouth Breeze** '95 thru '00
25020 **Full-size Front-Wheel Drive** '88 thru '93
K-Cars - *see DODGE Aries (30008)*
Laser - *see DODGE Daytona (30030)*
25025 **Chrysler LHS, Concorde, New Yorker, Dodge** Intrepid, **Eagle** Vision, '93 thru '97
25026 **Chrysler LHS, Concorde, 300M, Dodge** Intrepid, '98 thru '04
25027 **Chrysler 300** '05 thru '18, **Dodge Charger** '06 thru '18, **Magnum** '05 thru '08 **& Challenger** '08 thru '18
25030 **Chrysler & Plymouth Mid-size** front wheel drive '82 thru '95
Rear-wheel Drive - *see Dodge (30050)*
25035 **PT Cruiser** all models '01 thru '10
25040 **Chrysler Sebring** '95 thru '06, **Dodge Stratus** '01 thru '06 **& Dodge Avenger** '95 thru '00
25041 **Chrysler Sebring** '07 thru '10, **200** '11 thru '17 **Dodge Avenger** '08 thru '14

DATSUN

28005 **200SX** all models '80 thru '83
28012 **240Z, 260Z & 280Z** Coupe '70 thru '78
28014 **280ZX** Coupe & 2+2 '79 thru '83
300ZX - *see NISSAN (72010)*
28018 **510 & PL521 Pick-up** '68 thru '73
28020 **510** all models '78 thru '81
28022 **620 Series Pick-up** all models '73 thru '79
720 Series Pick-up - *see NISSAN (72030)*

DODGE

400 & 600 - *see CHRYSLER (25030)*
30008 **Aries & Plymouth Reliant** '81 thru '89
30010 **Caravan & Plymouth Voyager** '84 thru '95
30011 **Caravan & Plymouth Voyager** '96 thru '02
30012 **Challenger & Plymouth Sapporro** '78 thru '83
30013 **Caravan, Chrysler Voyager & Town & Country** '03 thru '07
30014 **Grand Caravan & Chrysler Town & Country** '08 thru '18
30016 **Colt & Plymouth Champ** '78 thru '87
30020 **Dakota Pick-ups** all models '87 thru '96
30021 **Durango** '98 & '99 **& Dakota** '97 thru '99
30022 **Durango** '00 thru '03 **& Dakota** '00 thru '04
30023 **Durango** '04 thru '09 **& Dakota** '05 thru '11
30025 **Dart, Demon, Plymouth Barracuda, Duster & Valiant** 6-cylinder models '67 thru '76
30030 **Daytona & Chrysler Laser** '84 thru '89
Intrepid - *see CHRYSLER (25025, 25026)*
30034 **Neon** all models '95 thru '99
30035 **Omni & Plymouth Horizon** '78 thru '90
30036 **Dodge & Plymouth Neon** '00 thru '05
30040 **Pick-ups** full-size models '74 thru '93
30042 **Pick-ups** full-size models '94 thru '08
30043 **Pick-ups** full-size models '09 thru '18
30045 **Ram 50/D50 Pick-ups & Raider and Plymouth Arrow Pick-ups** '79 thru '93
30050 **Dodge/Plymouth/Chrysler** RWD '71 thru '89
30055 **Shadow & Plymouth Sundance** '87 thru '94
30060 **Spirit & Plymouth Acclaim** '89 thru '95
30065 **Vans - Dodge & Plymouth** '71 thru '03

EAGLE

Talon - *see MITSUBISHI (68030, 68031)*
Vision - *see CHRYSLER (25025)*

FIAT

34010 **124 Sport Coupe & Spider** '68 thru '78
34025 **X1/9** all models '74 thru '80

FORD

10320 **Ford Engine Overhaul Manual**
10355 **Ford Automatic Transmission Overhaul**
11500 **Mustang '64-1/2 thru '70 Restoration Guide**
36004 **Aerostar Mini-vans** all models '86 thru '97
36006 **Contour & Mercury Mystique** '95 thru '00
36008 **Courier Pick-up** all models '72 thru '82
36012 **Crown Victoria & Mercury Grand Marquis** '88 thru '11
36014 **Edge** '07 thru '19 **& Lincoln MKX** '07 thru '18
36016 **Escort & Mercury Lynx** all models '81 thru '90
36020 **Escort & Mercury Tracer** '91 thru '02
36022 **Escape** '01 thru '17, **Mazda Tribute** '01 thru '11, **& Mercury Mariner** '05 thru '11
36024 **Explorer & Mazda Navajo** '91 thru '01
36025 **Explorer & Mercury Mountaineer** '02 thru '10
36026 **Explorer** '11 thru '17
36028 **Fairmont & Mercury Zephyr** '78 thru '83
36030 **Festiva & Aspire** '88 thru '97
36032 **Fiesta** all models '77 thru '80
36034 **Focus** all models '00 thru '11
36035 **Focus** '12 thru '14
36045 **Fusion** '06 thru '14 **& Mercury Milan** '06 thru '11
36048 **Mustang V8** all models '64-1/2 thru '73
36049 **Mustang II** 4-cylinder, V6 & V8 models '74 thru '78
36050 **Mustang & Mercury Capri** '79 thru '93
36051 **Mustang** all models '94 thru '04
36052 **Mustang** '05 thru '14
36054 **Pick-ups & Bronco** '73 thru '79
36058 **Pick-ups & Bronco** '80 thru '96
36059 **F-150** '97 thru '03, **Expedition** '97 thru '17, **F-250** '97 thru '99, **F-150 Heritage** '04 **& Lincoln Navigator** '98 thru '17
36060 **Super Duty Pick-ups & Excursion** '99 thru '10
36061 **F-150** full-size '04 thru '14
36062 **Pinto & Mercury Bobcat** '75 thru '80
36063 **F-150** full-size '15 thru '17
36064 **Super Duty Pick-ups** '11 thru '16
36066 **Probe** all models '89 thru '92
Probe '93 thru '97 - *see MAZDA 626 (61042)*
36070 **Ranger & Bronco II** gas models '83 thru '92
36071 **Ranger** '93 thru '11 **& Mazda Pick-ups** '94 thru '09
36074 **Taurus & Mercury Sable** '86 thru '95
36075 **Taurus & Mercury Sable** '96 thru '07
36076 **Taurus** '08 thru '14, **Five Hundred** '05 thru '07, **Mercury Montego** '05 thru '07 **& Sable** '08 thru '09
36078 **Tempo & Mercury Topaz** '84 thru '94
36082 **Thunderbird & Mercury Cougar** '83 thru '88
36086 **Thunderbird & Mercury Cougar** '89 thru '97
36090 **Vans** all V8 Econoline models '69 thru '91
36094 **Vans** full size '92 thru '14
36097 **Windstar** '95 thru '03, **Freestar & Mercury Monterey Mini-van** '04 thru '07

GENERAL MOTORS

10360 **GM Automatic Transmission Overhaul**
38001 **GMC Acadia** '07 thru '16, **Buick Enclave** '08 thru '17, **Saturn Outlook** '07 thru '10 **& Chevrolet Traverse** '09 thru '17
38005 **Buick Century, Chevrolet Celebrity, Oldsmobile Cutlass Ciera & Pontiac 6000** all models '82 thru '96
38010 **Buick Regal** '88 thru '04, **Chevrolet Lumina** '88 thru '04, **Oldsmobile Cutlass Supreme** '88 thru '97 **& Pontiac Grand Prix** '88 thru '07
38015 **Buick Skyhawk, Cadillac Cimarron, Chevrolet Cavalier, Oldsmobile Firenza, Pontiac J-2000 & Sunbird** '82 thru '94
38016 **Chevrolet Cavalier & Pontiac Sunfire** '95 thru '05
38017 **Chevrolet Cobalt** '05 thru '10, **HHR** '06 thru '11, **Pontiac G5** '07 thru '09, **Pursuit** '05 thru '06 **& Saturn ION** '03 thru '07
38020 **Buick Skylark, Chevrolet Citation, Oldsmobile Omega, Pontiac Phoenix** '80 thru '85
38025 **Buick Skylark** '86 thru '98, **Somerset** '85 thru '87, **Oldsmobile Achieva** '92 thru '98, **Calais** '85 thru '91, **& Pontiac Grand Am** all models '85 thru '98
38026 **Chevrolet Malibu** '97 thru '03, **Classic** '04 thru '05, **Oldsmobile Alero** '99 thru '03, **Cutlass** '97 thru '00, **& Pontiac Grand Am** '99 thru '03
38027 **Chevrolet Malibu** '04 thru '12, **Pontiac G6** '05 thru '10 **& Saturn Aura** '07 thru '10
38030 **Cadillac Eldorado, Seville, Oldsmobile Toronado & Buick Riviera** '71 thru '85
38031 **Cadillac Eldorado, Seville, DeVille, Fleetwood, Oldsmobile Toronado & Buick Riviera** '86 thru '93
38032 **Cadillac DeVille** '94 thru '05, **Seville** '92 thru '04 **& Cadillac DTS** '06 thru '10
38035 **Chevrolet Lumina APV, Oldsmobile Silhouette & Pontiac Trans Sport** all models '90 thru '96
38036 **Chevrolet Venture** '97 thru '05, **Oldsmobile Silhouette** '97 thru '04, **Pontiac Trans Sport** '97 thru '98 **& Montana** '99 thru '05
38040 **Chevrolet Equinox** '05 thru '17, **GMC Terrain** '10 thru '17 **& Pontiac Torrent** '06 thru '09

GEO

Metro - *see CHEVROLET Sprint (24075)*
Prizm - *'85 thru '92 see CHEVY (24060), '93 thru '02 see TOYOTA Corolla (92036)*
40030 **Storm** all models '90 thru '93
Tracker - *see SUZUKI Samurai (90010)*

(Continued on other side)

Haynes Automotive Manuals (continued)

NOTE: If you do not see a listing for your vehicle, please visit ***haynes.com*** *for the latest product information and check out our* ***Online Manuals!***

GMC
Acadia - *see GENERAL MOTORS (38001)*
Pick-ups - *see CHEVROLET (24027, 24068)*
Vans - *see CHEVROLET (24081)*

HONDA
42010 **Accord CVCC** all models '76 thru '83
42011 **Accord** all models '84 thru '89
42012 **Accord** all models '90 thru '93
42013 **Accord** all models '94 thru '97
42014 **Accord** all models '98 thru '02
42015 **Accord** '03 thru '12 **& Crosstour** '10 thru '14
42016 **Accord** '13 thru '17
42020 **Civic 1200** all models '73 thru '79
42021 **Civic 1300 & 1500 CVCC** '80 thru '83
42022 **Civic 1500 CVCC** all models '75 thru '79
42023 **Civic** all models '84 thru '91
42024 **Civic & del Sol** '92 thru '95
42025 **Civic** '96 thru '00, **CR-V** '97 thru '01 **& Acura Integra** '94 thru '00
42026 **Civic** '01 thru '11 **& CR-V** '02 thru '11
42027 **Civic** '12 thru '15 **& CR-V** '12 thru '16
42030 **Fit** '07 thru '13
42035 **Odyssey** all models '99 thru '10
Passport - *see ISUZU Rodeo (47017)*
42037 **Honda Pilot** '03 thru '08, **Ridgeline** '06 thru '14 **& Acura MDX** '01 thru '07
42040 **Prelude CVCC** all models '79 thru '89

HYUNDAI
43010 **Elantra** all models '96 thru '19
43015 **Excel & Accent** all models '86 thru '13
43050 **Santa Fe** all models '01 thru '12
43055 **Sonata** all models '99 thru '14

INFINITI
G35 '03 thru '08 - *see NISSAN 350Z (72011)*

ISUZU
Hombre - *see CHEVROLET S-10 (24071)*
47017 **Rodeo** '91 thru '02, **Amigo** '89 thru '94 & '98 thru '02 **& Honda Passport** '95 thru '02
47020 **Trooper** '84 thru '91 **& Pick-up** '81 thru '93

JAGUAR
49010 **XJ6** all 6-cylinder models '68 thru '86
49011 **XJ6** all models '88 thru '94
49015 **XJ12 & XJS** all 12-cylinder models '72 thru '85

JEEP
50010 **Cherokee, Comanche & Wagoneer Limited** all models '84 thru '01
50011 **Cherokee** '14 thru '19
50020 **CJ** all models '49 thru '86
50025 **Grand Cherokee** all models '93 thru '04
50026 **Grand Cherokee** '05 thru '19 **& Dodge Durango** '11 thru '19
50029 **Grand Wagoneer & Pick-up** '72 thru '91 Grand Wagoneer '84 thru '91, Cherokee & Wagoneer '72 thru '83, Pick-up '72 thru '88
50030 **Wrangler** all models '87 thru '17
50035 **Liberty** '02 thru '12 **& Dodge Nitro** '07 thru '11
50050 **Patriot & Compass** '07 thru '17

KIA
54050 **Optima** '01 thru '10
54060 **Sedona** '02 thru '14
54070 **Sephia** '94 thru '01, **Spectra** '00 thru '09, **Sportage** '05 thru '20
54077 **Sorento** '03 thru '13

LEXUS
ES 300/330 - *see TOYOTA Camry (92007, 92008)*
ES 350 - *see TOYOTA Camry (92009)*
RX 300/330/350 - *see TOYOTA Highlander (92095)*

LINCOLN
MKX - *see FORD (36014)*
Navigator - *see FORD Pick-up (36059)*
59010 **Rear-Wheel Drive Continental** '70 thru '87, **Mark Series** '70 thru '92 & **Town Car** '81 thru '10

MAZDA
61010 **GLC (rear-wheel drive)** '77 thru '83
61011 **GLC (front-wheel drive)** '81 thru '85
61012 **Mazda3** '04 thru '11
61015 **323 & Protogé** '90 thru '03
61016 **MX-5 Miata** '90 thru '14
61020 **MPV** all models '89 thru '98
Navajo - *see Ford Explorer (36024)*
61030 **Pick-ups** '72 thru '93
Pick-ups '94 thru '09 - *see Ford Ranger (36071)*
61035 **RX-7** all models '79 thru '85
61036 **RX-7** all models '86 thru '91
61040 **626 (rear-wheel drive)** all models '79 thru '82
61041 **626 & MX-6 (front-wheel drive)** '83 thru '92
61042 **626** '93 thru '01 **& MX-6/Ford Probe** '93 thru '02
61043 **Mazda6** '03 thru '13

MERCEDES-BENZ
63012 **123 Series Diesel** '76 thru '85
63015 **190 Series** 4-cylinder gas models '84 thru '88
63020 **230/250/280** 6-cylinder SOHC models '68 thru '72
63025 **280 123 Series** gas models '77 thru '81
63030 **350 & 450** all models '71 thru '80
63040 **C-Class:** C230/C240/C280/C320/C350 '01 thru '07

MERCURY
64200 **Villager & Nissan Quest** '93 thru '01
All other titles, see FORD Listing.

MG
66010 **MGB** Roadster & GT Coupe '62 thru '80
66015 **MG Midget, Austin Healey Sprite** '58 thru '80

MINI
67020 **Mini** '02 thru '13

MITSUBISHI
68020 **Cordia, Tredia, Galant, Precis & Mirage** '83 thru '93
68030 **Eclipse, Eagle Talon & Plymouth Laser** '90 thru '94
68031 **Eclipse** '95 thru '05 **& Eagle Talon** '95 thru '98
68035 **Galant** '94 thru '12
68040 **Pick-up** '83 thru '96 **& Montero** '83 thru '93

NISSAN
72010 **300ZX** all models including Turbo '84 thru '89
72011 **350Z & Infiniti G35** all models '03 thru '08
72015 **Altima** all models '93 thru '06
72016 **Altima** '07 thru '12
72020 **Maxima** all models '85 thru '92
72021 **Maxima** all models '93 thru '08
72025 **Murano** '03 thru '14
72030 **Pick-ups** '80 thru '97 **& Pathfinder** '87 thru '95
72031 **Frontier** '98 thru '04, **Xterra** '00 thru '04, **& Pathfinder** '96 thru '04
72032 **Frontier & Xterra** '05 thru '14
72037 **Pathfinder** '05 thru '14
72040 **Pulsar** all models '83 thru '86
72042 **Roque** all models '08 thru '20
72050 **Sentra** all models '82 thru '94
72051 **Sentra & 200SX** all models '95 thru '06
72060 **Stanza** all models '82 thru '90
72070 **Titan pick-ups** '04 thru '10, **Armada** '05 thru '10 **& Pathfinder Armada** '04
72080 **Versa** all models '07 thru '19

OLDSMOBILE
73015 **Cutlass** V6 & V8 gas models '74 thru '88
For other OLDSMOBILE titles, see BUICK, CHEVROLET or GENERAL MOTORS listings.

PLYMOUTH
For PLYMOUTH titles, see DODGE listing.

PONTIAC
79008 **Fiero** all models '84 thru '88
79018 **Firebird** V8 models except Turbo '70 thru '81
79019 **Firebird** all models '82 thru '92
79025 **G6** all models '05 thru '09
79040 **Mid-size Rear-wheel Drive** '70 thru '87
Vibe '03 thru '10 - *see TOYOTA Corolla (92037)*
For other PONTIAC titles, see BUICK, CHEVROLET or GENERAL MOTORS listings.

PORSCHE
80020 **911** Coupe & Targa models '65 thru '89
80025 **914** all 4-cylinder models '69 thru '76
80030 **924** all models including Turbo '76 thru '82
80035 **944** all models including Turbo '83 thru '89

RENAULT
Alliance & Encore - *see AMC (14025)*

SAAB
84010 **900** all models including Turbo '79 thru '88

SATURN
87010 **Saturn** all S-series models '91 thru '02
Saturn Ion '03 thru '07- *see GM (38017)*
Saturn Outlook - *see GM (38001)*
87020 **Saturn L-series** all models '00 thru '04
87040 **Saturn VUE** '02 thru '09

SUBARU
89002 **1100, 1300, 1400 & 1600** '71 thru '79
89003 **1600 & 1800** 2WD & 4WD '80 thru '94
89080 **Impreza** '02 thru '11, **WRX** '02 thru '14, **& WRX STI** '04 thru '14
89100 **Legacy** all models '90 thru '99
89101 **Legacy & Forester** '00 thru '09
89102 **Legacy** '10 thru '16 **& Forester** '12 thru '16

SUZUKI
90010 **Samurai/Sidekick & Geo Tracker** '86 thru '01

TOYOTA
92005 **Camry** all models '83 thru '91
92006 **Camry** '92 thru '96 **& Avalon** '95 thru '96
92007 **Camry, Avalon, Solara, Lexus ES 300** '97 thru '01
92008 **Camry, Avalon, Lexus ES 300/330** '02 thru '06 **& Solara** '02 thru '08
92009 **Camry, Avalon & Lexus ES 350** '07 thru '17
92015 **Celica Rear-wheel Drive** '71 thru '85
92020 **Celica Front-wheel Drive** '86 thru '99
92025 **Celica Supra** all models '79 thru '92
92030 **Corolla** all models '75 thru '79
92032 **Corolla** all rear-wheel drive models '80 thru '87
92035 **Corolla** all front-wheel drive models '84 thru '92
92036 **Corolla & Geo/Chevrolet Prizm** '93 thru '02
92037 **Corolla** '03 thru '19, **Matrix** '03 thru '14, **& Pontiac Vibe** '03 thru '10
92040 **Corolla Tercel** all models '80 thru '82
92045 **Corona** all models '74 thru '82
92050 **Cressida** all models '78 thru '82
92055 **Land Cruiser** FJ40, 43, 45, 55 '68 thru '82
92056 **Land Cruiser** FJ60, 62, 80, FZJ80 '80 thru '96
92060 **Matrix** '03 thru '11 **& Pontiac Vibe** '03 thru '10
92065 **MR2** all models '85 thru '87
92070 **Pick-up** all models '69 thru '78
92075 **Pick-up** all models '79 thru '95
92076 **Tacoma** '95 thru '04, **4Runner** '96 thru '02 **& T100** '93 thru '08
92077 **Tacoma** all models '05 thru '18
92078 **Tundra** '00 thru '06 **& Sequoia** '01 thru '07
92079 **4Runner** all models '03 thru '09
92080 **Previa** all models '91 thru '95
92081 **Prius** all models '01 thru '12
92082 **RAV4** all models '96 thru '12
92085 **Tercel** all models '87 thru '94
92090 **Sienna** all models '98 thru '10
92095 **Highlander** '01 thru '19 **& Lexus RX330/330/350** '99 thru '19
92179 **Tundra** '07 thru '19 **& Sequoia** '08 thru '19

TRIUMPH
94007 **Spitfire** all models '62 thru '81
94010 **TR7** all models '75 thru '81

VW
96008 **Beetle & Karmann Ghia** '54 thru '79
96009 **New Beetle** '98 thru '10
96016 **Rabbit, Jetta, Scirocco & Pick-up** gas models '75 thru '92 & Convertible '80 thru '92
96017 **Golf, GTI & Jetta** '93 thru '98, **Cabrio** '95 thru '02
96018 **Golf, GTI, Jetta** '99 thru '05
96019 **Jetta, Rabbit, GLI, GTI & Golf** '05 thru '11
96020 **Rabbit, Jetta & Pick-up** diesel '77 thru '84
96021 **Jetta** '11 thru '18 **& Golf** '15 thru '19
96023 **Passat** '98 thru '05 **& Audi A4** '96 thru '01
96030 **Transporter 1600** all models '68 thru '79
96035 **Transporter 1700, 1800 & 2000** '72 thru '79
96040 **Type 3 1500 & 1600** all models '63 thru '73
96045 **Vanagon Air-Cooled** all models '80 thru '83

VOLVO
97010 **120, 130 Series & 1800 Sports** '61 thru '73
97015 **140 Series** all models '66 thru '74
97020 **240 Series** all models '76 thru '93
97040 **740 & 760 Series** all models '82 thru '88
97050 **850 Series** all models '93 thru '97

TECHBOOK MANUALS
10205 **Automotive Computer Codes**
10206 **OBD-II & Electronic Engine Management**
10210 **Automotive Emissions Control Manual**
10215 **Fuel Injection Manual** '78 thru '85
10225 **Holley Carburetor Manual**
10230 **Rochester Carburetor Manual**
10305 **Chevrolet Engine Overhaul Manual**
10320 **Ford Engine Overhaul Manual**
10330 **GM and Ford Diesel Engine Repair Manual**
10331 **Duramax Diesel Engines** '01 thru '19
10332 **Cummins Diesel Engine Performance Manual**
10333 **GM, Ford & Chrysler Engine Performance Manual**
10334 **GM Engine Performance Manual**
10340 **Small Engine Repair Manual,** 5 HP & Less
10341 **Small Engine Repair Manual,** 5.5 thru 20 HP
10345 **Suspension, Steering & Driveline Manual**
10355 **Ford Automatic Transmission Overhaul**
10360 **GM Automatic Transmission Overhaul**
10405 **Automotive Body Repair & Painting**
10410 **Automotive Brake Manual**
10411 **Automotive Anti-lock Brake (ABS) Systems**
10420 **Automotive Electrical Manual**
10425 **Automotive Heating & Air Conditioning**
10435 **Automotive Tools Manual**
10445 **Welding Manual**
10450 **ATV Basics**

Over a 100 Haynes motorcycle manuals also available

10/22